Fluvial Processes: Theory and Applications

Roman S. Chalov

Fluvial Processes: Theory and Applications

Volume 1: Drivers and Conditions of River Channel Character and Change

Springer

Roman S. Chalov
Lomonosov Moscow State University
Moscow, Russia

ISBN 978-3-030-66182-3 ISBN 978-3-030-66183-0 (eBook)
https://doi.org/10.1007/978-3-030-66183-0

This Springer imprint is published by the registered company Springer Nature Switzerland AG
The registered company address is: Gewerbestrasse 11, 6330 Cham, Switzerland

Preface

The book presents an overview of the fluvial processes theory which was developed in former USSR at Lomonosov Moscow State University for the last 50 years since the famous book by Prof. Makkaveev was published in 1955. Even being published before purely in Russian language, the theory and the author of the book, Prof. Roman S. Chalov, are well known in the world. The book demonstrates basics of fluvial geomorphology and hydromorphology and its links to ecosystem approach in river analysis and management. Despite a strong tradition in fluvial geomorphology work in USSR, unfortunately little has been made of insights provided on the international stage. The fluvial processes theory accepted in USSR was separated from English-language World during the long period of development (since let 1955, when the first books in Russian appeared in this field) from the English-language literature. The studied rivers are mostly located in the domain of Northern Eurasia and much different compared to those described in classical English-language works by Schumm (1977), Knighton (1998) and located mostly in Europe and USA.

This 1st volume of classical book *Fluvial Processes: Theory and Applications* by Roman S. Chalov presents overview on the hierarchy and classification of the channel patterns. The book is based on the impressive empirical datasets on the most of large Russian rivers which have collected by the author since 1957 and are related to quantitative measures of channel patterns, channel evolution, sediment transport and adjustments of channel form, which are almost unknown to international reader. The book contains variety of classifications and illustration of various theoretical and field-based descriptions of river channels insights. Various hydrogeomorphological classification procedures have gained significant application in the river management arena characterization. In particular, the book in some parts relies on Ph.D. dissertations prepared by 39 Ph.D.-students supervised by Roman S. Chalov.

The book is dedicated to the memory of Prof. Nikolay I. Makkaveev—author's teacher and supervisor. The author thanks his colleagues Lomonosov Moscow State University, soil erosion and channel processes laboratory named after Prof. Makkaveev and Hydrology Department. The Russian version of the texts was carefully read by Prof. Nikolay Alexeevsky and Dr. Sergey Chalov. The preparation of the English version of the book was assisted by Vera Golovleva and coordinated by Sergey Chalov.

The author in particular thanks Springer Earth Sciences, Geography and Environment Editorial, and Dr. Alexis Vizcaino for helping with English-language translation, check, and tremendous support for publication of the book.

Moscow, Russia Roman S. Chalov

Contents

About the Author

Roman S. Chalov is the world's leading scientist in fluvial geomorphology and river fluvial processes.

He developed the widely recognized theory of channel processes and gave detailed description of channel patterns in various environments. His scientific activities have established integrated approach to study erosion and sedimentation. Chalov's system of stream classification and inventory which propose a unique system for linking variety of in-channel processes, bed forms, and physical and ecological processes was widely accepted by scientific and engineering communities. He has also conducted pioneer studies in a few scientific fields. His series of cornerstone paper has made a major contribution to crystallizing the modern concept and the strategy of the fluvial processes.

Results of research were widely used for solving a number of scientific, engineering, and social tasks in Russia and former Soviet Union, in particular for bank protection and hydrological hazards mitigation at largest rivers.

He graduated from the Faculty of Geography, Moscow State University in 1961. Doctor of Geographical Sciences (1978), Professor (1984). He served as a professor of the Hydrology Department, Faculty of Geography, Moscow State University . Head of the Makkaveev's Laboratory (institute) of Soil Erosion and Fluvial Processes (Faculty of Geography, Moscow State University).

His research interests are fluvial processes, sediment transport, river channel and erosion, river hydrology and hydroecology.

He was member of the WASER Council, Vice-president of the Russian Academy of the Water Resources Research Problems, corresponding member of the European Society for Soil Conservation (ESSC), and president of the Interinstitutional Scientific Coordinational Council on Erosion, River Channel and Mouth Processes Research Problems.

He supervised over 39 Ph.D. students who successfully defended Ph.d. thesis, among them are scientists from Russia, Ukraine, China, Mongolia. He has over two hundred academic descendants.

Chapter 1
Key Points and Development of Theory of Fluvial Processes

1.1 General and Geographical Studies: Subject of Study and the State-of-the Art in Science System

The scientific and technological progress of the second half of the twentieth century in the field of natural sciences is accompanied, firstly, by the further differentiation of science related to the deepening of knowledge, penetration into the essence of processes and phenomena, and the emergence of new directions of research in this regard. At the same time, "apart from the discovery of new phenomena of nature, which we cannot foresee, the main efforts of scientists are aimed at a deeper study of already discovered phenomena of nature, at solving methodological and applied problems" (Kapitsa 1977, p. 319). It is the in-depth study of natural phenomena that has become the main factor in the emergence of new branches of science with their own laws, objects and methods of research. Secondly, integration ties are being established, deepened and developed between the branches of science and scientific directions, first of all—related, but often quite distant. Such integration ties are particularly characteristic of new fields of knowledge, emerging and developing at the crossroads of science and, in essence, integrated. In this case, the emergence of a new industry may occur within a single scientific discipline, by virtue of the circumstances even narrower, but beyond its scope is conditioned by the use of research methods and achievements of related and fundamental sciences, and is accompanied by the absorption of the discipline—the ancestor, which completely or separately becomes part of its subsidiary discipline. In other words, the study and understanding of the laws of nature is accompanied, on the one hand, by an increasing differentiation of scientific knowledge, and, on the other hand, by the emergence of branches of scientific knowledge which, originating within one discipline, subsequently went beyond it through the use of methods and achievements of related sciences. The emergence of "transition of one science to another", i.e., along with differences and discreteness of separate sciences, the existing "connection, continuity" was revealed (Engels 1989, p. 218). This, in turn, has led to the

R. S. Chalov, *Fluvial Processes: Theory and Applications*,
https://doi.org/10.1007/978-3-030-66183-0_1

fact that some subjects or phenomena in nature have become the object of study of several different in nature sciences (for example, technical and natural sciences), or separate sections of science have become relatively independent disciplines with their own specific objectives and methods of research.

Something similar happened with the emergence and development of the doctrine of river channel, fluvial processes—riverbeds science.

Fluvial processes are the final link in hydrological processes and phenomena associated with the impact of constant water flows on the land surface. At the same time, water flows are the "motor spring" of fluvial processes (as figuratively expressed by Leliavsky (1904)), which is an active component in the process of interaction between the flow and the channel. From this point of view, the doctrine of fluvial processes is naturally part of the hydrological science cycle, being a special section of river hydrology (Mikhailov and Dobrovolsky 1991). It was as a branch of hydrology that N. I. Makkaveev considered the doctrine of fluvial processes (1955).

On the other hand, the impact of runoff on the land surface is one of the leading relief-forming processes. The result of this impact is the riverbed itself, the floodplain of the river, which is a derivative of the fluvial processes, and its relief. Due to channel changes, alluvium is accumulated and re-deposited, the floodplain is transformed into a floodplain terrace and, ultimately, the formation of the river valley itself. In this respect, fluvial processes are the most important geomorphological process, and from this point of view fluvial processes are the subject of study of geomorphology. N. E. Kondratyev (Kondratyev et al. 1959, p. 7) noted in this connection that "the fluvial process, taken as a whole, has its root cause of hydrological factors and in its final representation has a geomorphological character". In addition, as N. I. Makkaveev emphasized (1955, p. 3), "… fluvial processes cannot be considered as chains of phenomena, the development of which occurs in isolation from the geographical environment, without taking into account the specific features that characterize the landscape of the watershed. Flows and their catchment areas should be considered in close interconnection and interrelationship". The latter conclusion is particularly important, as it ultimately leads to two other principal provisions. Firstly, fluvial processes are the lower link in the chain of phenomena associated with the effects of runoff on the earth's surface: soil erosion (washout) caused by the activity of non-channel temporary streamflows flowing along the surface of slopes—gully erosion, produced by temporary flows in ravines and gullies—fluvial processes that develop in rivers. This ratio is the main content of the doctrine of a single erosion-accumulation process. Secondly, the fluvial processes develop under the influence of natural conditions, and the forms of their representation and the regime of reformation reflect a combination of physical-geographical and anthropogenic factors. Both of these provisions form the basis of the geographical direction in the study of fluvial processes, the main ideas of which were outlined in the classic work of N. I. Makkaveev "The river channel and erosion in its basin" (1955).

However, the stream and its structure represent an object of study for hydrodynamics (or river hydraulics in the case of natural water flow in the river channel).

Hence, the conclusion of Grishanin (1972, p. 4) that the theory of fluvial processes "reasonably combines geomorphological and hydrodynamic methods". This fully corresponds to the ideas about the mutual control of the flow and channel expressed by Velikanov (1949) and reflected in the law of limitation of Velikanov-Makkaveev's morphological complexes. Therefore, the study of riverbed development processes organically combines the study of the forms of channel relief and the channel itself, on the one hand, and the mechanisms of the influence of the flow, its structure on them on the other hand.

As a branch of knowledge, the doctrine of fluvial processes originated in the end of the nineteenth century, but it finally took shape in the middle of the third quarter of the twentieth century. However, for a long time the theory of fluvial processes has developed mainly as an applied science, a part of engineering hydrology, considering specific hydrotechnical problems. At the same time, having reached the fundamental level, it was considered as a division of the channel flows dynamics (Velikanov 1954, 1955; Karaushev 1960; Goncharov 1962; Grishanin 1969). As a natural-scientific (geographical) direction, the study of river channels for a long time was limited to the accumulation and systematization of actual material on morphology and dynamics of river channels. Only the synthesis of both directions—engineering (hydrodynamic) and geographical, the depth of study of problems associated with *channel processes*, theoretical generalizations and the selection of specific channel issues when solving applied problems led to the fact that at the end of the XX century it became possible to talk about an independent scientific discipline—learning about *channel* processes (this, in particular, is evidenced by numerous monographs and textbooks on *channel processes*, published since the 50s of XX century), the branch of knowledge, formed on the brink of continental hydrology, geomorphology, engineering hydrology and river hydraulics (fluvial dynamics). At the same time, a number of problems of fluvial dynamics were considered as part of the theory of fluvial processes. Essentially, this has also been recognized by those researchers who have remained in the position of considering fluvial processes within the framework of fluvial dynamics. In particular, Grishanin (1972, p. 6) wrote that "at a certain stage it became possible to speak about the occurrence of the theory of the fluvial process", which "is closely connected (not is part of it—*R.Ch.*) with the fluvial dynamics, but differs from it both by the methods used and by the content".

Until recently, there was no common name for a scientific discipline dealing with the study of fluvial processes. Usually in literature there are phrases "the theory of fluvial processes", "teaching about fluvial processes". The presence of the object (subject) of study, goals, objectives, and research methods (Chalov 1979; Kondratyev et al. 1982; Makkaveev and Chalov 1986; Baryshnikov and Popov 1988) allowed not only to state the fact of the existence of an independent branch of knowledge engaged in the study of fluvial processes, morphology and dynamics of river channels, but also to offer it a name (Chalov 1992)—riverbed studies (similar to those that have arisen in recent decades, erosion, mudflow, avalanche). Thus, as the branch of knowledge which is engaged in studying of one of the natural phenomena and its communications with natural and social and economic conditions, channel science is a geographical science (geographical channel science). The solution of applied aspects related to channel regulation and fluvial processes accounting in water management construction and design refers to the field of engineering channel science, which is included in the cycle of technical disciplines (Fig. 1.1).

In addition, it should be added that theory of fluvial processes (riverbed science) as a science has a system of laws, the totality of which in the most general way describes the phenomena studied by it. The most important of them were formulated by Makkaveev (1955, 1976), Velikanov (1958). Kondratyev (Kondratyev et al. 1959, 1982) and have already become a common property of scholars. This is evidenced, in particular, by the first experience of their systematization in a single list (Mirtskhulava and Snishchenko 1986; Chalov 1988).

Thus, riverbed science is a branch of knowledge that studies the conditions and processes of riverbed formation and develops methods and techniques for their regulation. The subject of the study of geographical riverbed science is the river channel and the processes that occur in it during the interaction of the flow with the soils that compose it—erosion, transportation and accumulation of sediment. The second part of the definition is the development of methods and techniques of riverbed control related to riverbed engineering. Both sections of riverbed science

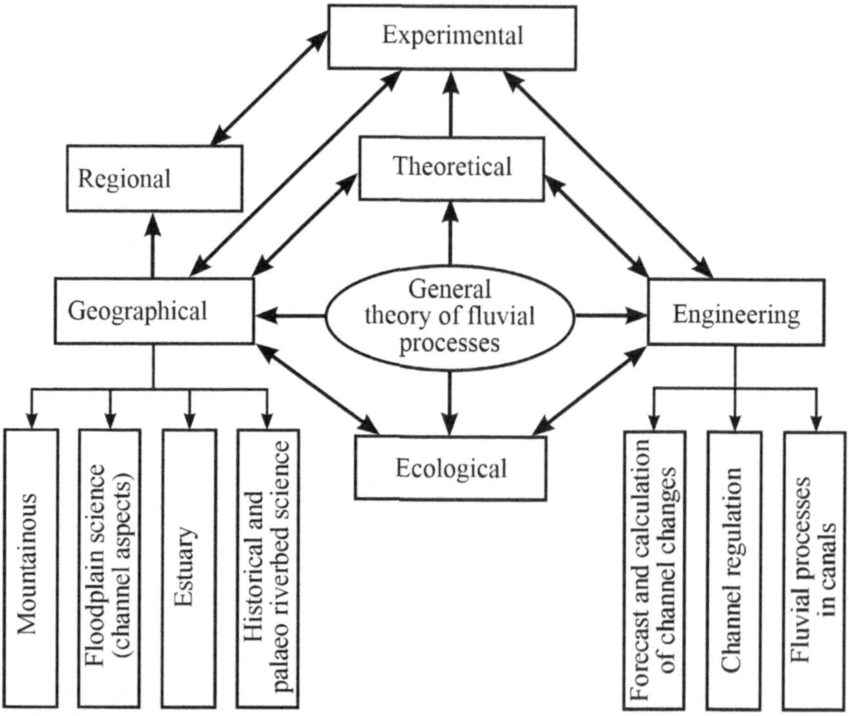

Fig. 1.1 Structure of the riverbed science

complement each other. Their interaction ensures the establishment of links between the river channel and the environment, the development of forecasts of its changes under the influence of economic activity, the identification of irreversible changes in the development of fluvial processes and their adverse effects on human life and activity, the justification of ways to optimize the measures to regulate the channels. The latter is already an object of ecological riverbed science, which, according to its objectives, considers the environmental aspects of fluvial processes —in their natural development, with changes in their direction and character as a result of economic activity, as well as in the case of such transformations of the river channel, which lead to the unfavorable condition of the rivers themselves, adjacent territories and river ecosystems (Berkovich et al. 2000).

Such interrelation of geographical, engineering and ecological riverbed science is determined by the fact that the doctrine of fluvial processes from the very beginning of its origin has developed in close contact with the production, when solving its specific problems. At the same time, the identified regularities were constantly being clarified, and the engineering applications made it possible to immediately carry out their inspection in practice. Lenin (1986, p. 195) wrote: "Practice is higher than (theoretical) cognition, because it has not only the dignity of universality, but also immediate reality". This, in turn, determined the rapid progress in the development of the fundamental positions of the young science sector. Even if we take only one geomorphological part of the doctrine of fluvial processes, which studies the mechanism of formation of channels as forms of relief, it is the interconnection with practice, according to Makkaveev (1960, p. 207), that has led to the fact that "now (mid-twentieth century—*R.Ch.*) is the most advanced branch of geomorphology in terms of the depth of study of the physical side of the process of formation and development of elementary forms of relief".

The content of science is always wider than the content of the object under study (Liamín 1986). Therefore, riverbed studies are not limited to the study of the riverbed alone. Including knowledge not only about its own object, but also its links with objects of other sciences, partially absorbing them, channel science relies on those sections of hydraulics, hydrology, geomorphology, soil science, hydrotechnics, engineering geology, which allow to reveal the mechanism of fluvial processes, the physical basis of channel changes, the nature of phenomena arising from the interaction of flow and channel, erosion, transportation and accumulation of sediment, as well as the interconnection with the natural environment as a whole and its individual components. In this regard, physical geography is the methodological basis for riverbed science, which allows the study of fluvial processes as a geographical phenomenon, and riverbeds—as elements of the landscape. On the other hand, the fluvial processes themselves are one of the landscape-forming factors both in the channel and within the valley bottom under their influence.

The most important section of the geographical riverbed science is regional riverbed science. It studies zonal, regional (basin) and local peculiarities of channel processes (in other words, representations in different natural conditions), substantiates regional schemes of fluvial processes development and along the entire length of the rivers, reveals interrelations with the processes in their basins, studies and describes the channels of specific rivers, develops scientific foundations of regional systems of channel regulation and fluvial processes management. The latter applies equally to the engineering field. Therefore, the synthesis of geographical and engineering approaches to solving specific tasks in the regional approach creates the conditions for ensuring environmental safety to the greatest extent. At the same time, if the problems of channel processes management to a greater extent now represent a task for the near future, the development of regional schemes of channel regulation in the last decades of the XX century in Russia has already acquired quite visible outlines. The founder of such an engineering and geographical approach was Makkaveev (1949), who was the first to link the solution of a specific engineering problem—the tracing of slots at the crossings of navigable rivers—with their channel regime.

One of the leading methods of regional riverbed management is the mapping of riverbed processes. The application of cartographic survey methods to the study of individual river sections, which are generally morphologically homogeneous (large scale), specific rivers over a significant distance (medium scale) and rivers within large regions or basins (small scale) makes it possible to identify at different levels of channel formation conditions, distribution of channel forms and channel relief, and to establish links between factors and forms of representation of fluvial processes. The greatest achievement in this respect was the creation of the first in the world practice small-scale maps of fluvial processes—"Fluvial processes on the rivers of the USSR" in scale 1:4,000,000 (1990), "Fluvial processes on the rivers of Altai Region" (scale 1:1,000,000) (1991), "Morphology and dynamics of river channels of the European part of Russia and neighboring countries" (scale 1:2,000,000) (1999), as well as special maps in the "Ecological Atlas of Russia" (2002).

The basis of general riverbed science is its theoretical direction, which develops systems of basic ideas and ideas about the regularities and essential links in the interaction of the flow and channel, determining the mechanisms of fluvial processes, the formation of river channels and channel relief and their relationship with natural and anthropogenic factors. At the same time, the theoretical coursework is to a greater or lesser extent accompanied by the development of mathematical models of channel processes. At the same time, natural material, which is usually the basis of the models, gives the researcher the results of fluvial processes in the form of various forms of channel and channel relief. About it, as a matter of fact, spoke Kondratyev (Kondratyev et al. 1959), specifying on geomorphological character of representation of fluvial processes. The task of theoretical riverbed science is to restore the mechanism of channel development by time slices, in which its certain state is fixed. In these conditions, the function of continuous monitoring of the development of processes and mechanisms of interaction between flow and channel in almost any scale of time, providing a deep penetration into the physical essence of the processes that determine the patterns of formation of river channels, takes on itself experimental riverbed studies. Its peculiarity is the study of channel processes on physical models, development of principles and methods of their modeling. At the same time, depending on the tasks to be solved, it gravitates towards geographical (free modeling method (Makkaveev et al. 1961) or engineering channel science. In the first case, the most general features of the formation of river channels are revealed, while in the second case the process is studied under strictly defined conditions on a model that often simulates a specific object on a certain scale, including the artificial impact on it.

The same role is played by computer modeling of channel processes, which allows imitating the flow, water and sediment yield, processes of riverbed reformation under given conditions, which determine them. It is based on the solution of Saint-Venant equations of motion and continuity together with the equation of changes (Saveliev and Zaitsev 2004).

Within the framework of geographical riverbed science, an important place is occupied by historical riverbed science and palaeoriverbed science, which, firstly, study the history of riverbed development in the process of environmental evolution, and, secondly, the changes taking place in the riverbeds and their relief at the historical stage, starting from the XVIII century (for the Russian rivers), when the first reliable documentary (mainly cartographic) information about the riverbeds appeared, including under the influence of economic activity. Palaeoriverbed science not only relies on palaeo-geographical methods of research, but also aims to explain the formation of river valleys and terraces over geological time intervals, alluvial deposits and their textures, to reconstruct climate change and, consequently, the palaeo-geographical situation through the interaction between the parameters of channels and the characteristics of water flow and sediment load. The problem of alluvial deposits formation as a product of river fluvial processes driver activity calls up with it.

Indeed, it is impossible to understand correctly the history of river valley development, terraces and floodplains formation, alluvial thick structure without knowledge of fluvial processes in all their representations: from the formation of ripples and local periodical lateral channel changes (development of bends and other forms of channel) to the directed general transformations of the longitudinal profile of the river. Otherwise, the researcher, dealing with the final result of the geological development of fluvial processes, creates speculative schemes devoid of physical basis. In this respect, Engels (1969, p. 30) pointed out that "in theoretical natural science, it is not possible to construct connections and enter them into facts, but rather to extract them from the facts and, finding, prove them as far as possible, experimentally".

The channel forms, being a conservative element of the "stream-channel" system, are the result not only of modern processes of interaction between the stream and the river bed, reflecting the physical and geographical conditions in the catchment area, but also of its historical development. Jasmund (1911) wrote in this connection: "Whoever wants to artificially promote the development of the flow, must know and understand, along with the modern forces of action, also the preceding development. Each flow has its own shape, and its uniqueness is determined by the geological development of its basin". In other words, "… cognition of fluvial processes is impossible without analysis of the history of each particular object, without taking into account the primary relief and those changes in fluvial processes driver conditions and channel forms, which the river has experienced at various stages of evolution" (Makkaveev 1955, p. 138).

The second part—historical riverbed science—is based on cartographic, archival and other sources, the analysis and comparison of which provides an opportunity to get information about the perennial reformation of river channels. Exactly in this area there is the most common, most applicable, and perhaps the most reliable method of the forecasting of channel changes. The latter is the one that is most closely associated with the ecological riverbed science.

Mining riverbed science, such specific section as "fluvial processes in estuaries" (estuary riverbed science or riverbed science of estuaries), floodplain science can be singled out by the objects of research. The first is due to the fact that the channels of mountain rivers constitute a special class of river channels, the formation of which is associated with a well-defined flow kinematics, specific conditions and forms of sediment transportation and, ultimately, special representations of fluvial processes (Chalov 1979, 1992). No less specific is the object of study of the wellhead channel science. In contrast to the estuary hydrology, which studies estuary processes based on the interaction and movement of river and sea waters with different hydrodynamic, physicochemical and biological properties, sedimentation and re-deposition of river and partially sea sediments (Mikhailov 1997a, b), the estuary riverbed science is focused on the formation and morphology of river channels in estuary regions and on the seashore under the influence of both river and sea factors.

Floodplain science, an integral part of the teachings about fluvial processes, is engaged in studying the processes of formation and evolution of river floodplains in the course of channel changes, the relief of floodplains and their geological structure, the hydrological regime of floodplains, the influence of floodplains on the development of river channels. At the same time, floodplain science, as well as the doctrine about river mouths, studies a wider range of issues beyond the scope of riverbed science (for example, formation of soil and vegetation cover on floodplains belonging to the field of soil science or geobotany, hydraulics of floodplains in the channels—the field of hydrodynamics, etc.). Not without reason the pioneers in the study of floodplains as a derivative of channel processes were geobotanists A. M. Dmitriev and R. A. Elenevsky, soil scientist V. R. Williams, geologist E. V. Schanzer; at present, hydraulics and hydrology of floodplains are successfully developing (Baryshnikov 1984).

The special sections of riverbed studies also include the "longitudinal profile theory" and the doctrine of riverbed movement, and "fluvial processes in artificial channels" (the latter already applies to the field of riverbed engineering).

In general, the objects of research of geographical riverbed science are: morphology of river channels and their relief; channel changes (vertical and lateral); factors of fluvial processes; erosion, transportation and accumulation of sediments in river channels; fluvial processes on specific rivers in different natural and natural-anthropogenic conditions; floodplains of rivers as derivatives of fluvial processes and their role in the development of river channels; zonal, regional and local representations of fluvial processes; specific forms of representation of fluvial processes (rocky channels, in conditions of active mudflow activity, in river mouths, in river junctions); longitudinal profiles of rivers and their transformation; channel processes as a component of erosion and channel systems; hazardousous representations of fluvial processes; siltation and degradation of small rivers; paleo and historical reconstruction of river channels; predictive assessments of channel changes in the event of fluctuations in the wave and global changes in the environment and climate.

In riverbed engineering, along with the problems of channel control and fluvial processes accounting in water management construction and design, a significant role is played by forecasts and calculations of channel changes, similar to hydrological calculations and forecasts in engineering hydrology. All these issues are solved on the basis of a wide natural science base, i.e. on the basis of general and regional riverbed management. Indeed, the effectiveness of channel control measures and, even more so, the management of fluvial processes, as well as the stability and reliability of engineering structures on the banks and in river channels, largely depend on the extent to which they correspond to the channel regime of the river, i.e. take into account the specifics of fluvial processes in a particular natural environment. Thus, there are relations between the main divisions of the field of riverbed studies, which determine the unity of the entire scientific discipline, and only in setting and solving problems can they be attributed to a particular (geographical or technical) cycle of disciplines.

Practical orientation (channel control) of the fluvial processes teachings predetermined its links with hydraulic engineering. But if hydro engineering helps to formulate and solve in general terms the problems of application of channel process theory to solving water management problems, then hydraulic engineering, which is engaged in the design, construction and operation of hydraulic structures, directly uses the results of channel research. In recent decades, methods of riverbed science have also penetrated into other applied fields of knowledge—prospecting geology in connection with the search and exploration of alluvial placers, water protection from pollution, agricultural land reclamation, etc. In this regard, it is more correct to talk about engineering and applied riverbed science as part of the general science of riverbeds.

Thus, engineering and applied riverbed science deals with the problems of river channel regulation and management of fluvial processes, development of methods of river recultivation and restoration, forecasts and calculations of channel changes in various types of river operation, use of water and river mineral resources, fluvial processes in canals and reclamation systems, taking into account fluvial processes for various sectors of the economy associated with rivers, issues of land reclamation and using of river floodplains, search and exploration of alluvial placers.

Thereafter, ecological riverbed science studies fluvial processes as a factor of ecological state of rivers and riverine territories, the effects of anthropogenic impact on fluvial processes and following changings of river channels, channel changing forecasts, optimization of economic regulation activities, assessment of vulnerability of channels to anthropogenic impact, assessment of secondary pollution of river waters with sediment erosion, and, finally, development of forecasts of riverbed processes upon termination, reduction or changes in anthropogenic loads.

Having originated and developed in the depths of hydrology and geomorphology, riverbed science naturally has the closest links with these two geographical sciences (Fig. 1.2). Due to the specifics of the object of research—the interaction between the flow and the channel (or soils that compose it)—it relies on river hydraulics and hydrophysics, on the one hand, and soil science and engineering geology, on the other. Finally, since erosion, transportation and accumulation of sediment are at the heart of riverbed processes, the doctrine of river load is the foundation upon which riverbed science rests. On the other hand, this doctrine is itself an integral part of riverbed science, since "channel phenomena ..., channel forms can be considered ... as one of the forms of transporting solid matter by flowing water" (Makkaveev 1955, p. 139), or that the same thing, "channel process is a form of transportation of sediment formed over the entire catchment area of the river" (Baryshnikov and Popov 1988, p. 5). The latter thesis determines the development of riverbed science in close contact with erosion science, forming a single doctrine on erosion and accumulation processes; soil erosion on slopes and ravine erosion are the most important sources of formation of river load.

1.2 Dynamics of Channel Flows and Its Connection with Fluvial Processes

Makkaveev (1955, p. 139) wrote: "Studies of the processes of channel relief development cannot be detached from the study of the flow structure", thus determining the organic connection between the theory of fluvial processes and hydrodynamics (fluvial dynamics). The river flow and the river channel form a

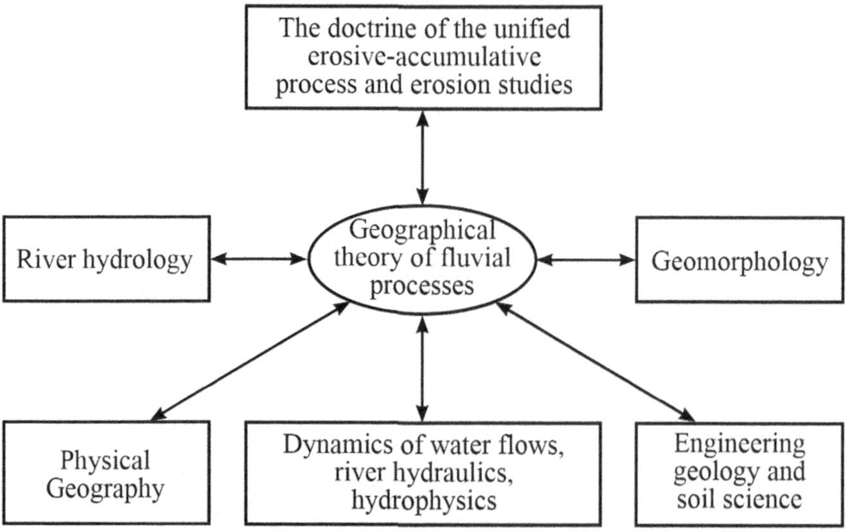

Fig. 1.2 Position of the riverbed science in a science system

single system in which the movement of various physical media takes place, the interaction of the flow—the element of the hydrosphere, which is in constant motion, with kinetic energy and power, and the soils that compose the bed of the rivers, which is an element of the lithosphere—the stationary part of the system. The latter, under the influence of the flow, are partially transformed into a new state, creating, together with the sediment flow from the catchment area into the flow of erosion products. Sediments transported by watercourses are already a special, intermediate medium between hydro- and lithosphere, in which individual elements are in different proportions both with the flow (in the suspended matter, distributing over its entire thickness, or concentrating at the bottom and making up the bedload) and with the surface of the river bottom. These ratios change over time due to seasonal fluctuations in water flow, which determines the discrete nature of both sediment transportation and its direct impact on the flow bed and the conditions under which the flow itself comes into contact with its bed. Sediment flow concentrated in the near-bottom area of the flowing water, depending on their concentration, can create a protective layer at the bottom, preventing direct contact of the water flow with the bed. At the same time, being in motion under its influence, this layer reflects the kinematic and turbulent structure of the flow, creating, depending on the size of the load, some form of accumulative relief.

In this complex multiphase interaction of different media, the active force is water flow; its presence determines both the modification of the lithogenic base (river bed) and the formation of the sediment flow. Emphasizing this fact, one of the founders of the doctrine of channel processes, N. S. Leliavsky, called his work "About river flows and formation of the river channel" (1893), emphasizing, on the one hand, the leading role of the flow in the "flow—channel" system, and, on the other hand, the continuity of interaction between the form of the river bed and its flow. Thus, the dialectical position of N. S. Leliavsky about the mutual connection of causes and consequences in the channel phenomena, in fact, became the basis for the modern theory of fluvial processes. However, the followers of N. S. Leliavsky in the first three decades of the twentieth century in their research and theoretical ideas focused on the study of only one side of the two-unit process—river currents. This led to the fact that the doctrine of fluvial processes (in its narrow sense) for a long time developed as a doctrine of channel flow, and channel processes were considered as a part of the dynamics of channel flows, which is reflected in the names and structure of a number of monographs and textbooks (Velikanov 1954, 1955; Karaushev 1960; Grishanin 1969) and still (early XXI century) retains its adherents. This point of view was defended, for example, by K. V. Grishanin in his last, unfortunately unpublished report at the 16th Planar Interuniversity Meeting on the Problem of Erosion, Fluvial and River Mouth Processes in October 2001.

On the other hand, M. A. Velikanov, who was the first to define the term "fluvial processes" already found in the literature (1946) and substantiated the position of the teachings on fluvial processes as one of the components of the fluvial dynamics (1954, 1955), being a representative of the engineering and hydrodynamic direction in their study, in his last book (1958) considered the main issues of fluvial dynamics (turbulence, flow kinematics, transverse circulation, etc.), as well as all the issues of sediment movement in the flows as sections of the general theory of channel processes. Grishanin (1972, p. 4), who wrote that "at a certain stage it became possible to talk about the emergence of the theory of the fluvial processes ... which not only generalizes, but also explains the observed phenomena", although at the same time and later in other works he subordinated it to the fluvial dynamics.

Fluvial dynamics deals with kinematics and structure of flows, their turbulence, distribution of velocities across the width and depth of river channels, transverse and whirl currents, wave phenomena in flows, water mixing, hydraulic resistances arising from the interaction of the flow with the channel, as well as ice cover, interaction of channel and floodplain flows, etc. The study of the hydrodynamics of sediment transportation has led to the development of the theory of suspended sediment transportation: diffusion (V. M. Makkaveev—A. V. Karaushev), gravity (M. A. Velikanov) and others. But they, as well as some other tasks, do not give a complete solution to the issues of fluvial processes. Thus, the dynamics of channel flows, while studying the flows, covers a much wider range of problems than it is necessary to solve the problems of formation of river channels. Therefore, the theory of fluvial processes "selects" those hydrodynamic phenomena and those sections of fluvial dynamics, which are necessary for understanding and cognition of the mechanisms of river channel formation.

Historically, the gap between engineering-hydrodynamical and natural history (morphodynamic or geomorphological) approaches to the study of fluvial processes has contributed to the emergence of schemes of riverbed development based on intuitive ideas or metaphysical views of their dynamics. This is, for example, the theory of the equilibrium profile, which is based on the idea of transforming the channel into a canal of water and sediment flow, with the default sources of their inflow, re-deposition, accumulation and outflow to the receiving basin (its critical analysis was given by Makkaveev (1955)), or exaggeration of the role of some hydrodynamic phenomena without taking into account the arising in the process of interaction between the flow and the channel of certain forms of the channel and the channel topography. The latter includes the conclusions of Losievsky (1934) about the role of transverse circulation in the flow in the formation of the channel. His ideas about the quasi-heterogeneity of the flow or about the presence of several dynamic axes in it have not lost their meaning so far. In this regard, it is necessary to recall the conclusion of N. S. Leliavsky about the mutual control of the flow and the channel, developed further by Velikanov (1958): "the flow controls the channel, the channel controls the flow" became a textbook). The same was pointed out by Makkaveev (1955, p. 137): "As soon as under the influence of this interaction (flow and channel—*R.Ch.*) some form of channel appears, the latter also becomes an important factor of fluvial processes, because it largely determines the hydraulic characteristics of the flow". Therefore, the study of riverbed development processes should organically combine the study of riverbed shape, structure and flow kinematics.

Such an approach, which was first mentioned in N. I. Makkaveev's book (1955), quickly became widespread in one form or another. Thus, Kondratyev (Kondratyev et al. 1959, p. 6) admitted that, although "the direct mechanism of interaction between the flow and the eroded bottom is quite determined by the laws of mechanics and hydrodynamics, … the possibilities of such a one-sided approach proved to be very limited". He further explained: "Channel forms arise under conditions of variable water regime, reflect the diversity of liquid and solid runoff characteristics, i.e. hydrological factors, and therefore cannot be explained by means of hydrodynamics only". Therefore, in the quoted monograph, its authors essentially consider channel processes as an object of independent research on the basis of analysis of river hydraulics and geomorphology, although it still retains an independent coexistence of the sections devoted to the fluvial processes themselves, the movement of sediments and the flow kinematics.

Later, Baryshnikov and Popov (1988), emphasizing the in relationship of channel processes as a natural phenomenon and recognizing the leading role of water flows and sediment transportation in the development of river channels, suggested that the relevant branch of knowledge should be called "the fluvial dynamics and fluvial processes". With such a name as academic discipline it was included in the curriculum of training of hydrologists, although at Moscow State University courses "Fluvial processes" and "Fluvial Dynamics" are read separately (Chalov 2002), and at RSHMU in a single course both its sections were conducted by different specialists. At the same time, the unified course program, as well as the above mentioned textbook by N. B. Baryshnikov and I. V. Popov, is designed in such a way that they include only those parts of the flow dynamics that determine the mechanisms of fluvial processes. Characteristically, at present, the Russian Hydrometeorological University has also divided a single course into two separate —"Fluvial Dynamics" and "Fluvial processes".

Snishchenko (2002, p. 7) came to the same conclusions concerning the correlation between the fluvial dynamics and fluvial processes: "While in the FD (fluvial dynamics abbreviation—*R.Ch.*) the issues of river flow and load (suspended) movement are solved with the help of the equation of hydrodynamics, it was impossible to obtain such solutions "from the position of forces" in the fluvial process. It required geomorphology and hydrology". However, his proposed hydrological structure chart considers fluvial processes (channel dynamics), along with the hydraulics of flows and the dynamics of river load, as one of the three sections of channel dynamics.

Such discrepancies are terminological, constitute the subject of discussion and are largely determined by the initial direction in the study of fluvial processes. In particular, the independent position of fluvial processes, for which the water flow is an active factor, is a "motor spring" (according to N. S. Leliavskiy), initially peculiar to the hydrologic and morphological approach to the study of phenomena related to the interaction of river flows with river channels. In the initial hydrodynamic approach to the theory of fluvial processes (even if its certain autonomy is recognized), its commonality with the dynamics of channel flows is preferred. But in any case, it is the interaction between the flow and the channel that constitutes the essence of the phenomenon itself—the fluvial processes.

1.3 Basic Concepts and Definitions

Interdisciplinary position of riverbed science and continuous parallel development of natural-historical (morphodynamic or geomorphological) and engineering (hydrodynamic) trends in the study of fluvial processes, which to a certain extent is still preserved today, the formation of riverbed science on the brink of natural and technical sciences, the comparative youth of the branch of knowledge, as well as the complexity of the object of research located and developing on the border of the division and the interaction of the two environments—hydro- and lithosphere, the first one of which is in a state of motion, with the formation of an intermediate phase—sediment flow—during this interaction, cause discrepancies in the use of certain basic concepts and terms and even the absence of a generally accepted definition of the concept of fluvial processes.

Rapid progress in the development of channel science in the second half of the twentieth century has led to the emergence of a large number of new terms, their active introduction in the scientific literature and in the recommendations for the consideration of channel processes in solving practical problems. A number of terms "migrated" from other fields of knowledge, including related ones, and their inaccurate application led to terminological confusion. At the same time, some terms became obsolete, "forgotten," but then reintroduced into the scientific vocabulary not always in the original sense.

One of the peculiarities of the development of riverbed science in Russia was also the wide application of terminology taken from the Russian folk language: the channel, the floodplain, the branch, the duct, the bend, the streamer, etc. The fact that the study of fluvial processes was stimulated by the solution of practical problems related to the development of waterways of communication, has led to the use in the theory of fluvial processes of many terms of track origin: the riffle, the riffle trough, side bar, medial bar, etc. Some of them have no analogues in English literature and sometimes it leads to ambiguous interpretation of terms in translations.

Thus, there is a problem of improvement and ordering of the riverbed science terminology. Proposing the ways of its solution, it follows from the criteria of standardization of terms, which are used in linguistics (Linguistic aspect …, 1993). Those are the criteria: (1) principle of unambiguity of the term; (2) conciseness; (3) correspondence of literal and actual meanings; (4) introduction; (5) undesirability of foreign-language borrowings; (6) internationalization; (7) regularity (i.e. the terms should be organically connected if possible); (8) linguistic correctness; (9) word forming ability; (10) simplicity and comprehensibility (memorability); (11) in relationship from context.

There are several definitions of the concept of "*fluvial processes*" used in both plural and singular. The first of them belongs to Velikanov (1946), according to which the fluvial processes consist in the continuous influence of the flow on the shape of the channel and the shape of the channel on the flow. A clearer formulation was given to them later (Velikanov 1949): the process of mutual control of the channel by flow, and flow by the channel is called fluvial. However, the term "fluvial processes" was used in scientific literature as early as in the 30s (Lvovich 1938a, b; Makkaveev and Sovetov 1940), although it was not yet defined.

Later, Velikanov (1955, p. 237) began to consider the definition given by him as the main content of fluvial processes, and the term itself formulated more narrowly: "...the whole complex of phenomena of both the initial formation of channel forms and their further changes is called fluvial process". In fact, he reduced the fluvial process to one of the forms of its representation—channel changes. This was practically acknowledged by the author himself, pointing out in the preface to the book that "the total effect of the transfer of sediment particles from one part of the channel to the other is represented in the form of channel changes, the combination of which forms a complex phenomenon of the fluvial process" (p. 5). In his last book (Velikanov 1958, p. 9), which he called "the result of a long, more than half a century of work on the study of the riverbed", M. A. Velikanov returns to his initial fundamental definition—"the interaction between the flow and the channel determines the very basis and the dynamic essence of the fluvial process" (p. 14).

The formulation given by Velikanov (1955) underlies the widespread definition of the State Hydrological Institute (SHI): "The fluvial process is understood as a modification of the morphological structure of the river channel, constantly occurring under the influence of the current water" (Kondratyev et al. 1959, p. 6) or (in the last edition): "the fluvial process is a change in the morphological structure of the river channel and the river floodplain, constantly occurring under the influence of the current water" (Kondratyev et al. 1982, p. 11). However, this limitation is compensated by the actual content of the studies (Kondratyev et al. 1959, 1982), formulated in the form of "basic postulates" of the hydromorphological theory of the fluvial process.

M. A. Velikanov's observation (1955) refers to the second most important component of fluvial processes—the movement of sediment, without which their definition is incomplete. Indeed, the essence of fluvial processes lies in the interaction between flow and channel, erosion, sediment accumulation and transportation. Water runoff is an active factor in the channel-flow system; it ensures the transportation of sediment from the catchment area to the channel, as well as through erosion of the bottom and banks of the channel itself.

Thus, in the interaction between the flow and the channel there is not just a flow, but a flow that transports the sediment. In this respect, the statements of N. I. Makkaveev and N. E. Kondratyev are indicative—the scientists who stood at the origins of the modern theory of fluvial processes, but headed different directions in their research. According to Makkaveev (1955, p. 137), fluvial processes "in the most general form can be defined as a "reflection" of the surface of a solid medium (i.e., soil composing the bed) of the peculiarities of the movement of water and its transported load". According to N. E. Kondratyev (Kondratyev et al. 1959, p. 11), "the transportation of sediment should be regarded as the content of the fluvial process and the morphological transformations as its external expression and form".

The unity of the processes of interaction between the flow and channel and the movement of sediment, in turn, determines the self-regulation of the "flow—

channel" system through the law of automatic leveling of the transport capacity (Makkaveev 1955), which represents itself in the development of the longitudinal profile of the river as a whole, in its individual sections, in the mechanism of movement of ripple forms of channel relief and the development of individual forms of the channel. Engels (1969, p. 199) wrote: "In order to understand individual phenomena, we must pull them out of the universal connection and consider them in isolation, and in this case, the successive movements appear before us—one as a reason, the other as an action". Therefore, a separate study of the processes of interaction between the flow and channel and the movement of sediment allows, under specific conditions, the identification of cause-effect relationships in the dynamics and development of the flow-bed system, i.e., the establishment of a flow control mechanism and the specificity of the reverse impact of the channel shape on the flow, on the one hand, and the determination of the impact of water runoff and sediment load on the development of the channel through changes in the transporting capacity of the flow, on the other hand.

Consequently, fluvial processes should be considered as a set of phenomena related to the interaction between the flow and the soils composing the river bed, erosion, transportation and accumulation of sediment, which determine the erosion of the river bed and banks, the development of various forms of channel beds and forms of channel relief, and the mode of their seasonal, perennial and secular changes (Makkaveev and Chalov 1986). This formulation takes into account the essence of fluvial processes (interaction of flow and channel, movement of sediment), their representations (channel and channel relief forms, channel changes), temporary variability (channel mode), and thus fully reflects the subject of study of the teachings about fluvial processes—channel science itself: fluvial processes (interaction of flow and channel; erosion, transportation and accumulation of sediment), riverbed morphology, channel changes, channel mode of the rivers.

When studying fluvial processes, such concepts as "*channel changes*" and its synonyms as "channel reshaping" and "channel reformation" are widely used to denote the forms of their representation. At the same time, even before the term "fluvial process" appeared, there was often another notion—"channel regime of rivers". These terms even appear in the titles of the books: "riverbed regime" (Makkaveev 1949), "change of river channels" (Popov 1965), although their definitions are absent in the literature. There are also other synonyms—"transformation of channels" (Faktorovich 1970), "fluvial processes driver" (Rzhanitsyn 1985), the use of which is apparently explained by the desire to avoid the unsuccessful application of the physical concept of "change". In physics (the theory of fluvial processes is based on the laws of physics) change is "a change in the position of points of a solid body, in which the distance between them changes as a result of external influences, and changes can be elastic and plastic, and in their form—stretching, compression, bending, twisting" (Soviet Encyclopedic Dictionary 1983, p. 385). In other words, the term applies to the body. In the same way, changes are treated in geology: "changes are understood as a change in the volume and shape of the body" (Koronovsky and Yakusheva 1991, p. 253).

The term "channel changes" was introduced into riverbed science literature at the beginning of the 60s, and its widespread use, apparently, began with the famous book by Popov (1965). Since then, he has received citizenship rights in scientific literature and began to be used together with the term "channel reformations". If we proceed from the above linguistic criteria, the concept of "channel changes" meets most of the requirements, including the word forming capacity (for example, "deformed channel"), while the russian version of this concept is less successful. Thus, despite the initial semantic inaccuracy of the term "channel changes", its use should be recognized as rooted and having the right to exist.

Sometimes the synonym for "channel changes" in the literature—"transformation of channels"—is more justified from the point of view of physics (Faktorovich 1970; Vexler and Donenberg 1981). In russian translation, the word "transformation" means "conversion". However, the term is less successful from the standpoint of other linguistic criteria, and therefore has not become universally accepted. At the same time, it is often used to assess channel changes in downstream waterworks, which is fair, because here, due to the regulation of water flow and sediment load, the channel is rebuilt radically, acquiring a completely different quality. Thus, the term "channel transformation" can be used to characterize channel changes in case of changes in fluvial processes.

Channel changes (changes of riverbeds) are the representation of fluvial processes. Therefore, they represent the most important object of research of riverbed science. Obviously, this is related to the frequent substitution with the term "channel changes" of a more general concept—"fluvial processes". Thus, in the "Hydrological dictionary" of Chebotarev (1964, p. 163), "channel changes" are defined as "changes in the size and position in the space of the river channel and individual channel formations, caused by the operation of the flow and associated with the re-deposition of sediment"; practically the same definition is given to the term "fluvial process": "constantly occurring changes in the morphological structure of the river channel and floodplain, caused by the action of the current water" (p. 162).

Channel changes are one of the components or forms of representation of fluvial processes. Based on the above definition of "fluvial processes" (Makkaveev and Chalov 1986), channel changes should be understood to mean changes in river channels caused by water flow, erosion, transportation and accumulation of sediment, resulting in increased or decreased bottom marks, movement of channel shapes or parts, scouring and bank build-up. With this approach, the measure of channel changes is their speed, i.e. changes in the linear, area or volume characteristics of the channel per unit of time.

To characterize the changes in time, starting with the monograph of Kondratyev et al. (1959), the adjectives "reversible" and "irreversible" are widely used. The latter term is more logical to apply when it comes to such anthropogenic impact on river channels, when under the influence of these impacts they acquire a new quality (they are transformed), including those that are not favorable for the condition of the river ecosystem. The term "reversible changes" is not successful from a dialectical point of view (in natural conditions channel changes never return to the original state), but it can be used in environmental assessments of riverbeds.

The use of the concepts of "directed" and "periodic" changes seems to be more reasonable (Chalov 1979). For the former, the sign of a change in the shape of the channel or its position in space remains unchanged over time, while the latter corresponds to the stage or cyclicality of the development of the channel and channel relief (for example, the development of the bend up to cutoff, after which a new bend begins to form, or the accumulation of sediment on the roll in one phase of the water regime, which is replaced by the erosion of the crest of the roll to another).

To assess the directionality of channel changes, their consequences can be compared with the direction of gravity (Chalov 1979). "Vertical" changes are a generic term for both the processes of river intrench and sediment accumulation at the bottom of the river valley, reflecting both directional changes in the longitudinal profile of the river and local scouring and shallowing of the channel, for example, within the limits of crossings. "Lateral" changes are synonymous (or close in meaning) with lateral erosion, which, however, determines only one side of the phenomenon—scouring the banks, while the lateral movement of the channel is accompanied by the accumulation of sediment in the other part of the channel, where the bank is formed. At the same time, lateral changes imply not only movement of the channel in space, but also through mutual control and interrelations in the flow-channel system, which also includes a change in depth (deepening of the pool cavity with the development of the bend, reshaping of branching are accompanied by deepening of one and shallowing of the other branches, etc.).

The other synonym is changes in plan. This frequently used concept reflects the change in channel position recorded when comparing multi-temporal maps (pilot, topographic, etc.), aerial and space images and channel plans based on the results of repeated surveys and measurements. In fact, the movement of the channel is accompanied by three-dimensional changes—the length of the front of erosion or accumulation, its width and height of the bank to be washed, counting from the bottom of the channel. On the other hand, the application of the word "in plan" to changes is devoid of physical meaning. The plan, which shows the riverbed, shows its position at a certain point, and only a comparison of multi-temporal plans and maps allows to judge about the channel changes. Characteristically, in related sciences (geology, geomorphology) similar changes in the position of objects are related to the lateral component of their overall changes.

The term "vertical channel changes" (Chalov 1979) was proposed instead of the term "deep channel changes". Obviously, the word taken from the concept of "deep erosion", which is widely spread in the geomorphological literature as a contradiction to "lateral erosion", is automatically transferred into it. And in both cases, the use of the word "deep" is unauthorized, because according to the "Explanatory dictionary of living Great Russian language" by Dal (1880, p. 357)—"depth—height, embroidered in the opposite sense: the extension of the plumb line from top to bottom from the surface to bottom. Deep—up to depth related", that is, deep changes are those changes that occur at depth. They can be correlated with bottom erosion, although the underwater part of the bank ledge is also being washed away. But they do not characterize the accumulation of sediment and the increase in bottom marks. In this respect, the term "vertical changes", which are carried out in the direction of gravity, vertically, is more consistent with the definition of V. I. Dal. On the other hand, it is preferable to use the term "river (flow) incision" or "bottom erosion" to denote vertical changes leading to lowering of the bottom.

Depending on the geological and geomorphological structure of the area where the rivers flow, channel changes may be free (in easily eroded rocks and wide floodplain valleys) or limited (in bed rocks, often floodplain valleys or floodplain widths, smaller channel widths). Conditions of limited development of channel changes do not coincide with the content of the concept of "limiting factors of fluvial processes", which is one of the most important in the hydromorphological theory of SGI (Kondratyev et al. 1959, 1982; Popov 1965; Baryshnikov and Popov 1988). These factors include "the general basis of erosion, the local basis of erosion, and exits in the channel and on the banks of non-destructible rocks" (Kondratyev et al. 1982, p. 12). If fluvial processes are identified with channel changes, this approach is quite reasonable—under certain geological conditions channel changes are limited; if we proceed from the definition of "fluvial processes" as a result of the interaction between the flow and the channel, and the movement of sediments, then they cannot be limited in the presence of the flow, only the effect of interaction changes.

The development of certain channel changes is determined by the combination of factors of fluvial processes and their variability; therefore, they are carried out in different ways in a particular natural environment. This difference determines the *channel regime of the rivers* as part of their hydrological regime. At the same time, the hydrological regime of the rivers itself acts as a leading factor in fluvial processes, as the flow is an active component of the "flow-channel" system, and its seasonal, long-term and secular changes cause constant transformation of the system itself. In other words, the features of the hydrological regime determine the specifics of the river channel regime in different natural conditions. The hydrological regime is "a set of characteristic changes in the condition of water bodies over time, mainly due to climatic features of the basin. The hydrological regime is represented in the form of perennial, seasonal and daily fluctuations of water level and flow rates, ice phenomena, water temperature, amount and composition of solid material transported by the flow, composition and concentration of dissolved substances, etc." (Geographical … 1988, p. 68). By analogy with this definition, the river channel regime is a set of characteristic changes in river channels under the influence of water flow over time. Accordingly, these changes can be seasonal, perennial, secular, etc.

To see the difference between the "channel mode" (or its type) and "channel changes" it is necessary to remind the statement of Makkaveev (1949, p. 15): "Flows with the same hydraulic characteristics can create different channel forms in different natural conditions and, in turn, the same channel formed can arise under the influence of completely different hydraulic processes", i.e. in case of morphological similarity of channel forms they can have a different mode of transformation. Therefore, it is necessary to speak about the types of channel regimes, considering them as one of the characteristic changes in the river conditions.

The term "riverbed regime", which was widely used in the 30s of XX century, was later undeservedly forgotten. Further progress in the development of geographical riverbed science, identification of zonal features and regional specifics of fluvial processes are possible only by establishing the types of channel regimes and their relations in space. At the same time, the general features of the channel regime of each river as a whole and at different structural levels—morphologically homogeneous sections, channel forms and their complexes (or series), crossings and other forms of channel relief—should be identified. This combination will allow not only to create a complete picture of the geography of fluvial processes, but also to develop regional schemes of channel control, accounting of fluvial processes in all the diversity of their representation in the economic use of rivers.

The revival of the term in its original meaning began with the publication of the book "The riverbed regime of Northern Eurasia" (1994). Earlier it was used by Rzhanitsyn (1985), giving it a completely different interpretation, understanding under it the types of lateral channel changes, to which "delta processes" and "deep incision" were added. It's obviously confusing terminology.

Naturally, the channel regime of rivers cannot be considered separately from channel changes: changes in channels over time due to fluctuations in the runoff rates of a river, the inflow of sediment into it and the influence of other factors, including economic activity, are superimposed on the change patterns inherent in a particular type of channel.

In the given definitions—fluvial processes, channel changes, channel mode of rivers—the *form of the channel* is understood as its morphology with its characteristic outlines in terms of plan (meandering; branched out into branches; relatively straight, single), width and depth. A riverbed is a form of relief of the river valley bottom, which is used for river runoff (water and sediment load, dissolved substances, etc.) into the low-water (in presence of a floodplain) phase of the water regime. Its formation of the channel is carried out as a result of erosion (washing out) effect of water flow on the underlying surface, movement and accumulation of sediments. Under certain conditions (when a river flows into a receiving basin—a sea, a lake, a water reservoir), the formation of a channel is associated solely with the accumulation of sediment at its flow rate and a decrease in its velocity.

During floods, the river flows out of its banks and floods *the floodplain of the river*, which is a lowland part of the river valley covered with vegetation, due to the origin of the river's erosion and accumulation activities, i.e. fluvial processes. Velikanov (1958) called the floodplain a "big channel", which allows all the increased water consumption to pass, in contrast to the "small channel", which is limited by the floodplain's edges. In other words, the floodplain, on the one hand, is a consequence of channel changes, and, on the other hand, an important condition for the formation of the channel itself. That's why the floodplain is one of the objects of research of riverbed science.

Channel topography forms arise as a consequence of the ripple movement of sediment, as well as due to local increases in the main flow bed, devoid of alluvium. The first ones are riffles, side bars, medial bars, and other accumulative formations —ripples of different sizes that determine changes in bottom markings, including time due to their movement, erosion, and neoplasms during the transport of bottom sediments. Sculptural forms of channel relief are rapids, longitudinal bends forming subvertical ledges with waterfalls, shivers and other outcrops of bed rocks that compose the river bed.

The relation of the channel to the straight, meandering and branched out channel corresponds to the separation of *channel patterns* (Chalov 1979). The definition of "channel pattern" reflects the unity of the channel shape (morphology), its lateral changes (dynamics of channel shapes), and the associated local changes in bottom marks (erosion and accumulation of sediment), which reflect the kinematic structure of the flow, as well as the peculiarities of its transport of sediment. Velikanov (1958, p. 28) noted that "each of … types of channel formation is a special form of channel, and it corresponds to a special type of the main flow". In this case, the concept of the "type of channel" characterizes the individual forms of the channel (straight, single-thread, bendy, branching) or a set of them (a series of bends, successive branches, straight area), which have a morphological similarity and a similar type of lateral channel changes.

In spatial terms (along the length of the river), extended areas within which the channel forms of a given pattern prevail and others do not have a continuous distribution, creating, as a rule, single formations, are called morphologically homogeneous (Popov 1965). In other words, *morphologically homogeneous areas* are a set of homogeneous channel forms characterized by the same morphology and changes. In most cases, they correspond to homogeneous local conditions of channel formation (river runoff rates, sediment load, geologically-geomorphological structure of the valley). Changes in these conditions (e.g., change in valley expansion by contraction or vice versa; merger with a large inflow, etc.) are accompanied by a change in the channel pattern or an increase (decrease) in its morphometric characteristics.

The concepts of "channel type" and "type of fluvial processes" have different meanings, since the latter defines a variety of mechanisms of interaction between the flow and channel, forms of transport of bed material load, determined by the kinematics of the flow. In the known classification of fluvial processes of State Hydrological Institute (SGI) (Kondratyev et al. 1982), including different types of channel meandering, channel anabranching and types of channels that differ in the form of movement of large ripple formations—side bar, medial bar, ripple-belt, these concepts are combined, which is associated with the same interpretation of the terms "fluvial processes" given by Kondratyev (Kondratyev et al. 1959, 1982), and "channel changes", placed in the "Hydrological dictionary" of Chebotarev (1964). In essence, it was recognized by one of its authors—I. V. Popov, who, describing the "classification of the fluvial process of SGI" also called it "classification of types of changes of river channels" (Baryshnikov and Popov 1988, pp. 278–279).

Thus, if we proceed from the main "velikanov`s" definition of the interaction between the flow and the channel, as well as the most important refinement related to the role of sediment transport (Makkaveev 1955; Kondratyev et al. 1959), the classification of fluvial processes should reflect the diversity of forms of sediment transport and the mechanisms of interaction between the flow and the channel caused by the unequal kinematics of the flow.

1.4 Dual Nature of Fluvial Processes

The essence of the fluvial processes, as shown above, is the interaction of the flow and the soils composing the riverbed, the mechanism of this interaction. According to Velikanov (1955, p. 237), "specific for the channel flow is the interaction between the velocity field formed under the influence of the boundary surfaces existing at the moment, and the channel, which shape is composed, in turn, under the influence of the velocity field. This interaction ultimately leads to the formation of special channel forms on the earth's surface". The quoted statement reflects the hydromechanical nature of fluvial processes: hydrodynamic features of flows through interaction with the river bed are reflected in its morphology, lead to bends of the channel, its branching, the formation of ripples of different sizes; the smaller the material (in case of its inconsistency) the channel is folded, the more the hydraulic structure of the flow is reflected in its structure. In particular, on sandy rivers, ripple forms of channel relief are developing, forming a hierarchy of ripples from the largest, comparable to the size (width, depth) of the flow and corresponding to the presence in it of the largest turbulent eddies, to the smallest, incommensurable with the parameters of the channel (Mikhailova 1966; Znamenskaya 1968).

Created under the influence of hydrodynamic phenomena channel forms have the opposite effect on the flow, causing the formation of inherent hydraulic disturbances in these forms. These are the circulation currents that occur at the bend of the channel or the eddy current—the roll in the basement of the ripple. Their appearance already as derivatives of the influence of the channel on the flow determines the dynamic stability of the channel forms themselves. "Channel forms are relatively stable and, consequently, typical if they excite processes that contribute to their renewal" (Makkaveev 1976, p. 12).

The enlargement of the particle size of the river bed leads to two possible discrepancies between the dynamic structure of the flow and the development of the hierarchy of channel formations. First of all, the size of turbulent eddies in the flow may be smaller than the diameter (particle size) of the particles. However, Mikhailova (1966) claims that the dimensions of the ripples (sand waves, in her terminology) at the bottom of the flow are determined only by the hydraulic characteristics of the flow and do not depend on the size of the sediment particles. In real life, rivers carry gravel, pebbles and even boulders, depending on the hydraulic characteristics of the flows. The diameter of these particles is more than

1 cm and reaches 100 cm and more, which exceeds the parameters of low-ranking eddy structures in the flows. Secondly, the increase in the size of the load is accompanied (on real rivers) by an increase in the heterogeneity of their composition and a decrease in the sorting of the load. This leads to the formation of alluvial detachment at the bottom of the flow, which is analogous to the indelible bottom. The immobility of particles lying on the bottom of the flow at its unconnected structure also makes it impossible to reflect the dynamic structure of the flow in the form of a channel, at least in the dry phase of the regime.

The impact of the flow to the bottom is characterized by frontal pressure on the particles that compose it, and the emergence of the lifting force associated with the vertical gradient of speed and pressure in the bottom area, turbulent ripple velocity. As a result, the particles break away from the bottom and move, forming a stream of bedded or suspended sediment. The mobility of particles is the reason for the formation of the bed ripple relief and channel forms corresponding to the velocity field, dynamic structures and circulation currents in the flow. Therefore, the enlargement of the composition of sediments leads to an increasingly lower probability of "mapping", according to Makkaveev (1955), the surface of the bottom (i.e., the soils that compose the river bed) of the features of the hydraulic structure of the flow.

Very fine sediment composition in the bed, on the contrary, results in rapid particle suspending. The water flow is saturated with them until the water flow itself changes in quality, until it becomes mudflow. This is usually the case for mountain rivers or ravine flows, where large quantities of loose material are collected from the catchment area at high flow rates. However, even on lowland rivers, the sediment content of the flow may exceed dozens of kilograms per 1 m^3: at the Huanghe River, after crossing the Loess Plateau, the average annual sediment concentration is 35.8 kg/m^3 with a maximum of 405 kg/m^3 (Chalov et al. 2000). Under these conditions, the channel shape reflects not so much the dynamic structure of the flow as the accumulation of sediment with a decrease in the hydraulic characteristics of the flow and a decrease in its runoff rates, although the location of the accumulation of sediment in the channel depends on the structure of the currents in this phase of the water regime.

Thus, the interaction between the flow and the channel from hydromechanical positions can be determined by the expression most clearly formulated by Velikanov (1958, p. 76): "the channel controls the flow, and the flow controls the channel" and Grishanin (1972, p. 151): "the channel $\leftrightarrow$ flow, i.e. the flow affects the channel, and the channel has the opposite effect on the flow" (Fig. 1.3). In the course of this interaction, a flow of sediment emerges which, on the one hand, influences the hydraulic characteristics of the flow, and, on the other hand, determines the development of the channel through channel change, influencing the development of the channel relief and the shape of the channel. Since part of the load flow (sometimes significant) is formed in the catchment area by soil erosion, gully erosion, and other denudation processes, this interaction is related to the physical and geographical characteristics of the river basins.

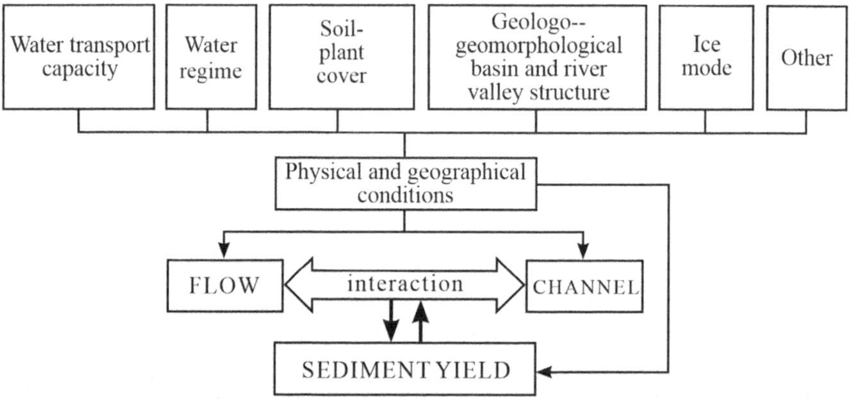

Fig. 1.3 Interactions between flow and river channels and associated environmental impacts

All of the above approaches to the interaction between the flow and the channel were based on the assumption that the channel is formed in unconnected, mainly alluvial or similar sediments. This thesis is represented in the recognition of "mobility of solid flow boundaries" (Grishanin 1972, p. 153) in the hydrodynamic solution of problems of fluvial processes and the separation of "limiting" (Kondratyev et al. 1959, 1982) fluvial processes factors. At the same time, the streams forming the channel in cohesive plastic and even more rocky soils generally fall out of the scope of research (the exception in this respect is the work of Karasev (1975) and Mirtskhulava (1988).

Cohesive plastic, semi-rocky and rocky soils composing the bed, as well as pebbles and boulders with alluvial detachment, determine the state of interaction in which the bed controls the flow. This was recognized by Velikanov (1955, p. 241), who noted the "directing influence of the channel on the flow … with relatively low mobility of solid particles that make up the channel". The need to take into account the second component of the interaction (the channel controls the flow) exists not only in terms of feedback (arising under the influence of the flow of form excites the corresponding dynamic structures in the flow), but also in its direct representation due to the specifics of the geological structure of the channels. This fact was emphasized by K. I. Rossinsky and I. A. Kuzmin (Gotlib et al. 1971), who proposed to divide the beds into three classes: I—folded by a homogeneous unbound material; II—by a complex structure with an inhomogeneous lithology of unbound material with outputs of the bonded material, up to and including rocks, both in the channel itself and on its banks; III—folded by a bonded soil (plastic, semi-rocky, rocky).

The same approach (formation of the channel in cohesive soils) is characteristic of the works of Karasev (1975) and Mirtskhulava (1988). They have shown that under these conditions fluvial processes are represented not only through the control of the channel flow, but also the emergence of specific forms of impact of the flows themselves on the channel, folded by a plastic or rocky soil.

Differences in the geological structure of the areas where the rivers flow and the lithology of the soils that make up their beds determine the arrival of sediment of varying size, strength and other physical and mechanical properties from the surface of the catchment area. As a result, with the same hydraulic characteristics, not only does the flow affect different soils, but it also transports unevenly sized sediments. Therefore, the interaction between the flow and the channel as a hydromechanical phenomenon is quite diverse in different natural conditions.

The geological-geomorphological structure ultimately determines the free or limited conditions for the development of channel changes. The width of the bottom of the valley, the channel itself and the floodplain depend on them. The first one determines the conditions of water flow in a flattened or cramped channel, its depth and, in feedback, determines the speed mode and dynamic structure of the flow. The floodplain causes the flow of the flow into the high-water phase of the water regime outside the channel (or its concentration in it), as well as the interaction at its flooding of the floodplain and channel flow, differing in speed, depth, roughness of the underlying surface, etc.

Natural channel flows are characterized in specific conditions by a certain water regime, depending on the climate (primarily, the distribution of precipitation and temperature in the year), different runoff rates, changing both along the length of the rivers and in the seasons of the year. Different conditions for the inflow of sediment from the catchment area into river channels depend not only on the geological structure but also on the topography and land cover (Fig. 1.3). The flow is also affected by cold weather ice, ice flow, in-stream and onbank vegetation development, wind influencing the water surface and flow rate.

Thus, fluvial processes are "in continuous and close connection with geological, geomorphological, climatic, soil and geobotanical conditions of the area and are part of the physical and geographical environment". (Velikanov 1955, p. 237). They "cannot be considered … isolated from the geographical environment, without taking into account the specific features characterizing the watershed landscape" (Makkaveev 1955, p. 3). At the same time, fluvial processes act as a landscape-forming factor, forming the channel, channel banks and floodplain, turning the latter into a floodplain terrace, as well as determining the evolution of landscapes at the bottom of river valleys, depending on the resulting channel changes of the ratio between water flows, channel formations and floodplain (Surkov 1999). Hence the duality of the nature of fluvial processes—they are a hydromechanical phenomenon that develops in a specific geographical setting, subject to natural laws.

1.5 Discreteness and Continuity of Fluvial Processes

One of the main theoretical positions of riverbed science is the idea of discreteness of fluvial processes. It was introduced into the conceptual machine in the early 1950s by Kondratyev (1953), although his ideas were only published in their most complete form in 2000. Summarizing the experience of N. E. Kondratyev's research on the mechanisms of interaction between the flow and the channel has shown that they are different and submit to different regularities at certain structural levels, corresponding to various forms of representation of channel processes—from a single sediment particle to the morphologically homogeneous section of the river or the longitudinal profile of the river as a whole.

N. E. Kondratyev singled out four structural levels—single particle, microforms, mesoforms and macroforms. He referred to micro- and mesoforms as rows of different sizes—small rows "covering usually all the bottom of the watercourse … and perceived as its roughness", and "large sand accumulations comparable to the width of the channel" (Kondratyev 2000, p. 237), to the macroforms—the forms of the channel itself and the floodplain ("an example of a macroform can be a river bend with the adjacent floodplain massif" (ibid., p. 238). Makkaveev (1955) also considered fluvial processes at four main independent levels: the longitudinal profile of the river; the floodplain; the channel shape; and the rolls as channel relief forms reflecting the transport of sediment. At the same time, he did not correlate them with the category of "discreteness", but actually studied the fluvial processes on each of them; he also has the term "the form of representation of fluvial processes" (Makkaveev 1974).

In the most complete form, the discreteness of fluvial processes can be represented in the form of a system of seven interrelated levels of interaction between the flow and the soils that compose the river channel, the forms of representation of fluvial processes and their combinations, which differ depending on the size of the river and the level of their change from source to mouth (Fig. 1.4). The first level—I—corresponds, as in the scheme of Kondratyev (1953, 2000), to the single particles of bottom sediments. Fluvial processes at this level represent themselves in the separation of soil particles (sediments), their movement and subsequent stoppage (accumulation). The mechanism of these phenomena—separation, movement and accumulation of particles—is hydromechanical, but their natural component is associated with the particle size (boulders, pebbles, gravel, sand, silt), the amount of incoming and transported sediment: geological-geomorphological structure of the valley, formation of the channel in alluvial, alluvial-delta and other loose sediments or in rocky, cohesive and other strong rocks, the nature of slopes and slope processes, forest cover or steppe of the basin area, etc. These are related to the conditions of transport of sediments in a confined or suspended state. At the next level of interaction between the flow and the channel, this is reflected in the unstructured (without ripple formation) or structural (rippled) transport of sediment.

The direct impact of the flow on the river bed represents a particular type of riverbed development. It is characterized by the specific features of the impact of

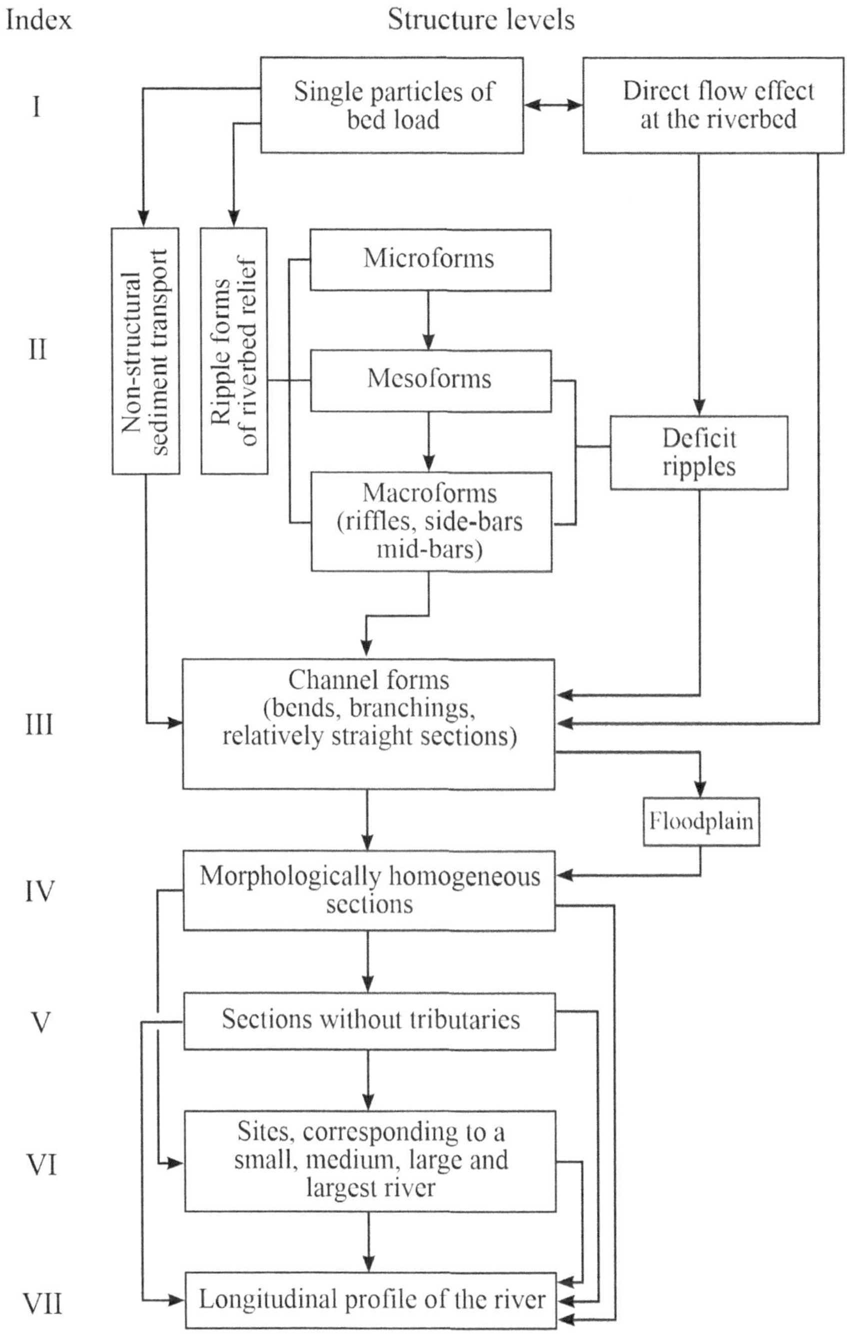

Fig. 1.4 Structural organization (discreteness) of the fluvial processes, forms of its representation and interaction of its levels

the flow on the bottom and the reaction of the latter to this impact. Its morphological representation is rocky beds and their analogues in bedrock semiscale and plastic (clay) soils. These conditions are characterised by a deficit of load and an absolute predominance of the transport capacity of the river flow over the load of river load $W_{tr} >> W$. This creates the preconditions for corrosive effects of the particles transported by the flow on the river bed.

The second (II) level of representation of fluvial processes is related to the mass movement of bed load in the form of ripples of different sizes, unstructured transport of sediment or the formation of scarce ripples. The latter arise under conditions where the transport capacity W_{tr} of the flow is much greater than the actual load carrying capacity of the river, with an absolute predominance of the bedload component, i.e. $W_{tr} >> W_G$, where W_G is the sediment yield. The deficit ripples appearing in these conditions (Debolskiy and Kotkov 1977) alternate with sections of the channel, where the flow directly contacts the root bed, or they are concentrated in the coastal parts of the channel, being absent in the core area.

The most common form of transportation of bed material load and the relief of the channel is the ripple form. This structural level can be represented by three sublevels—micro-, meso- and macroforms. Wherein, smaller ripples are formed and moved along the surface of larger ripples. As a result, both individual sediment particles and their aggregate amounts in portions in the form of small ripples arrive in the basements of the latter, causing a ripple in the velocities of their movement downstream. Thus, the entire range of ripples is an interconnected system in which each sublevel is characterized by its own specific interaction with the flow and influence on the morphology of the riverbed. The microforms of the riverbed relief are the smallest ripple formations, the sizes of which are incomparably small in relation to the flow dimensions (its width and depth). Sidorchuk (1993) calls them riffles, or the smallest ripples. Being associated with the ripple flow rate in the near-bottom area, the microforms have no effect on its structure. Covering the whole bottom of the stream, but without affecting the morphology of the channel as a whole, they are perceived as roughness of the channel (Kondratyev 2000, p. 236–237). In the case of an increased particle size (coarse pebbles, boulders), the sublevel of the microforms does not occur, since the height of the microforms that make up them varies from a few millimetres to the first centimetres, and the length from a few centimetres to a comparable depth of flow (Snishchenko 1983).

The mesoformations of the channel relief include three-dimensional ripples, the width of which is the tenth and hundredths of the channel width, and the length corresponds to the order of 10 depths of flow (Snishchenko 1983). They also determine the roughness of the channel, so that Kondratyev (Kondratyev et al. 1959; Kondratyev 2000) and Znamenskaya (1968) combine them into one class of microforms. In contrast to macroforms, these ripples, caused by the macroturbulence of the flow, in feedback excite vortices in it, covering the near-bottom region. Water level fluctuations and changes in the time of the hydraulic characteristics of the flow transform them, because the vortex structure of the flow is transformed, which is reflected in the ripples-mesoforms. That's why Sidorchuk (1992) distinguishes two stages of development of mesoforms (he calls them small and medium ripples): active, in which the ripples correspond to the turbulent structure of the flow, and passive, when the ripples appeared at the bottom of the ripple at a decrease in water levels no longer correspond to eddies in the flow of low runoff rates and are partially transformed under the influence of a new velocity field.

In the case of large sediment loads (boulders) and low runoff rates in rivers, mesoforms are also not formed due to the comparability of particle sizes with the depth of flow. As a result, at the sediment transport rate corresponding to the transport capacity of the flow ($W_{tr} \approx W$), the second level of representation of fluvial processes is related to the development of macroforms of the channel relief.

Macroforms include the largest ripples (the big ones, according to Sidorchuk (1992); mesoforms, according to Kondratyev (Kondratyev et al. 1959; Kondratyev 2000) and Znamenskaya (1968), riffles, according to Makkaveev (1955)), the height and width of which are commensurate with the depth and width of the channel. They determine the main shape of the riverbed's relief, causing changes in depth both along and across the river. Being connected by their origin with the largest turbulent whirls appearing in the flood flow, the macroforms to a great extent depend on its high-speed field and circulation currents appearing on the bends and in the branches of the channel. In the feedback of the ripple—macroforms influence the hydraulic structure of the flow, especially in the lowlands, when they partially dry out. They are related to the occurrence of secondary currents in the flow, the existence of which has a significant impact on the regime of seasonal channel changes. The ripples dried up in the lowlands of these ripples can turn into large elements of the channel forms (meander necks, islands, etc.), creating new areas of the floodplain, which belong to the next third—III—structural level of fluvial processes.

The separation of micro- (the smallest ripples), meso- (small and medium ripples) and macroforms (large ripples) inside the II structural level is justified by the constructed Sidorchuk (1992) graph of the connection between the flow depth kh (here k is the longitudinal wave number $2\pi/x$, x is the length of the river section) and the *Froude number Fr* (Fig. 1.5). There are four areas of existence of dynamically stable ripple forms of channel relief of different sizes, which are in different proportions with the size (depth) and kinematics (*Froude number*) of the flow. The microform distribution area (the smallest ripples) is located along the line, approximated by the relationship of the line $kh = 1.4Fr^{-1}$; the minimum sizes of the macroform (large ripples) are described by the equation $kh = 1.6\ g/C^2 exp(1.8Fr)$. Between them there is a wide spread of mesoforms—small and medium ripples, which Sidorchuk (1992) is divided into two regions, the boundary between which corresponds to the equation $kh = 1{,}2(2\ g/C^2)^{0{,}3}\exp\{-2{,}0Fr\}$.

No ripples are formed under certain hydraulic conditions and under the condition of $W_{tr} \approx W$. The non-structured transport of sediment during mass movement occurs in two cases: (1) at very high velocities in the rough flow (Froude number $Fr = \frac{V^2}{gh} > 2 - 3$, where V is the average flow rate, h is its depth, g is the acceleration of gravity), when the movement of load is carried out in a layer (in the smooth phase) without the formation of ripples (Kennedy 1963; Znamenskaya

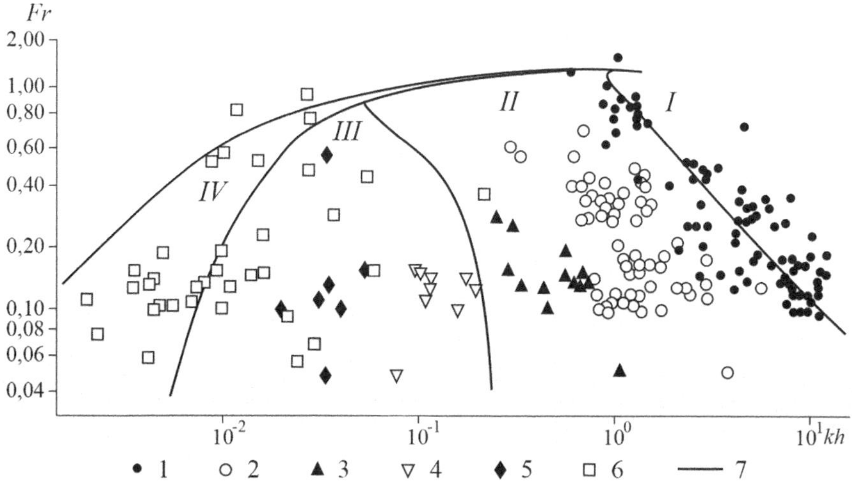

Fig. 1.5 Areas of formation of ripple`channel relief forms (according to Sidorchuk (1992)): I—microforms (1—riffles); II and III—mesoforms (2–5—ripples, in ascending order of size); IV—macroforms (6—side-bars)

1968). It's usually typical of mountain rivers with high slopes. Non-structured transport of sediment is also possible under the condition of $V > 2 - 2,5V_n$, where V_n *is a* non-scouring velocity of flow corresponding to the maximum state of rest of particles at the bottom of the flow, i.e. at fine sand and silt composition of bed material load (Amu Darya, Terek, Huang He). Usually, on such rivers, the transport capacity of the flow is realized through suspended sediment load, which is absolutely predominant in total sediment load and $W_R > W_{tr}$.

The third (III) structural level of fluvial processes representation is the formation and evolution of channel forms (according to Kondratyev (Kondratyev et al. 1959; Kondratyev 2000)). Their varieties are meanderin), branched out into branches and relatively straight, single-thread channels with their inherent partitioning between width, depth, velocity field of flow and circulation currents. The river channels are gradually reshaped under the influence of water flow, runoff and re-deposition of river load, both in the form of ripples and basically. The ripples make up the channel relief, which is transformed along the length of the channel shape depending on the peculiarities of changes in the flow structure and the conditions of sediment transport. During floods, the river flows out of its banks and inundates the floodplain. The presence of a floodplain section occupying the meander neck in the meandering channel, the island in a branched or stretched along a straight, unbroken section of the river—an indispensable element of the channel shape, formed in the conditions of free development of channel changes. During intensive incision of the river and formation of the incised channel the floodplain turns into a fluvial terrace, the fragments of which to some extent occupy the meander neck, the island or the coast of the straight channel. The release of water from the riverbed to the floodplain or its discharge from the floodplain to the riverbed causes various hydraulic phenomena in the stream, which have a significant impact on the development of the channel forms themselves. On the other hand, the floodplain regulates water flow and is an arena for the accumulation of predominantly suspended sediment. From the floodplain to the bed, clarified water is flowing into the bed. At the same time, the erosion of floodplain banks by the stream is a source of sediment load. Kondratyev (2000), emphasizing this peculiarity of channel forms, says that a new element in fluvial processes—the exchange of sediment between the channel and the floodplain—arises in the presence of the floodplain. According to Makkaveev (1955, p. 236), the floodplain, being an accumulation of alluvium, is "one of the forms in which the solid river flow is carried out".

Thus, the channel shape in relation to the ripple forms of the channel relief is a level of higher rank. Their development is due to longer fluctuations in water and sediment yield and is associated with the formation of the floodplain, overgrowth of macroforms of the channel relief by vegetation, and the influence of other factors that determine the diversity of representations of fluvial processes at this level.

Relatively long sections of rivers within which the shape of the riverbeds of this type prevail (bends, branchings, relatively straight sections) are morphologically homogeneous. Each of them is characterized by a certain set of factors of fluvial processes, which determine the known similarity of their forms throughout the whole length of such areas. In this case, the forms of the channel are conjugate, including those of different types (for example, branches of the channel are interrupted by separate bends or straight segments; meandering channel alternates with separate branches or straight segments, etc.). Their changes largely depend on the peculiarities of floodplain (floodplain massives) distribution, conditions of interaction between channel and floodplain flows within the limits of separate forms, changes in the general morphology of the valley (for example, during the transition from narrowing of the valley with a narrow floodplain to widening, due to which during floods or floodplains inundating the floodplain, there is an alternation of zones of support and decline of levels). Formation of morphologically homogeneous areas is the next IV structural level of fluvial processes representations. They correspond to the alternation of narrowings and widening of the valley bottom, change of the curbed by the wide floodplain, i.e. they are determined by the geological-geomorphological structure of the territory, etc. The boundaries of the sites may also include large tributaries, fractures in the longitudinal profile of the river, changes in the width of the valley bottom, the river's approach to the bedrock formed by bed rocks, or, vice versa, the location where the river departs from it, which determines changes in the conditions of interaction between the flow and the channel.

The confluence of rivers of the same or similar level rivers determines the transition to a higher level (V) of fluvial processes, as this doubles the river's runoff rates. This dramatically increases both the transport capacity of the flow and the flow itself, since $W_{G+R} = f(Q^m)$ and $W_{tr} = f(Q^m)$, where $m > 1$ or mainly $m \approx 2$–3 (Makkaveev 1955). In addition, the increase in the order of the river is usually accompanied by certain changes in its water regime, a decrease in the longitudinal slope, a significant increase in the partitioning between width and depth of the riverbed b_p/h and changes in other morphological, hydrological and hydraulic characteristics (Rzhanitsyn 1960, 1985). This is due to the change in the parameters of channel shapes, complication of branching or increase in the size of the bend, etc. In the most complete form, this is represented when comparing the morphological and dynamic characteristics of the channels of small, medium and large rivers (Pakhomova 2001), which gives reason to consider them as independent (IV) structural level of fluvial processes, along with the sections without tributaries.

The highest VII structural level is the development (evolution) of longitudinal river profiles. They integrate channel formation at lower structural levels, while at the same time representing a special form of fluvial processes. It is related to the influence of relief and tectonic processes on changes in the partitioning between $W \neq W_{tr}$ in the geological scale of time, the distribution of flow energy along the length of the river, changes in water availability, slope, flow and composition of bed material load, etc. These factors determine the shape of the longitudinal profile and the direction of its development due to vertical channel changes, which create a background on which the development of morphologically homogeneous sections, individual forms of the channel and macroforms of the channel relief takes place.

As a philosophical category "discontinuity (discreteness)—continuity" represents not only spatial, but also temporal delimitation and interrelation (interrelationship) of elements of the object state. In this respect, fluvial processes are discrete, as they represent themselves at different structural levels in different phases of the hydrological regime, with different intensities at the same level, with different directivity and a specific set of morphological formations with different parameters. This is the reason for the separation of channel-deforming water flows (as understood by Makkaveev (1955)), which are the most significant channel changes and correspond to one, two or three flow intervals, between which the changes are relatively attenuated, but at each of which they are represented at one or another structural level. The seasonal regime of the skips is characterized by an increase in the bottom marks (due to the accumulation of sediment) in the flood and the erosion of their ripples in the low water level, an inverse partitioning between shallowing and erosion in the low water level and high water level or more complex combinations of these processes, but in any case different in direction, intensity and latitude of its representation in different periods of time. In mountain rivers and rivers with a pebble-boulder composition of bed material load, their transport and channel changes are carried out only in the high-water phase of the regime and are completely attenuated in the low water period. Finally, significant parts of the channel within the limits of high parts of the secondary and a separated channels dry up at the flood recession and in the low water period, not contributing in the interaction with the flow in the low-water phase of the regime, and sometimes in low-water years. Small branches and ducts function only in floods, while others have different effects on sediment load in different phases of the water regime.

Developing the ideas of fluvial processes discreteness, Kondratyev (1953, 2000) at the same time noted that discrete properties of the "flow—channel" system are inseparable from the properties of a continuous medium, i.e. there is a continuity of this system. Indeed, its existence and development are ensured by the interrelation, interrelationship and conjugation of all forms of representation of fluvial processes, in which the forms of lower levels (ranks) are an integral part of the larger ones. Their development is subject to specific laws, but largely depends on the state of adjacent levels of representation of fluvial processes. Nevertheless, the ideas of discreteness turned out to be dominant in the theory and practice of riverbed science at a certain stage of development of the fluvial processes theory. Later, on the basis of the structural approach to the study of fluvial processes, more and more attention was paid to the issues of continuity of fluvial processes, channel forms and channel relief. This approach was formulated by Sidorchuk (1992, p. 6) as a principle of continuity of fluvial processes, which consists in "continuity of … channel flow and the field of marks of the riverbed bottom, interrelation and mutual transitions of separate forms of channels" (p. 6). One of the results of this approach was the method of calculating the bedload transport rate that takes into account the aggregate motion of ripples of different sizes (Alekseevsky 1998).

Continuity of fluvial processes is represented in the sequence of interrelationships between the formation and development of channels at different structural levels, and this relationship can be both direct (Fig. 1.4), directed from the lower structural levels to the higher and vice versa. The direct connection is represented in the consistent relationship of education and the evolution of the forms of representation of fluvial processes at lower levels. The formation of ripple forms of riverbed relief is impossible without the propulsion of individual particles of bottom sediments. The latter develop at three independent sublevels, which determine the formation and movement of smaller forms on the surface of larger ones, the size of which, in turn, affect the degree of development of the hierarchy and morphometry of smaller ripples. On the other hand, the movement of micro- and mesoforms of the ripple relief along the surface of macroforms and mesoforms causes their movement due to the portioned inflow of sediment into the basement of a larger ripple, causing its passive (Alekseevsky 1998), or indirect (Znamenskaya 1968), movement. The deposition in the basement of a ripple of any size of individual sediment particles moving along its surface determines the active movement of the ripple itself (Znamenskaya 1968; Alekseevsky 1998).

Macroforms of the channel relief, having an inverse effect on the velocity field and flow structure, especially in the dry phases of the regime, when they partially dry out, contribute to the transformation of the channel shape due to the erosion of the banks opposite to them, the emergence of deceleration and acceleration zones in the flow of current. As a result, macroforms, entering the deceleration zones of the current, stop shifting and grow in height due to the arrival of new load portions in the form of smaller ripples and, partially, the deposition of suspended loads. Fixed in these conditions by vegetation, they become the nuclei of the floodplain formation forming meander necks, islands or floodplains stretched along the straight channel. The conjugate development of individual forms of the channel is also represented in the development of adjacent forms of the channel within morphologically homogeneous areas.

The direct interrelationships of all structural levels (their continuity) are ultimately determined by the fact that the lower levels of interaction between the flow and channel are an integral part of the higher levels and, thus, the development of the latter depends to a greater or lesser extent on the nature and direction of the channel processes at the lower levels. In particular, the separation of individual sediment particles from the bottom or the direct impact of the flow on the river bed (involving corrosion, mechanical, biochemical and other types of underwater rock weathering) is ultimately an element of the evolution of the longitudinal profile of the river.

Morphological feedback is less significant and is often indirect. They are caused by the longitudinal movement of water and sediment, changes in flow energy at each structural level of fluvial processes. On the other hand, higher-order structural levels and their development are the background against which lower-order forms are formed and developed. The processes occurring in all hierarchy of levels, integrates among themselves, causing variety of forms of displays of fluvial processes as a whole.

Continuity of fluvial processes in time is the interrelation and interrelationship of all forms of their representation. It is most fully and visibly evident in those parts of the channel that are constantly under water in all phases of the water regime; hence, it provides a temporary continuity of interaction between the flow and the channel, although the forms of its representation and the intensity of the channel changes occurring at the same time change. On the other hand, in the macroscale of time, those parts of the channel that dry out in the dry phase of the regime, with the resumption of exposure to them flow continue to develop, maintaining a given direction. This is the case, for example, of a movement of the riffles during the flood, which occurs after a long (relative to the annual cycle) period of their drying up.

Identification of temporal discreteness and continuity of fluvial processes determines the channel regime of rivers, the analysis of which is no less important for understanding the regularities of river channel development than their consideration at different structural levels on the background of spatial continuity of their forms of representation. In the aggregate, the spatial and temporal discreteness (discontinuity) and continuity of the fluvial processes essentially represent one of the most general laws of their development.

1.6 Erosion-Accumulative and Fluvial Processes: Interconnections and Ratios in Erosion-Channel Systems

The fluvial processes constitute the lower link in the chain of phenomena related to the impact of runoff on the earth's surface, and their specificity reflects the formation of water and sediment yield within the river catchments. "The development of a river can only be properly understood if it is considered to be inextricably linked to the development of the upstream flows" (Makkaveev 1955, p. 32), including those represented by temporary watercourses in catchments. The impact of the processes taking place in the catchments is represented in the impact of river flows transporting sediment from the catchments on the underlying soils. The result is a new source of sediment—riverbed and bank erosion. At the same time, all water flows, starting from slope nonchannels and ending with rivers flowing directly into the seas and oceans (or lakes), carry out the destruction of the land surface, the movement of products of this destruction from the continents to the receiving water bodies, under certain conditions—their accumulation within the continents and, finally, repeated repositioning along the route of transportation. Ultimately, all this determines the formation of fluvial relief. The impact of runoff on the earth's surface is represented through erosion and accumulation processes, which are a set of erosion, transportation (movement) and accumulation of erosion and accumulation products, considered "in space and time, as well as in interaction with the determinants" (Makkaveev 1955, p. 83). Water runoff acts as an active factor in these processes, and the land surface, its lithogenic base, relief, land cover, represents the arena or factors controlling the activity of water flows. As a result, erosion and accumulation processes reflect the interaction of water flows and the earth's surface.

In terms of the volume of destruction, transport, re-deposition and delivery of solid matter to the world's oceans, erosion-accumulation processes are among the most powerful denudation agents that determine the formation of fluvial relief on Earth. According to various researchers, the amount of sediment delivered by rivers from the mainland to the ocean (Lopatin 1952; Makkaveev 1955; Losev 1989; Lisitsyn 1991) ranges from 12.7 to 26.7 billion tons per year. In addition, from 3.0 to 4.0 billion tons per year (Losev 1989; Walling and Webb 1987) is brought by rivers into the ocean in a dissolved state (as a result of chemical erosion of land by water flows). Air currents move 2.2–2.6 billion tonnes of particulate matter and sand particles to the ocean each year. The contribution of glaciers and sea shore abrasion to total solids inputs to the world's oceans is approximately the same—0.3–0.4 billion tons per year. From external sources, the Earth's surface annually reaches only 0.01 km^3 of space material (Makkaveev 1982). If we take into account that only 13.5–20% of the flushing products (soil erosion) reach the rivers (Golosov 2003), and the remaining 80–86.5% are deposited at the foot of slopes, in beams and other negative forms of terrain, and the sediment entering the rivers is accumulated partially on the floodplains or in the river beds themselves, the total share of substance transported by water flows on the earth's surface increases several times. For example, the total loss of soil due to erosion on Earth alone is estimated at 50 billion tons per year (Golosov 2003). In Ukraine, about 30% of the erosion-prone slopes are covered with reclaimed soils. In the North Caucasus, a short but very intense downpour was recorded, which led to the washing away of the upper 3–4 cm of soil, shifted a few tens of meters towards the bottom of the slope (Zaslavsky 1979).

The temporary streambeds at the bottom of the ravine quickly form a rather thick layer of spillage, which only comes into full motion in the event of catastrophic floods; in the event of less severe floods, it is only partially washed away. The spillage layer usually reaches its maximum capacity on the debris cones. The latter often form an almost solid strip along heavily contaminated steep slopes.

In rivers, the conjugation of erosive and accumulative processes is particularly evident. On any river you can find intensively washed out banks and recently, during the last flood, formed accumulative banks of the riverbed. The movement of huge amounts of solid material by rivers and their incision into the earth's surface is accompanied by the accumulation of alluvial layers. Most river islands move along the river due to the erosion of the headboard and the growth of the braid in the ear. The river bends are also displaced because the concave (outer) bank is washed away and the sediment accumulation at the convex (inner) bank is synchronized. Finally, the movement of bedload by river flows is carried out in the form of the movement of accumulation ripples, where the upper gentle slope is eroded and the lower steep slope is the place of their accumulation. The sediment deposited on the downhill slope is out of transit until it reaches the flush zone on the upper slope of the ripple.

As a result, lowland rivers carry to the sea only 7–10% of the products of soil removal from the catchments and erosion activities of the rivers themselves; for mountain rivers these values are much higher—not less than 50%. The rest of them remain within the river catchments (Golosov 2003).

The whole set of processes of soil and rock washout and erosion, movement, re-deposition and accumulation of sediment by water flows constitute a complex of erosion and accumulation processes, the individual links of which are closely connected and depend on each other. Sources of sediment transported by watercourses are: (1) products of erosion of the underlying surface (for rivers, including banks); (2) solids coming from slopes due to gravitational processes (landslides, screes, landslips), aeolian transfer, slow movement of soil on the slopes, solifluctions, etc. Under certain conditions, slope processes can play a decisive role in the formation of sediment load (e.g. on mountain rivers, especially in areas of increased mudflow activity); their share in total sediment load is generally low. For example, on the average Dniester River, 3.75 million tons of material per year arrive from the valley slopes to the riverbed as a result of debris and other slope processes (Kalinin 1987), with annual sediment load exceeding 6 million tons (Berkovich et al. 1992). The greater the water flow, the greater the absolute value and the smaller the relative share of non-erosional load. In the places where slope water flows originate, soil and soil erosion is practically the only source of sediment load at their rainfall genesis; at melt genesis, its role is difficult to separate from solifluction and other phenomena related to thawing of thawed soil.

According to Makkaveev (1955), the entire set of erosion-accumulation processes consists of three main interrelated links corresponding to certain types of water flows, which are the source of load for the next link, have their own geomorphological representation (Fig. 1.6), and are characterized by their unique development patterns, mechanisms of functioning, and spatial and temporal ratios of erosion, transport, and accumulation of load. These three links of the unified

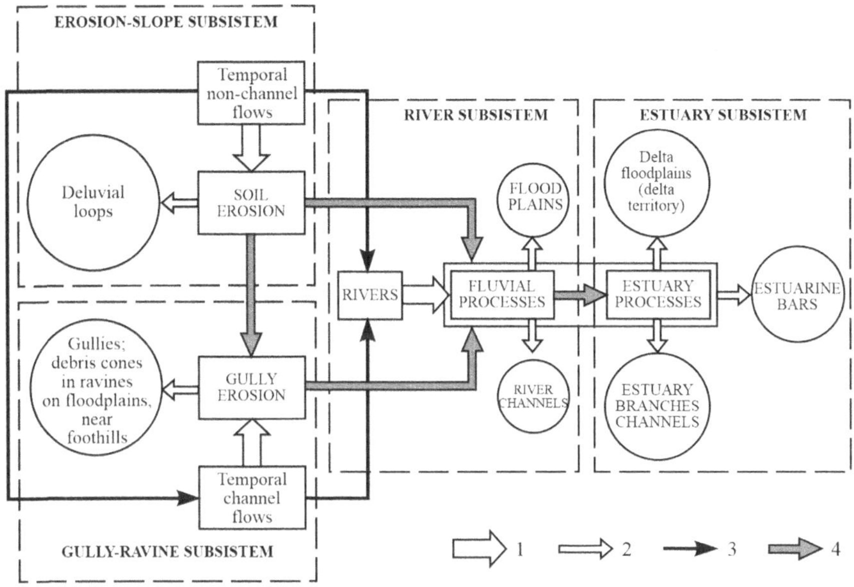

Fig. 1.6 Structure of erosional-channel system and erosional-accumulative processes; connection of types of water flows (1), the relief forms they create (2) and directed water moving (3) and solid material—sediments (4)

system of erosion and accumulation processes are: soil erosion, produced by temporary non-draught channel flows formed by rainfall and snow melting; ravine (linear) erosion, associated with the activity of temporary channel flows; fluvial processes as a set of phenomena caused by erosion and accumulation of rivers. This list can be supplemented by the wellhead processes that form the final link in a single system, which develops against the background of the directional accumulation of sediment during the flow into the receiving pond and the impact of marine phenomena on it (Mikhailov 1997a, b).

There are no clear distinctions between the types of water flows and the links of erosion and accumulation processes: the slope flow is divided into a network of streams, resulting in the formation of primary, although ephemeral, linear erosion forms on the slope—furrows and washouts; the ravine, crashing into the rock strata, reaches the aquifer and in this case has a constant water flow, small rivers dry up in dry periods, etc. Nevertheless, the water flows of each type differ in specific mechanisms of interaction with the underlying soils, the form and distance of sediment transport, and the peculiarity of accumulative savings. This makes it legitimate and practical to consider individual links in the erosion-accumulation system independently of each other, but taking into account their connection through the sediment load.

The evolution of the geographical environment, resulting in the transformation of runoff and change of land cover, determines changes in the characteristics of water flow, which determine their impact on the surface, the properties of the surface itself and, accordingly, its response to this impact. The surface properties (soil and ground composition, vegetation, slopes) determine the intensity of erosion and accumulation processes in the upper reaches of the watercourses network and the volume of material washed away by them, which is an important factor in the formation of sediment load in rivers. Other things being equal, the more soils are washed away and the more intensive the erosion in the catchments, the greater the river load, its basin component (Dedkov and Mozzherin 1984) and the share of suspended load in total sediment load. In the steppe zone, the latter is 80–90%, and on such a river as the Huang He (downstream), reaches 99.9%. In the Ob, in the steppe zone, where the river is cut into the Priobskoe plateau, composed of loesslike loams, to a depth of about 100 m and washes its benches, the share of suspended sediment load is almost 95%; in the south of the taiga zone, it decreases to 55%, and on other rivers in the forest zone, to 30–40% (Darbutas 1992; Chalov 1997).

Temporary non-channel flows carry washed away material over short distances equal to the length of the slope—from the first meters to hundreds of meters. There is a sufficiently clear spatial separation of the zones of predominant erosion, the transport of material coming from above and the accumulation of sediment (Fig. 1.7A) from the upper part of the slope to its foot, where the deluvial plume is formed. Non-channel flows produce dispersed (planar) erosion and dispersed accumulation. The part of the transferred substance that is not included in the composition of the accumulation formations within the slopes, goes to the next links of the system (ravines, gulls) or directly to the rivers. Fixed low level watercourses (first, second) are in direct contact with their catchments, taking a significant part of the material washed away from their areas. As the level of the river increases, this connection becomes more and more mediated. For example, in the channel and floodplain of small rivers in the Don basin, the annual layer of sediment accumulation, which is the product of flushing from ploughed catchments, ranges from 3 to 20–50 mm for rivers up to 10–25 km long and less than 1 mm for rivers over 100 km long. Gully erosion supplies solid material directly to the rivers where they wash away banks and valley slopes dissected by ravines. About 2 million m^3 of sediment is transported into the Don riverbed from ravines annually, which is comparable to 6 million m^3 of suspended sediment load from the river (Zorina et al. 1980).

These interconnections are especially evident in the development of anthropogenic (accelerated) soil erosion, which supplies excess sediment from catchments to small rivers (relative to the amount of sediment established during historical and geological periods of time). In lowland rivers, whose basins are heavily developed by economic activity, in different natural conditions, the activation of erosion processes results in an increase in the suspended sediment load modulus by a factor of 5–25 (Golosov 2003).

The naturally occurring load balance is disturbed by accelerated soil erosion, which leads to siltation and degradation of small rivers. This is greatly facilitated by the flow of chemically dissolved and biogenic substances into the rivers together with the sediments (Litvin and Kiryukhina 1995). As a result, river water salinity

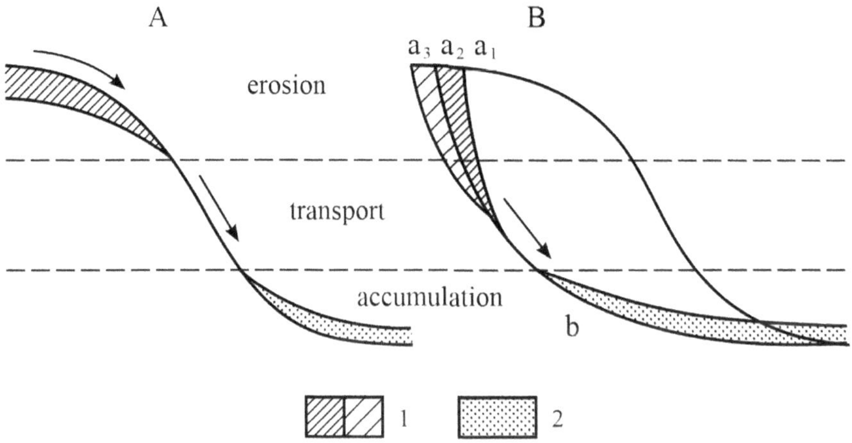

Fig. 1.7 Location of zones of erosion, transport and accumulation of sediments with temporal flows: slope non-channel (**A**), ravine (**B**); a_1b, a_2b, a_3b—successive positions of the longitudinal profile of the ravine): 1—erosion zone, 2—accumulation zone

increases, eutrophication intensifies and, as a result, aquatic vegetation actively develops, and riverbanks and coastal slopes are overgrown. This, in turn, contributes to the siltation of beds even if the natural background of sediment load from catchment areas is preserved.

A similar picture is observed in ravines (Fig. 1.7B), where erosion in the upper part of the ravine is even more pronounced due to the growth of the ravine top and waterfall effect. A debris cone is formed at the mouth of the ravine (if the ravine "opens" into the river, the debris cone can be destruct by the flow). In the central part of the area, the temporary water flow transits material from the upper reaches of the ravine, from the slopes of the ravine and from its catchment area. The distance of temporary channel flows reaches the first kilometers in the development of bottom ravines (along the bottoms of gulls). The length of ravines may slightly exceed the slope length (hundreds of meters, more than 1 km) due to regressive growth of their tops. Thus, temporary channel flows carry out linear erosion and concentrated accumulation (river transport). At the same time, the flow in a ravine at its periodic appearance should have such a transport capacity that it provides not only the transport of sediment formed during its existence, and the material entering the bottom in the absence of flow, but also the destruction of vegetation appearing here. Otherwise, the ravine is filled with sediment and transformed into a gull, becoming the arena for the development of accumulation processes (Makkaveev 1955).

In contrast to temporary rivers, permanent watercourses transport matter over distances commensurate with the width of the transverse (Amazon, 6670 km long) or half the transverse of the continents (the largest rivers in Europe, Asia and Africa). At the same time, due to the formation and development of channel forms, alluvial forms of channel relief and floodplains, the differentiation of erosion, transport and accumulation zones along the river length is not usually shown. The material being transported is re-deposited several times throughout the river. Part of the load is accumulated in the bed and on the floodplain, forming alluvial deposits. Systematic accumulation of sediment is characteristic of the lower reaches of rivers, where regressive accumulation is a consequence of their estuary elongation, sea level fluctuations (erosion base level) or tectonic deflections of the Earth's crust. Such accumulation of river load by rivers has been observed over centuries and geologically, and against this background, periodic channel changes occur, causing scouring and washdowns of riverbanks and river beds on a long-term, seasonal and current time scale (Alekseevsky and Chalov 1997). However, superimposed on directional vertical and lateral changes along the entire length of the rivers, these processes are accompanied by the preservation of some of the transported sediment flows forming alluvial deposits of floodplains and terraces. In the incised channel, where lateral channel changes are limited to bedrocks in which it is formed, such a "retention" of sediment is poorly defined, as it is mainly the transit of sediment, the amount of bottom sediment is small and the flow often comes into direct contact with the bedrock. In such conditions, erosion of floodplain and terrace banks can almost completely neutralize the accumulation of alluvial layers. This results in fragmentation, low capacity or absence of alluvium in the valleys of rivers with incised channel.

The rivers flowing in the spreading areas of easily eroded rocks form wide floodplain, weakly stable channels that are characterized by intense change; along with the transport of sediment, they are conserved in the forming floodplain massifs and new portions of them are brought into the stream when the floodplain and partly terrace banks are washed away. As a result, there is a constant mass transfer between the sediment flow, channel sediments and floodplain sediments (Fig. 1.8). Suspended load is deposited on the floodplain during flooding into the high-water phase of the water regime and forms the floodplain facies of alluvium; when the floodplain is washed away, the sediments of this facies are once again a source of suspended load. In the course of river intakes, floodplain sediments (channel and floodplain facies) are transferred to the terrace sediments, which genetically distinguishes them from the floodplain facies, which form as a result of lateral channel changes (channel facies) and accumulation of suspended sediment on its surface (floodplain facies). Sediments that make up high floodplains and terraces are a source of sediment when washed away by the flow of their ledges; at the same time, terraces can be deposited on the floodplain surface and in the channel due to the fragmentation of terraces by ravines, the basis for erosion in the development of which is both channel and floodplain.

The watercourses underneath the surface they affect and the erosion-accumulation processes resulting from the interaction of the two media form the erosion-river systems (ERS) that function within watersheds. The latter consist of ERS of different ranks—from the elementary catchment on the slope to the basin of the largest river. In general, ERS consist of individual subsystems (Fig. 1.6):

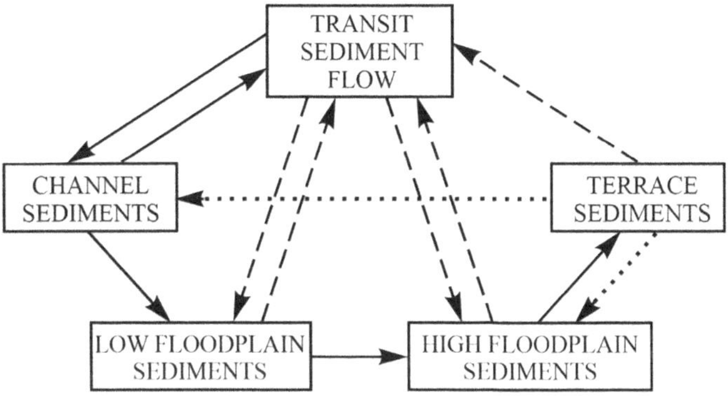

Fig. 1.8 The scheme of mass exchange in rivers between the flow and the river sediments (according to Alekseevsky (1998), modified and updated); continuous and dotted lines corresponds to more or less probability of mass exchange; dotted lines shows indirect mass exchange

(1) erosion slope, which has non-channel temporary flows on slopes, carrying out soil erosion;
(2) ravine-gully, corresponding to the activity of temporary flows, concentrated in linear erosion and other forms (channels) on the slopes (gully erosion), as a result of which ravines develop;
(3) a river, in which constant water flows operate, carrying out the transport of sediment from slopes and from the ravine-gully network, channel changes occur;
(4) the estuary.

Water flows are a dynamic component of erosion and channel systems (ERS), which underlay the surface (slopes, channel), i.e. static and erosion-accumulative processes, derived from their interaction (Fig. 1.9). Erosion-accumulation processes change the characteristics of water flows due to changes in the physical properties of water when it is saturated with suspended sediments and hydraulic resistances, for example, through the formation and constant transformation of ripples. This, in turn, is reflected in the mechanism of influence of streams on the underlying surface. On the other hand, the change in the characteristics of water flows (their runoff rates, speed) in the spatio-temporal relation changes the characteristics of erosion-accumulation processes, their intensity, forms of representation, etc. Spatial differentiation of erosion zones, transport, sediment accumulation or their simultaneous representation in ERS explains the use of various terms that determine the results of water flows in different links of the channel network: in the case of temporary non-channel and channel flows, erosion acts as a source of sediment for the next link and a destructive process that causes the reduction of land surface marks and its fragmentation. Hence the terms soil erosion and gully erosion. In the

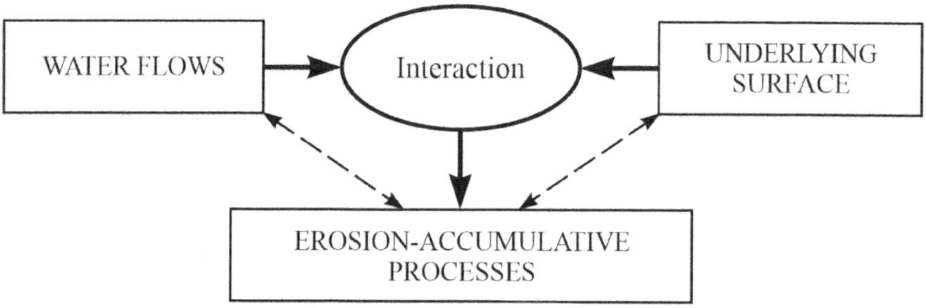

Fig. 1.9 Component structure of erosion-channel systems

case of rivers, the concept of "river erosion" limits the content of the erosion-accumulation process to one component and does not reflect its essence. Therefore, the work of river flows on erosion, transport and accumulation of sediment was called "fluvial processes".

On agricultural lands, soil erosion and ravine formation play an extremely negative role in the complex of living conditions, activities and life support of people, and their conservation requires various erosion control measures. Application of correct agricultural and other methods can completely prevent the development of these processes, or reduce them to permissible (from the point of view of soil conservation—environmental factor; reduction of costs of erosion control measures—economic factor) representations (Belotserkovsky 1996). The river subsystem focuses on the consideration of fluvial processes in the design, construction and regulatory work, including both the prevention of unfavourable erosion (erosion of the bottom and banks) and the accumulation of sediment, and the use of the river flow's most erosive and accumulative work in the desired direction (e.g., to increase depth at riffles of navigable rivers).

Interconnection of erosion and accumulation processes, developing in different links of the water flow network, determines the existence and functioning of erosion and channel systems (ERS) in nature, which are a set of interrelated forms of relief and processes due to the impact of water flows on the earth's surface. Baryshnikov and Samuseva (1999) refer to ERS as a self-regulating system "river basin—flow—channel". The processes in ERS represent a single chain of phenomena that combine erosion (particle separation by water flow), sediment transfer and accumulation. In different parts of ERS (slopes—soil erosion; ravines and gullies—gully erosion; river channels—fluvial processes; river mouths—delta formation processes) erosion and accumulation processes develop inextricably over time (without one cannot be the other), but are spatially either separated (on slopes, in ravines and gullies) or represented together (river channels). At the boundaries of the ERS (river basins), zones of absolute predominance of erosion (upper part of the separation slope) and accumulation of sediment (river mouth bar) are identified; the rest of the system (basin) is an area of predominant transport and re-deposition of sediment.

The interconnection between the ERS links can be direct and inverse (Fig. 1.10). At the same time, they are unidirectional for the active factor of erosive and accumulative processes—water flows—while between the corresponding processes there are direct or feedback connections. However, all other things being equal, direct connections are more significant and, as a rule, are direct, while the opposite are slower, often indirect and more complicated. The direct links are determined by the sediment flow formed during erosion. Therefore, the transport of sediment determines the nature of fluvial processes. Through the formation of sediment flow in river beds, soil erosion on slopes and gully erosion affect fluvial processes. On the other hand, channel changes as a form of representation of the latter are the most important factor in the formation of sediment flow in the channels, which determines the only significant feedback in the entire erosion and channel system.

The importance of the "soil erosion—gully erosion" interconnection is determined by the flow of flush products into ravines and the streamy nature of slope flows, so that any stream can be considered as a potential form of linear erosion and gully genesis. Feedback is represented in the development of slopes adjacent to the ravine and the formation of a ravine catchment area.

The development of estuaries depends on the sediment transport rate into the sea (lake) from the rivers and the fluvial processes in the delta channels. Feedback is

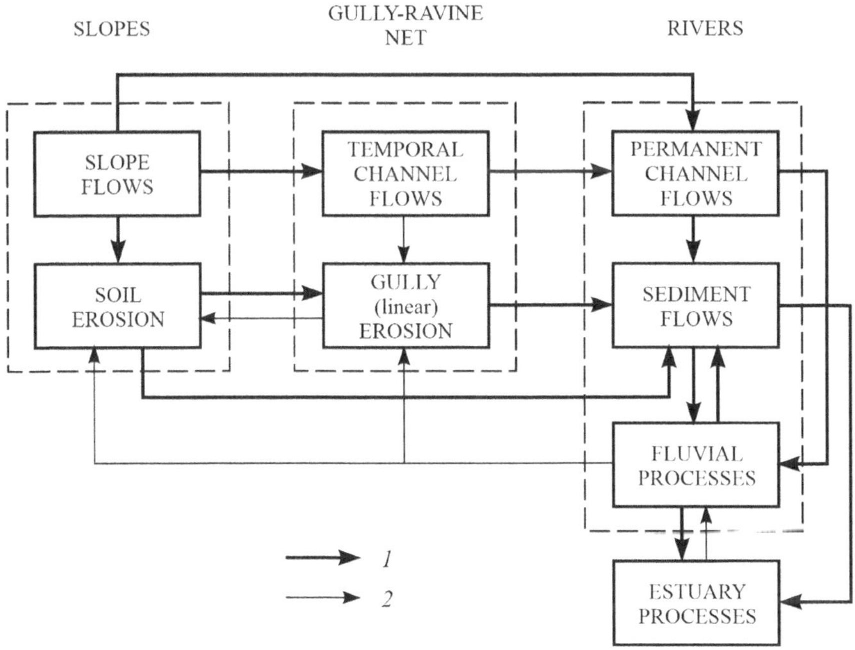

Fig. 1.10 Interconnection between structural components of erosion-channel systems (ECS): 1—permanently manifesting, direct and meaningful connections; 2—indirect, recurring and faintly manifested connections

insignificant and is mainly represented through the direction of vertical changes (regressive accumulation or erosion) in the lower reaches of rivers, caused by the delta's extension into the sea or erosion of its sea edge.

The significance of bonds in erosion-channel systems is conditioned by the exchange of substance and energy between the links of the system and, to a lesser extent, by changes in the functioning conditions of each link under the influence of the development or evolution of the entire system. The nature and even direction of these links depend largely on the landscape conditions (primarily relief and vegetation), the territorial and temporal changes of which may affect the "intra-system" relations in different ways. A striking example is the change in the share of slope runoff and runoff from the dry land network in the formation of the surface component of water runoff in the rivers of the European territory of Russia (ETR). In the forest zone of the ETR, the runoff from the slopes and bottoms of dry lands and river valleys is 25–30%, and in the steppe zone—more than 50% of the total surface runoff. In the steppe, in dry years, it becomes the main source of surface river runoff (Koronkevich 1990).

The role of anthropogenic landscape transformations is even more significant. Economic development, primarily the plowing of forest and pasture slopes, increasing the total intensity of erosion on the slopes many times (by several orders of magnitude) leads to a change in the sign of sediment balance in the channels of small rivers, resulting in their siltation. Moreover, the structure of ERS itself is changing—70% of ravines are of anthropogenic origin (Gully Erosion 1989).

Erosion-channel systems belong to the type of cascade systems with the prevalence of direct links coordinated with the "direction" of successive movement of matter from the slope to the mouth of the river over the feedback (Ecology … 2002). This means not so much the low physical intensity of the "reverse" effect as the rarity and locality of representations. For example, intensive lateral erosion of the river can destroy part of the valley slope, where a quasi-lateral, non-erosionally hazardous floodplain surface is formed, where the accumulation of sediment is absolutely dominant. The sediment loads come here with hollow waters flooding the floodplain, as well as in the form of deluvial plumes at the foot of the valley slopes and debris cones from the ravines resting on the rear parts of the floodplain. On the contrary, the increase in steepness and the destruction of vegetation on the washed out slope leads to local strengthening of soil erosion and activation of gully formation.

In general, in such systems, the strength and number of links decreases with the increase in the "system distance" of the links from each other, i.e. the influence of erosion and accumulation processes on the slopes gradually decreases in the row of the ravines—gullies—river—estuaries.

The territorial structure of the ERS and the interconnectedness of its individual elements are important for its interconnections. The general view of these characteristics for erosion and channel systems also helps to reduce the "downstream" effect on the system. For example, if a gully catchment area is 90–100% slopes, then as the order of channel flow increases, the relative area of slopes directly adjacent to the channel and supplying it with sediment decreases rapidly. At the estuary the extension delta has no contact with the slopes. Erosion processes on the slopes directly adjacent to the large river bed have a greater impact on the fluvial processes than those of the same intensity on the near-watershed slopes, where water and sediment load on the way to large rivers is transformed in the drywall network and small river beds.

Interactions within the erosion and channel systems are largely determined by the specific landscape situation and the degree of anthropogenic transformation of the landscape. The significance of changes in the genetic components of water and sediment load in rivers under the influence of these two groups of factors for the territory of Russia is shown in the works of Dedkov and Mozzherin (1984), Koronkevich (1990), and Alekseevsky (1998). For example, the share of slope load in mountain river load is significantly lower than that of flat rivers, and the intensification of soil-erosion processes has a much greater impact on flat rivers. Suspended sediments are usually of a basin-wide genesis, i.e., they are products of soil erosion, bed sediment—the result of channel changes. In lowland rivers, the ratio between the components of sediment load varies widely, depending on the specific natural conditions in which the ERS is located. Rivers in the forest zone are typically dominated by riverbed sediments, while steppe rivers are typically dominated by riverbed sediments. The channel genesis deposits are usually larger (sand, pebbles), coming from catchment areas—smaller (silt, clay). The average long-term content of fine particles in river waters naturally increases from north to south of Russia (Sediment yield … 1977), which is facilitated by the predominance of loess and loess-like sediments in the southern plain areas, which are the most easily eroded among all lithological complexes of rocks and sediments, as well as their confinement to the zones of the most active development of soil erosion.

Regional changes in the ratios of the basin and channel components of suspended sediment load are accompanied by the transformation of river load along the river systems. In southern rivers (with high suspended sediment loads, a predominance of their basin component and low sediment particle size), the share of bedload is low (within a few per cent and less than 1%). It increases downstream from south to north and decreases from north to south, from mountains to plains. Crossing of mountain massifs by rivers is accompanied by an increase in the share of bedload. In the lower reaches of rivers, where there is a directional accumulation of sediment, the share of bedload is reduced more intensively, as the massive material falls first and foremost out of the transit traffic. In rivers with high sediment load, this results in the accumulation of sediment in the channels and an increase in their bottom marks.

Small rivers react strongly and quickly to the intensification of erosion processes in the upper reaches of the erosion and channel systems to changes in water flow formation conditions. Their beds dry up due to the lowering of the groundwater level, become silted and degraded, which is especially characteristic of the steppe and forest-steppe zones. Small rivers in the humidal zone are high-water and less susceptible to siltation: on the contrary, in arid regions, the low runoff rates of rivers contribute to their degradation.

If melt water is the main source of sediment load into rivers, there is no noticeable siltation of the riverbeds, as the maximum sediment transport rate from the catchment coincides with the highest water discharge. Where rainfall is the main factor in soil erosion, the greatest amount of sediment washed away and discharged into the river network falls into the less high-water phase of the regime. Due to low current velocities and slow water exchange between the forests, the latter are silting, eutrophication processes are developing and shoals and coastal slopes are becoming overgrown. This causes an increase in channel roughness and a decrease in speeds during subsequent floods. As a result, the accumulation of sediment spreads over all phases of the water regime, causing siltation of rivers and degradation of the upper links of the channel network.

Sediment accumulation due to overflows from catchments and ravines sometimes extends to the middle rivers (Berkovich 1993). Large rivers are the most resistant to changes in water availability and sediment load. For example, a 30% decrease in the sediment transport rate from the Huang He River (China) due to erosion control measures in its basin has not led to changes in the characteristics of erosion-accumulation processes in its channel (Chalov et al. 2000).

1.7 General Laws of Erosion and Fluvial Processes: Self-Regulation of the Flow-Channel System

The unity of erosion, transfer and accumulation of sediments as components of erosion-accumulation processes, the interconnection and interrelationship of erosion-accumulation processes in various links of erosion-river systems constitute the two most general laws of the doctrine on a single erosion-accumulation process and its methodological basis. The essence of these laws was first revealed by N. I. Makkaveev in his monograph "The riverbed and erosion in its basin" (1955), whose very name reflected the content of one of them. Against the background of the representation of these most general laws that determine the functioning of erosion and channel systems and the essence of soil erosion, gully (linear) erosion and fluvial processes, N. I. Makkaveev has established five more universal laws both for the whole system and for the processes that make up them: nonlinearity of connections, factor relativity, mutual conditionality of the flow and the underlying surface, limitations of morphological complexes (the latter was first formulated by Velikanov (1946, 1949) as applied to fluvial processes), and automatic regulation of the transport capacity of flows (Makkaveev 1976; Chalov 1988).

The law of unity of erosion and accumulation processes is formulated as follows: erosion—movement (transfer)—accumulation of solid matter (sediments)—are obligatory components of a single process (Fig. 1.11). If soil erosion occurs, the particles entering the stream forming sediments will sooner or later be deposited forming alluvial, deluvial or proluvial accumulations. Sediment accumulation is impossible without erosion and the movement of washed away particles. The

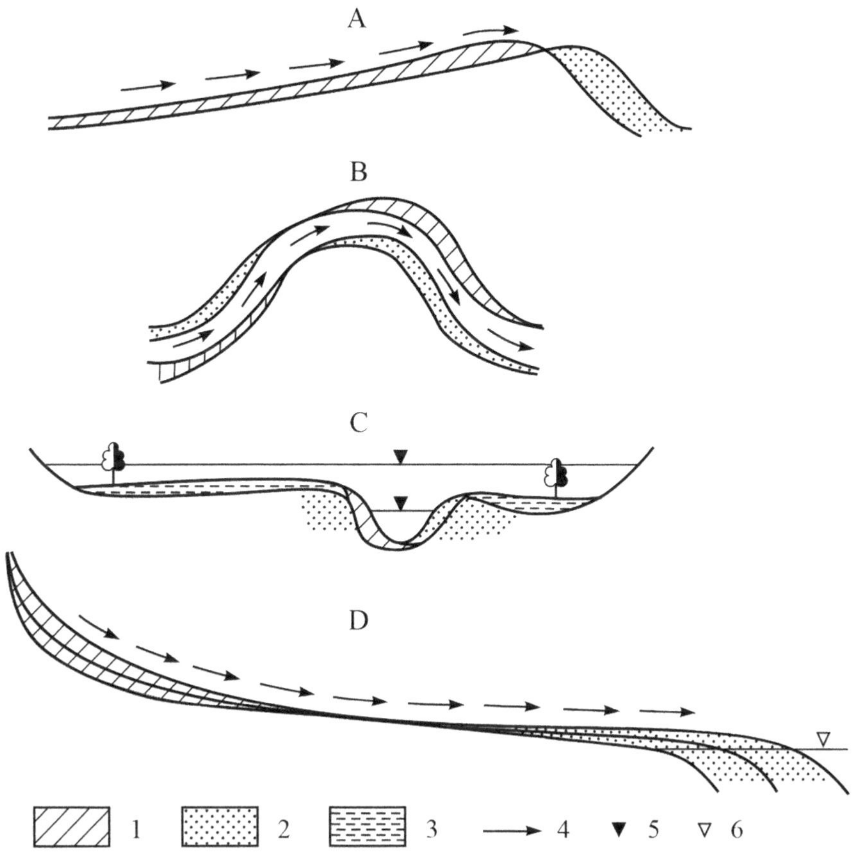

Fig. 1.11 Forms of representation of the law of the unity of erosional and accumulative processes: **A**—shifting of ripples in the channel; **B**—bend development; **C**—floodplain formation; **D**—longitudinal river profile development. 1—erosion of banks, channel bottom etc.; 2—accumulation of channel sediments; 3—accumulation of sediments on the floodplain surface; 4—movement of sediments by flow; 5—water levels in the river; 6—sea level

conditionally "pure" representation of erosion or accumulation can only be said to apply to the extreme points of the entire water flow system: only erosion occurs in the uppermost part of the slope—soil or ground washout; the accumulation of sediment is observed below the sea edge of the river delta. In general, the processes of erosion, sediment transfer and accumulation are so intertwined and interconnected in space and time that it is impossible and impractical to separate them from each other and to separate the areas of development. Only in some cases and with some degree of conditionally it is possible to allocate a site where prevalence erosion or accumulation is so great that the opposite process becomes subordinated: in the upper reaches of mountain rivers riverbeds is sometimes a rocky "tray", devoid of alluvial lodge; the river floodplain as an element of valley relief will look like an accumulative formation constantly filling with deposits except for erosion of it's banks by flow, as a result of which there a constant exchange of material between the channel and the floodplain; accumulation basically characterizes river deltas. The development of any form of channel is a set of washouts in some areas and the accumulation of sediment in others. Even the movement of ripple accumulative forms of channel relief by means of which channel sediment transport is carried out is a set of processes of erosion, transport, re-deposition and accumulation of particles (Makkaveev and Chalov 1986).

Interconnection and interrelationship of all types of erosion and accumulation processes, their conjugation in all links of the water flow network, as well as the relief forms and elements of non-stream forms of non-fluvial relief created by them, constitute the content of the second law of erosion and accumulation processes. It is most clearly represented in the partitioning between development of erosion processes in the catchment area and the processes of directional river intakes (Fig. 1.12). The river crossing and deepening of its valley are accompanied by an increase in the erosion dissection of the sides by the valleys of small rivers and the gully-ravine network, increased erosion of non-streams on the slopes (Borsuk and Chalov 1978). This leads to increased inflow of solid material into river channels from inflows, from slopes and from the catchment area as a whole (including due to non-fluvial processes, which are activated at lowering of valley bottom marks in relation to watersheds). As a result, the intensity of river intakes is slowing down. The development of erosion and denudation leads to the smoothing of slopes and the lowering of the catchment area relative to the river. A decrease in vertical terrain separation is accompanied by a decrease in the amount of material entering the river, and it starts crashing more intensively again. A new increase in the depth of the valley contributes to the activation of erosion and slope processes in the catchment area, and the river's incision is slowing down again. Naturally, these phenomena are observed synchronously, and changes in their ratio can be caused only by the change of physical and geographical situation or sign and intensity

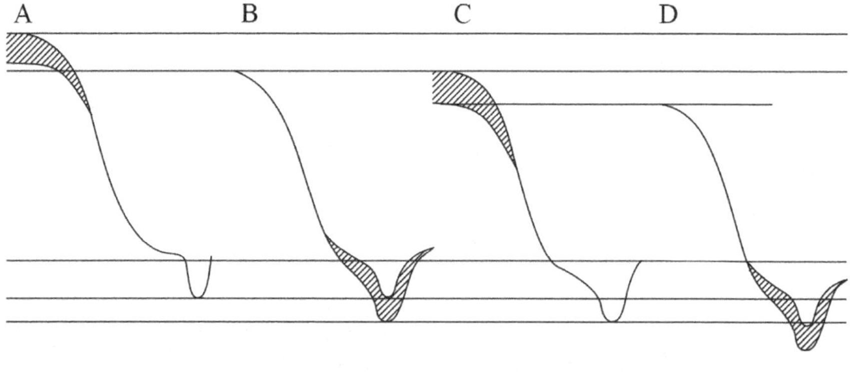

Fig. 1.12 Consecutive development of erosion of the catchment area surface (**A**, **C**) and river incision (**B**, **D**)

of tectonic movements. At relative stability of the latter and constancy in time of natural conditions interconnection and interrelationship of the whole complex of erosive and accumulative processes determine self-regulation in the development of the "river-watershed" system.

With a gradual slowdown in the rate of incision of the river, the vertical distance between the terraces is reduced with time. This is also facilitated by the reduction of longitudinal slope as the river incises (Makkaveev and Chalov 1986). At the same time, the lateral change (lateral erosion) of the rivers is reduced in scope, and as a result, the width of the valley floor decreases as the valley becomes deeper. This geomorphological effect of the ratio between erosion in the catchments and directed development of the river valley (river incision) is characteristic of the regions with ascending relief development, when the rate of elevation of the territory and incision of the river exceed the rate of denudation, resulting in $W_{tr} > W$.

The law of *automatic regulation of the sediment transport capacity of flows* determines the direction of fluvial processes. If the transport capacity of the flow is $W_{tr} > W$—the actual sediment transport rate, the flow erodes the underlying surface (ravines grow, rivers are dominated by incision—deep erosion), the sediment load increases, which is represented in the balance of solids on the slope (ravine, channel): $W_{i+1} = W_i + \Delta W$, where W_k is the volume of sediment load in the *i*-th and $(i + 1)$-sections, ΔW is the resulting balance; the «+» sign indicates the flow of material between the sections. If $W_{tr} < W$, the accumulation of excess sediment occurs; the sediment balance looks like $W_{i+1} = W_i - \Delta W$, where the «−» sign means a reduction in the amount of transported material due to its deposition at the bottom of the stream.

The increase in sediment load through the self-regulating flow-channel system leads to the formation of new forms of channel relief, the development of the floodplain, the shallowing of the channel as a whole, the intensification of its wandering, and the activation of bank erosion. Reduced sediment load leads to the termination of floodplain formation, incising of the riverbed, reduction in the intensity of change, and the disappearance of large forms of riverbed relief (side bars, medial bars). Changes in the water availability of the rivers, while maintaining constant sediment load, lead to similar effects. With $W = \text{const}$, the increase in Q_m is accompanied by an increase in W_{tr}, which may result in a condition $W_{tr} > W$, and the river will incise; with the decrease in Q_m, the condition $W_{tr} < W$ is created, and the river accumulates sediment.

Makkaveev (1955), considering the mechanism of formation of the developed longitudinal profile, has shown that it is based on the process of automatic regulation of the flow conveying capacity. "In those parts of the channel where the specific transport capacity of the flow is insufficient for the transit of sediment from the upstream section of the river bed, the bottom gradually rises until the slope of the river bed increases as a result of sediment deposits to a value that ensures the transit of sediment. The same areas where the river flow contains less sediment than it could carry are deepened by the greater the saturation deficit of the river flow. As the gradient in the section deepens, the slope decreases, which causes an increase in the slope in the upstream section of the river and an increase in the transport of sediment from the upper reaches" (p. 189). N. I. Makkaveev illustrated this by the example of possible changes in the longitudinal profile of the Chu river (Fig. 1.13) in the case of increasing the water flow module in its basin to a value corresponding to the climate of the forest-steppe of the Russian Plain (now the river crosses the semi-desert zone, ending in a "blind" mouth). An increase in transport capacity would result in the river crossing and dredging of the valley in the middle reaches of the river by almost 200 m.

Changes in sediment load are related to channel changes by the equation:

$$\frac{\partial W_{R+G}}{\partial x} + \rho\frac{\partial \omega}{\partial t} = 0 \qquad (1.1)$$

where W_{R+G} is the total sediment load (R—suspended, G—bed load); x—length of the section; ω—area of water section; t—time; ρ—density of load. According to Eq. (1.1), the rate of change in sediment load along the length of the river is equal to the rate of the area of water section of the flow over time, taken with the inverse sign.

Since the equation of sediment balance reflects the direction of erosion-accumulation processes in any form of their representation, the automatic alignment of the transport capacity of the flow should be considered as a mechanism for self-regulation of the "flow—underlying surface" system, and in relation to fluvial processes—the "flow—channel" system. Thus, in the upper part of the slope the flow is characterized by the highest transport capacity. Here, despite the low specific flow rate, the flows are least saturated with sediment; both the so-called additive turbulence and the wave movement of water flows favour erosion (Makkaveev 1971). Down the hill, as the flow of sediment is saturated and the degree of exposure to additive factors is reduced, the deficit of sediment is first reduced and replaced by excess sediment at the foot of the slope. If we proceed from these features of the mechanism of erosive and accumulative activity of

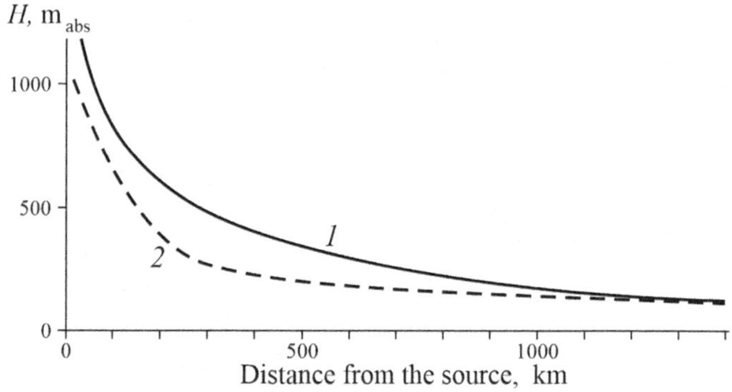

Fig. 1.13 Possible change of longitudinal profile of the Chu river during the increasing of water flow module ($M_{Q1} << M_{Q2}$) and sediment transport capacity of the flow ($W_{tr1} << W_{tr2}$). 1—current profile; 2—hypothetical profile (according to Makkaveev (1995))

non-channel water flow, the slope should eventually decrease. The limit of its alignment will be a lateral plane. However, the intensity of this process, as the slope's steepness decreases, becomes so weak that other factors begin to dominate the terrain and are able to maintain its roughness despite the erosion of the slope flows.

The effect of the law of automatic levelling of the transport capacity is represented in the development of the longitudinal profile of the river and the formation of terraces, thus having a geomorphological expression (Makkaveev 1955, 1970, 1971; Makkaveev et al. 1961; Makkaveev and Chalov 1986). It lies at the heart of the mechanism of movement of ripple forms of riverbed relief and in the development of forms of the channel itself. In the first case, it is related to changes in the length of the river and in the time of the ratio W_{tr} and W, which leads to its incision or accumulation of sediment, in the second—with changes in this ratio along the length of the ripple or in different parts of the channel within each of its forms (bend, branching, etc.).

Through the law of automatic levelling of the transport capacity of the flow, fluvial processes are affected by anthropogenic changes in the natural environment, leading to corresponding changes in water and sediment yield. As a consequence of these changes, the channel transformation or change in the rate of directional channel change takes place. The self-adjustment of the «flow-channel» system varies depending on the conditions in which the flow and the channel interaction: the flow controls the channel or the channel controls the flow. In real conditions one of these mechanisms of interaction in the flow-channel system becomes the leading one. In the first case, the processes proceed quickly, affecting the first years or decades, the channel is constantly changing under the influence of the flow. In the second case, the process is extremely slow (e.g. river incision caused by a shortage of sediment is represented in historical and geological scales of time). Correspondingly, anthropogenic changes in natural factors quickly affect the condition of the channels of lowland rivers under the condition of free development of channel changes. As a result, the rivers incise, for example, in tail-water of hydroelectric power plants, in the course of channel straightening or development of riverbed quarries, etc.), resulting in the transformation of the floodplain into a fluvial terrace; the accumulation of sediment as a consequence of anthropogenic soil erosion leads to the disappearance (degradation) of small river channels, their transformation into gulls, and above the reservoir—to the increase of bottom marks and increase inundation of floodplains. In the lower reaches of rivers, the regulation of runoff and the reduction of sediment load causes the deltas to stop advancing and even partially erode; artificial streambed straightening at the estuary causes a change in the accumulation of sediment by deep erosion. River crossings or accumulation of sediment will continue until the channel (width and depth), the volume of sediment and the flow of sediment are adequate to the transport capacity of the stream.

Self-regulation of the channel under anthropogenic impact on natural factors also causes changes in the development of channel forms. Their parameters are related to flow dimensions: e.g. the channel curvature radii are determined by the mean annual (or average maximum) flow rate. Changes in the water availability of a river result in a transformation of the channel pattern in the extreme case: for example, branching is replaced by meandering; more often there is a decrease in the size of bends, islands or their formation, dying of branches, and activation of bank erosion. Changes in the channel and especially the change in the type of its changes are accompanied by a corresponding transformation of the processes of floodplain formation.

Direct technogenic intervention in the channel (dredging slots, quarries of building materials, corrective and engineering structures, etc.) leads to a reverse impact on the flow and, ultimately, through the mechanisms of self-regulation of the system—to the restoration of equality $W = W_{tr}$. In this case, fluvial processes are changed in an upward and downward direction. In the event of a large number of human impacts, each of these impacts is cumulative or overlapping, so that the transformation of the channel is observed over a long stretch of time and is accompanied by consequences that meet the conditions of $W > W_{tr}$ or $W < W_{tr}$. These processes are equally evident in the wide floodplain sandy and pebbly incised beds.

The law of non-linearity of connections fixes disproportion between quantitative changes of active factors and intensity of representations of fluvial processes. Makkaveev (1976) considered that in the processes controlled by nonlinear connections, the cumulative effect can arise (from Latin word *cumule*—collect, accumulate) in the representation of fluvial processes in time and space for some changes in the active factor, caused by the effects in one direction. Time accumulation is most clearly represent through the presence of a square or cubic interconnection between the discharge Q and the transport capacity of the flows, which in the first approximation can be considered as a reflection of the total discharge of W_{R+G}:

$$W_{R+G} = f(Q^m), \tag{1.2}$$

where the degree m is greater than 1 and depends on the composition of the soils affected by the flow or bed material load.

The law explains the formation and development of the primary erosion furrow on the slope and its transformation first into a water gall and then into a ravine. Makkaveev (1971) called it a self-excited process. Due to the concentration of flowing water in the furrow, its flow rate as well as the depth of the stream increases. The increase in water discharge sharply increases the transport capacity of the flow, as the sediment load increases exponentially (1.2) compared to a linear increase in water discharge: when the discharge is doubled W_{tr} increases four or eight times depending on the m, as a result, it is significantly greater than the sediment load, i.e., $W_{tr} >> W_{R+G}$, which causes the erosion of the channel; an increase in the depth of the flow increases the flow rate in accordance with the Chezy formula.

$$V = C\sqrt{hI}, \tag{1.3}$$

where h is the depth, I is the slope of the flow, C is the coefficient depending on the roughness of the bottom. As a result, erosion is intensified and the primary furrow can be further deepened. The greater the plunging and depth of the furrow, the greater the runoff rates of the flow and the more active the erosion process. Self-development of the erosive form stimulates its deepening and transformation into a ravine. Growth of the ravine is therefore rapid until its increase in size matches the critical catchment area, when further increasing in water discharge stop. Its stabilization leads to the fact that $W_{tr} = W_{R+G}$, and the transformation of the ravine into a gull begins.

The transport capacity of the flows increases as the irregularity of the flow increases. This pattern is particularly evident in slope erosion: one heavy downpour erodes by an order of magnitude greater than a uniform, prolonged rainfall with the same amount of rainfall. Irregularity of runoff determines the predominant development of ravines in forest-steppe and steppe zones, where rainfall prevails. The total effect of linear erosion is greater for the same reason for rainwater than for meltwater. In rivers, the representation of this law is most clearly seen when comparing the "efficiency" of a short flood and a long low water. This seems to be one of the reasons why, in arid countries, where runoff is very uneven, each unit of river runoff usually carries much more sediment than in humid climates.

Cumulation in space is even more obvious. The fusion of flows in river systems of the same or similar order is usually accompanied by a much greater increase in river transport capacity than an increase in water availability. This leads to a decrease, all other things being equal, in the longitudinal profile slope downstream and its inflections being confined to the junctions with large inflows. Makkaveev (1971, p. 28) has shown that due to this "the rivers, merging, are almost never separated downstream". If there was no this pattern, the hydrographic network on our planet would have a completely different pattern: in particular, the formation of large valleys through the addition of smaller ones would be a rare phenomenon. The apparent violation of this pattern is observed only in the area of intensive accumulation of sediment by the rivers, on the removal cones in the foothills, in deltas, i.e., where the flow disperses and reduce its runoff rates along the length of the delta channels. The branching of channels into branches is related to the flow formation at a large width of the river and the formation of accumulative relief of the river channel. However, the very phenomenon of branching of channels by virtue of the law of nonlinearity of connections is reflected in the alternation of shallow (riffled) and deep (pool) sections along the length of the river, timed, respectively, to the nodes of division of the flow into branches and nodes of confluence of branches.

A peculiar representation of cumulation of the flow energy reflected in the process of channel formation takes place at the river bend (Makkaveev 1976). The bend of the channel causes a significant increase in the unevenness of longitudinal velocities in the living section of the flow, so that the conveying capacity of the flow at the bend increases by 1.5–2 times compared to the flow in a straight channel, which has the same average flow rate. For this reason, despite the increase in channel length and the corresponding increase in energy loss along the length, the flow in the bend competes (up to a certain limit of its development) in its transport capacity with the flow in the straightening branch, and the lateral scour pools in the bend of the channel have a greater depth than in the straightening branch. The same is true of branching, where the flow in the branch around the island forms a bend.

The law of *factor relativity* has recently found the most widespread use in research of fluvial processes. The essence of it consists in different and non-simultaneous reaction of erosive and accumulative processes in different links of water flows to changes in the geographical environment—climate, tectonic movements, fluctuations of erosion basis, etc. If water flows react to them almost instantly, then due to the stability of the underlying surface and river channels, their resistance to the impact of flows, erosion and accumulation processes and fluvial processes are characterized by greater or lesser inertia and at any given time morphological forms of their representation may not correspond to the characteristics of the flow. Makkaveev (1971) applied this law for the first time to the explanation of the mechanism of terrace formation and alluvium accumulation in river valleys, linking it to the directional changes in the longitudinal profile of rivers. He subsequently extended it to all the fluvial relief formation. When $W = W_{tr}$ according to the formula of the developed longitudinal profile QI = const erosion and accumulation activity of large rivers, characterized by low slope and high runoff rates, reacts sensitively to minor changes in slopes (for example, caused by fluctuations of base level). Small rivers, ravines and gullies and erosion processes caused by non-stream slopes are almost entirely controlled by climate and vegetation conditions. An example of the latter is the degradation and extinction of small rivers in the south of the forest and forest-steppe zones, associated with the reduction of forests, plowing of land, increasing the degree of unevenness of slope flow and, as a consequence, increasing the removal of products of soil erosion from the territory of catchments into the rivers, where their number is excessive compared to the carrying capacity of the flow.

The law of factor relativity is especially evident in case of anthropogenic impact on nature. Thus, deforestation contributes to a sharp increase in the unevenness of flow from small catchments. Maximum water discharge in gulls often increases by an order of magnitude. This is accompanied by the activation of erosion in the upper links of the erosion network, leading to intensive gully formation. The transformation of the flow regime in the lower links of the network is less dramatic; consequently, the increase in transport capacity is relatively small. As a result, erosion products brought from the upper links of the network accumulate in the channels of small rivers. Thus, the transformation of the flow regime in the catchment area has the opposite effect on the direction of fluvial processes in ravines and river channels.

An example of the fact that the length of the channel system in some cases, the transformation of the channel mode is extremely slow and stretches over time, can serve as an accumulation above large reservoirs. In the backwater wedging area appears the accumulation, which gradually, as the reservoir silts out, moves regressively, i.e. against the river flow. In the accumulation zone there is a noticeable change in the shape of the channel forms and the channel relief: islands and new riffles appear, etc. The front of such a change is moving on flat rivers at a rate of up to several hundred meters per year, and only in the rather distant future will it affect fluvial processes in the higher links of the network.

At the same time, the same anthropogenic factor—the creation of a reservoir—downstream of the river causes incision of the river, with the wave of erosion gradually spreading downstream. The products of this washout accumulate below the washout front, causing a shallowing of the channel, which gradually shifts downstream (Transport use … 1972; Babinski 2002).

Changes in the basic level (receiving basin level) significantly affect the lower links of the channel network. Their effects spread regressively, gradually fading over long periods of time. These changes have almost no effect on the erosion and storage processes in the upper network links. On the contrary, climate change first and foremost affects these processes in the upper reaches of the watercourse network, including small rivers.

The Law on *Mutually Conditionality of Flow and Channel* determines the extent to which banks and riverbeds influence the hydraulic structure of the flow and the simultaneous impact of the flow on the characteristics of the channel. It was first formulated by Velikanov (1958, p. 54) as "the principle of interaction between the flow and the channel": "The channel and the flow … become one organically bound complex, in which the channel reflects the form of the flow, and the flow reflects the form of the channel". However, the foundations of the law were laid much earlier. As early as 1896, N. S. Leliavsky wrote that the shape of the channel controls the flow, but at the same time, "the configuration of the bottom is in close relationship on the position of the current, which is the only natural engine that produces all the changes" (quoted from the edition of (1948, p. 156)).

N. I. Makkaveev also referred this law to fluvial processes. However, the course can be seen in a broader sense. Not only rivers have channels, but also temporary watercourses in gulls, ravines and washes. On the slopes, there are micro-river beds and a "boundless" bed of flat slope flow, the sides of which are the primary, often non-fluvial forms of micro relief of the slopes.

The active factor in erosion and accumulation (fluvial) processes is always water runoff, which is realized through water flows. Their characteristics (water availability, depth, speed, etc.) almost always change in the seasonal cycle and in the multi-year plan, and under the influence of changes in physical and geographical conditions in the historical and geological scales of time. This results in a continuous but slow change in the shape of the channel, which is reflected in the hydraulics of water, including the channel flows. Changing the shape of the channel requires time, which is more time-consuming, the more stable the channel is in relation to erosion, the more resistant the rocks that make it up. In relation to the development of river channels, this fact was expressed in the separation of free and limited development of channel changes (by the nature of the influence of the geological structure on the shape of the valley and the intensity of fluvial processes) (Chalov 1979; Makkaveev and Chalov 1986). In the first case, the weak stability of the rocks composing the river bed and the small channel alluvium determine the predominant role of the flow in the fluvial process (the flow controls the channel). Rivers have wide floodplains, within which their channels are displaced, branching into branches or meandering. These rivers are characterized by the most complete development of the whole complex of channel forms (channel forms and channel relief, the presence of terraces and thick thicknesses of alluvial deposits).

In the conditions of spreading rocks that resist erosion (rocky, cohesive), the flow of even significant energy is sometimes fully controlled by the channel. Limited conditions for the development of channel changes are represented in rivers incising through crystalline rocks. The intensity of channel changes here is so low that their geomorphological effect is revealed only in the geological scale of time. At many river reaches, the direction of the channel and valley follows tectonic disturbances. The valleys are often without floodplain, narrow, deeply embedded; the terraces are highly socle-rich or devoid of alluvial cover, and in many valley areas they are not preserved at all. Channels corresponding to such conditions are deep incised; the forms of their channel relief do not differ in diversity, often the sculptural forms of the channel relief prevail.

The law of mutual conditionality of flow and channel is closely related to the law of *morphological complexes*. Velikanov (1948, p. 13) formulated it in the following wording: "In nature, only a limited number of relatively stable forms of channel formations are carried out from the countless possible formations, which makes it possible to classify natural channel complexes geomorphologically". Subsequently, he connected this law with the mutual control of the flow and the channel, which "leads, as a result of all changes, to certain, most probable combinations between the morphometric characteristics of the channel and the hydraulic characteristics of the flow" (Velikanov 1958, p. 58). N. I. Makkaveev has extended the law of morphological complexes' limitation to all erosive and accumulative processes, defining its essence "by the presence of quite definite types of fluvial relief forms characterized by stability at stationary erosive and fluvial processes" (Erosion processes 1984, p. 27).

According to this law, the channel and relief forms associated with erosion processes are relatively stable and therefore typical if they excite in the flow of phenomena that contribute to their renewal. In the channel bend there is always a circulating current, which contributes to the preservation or growth of its curvature in the process of development and movement. The bottom ripple, moving along the channel, keeps asymmetrical shape in a quiet stream due to the eddy currents in the basement. The ravine on the slope contributes to the increasing concentration of melt and rainwater runoff, which, in turn, determines the preservation and further development of the ravine itself. There are many variations of forms corresponding to different stages of morphological complexes development caused by periodic or directed changes of natural factors and fluvial processes. However, the law of limited morphological complexes makes it possible to classify forms of fluvial relief and, in particular, riverbeds, as well as to identify the leading processes that determine the mechanisms of erosion and channel formation, to create basic schemes of their development.

If we consider the fluvial relief created by constant water flows, then the law of limited morphological complexes allows us to present all its forms in the form of a hierarchical staircase (Chalov 1983b): the river valley—floodplain—form of the channel—channel relief (macro-, meso- and microforms). It corresponds, with some variations, to the assumptions about the discreteness of fluvial processes and forms of their representation. The first form in this chain—the river valley—reflects in its morphology and structure in an integral form the changes in the complex of natural conditions in the river basin during the whole history of its development and has no visible connection with the modern water flow at the bottom of the valley. However, as long as there is a river, the constant water flow will transport sediment from the catchment area and as a result of erosion of the valley slopes and the river bed. This means the development of the valley as a whole as the largest form of fluvial relief. At the same time, valley and terrace slopes are no longer maintained in their development by the river (except in cases of slope scouring) and are exposed to numerous other factors, often altering or even destroying the primary relief created by the river. The floodplain terrace of the river, turning into a floodplain terrace, ceases to be a form of fluvial relief, which is constantly created by the river in the process of its wandering along the bottom of the valley. The relative stability of the channel form is disturbed, and the newly formed terrace can be destroyed by the river that created it in the course of lateral channel changes (wandering and bank erosion). The deeper the river valley and the deeper its depth, the more likely it is that the terraces will be destroyed by both the river itself and the gravitational processes. This leads to the loss of information about erosion and accumulation processes in the geological history of the valley.

River valleys, creating vertical topography and filled with various alluvial layers, influence the amount and composition of solid material entering the river; the shape of the valleys in the plan (alternation of contractions and extensions) determines the nature of flow, especially during floods, affecting the modern relief-forming activity of the river. Thus, river valleys, being (except for floodplains and channels) outside the sphere of direct influence of water flows, create one of the most important geomorphological conditions for the development of river channels.

The last two laws (mutual flow and channel conditioning and limited morphological complexes) fix the stationarity of erosion and accumulation processes due to the reflection of the underlying surface of the flow structure and the emergence of such structural elements in it, which are caused by the reverse effect on the flow of the underlying surface, the formation of a certain set of morphological forms as a result of the interaction between the flow and the underlying surface, reflecting the specifics of erosion, transport and accumulation of sediment.

1.8 History of Fluvial Process Research and Formation of the Riverbed Science

Engels (1969, p. 157) wrote that "… from the very beginning the emergence and development of sciences is conditioned by production", and then showed how the demands of practice and economic activity have led to the development of natural science from ancient times to our days, led to the emergence of new branches of science. These conclusions fully apply to the doctrine of fluvial processes—a branch of knowledge, which originated mainly at the turn of the XIX and XX centuries and was finally formed only in the 40–50 years of the twentieth century. The last date is adopted on the assumption that in 1946–1949, the period between the years 1946 and 1949 was marked by the fact that the period between the end of the year and the end of the year and the end of the year in question was marked by the fact that the period between 1946 and 1949 was marked by the end of the year. M. A. Velikanov was the first in the scientific literature to define the term "fluvial process", reflecting the interaction of water flow and the channel. Later Velikanov (1958, p. 7) noted that "the content of this concept arose in the middle of the last century mainly under the influence of the efforts of scientific and technical thought to rationally solve the problem of using flat rivers for navigation purposes". However, the first observations of the state of the river beds belong to the ancient times. Most of them were not documented, but without such observations it would have been impossible to carry out sufficiently large hydrotechnical works on rivers dating back to the 2nd millennium—VII-III centuries BC. (Degtyarev et al. 2001). The Assyrians, Phoenicians, Egyptians, and Babylonians also built canals and dams for irrigation, water supply, and defense. Shipping on rivers was already the main means of communication and trade. Therefore, the canals had a transport purpose, and ports were created on the rivers, the approaches to which required artificial depth maintenance. It is known that hand dredges were used in Ancient Egypt (Degtyarev et al. 2001, p. 4). Remains of hydraulic engineering systems in Ancient Egypt and Asia show that the builders of those times correctly found a place for the construction of head structures—at the tops of river bends, from the side of the concave bank, which provided the least inflow of sediment into the canals at the maximum flow of water.

The most ancient information about rivers, morphology and classification of their channels is contained in Chinese literary sources. This is obviously due to the fact that almost the entire history of China is connected with the systematic control of floods and riverbank erosion. The first information about the Yangtze is contained in the literary monument of the 2nd millennium BC—the book "Shanshu-Yugong". During the Western Han dynasty (from 260 BC to 24 AD), Jia Zhang applied the term "wandering" to the assessment of the reformation of the Huang He channel; he also organized the first measures to regulate it. During the period of the Three Kingdoms (220–265) Li Daoyuan wrote a book "Shuizinzhu" ("The importance of the hydrological regime"), which contains information on the movement of the Yangtze River bed, changes in its banks and the development of the islands (Chalov et al. 2000).

In the Middle Ages, Galileo Galilei consulted on the project of cutoff of the river Tiber designed to reduce flood levels in Rome. The same period also saw the beginning of regulatory work on the Rhine and other rivers in Western Europe in connection with flood control and the development of waterways: numerous bends were cutoff, dams were built and dredging was carried out. Even earlier, in Kievan Rus in the early XI century, near the fortress buildings were carried out works on the bank. In Russia, the development of rivers was almost universally associated with their use as transport routes. Therefore, the first data on the channels of Russian rivers were contained in chronicles. In historical documents of Russia, it has been fixed, that in XIII century artificial cutoff of the river Sukhona has been spent by building a 2 km long canal. In many Russian cities there was a pilot trade, which contributed to the passage of merchant caravans to the most difficult parts of the rivers, shallow or with stony fillings in the channels. The latter, for example, were extended to the Volga River near Ples city, where local princes used the difficulties of navigation and the need for pilotage of ships to organize the collection of duties from merchants using the waterway within their principality.

River channel regulation for navigation purposes was first undertaken under Peter the Great during the construction of the Vyshnevolotsk Shipping System. At the same time, Admiral Krujs took a picture of the Don channel (Fig. 1.14) to provide for the campaigns of Peter the Great to the Azovs, which became the basis for the first special cartographic image of the river channel—a prototype of future pilot maps of navigable rivers. Mass hydraulic engineering construction in the Peter the Great era was accompanied by the invitation of foreign specialists to Russia and the training of the first Russian hydraulic engineers abroad and in Moscow. In 1708, in Moscow, the first textbook on hydrotechnics, translated from French, was published. It contains initial information on river hydraulics, construction of canals, sluices, and dredging—"The book on the methods of free watercourse on rivers" (Fig. 1.15). The earliest documentary (including cartographic and planned) materials on the river channels are available for the Volga and some of its tributaries. They belong to the X, XVI, XVII centuries, and the German traveler Adam Alearius carried out the first survey of the Volga riverbed near the major cities of the Volga. The library of the Russian Academy of Sciences in St. Petersburg has preserved the originals of some of the eighteenth century channel shootings made on the Volga and Oka, and the archive of the Central Hydrographic Administration of the Navy has manuscript maps and plans for certain parts of the Volga River with the display of the depths and soils of the bottom, compiled by naval officers and navigators (Chalov et al. 2005).

There are three main periods of study of river channels: 1—informational (descriptive); 2—empirical; 3—scientific (analytical). The objectives of each of them, being related to economic development and the needs of society, have been maintained in the future, but their solution has already been based on new methods, the application of which has been determined by scientific and technological progress. The first period was characterized by the accumulation of information and

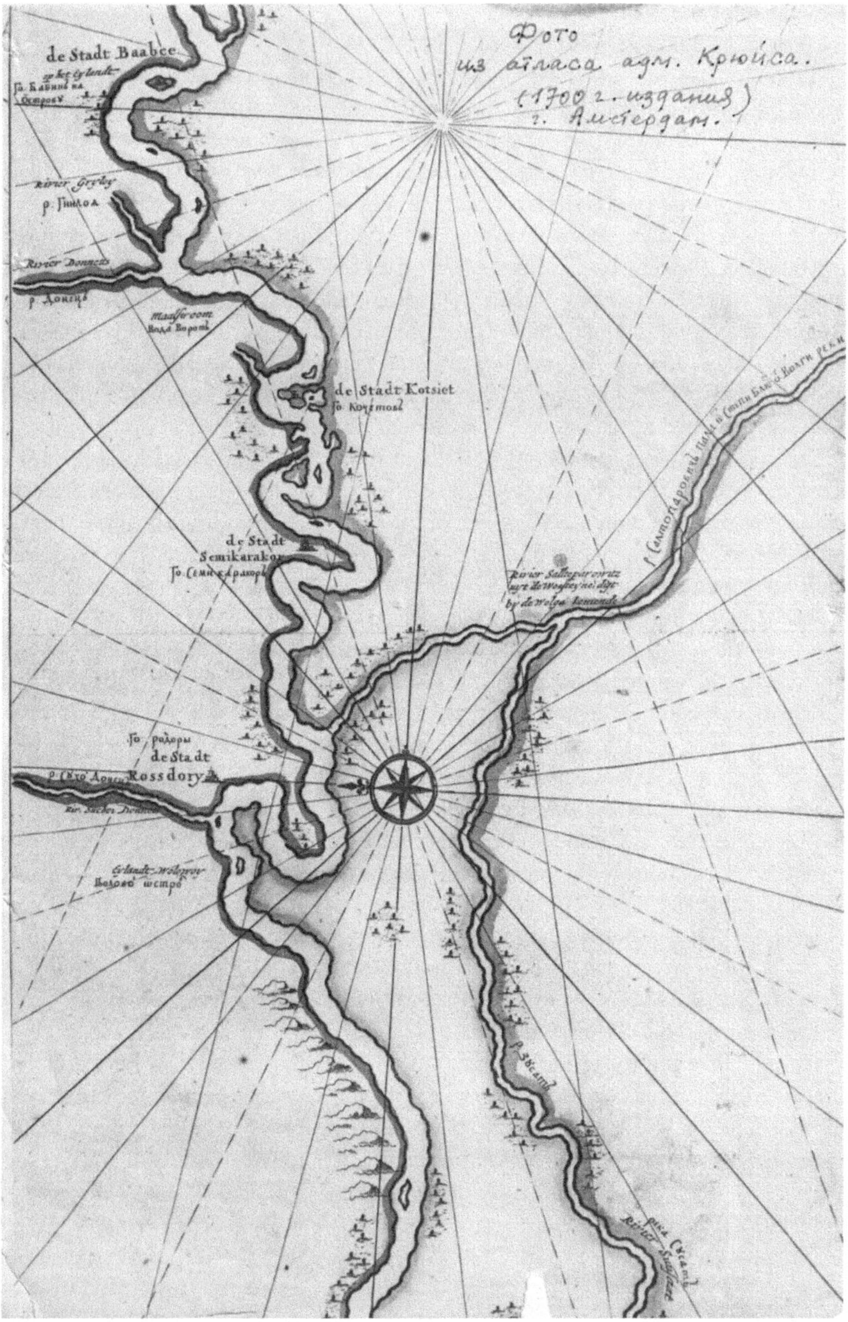

Fig. 1.14 Fragment of the map of the river Don (area of merging with Seversky Donets), made by admiral Cruys (1700, Amsterdam)

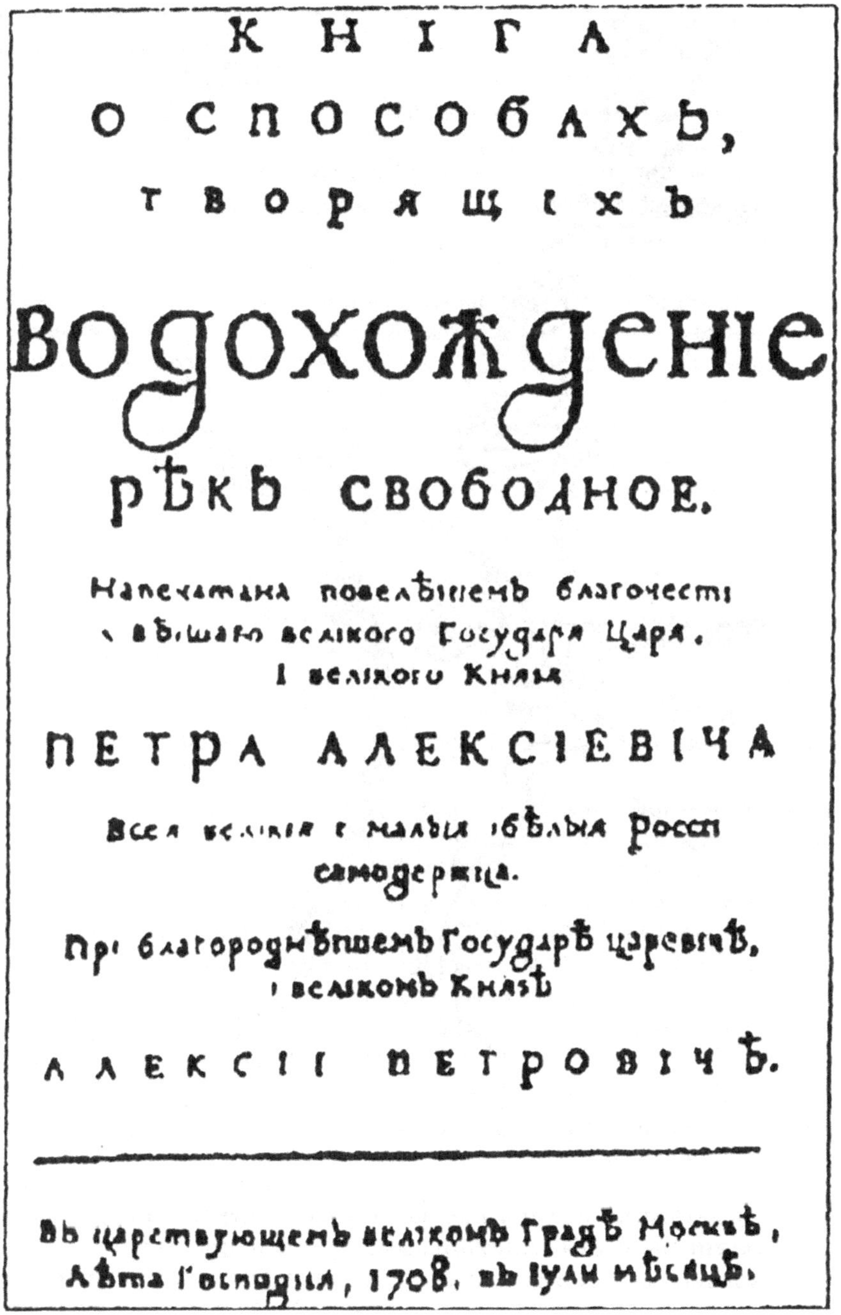

КНІГА
о способахъ,
творящіхъ
водохожденіе
рѣкъ свободное.
Напечатана повелѣніемъ благочести
вѣйшаго великого Государя Царя,
І великого Князя
ПЕТРА АЛЕКСІЕВІЧА
Всея великія и малыя и бѣлыя росси
самодержца.
При благороднѣйшемъ Государѣ царевичѣ,
і великомъ Князѣ
АЛЕКСІІ ПЕТРОВІЧѢ.
Въ царствующемъ великомъ градѣ Москвѣ,
лѣта Господня, 1708, въ Іулии мѣсяцѣ.

Fig. 1.15 Title page of the fist Russian version of the hydraulic engineering textbook published in 1708 in Moscow description of river channels in various literary sources, including manuscripts, the appearance of the first, and then the compilation of large-scale maps specifically devoted to the image of river channels in connection with the use of rivers as waterways. The beginning of the period refers to the Ancient World, during many centuries there was an unsystematized accumulation of data on individual rivers, although episodic. In Russia, the crowning feature of this period was the publication of "Ancient Russian Hydrography containing a description of the Moscow State of Rivers ..." (first half of the XVII century) and "Drawings of the Siberian Land" (second half of the XVII century), which describe and map the main rivers of Russia.

The second period was marked by the fact that observational data on channel reformation causing unfavorable living and working conditions of people on the banks of the rivers, associated, for example, with the hazard of their erosion by the water stream or, on the contrary, the river's "departure" from large settlements, with the need to ensure (in modern terms) the normal navigation conditions (deepening shallow areas—riffles), allowed us to come to an understanding of some general patterns of development of channels and lay the scientific foundations for their regulation to protect the coast from erosion and flood control, the creation of waterways, etc. The transition to this period did not mean that the accumulation of information ceased; on the contrary, the latter acquired a new meaning and moved from the description of the channels to the performance of special survey works, regular performance and improvement of their accuracy, organization of special survey expeditions, institutions and parties.

The importance of creating and further accumulating an information base has increased with the transition to the third analytical period. Its main feature was the development of the basis for the theory of fluvial processes and the branch of knowledge—riverbed science—owing to the generalizations and analysis of all available observation materials, surveys, network data. In essence, this meant that, along with the expansion of purely applied, private tasks, the fundamental science of the river channel began to form, while maintaining close ties with practice.

The beginning of the second period in Russia, obviously, correlates with the times of Peter the Great. After the survey of the Don in the course of two centuries, up to the end of the XIX century, a large number of works, including cartographic ones, appeared, which contained various information about the rivers, mainly in European Russia—the Volga, Don, Dniester, Oka, Northern Dvina, Vychegda, Kama. Many, mainly small and medium-sized rivers in the late XVIII century were reflected in the maps of the General Survey, which are stored in the Russian State Archive of Ancient Acts (Chalov et al. 2005). At the same time, attempts were made to generalize the available information about river channels and channel changes. For the first time M. V. Lomonosov drew attention to the latter, who wrote in his book "About the Earth's Layer": "Residents on the banks of large rivers are witnesses to that, if the great changes in the banks and stresses of their water flow, is the most damaging, causes. I don't mention the sands, which are washed away by spring and autumn, nor the meadows, which are quickly taken away from the front end, increases to the rear, but only the way the interior of the earth opens up, I imagine steep ravines, which sometimes great links with vegetable gardens and buildings settle and are refuted in the rivers, being washed away … Everything in the world miners won't dig through so much land, won't throw away stones in a hundred years, as one spring they will destroy the ice and the rapidity of unprecedented of the Russian waters".

In 1798, the Department of Water Communications was established, which began systematic work on the study and description of waterways, and these works covered not only large rivers, but also many of their tributaries (Tvertsa, Vazuza, Gzhat, Tsna, Vytegra, Sheksna, Mologa, Sura in the Volga basin). Their development was facilitated by the establishment of the Institute of Corps of Railway Engineers in 1810. The research included surveys and channel measurements (from the mouth of the Tom River to the confluence with the Irtysh River was taken at a scale of 1: 21,000), measurements of velocities, water levels, obtaining data on the floodplain, and sometimes—making a longitudinal profile. In the 50s of the XIX century the first pilot map of the Northern Dvina was drawn up, and since that time regular surveying works in the areas difficult for navigation have been started here. In 1861, the first atlas of the r. Volga from Tver to Tetyusha.

By the last quarter of the XIX century, surveys and measuring works were carried out on the Dnieper, Seversky Donets, Oka, Volkhov, Ob, Volga, Don, and Neva, and some of them, several decades later, were repeated. They resulted in maps and atlases of individual rivers, hydrographic maps, and the work of I. F. Shtukenberg on hydrography of Russia. Summarization of these materials in the form of information about channel changes on the rivers of Russia were published in 1854 in the "Shipping Road of European Russia". It noted that intensive changes in channels occur during floods; in the low water period they fade out, causing only local transformations of those forms of channel relief, which were created in spring. The most detailed description in this work describes the reformation of the Volga riverbed.

Data on the movement of the islands in the Volga channel were published in 1857 by K. M. Baer. He also proposed the "Bair's Law" explaining the movement of the main rivers of the Northern Hemisphere to the right and the Southern Hemisphere to the left as a consequence of the Earth's rotation. Thanks to this, the right banks of the rivers in the northern hemisphere are taller and steeper, while the left banks are more distant.

In 1875, navigation and description commissions were established under the Ministry of Railway Transport. A number of "described batches" are being organised at a later stage and have conducted detailed surveys of many navigable rivers. Based on the results of these works, 30 river atlases and 66 issues of "Materials for the description of Russian rivers and the history of improvement of their navigational conditions", separate pilot maps and notes were published. Since the end of the XIX century, pilot maps of European and a number of Siberian rivers have been published at intervals of 5–10 years. Among the authors of "Materials …" and notes were N. P. Puzyrevsky (Dniester, Seversky Donets, Oka), A. K. Staritsky (Selenga), E. V. Bliznyak (Yenisei). There are also generalized works on the river channels and their changes in time: V. M. Lohtin on the Dniester and

Chusovaya, V. I. Ragozin and N. A. Boguslavsky on the Volga, N. S. Lelyavsky on the Dnieper and Pripyat, N. P. Puzyrevsky on the Don, etc. Among them a special place is occupied by the report of N. S. Leliavskiy "On observations over the location of the jets on the Dnieper River near the Yekaterinoslav city", read in 1894, and his article "On the study of the movement of sand spits in Aleksandrovsk" (1896), in which for the first time the channel analysis of river sections was performed.

With the beginning of rapid development of water transport at the end of the XIX century—beginning of the XX century is associated with the laying of the foundations of the modern theory of fluvial processes, which marked the transition to the third, scientific (analytical), stage of the study of fluvial processes, which led to the formation of riverbed science as an independent branch of knowledge. In the works of travel engineers, which include the works of V. M. Lokhtin, N. S. Leliavskiy, V. G. Kleibor, N. N. Zhukovskiy, V. E. Timonov, V. M. Rodevich, S. P. Maksimov, were summed up the results of research of river beds, described the regularities of the processes occurring in them. An important role in the formation of the theory of fluvial processes played the congresses "Russian activists on waterways" and a number of international congresses on navigation. The decisions on standardization of river data were made at them, and the first scientific ideas about fluvial processes were developed in the discussions on ways to improve navigation conditions.

Theoretical foundations of the teachings on fluvial processes were laid in the classic works of V. M. Lohtin "On the mechanism of the river channel" (1897) and N. S. Leliavsky "On river flows and formation of the river channel" (1893). The latter came to the conclusion that the continuous interaction between the flow and the channel is the basis of the river channel development mechanism. In his report to the III Congress of Russian Waterways in 1896, he wrote: "The shape of the channel is in close mutual connection with the location and speeds of various jets, and therefore, for the full study of the movement of water in rivers with the submarine bed it is necessary, in parallel with the study of river jets, to do research on changes in the configuration of the channel" (quoted from the edition of (1948, p. 147)). The dialectical position of N. S. Leliavsky's substantiation of the mutual connection of causes and consequences in the channel phenomena is, in fact, fundamental in the modern theory of fluvial processes. However, the followers of N. S. Leliavsky in the first decades of the twentieth century in their research and theoretical concepts focused on the study of only one side of the two-unit process—river currents. This led to the fact that the doctrine of fluvial processes (in its narrow sense) for a long time developed as a doctrine of channel flow. Therefore, many researchers (A. V. Karaushev, V. N. Goncharov, K. V. Grishanin) considered fluvial processes as a part of channel flows dynamics. Only the author of the very concept of "fluvial process", M. A. Velikanov, being a representative of the engineering and hydrodynamic direction in general, in his last book (1958) referred the main issues of channel flow dynamics (turbulence, flow kinematics, transverse circulation, etc.) and sediment motion to the general theory of fluvial processes.

Another founder of the teachings on fluvial processes, V. M. Lohtin, for the first time expressed the idea of relationship of fluvial processes driver activity of rivers on natural conditions. Being an engineer-traveller, far from natural sciences, but being an inquisitive researcher and observer working on such different rivers as Chusovaya, Dniester, Volga and Neva, V. M. Lohtin came to the conclusion that the wording of which deserves to be completely given here (quoted from the edition (1948, p. 24)): "Any river is formed by a combination of three main independent elements, namely: (1) high runoff rates determined by atmospheric and soil conditions of precipitation on the river area and their flow into the river from tributaries, (2) slope, or steepness caused by the relief of the river crossing, and (3) greater or lesser dilution of the river bed, corresponding to the properties of its current layers of land. These three elements define the nature of the river, give it some specific features that distinguish it from other rivers, and all the hydraulic factors that we can observe at each individual cross section. … All of these factors constitute only the subsequent local representations of the interplay of the same three basic elements of nature. In addition, he introduced the notion of channel stability, proposing an indicator, "Lokhtin's number", which is still widely used today. He also substantiated the relationship between channel changes and sediment load, and showed that "sediment does not flow into the river due to bank erosion …; it is collected together with water flowing down from the entire basin, and more or less of it is an inevitable fact, independent of the condition of the channel itself" (1948, pp. 32–33), and also proved the relationship of channel changes on the hydrological regime of the river.

Among the foreign researchers of the late XIX—early XX centuries, the French engineer L. Farg stands out, who established a number of general patterns of formation of meandering river beds. Earlier French engineer G. Girardon created the first classification of river crossings. Studies of the processes of development of river bends, morphology and dynamics of rifts were conducted by R. Yasmund, F. Exner, M. Moller. However, their conclusions, based on rivers with relatively homogeneous conditions of formation, proved to be to a greater or lesser extent unacceptable for the rivers of the Russian Plain and Siberia, especially the largest, unparalleled in Europe. And although in some cases they anticipated the work of Russian hydraulic engineers on the physical validity of the conclusions, V. M. Lohtin, N. S. Leliavsky and other Russian engineers refused to mechanically transfer to the Russian rivers channel control methods developed for the rivers characterized by different conditions of channel formation in Western Europe, and laid the foundation for modern ideas about fluvial processes on the rivers of Russia.

In essence, in the works of V. M. Lohtin, N. S. Leliavsky and others, the principle of interaction between the flow and the channel was formulated for the first time, and thoughts were expressed about the relationship between the choice of methods of correction (regulation) of channels and the peculiarities of channel formation. This approach was adopted by Russian hydraulic engineers at the same time and has not lost its relevance to the present day. The essence of it is that the impact on the river channel and measures for its regulation should be based on the laws of natural processes of channel formation, thus causing their positive effect, which creates the prerequisites for the preservation of the river as a natural object and ensures environmental safety of these measures. However, the science of that time was not able to go beyond the limits of qualitative generalizations regarding the mechanism of river flows and explanation of channel change, although the research of N. N. Zhukovsky, relating to the beginning of the XX century, significantly enriched the knowledge about the structure of flows. Only in the 30s, due to laboratory modeling and field studies of river flows by A. I. Losievsky, A. Ya. Milovich, M. V. Potapov, N. F. Daneliya, A. A. Bolshakov, etc., the complex spatial nature of currents in river beds was studied quite well. In particular, a number of problems related to the peculiarities of the flow velocity field (A. Ya. Milovich), development of transverse circulation (M. V. Potapov) and others were solved. Later, on the basis of the data obtained and personal developments and research, M. A. Velikanov created the theoretical foundations of hydrodynamic direction in the theory of fluvial processes. V. N. Goncharov, V. M. Makkaveev, I. I. Levi, K. V. Grishanin and others made a great contribution to the development of the theoretical basis for the dynamics of channel flows. In the end, this direction became an independent scientific discipline in which, along with the issues of flow dynamics, the methods of calculation of sediment load motion and channel changes were considered (I. I. Levi, G. I. Shamov, V. N. Goncharov, V. G. Glushkov, M. A. Velikanov, K. V. Grishanin).

Of great importance for the development of the riverbed doctrine was the solution of problems related to large hydraulic engineering construction, which required the substantiation of forecasts of possible riverbed changes as a result of flow regulation. One of the first works on the channel regime of rivers with regulated flow was the publication of Polyakov (1933). Later, within the framework of these tasks, the attention of the researchers will be focused on the issues related to channel erosion in the lower reaches of the hydrosystems (works by K. I. Rossinsky and I. A. Kuzmin, I. L. Rozovsky, I. I. Levi, N. I. Makkaveev and B. G. Fedorov, V. S. Lapshenkov, A. B. Veksler and V. M. Donenberg).

Hydrodynamic direction in the 30–50s could not provide the fulfillment of all practical tasks. Therefore, studies of the morphology of river channels and their changes under the influence of water flows in the long-term and seasonal plans, taking into account the specifics of the hydrological regime of the rivers and the differences in sediment load, continued in parallel. These researches which have made a morphodynamic direction, continued in connection with the decision of questions of correction of river channels for improvement of conditions of navigation (works of A. I. Losievsky, N. I. Makkaveev, N. A. Rzhanitsyn, etc.), working out of forecasts of changes of morphology of river channels at regulation of a drain by hydrounits (N. I. Makkaveev, A. V. Serebryakov), their reformation in the zones of seepage in reservoirs (S. V. Rusakov, N. I. Makkaveev), consideration of fluvial processes in various types of water management use, construction of engineering facilities on the banks and river crossings (I. V. Popov, N. E. Kondratyev). These works were dominated by the engineering approach, which in turn was dominated by the water transport approach. Makkaveev and Sovetov (1940) generalized the experience of dredging on the rivers of the European part of the USSR and showed the relationship of the methods of improving the conditions of navigation on the peculiarities of the river channel regime. Later this direction was developed in the works of G. L. Sadovsky, A. I. Chekrenev, K. V. Grishanin, V. V. Degtyarev, F. M. Chernyshov. The advantage of this water transport approach was that it provided an insight into the fluvial processes in the rivers in the hinterland or in very long river sections. Therefore, it was within its framework that the concept of the riverbed regime appeared and the Geographical direction in the study of fluvial processes originated. His credo was formulated by V. M. Lohtin, a path engineer, in the form of "postulates" that determine the relationship of fluvial processes driver activity of a track on a set of natural conditions.

The issues of channel process accounting in irrigation construction were covered in the works, mainly of Asian channel workers. Studies of fluvial processes on Central Asian rivers, characterized by a number of specific features, and the development of methods to regulate their channels, are connected with the name of S. T. Altunin.

At the end of the XIX and beginning of the XX centuries, almost simultaneously with the development of the theory of riverbed formation, the doctrine of the floodplain, developed by meadow and soil scientists A. M. Dmitriev, V. R. Williams, and later R. A. Elenevsky, was born. Together with classical works by V. V. Dokuchaev and A. P. Pavlov on the formation of river valleys in their study initiated the geological-geomorphological direction in the study of fluvial processes. Issues of morphology of river floodplains in the pre-war period were covered in the works of R. A. Elenevsky, who described the floodplains of the Oka, Belaya, Volga and some other rivers of the European part of the USSR. In the book "Questions of studying and mastering floodplains" (1936), he summed up the results of many years of research and proposed a detailed genetic classification of floodplains. Many of the terms proposed by R. A. Elenevsky for the first time have been firmly included in the literature, and some of the later classification schemes, in fact, are the development and refinement of his ideas.

However, when explaining the floodplain morphology, the structure of alluvial strata and especially the mechanism of formation of river valleys and terraces, this direction of research was mostly based on intuitive ideas about the dynamics of river channels, rather than on the knowledge of its regularities. The same was demonstrated in geological-geomorphological works in connection with the study of river sediments, alluvial placers and paleogeographic reconstructions. However, hydrologists (for example, M. I. Lvovich) were involved in solving the problem of formation of alluvial deposits as well as the product of fluvial processes driver activity of rivers, and their contribution allowed revealing some physical features of the alluvial formation mechanism. For the first time, channel processes on mountain and semi-mountain rivers were studied. Summarization of the results of research conducted on small rivers in Eastern Siberia with reference to the formation of placers was made by Yu. A. Bilibin, in whose book "Fundamentals of placer geology" special chapters are devoted to the hydrological and channel regime of mountain rivers (the first edition was published in 1938). However, alluvial placers as derivatives of channel processes and objects of research of riverbed science began to be considered only in the 70–80s in the works of N. I. Makkaveev and N. V. Khmeleva.

While hydrotechnical engineers regarded fluvial processes almost exclusively as modern and “before our eyes”, geomorphologists and geologists, as a rule, dealt with such a form of their representation, which reflected the centuries-old, historical or even geological development of rivers. At the same time, the relatively independent development of hydrodynamic and morphodynamic (geomorphological) approaches to the study of fluvial processes is perceived by many researchers as a certain disadvantage, which should have been overcome. The logic of science development itself created objective conditions for combining hydrodynamic and morphodynamic (geomorphological) directions in a single branch of knowledge about fluvial processes. The parallel independent development of both directions led to the accumulation of extensive material, the theoretical understanding of which became impossible without the destruction of historical barriers between technical and natural disciplines, within which the theory of fluvial processes was formed. On the one hand, this resulted from the fact that the channel forms are the result not only of modern processes of interaction between the flow and the river bed, which ultimately reflect the entire complex of physical and geographical conditions in the catchment area, but also its historical development. On the other hand, it is impossible to understand correctly the history of the river valley development, the formation of terraces and floodplains, the structure of alluvial strata without knowledge of the fluvial processes in all their representations: from local periodic lateral channel changes (development of bends and other forms of channel), which are superimposed on the rippled movement of sediment, to the directed general transformations of the longitudinal profile of the river. Otherwise, the researcher, dealing with the final result of the geological development of fluvial processes, creates speculative schemes, often devoid of physical basis. Indeed, as Engels (1969, p. 30) pointed out, “in theoretical natural science, one cannot construct connections and enter them into facts, but must extract them from the facts and, finding them, prove them, as far as possible, experimentally”.

The rapprochement of both directions was observed in the 30–40s of the XX century, which was stated by Velikanov (1948, p. 14), who wrote that “now … the doctrine of fluvial processes … is formed into an independent division of hydrology, based on the synthesis of hydrodynamics with geomorphology”. To the fullest extent this was represented in the works of Makkaveev (1949, 1955), who created a geographical direction in the study of fluvial processes and proved the need to consider fluvial processes as part of a single erosion and accumulation process that unites in one chain all types of water flows (from slope to channel) in their interaction with the underlying surface. At the same time, geologist E. V. Schanzer (1951) made the first attempt to apply hydrodynamic representations of N. S. Leliavsky, A. I. Losievsky M. V. Potapov to explain the mechanisms of formation of floodplains and alluvial deposits, their structure and texture. However, E. V. Schanzer to a large extent used some hydrodynamic representations obtained in laboratory conditions and not confirmed by the analysis of fluvial processes on large rivers. Unfortunately, such an approach, which was justified in the late 40s and early 50s, was unconditionally accepted by its followers (Y. A. Lavrushin, A. A. Lazarenko) already in the 60s, entered without corrections into textbooks on general geology and geomorphology. In this respect, the work of Chistyakov (1978), devoted to the alluvium of mountain rivers and the regularities of its formation, which is based on modern ideas about the fluvial processes in mountain rivers, differs favorably from them.

Thus, "the synthesis of chemical hydrodynamics, with its complex experimental methods and theoretical generalizations, and geomorphology, which collects and analyzes the actual materials on the formation of river channels" (Velikanov 1948, p. 12) led to the intensive development in the 50–70s of the doctrine of fluvial processes, which eventually led to the formation of riverbed science as an independent science, developing at the junction of hydrodynamics, hydrology and geomorphology (Chalov 1992, 1997a, b). This became possible thanks to the general works of Makkaveev (1949, 1955), Velikanov (1954, 1955, 1958), Kondratyev et al. (1959), Rzhanitsyn (1960), Popov (1965), reflecting the dialectics of the development of science in general: "an infinite process of deepening the human cognition of things, phenomena, processes, etc. from phenomena to the essence and from a less profound to a deeper essence"; "from coexistence to causality and from one form of communication and interrelationship to another, deeper, more general" (Lenin 1986, p. 177). Summing up the stage of formation of riverbed science, Grishanin (1972, p. 4), a representative of the hydrodynamic direction in the study of fluvial processes noted that in the theory of fluvial process at the present stage (mid-2nd half of XX century) "reasonably combined geomorphological and hydrodynamic methods".

A special place in the formation of riverbed science is occupied by the book published in 1955 by N. I. Makkaveev "The riverbed and erosion in its basin", in which the development of fluvial processes was put in direct relationship on natural conditions and their place in the general system of erosion and accumulation processes was determined. N. I. Makkaveev has shown that river fluvial processes driver activity has geographical regularities and depends on the whole set of erosion processes within its basins. He also has the task of detailing and concretizing the ideas about the structure of flat river channels, developing the basic schemes for the development of meandering and branched channels, the formation of rolling stock, etc. The fluvial processes were considered by him as a result of the interaction between the flow and the channel, on the one hand, and sediment load, on the other. Thanks to N. I. Makkaveev's works, fluvial processes became a subject of study of geographical science. For the first time in one study of fluvial processes, geomorphological problems (formation of river valleys and terraces) and traditional issues of riverbed formation (riverbed regime) were considered.

Since the mid-50s, fluvial processes have been studied at the State Hydrological Institute. It has allowed to create the new direction which has received the name "the hydromorphological theory of fluvial process". The beginning of this direction was the famous monograph "The Fluvial Process" (Kondratyev et al. 1959). N. E. Kondratyev, I. V. Popov, B. F. Snishchenko, V. I. Antropovsky made a great contribution to the development of the riverbed, which, having shown a variety of morphological representations of fluvial processes, established quantitative links between channel types and factors of fluvial processes. At the same time SHI developed many aspects of the theory of fluvial processes. Thus, Znamenskaya (1968) proposed new methods for calculating channel changes based on the leading role of the ripple motion of sediment in the formation of channels. Under the direction of A. V. Karaushev (Sediment flow … 1977), the geographical patterns of river load formation were studied for the first time. Karasev (1975) considered the issues of channel formation depending on hydraulic resistance, sediment load and soil connectivity with which the flow interacts.

At the same time, various experimental studies were carried out on laboratory models of the processes of ripple formation and ripple motion of sediments (Znamenskaya 1968, 1992), as well as the development of longitudinal river profiles (Makkaveev et al. 1961), and the formation of bend radii (Experimental geomorphology 1969). In the 60s, for the first time, attention was paid to the specifics of fluvial processes on mountain rivers (Chalov 1968; Talmaza and Kroshkin 1968). From the standpoint of modern concepts of channel changes, Chernov (1983) considered the processes of floodplain formation. Baryshnikov (1978, 1984) analyzed the interaction of floodplain and channel flows, linking it to the peculiarities of morphology and formation of floodplains.

The current level of development of riverbed science is determined by a series of monographs and textbooks published from the beginning of the 60s to the beginning of the XXI century. These include works (listed in chronological order) by Popov (1961, 1965), Chalov (1979, 1997a, b), Kondratyev et al. (1982), Rzhanitsyn (1985), Makkaveev and Chalov (1986), Baryshnikov and Popov (1988), Mirtskhulava (1988), Sidorchuk (1992), Alekseevsky (1998), Chalov et al. (1998), Butakov (1999), Berkovich et al. (2000), Berkovich (2001), Obodovsky (2001). Some of them cover some aspects of the problem (riffles, sediment load), others open up new directions of riverbed science (ecological), are devoted to the mechanisms of interaction of the flow with the underlying soils or give an assessment of the problem as a whole. Along with them there were numerous articles devoted to the analysis of fluvial processes on specific rivers, and also generalization of regional researches of fluvial processes in the monograph "The channel mode of the rivers of Northern Eurasia" (1994) and maps of fluvial processes on territory of the former USSR, the European part of Russia and the contiguous countries, separate regions of Russia has been executed. The development of the cartographic method for the study of fluvial processes was preceded by the development and publication in the 60s of the XX century of a series of schemes for the distribution of riverbeds of different types (according to the SHI classification) for the separate regions of the former USSR, which together formed a single scheme for the entire country (Kondratyev et al. 1959; Pinkovsky 1961, 1966, 1967).

In the 90s of the XX century, a number of regional monographs were published, in which a comprehensive analysis of fluvial processes and related applied problems was given on the rivers of the Lena river basin (Waterways … 1995), the Ob

(Fluvial Processes and Waterways … 2001), the Amur (Soloviev 1967; Gusev 2002), the Lower Volga (Zaitsev et al. 2002), on the rivers of the Altai region (Fluvial processes … 1996), Perm region (Nazarov and Yegorkina 2004) and Udmurtia (Rysin and Petukhova 2006). In addition, a large number of articles have been published in journals and collections, which cover channel changes and morphology of the channels of individual rivers and rivers in different regions (Tatarstan, Primorye).

Among other researchers who have made a great contribution to the development of the doctrine of fluvial processes, we should mention K. I. Rossinsky and I. A. Kuzmin, who substantiated the first Russian classification of river channels, which now lies at the heart of almost all classification schemes available in the world literature, and developed methods for predicting the erosion of channels in the lower reaches of hydroelectric installations, O. V. Andreev and I. A. Yaroslavtsev, A. F. Kudryashov, V. S. Lapshenkov, V. S. Altunin, V. K. Debolsky, V. S. Borovkov, A. N. Butakov.

Abroad, in the second half of the XX century, the works of L. Leopold and M. Volman and E. Lane, dedicated to the formation of different types of channels and the search for quantitative estimates of their separation, became widely known. Further development of research in this area was carried out by R. Kollander, R. Griffiths, S. Shumm, who considered the problem of formation of river channels of different types from the point of view of their stability. Since the mid-1970s, foreign researchers have been paying great attention to mathematical models of riverbed formation (G. Parker, F. Engelund, I. Fredso) and the search for hydrologic and morphometric links for certain types of channels and their varieties (Z. Begin, H. Chang, T. Nakagawa, P. Ackers). In Germany, F. Nestmann and R. Kromer carried out research into the results of centuries-old riverbed regulation and justified the principles of natural recovery in order to improve the environmental situation. In Poland, Babiński (1982) gave a detailed analysis of the formation of the Vistula river channel in the lower reach of the Włocławek hydroelectric complex, and then (Babinski 2002) generalized the information on fluvial processes downstream of reservoirs in different parts of the world. The typology of mountain river beds was developed by Kzemen and Helmitsky (1998).

Major studies of fluvial processes have been carried out in China, where they are particularly relevant due to the systematic accumulation of sediment loads resulting from this threat of flooding and intensive scouring of coastlines. Their beginning dates back to the 50–60s of the XX century. At that time, specialized research institutes of water management of the Huang He River (in Zhengzhou) and the Yangtze River (in Wuhan) were established, which began to conduct a comprehensive study of river basins. In particular, since the 1950s, special attention has been paid to the development of the river load problem at the Huang He Plateau, which is the main source of river load. Based on them, in 1955, Juan Binwei drew up a map of soil erosion in the basin of the middle reaches of the Huang He River. It was the foundation for subsequent erosion control works, which led to a reduction in river load.

The largest studies of fluvial processes on the Huang He river belongs to Tsian Ning. He gave a detailed analysis of the conditions for the formation of the wandering channel in the lower reaches of the Huang He, its morphology, the development of bends, the movement of meso- and microforms of the ripple relief and other features of the dynamics of the channel, its spatial and temporal variability depending on the flow of water and sediment, the structure of the banks of the river, revealed the laws of erosion and accumulation in the channel and generally explained the reasons for the wandering of the channel in the lower reaches. He proposed, on the basis of statistical data processing for the Huang He, Weihe, Yundinghe, Fenhei and other rivers, a channel wandering indicator that is now widely used in China. All this is generalized Tsian Ning and Zhou Wenhao in the book "Fluvial processes in the lower reaches of the Huang He River" (1965). In 1987, Tsian Ning and his co-authors published the book "The Fluvial Process", which is a systematic presentation of the theory of fluvial processes in relation to rivers in China, which are characterized by very high sediment load.

Subsequent studies of fluvial processes at the Huang He River have been summarized in the book "Riverbed Geomorphology in the lower reaches of the Huang He River" (I. Tsingzao et al. 1990) and "Atlas of fluvial processes in the lower reaches of the Huang He River" (1985), compiled under the guidance of Zhao Ye'an. "Atlas …" Zhao Ye'an and accompanying monograph serve as the basis for the development of projects to regulate the lower reaches of the Huang He River to prevent bank erosion and flooding.

In the 1990s, the All-China Research Institute of Water and Hydropower (Beijing) and the Institute of Water Management of the Huang He River carried out studies on the impact on the fluvial processes of the Sanmensya Hydropower Plant, the construction of which led to the erosion of the channel and changes in the conditions of sediment accumulation, and the development of flood control methods. Their results were published in the book "Basic characteristics of fluvial processes in the lower reaches of the Huang He River" (I. Tsingzao et al. 1993).

They are carried out in conjunction with the study of the hydrological regime and river load and are closely related to practical tasks: the selection of locations for river ports, ripples in Wuhan and Nanjing, channel regulation, etc. At the same time, laboratory experiments were carried out, which together with the data of full studies made it possible to make reasonable forecasts of channel changes.

Major studies of the fluvial processes of the Yangtze River have been carried out due to shipping problems. Since 1972, the Institute of Geography of the Chinese Academy of Sciences and the Institute of Water Management of the Yangtze River have conducted a special study of the formation and regime of the branched channel in the middle and lower reaches of the Yangtze River. The results were published in the book "Characteristics of the channel and fluvial processes in the middle and lower reaches of the Yangtze River" (1985).

Literature

Alekseevsky, N.I.: Formation and Movement of River Load, p. 203. Moscow State University Publishing House, Moscow (1998)
Alekseevsky, N.I., Chalov, R.C.: Movement of Sediments and Fluvial Processes, p. 171. Moscow State University Publishing House, Moscow (1997)
Atlas of fluvial processes in the lower reaches of the Huang He River. Huang He River Water Management Commission, Zhengzhou, 114 l (1985) (in Chinese)
Babinski, Z.: Procesy korytowe Wisty ponizey Zaporu wodnei we Wloclawku. Dok. Geogr. Z. 1–2, 94 s (1982)
Babinski, Z.: Wplyw zapor na procesy kopytowe rzek abezialnych. Bydgoszcz, 186 s (2002)
Baryshnikov, N.B.: River floodplains, p. 152. Hydrometeoizdat, L (1978)
Baryshnikov, N.B.: Morphology, Hydrology and Hydraulics of Floodplains, p. 280. Hydrometeoisdat, L (1984)
Baryshnikov, N.B., Popov, I.V.: Dynamics of channel flows and fluvial processes, p. 456. Hydrometeoizdat, L (1988)
Baryshnikov, N.B., Samuseva, E.A.: Anthropogenic influence on the self-regulating system of a basin—river stream—channel, p. 220. Russian State Medical University, St. Petersburg (1999)
Belotserkovskiy, MYu.: Erosion-ecological condition of the arable lands of the European territory of Russia (in Russian). Problems of the Russian territory ecological tension estimation: factors, zoning, consequences, pp. 38–45. Moscow State University Publishing House, Moscow (1996)
Berkovich, K.M.: Modern transformation of the upper Oka longitudinal profile (in Russian). Geomorphology (1), 43–49 (1993)
Berkovich, K.M.: Geographical analysis of anthropogenic changes in channel processes, p. 164. GEOS, MOSCOW (2001)
Berkovich, K.M., Chalov, R.S., Chernov, A.V.: Ecological riverbed science, p. 332. GEOS, M (2000)
Berkovich, K.M., Zlotina, L.V., Ivanov, V.V., Nikitina, L.N., Ryazanov, P.I., Turykin, L.A., Chalov, R.S., Chernov, A.V.: Development of the lower and middle Dniester riverbed under the anthropogenic load (in Russian). Ecological problems of the soil erosion and fluvial processes, pp. 141–165. MSU Publications, Moscow (1992)
Borsouk, O.A., Chalov, R.S.: About the mechanism of formation of river terraces. Longitudinal profile of the rivers and their terraces, pp. 13–18. MF GS USSR, MOSCOW (1978)
Butakov, A.N.: Hydraulics of the river channel mesoform development, p. 216. RUDN Publishing House, Moscow (1999)
Chalov, R.S.: Some features of the mountain rivers channel regime (in Russian). Meteor. Hydrol. (4), 70–74 (1968)
Chalov, R.S.: Geographical investigations of fluvial processes, p. 232. MSU Publishing House, Moscow (1979)
Chalov, R.S.: Indicators of channel stability, their use for estimation of intensity of channel changes and ways of their improvement. Dynamics of channel flows, pp. 46–53. LPI Publishing House, L (1983a)
Chalov, R.S.: Factors of fluvial processes and hierarchy of channel forms. Geomorphology (2), 16–26 (1983b)
Chalov, R.S.: Laws of Fluvial Geomorphology and Problems of Theoretical Geomorphology, pp. 111–121. Nauka, Moscow (1988)
Chalov, R.S.: General, geographical and engineering investigations: subject of research and positions in the system of sciences. Vest. Moscow. Un. Ser. 5. Geography (in Russian) (6), 10–16 (1992)
Chalov, R.S.: General and geographical riverbed science, p. 112. MSU Publishing House, Moscow (1997a)
Chalov, R.S.: Fluvial processes as hydrological phenomena and their geomorphological reflection. Hydrology and geomorphology of river systems, pp. 23–32. Irkutsk (1997b)

Chalov, R.S.: Erosion and fluvial processes in training courses for hydrologists and geomorphologists in state (classical) universities. Problems of erosion and fluvial processes in training courses of universities, pp. 6–11. AGPI Publishing House, Arzamas (2002)
Chalov, R.S., Alabyan, A.M., Ivanov, V., Lodina, R.V., Panin, A.V.: Morphodynamics of the channels of plain rivers, p. 288. GEOS, M (1998)
Chalov, R.S., Liu, S., Alekseevskiy, N.I.: Sediment flow and fluvial processes on the big rivers of Russia and China, p. 216. MSU Publishing House, Moscow (2000)
Chalov, R.S., Kargapolova, I.N., Shtankova, N.N.: Study of fluvial processes on the Volga Basin rivers: history, condition, problems. Erosion and accumulative processes in the upper and middle Volga Basin, pp. 196–213. Udmurt Un, Izhevsk (2005)
Characteristics of channels and fluvial processes in the middle and lower reaches of the Yangtze River, p. 272. Science, Beijing (1985) (in Chinese)
Chebotarev, A.I.: Hydrologic dictionary, p. 224. Hydrometeoisdat, L (1964)
Chernov, A.V.: Geomorphology of floodplains of plain rivers, p. 198. MSU Publishing House, Moscow (1983)
Chistyakov, A.A.: Mountainous alluvium, p. 288. Nedra, Moscow (1978)
Dal, V.I.: Explanatory dictionary of living Great Russian language, vol. 1, p. 700. St. Petersburg. M.: Izd. of M.O. Wolf (1880)
Darbutas, A.A.: Fluvial processes of the R. Neman (Nemunas) and Influence of the Anthropogenic Factor on them: Abstract from the … Ph.D. geogr. sciences, p. 27. MSU Publishing House, Moscow (1992)
Debolsky, V.K., Kotkov, V.M.: Peculiarities of the deficit forms dynamics in the incoming flows (in Russian). Meteor. Hydrol. (10), 67–71 (1977)
Dedkov, A.P., Mozzherin, V.I.: Erosion and Sediment Sewage on Earth, p. 264. KazSU, Kazan (1984)
Degtyarev, V.V., Degtyareva, V.V., Mukhin, MYu.: The millennium of dredging on waterways (review of foreign and domestic periodicals), p. 28. NGAVT, Novosibirsk (2001)
Ecological atlas of Russia. ZAO "Karta", 128 l (2002)
Ecology of Erosion and Fluvial Systems in Russia, p. 163. MSU Publishing House, Moscow (2002)
Elenevskiy, R.A.: Questions of study and development of the river floodplains, p. 100. VASKHNIL Publishing House, Moscow (1936)
Engels, F.: Dialectics of Nature, p. 260. Gosolitizdat, Moscow (1969)
Erosion processes, p. 256. Thought, Moscow (1984)
Experimental Geomorphology, ed. 2, p. 173. MSU Publishing House, Moscow (1969)
Faktorovich, M.E.: Schematization of the channel formation processes and development of methods of calculation of channel changes (in Russian). Dynamics of deposits in open channels, pp. 32–37. Nauka, Moscow (1970)
Fluvial processes and waterways in the rivers of the Ob basin, p. 300. RIPEL-plus, Novosibirsk (2001)
Fluvial processes on the rivers of the Altai region, p. 243. MSU Publishing House, Moscow (1996)
Fluvial processes on the rivers of the Altai region. The scale is 1:1000000. GUGGK CSSR, 1 l (1991)
Fluvial processes on the rivers of the USSR. Scale 1:4000000. GUGGK USSR, 4 l (1990)
Geographical encyclopedic dictionary: Concepts and terms, p. 432. Soviet Encyclopedia, Moscow (1988)
Golosov, V.N.: Erosion-accumulative processes in the upper links of a fluvial network of the mastered plains of a moderate belt. In: Proceedings of Intern. Dr. geogr. of sciences, p. 45. MSU Publishing House, Moscow (2003)
Gotlib, Ya.L., Kuzmin, I.A., Razzorenov, F.F., Sokolnikov, N.M.: Natural hydrological studies in the design of HPP, p. 268. Gidrometeoizdat, L (1971)
Grishanin, K.V.: Dynamics of channel flows, p.488. Hydrometeoizdat, L (1969)
Grishanin, K.V.: Theory of the river fluvial process, p. 216. Transport, Moscow (1972)

Gusev, M.N.: Morphodynamics of the bottom of the upper Amur valley, p. 220. Dalnauka, Vladivostok (2002)
Jasmund, R.: Flissende Gewasser. Der Wasserbau. Bd. 111. Lerpzig (1911)
Kalinin, A.M.: Formation of Dniester Valley Slopes and Fluvial Processes. Regularities of Erosion and Fluvial Processes in Various Natural Conditions, pp. 311–312. MSU Publishing House, Moscow (1987)
Kapitsa, P.L.: Experiment, Theory, Practice, p. 352. Nauka, Moscow (1977)
Karasev, I.F.: Fluvial processes at transfer of the flow, p. 288. Hydrometeoizdat, L (1975)
Karaushev, A.V.: Problems of the natural water flow dynamics, p. 392. Gidrometeoizdat, L (1960)
Kennedy, J.F.: The mechanics of dunes and antidunes in erodible-bed channels. Fluid Mech. **16**, 521–544 (1963)
Kondratyev, N.E.: About discreteness of the fluvial processes (in Russian). In: Problem of the fluvial processes, pp. 34–42. Hydrometeoizdat, L (1953)
Kondratyev, N.E.: River fluvial processes and changes of reservoir banks, p. 258. Znak, SPb (2000)
Kondratyev, N.E., Lyapin, A.N., Popov, I.V., Pinkovsky, S.I., Fedorov, N.N., Yakunin, I.A.: Fluvial process, p. 372. Hydrometeoizdat, L (1959)
Kondratyev, N.E., Popov, I.V., Snischenko, B.F.: Basics of hydromorphological theory of the fluvial process, p. 272. Gidrometeoizdat, L (1982)
Koronkevich, N.I.: Water balance of the Russian Plain and its anthropogenic changes, p. 205. Science, M (1990)
Koronovsky, N.V., Yakusheva, A.F.: Principles of Geology, p. 414. Higher School, Moscow (1991)
Kzhemen, K., Helmitsky, V.: Application of numerical taxonomy to river bed typology (on the example of the Feshi River, Mount Cairngorm, Scotland). Soil erosion and fluvial processes, ed. 11, pp. 177–188. MSU Publishing House, Moscow (1998)
Leliavskiy, N.S.: About river flows and formation of the river channel. In: Proceedings of the 2nd congress of hydraulic engineers in 1893. St. Petersburg, 1893 (Problems of hydraulic engineering of free rivers), pp. 18–136. Rechizdat, M (1948)
Leliavsky, N.S.: On the study of the movement of sand spits near Alexandrovsk. In: Proceedings of the 3rd Congress of Russian Workers on Waterways in 1896, Part 2. St. Petersburg, 1896 (Issues of hydraulic engineering of free rivers), pp. 146–156. Rechizdat, M (1948)
Lelyavskiy, N.S.: About deepening of the big river niche (in Russian). 10th congress of Russian figures on waterways. Kiev, 1904 (Problems of hydraulic engineering of the free rivers), pp. 156–208. Rechizdat, M (1948)
Lenin, V.I.: Philosophical notebooks. Selected works, Part II, vol. 5, p. 672. Publishing House of the polit. lit., Moscow (1986)
Linguistic aspect of terminology standardization, p. 126. Nauka, Moscow (1993)
Lisitsyn, A.P.: Processes of territorial sedimentation in seas and oceans, p. 272. Nauka, Moscow (1991)
Litvin, L.F., Kiryukhina, Z.P.: Ecological aspects of soil erosion and contamination of surface waters with biogenic elements. In: Soil erosion and fluvial processes, ed. 10, pp. 5–16. MSU Publishing House, Moscow (1995)
Lohtin, V.I.: About the mechanism of the river bed. St. Petersburg, 1897 (Problems of hydraulic engineering of free rivers), pp. 23–59. Rechizdat, M (1948)
Lopatin, G.V.: Sediments of the USSR rivers, p. 368. Geografgiz, Moscow (1952)
Losev, K.C.: Water, p. 271. Hydrometeoizdat, L (1989)
Losievsky, A.I.: Laboratory research of riffle formation processes. In: Proceedings of TsNIIVT, ed. 36, p. 98 (1934)
Lvovich, M.I.: Experience of classification of the rivers of Russia. In: Proceedings of SHI, ed. 6, pp. 58–105 (1938a)
Lvovich, M.I.: Fluvial processes i gold content of a site of the river Sugor. In: Proceedings of trust "Zolotorazvedka" and inst. NIGRIZoloto, 8, pp. 169–185 (1938b)
Makkaveev, N.I.: Channel regime of rivers and tracing of slits, p. 202. Rechizdat, Moscow (1949)

Makkaveev, N.I.: River channel and erosion in its basin, p. 347. Published in the SA USSR, Moscow (1955)
Makkaveev, N.I.: Mutual connection of the hydrological and geomorphological investigations (in Russian). Methods of geographic investigations, pp. 206–208. Geografgiz, Moscow (1960)
Makkaveev, N.I.: Fluvial processes and their reflection in relief. Modern exogenous processes of relief formation, pp. 96–102. Nauka, Moscow (1970)
Makkaveev, N.I.: Flow and Stream Processes, p. 116. MSU Publishing House, Moscow (1971)
Makkaveev, N.I.: Forms of manifestation of fluvial processes. Educational materials for the VI session of 1974 International Hydrological Courses at MSU named after M.V. Lomonosov, pp. 17–21. MSU Publishing House, Moscow (1974)
Makkaveev, N.I.: General regularities of the erosion-fluvial processes (in Russian). In: Proceedings of IV All-Union hydrologic congress, vol. 10. Fluvial processes, pp. 8–12. Hydrometeoizdat, L (1976)
Makkaveev, N.I.: Denudation component of substance balance in the ocean-land system and its role in the formation of peneplates. Water resources (3), 147–155. River transport, Moscow (1982)
Makkaveev, N.I., Chalov, R.S.: Fluvial processes, p. 264. MSU Publishing House, Moscow (1986)
Makkaveev, N.I., Sovetov, V.S.: Trace of dredging slots on the riffles of the plain rivers of the European part of the USSR. In: Proceedings of the Central Research Institute of Russian Federation, ed. 3, p. 60. Issues of the way (1940)
Makkaveev, N.I., Khmeleva, N.V., Zaitov, I.R, Lebedeva, I.V.: Experimental geomorphology, p. 196. MSU Publishing House, Moscow (1961)
Mikhailov, V.N.: Hydrological processes in the river mouths, p. 176. GEOS, Moscow (1997a)
Mikhailov, V.N.: River mouths of Russia and neighboring countries: past, present and future, p. 413. GEOS, M (1997b)
Mikhailov, V.N., Dobrovolsky, A.D.: General Hydrology, p. 368. Higher School, Moscow (1991)
Mikhailova, N.A.: Transfer of solid particles by turbulent water flows, p. 236. Hydrometeoizdat, L (1966)
Mirtskhulava, C.E.: Basics of physics and mechanics of erosion of channels, p. 304. Hydrometeoizdat, L (1988)
Mirtskhulava, C.E., Snishchenko, B.F.: Estimation of the fluvial and erosion processes at the complex use and protection of the water resources (in Russian). General reports of the V All-Union Hydrologic Congress. V.II, pp. 88–106. Hydrometeoizdat, L (1986)
Morphology and dynamics of river channels of the European part of Russia and neighboring countries. Scale 1:2000000. Federal Geodesy and Cartography Service of the Russian Federation, 4 l (1999)
Nazarov, N.I., Yegorkina, S.S.: Rivers of Perm Territory: lateral channel changes, p. 156. Star, Perm (2004)
Obodovsky, A.G.: Hydrologic-ecological galvanization of fluvial processes (at the application level of Ukraine), p. 276. Nica-Center, Kyiv (2001)
Pakhomova, O.M.: Hydrologic-morphological characteristics of river beds and orderly structure of the river network. Autoref. diss. … Ph.D. geogr. sciences, p. 28. MSU Publishing House, Moscow (2001)
Polyakov, B.V.: Changes in river load and riffles of the Volga River due to construction of dams (in Russian). New Research Institute of Hydrotechnics **9**, 5–58 (1933)
Popov, I.V.: Methodic bases of the river fluvial process research, p. 208. Gidrometeoizdat, L (1961)
Popov, I.V.: River bed changes and hydraulic engineering, p. 328. Hydrometeoizdat, L (1965)
Ravine Erosion, p. 168. MSU Publishing House, Moscow (1989)
Rysin, I.I., Petukhova, L.N.: Fluvial Processes on the Rivers of Udmurtia, p. 176. Scientific Book, Izhevsk (2006)
Rzhanitsyn, I.A.: Morphological and hydrological regularities of the river network structure, p. 240. Gidrometeoizdat, L (1960)

Rzhanitsyn, N.A.: Channel-forming processes of the rivers, p. 264. Gidrometeoizdat, L (1985)
Saveliev, R.A., Zaitsev, A.A.: Definition of initial and boundary conditions of problems of computer modeling of fluvial processes. XIX plenary interuniversity coordination meeting on the problem of erosion, fluvial and estuary processes, pp. 178–180. Belgorod (2004)
Schanzer, E.V.: Alluvium of plain rivers of temperate belt and its importance for learning the regularities of structure and formation of alluvial formations. In: Proceedings of Geol, Inst. of the SA USSR, , ed. 135. Ser. geology (55), 275 (1951)
Sediment yield, its study of geographical distribution, p. 240. Gidrometeoizdat, L (1977)
Sidorchuk, A.Yu.: The structure of the relief of the river bed, p. 128. Hydrometeoizdat, SPb (1992)
Snishchenko, B.F.: Pair connections of parameters of a ripple and characteristics of a stream and a channel. In: Proceedings of SHI, ed. 288, pp. 15–25 (1983)
Snishchenko, B.F.: Development of theory of fluvial process in the XX c. In: Proceedings of SHI, ed. 361, pp. 5–26 (2002)
Solovyev, I.A.: Fluvial process and the waterways of the Lower Amur, p. 224. Dalknizhizdat, Vladivostok (1967)
Soviet Encyclopedic Dictionary, p. 1600. The Soviet Encyclopedia, Moscow (1983)
Surkov, V.V.: Dynamics of floodplain landscapes of the upper and middle Ob, p. 255. MSU Publishing House, Moscow (1999)
Talmaza, I.F., Kroshkin, A.N.: Hydromorphometric characteristics of mountain rivers, p. 204. Kyrgyzstan, Frunze (1968)
Transport use of water reservoirs, p. 224. Transport, Moscow (1972)
Tsian, N., Zhou, W.: Fluvial processes in the lower reaches of the Huang He River, p. 224. Nauka, Beijing (1965) (in Chinese)
Velikanov, M.A.: To statement of a problem of fluvial processes. Meteor. Hydrol. (3), 36–46 (1946)
Velikanov, M.A.: Russian channel hydraulic engineering and its role in development of teaching about the fluvial processes. Problems of hydraulic engineering of the free rivers, pp. 5–14. Rechizdat, Moscow (1948). (in Russian)
Velikanov, M.A.: Dynamic structure of the fluvial process. Izv. of the USSR SA. Ser. Geogr. and geophysics (6), 497–512 (1949) (in Russian)
Velikanov, M.A.: Dynamics of channel flows. V. 1. Flow structure, p. 324. Gostekhizdat, M (1954); V. 2. Sediments and channel, p. 324. Gostekhizdat, M (1955)
Velikanov, M.A.: Fluvial Process, p. 395. Gosfizmatizdat, Moscow (1958)
Vexler, A.B., Donenberg, V.M.: Influence of the flow regulation on the Yenisei channel transformation (in Russian). Regularities of the erosion and fluvial processes manifestation in different natural conditions, pp. 341–342. MSU Publishing House, Moscow (1981)
Walling D.E., Webb B.W.: Material transport by the world's river. JANS (164), 313–329 (1987)
Waterways of the Lena Basin, p. 600. MIKIS, M (1995)
Yang, H., Tan, Z.: Evolution of the Yangtze River bed in the middle reaches of the Tinjiang region, p. 245. Water Management and Hydropower, Beijing (1999) (in Chinese)
Zaitsev, A.A., Ivanov, V.V., Korotayev, V.N., Labutina, I.A., Lukyanova, S.A., Li, J., Rimsky-Korsakov, I.A., Rychagov, G.I., Scroll, A.A., Sidorchuk, A.Yu, Sychev, V.A., Chernov, A.V.: The Lower Volga: Geomorphology, Paleogeography and Channel Morphodynamics, p. 242. GEOS, M (2002)
Zaslavsky, M.N.: Soil erosion, p. 248. Mysl, Moscow (1979)
Znamenskaya, N.S.: Hydraulic modeling of fluvial processes, p. 240. Gidrometeoizdat, SPb (1992)
Znamenskaya, N.S.: Ripple Movement of Sediment, p. 188. Hydrometeoizdat, L (1968)
Zorina, E.F., Kosov, B.F., Prokhorova, S.D.: Experience of estimation of volume of ravine outcrops in the Don river basin. Vest. Moscow, Un. Ser. 5. Geography (3), 39–45 (1980)

Chapter 2
Natural Factors of Fluvial Processes

2.1 Fluvial Processes and Environment

The dual nature of fluvial processes determines, on the one hand, their development in accordance with the laws of hydromechanics: the shape of the channel and the shape of the channel relief in terms of its genesis are derived from the interaction of flow with the soils of the river bed, transport and accumulation of sediment. On the other hand, fluvial processes are one of the natural phenomena that arise and develop in a specific natural-historical situation. The forms of fluvial processes and their spatial and temporal (seasonal, perennial and secular) changes depend on the river water availability and their water regime, the volume and regime of sediment load, and through them on the climate, relief, geological structure and land cover.

Indeed, water flows, being the "motor spring" (according to N. S. Leliavskiy) of fluvial processes, arise in certain geographical conditions that determine the runoff rates, regime, longitudinal slope, particle size and composition of river load—all the main elements that determine the intensity and peculiarities of the development of fluvial processes in different natural conditions. Hence, they cannot be considered in isolation from the geographical environment without taking into account the specific features of the watershed landscape. Climate and runoff, geomorphological structure and properties of the territory's cover rocks, soil and vegetation cover, modern tectonic movements and fluctuations of base level marks are the main natural factors of fluvial processes. They determine the differences in form, intensity and direction of their representation in different physical and geographical conditions. These conditions, first of all, are reflected in the river flow, confirming the well-known formula of A. I. Voyeikov "rivers are the product of climate". Therefore, it is through the hydrological component of fluvial processes—water flows—that the physical and geographical conditions are most fully reflected in the formation of river channels. Depending on the natural conditions, it gives rise to natural combinations of forms of fluvial processes, so that in each region or zone certain types of channel patterns and forms of channel relief prevail as a reflection

R. S. Chalov, *Fluvial Processes: Theory and Applications*,
https://doi.org/10.1007/978-3-030-66183-0_2

of the different transport conditions of bed material load. At the same time, while the channel types are homogeneous in different natural conditions, the rivers differ in terms of their channel regime, which is characteristic of each geographic-hydrological zone (or region) and is determined by long-term and intra-annual peculiarities of water and sediment yield. As a result, with the general morphological similarity of the channel and channel relief forms in different natural conditions, they have a different regime of reformation. On this basis, we can talk about the types (subtypes, kinds) of channel regimes corresponding to the zonal, regional and local representations of the fluvial processes. In other words, since the river channels are an integral part of the physical and geographical environment, the fluvial processes are as zonal as other geographical processes (Makkaveev 1955), i.e., they are subject to the law of geographical zoning, or are characterized by regional and local peculiarities. The former are conditioned by the zonality of the leading factors of fluvial processes, i.e. water and sediment yield, while the latter are conditioned by the role of geological-geomorphological and other natural factors in the functioning of the "stream-river" system.

These peculiarities of the representation and development of fluvial processes were first emphasized by one of the founders of the doctrine of fluvial processes, V. M. Lohtin, and his position, known as the postulate of Lohtin, is the basis for the study of fluvial processes as part of the physical and geographical environment. The most common features of river fluvial process drivers are determined by the combination of climatic, soil, botanical and geomorphological conditions in the catchment area, the intensity of denudation processes, the products of which flow into the channel, and the geological structure of the territory (Chalov 1979). At the same time, the microform of the channel relief is a reflection of the turbulent structure of the flow, and therefore their connection with the natural conditions is mediated and manifested through the particle size of the load; however, the partitioning between latter with the kinematics of the flow, being a hydro-mechanical phenomenon, is associated with the slope of the channel, which, in turn, is, on the one hand, the most important hydraulic characteristic of the flow, and, on the other hand, is a consequence of the peculiarities of the relief of the territory where the river flows.

Geographically, the hydromechanical nature of formation and development of channel and channel relief forms determines the intrazonality of fluvial processes, i.e., their distribution in different zones, regions and areas with no visible connection to specific natural conditions. At the same time, forming a logical combination and differing in the commonality of changes in time, these forms are characterized by zonal features, the uniqueness of which on the rivers allows you to identify areas with appropriate types of channel regimes. Regional differences within the zones related to latitudinal or meridional changes in the channel regime, determined by the conditions of effective discharges, the emergence of new characteristic phases of the water regime in which channel changes take place, the influence of permafrost, freezing or drying up of rivers provide grounds for the identification of regions (or areas) to which the subtypes of channel regimes correspond. Similar regional differences and changes in the types of fluvial processes are associated with the spread of lowland, semi-mountainous, mountainous rivers and their types.

Local differences connected with geological-geomorphological factors determine the development of such forms of fluvial processes representation and specific features of the channel regime, which are beyond the connection with zonal characteristics and are azonal. The presence of the latter makes it possible to identify areas distinguished by a certain homogeneity of both zonal and azonal features of the channel regime of the rivers flowing within them.

Zonal, regional or local peculiarities of the river channel regime are reflected in the specifics of the mechanism of interaction of the flow with the soils composing their beds in specific natural conditions. This is due, firstly, to the different intra-annual distribution of flow, which leads to different changes in the timing of hydraulic flow characteristics, and, secondly, to different conditions of river load from the catchment area in different areas. As a result, apparently identical forms of channel and channel relief differ in their internal structure, hierarchy and combinations with each other on the rivers flowing in different geographical zones or geomorphological regions.

The degree of manifestation of each factor changes in time and depends on the size of the pool, the ratio with other factors. The larger the catchment area, the more even the river flow is (all other things being equal) and the less soil erosion and sediment transport work is carried out by the unit of water volume. The size of the catchment area and its form determine the number of landscape zones that fall within the boundaries of the basin. Most of the small basins are located within only one zone; while the channel regime of large rivers (including those with a strongly elongated meridional catchment) reflects the influence of a number of zones. This is particularly evident in the mountains, where the relatively small increase in the catchment area of the watersheds leads to their location in several high-altitude landscape zones with different types of weathering and denudation processes, which determines the change in their share of contribution in the formation of sediment composition and runoff. Accordingly, the degree of influence on the channel of natural factors depends on the scale of channel changes and the size of channel forms, on the one hand, and the size (runoff rates) of the river, on the other hand. Only the main forms of channel and channel relief create natural combinations, thanks to which channel types have a definite distribution in different regions. The connection of microforms of channel relief with natural conditions is mediated and often is not revealed at all. The only exception is the difference between steep slopes and flat rivers, as the former are characterized by rippleless transport of sediment or the formation of step-pool cascade channel.

The larger the river, the more its channel regime integrates the influence of a number of natural zones and differences in the geological-geomorphological structure of parts of the basin and valley. On the contrary, the channel regime of small rivers is fully determined by the local natural conditions in which they are formed. In this respect, the large rivers, and especially the largest rivers, are polyzonal in their channel regimes. Polyzonality is defined, on the one hand, by the transboundary transfer of regime features characteristic of other zones, and, on the other hand, by their transformation under the influence of natural conditions of a given zone (or region). The degree of transformation of the river channel regime is not the same, and not all transit rivers are polyzonal in this respect. These include Ob, Yenisei, Lena, Amur, Irtysh, which have a partial transformation of the channel regime, brought from other zones (regions) under the influence of natural conditions of the zone (region). The lower Dnieper, the lower Don, the lower Volga (in natural conditions—from the confluence with the Kama River), the Urals are characterized by a mismatch between their channel mode and the rivers they cross. Such rivers, being transit rivers, are azonal in their channel regime. At the same time, such rivers as the Amu Darya and the Syr Darya, which are in transit hydrologically, fully adapt the channel regime to the natural conditions of the deserts and semi-deserts of the Turan lowland.

Changes in the natural environment (moisture fluctuations, changes in vegetation cover, etc.) strengthen or weaken denudation processes, the most important of which is soil erosion. This causes sediment to accumulate in or be washed away from the river channels; each of these processes continues until there is a correlation between the amount of material entering the channel and the transport capacity of the flow. The same consequences, as well as the change of one type of channel to another, reduction or increase in the rate of channel change are caused by changes in the river runoff rates, intra-annual distribution of flow, its natural regulation. In natural conditions, these processes are slowed down and affect only the geological scales of time, represented in the morphology of river valleys and the structure of alluvial strata. When people interfere in the life of rivers (creating reservoirs, water intake, development of channel quarries, dredging and correction, as well as development of lands in catchments), they are accelerated, accompanied by changes in the morphology of the channel and affecting the river-adjacent areas.

The relationship of the river channel regime on the zonal and regional peculiarities of the factors of fluvial processes can be traced in general terms on the example of lowland rivers in Russia (Work of water flows 1987). In the tundra zone, the determining factors are the presence of permafrost soils and a high runoff coefficient, which reaches 0.7–0.9. The intensive development of slope processes (e.g. solifluction) in these conditions leads to an excessive inflow of sediment into small rivers, the channels of which "fill in" with solifluctation material exceeding the sediment transport capacity of flows on small rivers. At the same time, fluvial processes on large rivers, which are usually transit rivers for tundra, are developing very rapidly, which is facilitated by the high runoff rates of the rivers, the presence of through taliks under their beds, the significant flow of heavy loads and the significant role of ice congestions.

To the South of the tundra zone under flat conditions, the zonality of the fluvial processes is manifested through the peculiarities of the formation of sediment load, slopes and the degree of regulation of runoff, as well as the erosion resistance of the rocks composing their beds. Channel changes are slow in rivers with large lakes or rivers that cross the areas of distribution of erosion-resistant rocks. In areas where channel changes of rivers are not restrained by these factors, such forms of fluvial processes as erosion of banks, shifting of ropes and bends, development of channels, etc., are clearly visible. In the Russian Plain, the intensity of fluvial processes increases to the North of the main watershed by these signs. The reasons for this are the increasing flow rate in the rivers to the North and the spring ice drift accompanied by powerful congestions, which is typical for rivers flowing from South to North.

A similar change in the intensity of channel change is observed in the rivers of the Southern half of the Russian Plain, where, on the one hand, it is associated with an increase in the runoff rates of large rivers flowing down from the main watershed, and, on the other hand, with the crossing of the rivers of distribution areas of relatively easily eroded rocks. At the same time, in the Eastern regions of the South of the Russian Plain, fluvial processes develop under the conditions of periodic drying up of rivers, which has a significant impact on the shape of their channels.

In Western Siberia the development of channel changes in general does not differ from the rivers of the European part of Russia. However, here in the South there is a rather extensive zone, where the channels of large rivers are characterized by weak stability and very intensive reshaping. This is due to the fact that the rivers here, coming from the mountains of the South of Siberia, form vast "inner deltas" where, due to the change in the form of sediment transport (sandy load from the suspended load passes into the category of bed load), the rivers form complex branched channels, which are characterized by a complex mode of change; under the condition of $W_{tr} > W$, this is sometimes accompanied by a directed sediment accumulation.

The considered zonal changes of the forms and intensity of manifestation of the fluvial processes correspond quite clearly to the natural zones. For example, in the forest zone, the increase in heat and the presence of forest contributes to a decrease in the runoff factor (up to 0.4 and lower), but due to the significantly higher amount of precipitation, the flow rate here is increasing compared to that of tundra. Active representation of fluvial processes in the forest zone is favored by a number of conditions. Sufficiently high-water flow at low water intake from the catchment area of soil erosion products and gully erosion contribute to increasing the depth of river valleys and, as a consequence, the growth of groundwater inflow into the rivers. As a result, the rivers here are distinguished by a wide range of forms of representation of fluvial processes, intensive incision and very significant bank erosion.

In the steppe zone, low precipitation and increased evaporation lead to a significant reduction in the water runoff coefficient (up to 0.1) and flow rate, which affects the development of all links in water flows. A characteristic feature of the zone is the inverse relationship between the flow rate and the size of the catchment area; in river systems with small catchments the flow rate is higher than in systems with large catchments. The drier the climate, the more pronounced this pattern is. The influence of geomorphological factors and the character of the cover rocks on the flow of geomorphological factors increase significantly. Runoff rates increase with increasing slope steepness, the measure of relief dissection and watertightness of cover rocks. Winter precipitation is dominant in the formation of river runoff, but summer showers play a major role in the formation of slope runoff and runoff in the gully network. The unevenness of the flow increases sharply with the aridity of the climate. This applies to both the intra-annual distribution of runoff and the variability of its annual values. Hence, the "abnormally" large width of the river valley bottoms relative to the width of the inter-stream flow.

Due to the fact that the value of evaporation from the water surface is greater than the amount of precipitation, there is a possibility of existence of internal flow areas. Deep ground water occurrence causes a strong depression of the river network. The hydrographic network is characterised by interruptions in some places, which are most often observed in areas with flat topography. Erosion products and sometimes soil deflation are excessive in the river channels. Average sediment concentration of river water is significant, sometimes by orders of magnitude greater than in the forest zone. The abundance of fine material entering the rivers and the significant unevenness of flow contribute to the fact that per unit of flow volume there is several times the amount of sediment load. However, the sediment load flow rate are sometimes even slightly smaller than in the forest area, as the sediment load module is depleted to a greater extent than the sediment concentration of the river water increases. It is also important that a significant part of the load is deposited in local depressions and river valleys. Within small catchment areas, the amount of material transported into rivers from one area to another is much greater than in the forest zone.

In the desert and semi-desert zone, low precipitation and high evaporation cause insignificant runoff and its extreme variability. The amount of runoff can vary hundreds of times in some years. The river link of the hydrographic network in arid climate conditions practically falls out. The only exception is the transit rivers and rivers, which channel is formed in the exceptionally favorable geomorphological conditions. They form narrow deep valleys into which groundwater is discharged from deep horizons. The hydrographic network is often represented by small areas of internal runoff. Often, rivers coming from the mountains to the plain form "blind" estuaries, not reaching large transit rivers or receiving basins. By branching out into branches and accumulating sediment, they are either "lost" within the limits of the removal cones (Tejen, Murgab), or, when leaving their limits, their flow gradually decreases, and the channels decrease in size (Chu, Talas, Zeravshan, Emba) and gradually disappear. At the same time, they, as well as the rivers reaching the receiving basin (sea, lake), are characterized by extremely high sediment concentration and intensive accumulation of sediment, leading to the excess of the intermittent water level in the rivers over the marks and, as a consequence, to a systematic threat of flooding. This is often accompanied by a wandering channel and intensive erosion of banks composed of loess or loess-like sediments. As a consequence, the fight against this phenomenon in rivers of arid regions is no less relevant than flood prevention. These are the Terek in the lower reaches, the Amu Darya, on which the riverbank erosion was named bank scour, Huang He and Yangtze, where the threat of flooding is associated with the risk of destruction of flood dams due to erosion of the banks.

The intensity of fluvial processes in the mountains in general increases from highlands to foothills. While the upper and middle reaches of mountain rivers are usually cramped with rocks, in the foothills and plains, which are usually composed of easily eroded or loose sedimentary rocks, they are very unstable. In the same direction, the slope, the degree of kineticity of the flow, the shape of the sediment transport, the type of mountain bed and the nature of the channel changes change. Whereas in the upper reaches of mountain rivers the loads are transported in separate fragments during floods of rare recurrence, in the middle mountain belt the flow of bed material pebbles and boulders is carried out in the form of mass movement (layer). Low-lying areas and foothills correspond to the channels with developed alluvial forms of relief. Their changes consist not only in the movement (transportation) of sediment, but also in changes, sometimes very significant, in the channel itself. This scheme corresponds to mountain rivers with concave longitudinal profile; it is complicated by repeated changes of channel types along the length if the longitudinal profile has a stepped shape.

2.2 Classification of Fluvial Process Factors

Channel formation is a complex and multilateral process closely related to the natural features of the territory through which the river flows. The main factors of the natural fluvial process are water flow, and the geological structure of the area of sediment yield. At the same time, channel formation is also influenced by other factors of temporary or local character: ice phenomena, wind, vegetation, the existing shape of the channel and valleys, etc. All of these factors are not simply compounded effects on the fluvial processes; they are difficult to interact with and, in some cases, connected by causal links, i.e. they serve as both cause and effect in relation to each other. For example, sediment load depends to a large extent on the hydraulic characteristics of the flow (primarily flow velocity); at the same time, the particle size of the load affects the slope and roughness of the bed, i.e. the hydraulic resistance, and through them the flow velocity and other flow properties.

Anthropogenic (in a narrower sense—technogenic) impact on the nature changes the natural factors of fluvial processes, and only mechanical change of channel shape, creation of artificial base level (reservoir) can be attributed to the category of anthropogenic factors. But even in these cases, the primacy in the partitioning between anthropogenic and natural factors belongs to natural, because economic (engineering) activity only changes them and through them—the riverbeds. The main driving forces remain the water flow in the river, associated sediment load and the soils composing the riverbed and banks.

Table 2.1 divides all natural factors of fluvial processes into two groups depending on their role in fluvial processes. Water runoff and its time variability, which determine the channel formation process, are active. Discontinuation of flow entails the disappearance of the river and the elimination of the process itself. Under certain flow formation conditions and natural environment conditions, water flows have different runoff rates (in the range from elementary water flow—streams to the largest river) and different water regime. They differ in terms of their longitudinal slope, size and composition from the catchment area and directly at the time of

Table 2.1 Classification of fluvial process factors

Forms of representation	Factors	Type of impacts	Main characteristics
Active	Water runoff	Direct	Runoff rates
			Water regime
	Sediment yield	Direct	Sediment concentration, bedload and suspended sediment discharge
			Sediment yield regime
			Partitioning between suspended and bed load
Passive	*Basic*		
	Alluvial sediments that make up the riverbed and the floodplain	Direct	Grain sizes
			Sediment density
	Geological structure of the catchment area and valley	Direct	Lithology of riverbeds and banks and their erosion resistance
		Indirect	Lithology of river basin rocks and their erosion resistance
			Geological structure and tectonic movements
	Geomorphological structure of the catchment area and valley	Direct	Width and shape of the valley bottom
			Longitudinal river profile (slope distribution along the length)
			Valley morphology (bedrock banks, accumulative and basement terraces)
			Distribution, width and height of floodplain
	Riverbed morphology	Direct	Channel shapes and channel relief shapes
	Ice mode	Direct	Ice drift, congestion, ice jam
			Ice transport of solid material, ice accumulation of sediment
			Impacts of ice on banks and riverbeds
	Others		
	Soil-vegetation cover	Direct	Vegetation on the drying parts of the channel
			Vegetation on the banks
			Aquatic vegetation
		Indirect	Soils and vegetation on the floodplain
			Soils (erosion control) and vegetation in the river basin

(continued)

Table 2.1 (continued)

Forms of representation	Factors	Type of impacts	Main characteristics
	Meteorological	Direct	Wind mode, surging
			Spreading dried riverbed banks
		Indirect	Dune formation on the floodplain
			Precipitation and snowmelt mode
	Permafrost	Direct	Frozen soils on the bottom and banks
			Seasonal freezing of riverbed banks and shallow (frozen) channel areas
		Indirect	Permafrost in the river basin
	Slope and erosion processes	Direct	Direct supply of soil and gully erosion products to river channels
			Mudslide, talus, landslides on the banks.
			Mudflow
		Indirect	Erosion processes in the river basin
			Slope processes in the river basin
	Biogenic (activity of organisms)	Direct	Direct impact of living organisms on river channels (activities of beavers, fish, benthos, destruction of banks and bottoms at watering points of large animals, including domestic animals, etc.)
		Indirect	Biochemical (microbinal) underwater weathering

erosion of the banks and bottom of sediments and their concentration, the composition of underlying rocks (from loesses and sands of various sizes to rocks and dense bonded clays) and their erosion resistance, etc. They are significantly affected by the soil and vegetation cover, which regulates the flow regime and the conditions for the inflow of sediment into the river channels, as well as influencing the dynamics and structure of the flow (especially when flooding the floodplain) through the resulting hydraulic resistances.

Passive factors—external factors in relation to the flow, which determine the shape and steepness of the longitudinal profile, hydraulic resistances and through them the kinematics of the flow, its condition (open, under the ice, accompanied by ice flow, the formation of congestion and blockages), the saturation of sediments and their size, etc. Their flow modifies in the process of interaction with the natural environment or adapts to them, creating an appropriate form of manifestation of the passive factor internal structure. These also include the factors that determine the flow of sediment into the bed. The form of influence of passive factors on river channels and fluvial processes is twofold. It can be direct, corresponding to the conditions of their direct impact on the flow. Indirect form determines the indirect impact of factors on fluvial processes through intermediate links, types of impact or other factors. Such is, for example, the soil-vegetation cover at the catchment area, on which the development of erosion processes and the conditions for their products to enter the river network depend; the vegetation on the floodplain determines the hydraulics of floodplain flows arising during flooding of the floodplain, and, accordingly, the specifics of their interaction with the channel flow. The operation of the water flow is influenced by the lithology of rocks and alluvial sediments, which determines the inflow of sediment from the catchment area and from the erosion of the bottom and banks. The direction of modern tectonic movements leads to an increase or decrease in the area in which rivers flow, causing changes in their slopes. The distribution of permafrost in the river basin provides a peculiar regime for the inflow of sediment into the rivers: during spring floods, meltwater is formed on the frozen ground, and as a result, the sediment transport rate in the rivers is low; during summer floods, the active layer is washed away on a massive scale, and the sediment transport rate increases several times, despite the lower volumes of water flow (Tananaev 2002).

Sediment yield is the essence of fluvial processes, determining the formation and evolution of channel forms and channel relief forms. In the presence of a flow of sediment load, it is mandatory to carry out the discharge in one form or another, up to a chemically dissolved state, and the transport of sediment is carried out together with the flow, albeit at a lower rate (except for chemically dissolved substances). Particles moving together with the flow affect the river bed itself, corroding it, contributing to the separation of other resting particles and, getting into the flow column, changing its physical properties and, as a consequence, erosion and transporting (in relation to the bottom upstream sediments) ability (Makkaveev et al. 1970; Rossinskiy and Debolsky 1980).

On the other hand, sediment load is generated as a result of, firstly, the impact of the flow on the river bed and banks and, secondly, the inflow of sediment from the catchment area due to erosion, slope and other denudation processes. This creates the impression of a passive nature of sediment load as a factor in fluvial processes, since its occurrence is a derivative of the activity of water flows, including the fluvial processes themselves. Nevertheless, sediment load is an active factor of fluvial processes, in contrast to their composition, particle size and density, especially during the transition to bed sediments, which are essentially passive factors.

In contrast to the passive factors of fluvial processes, active factors, including sediment load, are only directed. It is they who are primarily subject to the laws of geographical zoning (Kuzin and Babkin 1979), which creates conditions for their manifestation in the fluvial process driver of rivers. Passive factors, even though they are zonal (land cover, climate), affect the channel only indirectly and therefore are not leading to fluvial processes. The contribution of such passive factors as the geological-geomorphological structure of the valley and the riverbed is more significant (first of all, direct). Being azonal in nature, they determine the regional specificity of channel processes.

2.3 Role of Water Runoff and Water Regime in River Channel Formation

Water runoff as the main active factor of fluvial processes determines the size of the channel, which is directly related to its size: for example, the width of channels in straight sections is proportional to the square root of the water consumption. The connection between the depths and the flow of water is not so close. Water consumption is mostly related to the relative depth of the channels. For large rivers, the relative depth of h/b_{ch} is less than for small rivers. Rzhanitsyn (1985) obtained a general relationship describing the changes in the relative depth of the channel on the average maximum flow rate $Q_{m.\max}$.

$$\frac{h}{b_{m.\max cp.Makc}^{-0.17}}, \tag{2.1}$$

where $A = 0.038\ K_{ch}^{0.5}$, K_{ch}—"the index of channel mode" of N. A. Rzhanitsyn, characterizing the change in the form of the living flow section in different phases of the water regime.

Since the channel width b_{ch}, depth h and ratio h/b_{ch} depend on the runoff rates of the river, the latter determines other morphometric and morphological parameters of the channels: water discharge (annual average or maximum average) is proportional to the size of the bend; the greater the water discharge, the lower the slope, which determines the concave shape of the longitudinal profiles of the rivers in humidal conditions and convex in arid (when the runoff rates of the rivers downstream due to evaporation losses).

Since the maximum discharge of $Q_{\max}$ rivers depends on the catchment area F (Sokolovsky 1968):

$$Q_{\max} = f(F) \tag{2.2a}$$

for rivers with a basin area of less than 50 km^2,

$$Q_{\max} = f(F^{0.84}) \tag{2.2.b}$$

– with a basin area of 50–100 km^2 and

$$Q_{\max} = f(F^{0.75}), \tag{2.2c}$$

– more than 100 km^2, i.e. in general $Q_{\max} = f(F^n)$, and the catchment area, in turn, is related to the length of the river (Makkaveev 1955)

$$L_r = f\left(\sqrt{F}\right), \tag{2.3}$$

then channel formation conditions, channel morphometric characteristics and channel shape parameters change along the river length. It is also known that small and medium rivers, under conditions of free development of channel changes, usually, meander, while large and, especially, the largest branching into branches. The growth of the catchment area and, consequently, the runoff rates of the river is mainly sudden, with the inflow of channels. Since the river load water flow is connected with each other by degree dependencies $W = f(Q^m)$ type, the confluence of the rivers causes a change in the transport capacity of the combined flow and, as a consequence, the direction of the river fluvial process driver. This is all the more significant if we take into account that the values of the degree m are close to 2 and even 3 depending on the size of the bed material load (Makkaveev 1955).

Where hydrological data are scarce, their water availability can only be determined using basin-specific calculations using formula $Q = f(F)$ or indirect methods, such as flow module maps. In this respect, in order to analyze changes occurring in the river channel regime from the upper reaches to the estuary under humidal climate conditions, it is important to identify the order of rivers that depend on the structure of the river network. The order of the river N is an individual characteristic that gives an average idea of the size of the river. The main models of the river network order structure, which allow assessing the impact of changes in river flow on the characteristics of the river channel and the river channel regime, are:

(1) R. E. Horton, A. N. Straler—V. P. Filosofov models and their modifications, assuming an increase of the order by one unit at the confluence of two first-order watercourses;
(2) the A. E. Scheidegger model (1964), in which the size of the river system is measured by the total number of elementary watercourses (1st order rivers); according to this model $N = \log_2(P) + 1$, where P is the number of first order tributaries. N. I. Alekseevskiy (Small Rivers … 1998) suggested that rivers of less than 10 km in length should be considered as first order tributaries, the number of which is given in hydrological publications;
(3) N. A. Rzhanitsyn model (1960), which is a transition between the first two.

The disadvantage of the R. E. Horton and A. N. Straler-V. P. Filosofov's models is a saltatory increase in the characteristics of rivers when their order changes. This contradicts the actual changes in hydrographic, hydrological and morphometric characteristics of the rivers, and, accordingly, does not allow their application in the analysis of fluvial processes and forms of their representation on the rivers. Rzhanitsyn (1960), having proposed a "transitional" version of the model of the order structure, used it to develop complex hydrological and hydrographic characteristics of the rivers in the different parts of the river flows. His studies have shown that a consistent change in the river order leads to natural changes in the average and average maximum flow rates, the ratios of typical water flow rates, the duration of spring floods and the morphometric characteristics of the channel.

The A. E. Scheidegger model in this respect provides more reliable estimates of changes in the hydrological and, consequently, channel characteristics of rivers, as it takes into account the gradual increase in the water availability of rivers as any inflow of any inflow occurs. In particular, a natural change in the mean perennial and typical water discharge along the length of zonal rivers has been revealed (Kositsky et al. 1999), which is described by the exponential equation (Fig. 2.1)

$$Q = ae^{bN} \tag{2.4}$$

where a and b are coefficients that have a regional meaning, differing in size between river basins. For polyzonal rivers that cross a number of natural zones, as the catchment area increases, the diversity of private basins that make up the catchment area affects the diversity of its components. For example, for the steppe rivers of the Don basin, the average long-term water discharge is much lower than for the rivers in the forest-steppe zone. At increase of N from 8 to 10 Q_0 at steppe rivers increases from 5 to 30 m^3/s, in a forest-steppe zone—from 10 to 40 m^3/s. This naturally affects the values of channel parameters, direction and rate of channel change. For example, the relationship between the channel width b_{ch} and the order of the rivers N in the basin of the Belaya (Chalov et al. 1999) is approximated by the equation (Fig. 2.2A)

$$b_{ch} = 1.87e^{0.34N}, \tag{2.5}$$

between channel depth h and river order (Fig. 2.2B)—by the equation

$$h = 0.22e^{0.16N} \tag{2.6}$$

and between the ratio b_{ch}/h and order of river (Fig. 2.2B)—by the equation

$$\frac{b_{ch}}{h} = 7.0e^{0.19N} \tag{2.7}$$

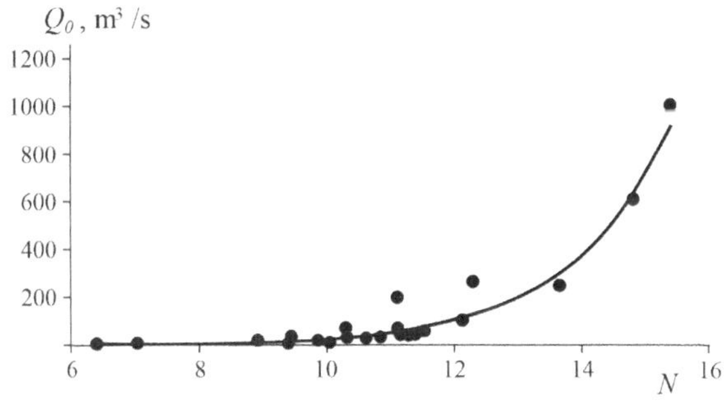

Fig. 2.1 Ratio of average perennial discharges Q_0 and river orders N in the basin of the Vychegda river (according to Kositsky et al. (1999))

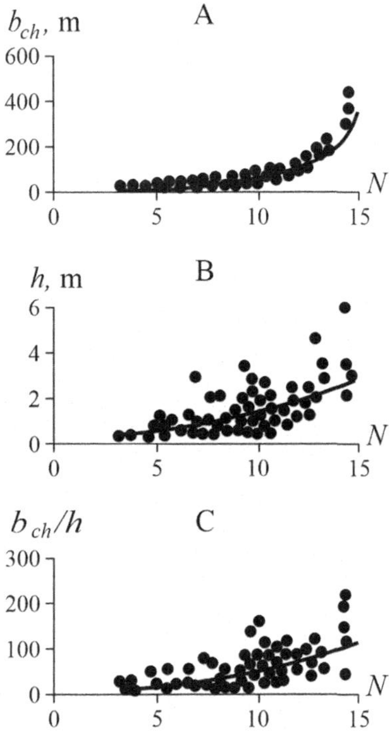

Fig. 2.2 Relationship between width b_{ch} (**A**), depth h of a channel (**B**) and the ratio b_{ch}/h (**C**) and order of the river N (Belalya river basin) (Chalov et al. 1999)

at correlation coefficients of 0.86, 0.6 and 0.61, respectively (on meandering rivers these relations are closer, correlation coefficients are higher: 0.94, 0.76, 0.90). At the same time, for all the links the inflections of curves fall on the rivers of 9–10 orders. This feature of increasing the intensity of morphometric changes can be used to determine the upper boundary of small rivers. It corresponds to the rivers with $N = 10$ (according to the model of A. E. Scheidegger). The boundary between medium and large rivers is of 14–15 order: at $N > 14$–15 the b_{ch}/h ratio, even at a slight increase in N, sharply increases, and the relationship $b_{ch}/h = f(N)$ becomes linear (at a significant change in width, under other equal conditions, the average depth of the flow changes insignificantly).

The dynamics of natural channel flows is strongly complicated by the irregularity of runoff in the seasonal and perennial term. The direction of the channel formation process depends largely on the intensity and amount of water flowing through the channel. As water discharge increases, the flow forms relatively larger bends, deepens the moulds, and forms rolls. When water discharge is reduced, the stream is unable to wash out deep moulds and deposits sediment in them. At the crossroads, where sediment deposition prevailed in the flood, the intermodal flow has increased slope, eroding their ripples, creating a relatively deepened furrow between the sides—the trough of the crossroads. The flood flow processes the forms of channel relief created in the low water period, and the inter-flow processes the forms created in the high water period. Thus, the unevenness of runoff is one of the reasons for channel changes.

All other things being equal, the lower the relative value of flow changes, the more stable the relief of the river channel is. Temporary reduction of channel change intensity is usually also observed in years with very low flood levels. High floods on rivers with unstable shallows increase the intensity of channel changes.

Such an impact of flow irregularity on channel transformation is also a consequence of the degree of relationship between water and sediment discharge $W_{R+G} = f(Q^m)$, where m = 2–3. In accordance with this, channel changes decay during the low water period and, conversely, are significantly activated in floods, since at this time the movement of sediment is intensified. The unequal influence on the transformation of the streambed into different phases of the water regime is associated with the allocation of effective water discharges, which "for a certain period of time (usually—a hydrological year) has the most significant impact on the channel compared to other observed during this time flow rates" (Makkaveev 1955, p. 182). In accordance with the peculiarities of the water regime, one—three water discharge intervals are distinguished (Chalov 1979), at which the impact of water flow on the channel is reflected in its form and in the morphology of the channel relief at various levels of representation of fluvial processes. In this sense, effective water discharges, having repeated repetitions during a hydrological year, determine the discreteness of fluvial processes depending on the specifics of water regime in morphology and transformation of river beds.

2.4 Geological and Geomorphological Factors

The geological structure, including rock lithology, and the relief of the area influence the shape of the valley, the longitudinal profile of the river, the composition of bed material load and the stability of the channel. When the river crosses mountain massifs, moraine and structural uplands, formed by relatively dilute rocks (rocky, ductile), the valley is narrowed, alluvium is enlarged and the channel becomes more stable. The longitudinal profile of the river thus takes on a stepped form due to the increase in the slope in the contractions. Raoids and shiver formation is associated with rock outcrops directly in the channels. Thus, on the middle Amur, which successively crosses the Zeya-Bureya plain, the Small Hingan Ripple and the Middle Amur lowland, the slopes in the corresponding parts of the channel are 0.085, 0.142, and 0.071‰, the wide floodplain channel is replaced by the incised and again wide floodplain channel (Zavadsky et al. 2000). Within the Khingan section, there are often local increases in bottom marks associated with rock outcrops forming a kind of spillway (the Union "riffle"). Similar rocky ledges are typical for the incised channel of the middle Yenisei, where they form the Kazachinsky (Fig. 2.3) and Osnovsky rapids; shivers—rocky outcrops blocking the channel from 0.5 to 0.8 its width are widespread in Angara, Vitim, and rocky raises of the bottom of the

Fig. 2.3 Kazachinsky step on the Middle Enisei. Photo by V. E. Sergutin

spillway type with a wide rapid—on the upper and middle Lena, upper Aldan. The steep slopes of the valleys in constrictions, especially those composed of rocks, are a source of coarse chipping material in the channels. As a result, the sediment here is pebbles and boulders. The same composition is characteristic of rivers crossing moraine uplands. For example, the Neman River within the Upper Neman lowland has a sandy free-meandering channel and a wide floodplain (Fig. 2.4A); downstream it crosses the Grodno moraine elevation, the floodplain is absent here, the channel becomes incised, composed of coarse sands, pebbles and boulders (Fig. 2.4B). Sediment loads are increasing here by a factor of 10–20.

Rivers originating in the mountains and on upstream hills are usually without floodplain, with an incised stable channel and a pebble-boulder composition of bed material load. The same rivers with constant flow regime, reaching the lowlands, folded by loose, easily eroded rocks, have a wide floodplain, lower slope of the longitudinal profile and less stable sandy channel.

The channels of the rivers flowing through the territories formed by loess or loess-like loam are especially unstable. These rivers are characterised by increased sediment load and a predominance of a suspended load component. Each flood changes the configuration of the channel, leads to its wandering, makes significant changes in the location of the rolls and sideways. An example of this is the river Huang He, crossing the Loess Plateau, and the Amu Darya in the middle and lower reaches. The forest sediments in the Huang He river basin are more than 250 m thick, and the Loess Plateau is a very strongly dissected gully area (the depth of vertical dissection reaches 100–150 m and more) (Fig. 2.5). The Loess Plateau is an arena of intense soil erosion, which is catastrophic in nature. It is the main source

Fig. 2.4 Sandy wide floodplain channel of the Neman river at the edge of Upper Neman lowland (**A**) and incised pebble-boulder channel crossing Grodno upland (**B**). Photo of the author

of suspended load entering the river in such quantity that $W_R \ggg W_{tr}$. This determines the regime of constant directional accumulation in the middle and especially in the lower reaches of the Huang He, where the river flows through the Great Plain of China. Being formed by the Huang He sediments, which occupy an extensive tectonic depression, experiencing constant submergence, the plain has the

Fig. 2.5 Loess Plateau dissected by the linear erosion (China) in the area of the Sanmenxia HPP on the Huang He river. Photo by author

form of very flat lowland, the surface of which is located below the interfluve water level in the river. This is the cause of flooding (when the river flows beyond its dams) and of repeated changes in the width of 600–800 km in the historical period of the river's position (Muranov 1957; Chalov et al. 2000).

The low stability of the channel and its intensive reshaping caused by geological reasons are also observed when the bottom of the valley is composed of more resistant to erosion than the banks. The flow "slides" along the roof of bedrocks without forming an erosion furrow in it. This is one of the main reasons for the instability of the Northern Dvina channel below the confluence with Vychegda, where the bottom of the valley is everywhere composed of moraine sediments of the Moscow glaciation time under the bed. Being covered by younger sandy sediments of alluvial genesis, in which the modern river bed is formed, moraine sediments create a kind of shielding horizon (Sladkopevtsev 1973), on which the bed wandering without crashing into it, washed away the banks and creating a very wide (up to 10–15 km) floodplain. After the retreat of the Moscow glacier along the valley of the Northern Dvina there was a sea bay, where the river flowed into formed a sandy delta (Sakharova 1960; Kostyaev 1960), morphologically expressed in the form of a wide hilly surface relative height (above the river) from 20 to 50–55 m. This surface approaches directly the river on the right bank, forming the Tolokonnaya Mountain (Fig. 2.6). Upstream at that time, there was an accumulation of alluvium from the high terraces. Sandy delta and alluvial deposits are now sources of large amounts of sediment in the bed, which is facilitated by wandering it on the roof of moraine loams. As a result, the most complicated morphologically and reshaped branched channel and rolling section are formed here.

Fig. 2.6 Tolokonnaya mountain near the Northern Dvina river—source of sandy sediments in the river. Photo by E. A. Lvovskaya

During the Valdai glaciation, the glacier supported the Northern Dvina and contributed to the formation of a very wide 11–14-m high sandy terrace upstream, which also serves as a source of abundant sediment load during the river washout. Downstream (up to the mouth of the Vaga River), the glacier left behind a moraine, which together with permetrias deposits causes the formation of an incised or confined channel here.

The role of the geomorphological factor in the formation of the channel is demonstrated by the Ob river below the Novosibirsk hydroelectric power station. Here the river flows through the marginal zones of two tectonic zones—the marginal part of the Kolyvan-Tomsk folded zone and the Kolpashevskaya depression (Matveevskaya and Ivanova 1960). Within the first of them (Novosibirsk—Baturino) sandstones, clayey and carbonaceous shales, siltstones of the upper Devonian and Lower Carboniferous silts and granite intrusions come out in the channel and along the coast. Protrusions of Paleozoic rocks and intrusive bodies determine the shape of the valley in the plan: its narrowing is connected with the places where the river crosses the granite massifs (areas of Novosibirsk, Kolyvan and Dubrovino). Between them, where the bedrock roof is submerged below the current river bed, the width of the floodplain increases significantly, although in both cases the channel remains wide floodplain (Fig. 2.7). At the same time, in the

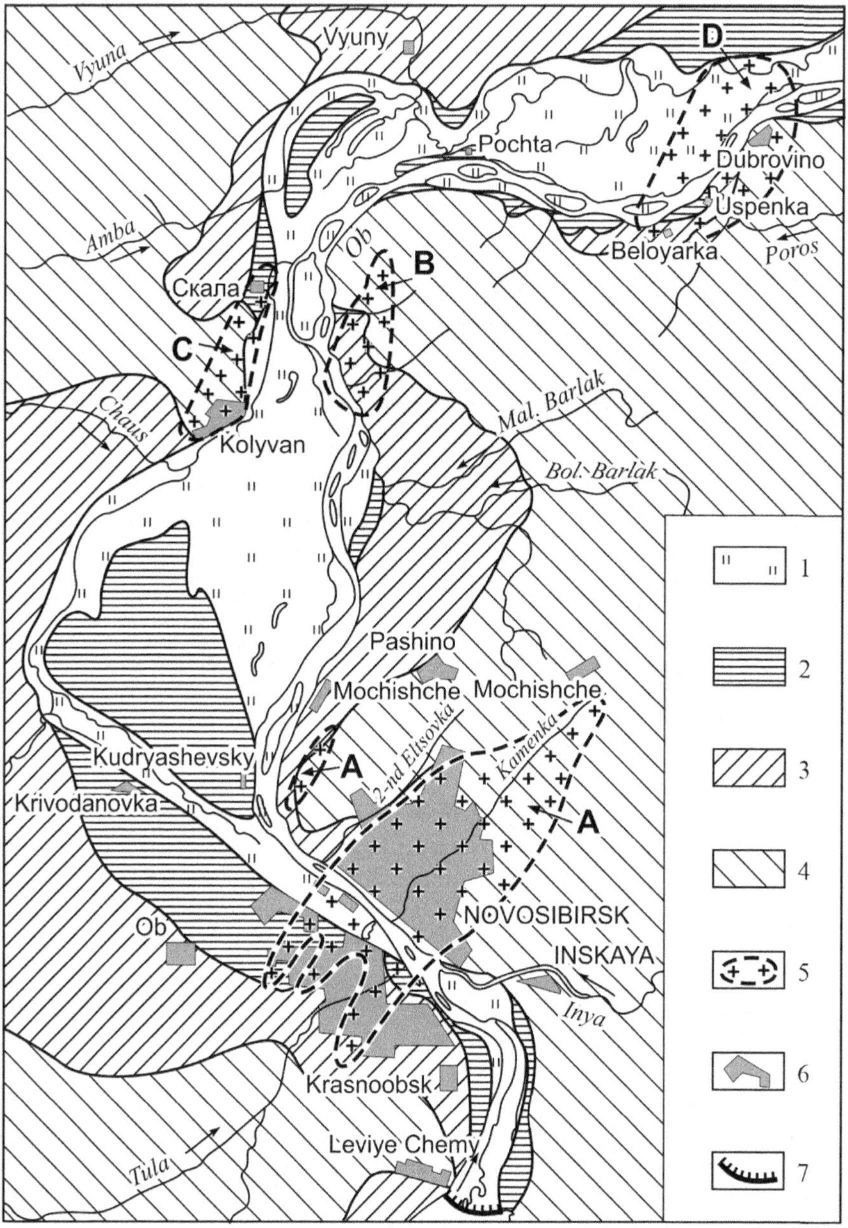

Fig. 2.7 Interchange of narrows and expands of the Ob river valley bed lower the Novosibirsk city, related with the crossing of granite massifs by the river; 1—floodplain; 2—first floodplain terrace; 3—high terraces; 4—Ob plateau and the right bank uplands; 5—granite massifs (**A**—Novosibirskiy, **B**—Barlakskiy, **C**—Kolyvanskiy, **D**—Dubrovinsky); 6—cities and populated areas; 7—Novosibirsk Hydroelectric Station

areas where the river crosses the boundary massifs, the main banks of the river have rocky capes, and in the channel itself, including in the middle of the river, there are numerous rocky ledges separating the pool valleys.

Lower the Baturino Paleozoic rocks are submerged under the thickness of lake-alluvial deposits of Pliocene and Pleistocene represented by clays and pebble-boulder deposits. The absence of rock outcrops causes the width of the valley floor to remain constant, although it is somewhat narrowed to the Tom river estuary due to the high position of the roof of the Lower Quaternary gravels (Feniksova 1966). The latter are the reason for the formation of bed sedges and the formation of pebbles in the sandy bed as a whole. Below the mouth of the Tom River, the Ob River enters the West Siberian lowland, where its channel and channels of almost all tributaries have a wide floodplain and sandy composition of bed material load.

Geomorphological factor often creates local conditions for channel formation. For example, the location of the channel along the high main banks of the river with an asymmetrical transverse profile of the valley bottom is the reason for its straightness. When the river washes away the sandy ledges of terraces, excess sediment enters the channel, which is the reason for the formation of directly downstream shallow water rifts and rolling areas. These are, for example, the Paechnye—Rubezh rifts below the Tolokonnaya mountain on the Northern Dvina. According to A. A. Zaitsev's calculations (1979), 1.7 million tons of loads enter the riverbed every year on the lower Viliui with the erosion of a concave sandy bank 25–30 m high. Their accumulation leads to the formation of a group of small rolls here.

In generalized form, the influence of the geological structure of the territory on the fluvial processes is represented in the free and limited conditions for the development of channel changes on the rivers (Chalov 1979). In the first case, the weak stability of the rocks composing the river bed and the shallow channel alluvium, determine the predominant role of the flow in the fluvial processes; on the contrary, in the conditions of the spread of rocks that resist erosion (rocky, cohesive), the flow, even with significant energy, is controlled by the channel. The latter case corresponds to the conditions of limited development of channel changes. The extent of channel changes limitation depends on the type of the rocks in contact with the flow. It is most fully manifested in the rivers that flow through crystalline rocks; the intensity of channel changes here is so low that their morphological effect is revealed only in the geological time scale.

In areas of free channel changes, rivers flowing among easily eroded loose rock have wide floodplain channels, where

$$B_f > (2 \div 3)\mathrm{b}_{\mathrm{ch}}, \tag{2.8}$$

where B_f is the width of the floodplain; b_{ch} is the width of the channel. Most often these are free-meandering (Fig. 2.8A) or branched out into branches. Lateral channel change velocities (movement of channel forms, bank erosion, etc.) are

Fig. 2.8 Forms of representation of fluvial processes in different geological-geomorphological conditions: **A**—free conditions of development of channel changes (wide floodplain channel of the Adycha river, aerial photo); **B**—limited conditions of development of channel changes (incised channel of the Aldan river). **A**—photo by P.N. Tersky; **B**—photo of the author

measured in meters—tens of meters per year, exceeding the intensity of vertical changes by several orders of magnitude. Directional channel movements are often obscured by complex periodic changes associated with the development of bends or branches of the channel, represented only in the historical and even geological time scales. Exceptions are only straight channels, for which the main type of reformation is directed shifts towards the main bank. However, even in this case, the straight shape of the channel often reflects a certain stage in the evolution of another type of channel: cutoffs along the main bank; shallowing and dying off of the branches passing through the floodplain, with the predominant development of the branch near the main bank. During channel change are constantly happening changes in the course of channel and the valley bottom is widening when the river washes out terraces and bedrock cliffs composed of easily eroded rocks. At the same time, a large amount of solid material is delivered into the channels, which usually has a higher erosion resistance than the erosion control stability of the soils composing the bottom and bank, and the size of the bed material load is lower than the critical one. Therefore, the rivers in these areas are characterized by the most complete development of the whole complex of channel forms. Particularly favorable conditions for channel wandering and bank erosion are created in the presence of shielding horizons represented by cohesive or rocky formations: moraine loams in the North of European Russia, traps—on the Middle Siberian Plateau. The rivers that have crossed into these sediments are already under conditions of limited channel change.

In areas of limited channel change, formed by bedrocks, channels are characterized by

$$B_f < b_{ch}, \tag{2.9}$$

or there's no floodplain at all. Lateral changes here have only a directional character and are comparable in tempo with vertical changes (plunging): the time of lateral movement of the channel at a distance corresponding to the size of the floodplain massif (for example, the longitudinal movement of the bend by the size of its pitch equal to the length of the floodplain segment) is less than the time required to turn the floodplain into a floodplain terrace.

In the conditions of limited development of channel changes, lateral changes are hindered by rocky formations that compose the banks and riverbeds; channels are often formed in floodplain, narrow, embedded valleys. The amount of bed material load is relatively small and even on flat rivers it is represented by pebbles or pebble-boulder material. The channels forming in this case are embedded, and the set of channel forms and forms of channel relief is minimal (Fig. 2.8B). The transport capacity of the flow usually far exceeds the actual amount of transported sediment, i.e. $W_{tr} >> W$, so that alluvial deposits are often absent at the bottom and the flow is in direct contact with the rocky or cohesive bedding of the bed. The destruction of the latter occurs due to tectonic fragmentation or rock fracture, chemical erosion, corrosion and biochemical erosion. Since all of these agents are most active on the root floor, the rate of deep erosion is higher than the intensity of bank erosion. Lateral erosion is also prevented by the ingress of coarse-clastic material from the slopes that form plumes and whips at the foot of the coastal slopes, where the size of the debris and boulders that compose them exceeds the critical size for the conditions of their movement and movement.

The most common on such rivers are incised bends and embedded straight channels (coincidence in terms of the shape of the channel and the shape of the valley), and among the branched ones—sculptural branching, where the base of the islands are the ledges of the rocky bed of the rivers. The shapes of the incised channels and their predominant distribution depend on whether the river is embedded into rocky or ductile formations (Matveev 1985). In rocky formations, the bed is fully confined to the geological structure and its forms are usually not related to the structure of the stream, although to some extent it is modeled by it. In ductile rocks, the flow actively simulates the primary forms of the channel, which often reflect their primal (initial) shape.

In transition from free to limited conditions of channel change, as well as in areas with frequent alternation of channel changes, intermediate channel types are formed. They develop in relatively narrow valleys or intermountain trenches, where lateral changes are limited to the main sides, but the channel itself is accompanied by a narrow floodplain. The criterion for selecting this intermediate channel type is

$$b_{ch} \leq B_f \leq (2 \div 3)b_{ch}. \tag{2.10}$$

Such channels can be called confined (Chalov 1997). They are usually found in border geomorphological areas. These are, for example, the Viliuy riverbed between the Central Siberian Plateau and the Central Yakut lowland (Bely et al. 1983) and the Dniester riverbed at the intersection of the South-Eastern slopes of the Volyn-Podolsk Upland (Berkovich et al. 1992). Similar channels are sometimes found in mountainous areas where river valleys inherit linearly elongated morphostructures (grabenne valleys, etc.). In such cases, the relatively narrow bottom of the valley is enclosed between steeply sloping rock walls, which restrict the development of lateral changes.

Another criterion of transition from free to limited conditions of formation of channels is the equality between the width of the bottom of the valley (floodplain) B_f and the transverse belt (relative to the axis of the valley) movement of the channel—meandering, branching $B_{m(br)}$

$$B_f \approx B_{m(br)} \tag{2.11}$$

on condition of optimal ratios between the channel length along the bend or main branch in branching l and the bend pitch or branching of the channel L—$l \approx 1.6$L (Makkaveev 1955; Makkaveev and Chalov 1986). The bend necks on meandering rivers and the islands in the branches should be floodplain accumulative formations that do not have bed ledges at the base. This criterion (2.11) cannot be applied to straight channels, for which only the B_{ch} and b_{ch} ratio remains valid.

Makkaveev (1955) took as an indirect sign of the prevalence of deep river erosion (incision) (2.9) or the development of lateral erosion—lateral channel change (2.8). The first case corresponds to the development of embedded channels, the second case corresponds to the development of wide floodplain channels.

To an even greater extent, geological and geomorphological factors determine the formation of lowland, semi-mountain and mountain rivers. Mountain rivers differ from lowland ones by their high steepness and variability in the length of slopes of longitudinal profiles, often their stepwise and undeveloped nature, pebble, pebble-boulder and boulder composition of bed material load, its insignificant underfloor capacity, the presence or predominance of rocky areas where the water flow directly comes into contact with the bedrock. All of this determines the specifics of fluvial processes which, on mountain rivers with high slopes, are rapid, represented in the originality of the forms of sediment transport and, as a consequence, in the morphology of the channels. In many cases, the fluvial processes here are controlled by debris flow activities. Mudflow outflows from tributaries often have a significant impact on the longitudinal profiles of even relatively large mountain rivers, on the composition of bed material load and on the type of bed.

However, not every river in the mountains is mountainous in nature of fluvial processes. A distinction should be made between mountain rivers and rivers located in mountains (Chalov 2002a, b). It is known that the longitudinal profile of the river under conditions close to depletion is described by the equation (Makkaveev 1955) $Q^m I = const$, where m is the coefficient approximately equal to 2/3. Other things being equal, the greater the length of the river, the area of its basin and runoff rates, the lower the slope and the greater the concave of the longitudinal profile. Therefore, the closer the watershed ripples are to the border of the mountainous country and the foothill plain area (Greater Caucasus, Zailiyskiy Alatau), the shorter and less water-intensive the rivers crossing the mountainous country, the more sloping they are, and all of them are mountainous in nature of fluvial processes. Only relatively large rivers (with a basin area of more than 100 km^2) have smaller slopes in the border zone (mountains—plains), and they even within the mountainous area become (in terms of the nature of fluvial processes) semi-mountainous (Mzymta, Bzyb, Kodori, Inguri—rivers starting on the slopes of the Greater Caucasus Range and flowing into the Black Sea).

In large and orographically complex mountainous countries (Altai, mountains of Eastern and Southern Siberia, Central Asia), the increase in runoff rates of the rivers (the area of the basins is more than 10,000 km^2) ensures the reduction of slopes to the values that are typical for semi-mountainous (Katun, Biya in the Altai) and lowland rivers (less than 0.30–0.50‰). For example, the upper Lena in the Kachug—Ust-Kut—Kirenga River mouth, flowing among the mountains of Baikal region, has slopes of 0.32–0.11‰ (basin area at Kachug—17.4 thousand km^2, below the mouth of Kirenga—140 thousand km^2; average annual water discharge, respectively,—87.6, 295 and 950 m^3/s). However, lowland rivers (in terms of gradients and the nature of fluvial processes) have a pebbly or pebbly-boulder bed in the mountains. These are the upper and middle Lena and its large channels (Kirenga, Vitim, Olekma, Aldan), Yenisei, flowing along the border of the Central Siberian Plateau, Tom within the Kuzbass, and the upper Amur. This is due to the direct flow of coarse debris into the rivers from the mountain slopes covered with stone runs and screes and the removal of pebbles and boulders from mountain tributaries.

This composition of channel alluvium on lowland rivers in mountainous areas creates an external resemblance to the actual mountain rivers (by fluvial processes). However, they differ from mountain rivers in terms of the mechanism of channel formation and transport of bedloads (Berkovich et al. 1985), and from sandy and sandy-pebbly plain rivers in terms of greater stability of the channels.

Mountain and semi-mountain rivers are mainly confined to areas of active upward development, which, as a rule, correspond to the limited conditions for the development of channel changes: in the mountainous countries of Northern Eurasia (the territory of the former USSR), these are 84% of the length of all mountain rivers and 62% of semi-mountain rivers. Only in intermountain trenches and intermountain depressions can mountain and semi-mountain rivers have wide floodplain channels formed in the conditions of free development of channel changes. There are no mountain river channel changes in flat areas with free conditions for the development of mountain rivers at all. However, here they can be found within elevated areas, especially in the upper reaches of rivers, where their channels are formed under conditions of limited development of channel changes. It was this that gave Makkaveev (1955, p. 79) a reason to conclude that «it is almost impossible to find a single significant flat river that would not have had any «mountainous» areas during its course». In addition to the upper reaches, it points to sections of the middle and even lower reaches of large rivers, where the channel becomes mountainous in lowland conditions. These were (before the construction of the Dnieper hydropower dam) the empty section on the Dnieper River or the section of the Kazachen rapids on the middle Yenisei.

The wide range of river gradients in the mountains and their natural reduction from highlands to lowlands and foothills cause a peculiar zonality in the distribution of mountain river beds of various types. In accordance with the increase in slopes, one type of mountain bed changes from foothills or large intermountain troughs, which are the basis for denudation of the surrounding mountains, to high mountains (Chalov 1985). Under certain conditions, fluvial processes on mountain rivers, especially in high mountains, are transformed into mudflows due to over-saturation of the flows with sediment and their transformation into structural ones. Mudflows on tributaries carry large volumes of solid material to the main rivers, exceeding the transport capacity of the channel flows. As a result, this material, accumulating in the river channel during repeated mudflows over long periods of time (sometimes millennia), creates kinks in the longitudinal profile. At the descent of especially large catastrophic in the form of mudflows, mudflows can create dams with the formation of temporary lakes (Bogomolov et al. 2002).

Landslides, crashes, scress on banks are also suppliers of solid material to rivers. Depending on the relief and its geological structure and lithology of rocks, gravity processes develop in the valleys of both mountainous and lowland rivers. They have a local impact on channel morphology and fluvial changes. Large landslides formed by cohesive rocks (clays) block part of the channel and have a directing effect on the flow for many years. As a result, the opposite banks are eroded and this can lead to the formation of bends, the development of floodplain branches or the formation of islands downstream. Narrowing channels, landslides, crashes, scress contribute to intensive local erosion of the river bed, up to the appearance of deep pits. Like mudslides, rock falls and landslides on the banks of lowland rivers sometimes create sub-pond lakes that have existed for centuries. In the mountains of a lake of landslide origin, sediment is introduced over time, causing a stepwise progression of the longitudinal profiles of the rivers (Berkovich and Chalov 1969). As the capacity of the lake is transferred and reduced during the overflow of water through the debris dam, the latter is eroded and the lake is lowered. At instant descent of the lake down the river there is a catastrophic mudflow (Kuznetsov and Chalov 1988).

In the plains of rivers, gully erosion along the banks of the rivers and gullies directly into the river beds play an important role in the formation of sediment load and channel morphology. For example, some researchers (Polyakov 1930) associated the shallowing of the Don riverbed in the early twentieth century with the intensification of gully erosion during this period. According to calculations by Zorina et al. (1980), the volume of ground removal from gullies increased steadily here until the beginning of the twentieth century and especially intensively at the end of the nineteenth century, when the area of arable land and, as a consequence, the number of growing gullies increased sharply. To date, the number of the latter has decreased due to the work on the protection of land from destruction; accordingly, the volume of removal from gullies to rivers has decreased (Table 2.2).

Where the gullies enter rivers typically riffles can be formed. Polyakov (1930) noted that there were cases when the current of the Don after heavy rain was blocked by a “dam” formed from gully outflows. When flood waters pass on the river, such “dams” are washed away and downstream of the gully mouths, riffles are formed with the products of removing from gullies deposited in the “high-speed shade” of debris cone that appears downstream, as well as in the upstream catchment area. Figure 2.9 shows one of the typical rifts of the Don—Olkhovatskiy, formed by outbursts from the beam of the Olkhovatka and a series of gullies on the right bank of the river. The channel plan clearly shows the riffle directly adjacent to the debris cone, the overall expansion and shallowing of the channel downstream. Such rifts are typical for rivers with deep valleys, the sides of which are dissected

Table 2.2 Changes in the volume of sediment transport from gullies to rivers upper Don basin (Zorina et al. 1980)

Years	Number of growing gullies, thousands	Removal volume, thousand m^3/year
1696	26.2	1422.7
1861	37.0	2002.4
1913	52.6	2839.5
1950	47.3	2444.3
1970	44.9	2419.1

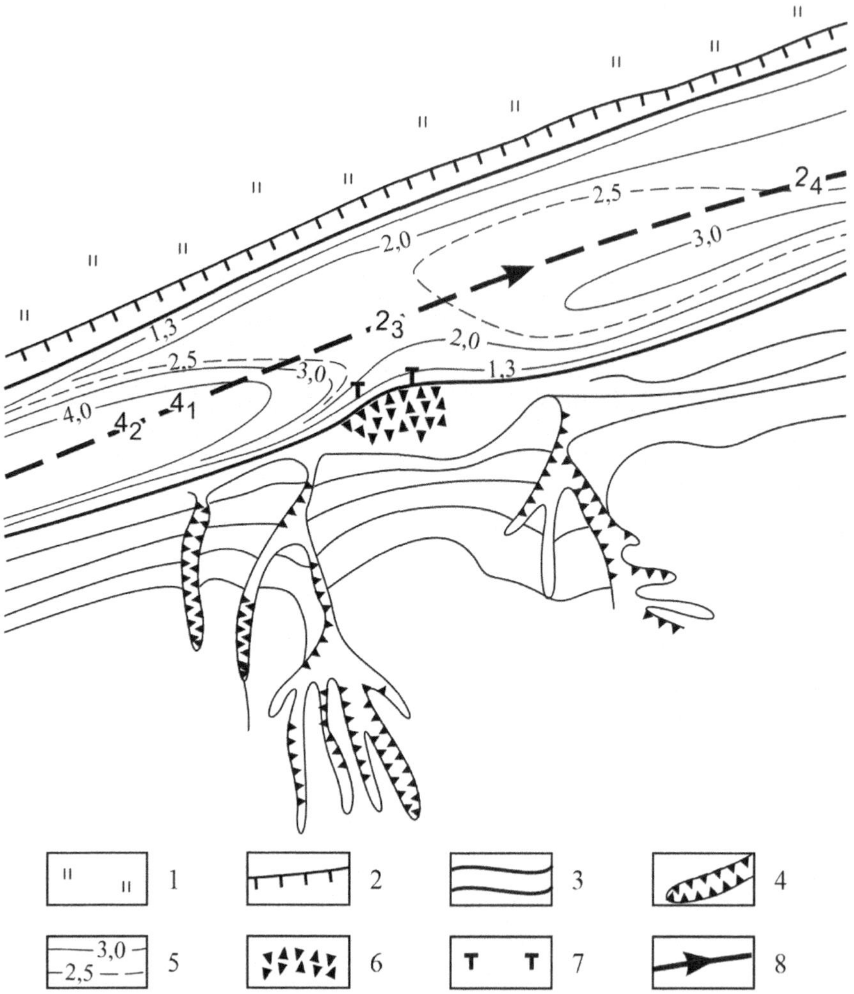

Fig. 2.9 Olkhovatsky riffle of the Don river, made by debris cones from gullies dissecting the right main bank: 1—floodplain; 2—floodplain edge; 3—horizontals of the main right bank of the valley; 4—gully; 5—isobates in the channel; 6—gully removal (surface); 7—rocky gully removal (underwater); 8—ship's passage

by the gully-beam network. They are quite widespread in the Don (Polyakov 1930; Engineering of ship's passages … 1964; Zorina et al. 2000) and the middle Dniester (Berkovich et al. 1992); in gullies on the banks of the Dniester River, mudflows are causing gully formation up to a depth of 150–200 m during downpours (Slastikhin 1987), which provide the most intensive growth of debris cones and shallowing of the associated riffles. In the 90s of the twentieth century, the mudflow along the Egorov Gully in the Protva River valley (the Oka River basin) fell off; the resulting debris cone deflected the river flow to the left, causing erosion of the opposite bank and formation of a bend in the riverbed.

2.5 Sediment Load, Its Components and Its Impact on Fluvial Processes

The physical essence of fluvial processes, along with the interaction of water flows with the channels and soils that compose them, is the transport of sediment. Therefore, many patterns of fluvial processes are associated with the magnitude and variability of sediment load and the mechanisms of their transport, and sediment load itself is an important factor in fluvial processes. This is the basic position of the theory of fluvial processes (called in Russian language as a «русловедение»—riverbed science). Nevertheless, sediment load, the partitioning between its longitudinal and weighted components, their relationship to the characteristics and indicators of the form of representation of fluvial processes, as compared with water flow, the structure of water flow, its turbulence, etc., remain the least developed issues in the theory of fluvial processes (the exception is the analysis of the morphology and dynamics of the ripple relief of river channels and balance relations to assess the direction of vertical change). This is obviously due, inter alia, to the lack of real data on the bedload transport rates and the high variability in sediment load characteristics along the length of the rivers. The dual nature of sediment load creates difficulties: it is a factor in fluvial processes, which is associated with the formation and development of channel forms and channel relief forms, and at the same time a derivative of them, since the load is carried into the stream during channel scour or from the catchment area. I. V. Popov (Kondratyev et al. 1982, p. 88) wrote: "Due to the lack of in situ information, it is not possible yet to construct quantitative links between the types of fluvial process and … the characteristics of solid runoff". His attempt to link the development of certain types of channel with sediment concentration of flows (their distribution over sediment concentration zones) did not reveal any connection between them. I. V. Popov, relying on G. I. Shamov's research (1959), rightly explained this by the predominance of non-rural-forming fractions in the composition of suspended load.

Recently, a number of publications have appeared in which the forms of fluvial processes in rivers of specific basins or large rivers are considered against the background of a detailed analysis of river load (Nekos and Chalov 1997; Babak et al. 2000; Chalov et al. 2000; Chalov and Shtankova 2000, 2003). This has made it possible, at least at a qualitative level, to establish the correspondence between the sediment load characteristics and the forms of fluvial processes. For example, for unregulated rivers in the Volga and Northern Dvina basins, differences have been identified in the predominance of certain channel patterns, depending on the size of sediment load and the partitioning between bedload and suspended components in the channel (Babak et al. 2000; Chalov and Shtankova 2003; Reznikov and Chalov

2005). The construction of diagrams "slope *I*—sediment concentration *s*", "flow power *QI*—sediment concentration *s*" (Chalov et al. 2000) and "flow power *QI* is a specific characteristic of the flow of bedloads" (Reznikov and Chalov 2005) made it possible to differentiate between meandering, branched out on branches and relatively straight, single-thread channels. It is fair to say that the Chinese scientific literature has always paid special attention to the problem of "riverbed sediments" (Alekseevsky and Chalov 1997), which seems to be related to the huge amount of sediment load, with an absolute predominance of a weighted component, the formation in these conditions of specific roaming varieties of channels, and an increased degree of hazard posed by the Chinese rivers due to the erosion of coasts and bunds, the constant threat of flooding due to the systematic accumulation of sediments and the increase in the amount of sediment load.

The impact of sediment load on fluvial processes is determined by: (1) the partitioning between its absolute value to the transport capacity of the flow; (2) the realization of the latter due to the suspended or bedload components; (3) the share of bed load in total river load; (4) the predominance of far-reaching or far-reaching or far-reaching components of river load in the composition of bed material load; (5) the particle size of bed material load; (6) the form of transportation of bed material load.

In general, sediment yield W affects on fluvial processes by changing the ratio between W and the sediment transport capacity W_{tr}. With W_{tr} = const, an increase in river load is accompanied by accumulation ($W > W_{tr}$), growth in the volume of river sediments, and a decrease ($W < W_{tr}$) in the volume of riverbed and bank erosion and a decrease in the volume of sediments (Alekseevsky and Chalov 1997). This causes vertical (erosion—accumulation, lowering—increasing of bottom marks) and lateral (erosion of banks in local growth areas of W_{tr}, formation of riverbed banks with local decrease of W_{tr}) channel changes. In the inland areas, most of the rivers have a deficit of sediment load relative to their transport capacity, which results in the absolute dominance of the incising rivers. This is due to the consistent increase in the runoff rates of the rivers downstream and the general tectonic upwelling of the land, which is rather slow in the plain-platform and quite noticeable in the mountainous areas). At the periphery of the continents and in areas of internal runoff (Caspian and Aral Sea basins), sediment accumulation is mainly facilitated by the longitudinal reduction in the runoff rates of rivers without tributaries, especially in arid conditions, by the tectonic movement of the negative sign (submergence of the territories) and the river mouth elongation. The intensity of this process depends on the absolute value of the sediment load. The implementation of the transport capacity of the flows due to suspended load causes very high flows, very high rates of sediment accumulation and increased intensity of lateral channel change (Huang He, Yangtze, Central Asian rivers, Lower Terek). With low suspended sediment yield and high bedload runoff, the rate of directional accumulation is relatively low (millimetre fraction in the lower reaches of Northern rivers), river intakes predominate and channels are more stable.

Sediment load comprises two components: suspended W_R and bed load W_G, i.e. its full value $W = W_R + W_G$. Weighted sediments are formed by small fractions (silt, dust, sand). Inhabited sediments also include gravel, pebbles and boulders. In various natural conditions, the ratio between the components of sediment load is highly variable, which is reflected in the form of fluvial processes, direction and intensity of channel changes. At low sediment concentration, suspended loads are mainly transit load, and bed formation is led by bed-ridden sediments. Under these conditions, $W_{tr} \sim W_G$, i.e. the transport capacity of the flow is mainly realized through the bedload transport rates. At high sediment concentration, suspended sediments become bed material load and $W_{tr} \sim W_R$.

The contribution of suspended and bed material in the formation of channels is represented in the comparison of cumulative curves of the particle size distribution of sediment (Fig. 2.10). The bed material load, regardless of their genesis, forms the channel. With their absolute predominance and large particle size (gravel-pebble, pebbles, and pebble-boulder), suspended loads are only transit and both cumulative curves overlap (Fig. 2.10A). These are mountain rivers and lowland rivers flowing in mountainous regions and along structural uplands. The opposite conditions correspond to the realization of the transport capacity through suspended load, which is absolutely dominant in total sediment load. In this case, the curves almost completely overlap each other, and only the thinnest curves in relatively small volumes are exclusively transit curves (Fig. 2.10B). It's characteristic of Huang He, Yangtze, Amu Darya, the Lower Terek. The most typical ratio is that the largest suspended loads are in the category of bed material load and the curves are only crossed in their outermost parts (Fig. 2.10B). It is characteristic of most sandy plain rivers.

Sediment yield is formed from solid material, which flows into the river with inflow waters, temporary streams, is carried by the wind, and is carried into the riverbed by collapses, debris and landslides on the banks. At the same time, sediments are discharged into the stream due to the erosion of the bed and riverbanks. Part of the material transported by the river forms ripple shapes of the riverbed, moving along the length of the river and forming the flow of bedload. These forms can be renewed if they are artificially destroyed by dredging, for example. The rate of form reconstruction, as well as the intensity of channel changes associated with uneven runoff, are greater than the greater the sediment load. In particular, the amplitude of seasonal changes in the ripple marks is directly related to sediment concentration of the flow. For example, in the Amu Darya, whose waters are very turbid, the depth during floods sometimes does not increase, but decreases, as the increase in the ripples of individual ropes exceeds the increase in water level. In rivers of the Russian Plain, which are characterized by significantly lower sediment concentration, the amplitude of changes in ripple marks usually does not exceed 1/3 of the level amplitude, and in rivers emanating from large lakes with clean water, floods do not practically affect the height of the ripples or contribute to their erosion.

Suspended sediment discharge depends on the mean annual discharge of Q_0 and sediment concentration s, i.e. $W_R = tsQ_0$, where t is the duration of the year (in seconds). Sediment concentration changes depending on zonal and azonal natural factors (Sediment yield … 1977; Dedkov and Mozzherin 1984), on which

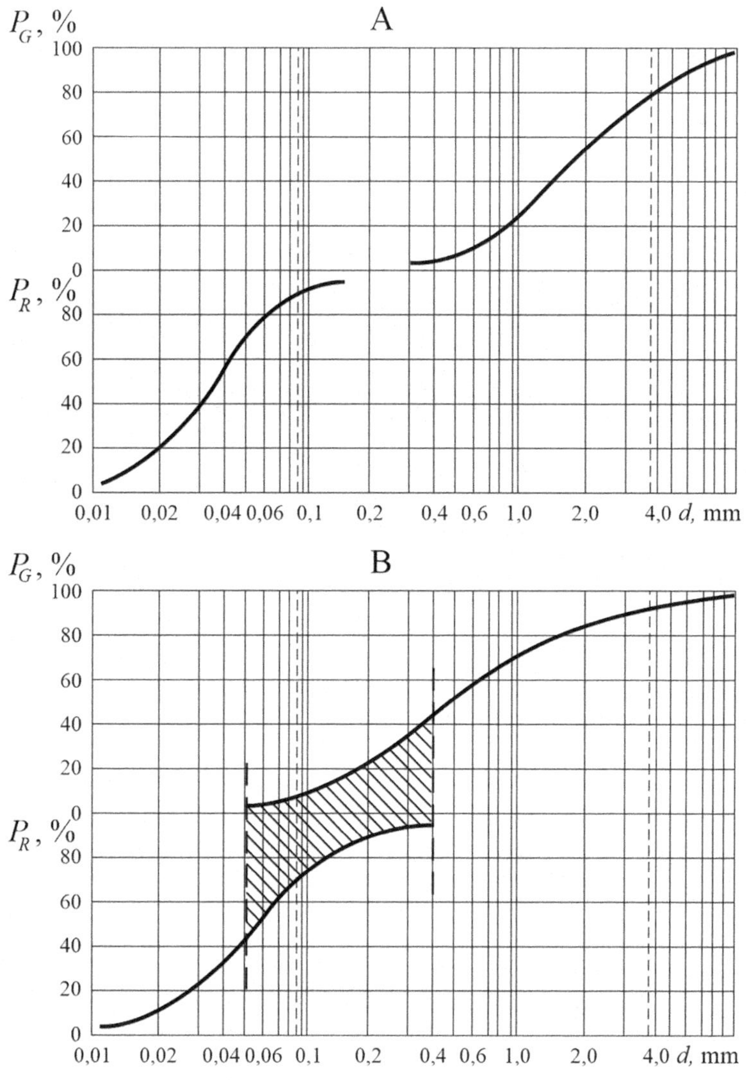

Fig. 2.10 Types of ratios of cumulative curves of particle size composition of suspended sediment and bed material load. **A**—rivers with pebble and pebble-boulder channel; **B**—rivers with sandy and sandy-pebble channel; **C**—rivers with sandy-silty and silty channel

anthropogenic loads are imposed on the river basin and river beds. With the same catchment area, there is a natural increase in water sediment concentration (even if the runoff rates of the rivers decrease) from humid to arid areas, from mountains to plains. This results in complex spatial changes in suspended sediment yield W_R, the runoff rates Q_0 of rivers and the sediment concentration s of the water in the rivers (Table 2.3). For example, the flow of Asian rivers separately into the Arctic, Pacific

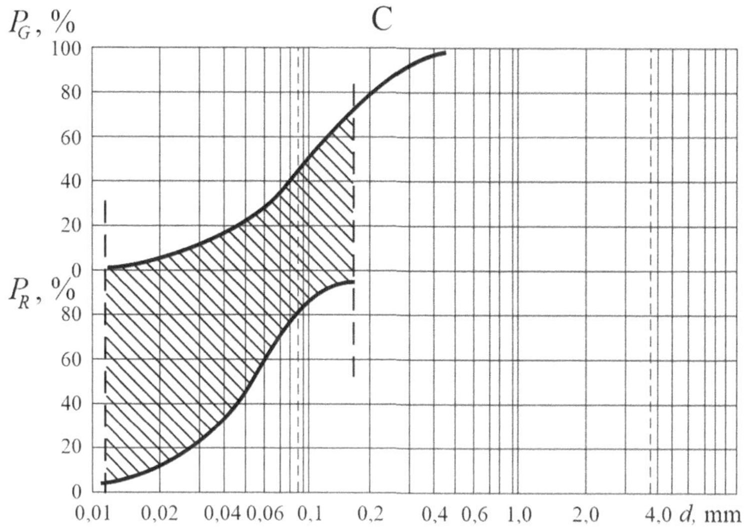

Fig. 2.10 (continued)

and Indian Oceans is about 1/3 of their total flow (with some excess of 1/3 of the flow of Pacific rivers and underestimation of the Northern rivers. Suspended sediment load from rivers flowing into the Pacific Ocean accounts for more than half of the load from Asian rivers. Sediments from rivers flowing into the Arctic Ocean account for only 1.8% of the total load from Asian rivers. Their differences (almost 50-times) in the value of the sediment load modulus are even greater (Liu et al. 2001).

The Yangtze has the largest water runoff between Asia's rivers. Then follow Ganges, Brahmaputra, Yenisei, Lena, Ob, Irawaddy, Mekong. The runoff rates of the Huang He at the time of its discharge into the sea is 1/20 of the Yangtze's water flow. At the same time, the Huang He (where the weighted component of the total load is absolutely dominant—more than 99.5%—prevails (Chalov et al. 2000), and ranks first not only in Asia, but also in the world. Suspended sediment load from each of the high-water rivers such as the Ganges, Brahmaputra, Indus, Yangtze, Irawaddy, Mekong and Shatt El-Arab is 2 or more times lower than that of the Huang He. The runoff rates of the Lena, Ob and Yenisei is 1/2–2/3 of the Yangtze runoff rates. Suspended sediment load from these rivers does not exceed 1/25 of that of the Yangtze. Exceeding the Huang He water flow by 7–10 times, the largest rivers of Siberia carry into the sea 60–80 times less suspended material than the Huang He.

Water sediment concentration in the estuaries of Northern Asia changes little (Fig. 2.11A), having a weak tendency to increase from West to East (from less than 0.05 kg/m^3 to more than 0.05 kg/m^3 with a maximum at the estuary of the Indigirka —0.213 kg/m^3). From North to South along the Pacific coast and Far West along the Indian Ocean coast, there is a double change in sediment concentration (from

Table 2.3 Average perennial water runoff and suspended load of main rivers in Asia flowing into the sea (except for the Atlantic seas)

Ocean	River (number in Fig. 2.11, name)	Basin area		Water runoff		The sediment concentration of the water, kg/m^3	Sediment yield		Data sources[a]
		Thousand km^2	%	Km^3 per year	%		Million tons per year	%	
Arctic Ocean	1. Ob	2990	28.4	402	19.7	0.033	13	15.6	7
	2. Nadym	64	0.6	18	0.9	0.022	0.4	0.5	7
	3. Taz	150	1.4	41.1	2.0	0.02	0.91	1.1	3, 7
	4. Pur	112	1.1	32.3	1.6	0,022	0.62	0.7	3, 7
	5. Yenisei	2580	24.5	537	26.4	0.022	13	15.6	7
	6. Pyasina	182	1.7	86	4.2	0.04	3.4	4.1	1–7
	7. Hatanga	364	3.4	105	5.2	0.05	5.2	6.2	7
	8. Anabar	100	0.9	25.2	1.2	0.024	0.4	0.5	7
	9. Olenyok	220	2.1	40	2.0	0.031	1.1	1.3	7
	10. Lena	2486	23.6	536	26.3	0.043	22.6	27.0	2, 11
	11. Omoloy	38.9	0.4	4.4	0.2	0.12	0.52	0.6	1
	12. Yana	238	2.3	29.9	1.5	0.093	3	3.6	5, 6, 7
	13. Indigirka	360	3.4	54.2	2.7	0.219	11.2	13.4	7
	14. Kolyma	647	6.2	122	6.1	0.067	8.2	9.8	7
	Totally	10531.9	100	2036.1	100	–	83.55	100	–
Pacific Ocean	15. Anadyr	191	2.8	67.9	2.8	0.092	3.6	0.2	7
	16. Kamchatka	56.9	0.8	32.6	1.3	0.095	3.1	0.1	7
	17. Penjina	73.5	1.1	24.8	1.0	0.040	1	0.03	7
	18. Amur	1855	27.6	355	14.5	0.065	14.9	0.6	5, 7
	19. Yalujiang	63.8	1.0	25.1	0.9	0.08	1.9	0.07	8
	20. Liao He	166.3	2.5	5.6	0.2	6.86	41	1.7	9
	21. Dalin He	23.2	0.3	2.1	0.1	21.9	36	1.5	9

(continued)

Table 2.3 (continued)

Ocean	River (number in Fig. 2.11, name)	Basin area		Water runoff		The sediment concentration of the water, kg/m^3	Sediment yield		Data sources[a]
		Thousand km^2	%	Km^3 per year	%		Million tons per year	%	
	22. Luan He	44.9	0.7	4.9	0.2	4.63	22.7	1.0	9
	23. Hai He (Yundinhee)	50.8	0.8	1.4	0.1	60.8	81	3.4	9
	24. Huang He	770	11.5	48.4	2.0	27.7	1320	55.5	11, 12
	25. Yangtze	1990	29.6	930	38.0	0.53	471	19.7	11, 12
	26. Jiangtanjiang	41.5	0.6	34.2	1.4	0.13	4.4	0.2	8
	27. Minjiang	61	0.9	58	2.4	0.13	7.5	0.3	12
	28. Julongjiang	15	0.2	15	0.6	0.21	3.1	0.1	12
	29. Hanshui (Hanjiang)	30.1	0.4	25.9	1.1	0.28	7.2	0.3	4
	30. Xijiang (Pearl)	329.7	4.9	227	9.3	0.34	71.8	3.0	10
	31. Hong Hungha (Red)	160	2.4	120	4.9	1.08	130	5.5	12
	32. Mekong	790	11.9	470	19.2	0.34	160	6.8	12
	Totally	6712.7	100	2447.9	100	–	2380.2	100	–
Indian Ocean	33. Irawaddy	430	5.7	430	20.7	0.61	260	12.2	12
	34. Brahmaputra	580	7.7	630	30.4	0.857	540	25.4	12
	35. Ganges	980	13.0	590	28.5	0.881	520	24.4	12
	36. Brahmani	28	0.4	16	0.8	1.25	20	0.9	12
	37. Godawari	310	4.1	92	4.4	1.85	170	8.0	12
	38. Krishna	250	3.3	32	1.5	0.125	4	0.2	12
	39. Pennaru	49	0.6	5.2	0.3	1.33	6.9	0.3	12
	40. Tapti	49	0.6	10	0.5	0.25	2.5	0.1	12
	41. Narmada	88	1.3	47	2.3	49	70	3.3	12
	42. Indus	970	12.9	175	8.4	2.49	435	20.4	12

(continued)

Table 2.3 (continued)

Ocean	River (number in Fig. 2.11, name)	Basin area		Water runoff		The sediment concentration of the water, kg/m^3	Sediment yield		Data sources[a]
		Thousand km^2	%	Km^3 per year	%		Million tons per year	%	
	43. Shutt-el-Arab	3800		46	2.2	2.17	100	4.8	12
	That's the total	7534	100	2073.2	100	–	2128.4	100	–

[a]1—Alekseevsky and Sidorchuk (1992); 2—Waterways … (1995); 3—Korotayev et al. (1978); 4—Liu (1986); 5—Mikhailov (1997); 6—Lower Yana … (1998); 7—Channel mode … (1994); 8—Physical geography of China (1981); 9—Ning et al. (1987); 10—Ning et al. (1993); 11—Chalov et al. (2000); 12—Milliman et al. (1995)

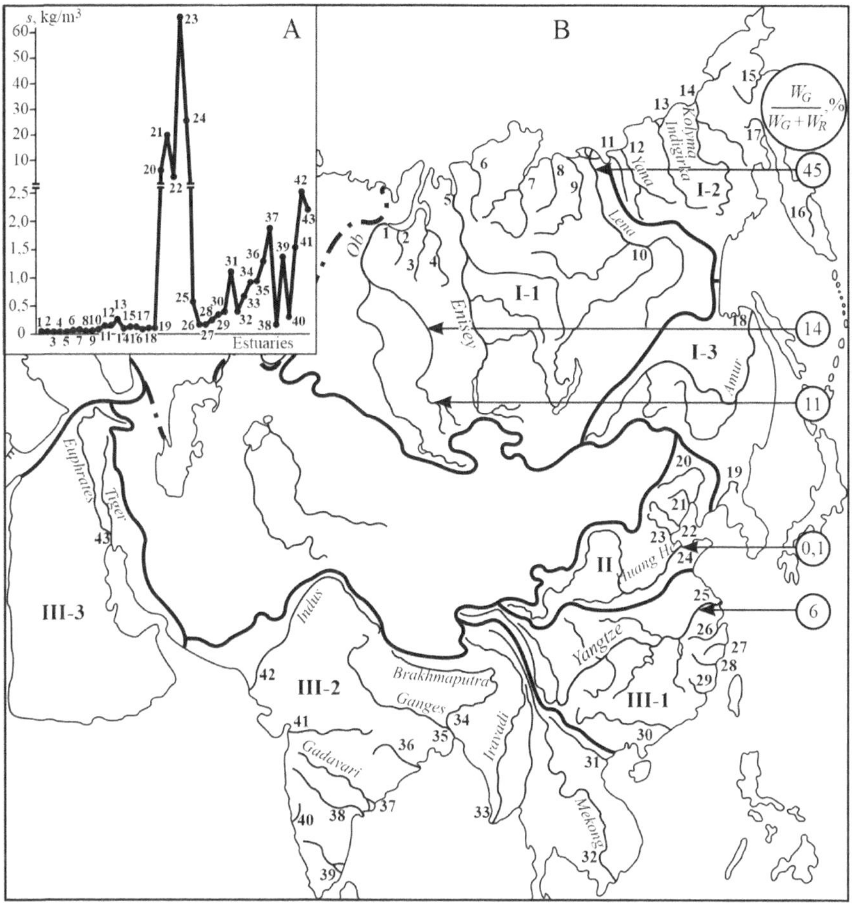

Fig. 2.11 The graph of changes of turbidity in estuaries of Asian rivers (**A**) and zoning of Asia by the conditions of forming and suspended sediment flow (**B**). 1 to 43—river numbers on the graph and scheme correspond to a Table 2.4. The right-hand side shows the typical ratios of bed load flow and total sediment load on rivers, %

small to large). Up to the Amur and Yalujiang it is close to the sediment concentration of the Northern rivers (no more than 0.07 kg/m^3). In Eastern China, it reaches a maximum of 60 kg/m^3 on the Haihe River (Yundinghee), 27.7 kg/m^3 on the Huang He River and 21.9 kg/m^3 on the Dalin He River. In South-Eastern and Southern China, sediment concentration is relatively low, although greater than the sediment concentration of Northern rivers by an order of magnitude. Among the rivers of South-East Asia, the Khonga River has the highest sediment concentration. It's much smaller at the mouth of the Mekong River. From the Mekong to the west, sediment concentration is gradually increasing, reaching a maximum (2.49 kg/m^3) at the mouth of the Indus River, which is less only compared to the Eastern Pacific rivers (from the Yangtze to the Liao He).

The entire territory of Asia can be divided into three districts in terms of sediment concentration of water in rivers (Fig. 2.11B). There are a number of subareas in areas I and III that vary in sediment concentration. Area I covers the Northern half of Asia and includes the basins of high-water rivers with the lowest (<0.1 kg/m^3) average annual sediment concentration and relatively low suspended sediment load (type 1). The main rivers of the region are the Ob, Yenisei, Lena, which flow into the Arctic Ocean, and the Amur River, which belongs to the Pacific Ocean basin.

If the proposed zoning scheme is to be extended to the European part, the Western extension of area I will obviously be the Pechora, Onega and Northern Dvina basins (for the latter at the closing post—Ust-Pinega village: annual water runoff of 104 km^3, sediment concentration of 0.037 kg/m^3, suspended load discharge of 3.86 million tons per year (Chalov et al. 2000)), the rivers of the Baltic Sea basin (for the Neman before the flow regulation of the Kaunas hydroelectric power station: annual water flow of 18.9 km^3, sediment concentration of 0.039 kg/m^3, suspended sediment flow of 0.4 million tons per year (Darbutas 1992)), the Black Sea (Dniester, Don), as well as the Volga (before the flow regulation).

Subareas I-1 and I-2 are located in the Northern part of the area. The sediment concentration of water from West to East, from the Ob to the Lena, changes insignificantly, and its absolute values are the lowest for all rivers in Asia. Sediment concentration then increases dramatically (although it remains lower than in other areas). Both subdistricts are characterized by a sharply continental climate, permafrost, and a location predominantly in the taiga zone (only the Southern part of the Ob and Yenisei basins is in the forest-steppe and steppe zones, but their influence on the formation of sediment load is almost completely obscured by the middle course of these rivers) (The channel regime … 1994; Chalov et al. 2000). However, in subarea I-1, the main water and sediment load flows during the spring floods; in subarea I-2, the spring and summer floods occur, and in summer the sediment flow is higher than in spring (Lower Yana … 1998). This is due, firstly, to the fact that the influence of rainfall from the Far East affects the formation of water runoff; secondly, the wide spread of permafrost causes that during the snowmelt period melt water quickly enters the channel without scouring the frozen ground. Summer rainfall floods occur when the soil thaws most, which determines its flushing and flow of washed material into the rivers. Besides, from West to East, the vertical dissection increases and the relief becomes mountainous. Therefore, the sediment concentration of water in the summer flood period on the rivers of the subarea is the highest. The relief and geological structure of the basins of subareas I-1 and I-2, and the predominance of physical, including frosty, weathering, lead to a high particle size of bed material load (The channel regime … 1994) and a high proportion of the flow of bed material in total sediment load—11 to 43% (Chalov et al. 2000).

Subarea I-3 is located in the South-Eastern part of district I, including only the Amur basin, which occupies more than 90% of its area. It is typically characterized by low sediment concentration of river water—less than 0.1 kg/m^3. Here, the forested mountainous terrain prevails. That's the main reason for the small amount of suspended sediment yield. A certain role is played by the directed accumulation of sediment, covering all the lower and partly middle course of the Amur and its tributaries. The total rate of sediment accumulation in the lower Amur is estimated at 0.6–0.8 mm per year (Makhinov et al. 1994). As a result, the annual sediment load at the estuary (14.9 million tons) is significantly lower than in the Khabarovsk area (26.1 million tons). The direction of the vertical channel changes in the middle and lower Amur allows us to draw a certain analogy between subarea I-3 and other areas of Eastern and South-Eastern Asia, where the rivers in the lower reaches are intensive (by an order of magnitude greater than on the Amur) accumulate sediment (Chalov et al. 2000), although it is related to the Northern region I in terms of its main features—low sediment concentration and low sediment load.

Area II is confined to East Asia, includes arid areas, has a relatively small area of river basins and water flow, maximum sediment concentration and sediment load. The main rivers are the Huang He, Liaohe and Haihe, which flow into the Pacific Ocean. The area of the region is 4.5% of the total area of the Asian river basins flowing into the sea, and the water flow is equal to 73 km^3/year—1.1% of the total flow of Asian rivers flowing into the sea.

Rivers have high sediment concentration and very high sediment load. The average sediment concentration of most rivers is over 6 kg/m^3, with the highest sediment concentration reaching 60 kg/m^3—the highest in the world. Sediment load is 1500.7 million tons per year, accounting for 32.7% of the total load of Asian rivers flowing into the sea. However, the bedload transport rate is very small, averaging less than 1% of the total load (Chalov et al. 2000). The amplitude of the intra-annual fluctuations in sediment concentration and sediment load is very high. The annual load variation coefficient C_v varies from 0.77 to 1.87, which is greater than the change in the flow variation coefficient.

At the Huang He River there is a hydrograph and the course of changes in sediment concentration of water generally coincide. Floods on this river, superimposed on each other, form a common wave with steep branches of rise and fall. However, flood peaks and sediment concentration peaks are not synchronous. The highest sediment concentration of water does not correspond to the highest water runoff and in most cases is observed 1–3 days or even a few hours before the highest flood peak. On some rivers of the district in some years the peaks of sediment concentration and flooding coincide or the first one even slightly lags behind the last one. Characteristically, the increased runoff rates in spring does not affect the sediment concentration of water in the rivers, which at that time is the lowest (less than 20 kg/m^3, although this value is an order of magnitude higher than the maximum sediment concentration of the rivers in other areas). Obviously, this is explained by the regulatory influence of reservoirs: in spring, when the flow increases, the load is delayed in them, while in the summer flood period of the Huang He reservoir, their bowl is washed out (Chalov et al. 2000). Since suspended sediments at the Huang He are bed material, such intra-annual fluctuations in their flow rate directly determine changes in the intensity of channel changes.

The reason for the very high sediment concentration of the rivers in the area is their crossing of the Loess Plateau, which is the main supplier of sediment to the rivers in the area. The average long-term soil washout within its limits is 5000–20,000 tons from 1 km^2.

The coastal part of the region (the Great Plain of China) is characterised by intensive directional accumulation of sediment, thanks to which the rate of riverbed rise reaches 7–9 cm/year. At the same time, a huge amount of load is being carried into the sea and the delta is rapidly advancing: at the Huang He, the riverbed at the mouth of the Huang He River lengthened by 13 km in just four years (1976–1980) (Mikhailova 1998). For comparison: the estuary elongation and advancement of deltas into the sea is so small that it is assessed only qualitatively (for the period of observations, their values are within the limits of survey accuracy for no more than 100 years) or for geological time intervals (Holocene, etc.).

Area III covers part of Eastern, South-Eastern and Southern Asia. The area includes large river basins that generate large flows of water and sediment, although sediment concentration is relatively low (0.5–2.5 kg/m^3). Annual sediment load is 2989.4 million tons (60.2% of the total load of Asian rivers), accounting for a significant proportion of the world's load. The area of river basins is 44% of the total area of Asian river basins flowing into the sea, and water runoff is 60.2% of the total water runoff. Most rivers flow from the highlands or mountainous countries of Central Asia. Measured and estimated bedload transport rate on the rivers of the region is no more than 100 million tons, which is only 1–2% of total load (Ning et al. 1987; Chalov et al. 2000).

The South-East Asian (III-1) sub-region is formed by the Yangtze, Jiang, Minjiang, Hanshuya and Xijiang river basins. The Yangtze, the longest river in Asia in terms of length and runoff rates, has the highest sediment concentration and crosses the subtropical monsoon area. Water runoff is more than 20 times higher than that of the Huang He, but the average sediment concentration is no more than 0.5 kg/m^3 (35 kg/m^3 in the Huang He).

In the Southern subdistrict (III-2), the rivers Hongkha, Mekong, Brahmaputra, Ganges and Indus flow into the open seas. Most of the large rivers flow from the Tibet Plateau and other mountainous areas of Central Asia, transporting huge amounts of sediment, partially accumulating it downstream and carrying it to sea in large volumes. Sediment concentration gradually increases from East to West and reaches the highest values in the Indus River—2.5 kg/m^3. At the mouths of the river, vast and actively growing deltas form; the rate of advance to the sea of the Mekong delta is 50–60 m/year, and the speed of advance into the sea of the Honghui (Red) delta is 80–100 m/year (Nguyen Van Ki 1990).

A special position in area III is occupied by the rivers Tigris and Euphrates, which flow in other natural conditions and flow into the Persian Gulf by the united river Shatt-El-Arab. Despite the great similarities (mainly rainwater feeding, high water runoff and high sediment concentration—sediment load), the isolation of their basin from other parts of South Asia gives grounds for its isolation into a special subarea III-3.

In close natural conditions (s = const), suspended sediment discharge is largely dependent on the size (order N) of the river. The larger the N, the larger the catchment area and, respectively, the greater the Q_0 and W_R. This pattern is often violated, especially on large rivers that cross several natural zones and territories with complex geological structures. Also relevant is the longitudinal variation in transport capacity associated with the reduction of river gradients with relatively little change in runoff rates. This is the reason of the accumulation of sediment ($W > W_{tr}$), resulting in a reduction of W downstream. For example, in the Ob, the annual mean sediment concentration in the area from the confluence of Biya and Katun to the city of Kamen-na-Obi increases 2.5 times (from 0.14 to 0.35 kg/m^3), and then decreases from 0.12 kg/m^3 in the center of Western Siberia (Kolpashevo) to 0.033 kg/m^3 in the estuary (Fluvial processes and waterways … 2001). Suspended sediment load increases from 5 million tons per year below the confluence of Biya and Katun ($N = 9.6$) to 17.8 million tons per year in Kamen-na-Obi ($N = 10.2$) and 16.6 million tons per year in Kolpashevo ($N = 12.4$). At the mouth of the Ob River, despite a significant increase in annual water runoff (more than 3 times), the sediment load is only 13 million tons per year. The decrease in load is due to the lower sediment concentration of the West Siberian tributaries of the Ob and the directional accumulation of load in the lower reaches of the river. This is also facilitated by the tectonic immersion of the coastal (marginal) areas and the river mouth elongation, with the higher intensity of the latter resulting in higher rates of accumulation. This is particularly evident in the different rates of directional accumulation of sediment in the lower reaches of the Huang He, Yangtze, Mekong, Hongkhi (Red), Terek, Amu Darya (in natural conditions), on the one hand, and Ob, Amur, Northern Dvina, on the other.

The rivers of the upper Ob basin, which cross the Pre-Altai steppe plateau formed by loess-like loam and the small rivers within the Biya-Chumysh Upland, have sediment concentration of 0.2–0.5 and more than 2 kg/m^3 respectively (Fluvial processes … 1996). Water sediment concentration is often reduced downstream (from more than 2 kg/m^3 on average to 0.12 kg/m^3 in the lower reaches of the Chumysh River, where $N = 6.5$), although suspended sediment load continues to increase due to increased water availability.

In the lower Amur, due to the accumulation of suspended load, the sediment flow decreases from 26.1 million tons per year (Khabarovsk) to 14.9 million tons (at the mouth) (Makhinov et al. 1994). In the lower reaches of the Huang He (Chalov et al. 2000), where the river crosses the Great Plain of China and has no tributaries (N = 9.4), sediment load decreases from 1736 to 1320 million tons/year.

In the Volga basin (Fig. 2.12) (Chalov and Shtankova 2003), five groups of regions are distinguished in terms of sediment concentration (kg/m^3): 1—<0.05; 2—0.05 to 0.1; 3—0.1 to 0.2; 4—0.2 to 0.5; 5—>0.5 (Shtankova 2003). Low sediment concentration (<0.05 kg/m^3) characterizes the rivers of the upper Volga basin up to the North-Western part of the Oka basin, the left-bank tributaries of the middle Volga, the middle reaches of the Vyatka river and the Northern and Eastern parts of the Kama basin. Thus, low sediment concentration of water is observed on the rivers of forest zone and distribution of podzolic soils, less susceptible to erosion, as well as on the rivers of mountainous area. Areas with a higher sediment concentration ($\gg$0.1 kg/m^3) belong to the Southern forest-steppe part of the Oka basin (Central Russian Upland), the central and Southern parts of the Kama basin

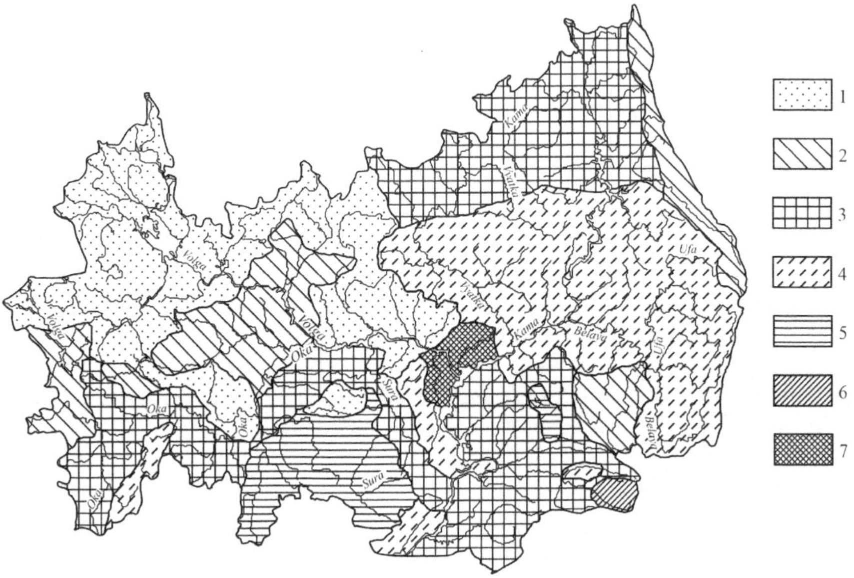

Fig. 2.12 Zoning of the Volga river basin by suspended sediment flow module M_R. Zones with M_R, t/(year*km^2): 1—<5; 2—5 to 10; 3—10 to 20; 4—20 to 30; 5—30 to 40; 6—40 to 60; 7—>60

(dissected areas of the Ufa plateau in the East, Sarapul upland, the Northern slopes of the Bugulma-Belebey and Southern Upper Kama Uplands, Vyatskiy Uval), which are characterized by a high degree of ploughing (40–60% and more) and a strong erosion hazard of lands (Map "Erosion hazard…" 1987).

The steppe Lower Volga region belongs to the region with higher sediment concentration (0.2–0.5 kg/m^3). The high sediment concentration of the rivers in this area is explained by the highest agricultural development of the steppe Bugulma-Belebey Upland, composed of loess-like loam, which is almost completely plowed and is an arena for intensive development of soil erosion. Against this background, the areas with the highest values (>0.5 kg/m^3), covering the basins of the Mesha River, which is located in the Volga-Kama interfluve, small rivers of the Volga Upland and the upper reaches of the Buzuluk, Samara, Tok, Big Kinel and Sok rivers, are distinguished.

The diversity of conditions for the formation of sediment load and the development of erosion processes in the Western part of the region (the Oka basin) has an impact on the longitudinal variation of the suspended sediment load modulus along the length of large rivers. At the Oka, they are reduced first by the confluence with the Zhizdra and Ugra rivers, where the rivers have low sediment concentration and less suspended sediment load than the Oka upstream, and then with the burgher rivers. At Moksha, a decrease in the suspended sediment load modulus to the central part of the Oka-Moksha Depression is observed, and then the same increase is observed below the confluence with Tsna (Oka-Tsna rampart). At Klyazma, the sediment load modulus is increased (river crossing of the erosion-hazardous areas of the Vladimir Opolye) to the middle reaches, but in the lower reaches it is reduced due to a decrease in flushing from the basin surface.

In the lower part of the Volga basin, some tributaries flow along the Volga Upland (right bank rivers), others along the Caspian Lowland (left tributaries). Correspondingly, the flow module is higher (20–30 tons/km^2—Tereshka River, middle reaches of the Sviyaga River) than that of the rivers on the left bank of the river. There is also an area of increased suspended sediment load (40–60 tons/year * km^2) at the head of the Samara River, which is apparently due to the local topography influence. This peak coincides with the belt of maximum sediment concentration (upper reaches of Samara, Big Kinel, Sok). The maximum modulus of suspended sediment discharge is noted in the upper reaches of the Sviyaga and Mesha (>60 tons/year * km^2), also coinciding with maximum sediment concentration.

The lowest values (<5 tons/km^2) of the модуль стока наносов are thus characteristic of the Upper Volga and central part of the Oka basin. In the Eastern part of the Volga basin (the Kama basin) such low values of the suspended sediment load modulus are not observed. The maximum values of the Suspended Sediment Flow Module (>40 tons/year * km^2) are confined to the steppe zone of the Lower Volga region.

Different conditions for the formation of suspended sediment load in different parts of river basins, and longitudinal regional changes in sediment concentration and in the sediment load module, often make each group of areas identified for sediment concentration consistent with its relationship on discharge (Fig. 2.13).

Suspended sediment discharge is determined by network observations. For unexplored rivers, its characteristics can be derived from known methods of hydrological calculations (Sediment yield … 1977). Determination of the discharge characteristics of bedloads is more difficult because there is no system of regular measurements and no reliable instruments for their carrying out, nor conventional

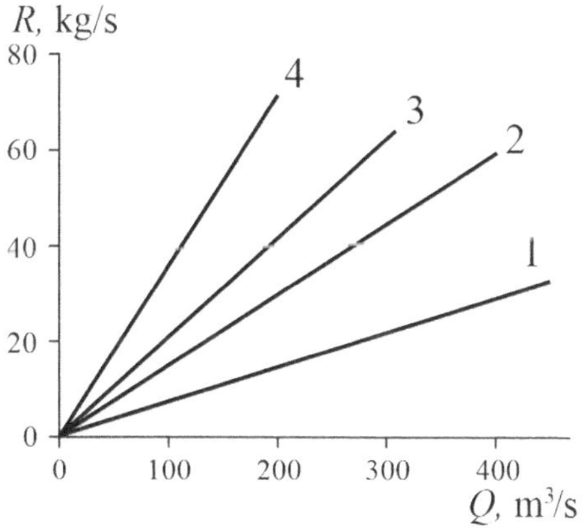

Fig. 2.13 Dependence of the bed material load discharge R from water discharge Q on the Don river basin rivers; numbers correspond to groups of areas with different water turbidity: 1—50 to 100 g/m^3; 2—100 to 250 g/m^3; 3—250 to 500 g/m^3; 4—>500 g/m^3

and reliable calculation methods. The differences between the actually observed bedload transport rates and the calculated discharge from formulas or standard instruments (e.g. the "Don" bathometer) are in the order of magnitude. Kopaliani and Kostiuchenko (2004, p. 26) named ten reasons for the imperfection of the methods for calculating the bedload transport rates, among which "insufficient account is taken of the structural (rippled—*R.Ch.*) forms of transport … which are the main form of transport of bottom canals in rivers". Therefore, the way out of the situation was practically found with the introduction of high-precision echo sounders in the practice of river measurement in the middle of XX century and the performance of special experimental studies in trays with glass walls with the use of photography and filming. This has made it possible to determine the flow rate of bed load sediment by the rate of movement of the ripples into which the loose (sand, sand-gravel) material is structured under the influence of the turbulent flow affecting it. On the basis of the data obtained on the ripple movement of sediments from the late 40s onwards, many Russian (Yu. M. Korchokha, M. A. Velikanov, A. F. Kudryashov, K. V. Grishanin, N. S. Znamenskaya, V. K. Debolsky, etc.) and foreign researchers have proposed forums or design dependencies for determining the bedload transport rate by the velocity of the ripple movement taking into account their size (subtraction). A thorough review of these formulas was made by Kopaliani and Kostiuchenko (2004), who compared the results of their calculations with the data of full-scale measurements and laboratory experiments. Errors in this case amounted to a minimum of −8 to 139% and a maximum of −38 to 1678%. This gave them reason to recommend the formula of Snishchenko and Kopaliani (1978), which gives the smallest discrepancies with the field data, for calculating the bedload on rivers with the ripple form of their movement:

$$g_{bl} = 0.11 h_r V Fr^3 \tag{2.12}$$

where g_{bl}—specific bedload transport rate, m^3/(second * m), and h_r is determined by the formulas of Snishchenko (1980)

$$h_r = 0.25h \text{ when } h < 1\text{m} \tag{2.13a}$$

$$\text{and } h_r = 0.20 + 0.1h \text{ when } h > 1\text{m} \tag{2.13b}$$

or the formula of Z.D. Kopaliani if data on flow rates and sediment size are available:

$$h_r = 0.39d \left(\frac{V}{V_0}\right)^{2.5} Fr^{-3.75}, \tag{2.14}$$

where $V_0 = 3h^{0.2}(d + 0.0014)^{0.3}$ is the flow velocity corresponding to the beginning of the movement of sediment particles. However, this approach does not take into account the hierarchy of ripples (only microforms are considered (Kondratyev et al. 1982)), their transformation under seasonal fluctuations in the runoff rates of rivers and the probability of the transfer of part of the load from the farmed to the suspended state and back, both in different phases of the regime and depending on changes in the hydraulic characteristics of the flow from the bedrock zone to the peripheral one. This inevitably leads to underestimation of the results of the calculations. Besides, in real conditions labour intensity of observations and the subsequent data processing has not allowed to use opened possibilities widely, first of all at network measurements, and they remain a destiny of specially put researches on concrete rivers, besides usually limited because of technical possibilities time of carrying out—as a rule, recession of a flood and an autumn. In any case, it is not known that such works on rivers should be carried out in the high-water phase of the regime (flood).

Solution of the problem in view of the expressed considerations became possible thanks to the calculation method developed by Alekseevskiy (1998), which is based on the relationship of parameters and velocity of movement of ripples from the order of the river, taking into account the differences in their formation in the main phases of the water regime (flood, low water level). Ripples differ in linear (height h_i) and dynamic (movement velocity C_i) characteristics, where i is the type of ripple (i—A, B, C, D, E). Ripples A and B are the largest channel formations—macroforms. The microforms of the channel relief include the D ripples. The C and D ripples—mesoforms—occupy an intermediate position.

Depending on the task at hand, one or another ripple classification can be used. N. I. Alekseevsky approach is convenient for calculating the bedload transport rate (separation of the ripples by their size, the separation of macro-, meso- and microforms when analyzing their relations with the flow structure and their position in the hierarchy of the channel relief forms); the morphological characteristics of the ripple should be given when assessing the channel relief and its reshaping.

In the lowland, sediment load is generated as a result of the active movement of microforms of the bedrock topography, and in the period of maximum runoff, the ripple of all five types is formed. Flow rate of bedload G_y for characteristic phases of the water regime (y—low water level or y—high water level)

$$G_y = k\rho_s b_a \Sigma(hiCi), \tag{2.15}$$

where k is the ripple shape coefficient (~ 0.6); ρ_s is the density of channel deposits (bed sediments); $b_a = 0.8\mathrm{b_l}$ is the active channel width depending on the width of low-water channel $\mathrm{b_l}$. The formula (2.15) is a modification of the well-known T. Tsubaki formula by N. I. Alexeevsky (Rossinsky and Debolsky 1980) $g_{bl} = k\gamma_s C_r h_r$ (here $\mathrm{g_{bl}}$ is the specific consumption of bedload, γ_s is the volume weight of sediment), due to the simultaneous movement of ripples of different sizes i and the introduction of the concept of active channel width. In this case, the bedload transport rate is proportional to the order N river, since

$$b_a = \varphi(N), h_i = \Psi_i(N), C_i = \beta_i(V_{l,f}N), \tag{2.16}$$

where $V_{l,f}$ is the flow rate (indexes l are low-water, f is flood). Table 2.4 shows the dependencies for the calculation of ripple heights and velocities of their movement separately for low and high water (flood).

The effectiveness of the method for calculating the average bedload transport rate during low-water and flood periods proposed by N. I. Alekseevsky has been confirmed by special studies on the Ob, Lena, and Dniester (Alekseevsky and Gaikovich 1987; Alabyan et al. 1992; Alekseevsky et al. 1996) and other rivers (Fluvial Processes … 1996; Darbutas 1992).

The data of longitudinal echo sounding of the Northern Dvina channel in the river section between the mouths of the Vychegda and Vaga rivers, performed in the summer of 1997, showed a good convergence of linear characteristics of the natural ripples and the calculations (Table 2.5). Error in calculating the height of all five types of ripples ranges from 27.3 to 7.3% (Chalov et al. 2000).

Testing of the methodology on rivers with very high sediment load and absolute predominance of suspended sediment (Huang He, Yangtze) has been carried out based on empirical data obtained by Ning et al. (1987), represented in the form of combined longitudinal profiles along short sections of these rivers (Fig. 2.14). The comparison has shown that for Chinese rivers the calculation method gives overestimated and underestimated values of characteristics (Tables 2.6 and 2.7). For the Huang He, the errors in determining the height of moving channel shapes range from 32% (Ripple A) to 11% (Ripple Г), for the Yangtze River from 32% (Ripple Д) to 16% (Ripple Г). Therefore, it is necessary to enter the correction factor $k_h \sim 0.8$ in order to calculate the ripple height. The actual movement velocities of B and Б on Huang He, Г and B on Yangtze are 50 and 5, 19.8 and 26.0 m/day,

Table 2.4 Correlation relationships for calculating altitude and velocities of ripples movement from the order of rivers (according to Alekseevsky (1998))

Mode phase	h_i	C_i
Low-water	$h_Г = 0.005 * N^{2.0}$	$C_{Г,Д} = 600 * V_l N^{-2.0}$
	$h_Д = 0.003 * N^{2.0}$	
Flood	$h_A = 0.046 * N^{1.95}$	$C_A = 1.1 \times 10^{-6} * V_f N^{5.4}$
	$h_Б = 0.018 * N^{2.0}$	$C_Б = 3.1 \times 10^{2} * V_f N^{-2.6}$
	$h_В = 0.032 * N^{1.5}$	$C_В = 3.5 \times 10^{-4} * V_f N^{4.6}$
	$h_Г = 0.023 * N^{1.4}$	$C_{Г,Д} = 2.2 \times 10^{4} * V_l N^{-3.5}$
	$h_Д = 0.005 * N^{1.9}$	

Table 2.5 Actual and estimated heights of the ripples at the Northern Dvina

Type of ripple	А	Б	В	Г	Д
Estimated data	4.60	2.03	1.10	0.63	0.44
Field data	4.09	2.19	1.15	0.57	0.32
Deviation $\frac{h_{i_{field}} - h_{i_{estimated}}}{h_{i_{estimated}}} \times 100\%$	−11.1	7.9	4.6	−9.5	−27.3

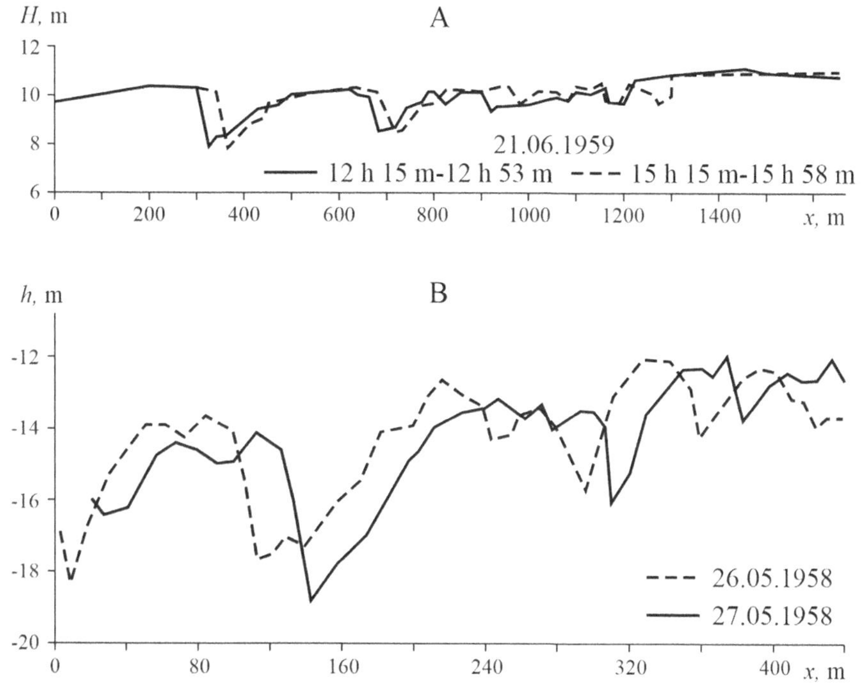

Fig. 2.14 Combined longitudinal profiles of ripples in Huang He river channel near the city Huajuankou (**A**) and in Yangtze river channel near the city Datun (**B**): H—water level; h—channel depth (according to Ning et al. (1987))

respectively. Estimated data on Huang He is 2.8 times less, on Yangtze is 2.1 times less. To reduce the errors, a correction factor $k_c = 2.45$ is introduced into the calculation equation for determining the velocity of the ripple movement. The total effect of using both coefficients will take the form of an increase in the calculated G_y values by 1.96 times.

These differences in the design and actual characteristics of the ripple seem to be due to the fact that the Alekseevsky method was developed on the basis of data from Russian rivers with a sandy composition of bed material load, which differ significantly from those in China (Fig. 2.11, areas II and III in terms of sediment concentration and suspended sediment load), the ratio between the particle size distribution of suspended sediments and bed load, and the absolute predominance of a suspended component in sediment load and bed material load. Under these conditions, lower-altitude ripples are formed than is typical of rivers with relatively low suspended sediment load and larger (medium and coarse sand) bed material load, the former moving almost exclusively in suspension and the latter in ripple form. This determines the mismatch between the sign of inequality $W \neq W_{tr}$ on the rivers of Russia and China, resulting in $k_h < 1$. On the other hand, the correspondence between the particle size distribution of suspended and fractional sediments

Table 2.6 Comparison of actual (Ning et al. 1987) and calculated height of the ripple (Liu 1998; Chalov et al. 2000) on the Huang He and Yangtze Rivers

River—hydrological control station	Ripple height, m										Deviation from the calculated value, %				
	Actual					Calculated									
	А	Б	В	Г	Д	А	Б	В	Г	Д	А	Б	В	Г	Д
Huang He-Huayuankou	–	1.31	0.84	0.49	0.32	–	1.68	0.95	0.55	0.37	–	−22	−12	−11	−14
Yangtze—Nanjing	–	2.63	1.26	0.76	0.50	–	3.42	1.64	0.91	0.73	–	−23	−23	−16	−32

Table 2.7 Comparison of actual (Ning et al. 1987) and calculated velocities of ripple movement (Liu 1998; Chalov et al. 2000) on the Huang He and Yangtze Rivers

River—hydrological control station	Ripple movement velocities, m/day									
	Actual					Calculated				
	$C_Д$	$C_Г$	$C_В$	$C_б$	$C_а$	$C_Д$	$C_Г$	$C_В$	$C_б$	$C_а$
Huang He-Huayuankou	–	–	50	5	–	14.34	14.34	21.53	1.55	0.41
Yangtze—Nanjing	–	19.8	26.0	–	–	3.02	3.02	82.15	0.45	2.11

and the large role of the former in the formation of channel sediments (bed material load) causes abnormally high velocities of ripple movement, which results in the $k_c > 1$. It is typical that Alekseevsky (1998) has shown that on rivers with a pebble composition of bed material load, there is an inverse correlation between the errors in the calculation of the height and the rate of movement of the ripples, which leads to a corresponding underestimation of the value of G_y and the bedload transport rate W_G. Therefore, they were offered a modification of the method for calculating parameters and velocities of ripple movement for sandy-pebble and pebbly-boulder beds (Alekseevskiy and Melnik 1991). Deviations in the design characteristics of the bedload discharge from the actual discharge towards the overestimation are also possible in the event of freezing of shallow water in the lower reaches of the channel and the closure of the seasonal freezing to permafrost. As a result, the heavy flood flow is not able to move frozen sediment particles until the thawing of the ground and the formation of an active layer, which usually occurs already during the recession of the flood and in the low water period. These are, for example, the conditions of transport of bedload on the middle and lower Lena (Alekseevsky et al. 1996; Tananaev 2004). However, to a certain extent the same factor (seasonal freezing of shallow waters) limits the flow of bedload in the flood waters of the Northern Dvina, Ob, Pechora and middle rivers flowing outside the permafrost zone.

Ultimately, the bedload transport rates is calculated by the formula of N. I. Alekseevsky

$$W_G = W_{G_1} + W_{G_2} = (365 - T_n)\sum_{i=Д}^{i=Г} G_1 + T_f \sum_{i=Д}^{i=A} G_2\,, \tag{2.17}$$

where W_{G_1} and W_{G_2}—runoff of bedload, respectively, in the low water period and in the high water period; T_f—duration of the flood period; $G_{1,2}$—parts of bedload discharge determined by the formula (2.15) for the ripples А-Д (G_2) for flood conditions and G_I—for low-water conditions.

Calculation of the bedload transport rates from the data on the movement of ripples of different sizes can be carried out for conditions of mass transport of bedloads where the flow rates exceed the scouring velocities V_s for sediments and soils of a given size that compose the bed, and $V_s = 1.4V_n$ (Mirtskhulava 1988), where V_n is the non-destructive velocity at which the soil particles are stationary. If $V_f < V_n$, the consumption of bed sediment G =0; if $V_f > Vn$, then G > 0.

The ripple transport of bedloads consists in the movement of the forms of channel relief, the height of which is $h_r \ggg d_{50}$. This determines the reduced hierarchy of ripple shapes on rivers with pebble-boulder composition of bed material load, where at $d_{50} > 50$ cm the A and Б-types of ripples are formed (according to N. I. Alekseevsky), (Fig. 2.15A); and the condition $V > V_s$ is observed only during floods, and in the lowland the transport of large loads (pebbles, boulders) does not occur.

Variations in the conditions of interaction between flow and channel sediments cause changes in the ripple parameters (Znamenskaya 1968, 1976). The increase in water flow rate, leads to an increase in their length l_r and height h_r. At the level decrease, the emerged elements of the channel relief do not always have time to go through a full cycle of opposite changes, and in the channel relief both small ripples corresponding to the hydraulic conditions of the low water level and large ones, preserved from the previous high-water phase of the water regime, are expressed (Fig. 2.15B). Ranges of different sizes, creating a hierarchy of channel forms, move actively or passively (Znamenskaya 1968). Active movement is related to the transport of particles at a distance of $x = l_{r_i}$, where i is the type of ripple. Passive

Fig. 2.15 Different forms of sediment transport on rivers: **A**—large ripples on the Timpton river (tributary of Aldan river) with pebble-boulder sediment composition (photo of the author); **B**—combination of sandy ripples-macroforms (type A, according to N. I. Alekseevsky) and microforms (type E) on the bank shoal of the Vychegda river (photo of the author); **C**—flat, non-ripple sandy-silty shoal in lower current of the Huang He river (photo by Liu Shuguang)

movement characterizes the movement of large shapes by bringing smaller, actively displaced ripples into their basement.

Constant changes in ripple parameters, the destruction of small ripples and the formation of larger ones with increasing flow rates, and different conditions for their development in the bedrock and peripheral zones of the channel lead to an opinion that the total bedload transport rate is greater than that which is carried out in the form of the ripple G_r: according to Rossinsky and Debolsky (1980), $G_r = 0.7G$. The aspect ratio depends on the diameter of channel deposits and the relative depth of the channel h/b_{ch} (Grishanin 1969). However, Znamenskaya (1968), Simons et al. (1966) believe that all bottom sediment load is carried out in the form of a ripple. On the other hand, the differences between the definitions of total bedload discharge and the 30% of the measured ripple movement appear to be within the limits of accurate calculations and can be neglected.

If the flow velocity $V > 2.5V_n$, the ripples are either destroyed or not formed at all, as in the lower reaches of the Huang He (Fig. 2.15C), in the Amu Darya, lower Terek, and the load moves only in the smooth phase (Znamenskaya 1968). The same thing happens in the turbulent flows of mountain rivers with pebble-boulder deposits (if $Fr > 2$).

The results of calculations of the bedload transport rate are presented in Table 2.8. They characterize the variability of W_G in rivers flowing in a variety of natural conditions. The relationship of W_G on natural conditions is less clear than W_R. It's mostly due to the size (order) of the river, the duration of the flood (or flood period), the composition of the channel sediments, and the partitioning between basin and channel components in the sediment load. This is also characteristic of the partitioning between bedload transport rates in the phases of increased (flood) and decreased (low-water) runoff rates, which varies widely. In general, it increases with the growth of the river order and also depends on local conditions. For example, it is several times bigger on the Yangtze than on Lena (Tabaga village).

The impact of natural conditions also has an impact on partitioning between bedload and suspended load. It is generally accepted that the consumption of bedload is not more than 1% of the consumption of suspended sediment (with a particle diameter of <0.25 mm), and with larger particles (pebbles and pebble-boulder sediments) in the active bed layer and $d/h \geq 0.01$ (d—particle diameter, h—flow depth) the R/G ratio increases to 5–10% (Karasev 1975). According to a generalization made by Kopaliani (1985), in lowland rivers the partitioning between bottom load to river load is usually $\delta = \frac{G}{G+R} \times 100 = 1 - 3\%$ (rarely 10%), and only in mountain rivers does δ change from 15 to 70%. These estimates often do not coincide with the data obtained from the analysis of the dynamics of the rolled strips, side-shifting, dredging slots and channel quarries. According to the latest data (Alabyan et al. 1992; Fluvial processes … 1996; Alekseevsky et al. 1996), the share of bedloads may vary considerably wider. Similar conclusions were also reached by Babinski (1982) in the course of research on riverbed processes in the Vistula (Poland).

Table 2.8 Bedload sediment yield on large Russian and Chinese rivers and their tributaries in natural conditions (Liu 1998; Chalov et al. 2000; Babak et al. 2000; Shtankova 2002; Reznikov and Chalov 2005)

River basin	River—hydrological control station	Average annual discharge Q, m^3/s	The order of the river N	Bed sediment yield, million tons			Average annual bedload transport rate G, kg/s
				Low-water	Flood	Average annual	
Northern Dvina	Suhona—Kalikino	431	8.5	0.21	0.19	0.40	12.7
	Ug—Gavrino	289	8.0	0.07	0.11	0.18	5.71
	Vychegda—Fedyakovo	1010	9.7	0.24	0.30	0.54	17.1
	Uftuga—Yakuhino	35.3	5.8	0.03	0.04	0.07	2.22
	Northern Dvina—Abramkovo	1930	10.6	0.26	0.45	0.71	22.5
	Yerga—Yurikinskaya	4.85	4.35	0.004	0.010	0.014	0.44
	Lower Toima—Metil	18	5.25	0.023	0.044	0.067	2.13
	Codima—Shev	2.97	4.0	0.008	0.027	0.035	1.11
	Waga—Ust-Syuma	372	8.3	0.11	0.13	0.24	7.62
	Northern Dvina—Zvoz	2570	11.0	0.17	0.44	0.61	19.4
	Northern Dvina—Ust-Pinega	3310	11.5	0.16	0.49	0.65	20.6
	Pinega—Culogory	372	8.3	0.18	0.20	0.38	12.1
Ob	Ob—Fominskoye	1150	9.6	0.19	0.46	0.65	20.6
	Charysh—Charyshsky	189	6.9	0.05	0.17	0.22	7.00
	Ob—Barnaul	1470	10.0	0.13	0.44	0.57	18.1
	Chumysh—Talmenka	138	6.5	0.06	0.13	0.19	6.03
	Ob—Kamen-na-Obi	1570	10.2	0.22	0.77	0.99	31.4
	Ob—Novosibirsk	1742	10.4	0.43	0.94	1.37	43.5
	Tom—Tomsk	1080	9.3	0.18	0.25	0.43	13.6
	Chulym—Baturino	779	8.9	0.17	0.25	0.42	13.3
	Ob—Kolpashevo	4260	12.4	0.40	2.49	2.89	91.8

(continued)

Table 2.8 (continued)

River basin	River—hydrological control station	Average annual discharge Q, m^3/s	The order of the river N	Bed sediment yield, million tons			Average annual bedload transport rate G, kg/s
				Low-water	Flood	Average annual	
Lena	Lena—Tabaga	7150	13.9	0.85	4.58	5.43	172.4
	Aldan—Verkhoyansk Perevoz	5190	13.0	0.48	1.87	2.35	74.6
	Viliui—Khatyryk-Homo	1430	10.3	0.16	0.27	0.43	13.6
	Lena—Kyusur	16,700	16	1.06	16.39	17.45	554.0
Huang He	Huang He—Huayuankou	1592	9.65	1.38	3.48	4.86	77.1
	Huang He—Jiahetan	1530	9.40	1.22	3.26	4.48	71.1
	Huang He—Gaotsun	1538	9.42	1.00	2.26	3.36	53.3
	Huang He—Sunkou	1521	9.40	0.90	2.56	3.46	54.9
	Huang He—Aishan	1544	9.42	0.72	1.88	2.60	41.3
	Huang He—Lekou	1539	9.42	0.38	0.98	1.36	21.6
	Huang He—Lijin	1523	9.40	0.48	1.44	1.92	30.5
Yangtze	Yangtze—Yichang	14,300	15.5	0.80	12.28	13.08	207.6
	Yangtze—Loshang	20,300	16.4	1.96	23.24	25.20	400.0
	Yangtze—Hankou	23,400	16.7	1.92	26.34	28.26	448.6
	Yangtze—Datong	28,900	17.1	1.70	32.68	34.38	545.7
Volga	Volga—Rzhev	93.5	10.7	0.03	0.108	0.13	0.41
	Volga—Staritsa	156	11.4	0.05	0.193	0.24	0.76
	Unja—Kologriv	95	10.7	0.03	0.110	0.14	0.44
	Unja—Manturovo	135	11.1	0.04	0.159	0.20	0.63
	Unja—Makaryev	157	11.2	0.03	0.133	0.16	0.51
	Sura—Penza	53	11.0	0.02	0.080	0.10	0.31
	Sura—Kadyshevo	103	11.7	0.02	0.066	0.08	0.25
	Sura—Poretskoye	232	12.7	0.04	0.151	0.19	0.60

(continued)

Table 2.8 (continued)

River basin	River—hydrological control station	Average annual discharge Q, m^3/s	The order of the river N	Bed sediment yield, million tons			Average annual bedload transport rate G, kg/s
				Low-water	Flood	Average annual	
	Vetluga—Vetluga	163	11.4	0.03	0.134	0.17	0.54
	Vetlug—Vetlutsky	191	11.7	0.02	0.108	0.13	0.41
	Oka—Vendarovo	2.35	6.9	0.002	0.04	0.042	0.13
	Oka—Kaluga	291	12.5	0.04	0.09	0.13	0.41
	Oka—Murom	916	14.0	0.12	0.28	0.40	1.27
	Zusha—Mtsensk	28	9.9	0.11	0.03	0.04	0.13
	Moksha—Shevelov Maidan	107	11.4	0.02	0.07	0.09	0.28
	Klyazma—Pavlov Posad	32.6	9.5	0.01	0.04	0.05	0.16
	Klyazma—Vladimir	80.3	10.9	0.02	0.03	0.05	0.16
	Klyazma—Kovrov	137	11.6	0.04	0.07	0.11	0.35

It should be recalled that Lopatin (1952, p. 307) emphasized that "the data ... of measurements not only considerably underestimate the value of the bedload transport rate, especially during floods, but also do not give an idea of the regime of the bottom process. These remarks are more relevant for large rivers (with significant depths and flow velocities)". G. V. Lopatin has estimated the partitioning between annual runoff of bedload to suspended load of up to 10%, which he obtained from "general considerations of sediment transport". Maximov (1905) made the first attempt to determine the bedload transport rates from the data on the dynamics of the riverbed terrain, proposing to use sequential bed plans for this purpose on the ropes obtained by survey batches of waterway administrations. Subsequently, this approach was applied by Makkaveev (1955) and Chalov (1964). Calculations for Vychegda (Chalov 1964) on the movement of side ropes gave a value of annual runoff of bedload equal to 0.41 million tons, which is 25% less than the method of N. I. Alekseevsky for the same river section. This is obviously due to the fact that the determination of sediment load according to the compared plans gives underestimated results, especially in the intermittent period, since it takes into account the volume of load that composes only the macroforms of the bedrock relief—side bars (the A-ripples according to N. I. Alekseevsky) and possibly the B- and C-type ripples; smaller ripples have practically no effect on the plans.

This is also indicated by the data on suspended and coarse sediment load (δ) for some of the large rivers in Eurasia (Table 2.9), which are summarised in Fig. 2.16. In rivers with high suspended sediment load and low overall sediment deposition, the share of coarse sediments in total sediment load is significantly lower. In Huang He, it's only 0.1–0.3%. This proportion is more than halved at the mouth of the river due to a reduction in transport capacity and the average diameter of channel sediments. On the Yangtze, the bedload transport rates already forms from 2.4 to 7% of the total discharge. The share W_G is increasing in the Loshan-Datong area, where the Yangtze crosses the mountain range. At the Yangtze, an increase in the volume of bedload and δ is observed against the background of the directional accumulation of load in the area of the Lianghu Plain. Here, the water flow is regulated by the Dantinghu lake system, which accumulates suspended sediment. Downstream, the river is weakly incised, changing the components of sediment load. The process of river load accumulation dominates in the lower reaches. The share of bedload in the upper Ob is even higher: here it varies from 5.2 to 14.8%, decreasing downstream from the confluence of Biya and Katun. A decrease in the share of transport of heavy loads is observed up to the town Kamen-na-Obi, which is due to a relative increase in suspended sediment load due to the erosion by the riverbeds of the Pre-Altai steppe plateau, composed of loess-like loam. According to the data of full-scale measurements in the high lowlands (Berkovich et al. 1989), at the confluence of Biya and Katun the sediment concentration of water in different parts of the channel is from 78 to 200 g/m^3, 90 km downstream—50–60 g/m^3, leveling out the width of the channel. Below the confluence with Charysh due to the inflow of sediment from plateau's scarps and removal from the left-bank tributaries and gullies that separate it, the intermittent sediment concentration reaches 300–500 g/m^3 in the area of the village Vyatkino (50 km below the river approach to the

Table 2.9 Correlation between suspended W_R (million tons) and bed W_G (million tons) sediment yield on some major rivers in Eurasia

River	Hydrological control station	The order of the river N	W_R	W_G	$W = W_R + W_G$	$\delta = W_G/W$, %
Lena	Tabaga	13.9	8.99	5.43	0.1440	37.6
	Kyusur	16.0	22.66	17.45	0.4010	43.5
Northern Dvina	Abramkovo	10.6	13.2	0.71	0.0200	35.0
	Zvoz	11.0	1.77	0.61	0.0240	25.6
	Ust-Pinega	11.5	3.86	0.65	0.0450	14.4
Ob	Fominskoye	9.6	5.00	0.65	0.0560	11.5
	Barnaul	10.0	8.10	0.57	0.0860	6.6
	Kamen-na-Obi	10.2	17.86	0.99	0.1880	5.2
	Novosibirsk	10.4	13.38	1.37	0.1480	9.3
	Kolpashevo	12.4	16.65	2.89	0.1950	14.8
Yangtze	Loshang	16.5	425	25.20	4.50	5.56
	Hankou	16.8	429	28.26	4.57	6.13
	Datun	17.5	471	34.38	5.05	6.73
Huang He	Huayuankou	9.6	1500	4.86	15.0	0.32
	Jiahetan	9.4	1736	4.48	17.4	0.26
	Gaotsun	9.4	1423	3.36	14.2	0.23
	Aishan	9.2	1317	2.60	13.2	0.20
	Lakou	9.2	1227	1.36	12.3	0.11
	Licensee	9.2	1320	19.2	13.2	0.14
Volga	Rzhev	10.7	0.04	0.13	0.17	76.6
	Staritsa	11.4	0.18	0.24	0.42	57.3
Oka	Wenderovo	6.9	0.007	0.04	0.047	85.2
	Kaluga	12.5	0.64	0.13	0.77	16.7
	Murom	14.0	1.54	0.40	1.94	20.5
Kama	Volosnitskoye	10.5	0.16	0.07	0.23	31.7
	Gainy	11.7	0.26	0.25	0.51	48.5
Belaya	Arsky Kamen	8.7	0.016	0.077	0.093	82.2
	Sterlitamac	11.4	0.63	0.13	0.76	17.1
	Ufa	13.3	2.15	0.45	2.60	17.4
	Birsk	13.5	2.62	0.54	3.16	17.1
Vyatka	Kirov	12.4	0.50	0.54	1.04	52.0
	Vyatskiye Polyany	13.5	3.09	0.38	3.47	10.9

cliffs of the Priobsky plateau) and 1000–1300 g/m^3 (at the mouth of the Kalmanka River, another 50 km lower). Downstream, the bedload transport rates increase to a greater extent than that of suspended sediment.

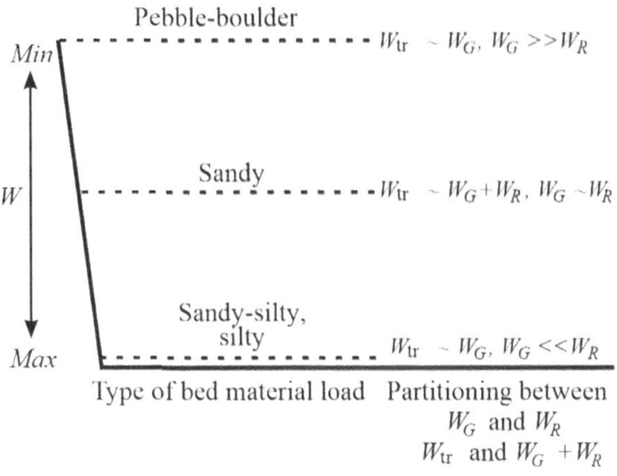

Fig. 2.16 Ratio of suspended sediment flow (W_R) and bed load flow (W_G), total sediment flow (W) and sediment transport capacity (W_{tr}) on rivers with different types of bed material load and different volume of sediment runoff

On the Lena River, the share of bedload is tens of percent, reaching a maximum in the lower reaches of the Lena River (43.5%). Its longitudinal change is characterized by a relative increase of W_G, as in the lower reaches of the river flows along the necks of the Verkhoyansk range, and then crosses the mountain massif—the Haraulakh mountains and the Chekanovskiy ripple. On the Northern Dvina, below the Vaga mouth, there is a directional accumulation of sediments due to the fact that the proportion of bedload transport rates downstream is reduced significantly (by almost 2 times).

The smallest (less than 1%) δ value is typical for Southern rivers (the Huang He River and others). This is due to the formation of thick corrugations, the predominant distribution of loess and loess-like rocks on the plains, as well as the highest soil erosion rate in the world in natural and anthropogenic disturbed steppe and semi-desert landscapes.

In the upper part of the Volga basin (forest zone) in the upper part of the Oka basin, which is located within the small part of the Central Russian Upland, and in the East of the Ural Mountains, the share of bedrock material in total river load exceeds 50% (Fig. 2.17). In some rivers, ripple transport of sandy load forms a dominant ($\delta = 90\%$) part of the total sediment load. The smaller the river and the greater the forest cover of its basin, the greater the amount of bedload and δ. These characteristics also increase where rivers cross flat, sand-covered lowlands (Moloko-Sheksna, low Volga region) and moraine uplands (Valday). The lowest share of bedload ($\delta < 25\%$) is characteristic of the Southern part of the Volga basin. In the steppe regions it decreases (up to 9–10%) and increases within the Central Russian, Volga and Bugulma-Belebey uplands. In general, the ripples movement contribute to 50% of total sediment load in the Northern humid areas.

Of sediment load in the Northern humidal areas of the basin. The rivers in this region are characterised by low suspended sediment loads. Thus, all other things being equal, the share of bedload in the rivers of the Volga basin increases from South to North and from West to East, reaching the highest values on the rivers of mountains and foothills of the Urals, sandy lowlands and uplands. The smallest

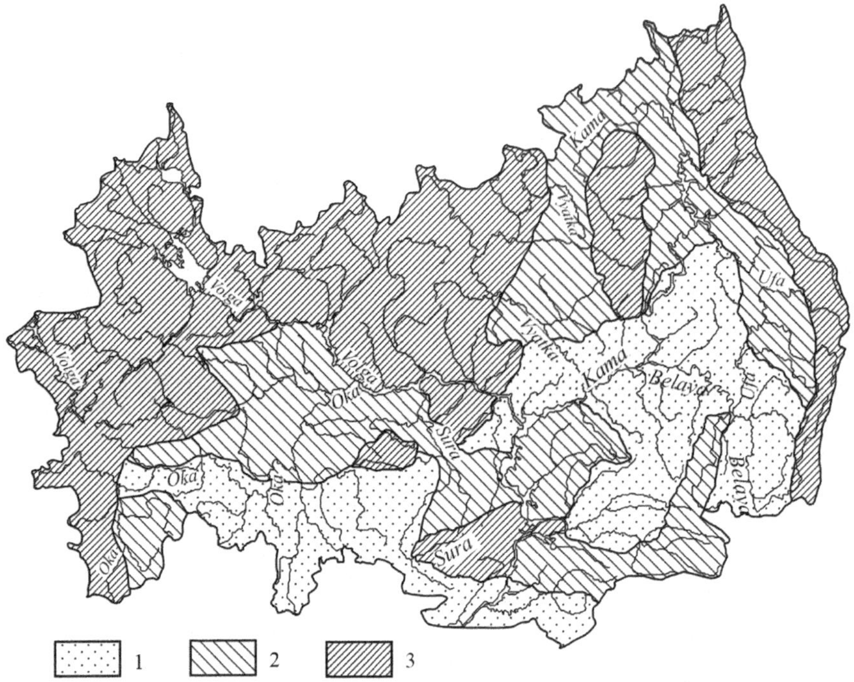

Fig. 2.17 Zoning of the Volga river basin due to proportion of bed load runoff and total sediment runoff: 1—<25%; 2—25 to 50%; 3—>50% (Chalov and Shtankova 2003)

share of bedload is characteristic of the steppe and forest-steppe plain part of the Kama basin, where the suspended load component, which has a predominantly basin genesis, is absolutely dominant. In the Kama basin, the share of heavy load is higher than in the Oka basin, which is explained by its partial confinement to the Ural Mountains, which are the main suppliers of debris to the rivers.

From North to South (on a global scale), there has been an overall increase in suspended sediment load. Against this background, there is a decrease in the share of farmed sediment load, from a few tens of percent in the North to a few percent in the South. The predominance of a weighted component in bed material load results in high rates of channel change and instability of river channels: the corresponding figures have the lowest values here (e.g., the Lokhtin's code $L \leq 1$–0.5). In the inland areas, the corresponding river length patterns are identified, but the direction of changes in the partitioning between sediment load components, stability and rate of change depends on the flow of the rivers from South to North or North to South (e.g., the Ob and Volga).

The partitioning between suspended and bed component in sediment load determines, to a certain extent under specific natural conditions, the development of river channels of different patterns and the intensity of their transformation. On rivers of 1–2 orders, with low overall river runoff, rectilinear, unbroken channels prevail (Fig. 2.18), which are relatively stable. Small and medium rivers that mainly meander, characterized by segmental, omega-like or bursting bend, and suspended sinusoidal rivers with predominance of bedloads.

On large rivers, the greater the amount of bedload, the more complex their branching is. This can be seen in the comparison of branches on the Yangtze River, where δ = 5–6% (with the predominance of thin and small bed material load) and on the Lena River, where δ = 40% (with the predominance of medium and large sands) (Fig. 2.19). On rivers with high sediment load, an absolute predominance of the suspended component and their directional accumulation of branching do not

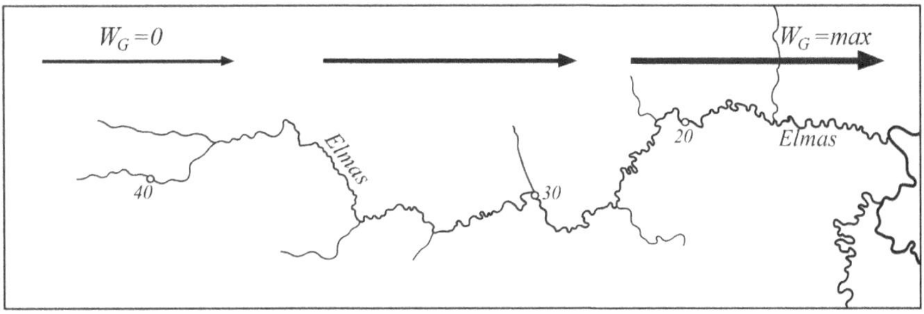

Fig. 2.18 Changes of channel pattern from straight to meandering downstream from the source due to increasing of bed load runoff (left-bank part of Northern Dvina river basin)

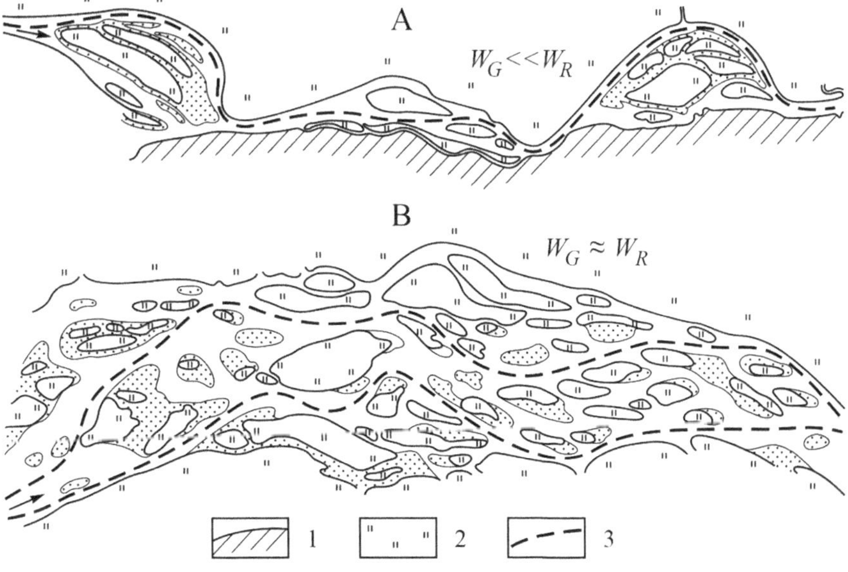

Fig. 2.19 Typical braided reaches of Yangtze river (**A**) and Lower Lena river (**B**), formed at different ratio of bed load (W_G) and suspended sediment (W_R) runoff

form or only their simple varieties are present. One of the reasons for this is the "mudding" of the ducts and branches that arise as a result of the sediment load entering them, their rapid shallowing and the attachment of sediments or islands to the bank. At the same time, with relatively less sediment load and a high δ value, as on the Lena, branching is the most common type of channel, including in the intermittent rivers, for which inequality $W < W_{tr}$ is true.

To a certain extent, the predominance of suspended or bed sediments in total load depends on the conditions in which they are generated. The sediments have channel genesis if they are a consequence of riverbed and bank erosion, provided that the actual flow velocities V_f exceed the non-destructive velocities for the soils (sediments) composing them, i.e. when $V_f > V_n$ (Fig. 2.20). At the same time, river load comes from the catchment area and the gully-beam network, i.e. it has a basin genesis. River basin sediments are mainly responsible for suspended sediment load. The riverbed sediments form the bedload transport rate, although they may enter the slurry if the finely dispersed sediments of the floodplain facies of alluvium, facies of swirls, etc. are washed away. At the same time, the gully outbursts, mudflows and curums on mountain slopes are a source of bedrock material in the rivers.

Bed and some suspended sediments are bed material load, which are "predominantly contained in sedimentary sediments" when transport ceases (Karasev 1975, p. 146). The proportion of suspended load in the bed material load, as well as the partitioning between suspended W_R runoff to the total runoff of exposed W_G loads, varies widely, depending on the geographical patterns of sediment load formation as a whole, the conditions for the realization of the transport capacity of W_{tr} and the spatial and temporal changes in the hydraulic characteristics of the flow (from the bedrock to the periphery of the channel, to the high-water and low-water phases of the regime, etc.). It is larger at low sediment particle size (sandy-muddy), which coincides with an increase in total load of W_{R+G} (Table 2.9), which increases in Eurasia as a whole from North to South and reaches its maximum on the rivers of South-East Asia. Relationship on the specific conditions of the formation of load is manifested in the increased values of bedload flow in the rivers flowing through the sandy lowlands, while the rivers of neighbouring areas, which have the same size (order) but divide the uplands, are characterized by values 2.5–4 times smaller.

The source of sediment from Southern rivers is formed by loess or loess-like strata and crust of weathering on plateau. As a result, river load is dominated here by sediment loads of basin origin and an absolutely dominant suspended component, which ensures the maximum load of load from the load of sediment ($W_{tr} \approx W_{R+G}$). Changes in W_{tr} and hydraulic flow characteristics due to intra-annual flow irregularities ensure that the runoff rates of the river is reduced and that suspended sediments are transferred to the category of bedridden. As in this case $V_f >> V_n$ their movement occurs without formation of ripples; They accumulate in river channels, forming continuous shallow waters, stop shifting in the lowland, becoming channel sediments, dried up to a large extent, and transforming the channels into silt-sandy fields along the rivers, among which the stream either forms curves (sediment accumulation is similar to that of sideways) or breaks into a network of ducts (sediment accumulation is similar to that of sediments).

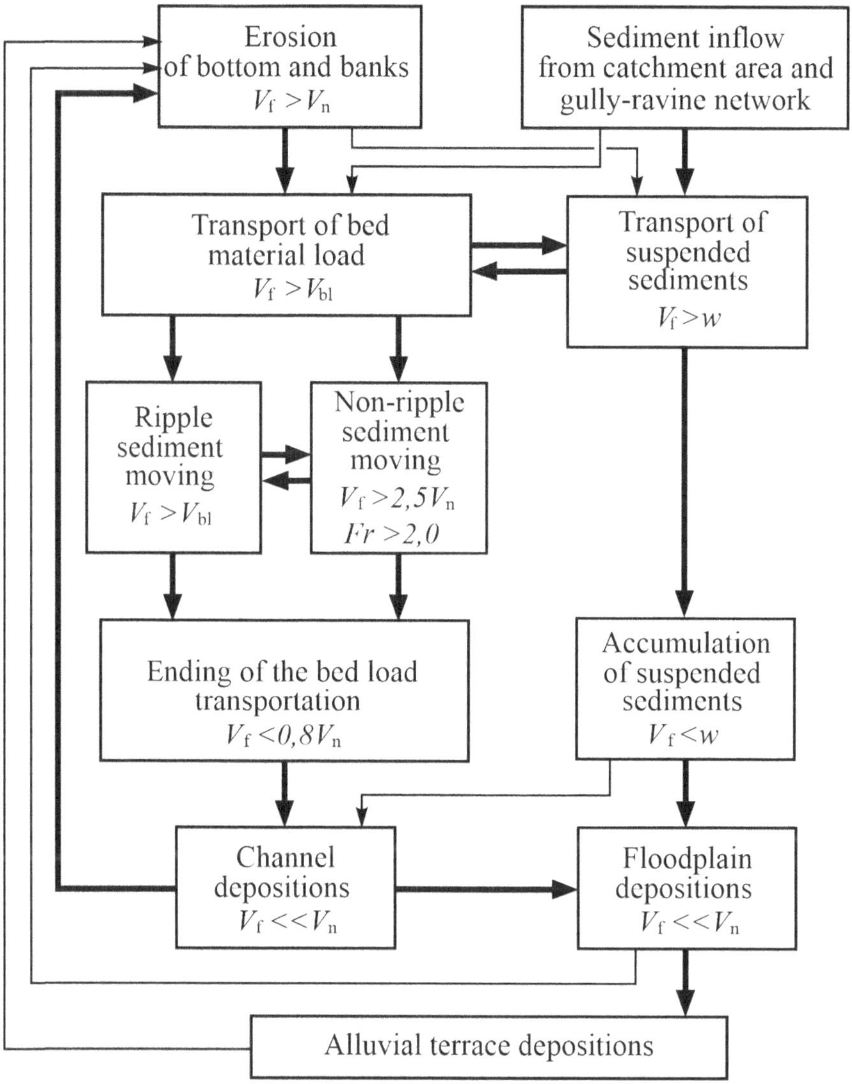

Fig. 2.20 Scheme of the relationship of channel erosion, forms of sediment transport and sediment accumulation in fluvial processes (thickness of arrows shows main and secondary role in mass transfer): V_f—average factual flow velocity; V_n—non-washout velocity; V_{cr}—critical share stress; *w*—hydraulic grain size; Fr—Froude number

Their stirring-up into the next high-water phase causes the wandering of channels, which are absolutely unstable in terms of the intensity of change (Huang He, Amu Darya) (Chalov et al. 2000). This, in turn, leads to the preservation of a straight form of the riverbed, as numerous dried up in the depths of sediment accumulation (analogues of side and axillary sediments) do not have time to fix vegetation and turn into a floodplain. The condition $W_{tr} < W_{R+G}$ determines the directed accumulation of sediment in the channel, i.e. the conservation of sediment in the form of channel sediments, leading to a systematic increase in the bottom marks: at the lower reaches of the Huang He it occurs at a rate of 4.5–7.7 cm/year (Chalov et al. 2000), on the Amu Darya—3.4 cm/year (Tsvetkova 1963), on Brahmaputra—3.4 cm/year (Gusarov 2002). As a result, the rivers have bunded channels, where the water level in the lowlands exceeds the level of adjacent territories.

With larger volumes of load entering river channels (sand, sand-pebble), a reduction in the basin component of sediment load and an increase in the share of bedload, a hierarchy of ripple forms of riverbed relief emerges. Its structure is transformed depending on the river's runoff rates during the change of hydrological regime phases and depending on the position in the channel (spindle or peripheral zones). In the flood period, all rows of all generations—from macro- to ultra-microforms of the channel relief—move across the entire width of the riverbed in floodplains; in the low water, when the parts of macro- and mesoforms (side bars, middle bars), which have not been washed away by the water level, dry out or have not been washed away, the ripples—mainly microforms and partly mesoforms—move only between them, and the width of the strip of sediment transport in the form of ripples in the lowlands up to 2–3 times can already be so at the maximum discharge of water.

Each ripple is essentially an accumulative formation in which more or less of the sediment has temporarily stopped moving, forming a stationary core. However, as sediments on the upper (pressure) slope of the ripple erode, sediments return to motion, settling and returning to the sediment category on the lower steep slope (basement). Thus, the formation of ripples as accumulative formations in the channel is not yet evidence of the formation of alluvial strata, but is a temporary stop (cessation) of transport of sediment particles, which resumes as the ripple shifts. Nevertheless, each ripple, especially those that dry out in the lowlands, has the potential to transform, in part or in full, such seasonal deposits into alluvial layers. This is due to the reverse effect of the ripples—macroforms on the kinematic structure of the flow, erosion of the banks opposite to the drying ripples, appearance of vegetation ripples on the surface and cessation of their further movement. The vegetation fixation of ripple macroforms is accompanied by the deposition of suspended load on their surface, which has a hydraulic grain size $w > V_f$, and the formation of a floodplain filler. Suspended sediment accumulation is a process that dominates the floodplain surface and results in the formation of a floodplain loamy alluvium facies. The greater the sediment concentration of the water and the greater the sediment load, the more intense the sediment accumulation on the floodplain and the greater the thickness of the actual floodplain sediments. As a result, the primary dynamic axis flow curves near side bars are transformed into channel bendings, and the branches of the flow around the axes—into branches of the channel (side bars turn into meander necks, the mid bars—into islands). As a result, there is a channel of one or another pattern in which further movement of macroforms is accompanied by the development of bends, redistribution of water runoff, deepening and shallowing of branches, reshaping of relatively straight single-thread channels, i.e. determines the lateral channel changes.

A condition for this transformation of ripple forms of sediment transport into channel forms is the greater stability of sandy, much less pebbly sandy beds (usually *Lokhtin's code* $L > 2–3$) as compared to sandy-silty beds. The bed remains straight with low bedload transport rate, as there are no relatively slow shifting ripple macroforms of the channel topography. This is characteristic of first order watercourses that have not yet received a sufficient amount of sediment from interaction with the bottom and banks to form a ripple of sediment, as well as for large rivers with weakly stable or unstable channel, which are characterized by high mobility of macroforms of ripple relief, due to which they do not have time to consolidate vegetation (including those rivers, where bed material are mainly suspended loads) (Chalov et al. 1998).

Sandy and pebbly-sandy rivers are characterized by $W_{R+G} \approx W_{tr}$ ratios, and usually W_{tr} little more than W_{R+G}, so they incise very slowly (fractions of millimeters per year). In historical or geological periods of time, this lead, for example, to the formation of stepped floodplains, their transformation into floodplains. The inverse ratio is only present in the lower reaches and estuaries of rivers, but due to the insignificance of W_{R+G} over W_{tr}, the rate of directional load accumulation is also low and is compensated for by the accumulation of sediment on the floodplain. As a result, there are no bunded floodplains here, but superimposed floodplains (with periodic change of incision by accumulation) or one-tier floodplains (with directed accumulation of sediment) appear.

Together, the values of sediment load, its bed load and suspended components, their ratios and the realization of the transport capacity of the flow determine the complexity of channel shapes, their stability, the intensity of channel changes, the degree of shallow water in the channel, etc. (Figure 2.21). On the one hand, the latter are a function of sediment load, i.e. an integral characteristic of the complexity of channel morphology and the intensity of channel changes $\psi = f(W)$: the larger it is at $W > W_{tr}$, the more morphologically complex the channel is, the less stable it is, the shallower it is, and the more intensive the reshaping is. On the other hand, these characteristics increase with the increase in bed load only up to certain limits, and their maximum is achieved when the transport capacity of river load is realized at the expense of suspended sediment load of $W_{tr} \approx W_R$.

The formation and movement of ripples on rivers with a pebble-boulder composition of bed material load is distinguished by their specific features. The suspended component does not participate in their formation at all. In the low-water period, when $V_f < V_n$, the sediment transport stops. In the high-water phases of the regime, the mass transport of sediments begins only after the decay of the armouring layer, when $V_f > V_{n\ arm\ 1}$. At this point, for alluvium lying beneath the bripple, $V_f \gg$, $V_{bl_{ai}}$ which causcs a rapid, often avalanche-like growth in the bedload transport rates due to the almost instantaneous involvement of smaller material in motion and, consequently, a decrease in V_n when the flow is saturated with fine material (Makkaveev et al. 1970). During flood the first thing that comes out of transportation is the large particles of bedded sediment that form the scaffolding and bury granulometrically heterogeneous loads. Due to the large size of the

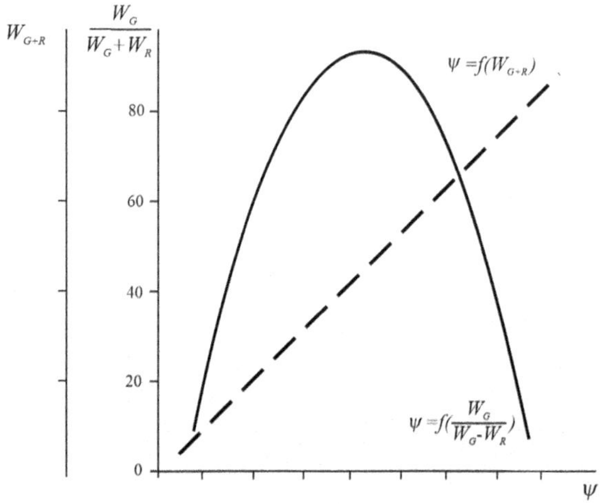

Fig. 2.21 Connection of integral index of channel morphology and intensity of its changes ψ and total sediment runoff W_{R+G} and share of bed load runoff in it

load and the presence of a scaffold, the ripple topography hierarchy has been reduced. It presents only macro- and partly mesoforms, the ripples of which are not eroded at the flood recession and in the low water level. Therefore, the pebble-boulder beds are stable and absolutely stable ($L > 10$–50).

Usually for pebble-boulder rivers $W_{R+G} \ll W_{tr}$. Under these conditions, there is a shortage of sediment and rocky beds are formed (Chalov 2003). When changing lithology of rocks (magmatic rocks to softer sedimentary rocks, rocky formations to semi-rocky formations, etc.)) The width of the channels increases, W_{tr} decreases, more rust and denudation of slopes enters the channels, and they, coming out of narrow canyons, where the stream was constrained by rocky banks, melt and often become branched out. This is especially true for mountain rivers that come out of the mountains into the foothills or intermountain hollows, where the water temperature decreases due to the reduction of slopes, among other things. As a result, the mass unriddled transport of debris into the high-water phases of the regime is replaced by the formation and movement of ripples that create extensive pebble-boulder fields (Figs. 2.22A), among which the river is divided into numerous channels (there is a known analogy with sandy-silty fields on the wandering rivers with an absolute predominance of suspended sediment in the bed material structure). Similar pebble-boulder fields (outwash plains) occur in the glacial zone when water flows accumulate fluid glacial fluvioglacial material on the flat bottom of path valleys (Fig. 2.22B).

In mountain rivers, due to the high flow velocities, the transport of sediment is carried out with the formation of an antidune ripple when Fr values from 0.8 to 2.0 and relatively small gradients (channels with developed alluvial forms) or a continuous layer (unstructured transport) when $Fr > 2.0$. At very high gradients on mountain rivers, large fragments form porous waterfalls (Makkaveev and Chalov 1986).

Bank erosion results in the return of some of the alluvial deposits that make up the floodplains and floodplain terraces to sediment transport. Thus, floodplain

Fig. 2.22 Pebble-boulder fields on mountain river: **A**—in the intermountain hollow (photo from the archive of N.I. Makkaveev); **B**—in the glacier area of the Djankuat glacier (Caucasus). Photo by N.L. Frolova

massifs are essentially one of the forms in which sediment load is carried out (Makkaveev 1955). The wider the floodplain and the more stable the channel, the greater the part of the load is preserved in the floodplain; the terraces, the surfaces of which have already risen from the floodplain level, serve only as sources of load entering the channels, but only when the flow of their benches to the river washes away. At intensive erosion of sandy terraces downstream, shallow rolling areas are usually formed, represented by random clusters of ripples, or the most complicated morphology and change regime by branching. These are the channels of the Northern Dvina river below the Tolokonnaya Mountain (Chalov et al. 2000) and the Zeya river below the White Mountain (Vinogradov et al. 2003).

Thus, sediment flow, its value and characteristics are determined by the physical and geographical conditions of the river basins. Aquifer capacity and the water regime of rivers are equally important. In different natural conditions, different rivers of the same order have different sediment flow characteristics. The nature of the intra-annual distribution of runoff, the partitioning between particle size distribution of suspended and bed load, channel sediments, the actual flow rate of river load and the transport capacity of the flow are important for their spatio-temporal variability. All of these factors, in their complex interrelationships, influence the partitioning between suspended and bed sediment load in the structure of the overall transport of lithogenic material. In turn, the differences in sediment load determine the formation of alluvial forms of riverbed relief and, through them, the formation of beds of a particular pattern and the mode of its seasonal and perennial changes.

New data on the proportion of suspended and bed components of total sediment load raises the question of determining the ***sediment transport capacity of the flow***, which is commonly understood to be the maximum amount of load that can be carried with these hydraulic characteristics (Baryshnikov and Popov 1988). In practice, the average sediment concentration of flow s is taken as an indicator of transport capacity. The empirical formulas of E. A. Zamarin were most widely used in practice (1951):

$$s_{tr} = 0.022\left(\frac{V_{medium}}{w}\right)^{\frac{2}{3}}\sqrt{RI} \tag{2.18}$$

when $0.002 \leq w \leq 0.008$ m/s, and

$$s_{tr} = 11_{medium}\sqrt{\frac{RV_{medium}I}{w}} \tag{2.19}$$

at $0.004 \leq w \leq 0.002$ m/s (here R is the hydraulic mean depth).

Specifying this approach, Rossinsky and Kuzmin (1964) proposed to consider sediment concentration corresponding to the transporting capacity of the flow as a state of its saturation limit, which is determined by the upper envelope relationship

$$s_{tr} = 0.24\frac{V^3}{hw} \tag{2.20}$$

There is also a fairly large number of empirical formulas and formulas proposed by other researchers based on the discovery of the physical nature of the sediment transport process, mainly in a suspended state. A detailed analysis of them, as well as justification of the new formulas of transporting capacity of the flow was performed by Karasev (1975) and A. V. Karaushev (1977). They are all based on the provision for large differences in sediment discharge between bottom and suspended sediment. According to Karasev (1975), the ratio between them (β) for small fractions (d < 0.25 mm) is expressed by the relationship

$$\beta = 30\frac{d}{h} \tag{2.21}$$

at relatively shallow depths of H (about 1 m) does not exceed 1%. For larger loads ($d > 1.5$ mm)

$$\beta = \frac{1300\left(\frac{V_{medium}}{V_n} - 1\right)}{\left(\frac{V_{medium}}{V_n}\right)^2 - 0.5\frac{V_{medium}}{V_n}} \cdot \frac{d}{h} \tag{2.22}$$

and its value increases on mountain rivers up to 10–20%.

Since, all other things being equal, suspended sediment loads are related to each other, sediment concentration can be used as an indirect indicator of the transport capacity of the flow.

The above data on the partitioning between suspended and distant components of sediment load indicate from a modern standpoint that the issue of determining the transport capacity of the flow is much more complex. The uncertainty and physical invalidity of the current interpretation of W_{tr} as the ultimate "saturation of the flow" was pointed out by N. E. Kondratyev (Kondratyev et al. 1959), who put forward almost 50 years ago the problem of developing the W_{tr}. The realization of the W_{tr} is indeed possible due to the saturation of the flow with suspended sediments, the hydraulic grain size of which is $w \le V^*$, where V^* is the vertical component of the turbulent flow rate. At the same time, the W_{tr} can be realized by movement of particles and fragments of various particle sizes up to large boulders and boulders in a faraway state, and the flow energy consumption for their movement is much higher (by orders of magnitude) than for the weighing of fine mud and fine sand particles. The solution to the problem of determining the transport capacity W_{tr} of the stream should therefore be to assess the loss of shear pressure, movement, suspending and transport of bed and suspended sediments. This corresponds to an integral assessment of the transport capacity of the flow as the maximum annual runoff of river load under conditions of changes in water availability and hydraulic flow characteristics. With this approach, changes in transport capacity result in excess or, conversely, a deficit of sediment load entering the flow in a given section of the river. In the former case, a part of the bedload is stopped in the form of suspended sediment and some of the suspended sediment is transferred to the category of bed load; as a result, the flow is cleared, the sediment concentration decreases and the bottom marks increase. In the second case, the opposite process takes place: the smallest fractions of the exposed sediment are suspended and the particles lying on the bottom begin to move; sediment concentration of the flow increases.

The last proposal for determining the transport capacity (which was not included in the above-mentioned reviews, respectively) belongs to Zorina (2000), who, studying the detachment, suspending and movement of soil (highly saturated mixtures, mash) by pressure flows in the pipelines of river dredgers, got the expression

$$W_{\text{tr}} = Q \cdot \frac{A\left(V^{2.7} - V_{bl}^{2.7}\right)}{0.2 + 0.65A\left(V^{2.7} - V_{w}^{2.7}\right)}, \qquad (2.23)$$

where Q—average water flow rate for the calculated period of time; V—flow rate, m/sec; V_{cr}—critical share stress, m/sec; $A = \frac{n^2}{V^{0.7}h^{1/3}}$ h—flow depth; n—flow bed roughness coefficient. Given the object of study of sediment transport (pressure flows in pipes), the ultimate saturation of the flow is due to both suspended and excavated material, which distinguishes this approach from the rest, which is based on the sediment concentration of the flow (i.e. filling it with suspended sediment) and the hydraulic particle size of the load.

Saturation of the flow of sediment cannot occur instantly; with a sharp increase in flow velocity, the flow at some distance has a “saturation deficit”, which is extinguished by increased erosion of the river bed. Erosion occurs until the longitudinal profile of the watercourse becomes flatter than in the adjacent areas, and until non-destructible rocks are exposed or alluvial decay is formed. As in any system of water flows its components have different hydraulic characteristics, at their confluence of water of the incorporated stream appear unsaturated with sediment, as at change of hydraulic conditions as a result of merge also the general carrying capacity of a stream changes. In other words, a deficit of load arises if $W_2 - W_1 = \Delta W > 0$ and $W < W_{tr}$, where the indices 1 and 2 indicate the position of the sediment flow measurement, respectively, downstream and upstream.

Other common cases of load deficits in flows are a sharp increase in the runoff rates of the river during the rise in water levels during flooding, clarification of the water supplied to the downstream section of the dam from the reservoir, etc. The effect of changes in the hydraulic characteristics of the flow on its transport capacity can be illustrated by the following example: when levels rise rapidly during a flood at the front of its wave, there is a large longitudinal slope, the flow velocities are high and there is a strong saturation of the sediment flow; on the contrary, when levels rise slowly at the front of the wave of the flood insignificantly increases, the rate increases slightly and significantly the rate of channel erosion and saturation of the flow of sediment does not occur.

Excess sediment in the flow, i.e. the resulting balance of $W_2 - W_1 = \Delta W < 0$, is observed in the event of a flood recession or in the rear of a single movement wave, in the event of flow backflow, etc.

In case of seasonal changes in water runoff, as well as in case of greater difference in the slopes of the free surface of the flow in the flood and in the low water period, the total value of the sediment load (suspended and bed) transporting capacity of the flow in the annual section is determined by the relationship of Makkaveev (1955):

$$W = A_{er} I Q^m \tag{2.24}$$

In formula (2.24), the A_{er} coefficient depends on the degree of irregularity of flow, the nature of the rocks composing the river bed, as well as on the mechanical composition of the load supplied by inflows, melt and rainwater and slope processes. In particular, the flow at a given gradient and water flow rate can carry more solid material the less it is deep. As a result, the value of the A_{er} coefficient can vary significantly depending on local denudation patterns in the catchment area. Therefore, it is called the “erosion” coefficient by N. I. Makkaveev, the empirical values of which for a number of rivers are given in Table 2.10.

In accordance with the degree m, the connection between water and sediment discharge in flat rivers is usually close to a quadratic one, and in mountainous rivers it is close to a cubic one. The increased importance of m in mountain rivers is due to a number of reasons: the movement of load on mountain rivers at high gradients is not only related to the action of the flow, but also to the slope of the bottom;

alluvium in mountain rivers is characterized by a much greater heterogeneity of the mechanical composition; at the bottom of the flows there is formed a scaffold that prevents the movement of sediment between the lowlands; the velocity of flow on mountain rivers decreases to a greater extent to the lowland than that of flat rivers due to the increased, due to the particle size of the alluvium, roughness of the bed.

The relationship (2.24) for lowland rivers, if we assume $m = 2$ and replace the speed value of the Chezy formula (1.3), shows that the maximum total sediment transport rate is approximately proportional to the flow rate in the sixth degree:

$$W_{\mathrm{tr}} = f(V_{\mathrm{av}})^6. \tag{2.25}$$

The links between water and sediment discharge, flow rate and transport capacity of the flow have a number of important implications for the understanding of erosion and storage patterns in rivers. It follows from them that the unevenness of the runoff contributes to the increase in the transport capacity. This is the only factor that increases the total annual sediment load (and therefore the average annual sediment concentration of water) by several times. If, for example, the flow is assumed to have a constant flow rate of 5 m^3/s and a sediment concentration of 0.05 kg/m^3, then the sediment transport rate is 0.25 kg/s and the annual sediment transport rate is about 800 tons. Having released the same annual amount of water within one month, sediment concentration can be expected to increase to 600 g/m^3, sediment discharge to 36 kg/s (during flooding) and total sediment load to 9600 tons, although most of the year the flow will be dry and carry very little sediment.

The merging of flows results in an increase in the transport capacity of the combined flow to a greater extent than the increase in water consumption. If $Q = Q_1 + Q_2$, then $W >> W_1 + W_2$, where Q and W are, respectively, the values of water runoff and sediment load of the combined flow, and the rest of the indicators —its components. Hence, with the increasing area of the basin, the transport

Table 2.10 Values of erosion coefficient A_{er} in the formula (2.24) (Makkaveev 1955)

River—Hydrological control station	Average slope, ‰	Square of catchment area, thousand km^2	Erosion coefficient A_{er}
Vyatka—Vyatskiye Polyany	0.06	124.0	1.13
Kama—Perm	0.05	167.8	0.31
Don—Hovansky	0.06	168.8	2.94
Volga—Kineshma	0.03	187.8	0.51
Oka—Novinki	0.03	244.8	0.97
Ob—Novosibirsk	0.08	246.2	0.32
Yenisei—Bazaika	0.23	299.4	0.53
Kama—Tarlovka	0.04	367.0	0.24
Don—Razdorskaya	0.06	378.0	2.50
Ob—Kolpashevo	0.04	481.0	0.39

capacity of the flow increases to a greater extent than the flow rate along the length of the river.

In different phases of the hydrological regime at the same flow rate, the transport capacity of the flow on most rivers is different, which is represented in the ambiguous connection between the flow rate of water and sediment at the rise and fall of floods (Fig. 2.23A). This connection takes the form of a loopy curve, the right branch of which corresponds to the period of flood rise and the left branch to the recession of the flood. At the rise of the flood wave front, the slope of the flood wave is increased; the flow rate also increases due to the velocity of propagation downstream of the wave itself. According to the relationship (2.19), the transport capacity of the flow sharply increases. In the rear part of the wave, the flow velocities drop, the sediment transport capacity becomes smaller, sediment concentration and the consumption of sediment decreases. Only on rivers with a flood regime, small and rather homogeneous composition of alluvium (Amu Darya, Syr Darya) a single-valued curvilinear relationship $(R + G) = f(Q)$ is formed (Fig. 2.23B).

In large rivers, the maximum sediment transport rate is often higher than the flood peak, which is due to the specifics of the flow of material washed away from the catchment (Shamov 1959). The main source of suspended sediment during the flood rise is slope flows. The peak of flooding occurs when erosion processes in the catchment area have subsided and the introduction of sediment into the rivers comes from the gully network and from tributaries carrying larger material. At the same time, it begins to erode the riverbed itself. The increase in the size of the load transported by the flow of sediment leads to an increase in the energy losses of the flow and to an overall reduction in sediment concentration. In addition, the floodplain of the river is usually flooded at the peak of the floodplain, where part of the load accumulates. At the same time, the release of water to the floodplain is accompanied, all other things being equal, by a decrease in the carrying capacity of the flow due to an increase in the cross-section of the "big channel" (the channel together with the floodplain—as defined by Velikanov (1958)), a decrease in its average depth and an increase in the roughness of the underlying surface due to vegetation on the floodplain. As a result, at maximum water discharge due to the kinematic effect, the transport capacity decreases (Zheleznyakov et al. 1970; Baryshnikov 2005).

The intensity of the transport of bottom sediments and, consequently, the total transport capacity of the flow increases as the concentration in the flow of finer material increases (up to certain limits). Figure 2.24A shows the relationship between the sediment concentration of the Amu Darya flow, which is smaller than 0.05 mm, and the content of moving sediment particles with a diameter of 0.2–0.5 mm: as sediment concentration increases, the concentration of coarse bottom sediments increases and then begins to decrease when the suspended solids saturation of the flow exceeds 1000 g/m^3 despite the increase in the flow rate. This effect of sediment concentration on the flow rate of bedloads is represented in their flow rate (Fig. 2.24B), which first increases as the sediment concentration increases and then decreases (Rossinsky and Debolsky 1980).

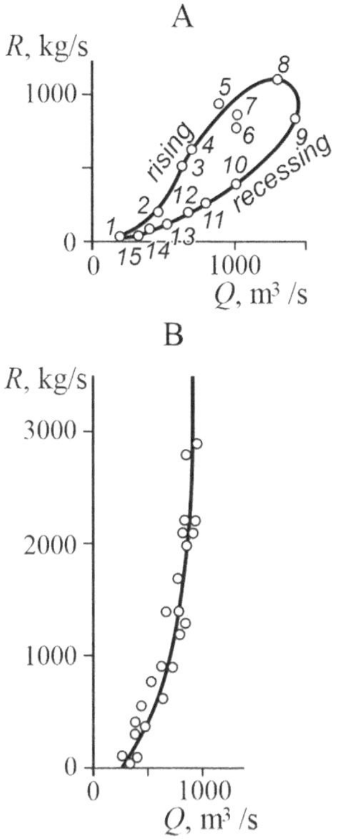

Fig. 2.23 Dependence of suspended sediment discharge R and water discharge Q: **A**—loopy, on rivers with a pronounced flood and heterogenous bed material load composition (Ural river, Topoli village); **B**—definite, on rivers with non-pronounced flood and fine-grained homogenous bed material load composition (Syr Darya river, city Kazalinsk) (according to Karaushev (1960))

The reason for the increase in the mobility of bottom sediments and their involvement in the transport of sediment is due to the fact that the hydrodynamic pressure on the soil particle is proportional to the density of the liquid. The presence of suspended solids in the water, which reaches a maximum at the bottom, increases the density of the water in the bottom layers of the flow and, consequently, increases the pressure on the particles. This explains, in particular, the movement of some lowland rivers flowing in mountainous areas (the upper Lena and Aldan) and mountain rivers of pebbly-boulder sediment of increased size, for the movement of which in other conditions requires speeds that are significantly higher than the actual ones (Makkaveev et al. 1970). Due to the large heterogeneity of alluvium, lying under the decanter, when the latter is destroyed, the flow is saturated with fine material and acquires a structural character, so that the mechanism of movement of pebbles and boulders in the rivers resembles water-stone mudflows.

In addition, as the concentration of bedload increases, the probability and frequency of impact between particles and their impact on stationary particles increases. The latter contributes to the movement of particles lying at the bottom. Collisions also transmit the amount of motion from particles moving at higher speeds, i.e. smaller sizes, to larger and slower particles. As a result, the average speed of bottom sediments increases. The impact of impacts from smaller particles

Fig. 2.24 Influence of water turbidity on bed material load transport: **A**—connection between turbidity s of Amu Daria river (average particle size d < 0.05 mm) and bed material particle content s_d = 0.2–0.5 mm; **B**—connection between average turbidity *s* and bed material load discharge G (according to experimental data (Rossinsky and Debolsky 1980)

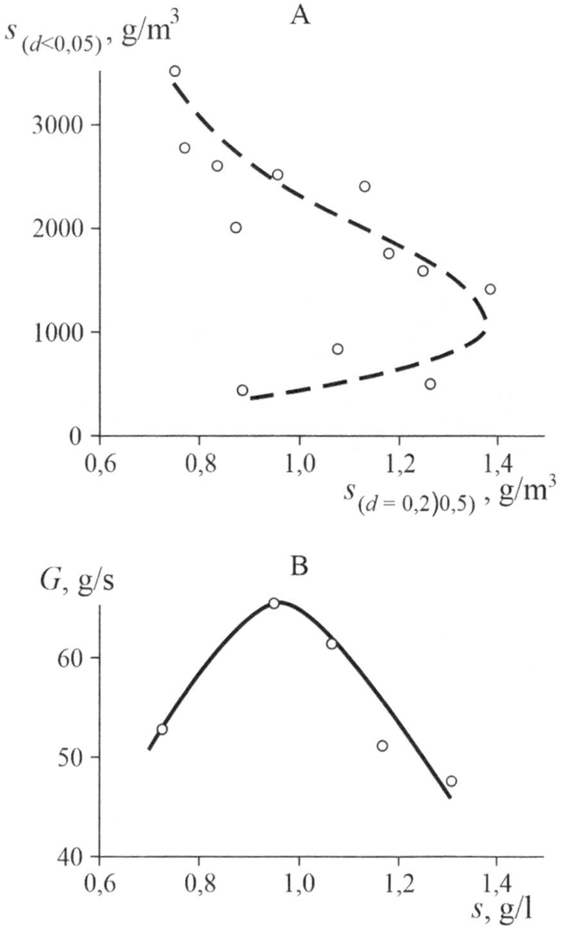

on the transport of bottom sediments is particularly evident in the pebble-boulder composition of bottom sediments. Due to this phenomenon, the inflow of fine pebbles into the stream has a more significant impact on the transport of boulders than the inflow of sand.

The inverse effect of the further increase in sediment concentration on the transport of exposed sediments is due, firstly, to the possible collation of the bottom ground by suspended sediments, which fill in the unevenness of the bottom and the pores between coarse particles and level its surface. The roughness of the bottom decreases, and as a result, the ripples of the bottom flow rate decrease. Secondly, with a high saturation of the flow of sediment, the bottom of the river is covered by smaller particles, the concentration of which in the flow increases. As a result, coarse debris lying on the bottom is protected from direct exposure to the flow of smaller particles in motion.

2.6 Channel Shape and Channel Relief as Factors of Fluvial Processes

The forms of the channel and channel relief in feedback arising from the interaction of the flow with soils and alluvial sediments and in the process of transport of riverbed sediments through the law of mutual conditionality of the flow and channel of M. A. Velikanov-N. I. Makkaveev act as a factor of fluvial processes. Although the channel shape is created mainly by the flow activity, its subsequent changes under the influence of the flow are inertial and delayed in comparison with changes in the flow characteristics: if the hydraulic characteristics of the flow change with changes in the interaction conditions almost instantly, the transformation of the channel shape takes place over a period of time determined by the partitioning between soil size and mobility, on the one hand, and the flow rate, its kinetic energy, on the other hand.

Continuous influence of the flow on the channel and channels on the flow causes the appearance of such kinematic structure (circulation currents, velocity fields, etc.) in the flow under the influence of its shape and the channel relief forms, which contributes to its constant renewal and relative stability of the forms themselves. For example, the preservation of the sand ripple is provided by the formation of a whirling current (roller), which appears in its basement. The bottom branch of this current is directed against the main stream and holds the sediment on the lower steep slope of the ripple (Fig. 2.25A, B). A. N. Lyapin (Kondratyev et al. 1959) draws an analogy with the separation of vortices (rollers) from the edges of the streamlined body, comparable to the geometric dimensions of the streamlined body itself and the flow (Fig. 2.25B). A decrease in the depth of flow over the ripple, similar to its compression, leads to a decrease in the level above its ripple (Fig. 2.25A, B) and an increase in speed here. Due to this, particles are washed off and moved along the pressure slope of the ripple, and it occurs at a negative (against the current) slope of the bottom.

The circulation currents and the structure of the high-speed flow field at the bend of the channel, being a derivative of the bend of the channel, provides stability and further development of the bend itself through the formation of corresponding zones of sediment accumulation and erosion of the banks. Analysis of the flow structure on the bend radii shows that these channel forms exist not only because of the bending circulation currents, but also because its transporting capacity increases significantly at the bend of the flow and (up to a certain limit of the bend) and turns out to be greater than W_{tr} at the straight sections of rivers. Therefore, when river channel shapes are formed, those that contribute to the greatest erosion and transport performance of the flow are developed in preference.

At a given point in time, the forms of the channel or channel relief significantly affects the hydraulic characteristics of the flow, transporting capacity and conditions of sediment transport. As a result, each channel formation (channel form and channel relief form) affects to some extent channel changes within adjacent forms and determines their development at a lower structural level. The first represents

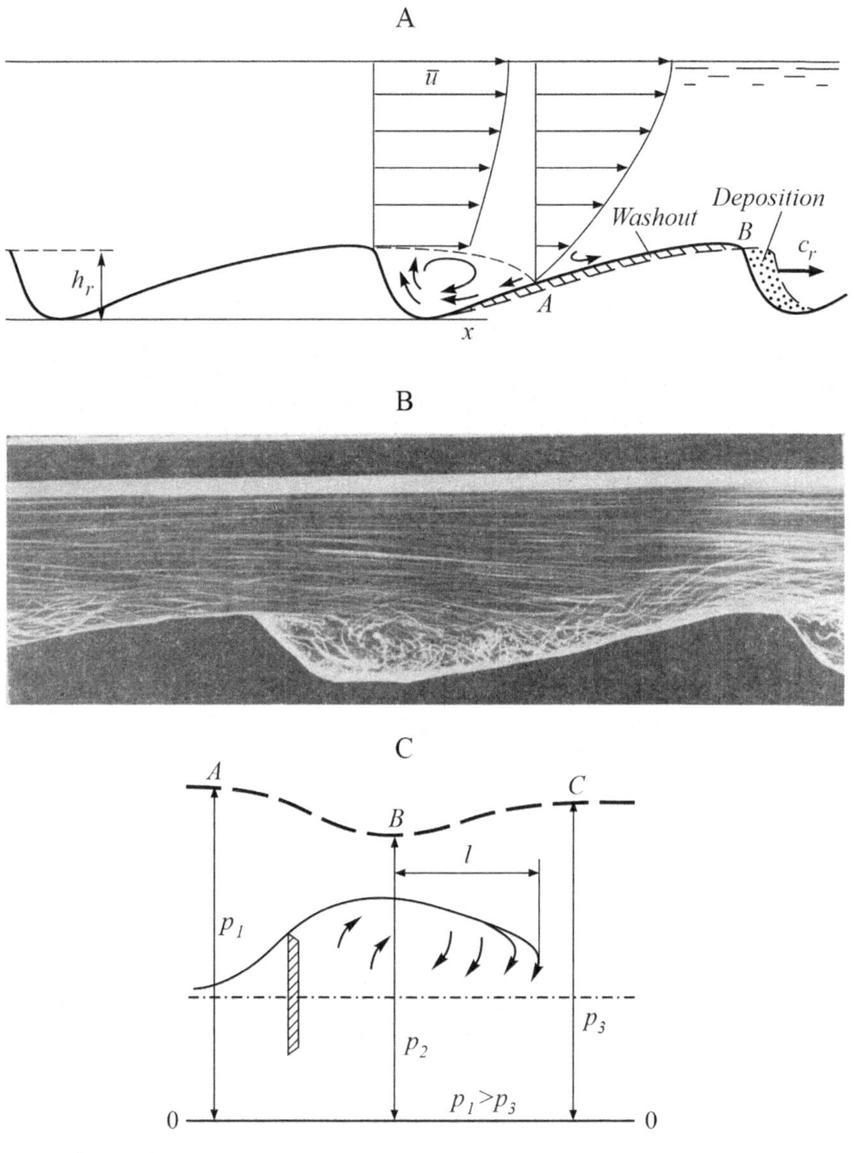

Fig. 2.25 Formation of a vortex with a horizontal axis in a scour hole of a ripple (**A**—scheme according to Grishanin (1972), simplified, with additions, according to Kondratyev (1964)) and kinematic structure of the flow within the ripple bottom relief in a glass tray (**B**—photo by N. A. Yarnykh (Znamenskaya 1968)) and beyond the streamlined solid (C—according to A. N. Lyapin (Kondratyev et al. 1959))

itself in the conjugate development of bends and branches in branchings formed by successive islands. Cutoff of one of the bends in the series leads to the rearrangement of the channel both upstream and downstream due to the arising mismatch of the kinematic structure of the flow in the new channel of the previously formed forms. The same thing happens in branching when one of the branches dies out and the island is divided into two parts. At the conjugate development of a bend and branching consecutive increase in a curvature of a channel on a bend promotes concentration of the most part of the expense of water in a stream in a branch at opposite coast and its priority development; the second branch, having appeared behind a shoulder of the concave coast, shallows down and, finally, can freeze. When the bend is straightened, the flow is redistributed between the branches and the previously dry or dying branch is activated.

Different kinematic structures and changes in the position of the dynamical axis of a watercourse at different water levels have a great influence on the formation of riffles and branches. One of the reasons for the formation of the riffles is the significant local divergence of the flow directions in the flood and low water period (Makkaveev 1949, 1955). Relative stability of many branches is connected with the fact that the main stream of the river moves from one branch to another depending on water availability of the stream, as a result of which one branch gets priority development in the high-water phase of the regime or in high-water years, and in low-water years—another.

Changes in channel width cause the formation of local zones of drop in levels and flow support, its constriction and spreading, which leads to the formation of pool hollows, in the first case, and riffles of the "inner bar" type (Makkaveev 1955), in the second case. Formation of the channel bend is accompanied by a natural arrangement of pool hollows at concave banks, banks (sideways) with developed ripple relief and riffles at the bend between the bends. The formation of the bend is accompanied by a narrowing of the channel, deepening of the pool hollow, increase in the length of the channel and growth of hydraulic resistance. The branching of the channel into branches results in a local decrease in transport capacity, since $W_{R+G} = f(Q_m)$, where $m > 1$, so that at $Q = Q_1 + Q_2$ (here the indexes mean the characteristic belongs to the branches) $W < W_1 + W_2$. Therefore, the division of the flow into branches is accompanied by the formation of riffles at the entrance to them, similar to the formation of shallow water (riffle) at the bends between adjacent bends. However, in the branching itself, the bending of the stream near the island causes hydrodynamic phenomena similar to those on the bend, which results in the growth of the W_{tr}, providing for the development of branching as a form of channel and transport of sediment. In both cases, due to the law of self-regulation of the "flow-river" system, the bending of the channel on the bend or near the island in the branching is accompanied by such a change in the structure of the flow, which causes an increase in the transport capacity of the flow, compensating for the loss of pressure due to the growth of hydraulic resistances and the dispersal of the flow on the hoses.

These examples illustrate the impact of channel development (bend, branch) on the formation of large forms of channel relief (riffles, pool hollows). Even more significant is the role of the floodplain, which is the result of the development of bending, branching, and movement of the rectilinear single-thread channel, i.e. it is a derivative of channel changes. During floods, when the floodplain is flooded, a complex interaction of floodplain and channel flows occurs (Baryshnikov 1978, 1984), which is determined by the mutual location of the floodplain massifs and the channel and affects the conditions of movement of the channel flow, its hydraulic characteristics and, ultimately, the carrying capacity of the flow (Baryshnikov 2005). This, in turn, leads to the formation of a series of riffles (riffle sections) in some parts of the river, and in other parts of the river—the formation of extended pools (pool sections). The influence of the floodplain in areas of concentrated outflow of water from the channel to the floodplain at the beginning of the floodplain massif and at the time when water is discharged from the floodplain into the channel at its end, at the approach of the river to the root bank, is particularly noticeable. In the first case, the rolls are formed "on a breakaway", "described by Leliavskiy (1896) (the term is also proposed by N. S. Leliavskiy), associated with the reduction of runoff rates in the channel flow. In the second case, there is a concentrated discharge of clarified water from the floodplain into the channel, resulting in a deficit of sediment in the stream, which contributes to the erosion of the channel and the formation of a pool hollow near the main bank of the river bed, beginning with the local rein deepening of the channel (usually sturgeon and sterlet holes). Above the confluence of floodplain and riverbed flows there is a pressure and a local decline in the riverbed, which in turn leads to the formation of riffles and branches here. A similar phenomenon is also characteristic of the channel section above the general narrowing of the floodplain, where, along with the discharge of water from the floodplain, there is a rise in levels due to a decrease in the specific width of the flood flow.

2.7 Climatic and Meteorological Factors

Climatic conditions of river basins, regions and natural zones, in which they are located or which cross transit rivers, determine the specificity of river flow and its regime as the main, active factor of fluvial processes. At the same time, some elements of the climate and meteorological features of the areas affect the development of channels indirectly, through their impact on the flow, the soils that compose the river channels, and the resulting phenomena.

The wind has a significant impact on fluvial processes in the case of large rivers. Wind power increases markedly over large water areas; the long acceleration length contributes to the development of considerable excitement. At very strong winds (up to 20 m/s), by analogy with reservoirs, since the corresponding observations on the rivers are not carried out, on the largest rivers (Lena, Yenisei, Ob) at free water area in width of several kilometers the wave height reaches 2.5–3.0 m; at presence of numerous islands and shoals it is essentially lower—1.0–1.2 m (Fluvial processes and waterways … 2001). At the same time, wind currents with velocities of up to 0.2–0.3 m/s are formed.

On smaller rivers and rivers with branched channels, such wind wave conditions occur in the high-water phases of the regime (Amur), when islands and wide floodplains are under water. On some rivers, especially those flowing in constrictions between mountain ranges (e.g. the lower Lena section between the Chekanovsky and Haraulakh mountains, called the "Lena pipe"), the influence of wind on the flow state is maximal when its vector coincides with the direction of the valley. The valley of the Lena River near the mouth of the Viliuy River is one of the strongest winds in central Yakutia, oriented from the South-East to the North-West, which coincides with the two prevailing wind directions in this area (SE—46% and NW—33%). Average wind speeds near the Viliuy river mouth are: 3.9–5.8 m/s for SE and 4.6–5.1 m/s for NW. The duration of the first reaches 13–15 days, the second—10 days. Other parts of the river do not have such a complete coincidence between the prevailing winds and the valley, and the frequency of the prevailing winds is lower there.

The waves that appear on the rivers are destroying the banks. Wind currents coinciding with the river flow contribute to the activation of bank erosion. In the upper horizons of the flow under the influence of wind the velocities increase, and the profile image becomes hypertrophied (Fig. 2.26A), which increases the erosion ability of the flow. Field observations indicate that the rate of erosion of sandy floodplain banks in such conditions increases up to 1.5 times. On the contrary, the oncoming winds (such as the North-West winds on the lower Lena near the Viliuy estuary) slow down the flow rate in the upper horizons (Figs. 2.26B), and the rate of erosion of sandy banks, all other things being equal, is reduced by 1.5 times.

At transverse position of prevailing winds in relation to the direction of the river flow, the formation of the channel is affected by runoff towards the windward bank. This causes an increase in the level near them (by analogy with flat reservoirs—up to 70 cm) (Fluvial processes and waterways … 2001) and the formation of transverse slopes of the water surface. In the near-bottom layers of the stream, currents are generated that are directed to the opposite leeward bank and carry sediment to it. As a result, the channel is shifting towards the windward bank, while the leeward part is shallow, the banks are located here and the floodplain is formed.

This process is disturbed if the wind intensively transfers dried up in the lowland sands in stormy days, after which the depths at the crossroads often decrease. In the upper Ob, the systematic transport of sand by the prevailing South-Western winds is a significant source of load to the right channels. As a result, the river is washed by the leeward left bank of the Ob steppe plateau, represented by an 80–100-m scarp, while along the right floodplain windward slopes, islands and dry branches are located. With strong storm winds in the low winds, when dried numerous side bars and middle bars are waved, rising above low water levels up to 2–3 m or more, the air above the river is so saturated with sand and dust that visibility is reduced to 50 100 m (Makkaveev and Chalov 1964).

The most active riverbed sandbanks are fluttering where the riverbed follows directly along the scarp of the Altai steppe plateau, which is facilitated by specific aerodynamic phenomena arising from wind failure from a 100-m scarp. With the prevailing South-West wind, strong eddies occur on the leeward side of the ledge. In winter, when the strongest winds are observed, a downward flow caused by cooling of the air above the plateau surface and higher air temperatures above the

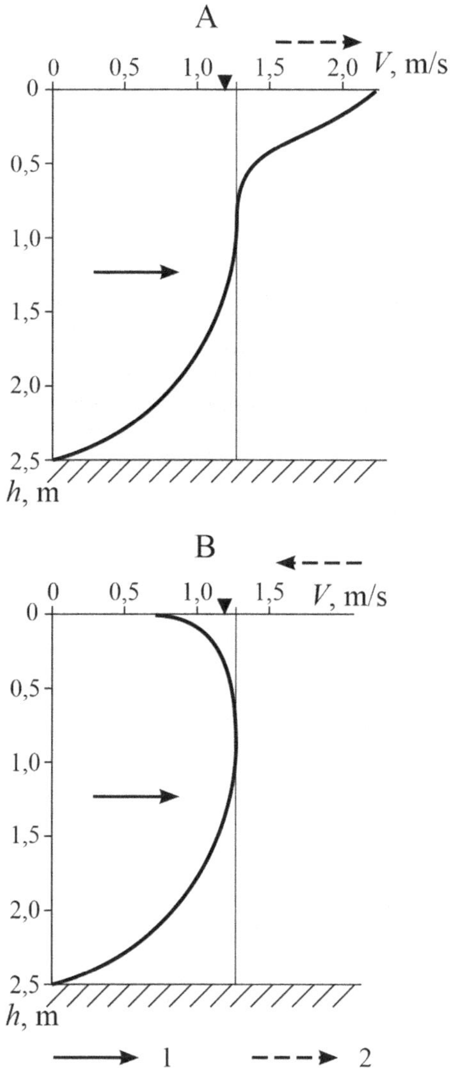

Fig. 2.26 Transformation of the epure of flow velocities: **A**—in coincidence, **B**—on the contrary to flow vectors (1) and wind (2)

river is added. According to Makkaveev (1955), the wind speed at the river increases by 30–50% compared to the watershed areas. In addition to the riverbed sand of channel banks carried across the river in summer, during dust storms or storm winds above the plateau, the surface of which is almost completely plowed out, loess material is added to it.

Scattering of riverbed banks and crosswind transport of sand material in relation to the direction of the river flow leads to the fact that the banks located in the right bank (floodplain) part of the channel rises above the river in the lowland by more than 3 m, while the left bank (under the scarp of the plateau) is more than 1–1.5 m high. Systematic loss of bed material in the left part of the channel and its accumulation in the right part of the channel affects the directional movement of the entire riverbed to the left, towards the leeward bank. This is also favored by a rapid increase in height (due to aeolian accumulation) of the riverbed on the right bank of the floodplain and the formation of inundated dunes on it. This leads to the redistribution of specific water discharge towards their increase near the leasable left bank (Makkaveev et al. 1966).

Aeolian material from the adjacent territories enters the rivers from deserts and semi-deserts of Central Asia. In the Amu Darya during dust storms, shipping stopped for 3–5 days and the Termez port stopped working; after it was stopped, changes in the position and shallowing of the fairway were observed at the crossroads. In some places, the leeward slopes of dunes descend directly to the river. A similar phenomenon is observed in the sandy massifs of the Central Yakutia Lowland—the Kukulans. On the average Viliuy, on the Lena above Olekminsk, modern dunes descend to the river and are washed away, being a local powerful source of sediment load.

Dune ripple formation along the windward floodplain banks is associated with the erosion of riverbed sands (Fig. 2.27). Here, under the influence of floodplain vegetation in the strip up to 200–300 m in the lower layers of the air flow, the most saturated with sand, there is a slowdown in wind speed by 3–4 times. This causes aeolian material to accumulate in the coastal strip. On the upper Ob dunes form solid shafts up to 50–100 m wide and 3–6 to 10–12 m high above the floodplain surface, which is higher than the maximum water level during floodplain flooding. Moving deep into the floodplain, the dunes capture a strip of 200 to 700–800 m wide, stretching along the coast for several tens of kilometers (Makkaveev et al. 1966). The largest massifs of freshly crossed dunes are observed where the river follows along the plateau cliff and has a clearly expressed one-sided right-bank floodplain. As the rivers move away from the scarp of the plateau and the left bank floodplain appears, the dunes are overgrown and, as a result of the erosion of the banks in the course of channel changes, partially destroyed, and the dune ripple, already fixed by vegetation, becomes interrupted.

The aeolian forms on the Lena River near the mouth of the Viliuy River are usually 4–6 m high, 12 m long and 20–25 m long, and 60 m long at the maximum (Borsuk et al. 1975). Both rivers have a close link between the formation of dunes and the development of sideline dunes. Dunes are formed at the edge of the rear of the side and floodplain. As the aeolian material is displaced, the aeolian material flow to the dune decreases or stops, and the dune eventually becomes overgrown and becomes a natural "dam" stretched along the floodplain's edge (provided that it is not destroyed by riverbank erosion). The formation of a new sideline leads to the formation of a new dune ripple, which is sub-parallel to the existing one. As a result, in some places there is a kind of maneuverable dune relief, represented by a system of 3–5 ripples. The dying out of branches and the unification of islands leads to the separation of aeolian forms from food sources (bed sands) and their consolidation.

Fig. 2.27 Dune on a right-bank floodplain of Upper Ob downstream from the confluence with Charysh river. Photo of the author

In the upper Ob (from the confluence of Biya and Katun) to the Novosibirsk Reservoir, modern dunes, which form the floodplains stretched along the riverbed, are mainly developed in the meridional part of the valley between the mouth of the river Charysh and Barnaul (Fig. 2.28), where the river passes directly near the cliffs of the Ob steppe plateau. At the same time, rising above the water even during the maximum spills, these ripples separate the channel and floodplain flood streams, which move in parallel to each other, without interacting with each other. On the lower Lena, where strong winds, being "lower", North-Western, directed up the valley, and South-Eastern (down the valley), contribute to the formation of dunes, in addition to the coastal areas of the floodplain with their appropriate orientation, in the downstream ends and on the tips of the islands.

Another form of wind impact on fluvial processes is inherent in the estuary areas of rivers, where overtaking and overrunning of levels determines the periodic formation of curves of support and decline of levels, which directly affects the peculiarities of channel formation. On the other hand, the delineation of the estuary area of the river itself is made by the distance of upstream overtaking and overrunning.

The ice regime has a noticeable impact on the fluvial processes in the rivers flowing from South to North, and in general on the rivers where thick ice cover is formed in winter. They are characterized by frequent spring congestion, the places of formation of which are confined to steep bends, branches of the channel, junctions with inflows, riffles. The fluvial processes themselves and the morphology of

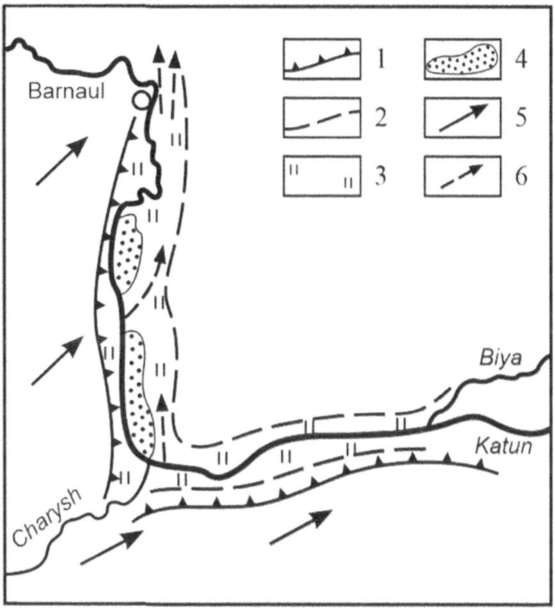

Fig. 2.28 Distribution of dune ripples on the Upper Ob floodplain: 1—the ledge of Altay steppe plateau; 2—terrace boundary; 3—floodplain; 4—young dunes; 5—prevailing wind direction; 6—floodplain flows during floods

the channel through its capacity and its longitudinal and temporal changes are factors in the formation of ice conjuctions. Longitudinal change of channel transport capacity is proportional to its relative width b_{l-w}/b_f. In the narrow channel, where its low-water width $b_{l\text{-}w}$ is close to the channel width during the flood period b_f, the ice fields are less closely connected with the channel sediments. In the melted bed $b_{l-w} < b_{f,}$ *the* riverbed banks have a slowing effect on the movement of ice fields, which creates prerequisites for ice rushing and congestion. For the Northern Dvina below the confluence of the rivers Yug and Sukhona (the area of Velikiy Ustyug, where floods are connected with congestion and the city is overflowing with water), the condition $b_{l\text{-}w}/b_f < 0.4$ corresponds to the maximum probability of formation of the congestion lock (Fig. 2.29). If the value of $b_{l-w}/b_f > 0.6$ congestion does not occur (Alabyan et al. 2003).

A more significant factor is the directional change in depths at the riffles. In the 50–80s of the twentieth century, in connection with the creation of a modern ship's passage route at the Northern Dvina riffles, dredging slots were developed annually to ensure the maintenance of depths of 1.7 m (from the design level of 90 cm at the Medvedka hydrological station) in the low water depths (before freezing). Since the stoppage in end of the 80s of the XX century of this type of regulatory work on the Northern Dvina riffles in the period 1985–1997 has increased by an average of 15–40 cm and a consistent decrease in the minimum depth, and at the design level they decreased to 1.3 m, and at some riffles—to 1.1 m. This, in turn, led to an increase in the water level by 25–30 cm. As a result, the flow through the shallows increased and the b_{l-w}/b_f decreased, which corresponds to a decrease in the channel transport

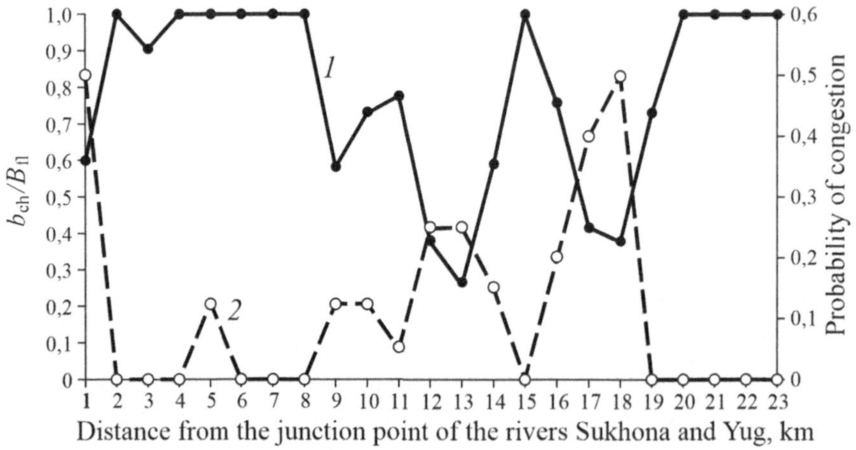

Fig. 2.29 Ratio of relative channel width (1) and ice jams formation possibility (2) on a Small Northern Dvina river (Alabyan et al. 2003)

capacity during the ice drift and increases the probability of congestion (Alabyan et al. 2003).

On the Lena River, congestions that caused a series of catastrophic floods in the city of Lensk occur at the head of Batamai Island, near which both branches make large bends (Zaitsev and Militeyev 2003). In the wide floodplain channel, the flow bypassing the icy dam washes out the side branches of the channel, which can eventually become major branches of the channel. On the Lena River (between the city of Yakutsk and the mouth of the Aldan River), in the Khaitialaakh branching in the 40s of the twentieth century, the left branch—the Khaitialaakh channel—was dry, narrow and shallow, and the water flow was distributed mainly between the Pryamoy and At-Ary channels. Its development up to the state of the main shipping channel, where it passes in different phases of the regime from 30 to 42% of water consumption in the river, has occurred due to the systematic formation of congestions that block almost all the channel, bypassing which the flow rushes into the transverse channel between the islands to the left bank of the river. The development of the Khaialaakh channel led to the complication of the entire branching, which turned from two-legged (Pryamoy and At-Ary channels) to three-legged (Fig. 2.30).

In the Northern Dvina before its merging with Vychegda, the formation of branched channels and floodplains is also a consequence of the systematic occurrence of ice jams here. Zaitsev (2003) described the formation of a transverse channel between the branches across an island several hundred meters wide on the river Kirengue, which has a pebble channel, due to congestion in one of the branches. In a meandering channel, the congestions help to straight the meanders, through the necks of which almost all the water discharge goes when the channel is blocked by ice.

When the congestion breaks out, a powerful, fast-moving wave is formed, which has a huge destructive force and produces streambed scouring. In rivers in the North-West, where winter is characterized by frequent thaws, and in areas of rivers below thresholds, large reservoirs and lakes, the formation of blockages is common, blocking the channel with its loose spongy mass and is also one of the causes of lateral rejection, the development of shallow ducts near sedges, and sometimes the formation of branches.

During ice drift, floating ice floes and interactions with the banks cause their destruction; traces of "ice erosion" in the form of cut bushes and trees are usually well visible on the tips of islands, concave banks of channel bends and branches, i.e., where the flow is pressed against the banks and contributes to the accumulation of ice floes on them. On the rivers with a boulder bed, the floating ice, passing over the flooded scourges, compacts the outcrop and turns it into a kind of "cobblestone bripple" (Lodina and Chalov 1994; Zaitsev 2003), which is a factor in stabilizing the banks. At the same time, ice floes sometimes plough the surface of such whips, destroying the sprout.

On rivers with pebbly-boulder alluvium, rocky banks, high and steep mountain slopes with developed on them landslides, rock streams, river ice screes carry a large amount of coarse debris. It arrives directly from the slopes, breaks through the ice when it freezes during freezing with rocky banks, boulder whips and shallows. Non-rolled debris often has dimensions of more than 1 m^3, sometimes reaching 10–12 m in length and up to 2 m in cross-section (Zaitsev 2003). Moving with the ice, they are dumped into the channel by congestion, rutting and ice floes on the bank. In the channel and on the banks of the riverbed, they form «lonely» stones

Fig. 2.30 Haityalaakhskaya branch (shipping branch) in three-branch channel of the Lena river, caused by ice jams, blocked other branches. Photo by A. A. Zaytsev

(Fig. 2.31), which are usually confined to places of systematic congestion. Often the petrographic composition of the stones does not correspond to the geological structure of the river valley in the areas where they are distributed in the channel. For example, on Aldan, Vitim, and Kireng, «lonely» stones are represented by granites or gneisses where the valley is cut into limestone and granites and gneisses are spread hundreds of kilometers upstream.

In extreme climatic conditions, rivers dry up or freeze. ***Drying out*** in the dry period is typical for small rivers in arid regions. Here, only transit rivers and rivers with deep incised valleys and groundwater feeding have a constant flow. The beds of drying up rivers are characterized by relatively deepened moulds (up to 4–6 m), which are not filled with sediment in the intermittent period of time when the river flow is interrupted and the ropes are not washed away; at the same time, there is an accumulation of organic remainders due to the extensive development of aquatic and coastal vegetation. Alluvium of such rivers is characterized by increased plastering (Sladkopevtsev 1963). Riffles during the low-water period are drying bripples between lake-shaped moulds covered with gypsum crust, preventing their subsequent erosion during the renewal of the river flow.

Transit rivers of arid regions have such a significant water flow that they cross the driest areas, losing only part of the water, but not completely dried up (Nile, Amu Darya, Syr Darya, Ili, etc.). For example, the evaporation layer from the water surface of the Amu Darya in its middle reaches is 162 cm per year, and within the delta evaporates about 22% of the annual runoff. Nevertheless, since its main sources of supply are glacial (44% of runoff) and snow (37%) water, the river

Fig. 2.31 Bedrocks on the bank shoal (mid-bar) in the area of Tulmudur on the Upper Aldan. Photo of the author

reached the Aral Sea (prior to the dismantling of runoff for irrigation), crossing the anhydrous deserts of the plains of Central Asia. Here, due to the huge evaporation losses, the transport capacity of the flow decreases and the river accumulates sediment, often in natural dams above the surrounding area. Smaller rivers in these conditions do not reach the main rivers or receiving basins, forming “blind mouths” (Tejen, Murgab, Zeravshan, Chu, Talas).

Freezing of rivers to the bottom leads to the fact that the soil of the bed and alluvial deposits are cemented, turning into analogues of rocks by their physical properties. The result is a deep cracking of frozen soil. During the formation of ice, debris breakups have been observed to form accumulations of large debris of various sizes, including frozen alluvial sand, covering all the valleys of such rivers. As a result, watercourses in the lowland are buried under the debris strata, lose their channel shape, but then in the process of evolution of the ice glades and as the roaming flows of the ice bed are processed, a lot of shallow watercourses are formed, which connect at the end of the ice glade (Alexeev 1997). On the 44-km long Levtyrinivayam River (North of Kamchatka) with a valley bottom width of 320 m, the ice glade is 800 m long, and the channel itself is represented by a network of streams (more than 100 in number) separated by unvegetated sedges (floodplains, respectively) of different sizes and configurations (Chalov 2005). Such ramifications are sometimes called the icy multiharm (Prokacheva et al. 1982), although we are not talking about branches, but about ducts among the midbars. The peculiarity of their development is the instability of branches and islands, the formation of new branches and ducts in floods, including due to thermal erosion processes, melting of recooled ice and other subsurface ice formation.

Freezing to the bottom is characteristic even of such a large river as the Yana in the middle reaches: at the hydrological station of Jangky in January-February there is no water runoff. Downstream this leads to sagging of ice and due to insufficient runoff rates in the estuary area of the river, salty sea water penetrates to the top of the delta.

The permafrosty soils common in the cryolithic zone rivers where their beds freeze to the bottom have a limiting effect on channel changes. The permafrost cements alluvium, which resists erosion like rock. This leads to a large width of the rivers with completely freezing watercourses in comparison with the same runoff rates in the temperate climate zone (Molchanov 1971). Often, only sandbanks dried up in the lowland and above water level in the pre-frost period are fenced with frozen frost. However, if the channel is pebbly, the processes of heat exchange of channel flow and groundwater due to the permeability of pebble alluvium ensure the formation of talik under the whole channel, including the banks (Mikhailov 1998). On rivers with sandy alluvium, permafrost is formed not only on dried banks, but also on adjacent parts of the channel. This is due to the significant moisture content of the sands and the continuing lowering of the water level in the rivers during the winter lowering season (up to 2–2.5 m on such large rivers as the Lena), which means that by the end of winter up to half and more of the riverbed area is drying up. With an ice thickness of 1.8–2.5 m, the contact of the river ice with the bottom also extends to the part of the channel where the depth from the lowest winter level is about 2 m. Besides, outside the zone of freezing of ice with the bottom between the bottom and the bottom surface of the ice there is a layer of water of increased mineralization (the upstream water coming into the channel flow, has a mineralization of 5–2 g/l with water salinity in the Lena 0.04–0.08 g/l) and subcooled to −4–5 °C, which also causes freezing of bottom soils (Waterways … 1995). In such conditions, permafrost is formed on most of the channel area, except for the narrow stripe near the spindle, where the depths significantly exceed the thickness of the ice cover. In the middle and lower reaches of the Lena River, the talik occupies only 10–15% of the total area of the channel (Tananaev 2004). It is in this narrow strip in the core zone of the interflow that the most intense changes are localized, without capturing the entire channel as a whole (Fig. 2.32).

In the erosion of coastlines formed by frozen rocks, the thermo-abrasion effect is of great importance, which determines specific forms of their dynamics in the process of flow erosion. Typically, niches are formed on rivers at the intertidal water level, resulting in large blocks of collapsing banks temporarily until they are completely thawed and washed away, protecting the bank from flow, although the average rate of bank retreat is the same as in the permafrost zone.

The indirect impact of permafrost on fluvial processes is represented through the formation of the river basin component of sediment load. On the graphs of connection $R = f(Q)$, constructed for the rivers fully flowing in the cryolithic zone (Yana, Indigirka, Olenyok) (Tananaev 2002), the flood loop significantly exceeds the flood loop with equal water discharge (Fig. 2.33). Flooding takes place at a time when the soil and grounds in the basin are frozen, the active layer has not yet formed and melt water flows down the permafrost. The sediment concentration of the water at this time is relatively low. On the contrary, in summer rainfall floods, runoff is generated when the frozen ground in the basin thaws out the most, and the water in the rivers has increased sediment concentration (the local population calls this phase of the "black water" regime). For example, according to full-scale measurements at the Yana estuary during the flood of 15–25 August 1973, the

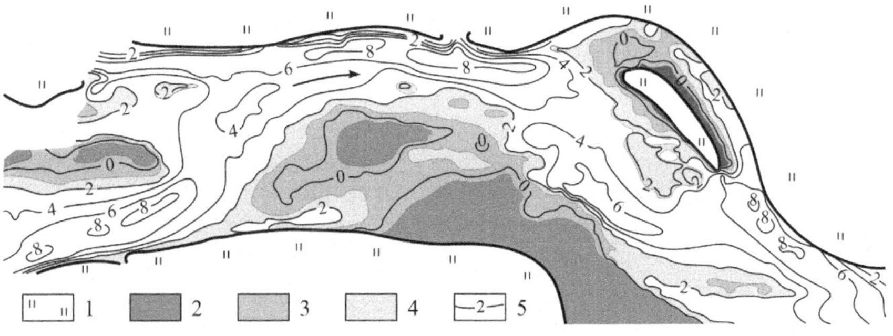

Fig. 2.32 Distribution of frozen ground in the Lena riverbed near Yakutsk (winter period 2001–2002): 1—floodplain; 2—areas of the channel freezing from the surface immediately after the air temperature passes over 0 °C; 3—the same as a result of water level decreasing in winter; 4—when the bed is freezing to the ground; 5—taliks and isobaths (N. I. Tananaev's data)

average sediment concentration of the flow ranged from 191 to 435 g/m^3, while in the flood at the same flow rate it was 25–140 g/m^3 (Korotayev et al. 1978). As a result, the riffle ripples are eroded during floods, while they become shallower during floods.

2.8 Influence of Vegetation on Fluvial Processes

Vegetation has a direct and indirect impact on fluvial processes (Table 2.11). Indirectly, it is the regulatory influence of vegetation on runoff and sediment transport rate into the river. *Vegetation in the catchment area*, especially forest *vegetation*, contributes to the reduction of maximum flood and flood discharge, as well as to a certain increase in the runoff rates of the lowland. Accordingly, forest clearing leads to a redistribution of runoff and, as a consequence, to an increase in the erosion and transport capacity of channel flows in the flood. However, the same, as well as the plowing of land, leads to a decrease in the level of groundwater, a reduction in the underground feeding and a decrease in the inter-course flow up to the drying up of small rivers. According to the data of Dedkov and Mozzherin (1996), due to the disappearance of sources, the number of 1–2 orders of rivers in the forest zone decreased by 2.2 times.

The influence of vegetation on the flow of sediment into rivers is even more noticeable. According to G.V. Lopatin (1952), the difference in the saturation of river waters by sediments with forested and non-forested catchments reaches an order of magnitude. The removal of natural vegetation, plowing of land and associated soil erosion cause excessive (in relation to the transport capacity of the flow) inflow of sediment into the rivers, which leads to siltation and degradation of small river beds.

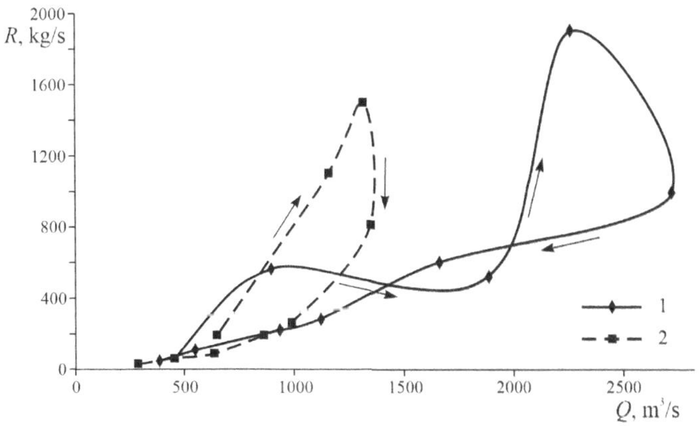

Fig. 2.33 Connection curves of suspended sediment discharge R and water discharge Q on the Yana river (Verkhoyansk city) in 1967: 1—flood; 2—summer rain floods (Tananaev 2002). Arrows shows chronological orientation of changes

The intensification of soil erosion at the turn of the XIX-XX centuries affected the shallowing of some rivers even in the middle reaches—the Oka (Berkovich 1993, Don (Polyakov 1930). In the steppe zone, as a result of siltation, small rivers of the first order disappeared completely, up to the fourth order, resulting in a total reduction of the river network by 30% (Kovalchuk et al. 1996).

The type of *vegetation on the floodplain* (forest, shrub and meadow) determines the roughness of the floodplain flow occurring during flooding, the regulation of the floodplain runoff into the spill phase and the conditions for deposition of suspended sediment load on the floodplain. If the roughness coefficient n in the Chezy-Manning formula for lowland river beds

$$n = \frac{1}{V} R^{2/3} I^{1/2} \tag{2.26}$$

(here R—hydraulic radius, I—slope) fluctuate within the limits of 0.0025–0.040, for meadow floodplains with rare shrubs and leveled surface n = 0.050, for floodplains with complex relief, shrubs and trees n = 0.100, and for floodplains overgrown with dense forest, n = 0.200 (Sribnyi 1960). Sokolov (1988, 1991) proposed to determine the floodplain roughness caused by vegetation by the following formula

$$n_v = 0.25\left(P, \frac{l}{d}\right), \tag{2.27}$$

Table 2.11 Impact of vegetation on fluvial processes

Type of impact	Location of vegetation	Type of impact	The form of manifestation in the riverbeds processes
Direct	On the banks	Bank protection function Woody debris Creation of creases, backwater and drop in levels	Reduced or increased intensity of bank erosion Destruction of riverbanks Sediment accumulation above creases, lower bed erosion and erosion during logjam destruction
	In the channel	Increased roughness of the channel	Sediment accumulation and siltation of beds
Indirect	In the river basin	Flow regulation of small rivers Regulation of soil erosion products inflow	The drying up of small rivers Siltation and degradation of small river beds Shallowing of small and medium river beds
	On the floodplain	Floodplain flow velocity mode	The development of the anabranching channel Cutoff of the bend Sediment accumulation and clarification of the floodplain flow

where P is a vegetation parameter defined by the formula

$$P = \frac{\sum_{i=1}^{m}(S_i h_i \delta_i)}{a^2 h_f}, \tag{2.28}$$

S_i—lateral projection of vegetation, m^2; h_i—height of vegetation flooded with water, m; δ_i—parameter taking into account the degree of vertical complexity of vegetation (for bushes δ_i = 0.05–1.0, for trees δ_i = 1.0, for grass δ_i = 0.1–1.0); a—linear size of the floodplain area; h_f—average depth of flow in the given floodplain area; indices i = 1, 2, 3 … corresponds to vegetation types, the number of which is m; l—distance between vegetation elements, m.

The flow velocity on a forested floodplain is much lower than on a meadow one. If the floodplain is forested, then under other equal conditions the anabranching channel is poorly developed, as weak currents do not favour the maintenance of the flow of the branches which are separated from the river. On the contrary, in meadow vegetation, if the effective water discharge passes through the floodplain, the anabranching channel is most developed. Under the same conditions, the meandering rivers form broken bends, and the loopy bend is straightened by washing through the neck of the trough until their wings come closer due to counter-erosion of the banks. When the floodplain is forested, broken bends (incomplete meandering, according to the SHI (State Hydrological Institute) classification (Kondratyev et al. 1982)) are not formed, and the cutoff of loopy ones is possible only as a result of counter-erosion of the banks on their wings.

On the surface of the forested floodplain, erosion activity of the floodplain flooding is practically impossible. On the contrary, the formation of erosional forms is possible in meadows due to whirlpools caused by uneven terrain near stand-alone trees or bushes. In places of water overflow from the riverbed to the floodplain, forest and shrub vegetation contributes to a sharp damping of the flow rate and concentrated accumulation of sediment along the floodplain edge. The width of the strip in which sand deposits are deposited does not usually exceed 50 m (Rubtsov 1982). On the meadow floodplain, sediment is distributed throughout the riverbed, causing it to grow in height. Accordingly, forested floodplains provide more lightening of the flow than grassland floodplains, which affects the transport capacity of the channel flow below the point where water is drained from the floodplains into the channel.

The impact of *vegetation on riverbanks* on fluvial processes is more complex and diverse. According to Rubtsov and Deryugin (1981), the shift of banks covered with deciduous forest at Vychegda and Vyma is 1.5–2 times less than that of grassland; at the same time, the erosion of banks covered with spruce is 25% greater than that of meadow floodplain. The latter is explained by the peculiarities of the root system of trees and its shallow depth of penetration into the ground, so that the erosion of the bank often goes below the main mass of roots. Rootstrapping and floodplain overgrowth stabilize the channels of small rivers, as the root system penetrates most or all of the terrain at lower floodplain heights, protecting it from erosion.

In tundra, the mossy or shrub cover of the tundra covers the bank scarp in a raincoat-shaped manner and prevents it from retreating.

When the river erodes floodplain banks, fallen trees lie on their ledges, accumulate along them, and are intertwined with root systems and crowns. This also, to some extent, protects the banks from erosion. However, during floods, trees that have fallen into the river are floating, and when the flow is pressed against the bank, they have a damaging effect similar to that of ice drift. In the middle and especially small rivers of the taiga zone, floating trees form a woody debris, which is an important factor in the destruction of banks. On bends and in branches of the channel, on ropes with the width of the channel up to 50 m forest creases are formed, similar to ice congestion, but in contrast to the latter they form almost permanent dams that block the channel in areas up to 1.5–2 km long and rise above the intertidal level up to 10–12 m. Small rivers sometimes have up to 20–50 creases (Zemtsov 1976). Such debris from forested areas contributes to the shallowing of rivers, as it completely disrupts the transport of sediment flows, causing them to accumulate. On larger rivers, the creases occupy part of the channel, deflecting the dynamic axis of the stream to the opposite bank and activating its erosion.

Macrophytes directly affect the *fluvial* processes of small rivers. Aquatic vegetation covers most of the bottom area, occupying all the shallow-water areas of the channel, the roughness of the channel increases by 3–4 times, and the diagram of velocities is transformed in such a way that only in the upper layers of the stream there is a noticeable current. In the rest the speed diagrams are close to the bottom speed. Under these conditions, small rivers form swampy channel represented by alternation of swampy meadows—lake-like extensions, connected by narrow shallow ducts among the thickets. The main process is the accumulation of sediment and organic residues, which further contributes to the further overgrowth of the riverbeds. Shallow water parts of the channel sometimes overgrowth is typical for large rivers: for example, the Angara riverbed below the Ust-Ilimskaya HPP.

2.9 Biogenic Factors

Biogenic factors (activity of living organisms and bacterioflora) are secondary factors of fluvial processes, but sometimes their role is quite noticeable. Bolysov (2003) singles out both indirect influence of organisms on geomorphological, including fluvial, processes and direct zoogenic relief formation, which also includes mechanical changes of channels by living organisms. These are the places of watering places for cattle, loosening sediments on the banks of the riverbeds, excreting sediments on the adjacent shallow river channels, smoothing and destroying the coastal slopes.

On small rivers in the forest zone, the morphology of the channels is sometimes completely transformed by the "hydrotechnical activity" of beavers (Sinitsyn and Rusakov 1990). Cascades of beaver dams together with beaver ponds (Fig. 2.34A) reach a length of 0.7–1 km.

The dams of the latter straightenings are also contributed to by the creation of dams by beavers (Fig. 2.34B). In the places of mass settlements, beavers break through the channels through the most stern areas of the floodplain leading to significant changes in the position of the channel, sometimes to their elongation; such channels are usually found between beaver ponds that do not form cascades and are separated by areas of the natural channel (Fig. 2.34B).

Swallow's nests in the upper part of the ledges of floodplain banks, dividing sufficiently compacted soil into separate blocks, contribute to its erosion at high water levels. Sometimes the number of nests reaches 20–25 per 1 m^2.

In rivers of spawning importance, bottom sediments are intensively cleared of silt and other small fractions by fish, resulting in changes in their grain sizes. This is due to the spawning mechanism, i.e. the introduction of eggs by the fish into the ground, during which it excites itself. When several tens of thousands of salmon spawn, up to 1000 tons of silt are carried from the river to the receiving basin (Levanidov 1968). On the other hand, the layers of the bottom ground in which the eggs are located are not subsequently silted as the eggs increase the porosity of the deposits and are therefore well washed.

The activity of microorganisms plays a special role in fluvial processes. According to the data of Dobrovolskaya (2001), in the waters of the Lower Lena, where the anthropogenic impact on the formation of microbial river runoff is minimal, the number of bacteria in the water layers, depending on the morphology of the channel varies from 2–6 to 20–30 thousand. KOE/ml (KOE is a positive unit per 1 ml or 1 g of substance); in the bottom sediments of quartz sand it contains from 264 thousand to 1558 thousand KOE; on limestone, coming out on the river bottom in the coastal zone of the channel, on the average 640 thousand KOE/g, and on guarantors in other natural conditions, the number of microorganisms exceeds several million per 1 g of rock (Dobrovolskaya 1980). The concentration of bacterial communities in small rivers in the Southern taiga zone of the UTR ranges from 0.39 to 3.2 million cells in a millimeter of water, and in the Protva River (Southern forest zone) it reaches 20 million cells/ml (Dobrovolskaya 2003).

The impact of microorganisms on rocks and minerals can be direct and indirect, and the effectiveness of the latter, as a rule, is much higher. The result of direct exposure is the destruction of rock under the influence of microbial mucus and enzymatic destruction. Indirect effects are carried out by biochemical compounds produced by microorganisms in the process of metabolism. Mineral and organic acids, biogenic alkalis, and kelatogenesis, being active chemical reagents and possessing reducing properties, are capable of chemical reactions with practically any mineral (Dobrovolskaya 1980). Therefore, the presence of microorganisms in running water, on rocks and on sediment particles causes their destruction by biochemical weathering processes. Products of microorganisms (metabolites), both liquid and gaseous, cause disintegration of rocks. The intensity of their destruction depends on the qualitative and quantitative composition of microorganisms living in the water flow, pores and cracks of rocks. As a result, even such a stable mineral as quartz has an output of SiO_2 molecules under the influence of microorganisms of 18.8 mg/l for 100 days. At the same time, the water flow washes away both the

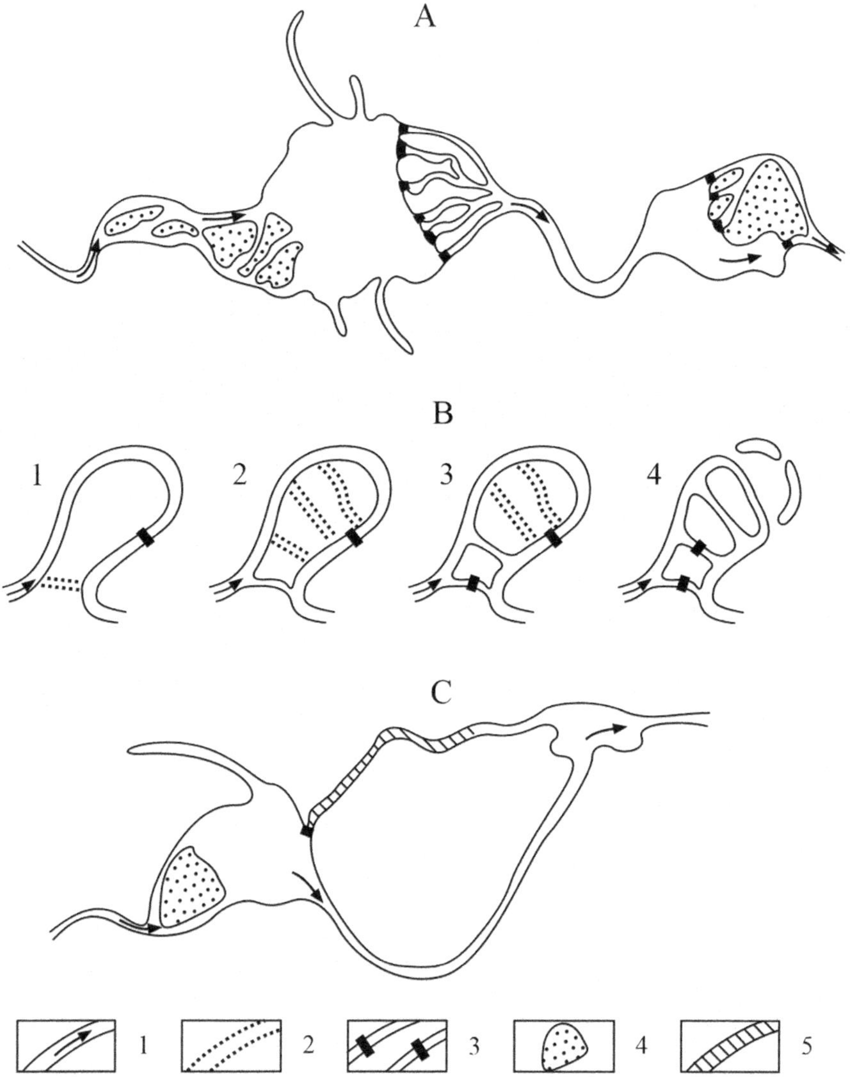

Fig. 2.34 Changes in the morphology of channels of small rivers in forest zone in response to river beaver activity: **A**—cascade of beaver dams and ponds; **B**—bends straightening along the beavers' trails; **C**—channel position change after beaver channel creation. 1—channels and ponds; 2—beaver's trails on the floodplain; 3—beaver dams; 4—alluvial depositions; 5—dried up old channel (Sinitsyn and Rusakov 1990)

products of biochemical weathering and the products of the bacteria themselves, constantly exposes the rock for insemination with new microorganisms. As a result, there is a constant destruction of the rocks, which is formed with chemical weathering and other effects on the rocks, causes the incision of the rivers in the strongest rocks.

In sub-aerial conditions the products of destruction and vital activity accumulate, leading to death of microorganisms themselves and, finally, to decrease of intensity and termination of their destruction under the influence of this factor.

Similar processes of biochemical weathering occur in the sediment strata transported by the water flow. Their intensity is closely related to the mechanical abrasion of particles. During their transportation by sliding, drawing, rolling and impact of particles, the formation of finer fractions and exposure of fresh surfaces due to the continuous removal of weathering products by the flow. Particle surface renewal contributes to the biochemical activity of microorganisms, because together with the upper destroyed layer, metabolic products and weathered layers combined with nutrients are removed from the particles. Thus, the presence of microorganisms in the water and on the surface of the sediment particles contributes to the change of their grain sizes.

2.10 Slope Processes

Landslides, collapses, and mudflows are primarily sources of sediment inputs to rivers. They play a particularly important role in the formation of sediment in mountainous areas. In the basin of the Bzyb river (Western Caucasus), in the geological structure of which about 70% of its are rocky (Makkaveev et al. 1968), in the large ripples descending to the rivers, about 50 thousand m^3 of debris material is concentrated. According to Khmeleva et al. (2000) up to 27,000 m^3 of debris enters the riverbeds each year through direct contact with debris. If we assume that the annual bedload transport rates at the estuary is 205,000 tons (Khmaladze 1978), which corresponds to about 100,000 m^3, the debris is a supplier of almost 25% of the bedload flow.

On the upper Aldan, in the Aldan plateau (Tommot Mountain—the mouth of the Uchur River), the source of pebbly-boulder sediment is collapses, mudflows and stone runs. According to the estimates of Kalinin and Kirik (1983), landslides and debris are supplied in its course annually in an average of 75 thousand m^3, and kurums—30 thousand m^3 per year. Landslides and embankments cause the inflow of debris from the steep slopes of the Volyn-Podillya Upland into the Dniester River channel in the area between the Dniester and Dubasari reservoirs in average 19.5 thousand m^3 per year (Kalinin 1987). This results in a pebble composition of the load, the particles being poorly pelletized.

Collapses, especially on mountain slopes, cause the formation of debris lakes. Most of them exist for a very short time, as the dam that creates them is washed away by the overflow and the lake descends. Sometimes this is accompanied by the formation of mudflow due to the almost instantaneous descent of the lake. This was the case, for example, with Lake Issyk in the Zailiyskiy Alatau, which existed for less than a century (Lake Sarezskoye in the Pamirs and Lake Amtkeli in the Caucasus belong in this respect to the "long-livers"). Other debris lakes have been preserved for many centuries and are gradually filled with lakes and cause a specific shape of the longitudinal profile of the river (Berkovich and Chalov 1969) with alternation of sections with slopes of channels from 1–2‰ and less (within the former lake basin) to 50–70‰ on the lower slope of the debris dam.

Large landslides, formed by cohesive plastic rocks, block some of the channel and for many years have a directing effect on the flow until it is washed away by the latter. This leads to the erosion of the opposite, including the floodplain coast, the formation of local expansion of the channel and the formation of single branching. Thus, the riverbed of the Kuban River was branched (Fig. 2.35). A similar impact of landslides on the riverbed is also present in the loess-like structure of indigenous landslide banks. In these cases, it is determined by the huge masses of soil that crawled into the river. During their scouring, the channel can undergo significant changes that affect its morphology and river channel regime, sometimes over many years. For example, on the Ob River below the mouth of the Charysh River in the 40s of XX century, a huge landslide on the left bank of the river about 100 m high blocked the channel more than a third of its width. Downstream, the riverbed was shallowed by landslide erosion products, and the stream developed a right, previously dry branch behind Belovsky Island, which had been navigable for more than 10 years. In the early 1980s, a major landslide blocked the Prut riverbed in Moldova. This led to its shallowing upstream, where a several kilometers long retaining zone was formed, and downstream, where the products of landslide erosion came in.

The influence of other slope gravitational processes is characterized by different intensity in different natural conditions and, accordingly, has different effects on the formation of sediment and in the channel mode of rivers. The size of the rivers also matters: what is very important for small rivers is almost invisible for large ones. In regions with a pronounced continental climate, slope processes have a decisive impact on the channels of small rivers. Simonov (1972) speaks even about "oppression of fluvial processes and water flow suppression", singling out in the Far East and Baikal region the valleys-stone-runs and valleys-mari. Here, first order rivers often do not have clear channels, the water flow appears only during floods and constantly changes its position on the flat bottom of the valley. In the mountainous areas, large debris flows to the bottom of such valleys, and at high gradients, the channel is a linear stone-run represented by weakly rolled boulders and sharp-coal fragments that bury the stream. When developing on the slopes of solifluction, a large amount of fine-grained soil enters the river channels. In such cases, wide mari valleys are formed in which the channel forms a narrow, in some places rectilinear channel 0.5 m deep and no more than 1 m wide, often overgrown with grass. Filling the channels of small rivers (of first order) with solifluxation material is also characteristic of the tundra zone, where the presence of permafrost contributes to this.

Fig. 2.35 Channel branching on the Dniestr river near Kriulyany village, formed in the expansion after a big confluence. Photo of the author

2.11 Human Impact on Fluvial Processes

River management activities are multifaceted, and different types of activities have varying impacts on channel processes. However, the majority of them, to a greater or lesser degree, breaks the existing natural balance in the system "basin-flow-riverbed" and thus leads to changes in channels, directionality and intensity of channel changes, the risk of emergence of new anthropogenic hazardous representations of fluvial processes and the need to protect banks and engineering objects from their destruction, deepening of rivers for their transport use, ensuring the operation of water intakes, etc. At the same time, anthropogenic changes in fluvial processes are often related to the objective need to use rivers and their resources to ensure the life and livelihood of people.

The variety of types of interaction between economic activities and engineering structures and fluvial processes makes it necessary to type them, which in turn allows to assess the consequences of this interaction. For the first time, B. F. Snishchenko proposed a substantiated classification of the impact of engineering structures on river channels (Kondratyev et al. 1982); it is based on the separation of two classes—active and passive. The former modifies river channels themselves (e.g. quarries, dredges, etc.), while the latter influence river channels passively, by their presence (e.g. bridge piers, quayside walls, etc.). However, along with this approach, the classification of types of impacts of economic activities can be approached from other positions, based on the nature of interaction between fluvial processes and economic activities. Economic activity can change both the channel itself, the shape of which and the channel topography are factors of fluvial processes, and the factors of fluvial processes (e.g., water and sediment yield). In the latter case, the channel is transformed, adapting to the new conditions. On the other hand, structures not only influence fluvial processes; their functioning and safety depend on channel changes, their correct accounting and forecasting. This approach makes it possible not only to classify the types of economic activities, but also to determine the objectives of fluvial processes assessment in environmental management projects.

Typically, the impacts on river channels and fluvial processes are combined by a common concept of "anthropogenic factors". For the sake of brevity, it will be applied even lower, although in essence it is not, since the effects of economic activity have an effect directly or indirectly on the transformation of natural factors —water and sediment yield, as well as on the shape of the channel itself and the riverbed terrain. Since it is now difficult to find rivers for which the factors of fluvial processes have not been changed to some extent, the latter should be called natural-anthropogenic.

Table 2.12 provides a classification of economic activities in basins and on rivers by their interaction with fluvial processes, their impact on them, and the nature of tasks in the design and construction of structures, as well as hydrotechnical and water management activities. The first type of impacts are those measures and structures that affect the factors of fluvial processes (change the flow of water, its seasonal distribution, increase or decrease the inflow of sediment into the river). These include waterworks, industrial, municipal and irrigation water intakes, reclamation and forestry activities in the basin or river valley. The role of the waterworks is not limited to the influence on the factors of fluvial processes. The reservoir under the hydroelectric complex is an artificial basis of erosion for the upstream section of the river and represents a reservoir where river load is accumulated. As a result, sediment load is interrupted on the river and an area above the reservoir is created in which fluvial processes develop under the influence of alternating pressure and regressive accumulation of sediment.

The same type of impacts includes artificial collapse of the channel, which separates the floodplain from the river, prevents its flooding in the floodplain, increases the flow of water in the river channel in this phase of the regime, changing the conditions of its impact on the channel.

Berkovich (2001) proposed a three-level classification of engineering structures and measures, considering the types of economic activities to spread the effect of their impact on fluvial processes (regional—on the entire river system, the river or its long section; local—on a separate, relatively short section of the river; on the only one profile.

The duration of the impact and the sign of changes in orientation of the channel changes caused by them—erosion of the channel or accumulation of sediment (Table 2.13). At all levels of each classification attribute is related to the direct or indirect impact of the structure (or measure) on the channel. Directly related to direct, usually mechanical change of channel, indirect influence on the channel

Table 2.12 Types of economic activities and their connection with fluvial processes

Types of interaction of engineering and water management facilities and measures with fluvial processes	Types of events and buildings	Tasks of studying fluvial processes			
		Limitation of channel changes	Accounting for channel changes	Forecast of channel changes	Environmental impact assessment
Changing factors in fluvial processes	Hydrosystems	–	–	+	+
	Land reclamation and forestry measures in the river basin	–	–	+	+
	Artificial collapse of the bed	+	–	+	+
Influencing channel morphology and changes	Dredging and correction works on navigable rivers	–	+	+	+
	Building materials quarries	–	+	+	+
	Embankment construction	+	-	+	+
	Channel regulation in urban areas and in areas of large industrial enterprises	–	+	+	+
Influenced by channel changes	Bridge crossings	–	+	+	–
	Water intakes	+	+	–	–
	Underwater crossings	–	+	–	–
	Onbank Engineering Facilities	+	–	–	–
	Quay walls	–	+	–	–
	Bank protection from erosion	+	–	–	–
	Recreational zones near the river	–	+	–	–

through the change of external (in relation to the channel as a form of relief) factors of fluvial processes. On the other hand, objects and structures on rivers are affected or dependent on fluvial processes, and for their protection and normal functioning require the construction of regulatory and bank protection structures, special measures, i.e. direct technogenic impact on the channels.

Direct impact on the channel leads to artificial transformation of the cross-section, redistribution of water flow rates and flow rates across the width, creation of anthropogenic forms of channel relief. At the same time, the effect of the impact on the fluvial processes of engineering structures, measures, technical means, etc. depends on the specific conditions in which the river channel is formed, the size of the river and the relationship between the types of anthropogenic impacts, while simultaneously exhibiting several of them. Thus, direct and indirect impacts on the factors of fluvial processes, interacting with each other (Fig. 2.36), create conditions of "double action". If direct impacts directly change the channel and fluvial processes, and indirect ones cause changes in the natural conditions of channel formation, the "double action" changes the nature of fluvial processes, both directly and by changing natural factors. The peculiarities of the same influences on river channels and floodplains are often the opposite. For example, by altering the channel by creating reservoirs, the fluvial processes are influenced mainly indirectly by the regulation of water and sediment flow (in the case of the latter, its full or partial interception by the reservoirs) for the lower tailrace, which results in the creation of a condition $W_{tr} >> W$ and the emergence of an erosion basis for the section above the reservoir. At the same time, in the lower tailrace, the unsteady flow regime, which is one of the reasons for channel erosion, should be attributed to the direct consequences of the hydraulic system operation. On the other hand, the multidirectionality of the impact of the dam on the fluvial processes is manifested not only through the differences in changes in their factors in the downstream (flow regulation) and upstream of the reservoir (creation of the base level). Thus, streambed scour in the lower tailings of the hydroelectric complexes is usually accompanied by a "wave of accumulation", passing before the front of the scour and caused by the inflow of erosion products to the downstream part of the channel. Thus, accumulation is observed both above the reservoir and below the dam, although at some distance from it, but the genesis of sediment accumulation in each case is fundamentally different.

Anthropogenic changes in channels related to economic activity in catchments are regional, as they cover all small rivers within the regions or large rivers over a considerable length of time. They are indirect, as they are determined by the transformation of the water flow regime and the conditions for the formation and flow of sediment into rivers. The same characteristics are also present.

The consequences of large hydrotechnical construction, which disrupts the natural hydrological factors of fluvial processes and interrupts the transit flow of sediment: transgressive deep erosion in the lower tailrace and regressive accumulation of sediment above reservoirs.

Bunding dams, which are built along the riverbeds over a long period of time to prevent flooding of floodplains when used in agricultural production (rivers in the

Table 2.13 Classification of engineering structures and measures by duration of impact, direction of changes in channel changes caused by them and impact scales (according to Berkovich (2001))

Type of impact (activity)	Duration of impact			Direction of channel changes			Impact scale		
	Constant	Longstanding	Temporary	Erosion	Accumulation	Neutral	Regional	Local	In a river profile
Direct									
Dams of waterworks (reservoirs)	+			+	+		+		
Correction structures (continuous correction)	+			+	+			+	
Bridge crossings					+				
Bunding dams	+			+					+
Straightening the channel	+					+	+	+	
Dredging (capital)		+		+				+	
Channel Careers		+		+			+	+	
(a) solid		+		+			+		
(b) single									
Laying underwater utilities			+	+				+	
Individual buildings (elective rectification)			+			+			+
Quay walls, bank protection structures		+	+						+
Dredging (development of slots along the ship's routes)	+					+			+
			+	+					+

(continued)

Table 2.13 (continued)

Type of impact (activity)	Duration of impact			Direction of channel changes			Impact scale		
	Constant	Longstanding	Temporary	Erosion	Accumulation	Neutral	Regional	Local	In a river profile
Development of alluvial mineral deposits									
Indirect									
Dams of waterworks (reservoirs)	+			Downstream side	Above the reservoirs		+		
Agrotechnical activities		+			On small rivers	On large rivers	+		
Deforestation		+			On small rivers	On large rivers	+		
Drainage withdrawal					+			+	
Runoff increase	+			+				+	
Dams	+			+				+	
Channel Careers	+		+	+				+	

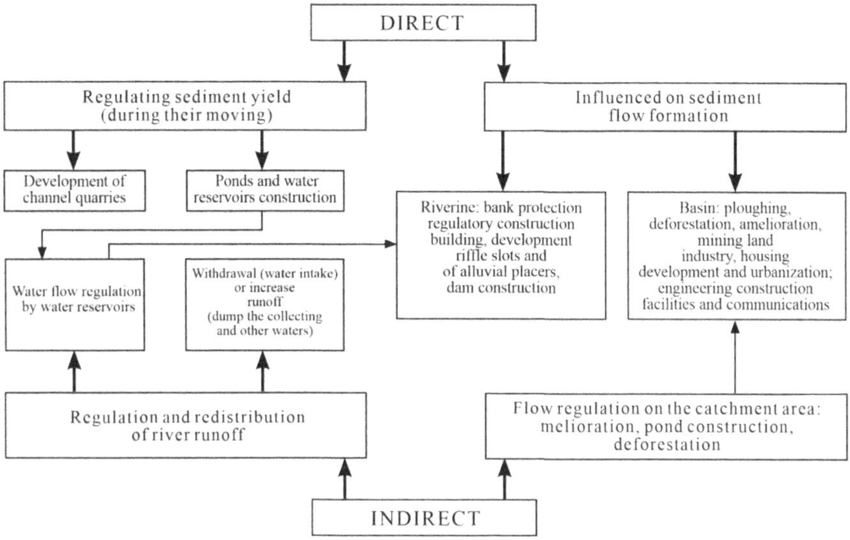

Fig. 2.36 Interaction of direct and indirect anthropogenic factors of fluvial processes

South of Russia, Ukraine, and Moldova), to control floods (Terek, rivers in China—Huang He, Yangtze in the lower reaches) also have a regional impact on fluvial processes. The total length of dams along the Huanghe River bed in the lower reaches exceeds 1320 km. At the Yangtze River, the bunding system (Fig. 2.37) is, as at the Huang He River, a complex of structures that create a peculiar cellular (parallel transverse) structure in the riverside zone to improve the safety of the territory in the event of the destruction of the coastal dam. Their total length is 36,000 km. Dams 9000 km long are also erected along the tributaries of the Yangtze (Chalov et al. 2000). Dams contribute to the concentration of maximum water discharge in the channels, which changes the nature of fluvial processes, although the shape of the channel and its relief do not change radically. Bunding is usually a way of pushing the channel into the watercourses. At the same time, due to the concentration of all water consumption during the "shutdown" of the floodplain in the channel itself, the size of the channel and its width are increasing; the latter is accompanied by an intensification of riverbed scouring and threatens to destroy the dams themselves, necessitating the need to combine the construction of dams with bank protection structures. In conditions of very high sediment load, when $W \gg W_{tr}$ bedding by collapse dams, causing insufficient transport capacity for flood discharges, does not lead to bed incision due to an increase in the specific flow rate of water, but, conversely, contributes to the accumulation of sediment (Chalov et al. 2000). This manifests the ambiguity of the influence of the same anthropogenic impact depending on other natural factors and forms of manifestation of fluvial processes.

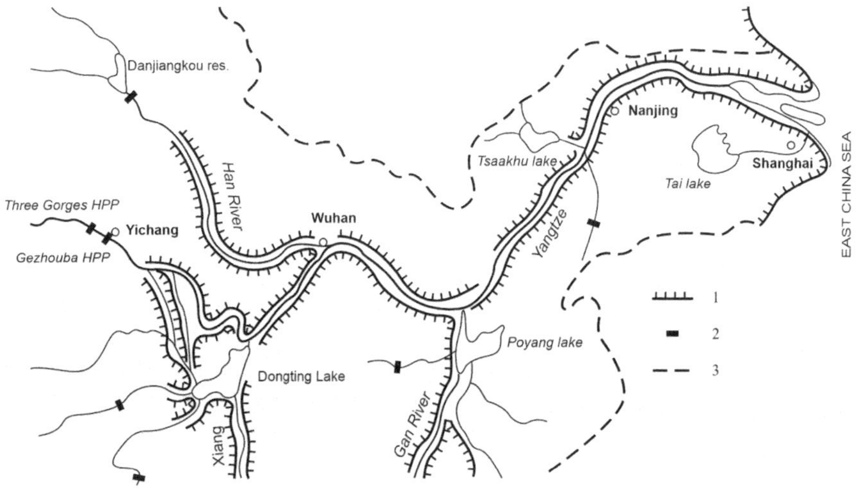

Fig. 2.37 Leeve dam system in the middle and lower current of the Yangtze river: 1—dams; 2—hydrosystems; 3—basin boundaries

In many cases, dams protect small areas from river spills, usually in urban areas. In this case, the effect of their impact on the channel is local.

Significant regional impact on fluvial processes is caused by runoff diversion to large reclamation canals, as well as for industrial and municipal needs. This is usually accompanied by accumulation of sediment and general riverbed shallowing, including the transformation of rivers into small watercourses (Amu Darya, Syr Darya); in China, even such a large river as the Huang He due to water withdrawal and the operation of reservoirs (they are filled in the winter lowlands) in the lower reaches of the river completely dries up in winter, not reaching the sea. Shallowing of the rivers and intensive accumulation of sediment below the water intake are characteristic of many lowland rivers in the North Caucasus (Kuban, Lower Terek) and Central Asia (Ili, Chu, Zarafshan). Changes in the channel relief depend not so much on the size of the river as on the volume of water intake and the amount of sediment load. In 1962–1968, for example, the Amu Darya bed markers below the headworks of the Karakum canal increased by 1 m; at the first 15 km below the structure, according to estimates by Shapiro (1973), about 8 million m^3 of sediment was deposited; water consumption decreased by 120–300 m^3/s.

At the same time, small volumes of water withdrawals from rivers, which constitute the first percent of their annual flow, especially from reservoirs, do not have a significant impact on fluvial processes.

Rivers receiving water from irrigation canals or inter-basin flow redistributions when solving the problem of water supply to mega-cities (the Moscow River) and urban agglomerations, rivers whose channels are used as irrigation or flow transfer canals, are characterized by an increase in the carrying capacity of flows due to a sharp increase in water discharge. The natural channels of such rivers are unable to pass the increased water discharge. Therefore, deep and lateral erosion is activated in such channels, the rivers cut in, and their channels turn into canyons. Thus, the Kalaus River after its connection with the Big Stavropol Canal crashed almost 5 m, the Egorlyk River after the beginning of releases from the Nevinomyssk Canal—10 m (Karasev 1975). Following the incision of the main river, even low-water inflows into it are intensively cut in.

Many economic activities have a local impact on fluvial processes; however, with their abundance and continuous location along the river, the impact of anthropogenic loads is so strong that changes in river channels become regional in nature. These impacts include bed quarries where sand and gravel are extracted as construction material. The development of individual quarries has a local impact on the channel, which, when properly located, is not accompanied by changes in the river channel regime, and the quarries themselves are quickly (over several years) filled with sediment.

Mass development of channel quarries results in the irrevocable removal of such an amount of alluvium from the bed, which significantly exceeds the annual flow of bed material load, and often extracts ancient alluvium, which underlies the modern bed. In this case, the impact of the quarries on the channel has a regional character—it is represented in a significant change in the morphometric and hydraulic characteristics of the channel and flow, channel scour, which are higher and lower quarries spread transgressively or regressively along the river. This results in lower water levels in the river, an end to floodplain flooding and the concentration of flood flow in the river channel.

Development of alluvial mineral deposits in small semi-mountainous and mountainous rivers changes beds and floodplains in no less and often on a large scale (Khmeleva et al. 1995; Chalov 2005; Makhinov and Makhinova 2006). In this case there is a mechanical change not only in the riverbed, but also in the floodplain, and even low terraces, up to their complete destruction. Sediment concentration of the water sharply increases along the whole length of the river: on the rivers near the Sea of Okhotsk it reaches 2 kg/m^3 in floods, which is hundreds of times more than the background values (Makhinov and Makhinova 2006). The increase in sediment concentration occurs even on large rivers, starting from inflows of tributaries flowing through mining areas: for example, on the Aldan above and below the mouth of the Tommot river (Aldan gold mining district) during the low-water period the sediment concentration of water, respectively, is close to 0 and reaches several tens of g/m^3. At the site of the dredge workings in rivers and floodplains, anthropogenic water bodies are formed, which reduce the average annual and especially minimum water consumption; when the flow of alluvium products is washed away and eroded by the stream of alluvium products, the composition of alluvium is enlarged.

Track works at continuous rectification of rivers on extended sections change the shape of the cross-section and the relief of the channels on the rolls. This is achieved by increasing the average depth, completeness of the living section, reducing the width of the channel, the destruction or rejection of side channels, the formation of artificial forms of the relief of the channel at the site of dumps. At the same time, the coarseness of bed material load is increasing, which helps to increase the stability of channels. Capital remedial works include deepening of the branches, which are perspective with respect to channel changes and ship's route, cutoff of bend, narrowing of the channels by the systems of dams and semi-ponds.

The volume of impact of dredging and remedial works on fluvial processes depends to a large extent on the extent to which their design takes into account the natural trends in channel development. Therefore, the design of such works is usually carried out on the basis of the positive (in respect of the finished result) effect of the water flow itself, its fluvial process driver (Engineering of ship passages … 1964; Berkovich and Chalov 1993).

Capital corrective works lead to stabilization of channels, increase of depths on rivers, fixation of the main stream of rivers along which the fairway runs. Dredging and correction work does not result in irrevocable removal of alluvium from the bed, but, as a rule, takes into account the natural regularities of channel changes. Positive role of track works on the rivers was especially evident in the 90s of the XX century, when the dredging was sharply reduced or, on some rivers of Russia, completely stopped. This has led to an increase in bank erosion, shallowing of the channel and, as a consequence, an increase in the likelihood of ice congestions and associated floods, siltation of spawning grounds and wintering holes, etc.

However, in case of violation of the technology of track dredging, excessive deepening of channels, elimination of rolled products regulating the level regime, performance of continuous correction of channels without taking into account the orientation of channel changes, aimed at their canalization out of connection with them, there is an artificial transformation of the channel, which acquires a new morphological shape and mode of change.

Structures and measures that have a local impact on river channels do not change their shape and do not affect channel changes. But they themselves are more affected by channel changes than others and require that they be accounted for, predicted and preventively managed. Incorrect decisions in these cases lead to the emergency state of engineering facilities, difficulties or termination of their operation and, as a consequence, the need to take measures that are objectively directed against the operation of the river itself.

Rivers in urban areas occupy a special place in the interaction of fluvial processes with economic activity. Here, along with the concentration of structures and various activities, there are a number of specific phenomena (Borovkov 1989) associated with the influence of wastewater, contamination of channels and their overgrowth. Often these processes are of a physicochemical nature, which results in a new type of anthropogenic river sediments and new conditions of sediment transport. The interaction of all types of structures, activities and phenomena leads to unforeseen and difficult to predict consequences, exacerbated by the fact that the design of engineering structures is conducted without taking into account the mutual influence of economic activities on the fluvial processes.

Assessment of the degree of impact of urbanization on river channels and fluvial processes on a 5-point scale is presented in Table 2.14. At the same time, 0 points characterize the absence of changes in river channels under the influence of the city,

1 point—the emergence of individual changes associated with bank protection, urban water intake, dredging to ensure water approaches to the pier, the channel remains in its natural state. A 5-point assessment is responsible for the elimination of the river ecosystem, with channel changes often spreading far beyond cities downstream. For example, hail industry siltation of the Insara River is observed for about 100 km from the city of Saransk to the inflow into the Alatyr River (Berkovich et al. 1998a, b).

Especially strong anthropogenic press in cities is exposed to the channels of small rivers, which are either buried and climbed into pipes, or turned into sewers, concrete trays, etc. Mechanical changes in the channels of small rivers, their transformation into dumps of household waste and industrial waste is also typical for rural settlements and urban-type settlements and timber farms. Large rivers are also significantly affected by cities, especially the largest and largest ones. Their beds often lose their natural morphological appearance. These are Yenisei in Krasnoyarsk, Ob in Novosibirsk, Irtysh in Omsk. Nevertheless, their resistance to anthropogenic loads is quite high, and as a result, as a rule, complete transformation of fluvial processes on large rivers does not occur. For example, at the Ob, in the area of Barnaul, despite two bridge crossings, a port, ditches, several dozen different engineering structures, including water intakes, quarries of building materials and dredging works, the channel has practically retained its pattern, and its changes have received specific development only near the bridge crossings and under the influence of channel quarries (in both cases it is a question of deep erosion that led to the reduction of levels).

On the largest rivers, the influence of large cities on the fluvial processes is usually relatively small (Lena near Yakutsk, Amur near Khabarovsk, Volga near Astrakhan and Nizhny Novgorod). But even here it is possible to significantly change the direction of channel changes under the influence of activities related to the functioning of the city on the river. On the Lena River near Yakutsk (Fig. 2.38), the design of floodplain reclamation (increasing its marks) to expand the development area in the area adjacent to the center of the city, it was assumed to develop quarries in the left dying river branch. The volumes of excavated soil were planned to be so large that it could, according to the forecast, lead to the development of this branch and a complete restructuring of the branched channel. At the same time, the area to be constructed in the reclamation area will be subject to erosion and will require bank protection.

Many medium-sized and small towns do not have a noticeable impact on the largest rivers, while many large rivers are usually not significantly affected. Exceptions are those cases when river channels are fundamentally altered due to hypertrophied influence of one or a small number of anthropogenic factors (quarries of building materials, tail-water of hydroelectric power plants, their joint influence). Such is the complete transformation of the Tomsk riverbed in Tomsk under the influence of channel quarries (Berkovich et al. 1998a), the Ob channel in the lower reach of the Novosibirsk hydroelectric power station, which affected both the influence of the hydroelectric installation and the development of quarries, strengthening of banks, dredging, corrective structures, bridge crossings, etc.

Table 2.14 Assessment (in points) of the impact of urbanized areas on river channels and fluvial processes (Chalov and Ruleva 2001)

Points	Influence of the city on the channel
0	No influence
1	Separate regulatory structures, bank reinforcement, water intakes, shipboard dredging. The riverbed is kept in its natural state
2	Embankment in the central part of the city, dams, dikes, berthing buildings; bridge crossing; water intakes; bank protection, capital dredging in the channel to ensure water approaches. The banks have partially lost their natural appearance, but the channel has not been changed
3	Embankments or bank reinforcement along 20–60% of the coastline within the city, partial floodplain reclamation for construction or bunding, changes in floodplain landscapes due to floodplains overlapping with dams, continuous straightening of the channel for navigation, water intakes, bripples and underwater crossings, in-stream quarries, barrages and dams, overlapping of side branches, occurrence of anthropogenic sediments in the channel. Partial changes in the morphology of the channel and its relief, loss of natural appearance of the river, disruption of the river ecosystem
4	Solid embankments and bank reinforcement, mass quarrying, loss of recreational qualities in the city by the river, pollution of bottom sediments, bripples, water intakes, underwater crossings, mass correction facilities in the channel, dams, etc. Total loss of natural appearance of the riverbed, partial elimination of the river ecosystem
5	Complete channel canalization, river intake into pipes, continuous location of engineering structures along the banks, mass bripples and underwater crossings, dams regulating the flow, accumulation of anthropogenic sludge, urban and industrial development of the floodplain, its bunding or landfill. Complete elimination of the river ecosystem, spreading the influence of the city downstream

Active impacts indirectly, through changes in the fluvial processes of mainly small rivers, include deforestation and plowing of catchment areas. At the same time, the most common phenomenon in the steppe and forest-steppe zones is the siltation of small rivers as a result of soil erosion in the ploughed catchments. The rate of sediment accumulation in the channels is 1.5–3.0 cm per year. As a result, many channels of the first (streams) and second order were generally buried under a layer of sediment; on the larger rivers the runoff remained, but the channels themselves turned into swampy ones, consisting of a chain of small lakes-swampy meadows and narrow overgrown streams between them. Pollution of riverbeds with nutrients (primarily phosphorus), as a result of mineral fertilizers being applied to soils and their subsequent entry into rivers, contributes to the development of aquatic vegetation, strengthening of the processes of siltation of beds and reduction of water runoff. Because of this, many small rivers have ceased to exist.

In the forest zone, deforestation leads to a decrease in the level of groundwater, drying up of springs and drying up of rivers in the low-water. This phenomenon is also accompanied by the accumulation of sediment in the channels and floodplains of small rivers, which increases their degradation.

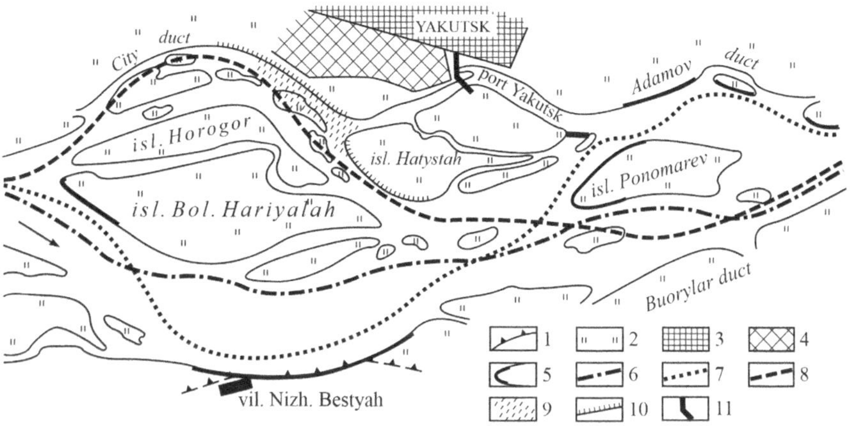

Fig. 2.38 The scheme of Yakutsk branching of the Lena river: 1—root bank; 2—floodplain; 3—Yakutsk city territory; 4—city district on accumulation territories; 5—nowadays erosion of river banks; 6—possible position of the main current of the river in the natural development of the channel; 7—current situation of the main river current supported by dredging; 8—possible position of the main river flow during quarrying in the Gorodskaya branch; 9—position of the planned quarry in the Gorodskaya branch for the accumulation on the territory; 10—areas of potential bank erosion in the Gorodskaya branch during quarrying in it; 11—existing dam protecting the river port

Reclamation of the territories adjacent to the rivers has a controversial effect. Drainage reclamation is carried out mainly on floodplains and low floodplains. Reclamation canals reduce the level of groundwater, drains bogs, and introduces them into agricultural circulation. However, the drainage of wetlands reduces flow regulation, and many small rivers dry up in the low-water period. Secondly, in drainage systems, small rivers are often used as main canals for which they are straightened. At the same time, the rivers begin to run into each other, which leads to the incision of the lower reaches of the reclamation canals tied to them. Flushing products are deposited in the channels of small rivers and canals, which leads to their siltation (Chemeris 1992).

Regulation of small rivers by pond cascades for irrigation or small hydropower generation due to the accumulation of flood runoff and water discharge in the low water phase does not allow the rivers to dry up in the dry phase of the regime in conditions of excess sediment load—products of accelerated soil erosion. On the other hand, the deaf earth dams, which in the steppe zone are used to block small rivers in mass quantities in order to accumulate water for the period of arid low water level, completely stop the interflow between the dams. In extreme floods and floods, dams often break through, and the mass of scour products is accumulated downstream, contributing to the silting of beds and floodplains.

Literature

Alabyan, A.M., Alekseevsky, N.I., Berkovitch, K.M.: Sediment transport in the downstream area of Novosibirsk HES (in Russian). In: Proceedings of WestSibNIGMI. 98, pp. 79–82 (1992)

Alabyan, A.M., Alekseevsky, N.I., Evseeva, L.S., Zhuk, V.A., Ivanov, V.V., Surkov, V.V., Frolova, N.L., Chalov, R.S., Chernov, A.V.: Genetic analysis of the spring flooding causes of the Small Northern Dvina Valley in the area of Velikiy Ustyug city (in Russian). Soil erosion and fluvial processes, Ep. 14, pp. 104–130. Moscow State University Publishing House, Moscow (2003)

Alekseev, V.R.: Morphodynamic features of glacial sections of river valleys. Hydrology and geomorphology of river systems, pp. 207–208. SO RAS, Irkutsk (1997)

Alekseevskiy, N.I., Gaikovich, A.B.: To calculate the flow of incoming load on the unexplored rivers in the period between periods. Meteor. Hydrol. (8), 96–102 (1987)

Alekseevskiy, N.I., Mel'nik, O.V.: Structure and peculiarities of the relief development of the gravel-pebble and boulder channels (in Russian). Geomorphology (3), 59–65 (1991)

Alekseevskiy, N.I., Zaitsev, A.A., Chalov, R.S.: Balance of river load, channel changes and possibilities of regulation of the branched-out channel of the largest river (on an example of the river Lena near Yakutsk). In: Proceedings of AVN, 3, pp. 90–108. M (1996)

Alekseevsky, N.I.: Formation and movement of river load, p. 203. Moscow State University Publishing House, Moscow (1998)

Alekseevsky, N.I., Chalov, R.S.: Movement of Sediments and Fluvial Processes, p. 171. Moscow State University Publishing House, Moscow (1997)

Alekseevsky, N.I., Sidorchuk, Yu.A.: Accelerated erosion in landscapes disturbed by mining activities (by the example of the Omoloy and Yana river basins). Ecological problems of soil erosion and fluvial processes, pp. 187–198. Moscow State University Publishing House, Moscow (1992)

Babak, Yu.V., Turikin, L.A., Chalov, R.S.: Sediment flow, effective water discharge and channel patterns in the Oka basin. Geomorphology (4), 57–71 (2000)

Babinski, Z.: Procesy korytowe Wisty ponizey Zaporu wodnei we Wloclawku. Dok. Geogr. Z. 1–2. 94 s (1982)

Baryshnikov, N.B.: River floodplains, p. 152. Hydrometeoizdat, L (1978)

Baryshnikov, N.B.: Morphology, Hydrology and Hydraulics of Floodplains, p. 280. Hydrometeoisdat, L (1984)

Baryshnikov, N.B.: Influence of Effect of Interaction of Floodplain and Channel Flows on Transport Capacity of Channel Flow. Erosion and Fluvial Processes, 4, pp. 141–151. MSU Publishing House, Moscow (2005)

Baryshnikov, N.B., Popov, I.V.: Dynamics of channel flows and fluvial processes, p. 456. Hydrometeoizdat, L (1988)

Beliy, B.V., Borsouk, O.A., Ivanov, V.V., Lodina, R.V., Matveev, B.V., Nikitina, L.N., Chalov, R.S., Chernov, A.V.: Conditions of formation, morphology, dynamics and regulation of the middle and lower Vilyui riverbeds (in Russian). Soil erosion and fluvial processes, 9, pp. 146–175. MSU Publishing House, Moscow (1983)

Berkovich, K.M.: Modern transformation of the upper Oka longitudinal profile (in Russian). Geomorphology (1), 43–49 (1993)

Berkovich, K.M.: Geographical analysis of anthropogenic changes in channel processes, p. 164. GEOS, MOSCOW (2001)

Berkovich, K.M., Chalov, R.S.: Traces of the ancient lakes in Caucasus mountains (in Russian). Notes of Transbaik. br. of. GS USSR. V. XXXI, pp. 48–53 (1969)

Berkovich, K.M., Chalov, R.S.: River channel mode and principles of its regulation at water transport development (in Russian). Geogr. Nat. Resour. (1), 10–17 (1993)

Berkovich, K.M., Zaitsev, A.A., Lodina, R.V., Chalov, R.S.: Fluvial processes on the large East Siberian rivers with a pebble-boulder alluvium and features of their regulation (in Russian). Vest. Moscow, Un. Ser 5. Geography (in Russian) (3), 35–41 (1985)

Berkovich, K.M., Zlotina, L.V., Ivanov, V.V., Nikitina, L.N., Ryazanov, P.I., Turykin, L.A., Chalov, R.S., Chernov, A.V.: Development of the lower and middle Dniester riverbed under the anthropogenic load (in Russian). Ecological problems of the soil erosion and fluvial processes, pp. 141–165. MSU Publications, Moscow (1992)

Berkovich, K.M., Ruleva, S.N., Chalov, R.S.: Riverbed regime of the Lower Ob. Geogr. Nat. Resour. (4), 54–61 (1989)

Berkovich, K.M., Rouleva, S.I., Surkov, V.V., Chalov, R.S.: Fluvial processes and anthropogenic reformations of the lower Tom riverbed and their influence on the floodplain landscapes (in Russian). Soil erosion and fluvial processes, Ep. 11, pp. 215–240. MSU Publishing House, Moscow (1998a)

Berkovich, K.M., Zlotyna, L.V., Chalov, R.S.: Fluvial processes and urban industrial siltation of the Insar riverbed in Mordovia. Geogr. Nat. Resour. (2), 97–101 (1998b)

Bogomolov, A.L., Vinogradova, N.I., Vlasov, B.N., Krylenko, I.V., Chalov, R.S.: Influence of the catastrophic mudflows in Tyrnyauz on the Baksan River bed. Geomorphology (1), 65–74 (2002)

Bolysov, S.I.: Biogenic relief formation on land: Autoref. Diss. Dr. Geogr. Sciences, p. 41. MSU Publishing House, Moscow (2003)

Borovkov, V.S.: Fluvial processes and dynamics of the river flows in the urbanized territories, p. 288. Gidrometeoizdat, L (1989)

Borsouk, O.A., Gornak, A.A., Kirik, O.M., Chalov, R.S.: Eolian processes in the Lena r. valley. Bull. MPEI, Department of Geologist, 2, pp. 151–152 (1975)

Chalov, R.S.: Determination of sediment discharge by the rate of displacement of accumulative forms of the river channel relief (in Russian). Quantitative methods in geography, pp. 19–22. MSU Publishing House, Moscow (1964)

Chalov, R.S.: Geographical investigations of fluvial processes, p. 232. MSU Publishing House, Moscow (1979)

Chalov, R.S.: Vertical zoning in development of the fluvial processes on the mountain rivers (in Russian). Study of natural conditions and its applied aspects, pp. 70–76. Nauka, Moscow (1985)

Chalov, R.S.: Alpine rivers and rivers in mountains: longitudinal profile, morphology and channel dynamics. Geomorphology (3), 26–40 (2002a)

Chalov, R.S.: Erosion and fluvial processes in training courses for hydrologists and geomorphologists in state (classical) universities. Problems of erosion and fluvial processes in training courses of universities, pp. 6–11. AGPI Publishing House, Arzamas (2002b)

Chalov, R.S.: Rocky channels: spreading, morphology, specificity and conditions of deformations. In: Eighteenth plenary interuniversity coordination meeting on the problem of erosion, fluvial and etuary processes, pp. 211–212. Kursk (2003)

Chalov, S.R.: Fluvial processes on small rivers of Kamchatka and their technological changes. Soil erosion and fluvial processes, ed. 15, pp. 164–175. MSU Publishing House, Moscow (2005)

Chalov, R.S.: General and geographic riverbed science, p. 112. MSU Publishing House, Moscow (1997)

Chalov, R.S., Ruleva, S.N.: Changes of riverbeds and dangerous displays of fluvial processes on urbanized territories. Geogr. Nat. Resour. (4), 17–23 (2001)

Chalov, R.S., Shtankova, N.N.: Sediment runoff, effective water discharge and morphodynamic types of river beds of the Kama basin. Problems of physical geography and geoecology of the Urals, pp. 99–116. Perm State University, Perm (2000)

Chalov, R.S., Shtankova, N.N.: Sediment flow, the share of river load in it and their reflection in forms of manifestation of fluvial processes on the rivers of the Volga basin (in Russian). In: Proceedings of AVN, vol. 9, pp. 195–205 (2003)

Chalov, R.S., Alabyan, A.M., Ivanov, V., Lodina, R.V., Panin, A.V.: Morphodynamics of the channels of plain rivers, p. 288. GEOS, M (1998)

Chalov, R.S., Liu, S., Alekseevskiy, N.I.: Sediment flow and fluvial processes on the big rivers of Russia and China, p. 216. MSU Publishing House, Moscow (2000)

Chalov, R.S., Zavadsky, A.S., Pakhomova, O.M.: Natural and anthropogenic manifestations of the fluvial processes in the different parts of the river network (in Russian). Problems of hydrology and hydroecology, ed. 1, pp. 203–216. MSU Publishing House, Moscow (1999)

Channel regime in Northern Eurasia, p. 336. MSU Publishing House, Moscow (1994)

Chemeris, M.P.: Estimation of the scale of anthropogenic transformations of water basins and channels of small rivers in Ukrainian Polesie. Problems of erosion and fluvial processes, pp. 99–101. Izhevsk (1992)

Darbutas, A.A.: Fluvial processes of the R. Neman (Nemunas) and Influence of the Anthropogenic Factor on them: Abstract from the … Ph.D. Geogr. Sciences, p. 27. MSU Publishing House, Moscow (1992)

Dedkov, A.P., Mozzherin, V.I.: Erosion and Sediment Sewage on Earth, p. 264. KazSU, Kazan (1984)

Dedkov, A.P., Mozzherin, V.I.: Basic approaches to the study of changes in the flow regime and their geomorphological consequences. Causes and mechanism of drying of small rivers, pp. 9–26. KazSU, Kazan (1996)

Dobrovolskaya, N.G.: Features of subaqueous weathering on plain and mountain rivers (on the example of Lena and Kara-Bau, Western Tien Shan). Vest. Moscow, Un. Ser. 5. Geogr. (2), pp. 71–76 (1980)

Dobrovolskaya, N.G.: Microorganisms as a component of river geosystems. Soil erosion and fluvial processes, ed. 13, pp. 133–144. MSU Publishing House, Moscow (2001)

Dobrovolskaya, N.G.: Some features of microbial runoff of the small rivers (in Russian). In: XVIII interuniversity meeting on the problem of erosion, fluvial and estuary processes, pp. 108–109. Kursk (2003)

Engineering of ship passages on free rivers. TsNIIEVT works, ed. 36, p. 262. Transport, Moscow (1964)

Feniksova, V.V.: Anthropogenic deposits of the South-East of Western Siberia. Problems of Geology and Paleogeography of Anthropogene, pp. 185–211. MSU Publishing House, Moscow (1966)

Fluvial processes and waterways in the rivers of the Ob basin, p. 300. RIPEL-plus, Novosibirsk (2001)

Fluvial processes on the rivers of the Altai region, p. 243. MSU Publishing House, Moscow (1996)

Grishanin, K.V.: Dynamics of channel flows, p. 488. Hydrometeoizdat, L (1969)

Grishanin, K.V.: Theory of the fluvial process, p. 216. Transport, M (1972)

Gusarov, A.V.: Trends of erosion and suspended sediment flow in Asia in the second half of the XX century (in Russian). Geomorphology (4), 70–87 (2002)

Kalinin, A.M.: Formation of Dniester Valley Slopes and Fluvial Processes. Regularities of Erosion and Fluvial Processes in Various Natural Conditions, pp. 311–312. MSU Publishing House, Moscow (1987)

Kalinin, A.M., Kirik, L.S.: Quantitative estimation of a role of slope processes of a valley of the top Aldan as the factor of channel formation. Research of fluvial processes for practice of a national economy, pp. 229–230. MSU Publishing House, Moscow (1983)

Karasev, I.F.: Fluvial processes at transfer of the flow, p. 288. Hydrometeoizdat, L (1975)

Karaushev, A.V.: Problems of the dynamics of natural water flows, p. 392. Hydrometeoisdat, L (1960)

Karaushev, A.V.: River hydraulics, p. 416. Hydrometeoisdat, L (1975)

Karaushev, A.V.: Theory and methods of river sediment calculation, p. 272. Hydrometeoizdat, L (1977)

Khmaladze, G.N.: Sediment loads by the rivers of the Black Sea coast of Caucasus, p. 168. Gidrometeoizdat, L (1978)

Khmeleva, N.V., Vinogradova, O.V., Sysoeva, S.M.: Influence of development of alluvial rashes on fluvial processes of alpine and one and a half rivers in Eastern Siberia. Soil erosion and fluvial processes, ed. 10, pp. 121–132. MSU Publishing House, Moscow (1995)

Khmeleva, N.V., Vinogradova, N.N., Samoilova, A.A., Shevchenko, B.F.: Mountain river basin and exogenous processes within it (results of stationary studies), p. 187. MSU (2000)

Kondratyev, N.E., Lyapin, A.N., Popov, I.V., Pinkovsky, S.I., Fedorov, N.N., Yakunin, I.A.: Fluvial process, p. 372. Hydrometeoizdat, L (1959)
Kondratyev, N.E., Popov, I.V., Snischenko, B.F.: Basics of hydromorphological theory of the fluvial process, p. 272. Gidrometeoizdat, L (1982)
Kondratyev, N.E.: Kinematic flow structure with a ripple bottom structure. In: Proceedings of SHI. Iss. 116, pp. 3–18 (1964)
Kopaliani, Z.D.: About the ratio of bottom and suspended sediment discharge in rivers. Hydrophysical processes in rivers and reservoirs, pp. 143–147. Nauka, Moscow (1985)
Kopaliani, Z.D., Kostiuchenko, A.A.: Calculations of the bottom sediment discharge in the rivers. Coll. of works on hydrology (27), 25–40 (2004)
Korotayev, V.N., Lodina, R.V., Miloshevich, V.A., Sidorchuk, A.Yu., Chalov, R.S.: Formation of the Yana river delta and development of its estuary bars (in Russian). Soil erosion and fluvial processes, ed. 6, pp. 123–159. MSU Publishing House, Moscow (1978)
Kositskiy A.G., Reteyum K.F., Chutkina L.P., Shenberg, N.V.: Scale effects of flow changes along the length of the river systems (in Russian). Problems of hydrology and hydroecology, ed. 1, pp. 141–156. MSU Publishing House, Moscow (1999)
Kostyaev, A.G.: Periglacial deposits and structure of the low terraces of the Walday age in the Northern Dvina valley (in Russian). Periglacial phenomena on the territory of the USSR, pp. 188–200. MSU Publishing House, Moscow (1960)
Kovalchuk, I.P., Volos, S.I., Kholodko, L.P.: River systems of the West of Ukraine: scales and tendencies of transformation of structure, mechanism of change of a condition in XIX-XX centuries And Reasons and mechanism of drying of small rivers, pp. 43–56. KazSU, Kazan (1996)
Kuznetsov, K.L., Chalov. R.S.: Fluvial processes and morphology of the mountain river beds in conditions of active mudflow activity (on an example of the rivers of the Northern slope of Zailijsky Alatau). Geomorphology (2), 71–78 (1988)
Kuzin, P.S., Babkin, V.P.: Пeographical patterns of the hydrological regime of rivers, p. 200. Hydrometeoizdat, L (1979)
Leliavsky, N.S.: On the study of the movement of sand spits near Alexandrovsk. In: Proceedings of the 3rd Congress of Russian Workers on Waterways in 1896. Part 2. St. Petersburg, 1896 (Issues of hydraulic engineering of free rivers), pp. 146–156. Rechizdat, M (1948)
Levanidov, V.Ya.: On the hydrological regime of spawning grounds for chum salmon and pink salmon (in Russian). Izv. TIN-RO. **64**, 101–125 (1968)
Liu, D.: Physical Geography of the World. High Education, Beijing (1986) (on Chineeze)
Liu, S.: Comparative Analysis of Channel Processes on Large Rivers of Russia and China: Author's Disc. Ph.D. Geogr. Sciences, p. 23. MSU Publishing House, Moscow (1998)
Liu, S., Chalov, R.S., Ding, J., Ivanov, V.V., Korotayev, V.N., Li, J., Yang, H., Yang, S.: Regional variations of the Asian rivers suspended sediment flow in estuaries (in Russian). Vest. Moscow univ. Ser 5. Geography (3), 44–51 (2001)
Lodina, R.V., Chalov, R.S.: Cobblestone bridge on the big rivers (in Russian). Nature (7), 57–62 (1994)
Lopatin, G.V.: Sediments of the USSR rivers, p. 368. Geografgiz, Moscow (1952)
Lower Yana: estuary and fluvial processes, p. 212. GEOS, M (1998)
Makhinov, A.N., Makhinova, A.F.: Transformation of the anthropogenic relief in the area of alluvial deposits development (North of Khabarovsk Territory). Geomorphology (1), 43–49 (2006)
Makhinov, A.N., Chalov, R.S., Chernov, A.V.: Directed sediment accumulation and morphology of the lower Amur bed (in Russian). Geomorphology (3), 70–78 (1994)
Makkaveev, N.I.: Channel regime of rivers and tracing of slits, p. 202. Rechizdat, Moscow (1949)
Makkaveev, N.I.: River channel and erosion in its basin, p. 347. Published in the SA USSR, Moscow (1955)
Makkaveev, N.I., Chalov, R.S.: On the development of the surface topography of river terraces and signs of deep erosion by the example of the Upper Ob. New. Ser. Geogr. (4), 120–125 (1964)

Makkaveev, N.I., Chalov, R.S.: Some peculiarities of the big river valley floor connected with periodic changes of the flow norm (in Russian). Geography issues, Col. 79, pp. 156–167. Geografgiz, Moscow (1970)
Makkaveev, N.I., Chalov, R.S.: Fluvial processes, p. 264. MSU Publishing House, Moscow (1986)
Makkaveev, N.I., Belinovich, I.V., Vyshlov, K.P.: About the influence of the Caspian Sea level decline on the navigable conditions in the lower Volga and Ural. Fluvial Investigations to Improve Shipping Conditions, pp. 24–28. River Transport, Moscow (1958)
Makkaveev, N.I., Sakharova, E.I., Chalov, R.S.: Modern aeolian processes in the upper Ob valley (in Russian). Vest. Moscow Un. Ser 5. Geography (2), 49–53 (1966)
Makkaveev, N.I.: Mandych, A.F., Chalov, R.S.: Influence of upward landform development on deep erosion and sediment load of rivers in West Georgia. Vest. MSU Ser. 5. Geography. (4), 52–58 (1968)
Makkaveev, N.I., Litvin, L.F., Khmeleva, N.V.: Using the transport capacity of the river flow for practical purposes. Vest. MSU. Ser. 5. Geography. (2), 82–89 (1970)
Map "Erosion hazard of agricultural lands of European part of USSR". The scale is 1:2000000. MSU Publishing House, Moscow (1987)
Matveev, B.V.: Influence of the geological-geomorphological factors on formation and morphology of the river bend (in Russian). Geomorphology (3), 51–57 (1985)
Matveevskaya, A.L., Ivanova, E.F.: Geological structure of the Southern part of the West Siberian Plain in connection with oil and gas bearing issues, p. 264. Publishing house of the SA USSR, M, L (1960)
Maximov, S.P.: On study of river sediment movement (in Russian). Questions of river life. St. Petersburg, pp. 243–254 (1905)
Mikhailov, V.M.: Estuaries of Rivers of Russia and Neighboring Countries: Past, Present, and Future, p. 413. GEOS, M (1997)
Mikhailov, V.M.: Talik development in the Kolyma river valley and the river water temperature (in Russian). Geoecology (6), 100–110 (1998)
Mikhailova, M.V.: Hydrologic regime and dynamics of the Huang He estuary hydrographic network (in Russian). Water Resour. **28**(1), 105–117 (1998)
Milliman, J.D., Rutkowski, C., Meybeck, M.: River discharge to the sea. A global river index (GLORY), LOICZ reports and studies. Texel, The Netherlands. (1995)
Mirtskhulava, C.E.: Basics of physics and mechanics of erosion of channels, p. 304. Hydrometeoizdat, L (1988)
Molchanov, A.K.: Research of dependence of the morphometric characteristics of the riverbeds on the factors determining them: Authors' dis… Ph.D. geogr. sciences., p. 16. MSU Publishing House, Moscow (1971)
Muranov, A.P.: Huang He River, p. 88. Gidrometeoizdat, L (1957)
Nekos, S.V., Chalov, R.S.: Sediment runoff and fluvial processes on rivers of the Don basin. Geomorphology (2), 60–71 (1997)
Nguyen Van Ki.: Dynamics of the estuary areas of the rivers of the Socialist Republic of Vietnam: Abstract of Diss. Doc. geogr. sciences, p. 53. MSU Publishing House, Moscow (1990)
Ning, Q., Zhang, Z., Zhou, Z.: Fluvial Process, p. 584. Science, Beijing (1987) (in Chinese)
Ning, Q., Tsintsao, I., Venhao, C.: Changes in water flow and river sediment in the main stream of the Huang He River and fluvial processes. Beijing (1993) (in Chineeze)
Physical geography of China: Land waters. Science, Beijing (1981) (in Chineeze)
Polyakov, B.V.: Hydrology of the Don river basin, p. 328. Rostov-na-Donu (1930)
Prokacheva, B.G., Snischenko, D.V., Usachev, V.F.: Remote methods of hydrological study of BAM zone, p. 224. Gidrometeoizdat, L (1982)
Reznikov, P.N., Chalov, R.S.: Sediment flow and conditions of channel formation on the rivers of the Basin of the Northern Dvina (in Russian). Geomorphology (2), 73–85 (2005)
Rossinskiy, K.I., Debolsky, V.K.: River Sediments, p. 216. Nauka, Moscow (1980)
Rossinskiy, K.I., Kuzmin, I.A.: Balance method for calculation of changes of the flow bottom. In: Proceedings of Hydroproject. Col. 2, pp. 265–271 (1964)

Rubtsov, M.V.: Scientific bases of the forestry measures in the protective and water-protective forests along the rivers of the European North: Author's thesis … Dr. Techn. Sci. L, p. 37 (1982)
Rubtsov, M.V., Deryugin, A.A.: Influence of forest on channel of banks of taiga rivers. Laws of manifestation of erosion and fluvial processes in various natural conditions, pp. 264–265. MSU Publishing House, Moscow (1981)
Rzhanitsyn, I.A.: Morphological and hydrological regularities of the river network structure, p. 240. Gidrometeoizdat, L (1960)
Rzhanitsyn, N.A.: Channel-forming processes of the rivers, p. 264. Gidrometeoizdat, L (1985)
Sakharova, E.I.: Geomorphological research in the study of fluvial processes in the Northern Dvina. Methods of geographic research, pp. 101–109. Geografgiz, Moscow (1960)
Scheidegger, A.E.: Theoretical geomorphology, p. 452. Progress, Moscow (1964)
Sediment yield, its study of geographical distribution. L.: Gidrometeoizdat, 1977. 240 p
Shamov, G.I.: River sediment, p. 380. Hydrometeoizdat, L (1959)
Shapiro, H.Sh.: Regulation of Solid Flow and Fluvial Processes at Water Intake in Irrigation Systems: Authors' Diss. Doct. Sci. (Techn.), p. 52. MGMI, M (1973)
Shtankova, N.N.: Sediment flow on rivers of Volga basin (in Russian). Dynamics of gully-beam forms and fluvial processes, pp. 112–119. MSU Publishing House, Moscow (2002)
Simonov, YuG: Regional geomorphological analysis, p. 264. MSU Publishing House, Moscow (1972)
Simons, D.B., Richardson, E.V., Nordin, C.F.: Sedimentary structures generated by flow in alluvial channels. SEPM div. AAPS, special publication, 12, pp. 48–67 (1966)
Sinitsyn, M.G., Rusakov, A.B.: Influence of river beaver activity on relief of valleys and river beds of small rivers of Vetluzhsko-Unzhenskoye Polesiye. Geomorphology (1), 85–91 (1990)
Sladkopevtsev, S.A.: On the structure of forests in the valleys of the temporary waterways of Central Kazakhstan (in Russian). New. VGO. **95**(5), 449–450
Sladkopevtsev, S.A.: Development of the river valleys and neotectonics, p. 184. Nedra, Moscow (1973)
Slastikhin, V.V.: Process of erosion on the selective watersheds in Moldova. Regularities of erosion and fluvial processes in different natural conditions, pp. 136–137. MSU Publishing House, Moscow (1987)
Small rivers in the Volga Basin, p. 335. MSU Publishing House, Moscow (1998)
Snishchenko, B.F.: About the connection of the sand ripples height with the parameters of the river flow and channel (in Russian). Meteorol. Hydrol. (6), 84–91 (1980)
Snishchenko, B.F., Kopaliani, Z.D.: About speed of a ridge movement in the rivers and laboratory conditions (in Russian). In: Proceedings of SHI, ed. 252, pp. 30–37 (1978)
Sokolov, Yu.I.: Transport capacity and changes of the river floodplains (in Russian). In: Proceedings of V All-Union hydrological congress, vol. 10, Kn. 2, pp. 196–203. Hydrometeoizdat, L (1988)
Sokolov, Yu.N.: Hydraulic Resistances on the Floodplains of Plains Rivers: Author's Disc. Dr. Sci. (Techn.), p. 39. YVP AN SSSR, M (1991)
Sokolovsky, D.L.: River flow, p. 540. Hydrometeoisdat, L (1968)
Sribnyi, M.F.: Formula of average speed of the rivers flow and their hydraulic classification by resistance to movement. Research and complex use of water resources, pp. 204–221. Publishing House of the SA USSR, Moscow (1960)
Tananaev, N.I.: Influence of perennially frozen soils on suspended sediment flow and fluvial processes. Dynamics of gully-beam forms and fluvial processes, pp. 107–111. MSU Publishing House, Moscow (2002)
Tananaev, N.I.: Seasonal and long-term freezing of the cryolithozone river beds and its influence on channel changes (on the example of average Lena). Erosional, fluvial processes and hydroecology problems, pp. 195–201. Geogr. ft. MSU, Moscow (2004)
Tsvetkova, N.A.: Mode of Amudarya river sediments (in Russian). Questions of hidrotechnik, ed. 13, pp. 5–86. Tashkent (1963)
Velikanov, M.A.: Fluvial Process, p. 395. Gosfizmatizdat, Moscow (1958)

Vinogradov, V.A., Klaven, A.B., Nikitin, V.I., Turutina, T.V.: Non-equilibrium processes in river bed formation, In: Proceedings of AVN, vol. 9, pp. 55–68 (2003)
Waterways of the Lena Basin, p. 600. MIKIS, M (1995)
Work of water flows, p. 194. MSU Publishing House, Moscow (1987)
Zaitsev, A.A.: Formation of the free bends on the plain rivers: Author's thesis … Ph.D. geogr. sciences, p. 24. MSU Publishing House, Moscow (1979)
Zaitsev, A.A.: The role of ice phenomena in formation of beds and banks of channels of large rivers of Lena basin. Safety of power structures. Exhibit. 11, pp. 211–223. NIYEZ, M (2003)
Zaitsev, A.A., Militeyev, A.I.: Ice jams and fluvial processes on the Lena river. Security of power structures, ed. 11, pp. 224–237. NIYEZ, Moscow (2003)
Zamarin, E.A.: Transport capacity and allowable flow speeds in channels, p. 32. Gostransizdat, Moscow, L (1951)
Zavadsky, A.S., Makhinov, A.N., Chalov, R.S.: Formation of the Middle Amur channel and its morphodynamic types. Water Resour. **27**(2), 133–140 (2000)
Zemtsov, A.A.: Peculiarities of fluvial processes development on taiga rivers. Regularities of erosion and fluvial processes manifestation in different natural conditions, pp. 258–259. MSU Publishing House, Moscow (1976)
Zheleznyakov, G.V., Baryshnikov, N.B., Altunin, V.C.: Impact of kinematical effect of the non-pressure flow on sediment transport (in Russian). Sediment movement in open channels, pp. 19–24. Nauka, Moscow (1970)
Znamenskaya, N.S.: Ripple movement of sediment, p. 188. Hydrometeoizdat, L (1968)
Znamenskaya, N.S.: Bottom sediments and fluvial processes, p. 142. Hydrometeoisdat, L (1976)
Zorina, E.F.: Gullies, gully formation and development potential. Soil erosion and fluvial processes, ed. 12, pp. 72–95. MSU Publishing House, Moscow (2000)
Zorina, E.F., Kosov, B.F., Prokhorova, S.D.: Experience of estimation of volume of gully outcrops in the Don river basin. Vest. Moscow, Un. Ser. 5. Geography (3), 39–45 (1980)

Chapter 3
Conditions of River Channel Formation and Their Hydrology and Morphology Analysis

3.1 Conditions of River Channel Formation and Their Connection with Factors of Fluvial Processes

Factors of fluvial processes, in accordance with the generally accepted definition of the concept of "factor" (Soviet…, 1982), are defined as main drivers which determine character and individual features of the processs and phenomena. The main factors of fluvial processes are water runoff, the lithogenic surface of the territory and its relief, with which water flows interact, and sediment yield, which is, on the one hand, a derivative of this interaction, and, on the other hand, determines the fluvial processes themselves. From this point of view, other factors—vegetation, permafrost, ice phenomena, wind—essentially represent the conditions under which the interaction of the flow and channel is implemented and on which the shape, intensity and other features of the fluvial processes depend. Nevertheless, they are traditionally considered to be the factors of fluvial processes, and therefore they are considered as such, but with their classification as indirect and secondary. However, this logical violation is to some extent justified by the fact that, for example, the water flow interacting with the lithogenic base (the bottom and banks of the river), which can be modified by the presence of aquatic vegetation, root systems of plants, etc., as a result of which the interaction of the flow occurs with the environment that integrates both litho- and biogenic base. The same can be said about the role of permafrost, which cements soil and changes its physical properties, meteorological conditions, etc.

Geological-geomorphological factors can be considered as a condition determining the nature and forms of representation of fluvial processes. This is due to the passive form of influence of these factors on the fluvial processes, which cannot be considered as a literally driving force of the process. However, the forces that resist the effects of the flow are associated with them and, depending on them, the interaction of the flow with the riverbed sediments is implemented. From this point of view, the relief and the rocks and sediments composing it are factors of fluvial

R. S. Chalov, *Fluvial Processes: Theory and Applications*,
https://doi.org/10.1007/978-3-030-66183-0_3

processes. At the same time, they also create the background against which fluvial processes develop, representing certain properties of the underlying surface and beds of the channel flows, which are favourable or unfavourable for the separation of soil particles from them and the formation of sediment flow, the composition of which, in turn, is determined by the physical properties of the surface itself.

The same approach can be applied to the assessment of sediment load. On the one hand, this is the most important active factor in fluvial processes. On the other hand, river sediments and the deposits they form are the background against which fluvial processes develop. This refers to their composition (grain size, sorting, pelletizing, etc.), which depends on the geological and geomorphological situation (composition of the basin, river valley and bedrock bed, river slope, determined by mountainous or flat terrain, etc.) and itself determines such a major condition for the formation of channels as their stability, i.e., the property with one or another intensity to be exposed to the impact of the flow and deformed due to it. At the same time, the stability of the channel depends on the geomorphological situation in which sediment formation takes place. The slope of the river, which is one of the hydraulic characteristics of the channel flow itself, is also connected with this situation. Being by its physical nature a reflection of the energy losses of the flow to overcome the resistance, the slope is nevertheless created by the difference in altitude of the terrain on which the river flows. The partitioning between sediment particle size (the composition of the bed soil) to the gradient that determines the flow rate and its power, determines the intensity of channel changes and, consequently, the stability of the bed as a condition for its formation.

Water runoff is realized as a factor of fluvial processes, on the one hand, through its capacity, and, on the other hand, through effective discharge rates. Transport capacity characterizes the conditions that determine the interaction between the flow and the channel, i.e. fluvial processes, and ultimately the formation of a channel pattern in each given river section through changes in the runoff rates and slope of the flow. The effective discharge is associated with long-term and seasonal fluctuations in water content, which determine a different role of each phase of the water regime or each actual water discharge in the impact of the flow on the channel and, accordingly, in channel changes.

On the Fig. 3.1 the scheme of influence of factors and conditions of formation on the development of channels of this or that pattern is presented. Both direct natural factors (runoff rates of the river, expressed through the flow rate; geological-geomorphological; sediment load and composition) and the influence of other factors and conditions of fluvial processes, including those that are integral characteristics of the conditions that determine the formation of channels, are distinguished. It follows from it that the channel pattern is a reflection of the influence of all leading factors of fluvial processes and the conditions in which they are represented in their complex relationships. At the same time, the influence of anthropogenic conditions on the formation of channels is not included in the scheme, because they are manifested through changes in natural factors and the transformation of conditions for the development of fluvial processes.

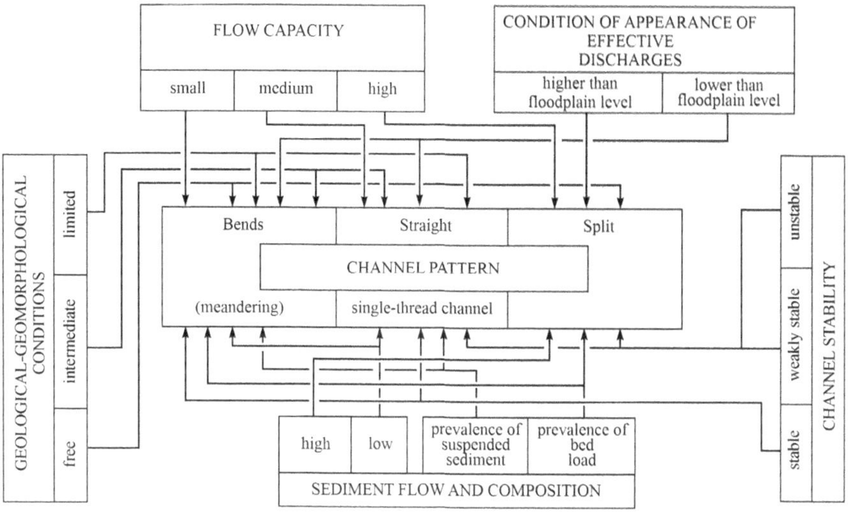

Fig. 3.1 Integrated scheme of natural conditions influences on the formation of different channel patterns

The proposed scheme allows to model possible ways of river channel transformation in case of changes in some or other conditions on catchments and on the rivers themselves, including as a result of anthropogenic interference, which inevitably leads to changes in the characteristics of factors and conditions of channel formation. Of these, the most conservative with respect to anthropogenic impacts are geological and geomorphological conditions, changes in which in the natural environment affect only in the geological scale of time, and with anthropogenic intervention—only in the case of major technological changes in the relief, such as the destruction of mountain ranges. The same allows to solve the opposite problem: to restore natural characteristics in the river basin, especially at paleo-hydrological constructions, according to the existing or existing type of channel (for example, reconstructed according to the relief of the floodplain). Reconstruction of the channel position by dating the age of alluvial deposits allows to characterize climatic and geomorphological conditions, remote for the whole geological epochs.

The channel pattern is a reflection of lateral channel changes. Similar schemes can be made for vertical channel changes and channel changes caused by the ripple sediment motion. However, their graphic representations are much more complex, despite the simpler form of their representation. This is due, for example, to the fact that the conditions for the development of vertical channel changes, on which their orientation depends, are the relief and characteristics of the lithogenic basis, the sign, direction and intensity of tectonic movements, changes (positive or negative) in the base level, the rate of estuary elongation, in turn, associated with the value of sediment load; the latter, being in some proportion to the transport capacity of the flow, in turn, or determines, often regardless of the ascending or descending flow rate, or determines, regardless of whether it is ascending or descending. At the same time, they depend on the water content of the rivers and its changes associated with climatic fluctuations of different periods and directions, which is superimposed by constant fluctuations in the amount of load from the catchment area. As a result, the same factor determining the development of vertical changes turns out to be in very different conditions and thus leads in each case to the opposite representation of the processes.

This complexity of the conditions is even greater in the analysis of channel changes associated with the flow of sediment, although the external representations of the conditions in which they develop are even more simplified.

3.2 Bed Material Load

Bed load and some suspended sediments are bed material load, which are "predominantly contained in bottom sediments" when transport ceases (Karasev 1975, p. 146). In this sense, the concept of "bed material load" is synonymous with "bottom (channel, river) sediments". Bed loads are always part of the bed material load; the proportion of suspended load in it varies widely, depending on the geographical conditions for the formation of river load, the ratio between the suspended W_R and the bed load W_G in total river load, as well as spatial and temporal changes in the hydraulic characteristics of the flows (from the bedrock to the peripheral part of the channel; in the high-water and low-water phases of the regime, etc.). In general, the larger the composition of bed material load, the smaller the proportion of suspended sediment in it. The application of the graphical method for dividing the load into suspended and bed sediments proposed by V. Kresser (Baryshnikov and Popov 1988) has shown (see Fig. 2.10) that in pebble-boulder alluvium suspended loads do not practically participate in its structure (these are the bed material loads on mountainous and pebbly-boulder plain rivers), while rivers with sandy-silty alluvium are characterized by an absolute predominance in suspended sediment deposits that settle on the flood recession and in the low water period. In most sandy rivers, depending on the proportion of suspended sediment in total sediment load, more or less suspended sediment is present in river sediments, which determines the partial overlap of both cumulative curves.

The composition of bed material load is an essential condition for the development of channel changes. They are formed as a result of erosion by the flow of its bottom and banks, removal of solid material from tributaries and partial inflow from the catchment area. As a result, bed material load reflects the entire complex of erosion-accumulation and denudation processes in the river basin. Therefore, their composition is determined by the geological structure of the territory along which the river flows and the history of its geomorphological development. Accumulating in river valleys, bed material load form alluvial layers that witness geological activity of the rivers and at the same time serve as a source of alluvial material inputs to the stream during the erosion of banks and riverbeds. At the same time, each ripple as a form of sediment transport is an accumulative formation composed of bed material load that have temporarily escaped from the transport state, forming its fixed core. As sediments on the upper (pressure) slope of the ripple are washed away, the sediments return to motion, settling and returning to the category of sediments on the lower steep slope (in the basement). Thus, the formation of ripples as accumulative formations in the channel is not yet evidence of the formation of alluvial strata, although it contains, especially when dried up in the lowlands, the potential for the transformation, partially or completely, of such accumulations into alluvial deposits. This is due to the inverse effect of the ripples—macroforms—on the kinematic structure of the flow, the washout of opposite drying ripples (side bars, mid bars) of the banks, the appearance of vegetation on them and an increase in the proportion of suspended material in the composition of the sediment forming them, which is deposited during subsequent flooding in the high-water phase of the regime. The distribution of sediment bed widths of different sizes reflects the changes in the position of the dynamic axis of a water-course in different phases of the regime and the redistribution of water discharge along the channels (in the braided channel), indicates the tendency of the development of some and the dying out of others, and captures mainly the zones of erosion and accumulation of sediment in the channel.

The composition of bottom sediments is formed mainly during floods or high floods, i.e. when effective discharges are passed. Therefore, in the intermittent period, the average diameter of the sediment particles lining the riverbed, even in the tributary flow zone, may be larger than the critical size at these flow velocities, since $V_{average} < V_n$. This is particularly evident in rivers where pebbles or pebbles and boulders are the bed material load. A large difference (by an order of magnitude) between the critical rates of erosion of such soil and the actual flow rates in the low water causes a sharp decrease in sediment concentration, up to a complete clarification of the flow and preservation of the bottom of the channel for the entire period between. In rivers with a sandy bed, the movement of sediment, particle suspending, and erosion of the seabed and banks continue into the low water table, but the area in which they occur is greatly reduced to a narrow strip along the streambed. There is virtually no change in the mechanical composition of the bottom sediments, since the movement of sediments is carried out within the range of the soil of one size or another. Only in those cases, when on the sites, lined with small soils, the ripples, folded by larger material, are coming, local consolidation of the composition of bottom sediments is observed. On the contrary, in places where the current slows down (in the basements of large ripples, in the ears of side and abysses, at convex banks of bends, etc.), it is possible to expect a partial overlap of the smaller sands or silt of the areas of large sediment transported here during the flood. However, these local changes lead to an insignificant redistribution of sediments in the channel (bed material load) in terms of channel size, causing a deviation from the distribution of sediment at a effective discharge of 15–20%. The average particle size of load varies even less noticeably; as the size of the river section for which averaging is carried out increases, the deviations become smaller and do not exceed the accuracy of the measurements. Therefore, the study of the composition of bottom sediments, which is usually carried out in the lowlands, provides a picture of the distribution of the bedload ranges and sediments of different sizes, which integrates the conditions for the passage of effective discharge during the previous hydrological year.

Classifications of bed material load are used to determine the distribution of bedload in the river channel. For lowland rivers with sandy or muddy alluvium, the type of load is determined by the content of the individual fractions. There are a large number of different classification schemes, which differ mainly in the fractionality of the selection of individual types. Table 3.1 shows the classification developed in the research laboratory of soil erosion and fluvial processes of Moscow State University (Chalov 1995).

On rivers with sandy-silty sediments and in the estuaries where muddy sediments predominate, this classification is supplemented by a modified classification of the Institute of Oceanology of the Russian Academy of Sciences (Budanov 1964) (Table 3.2).

The larger the sandy bed material load that form river sediments in the lowlands, the worse their sorting is. On plain rivers, fine-medium-grained sands often have an admixture of gravel particles (up to 5%); large-grained sands are seldom well sorted at all, and the admixture of gravel in them is quite significant (up to 20% on average). The presence of gravel particles in sandy sediments increases their average particle size, but is not an important fluvial processes driver. The

Table 3.1 Classification of sandy bed material load of plain rivers (developed by N.V. Lebedeva, R.V. Lodina and R.S. Chalov)

Type of sandy sediments	Content of characteristic fractions	Average particle diameter, mm
Silt	Fraction particles <0.1 mm more than 80%	0.10–0.12
Thin-grained	Fraction particles <0.25 mm more than 95%	0.12–0.16
Ultrarine sand	Fraction particles <0.25 mm more than 90%; fraction 0.25–0.1 mm (about 80%) predominates	0.16–0.18
Fine sand	Fraction particles of 0.25–0.1 mm more than 75%	0.18–0.20
Fine to medium dan	0.5–0.1 mm fraction particles more than 90%; 0.25–0.1 mm fraction prevails	0.20–0.30
Medium sand	Fraction particles of 0.5–0.25 mm more than 60%	0.30–0.40
Course to medium sand	Fraction particles of 1.0–0.25 mm about 70–75%; fraction of 0.5–0.25 mm prevails	0.40–0.60
Course sand	Fraction particles of 1.0–0.25 mm more than 80%; fraction of 1.0–0.5 mm is about 30%	0.60–0.90
Sandy-gravel	Gravel particles with a size of 1 mm about 30–40%	0.90–1.30
Gravel	Content gravel particles over 50%	1.3–2.0

Table 3.2 Classification of muddy bed material load

Type of sediment load	Content characteristic fractions	Average diamete particles, mm
Large silhouette	Fraction particles 0.1–0.05 is >50%	0.05–0.1
Small-silhouette silt	Fraction particles <0.05 mm more than 50%	0.05–0.01

movement of such particles is usually influenced by the abluvial effect in its microexpression.

In rivers with pebbles and boulders, the characteristic of bed material load is usually limited to determining the size of the debris and particles and then separating them into fractions (Table 3.3). This, however, does not allow the type of sediment to be defined as a set of particles and debris of different sizes. Therefore, R.V. Lodina (Chalov 1995) proposed a classification of pebble-boulder bed sediments, in which their types are distinguished by the content of individual fractions of boulders and pebbles (Table 3.4). This is all the more important because the sorting of large deposits is very low; at the same time, thanks to it, pebbles and boulders in the lowlands form a bedding of the largest material, which provides stabilization of the channels in this phase of the regime.

In the process of sediment transport, the size of the particles involved decreases and the river length is differentiated according to particle size and shape, petrographic or mineralogical composition, etc. Against the background of mechanical particle size reduction and hydraulic sorting of sediment particles from the basin area, and as a result of the erosion of the banks and bottoms, new portions of solid material of various textures are constantly entering the river beds, making the picture of the distribution of bed material load along the length of the river very complex.

As the flow of each individual sediment particle moves, its volume and weight gradually decrease. The reasons for fining are: the abrasion of particles in contact with other moving particles and soils of the bottom, impacts on the solid surface of the river bed or mutual impact of moving particles, resulting in their fragmentation, the gradual dissolution of solid matter particle with water and its biochemical

Table 3.3 Classification of gravel, pebbles and boulders by size

Name of sediment by particle size	Particle size, mm	Name of sediment by particle size	Particle size, mm
Rocks	>1000	Medium pebbles	50–25
Large boulders	1000–500	Small pebbles	25–10
Medium boulders	500–250	Large gravel	10–5
Small boulders	250–100	Medium gravel	5–2
Large pebbles	100–50	Small gravel	2–1

Table 3.4 Classification of boulder and pebble bed material load [according to R.V. Lodina (Chalov 1995)]

Type of riverbed sediments	Content of characteristic fractions	Average particle diameter, mm
Large boulders	Boulders—about 90%, including 500–1000 mm fraction—about 40%	500
Medium boulder	Boulders—about 70%, including 500–250 mm fraction—about 50%	250–300
Petty boulder	100–250 mm fraction more than 40%	110–140
Large pebbles	100–50 mm fraction—more than 70%; 100–75 mm fraction prevails	70–90
Mid-large pebbles	100–50 mm fraction—from 50 to 70%; 75–50 mm fraction prevails	60–70
Medium pebbles	fraction 75–25 mm—about 80%; fraction 75–50 mm prevails	45–55
Small-middle pebbles	50–10 mm fraction—80%; 50–25 mm fraction prevails	40–45
Small pebbles	50–10 mm fraction—90%; 25–10 mm fraction prevails	25–30

destruction. The weight loss due to abrasion is proportional to the particle's weight P and the distance x to which it has moved:

$$dP = -aP_0 dx \tag{3.1.}$$

or, after the integration.

$$P = P_0 e^{-ax} \tag{3.2.}$$

In this expression, known as A. Sternberg's equation, P_0 is the initial weight of the particle, and a is the coefficient of performance depending on the strength of the rock (mineral) that makes up the particle, the flow rate and the hardness of the ground along which the particle moves. If the weight of particles is replaced by their diameter, Eq. (3.1.) takes the form

$$d_x = d_0 e^{-ax} \tag{3.3}$$

According to A. Sternberg's Eq. (3.2.), granite pebbles weighing 100–150 g are transformed into sand at a distance of 400 km, limestone—200 km. When the particles reach the size of the grain size, their further mechanical crushing becomes almost imperceptible. The main reason for the "lower threshold" of particle crushing is the latent micro-fracture of rocks (minerals). The average cross section of the compartments is 1–2 mm. Particles change very slowly when they become monolithic.

Razumikhin's experiments (1965) on abrasion and crushing of different fractions during their transportation (Fig. 3.2A) have established that the maximum crushing and abrasion is carried out for the majority of rocks at the beginning of the journey (not more than 100 km). Crushing prevails for pebbles, abrasion prevails for gravel. Large particles wear out more intensively than fine particles due to differences in the way they are transported: large particles are characterized by dragging, which causes collision and abrasion, while small particles are mainly suspended. Even with sand abrasion, the greatest loss of weight is inherent in large sand differences. For gravel particles, in contrast to pebbles and sand, fine gravel (2–4 mm) is subject to the greatest wear. Thus, the fractions of medium and large gravel are more resistant to crushing and abrasion than pebbles and sand. N.V. Razumikhin established a certain sequence in the abrasion of particles of different mineral compositions, placing the formations in order of decreasing stability: chalcedony, quartz, quartzite, diabase, sandstone, limestone. A similar relationship of the abrasion intensity of rock particles was obtained for the rivers of Eastern Siberia (Borsouk and Lodina 1987), on the basis of which a number of abrasive rock stability was proposed: granites, gneisses, quartz sandstones, siltstones, porphyries, porphyrites, quartz.

Analyzing the results of the experiment conducted by N.V. Razumikhin, we can come to the conclusion that the fraction of fine gravel is subject to the greatest mechanical abrasion, and the maximum crushing is inherent in fine pebbles (with the initial grain size in the experiment 20–30 mm—fine pebbles). Middle and large gravel fractions are the least crushed. Thus, according to the results of N.V. Razumikhin's experiments, it would be possible to suggest that gravel beds should be widely spread in nature. In reality, however, this is not the case, and therefore the consideration of a single mechanical abrasion and crushing does not allow us to

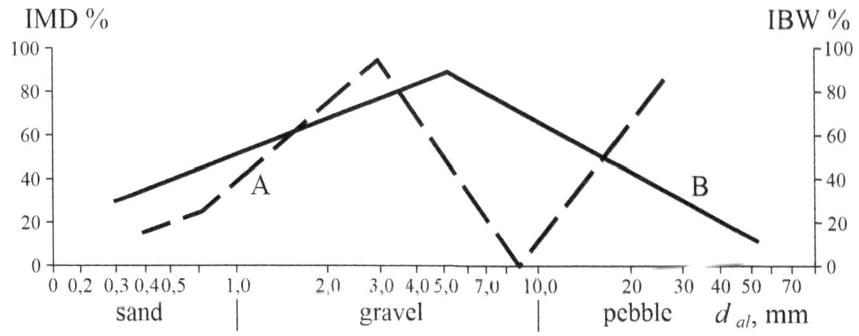

Fig. 3.2 Ratio of the intensity of mechanical (in % of shredding and abrasion at a distance of 150 km) destruction IMD (**A**) and biochemical (in % of production of Ca^{++} mg/l at 1 g of a formation) weathering IBW (**B**) of pebble, gravel and sand in channel flows [based on data from Razumikhin materials (1965), biochemical weathering data—from materials of N.G. Dobrovolskaya (1980, 2001)]

explain the real change in the particle size of sediment during its transport (Dobrovolskaya et al. 1991).

The role of subaquatic biochemical weathering (Fig. 3.2B) under the influence of microorganisms is no less important in the crushing of rock fragments and particles that make up the bottom sediments (bed material load). Laboratory and field studies (Borsouk et al. 1981; Glazovskaya and Dobrovolskaya 1984) have shown that, all other things being equal, the rate of biochemical weathering depends on the strength of the crystal lattice of minerals, the conditions of the environment in which they dissolve, the nature of the rock-destroying agent, and the petrographic and fractional composition of deposits. Microbacteria, on the one hand, contribute to the destruction of minerals and the complete dissolution of individual areas of mineral grains, on the other hand, stimulate the formation of large, stable aggregates (through various excreta), which indicates the great influence of biochemical processes on the change in the grain size of the mineral substrate.

The intensity of biochemical weathering is closely related to the mechanical abrasion of particles of bed material load. As they are transported by sliding, drawing, rolling, they collide and crush into smaller fractions. Exposure of fresh surfaces due to the continuous removal of the flow of weathering products contributes to an increase in the biochemical activity of microorganisms, because together with the upper layer removed metabolic products and weathered layers, depleted nutrients. Thus, mechanical abrasion and crushing stimulate the process of biochemical weathering. The results of experiments with limestone particles indicate a very active dissolution under the influence of microorganisms, primarily gravel fraction particles (4–6 mm). The most stable were particles from 0.5 to 1 mm. Dissolution of quartz under the influence of microorganisms is much slower, which indicates selective erosion (predominant weathering of some rocks compared to other rocks), although a clear trend towards maximum weathering is also observed for the gravel fraction: the intensity of destruction of quartz gravel is 4–8 times higher than that of pebbles and 2 times higher than that of sand, which is 6–8 times lower than the intensity of limestone destruction.

Thus, the gravel fraction is the most resistant to mechanical shredding in the process of channel transport (Razumikhin 1965), is the least resistant to microorganisms. Perhaps this explains the rare occurrence of gravel channels on the rivers (Dobrovolskaya et al. 1991), and only a few relatively short sections or parts of the channel (in the cross-section) can be lined with gravel, but due to the hydraulic sorting of alluvium, which is heterogeneous in terms of grain size. At the same time, limestone is very unstable in relation to any grinding agents: mechanical (abrasion and crushing), biochemical and dissolution. When it passes the gravel stage, other processes, including chemical ones, come to the fore in its destruction. The final stage of limestone grinding is limestone flour, which follows immediately after the stage of fine gravel—coarse-grained sand. Hence the absence of sandy limestone channels and the very rapid disappearance of limestone from the petrographic spectrum of deposits as soon as the source of their flow into the rivers disappears.

The impact of microorganisms on sediment particles of different petrographic composition determines the selective nature of erosion products washing off by water flow. According to N.G. Dobrovolskaya (Chalov and Dobrovolskaya 1993), this may be due to the highest activity of microorganisms on rocks containing quartz. Thus, the channel sediments of the Sukokskaya River (Western Tien Shan), consisting mainly of particles of limestone, contain 2–10 times more microorganisms than the andesite porphyrites in the alluvium of these rivers. The maximum activity of microorganisms is characteristic of granite particles. This leads to a more rapid disappearance of particles from the petrographic spectrum of sediment, which are least resistant to biochemical destruction and relative uniformity of the mineral and petrographic composition of channel alluvium at a distance from local sources of its supply by debris material.

Thus, the most resistant to mechanical shredding were medium and coarse gravel, medium and fine sand. Despite this, it is the gravel fraction that has a much higher intensity of biochemical weathering under the influence of microorganisms than pebbles and sand. This results in the deposition of gravel as a type of bed material load due to continuous subaqueous biochemical weathering. This determines the fact that almost all rivers do not have a gravel bed, which is either in the form of impurities in the sand or pebbles, or composes separate forms of the bed or parts of it due to hydraulic sorting, and in the transition area from mountains to plains for this latter reason—lays relatively short sections of the bed.

The analysis of changes in the composition of bed material load from rivers, which are mountainous in the upper reaches of the river and otherwise flat, and which, accordingly, have a pebbly-boulder bed and then a sand bed, shows that the transition from a pebbly bed to a sand bed is either gradual (through a pebbly-sand bed) or abruptly, bypassing the gravel beds in all cases (Table 3.5). The Terek River (Lodina et al. 1987) gives an example of a sharp transition from the pebble bed to the sand bed, where an abrupt decrease in the average size of the channel alluvium is observed (approximately 50 times). At 200–250 km from the source, where the river extends beyond the foothill ripples to the Terek-Kuma plain, pebble deposits (average particle size of 80 mm) are replaced by sandy ones (average particle size of 0.55 mm). Upstream and downstream, the change in the particle size of bed material load, depending on the flow rate, occurs according to the lognormal curve (Fig. 3.3). Between the individual branches of the graph, there is a gap in the relationship $d_{average} = f(\rho gIQ)$. The lack of pebbles from tributaries and valley sides in the transition zone between the mountains and the plain causes a mismatch between the specific flow rate of ρgIQ and the size of bed material load. This abrupt transition from the pebble bed to the sand bed is typical of areas with a pronounced boundary between mountains and foothills, which is usually accompanied by a sharp change in slopes.

In the gradual transition from the pebble bed to the sandy bed, which is typical for the middle reaches of the Lena River, the large river bed is divided into two parallel branches in the section of several hundred kilometers in length, of which in the right river bed, which runs along the cliffs of the Lena-Aldan plateau, the bed material load sediment is pebbly-boulder with a gradually decreasing particle size

Table 3.5 Changes in the grain size distribution of bed material load on the Lena and Terek rivers

River	Distance from mouth, km	Type and shape of channel	Fraction content, %		
			Boulders, pebbles	Gravel	Sand
Lena	Incised				
	2423	Bend	100.0		
	2110	Single branching	89.7	10.3	
	1965	One-sided branching	70.0	16.3	13.7
	1860–1850	Parallel branch branching	28.0		72.0
	Wide floodplain				
	1742–1732	Conjugate branching	25.7		74.3
	1630	Conjugate branching	0.2	0.1	99.7
Terek	Incised				
	598	Straight line section	75.6	14.4	10.0
	393	Complicated branching	94.1	3.1	2.8
	250	Braided, meandering section			100.0

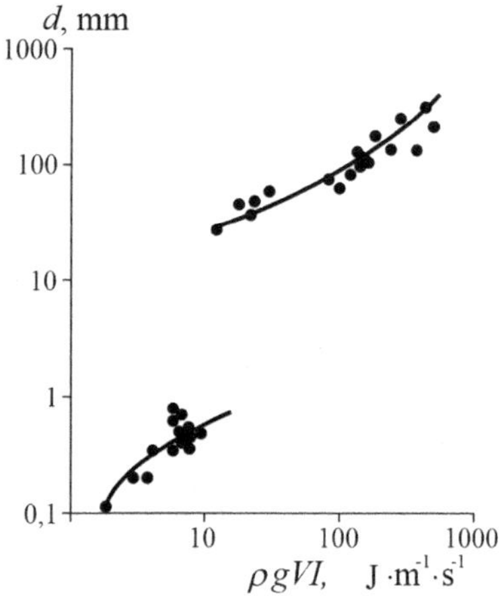

Fig. 3.3 Relationship between weighted average diameter of channel sediments (d, mm) of Terek river with the specific flow capacity [according to Lodina et al. (1987)]: V—flow velocity, I—water surface slope, ρ—water density

downstream to pebbly-sandy, and the left river bed is sandy for almost the entire length of the river. Sand spreading in the left-bank part of the channel is connected with the wide development of sandy terraces on the left bank of the riverbed, where significant amount of sand enters the channel as a result of washing up and sand inflow from the left-bank tributaries cutting through terraces and originating within the sandy Central Yakutsk lowland. These types of crossings occur on rivers flowing in the border area along the boundary of mountains and plains. If the rivers cross this area in a normal way, and the area itself is represented by a fairly wide zone of low-mountainous foothills, the pebble channel is replaced by pebbly and sandy, then sand and pebble and sandy with a gradual decrease in slopes, and depending on the hydraulic conditions, different parts of the channel are composed of either sand (side bars, shoals) or pebbles (pools, spindles) material with a gradual increase in the area of distribution of some and the reduction of others.

The natural reduction in the particle size of bed material load along the length of the river in accordance with A. Sternberg's Eq. (3.3) is usually most clearly visible in mountain rivers, but with no change in gradient, runoff rates or other flow characteristics, i.e. with no longitudinal hydraulic sorting. In particular, since the particle size of the load is correlated with the change in the thickness of the water flows ρgVI (here V is the average flow rate; I is the slope of the water surface; ρ is the density of the water; g is the acceleration of free fall), the decrease in I and V is accompanied by a decrease in the thickness of the flow and the average diameter of the particles that make up the channel sediments. At the upper 100 km of the Terek channel, abrasion of brown porphyrite fragments causes a 30% decrease in their initial weight (Lodina et al. 1987). For the next 100 km, it explains the reduction of $d_{average}$ by only 3%; the main factor in the change in particle size in the active bed layer here is the decrease in the sediment transport capacity of the flow, rather than the abrasion of the particles.

The A. Sternberg statute checkup was made by Sakharova and Lebedeva (1968) on the example of the river Mzymta (Western Caucasus). They found that the relative strength of rock fragments during their transportation by stream changes with distance from the source of inflow to the rivers and the reduction of particle size depending on the petrographic composition of the rocks. In the course of the transfer, low-stable rocks disappear in the deposition at a small distance from the feeding zone. At the first kilometers of transport, the rocks collide and rub against each other, and they are also washed away by a thin layer of rock prepared by weathering. Particles of the strongest rocks—basalts, some types of quartzites, effusives, flint—travel hundreds of kilometers, with little change in diameter. Therefore, the partitioning between rocks of different petrographic composition in pebble-boulder fractions of channel alluvium is determined not so much by the degree of their distribution in the basin as by their resistance to destruction. As the size of the river increases, the selection of the material to be transported begins to play an increasingly important role in terms of its abrasion resistance. Determination of the impact of changes in the length of the river on the hydraulic characteristics of the flow: all other things being equal, in rocky gorges, where the flow is heavily constrained, the shredding of debris, especially the large ones, is much more intense than in adjacent areas with "normal" channel. As a result, E.I. Sakharov and N.V. Lebedev proposed a modification of A. Sternberg's equation for the main groups of erupted and sedimentary rocks developed in the Mzymta river basin (Table 3.6); taking into account the specifics of abrasion and crushing of debris particles (boulders, pebbles) in the process of their transportation by flow, Eq. (3.3) took the form

Table 3.6 Changes in the average diameter of pebbles and boulders of different petrographic composition from the length x of their transfer by the flow [according to Sakharova and Lebedeva (1968)]

Rock formation	Formula	Largeness (average diameter, cm)		Error, %
		d_0	d_x	
Quartzites, quartzite sandstones	$d_x = d_0 - 2.6x^{0.6}$	29.5	9.2	9–22
Tuffs, tupphobreccia, diabase	$d_x = d_0 - 2.2x^{0.6}$	23.8	7.9	1–2.5
Tupphobreccia, porphyrites	$d_x = d_0 - 1.7x^{0.6}$	28.1	8.9	2.5–12
Limestone	$d_x = d - 3.5x^{0.6}$	26.5	6.6	8–31
Granites, granite-gneisses, granodiorites, crystalline shale	–	19.9	11.1	–

$$d_x = d_0 - ax, \tag{3.4}$$

where a is the coefficient depending on the petrographic composition of rocks by groups in Table 3.6.

The A. Sternberg Eq. (3.3) and its modification (3.4) apply only to pebbles and boulders of a certain petrographic composition. In fact, on rivers in mountainous countries, bed material load are represented by a fairly diverse range of rocks, and changes in their average particle size are determined both by the hydraulic sorting of debris and by the continuous flow of material from slopes or tributaries. The most important condition for the hydraulic sorting of bed material load is the slope of the river. According to the equation

$$d_{\text{av}} = f(I), \tag{3.5}$$

the size of the bed material load particles is proportional to the slope of the free flow surface. This ratio on the basis of field observations was confirmed by many researchers. For mountain rivers in Central Asia (Altunin 1962), the relationship between the average diameter of channel sediments and the slope of the channel is as follows

$$d_{\text{medium}} = 4.5I^{0.9} \tag{3.6}$$

Litvin (1971) established a linear link between the slope of rivers and the average diameter of bed material load on the rivers of the Caucasus.

Since the formula (3.6) is empirical and does not meet the dimensional conditions, A.N. Kroshkin (Talmaza and Kroshkin 1968) proposed the expression

$$d_{\text{medium}} = 1.6I^{0.9}\left(\frac{Q_{\text{f}}}{\sqrt{g}}\right), \tag{3.7}$$

The amount of bed material load linking the particle size of the bed material load depends on the slope of the river and the size of the effective discharge Q_f. An increase in the runoff rates of the river (flow rate) during a flood results in both an increase in water sediment concentration and sediment discharge (2.18) and an increase in the consolidation of bed material load. The physical meaning of formula (3.7) depends on the size of the sediment load on the flow rate and its transport capacity, which is determined by the runoff rates of the river.

Since, according to the Chezy formula, the slope and flow rate are related, the relationship of the sediment particle size on the flow parameters can be obtained from the assumption that changes in flow rate, i.e. changes in the flow rate, are the leading factor in their hydraulic sorting

$$d_{av} = f(V). \tag{3.8}$$

Replacing the depth h in the Chezy-Manning formula with the expression $h = \frac{Q}{V_p} h$, we obtain

$$V = n^{0.6} Q^{0.4} b_{av}^{-0.4} I^{0.6} \tag{3.9}$$

where b_{ch}—channel width; $Q = V\omega$—water consumption; ω—living area; n—roughness coefficient. For mountain rivers, it can be assumed (Altunin 1962; Talmaza and Kroshkin 1968) that the coefficient n is mainly determined by the granular roughness: $n = f(d)$. Then

$$d_{medium} = kQ^{0.4} b_{ch}^{-0.4} I^{0.3}. \tag{3.10}$$

Since the river's runoff rates (average annual discharge Q) is proportional to its catchment area F, the

$$d_{medium} = f\left(F^{0.4} b_{ch}^{-0.4} I^{0.3}\right). \tag{3.11}$$

The necessary data for the construction of this relationship were obtained during the study of the Western Tien Shan rivers (Khakimov 1992; Lodina et al. 1994). The range of points in the graph (Fig. 3.4) is very large, which does not allow us to speak about a stable connection between the considered values. This is explained by the action of factors external to the flow that determine the flow of sediment into the channel: debris flow and slope processes. In mountainous terrain, their influence on the composition of bed material load is so great that it obscures the role of the hydraulic factors themselves. Nevertheless, the influence of hydraulic sorting can be traced along the upper point envelope field, which limits the maximum possible average particle size of bed material load.

The distribution of bed material load along the length of the rivers of the Western Tien Shan tends to reduce their average diameter downstream due to the reduction of slopes and the abrasion and crushing of rock fragments. However, mudflow inflow outbursts and alternation of narrowings and valley extensions break

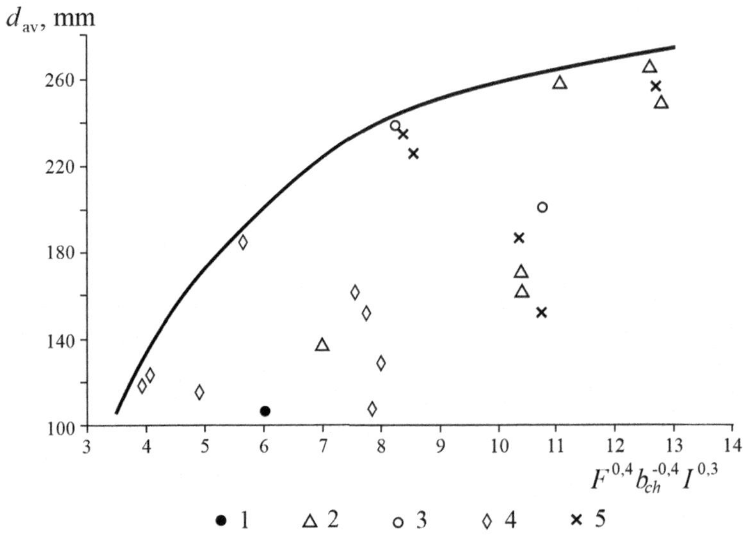

Fig. 3.4 Relationship between average size of sediments in armouring layer ($d_{av.arm.}$) with index (3.11) $F^{0.4}$ $b_{ch}^{-0.4}$ $I^{0.3}$ on rivers of West Tien Shan; 1—Oygaing river; 2—Pskem river; 3—Chatkal river; 4—Ugam river; 5—Akbulak river

this regularity, forming sharp maximums on the graphs of average particle size distribution along the length of the rivers. For example, the graph of the change in the average diameter of bed material load along the length of the Ahangaran river (Fig. 3.5A) shows a general decrease in the average diameter of sediment particles in the downstream section of the ripple: 402 mm in the upper section, 176 mm in the middle section and 149 mm in the foothill area near Angren town. Against this background, a significant increase in load is observed at 32 km from the source ($d_{average}$ = 289 mm) due to mudflows from the left-bank tributary of the Kengkul and in the middle reaches (43 km), where a sharp increase in the average diameter of the slipway is due to the constriction of the channel in the gorge and a significant increase in flow velocities, which ensures the transit of smaller material outside the area. Above the gorge, where the channel is wide floodplain and has an impact on the pressure from constriction, the particle size of the load is reduced to a minimum. The alternation of contractions and expansions determines the changes in the hydraulic and morphological characteristics of the flow—slope, channel width, depth and flow rate (Table 3.7), which results in an increase or decrease in the size of bed material load.

As a result, only in the foothill zone, where there are no sharp changes in the width of the valley floor, a gradual decrease in the average diameter of sediment is observed. This is also due to the fact that there are no sources of sediment inputs into the river bed.

On the rivers Pskema, Ugama, Akbulak, the tendency to decrease the average size of the material in the paving is less clearly expressed (Fig. 3.5B), which is

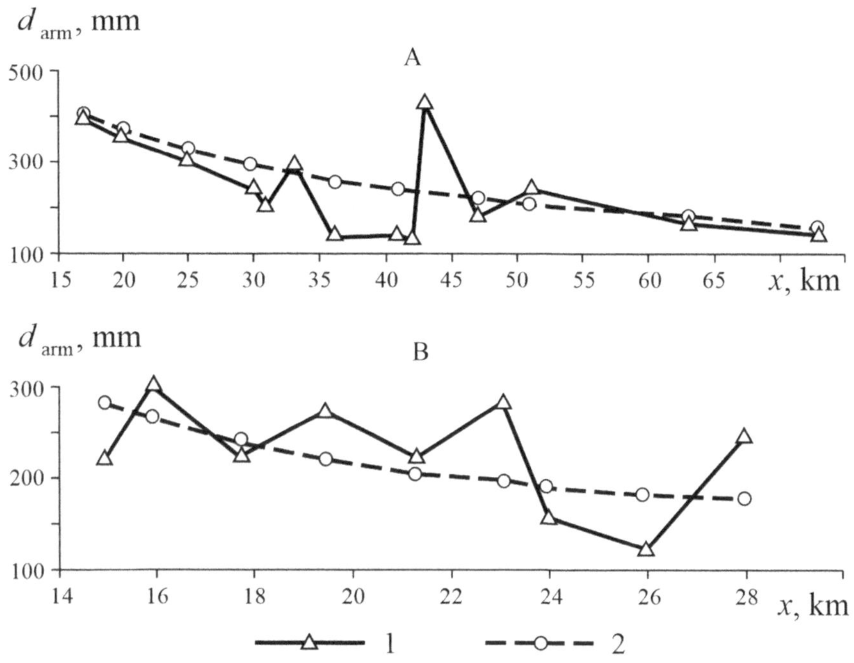

Fig. 3.5 Change of average size of pebble-boulder sediments in armouring layer along the rivers: **A** Akhangaran; **B** Akbulak (Western Tien Shan). 1—average diameter of armouring layer material; 2—longitudinal profile of the river [according to R.V. Lodina et al. (1994)]

explained by the more active influence of slope and mudflow processes in the formation of bed material load in comparison with the valley of Ahangaran.

Typically, the formulas (3.10) and (3.11) take into account to a greater extent the factors of hydraulic sorting of sediments: along with the slope, they also include the width of the channel, the changes in which cause the flow to become constricted or to melt, which is accompanied, respectively, by an increase and decrease in the average particle size of sediments.

The runoff rates of rivers as a factor determining the particle size of bed material load was used by G.A. Petukhova with respect to rivers in the large plain region (Sediment load … 1977). It revealed the relationship between the grain-size composition of bed material load on rivers in Eastern Europe (for its characteristic, the particle size of the sediment corresponding to a 50% probability on the distribution curve of $d_{50\%}$) and the mean multiyear water flow module M_Q (Fig. 3.6). This relationship is differentiated by different geomorphological regions, which differ in terms of the conditions of sediment formation: I—north of the European part of Russia and the Baltic States; II—area of uplands forming the main watershed of the ETR; III—water-glacial and alluvial plains along large rivers of steppe and forest-steppe zones (Dnieper-Donetsk, Oka-Don and Middle Volga regions); IV—uplands of forest-steppe and steppe zones (Central Russian, Volga, Obshchy

Table 3.7 Riverbed sediment size (in the armouring layer) and hydraulic flow characteristics in constrictions and enlargements of some river valleys Western Tien Shan (Lodina et al. 1994)

River–gauging station	Valley morphology	Flow velocity in low water, m/s	Slope I, ‰	Sediment diameter	
				$d_{average}$, mm	d_{max}, mm
Ahangaran—the mouth of Kamchiksay	Enlargement	1.31	6.31	131	256
Ahangaran—0.5 km lower Irtashsay's mouth	Constriction	2.43	13.48	423	716
Ugam—lower the resthouse "Algorithm"	Enlargement	1.85	6.70	97	301
Ugam—upper the mouth of Karakuzi	Constriction	2.86	10.22	153	353
Oygaing—lower the mouth of Koksu	Enlargement	2.01	14.19	85	261
Oygaing—8 km from the mouth	Constriction	2.56	19.11	103	296
Akbulak—lower the mouth Terekli	Enlargement	1.45	13.20	154	343
Akbulak—3 km from the mouth	Constriction	2.66	23.10	324	688

Syrt, Volyn-Podolsk, Dnieper, Donets Ridge); V—Black Sea and Caspian lowlands; VI—foothills of the Caucasus.

Lodina (1974), having accepted that the average particle size of alluvium (d_{av}) on mountain rivers is a function of the river slope (I) and water discharge, the indirect measure of which is the length of the river from the source (l), i.e. $Q = f(x)$, found for the rivers of the Western Transcaucasia that

$$d_{av} = AIx, \tag{3.12}$$

where coefficient A depends on the predominant distribution of rocky formations in river basins of one or another lithological composition: $A = 0.001\ F'$—for river basins predominantly composed of sedimentary rocks (limestone, sandstones); $A = 0.0025\ F'$—for basins composed mainly of erupted rocks; $A = 0.01\ F'$—for basins composed mainly of rock complexes (here F'—the share of the basin area composed of rocks of different composition, %).

The natural reduction in the size of bed material load on mountain rivers is characteristic of the relatively simple orographic mountainous countries that form a single arch (the Greater Caucasus, the Ural mountains); the main rivers, originating in the axial zone, normally cross the mountain system and enter the foothill or plain zone. A similar picture is observed in the complex orography of mountainous countries, but in the rivers flowing down from the edge of the ripples. The rivers flowing through such mountainous countries on a large length, crossing repeatedly

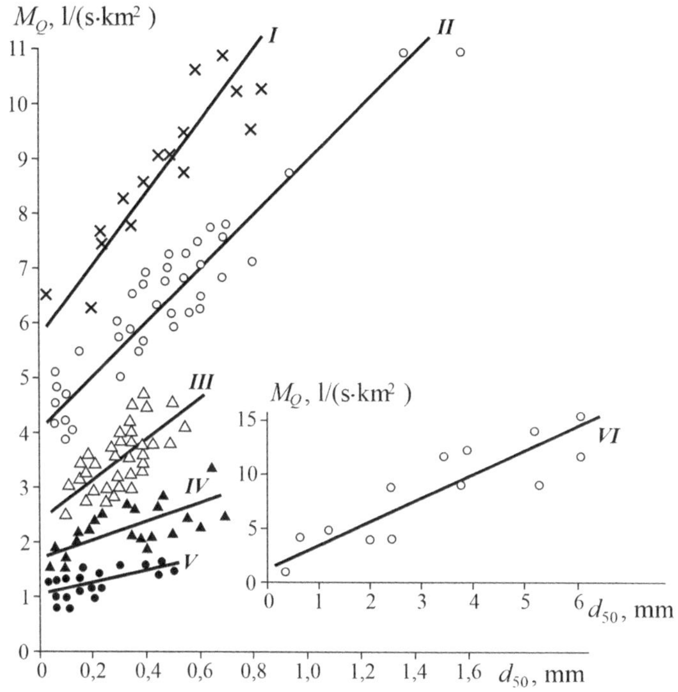

Fig. 3.6 Dependence of grain size composition of bed material load d_{50} from average water runoff module M_Q: I to VI—numbers of areas (see text) [according to G.A. Petukhova (Sediment runoff … 1977)]

mountain ranges and intermountain troughs, composed of rocks of different strength and petrographic composition, are distinguished by the lack of connection between the d_{av} and the slopes of the rivers. On the Fig. 3.7 longitudinal profiles and graphs of changes in the mean diameter of pebbly-boulder sediment at the Kodori River in the Western Caucasus (A) and the Katun and Chuy Rivers in the Altai (B) are presented for comparison. In the first case, the rapid decrease of slopes in the high-mountainous area is accompanied by a relatively slow decrease of d_{av} in the upper course; then there is a synchronous sequential decrease of both d_{av} and I with a slightly higher intensity for d_{av}. In the Altai rivers, differences in the shape of longitudinal profiles and the complexity of the relief and geological structure of the basins result in different changes in the size of bed material load along the length of the rivers (Demin et al. 1991). The only common factor is its reduction within the limits of the intermountain troughs, especially at Katun, and when the river comes out of the mountains into the foothills. Below the Katun depression, there are clearly marked maximums of increased sediment particle size against the general background of its consistent decline to the mouth of the river. The peak of grain size at 240 km is confined to the gorges between the Kuminsk and Seminsk ridges.

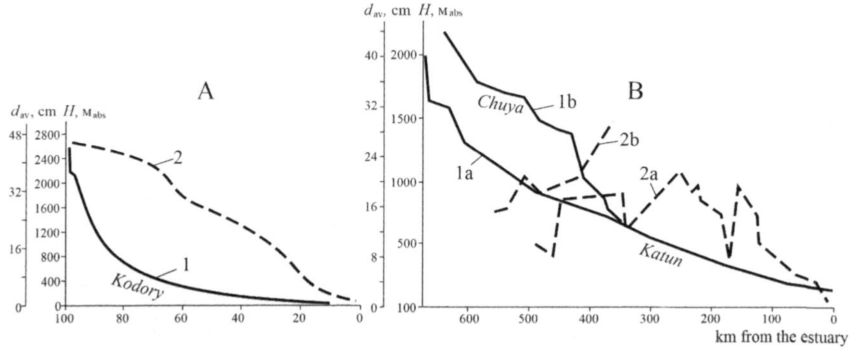

Fig. 3.7 Longitudinal profiles (1) and change of average diameter of pebble-boulder sediments (in the armouring layer) (2) along the rivers: Kodori in the Western Caucasus (**A**), Katun (a) and Chuya (b) in the Altai (**B**)

The change in the average size of the bed material load in Chuy, the largest right tributary of the Katun river in the mountainous zone, shows a general trend towards their consolidation downstream. Such an anomalous distribution of mean size along the length reflects the peculiarities of valley morphology, longitudinal profile and changes in channel types. In the highland Chuy basin, the river has a flat channel with low slopes and, consequently, a relatively small composition of alluvium. The crossing of the Kurai Ridge, where the longitudinal profile of the Chuy has a steep step, is accompanied by an enlargement of the average diameter of load. Below the Kurai steppe, the river, flowing in gorges, increases the slope by almost 20 times compared to the upstream section, which causes more than a twofold increase in the average particle size of bed material load in the lower reaches of the river.

The fact that on the Chuy in the upper and middle reaches of the riverbeds the mountain types of the channel are replaced within the depressions by the plain ones, determines a discontinuous character in the distribution of pebbly-boulder sediment. At Katun, pebble and boulder deposits are not interrupted anywhere despite fluctuations in their size along the length of the river. With the same specific flow rate, the sediment load on Katun is 2–2.5 times larger than on Chuy (Fig. 3.8); only with small values of the product I_{av} V is the size of bed material load on both rivers identical (fine pebbles).

The smaller the particle size of the load, the greater the sorting, i.e. the homogeneity of the particle content of one or similar fractions. On sandy rivers, medium-grained sands often contain an admixture of gravel and are almost never mono-fractional. Pebble and pebble-boulder deposits are a multifraction mixture in which only the particles of the fractions that define their names predominate. The larger the particle size and the less sorted the particles, the more likely it is to form an alluvial alluvial sinking of the alluvium armouring thickness from a multifraction material. On large flat rivers with a pebbly-boulder composition of sediments, the bedding size is usually 30–40% larger than that of the sediment underneath it (Fig. 3.9A), and on mountain rivers it is up to 1.5–2 times larger, although more

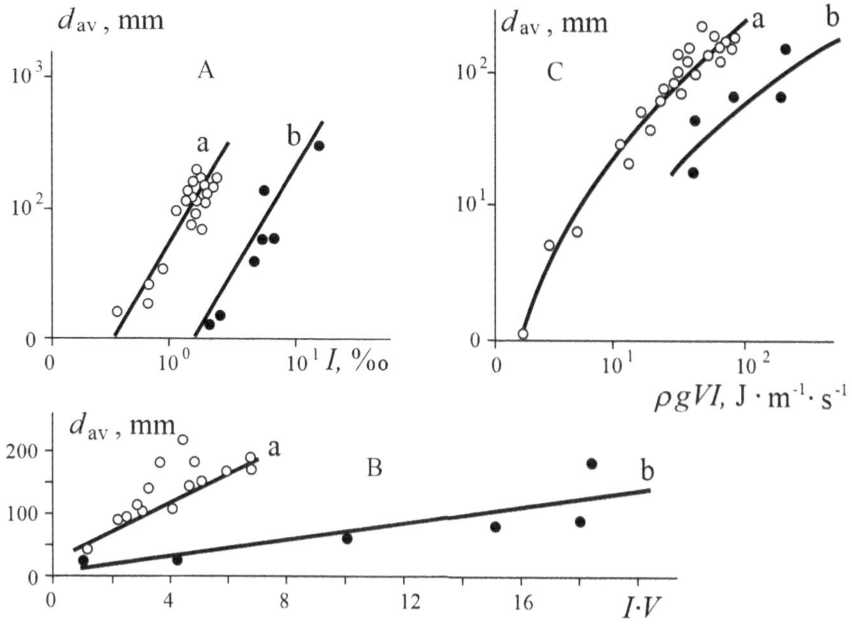

Fig. 3.8 Dependence of weighted average diameter of bed material alluvium d_{av} of Katun river (a) and Chuya river (b) from average slope I (**A**), composition of the slope and flow velocity during low-water IV (**B**) and specific flow capacity ρgVI (**C**)

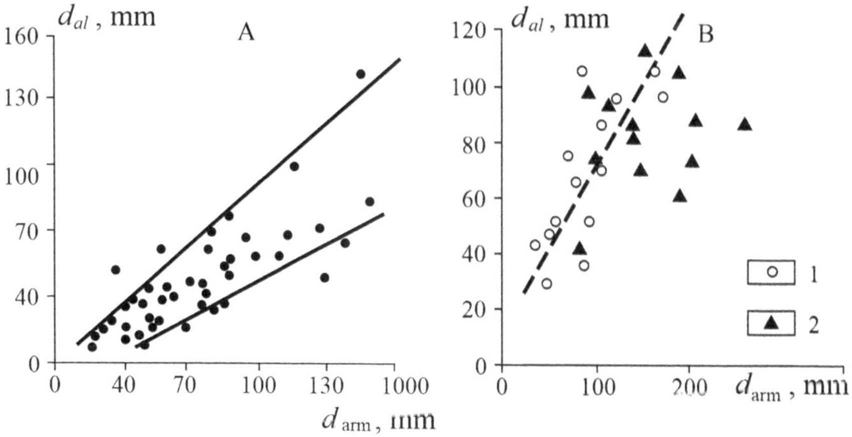

Fig. 3.9 Connection between average size of bed material (d_{al}) and average diameter of the material of alluvial armouring layer ($d_{arm.}$): **A** on the lowland river with pebble-boulder channel (Aldan river) [according to Berkovitch et al. (1985)]; **B** on mountain rivers of Western Tien Shan [according to Lodina et al. (1994)]: 1—Ugam river; 2—Akbulak river

significant differences are possible (Fig. 3.9B). On the upper Yenisei, the average particle size of the sill is 64–128 mm (maximum dimensions—up to 266 mm); the thickness of channel sediments under the sill is represented by the average pebbles, its average diameter is 37.4 mm. On rivers with a pebbly-boulder bed, the bluff is a very characteristic phenomenon. In the low-water, a destructive of such a channel practically stops. In floods, when the wreckage components come into motion, a large amount of gravel and sand material, lying under the ripple, is involved in the flow, and its sediment concentration increases by tens and hundreds of times. On the flat rivers formed in sandy sediments, there is no detachment.

The formation of the armouring layer can happen in two ways. Firstly, due to the different impact of the flow on different particle sizes, selective erosion of the bottom occurs, due to which the predominant removal of relatively small particles takes place, while the larger ones remain in place. This process may eventually lead to the formation of an island. Such sedimentation, if it has formed, practically means stopping change associated with bed scour and the involvement of sedimentary particles in transport. Usually it is typical for rivers, the bottom of which is underlain by non-alluvial (moraine, fluvioglacial) or ancient alluvial pebble-boulder deposits. In this case, the armouring layer is formed from the coarse pebbles and boulders—perluvium—remaining in place after the erosion of the main sediments or ancient alluvium (Lamakin 1943).

Secondly, the stoppage of sediment load at the flood recession occurs in conditions where the hydraulic characteristics of the flow (speed, depth) are reduced. This results in a sharp stop of the particles of the largest sediment fractions (coarse pebbles, boulders). As a result, there is a scaffold covering a variety of material, in which large particles (pebbles, boulders) were moved. Only a fraction of the finer loads (fine pebbles, gravel, sand), which at the time of detachment were moved as part of the suspension or saltation, are carried outside the area, causing a longitudinal differentiation of the sediment in the detachment in terms of particle size. Such armouring layer is a temporary formation that occurs at the flood recession and in the low water and collapses at a new rise of levels. It is characteristic of bed material pebble-boulder deposits, both mountainous and plain, flowing in the mountains of rivers.

These formulas (3.3), (3.4), (3.7), (3.10) characterize the grinding intensity of coarse sediment (boulders, pebbles, gravel). When the size of the particles becomes less than 1 mm (sand fraction), their further abrasion under the influence of mechanical agents is practically stopped. Therefore, on flat rivers with sand-forming bed material load, the natural change in alluvium particle size along its length appears to be strongly shaded under the influence of local factors and hydraulic (depending on changes in the hydraulic characteristics of the flow) sorting of transported sediment particles. At the same time, it is often not the slopes, which in general do not change much along the length of lowland rivers, that come to the foreground, but the changes in the width of the channel and the bottom of the valley (floodplain). The role of local sources of sediment inputs is also increasing significantly, such as sediment transport from tributaries and ravines, erosion of banks with different geological structures, crossing of structural or moraine uplands by rivers, etc. As a result, the overall reduction in the size of bed material load on flat rivers is only evident at the macro level, from the upper reaches to the mouth of the river, along the entire length of the river. Therefore, some researchers, studying the composition of load only over a limited, although sometimes quite long, area, conclude that there is no natural change in the size of load on flat rivers at all (Lebedeva 1966).

The alternation of narrowing and widening of the valley bottom causes concentration or flow spreading in the flood. In constrictions, the flow, concentrating in a single channel, significantly increases its transport capacity. As a result, the average diameter of bed material load here increases compared to the adjacent sections of the channel. Large floodplain expansions are characterized by a 2–2,5-fold decrease in the average weighted diameter of the load, which is associated with a decrease in the specific values of effective water discharge. Within each major expansion of the valley floor, the change in the size of channel sediments from the beginning to the end of the expansion is clearly visible. The local increase in speeds before and at the beginning of expansion causes bed scouring and the removal of smaller material to the downstream areas. In the expansion itself, by reducing the specific water consumption, the size of sediments is significantly reduced. In the lower reaches rate expansion due to the support from the next contraction decreases even more, and here the smallest material is deposited. For example, on the lower 250 km of the river. The average diameter of pebbles and sand deposits varies from 78 to 25 mm (Belyi et al. 1979). Against the background of a general decrease in alluvium particle size, the bead-shaped structure of the valley floor results in a wave-shaped change in the average diameter of load (Fig. 3.10): maximums correspond to the valley floor contractions and minimums to expansions. The average alluvium particle size determined for expansion and contraction tends to decrease downstream, as the main factor in its distribution of load on particle size is hydraulic sorting as a result of the reduction of downstream gradients. Other factors (increase or decrease of specific effective discharge during floods, inflow of sediment as a result of screes and collapses from slopes or at erosion of terraces, tributary inflows) determine changes in d_{av} through fluctuations in the transport

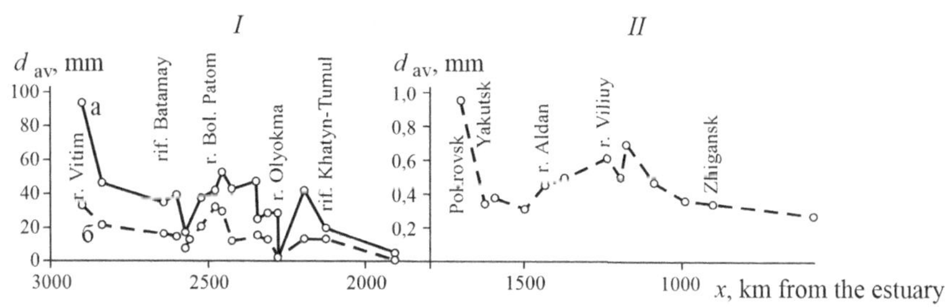

Fig. 3.10 Changes of size (d_{av}) of bed material load along the river on the Middle and Lower Ob: **I**—for pebble-boulder and sandy-pebble channel (a—armouring layer, b—alluvium layer), **II**—for sandy channel

capacity of the flow or have a local impact on the composition of the load. The first circumstance leads to the fact that within the floodplain logs there is an inverse relationship of the sediment load size on the average channel width (b_{ch})

$$d_{av} = kb_{ch}^{-a}, \quad (3.13)$$

It is traced against the background of changes in the particle size distribution of the load due to alternating contractions and valley bottom extensions. In the Ob, the coefficient values in the formula (3.13) are the same, at least from the confluence of Bia and Katun to the mouth of the Tom river: k = 1.42, and = 1.17 (Berkovich et al. 1972); on the middle and lower Lena, the degree a depends on the distance of the river section from the main source of inflow to the channel of relatively large material: at a distance of up to 50 km a = 1.8, from 50 to 100 km a = 1.2, from 100 to 150 km a = 1.0 and more than 200 km a = 0.7 (Berkovich et al. 1981).

On rivers with a complex braided and weakly stable channel, this relationship is complicated by the dispersion of the flow between the main and secondary branches. If we take as an indicator of the branching of the channel the number of branches n_{sl} in which the flow rate $Q_{br} \leq Q_0$, where Q_0—the total flow rate in the river, the connection between d_{av} and the n_{br} looks like

$$d_{medium} = kn_{br} + c, \quad (3.14)$$

where k and c are empirical coefficients equal to 0.03 and 1.2 on the middle and 0.007 and 0.8 on the lower Lena. Local increases and decreases in the average particle size of bed material load are also due to changes in the velocity field at the bend of the channel, at branching of the channel into branches or at the confluence of branches, in pool valleys and at riffles. The inverse relationship between the mean particle size of the load (d_{av}) and the radius of curvature of the bend (r) is characteristic of free radii. This is due to the local increase in currents at the bend. In the case of entrenched bending, this connection is obscured by the influence of the main bank. The reduction in the curvature radii of the three adjacent free radii on the Vychegda River (over the period 1910–1994), sequentially from 1.5 to 1.4 km, from 1.25 to 1 km and from 3.5 to 1.1 km, has led to an increase in the average diameter of bed material load from 0.37 mm (Lebedeva 1966) to 0.60 mm (Beliy et al. 2001). The upper wings of the upper Ob bend are lined with coarse, medium-large and medium-grained sands; the lower wings of the same bend are composed of finely medium, fine and fine-grained sands. Local increases in the particle size of the load are observed in the straightening branches of the bend. Thus, in the straightening Izvekovskaya channel (Ob River) the average diameter of alluvium is 0.36 mm, while the average diameter of this river section is 0.23 mm.

The distribution of sediment load in the branches is related to their relative runoff rates: branches with higher runoff rates are lined with larger sediments (Table 3.8).

Hydraulic sorting of bed material load is observed along the width of the riverbed according to the distribution of flow velocities: the bed of the channel is usually lined up along the streambed with larger material, and in coastal zones with

Table 3.8 Ratio between the bed material size in the branches of the Ob River and their relative runoff rates

Title of branches and branches	Relative runoff rates, %	Average diameter, mm
Voronovskoye		
Kashtakovskaya cannel (right)	64	0.46
Voronovskaya cannel (left)	13	0.34
Boroshnoye		
Right	64	0.74
Boroshnaya cannel (left)	19	0.41
Monastery		
Left	57	0.59
Right	42	0.38
Ust-Tomskoye		
Braginskaya cannel (left)	60	0.91
Bitenevskaya cannel (right)	19	0.35

smaller material. The drifting of the streambed with level fluctuations sometimes results in the fact that during the low-water period the dried bedrock banks turn out to be composed of larger sediment loads than the bedrock zone in this regime phase. A similar phenomenon is also observed in the braided channel, where low-water hoses in the low water level can be characterized by the increased size of bed material load, reflecting the hydraulic characteristics of the flood flow and runoff rates of the hose at high levels. In particular, the more coarse loads in asymmetrical branches are characterized by more straight branches, because in them the dynamic axis of the flow shifts into the high-water phase of the regime and its runoff rates increases. For example, the Kuchukskaya channel on the upper Ob is lined with sands of average size of 0.42 mm, which is 1.5 times larger than the size of the load in the left curved and longer main channel, but with less runoff rates.

All these regularities are disturbed due to the heterogeneity of the geological structure of the riverbed and the indigenous banks, as well as the composition of alluvial deposits that compose the terraces and floodplains. In this respect, the change in the size of the load along the length of the Lena River, which crosses various geomorphological regions of Eastern Siberia, and the typical river of the plains, the Neman, is indicative. At the upper Lena (Borsouk and Lodina 1974), the valley of which dissects the Lena plateau in the absence of significant tributaries downstream of Ust-Kut, pebbles are rapidly crushed and 250 km before merging with the Kirenga River, the bed is sandy. Downstream, as numerous small mountain rivers flow into the river, the sediment becomes pebbly again, but sandy-pebbly before it merges with Vitim. On the middle and lower Lena (Waterways … 1995) on the background of transition from the pebble-boulder bed to the pebbly-sand bed in the area from the mouth of the Nui River to the Pokrovsk, where the river enters the Central Yakutia Lowland and the sandy downstream and sandy-silty in the delta branches, there are several peaks of consolidation of bed material load associated with changes in the geomorphological structure of the basin and the removal of larger loads from the tributaries (Fig. 3.10). In the middle of the Lena River, sediment accumulation occurs where the river flows along the Baikal-Patom Plateau. Above the mouth of the Olekma River in the Killah expansion channel is sandy. Sediment removal from Olekma makes the channel pebbly-boulder again. Downstream, the channel is characterized by two stripes of different types of alluvium: the left-bank part of the channel, including all the left-bank branches and channels, is lined with sandy deposits, while along the right "mountain" bank stretches pebbly-boulder deposits. Below the village of Pokrovsk in the sandy bed, the borrowed consolidation of load occurs between the mouths of the Aldan and Vilyui, due to, firstly, the arrival of larger loads from the Aldan, and secondly, the location of the river along the spurs of the Verkhoyansk mountainous country. In the lower reaches before entering the delta, the river flows in the section of about 200 km in the Lena pipe—a gorge between the Kharaulakh Mountains and the Chekanovsky ridge. This again leads to a sharp increase in load due to local sources of replenishment; transiting sands from above are transported through this area to the delta.

In terms of the composition of bed material loads, the Neman (Darbutas et al. 1993) can be divided into several sections (Fig. 3.11). From the confluence of the Ussa and Losha rivers to the mouth of the Shchary River, where fine and medium sands predominate. Here, the Neman flows along alluvial and water-glacial lowland plains composed of sandy and sandy-clay sediments (Voznyachuk and Valchik 1978) and has a wide floodplain meandering channel. From the estuary of the Shchary River to the estuary of the Kotra River, fine and coarse-grained sands are widespread in the valley extensions along with medium-grained sands, and in the

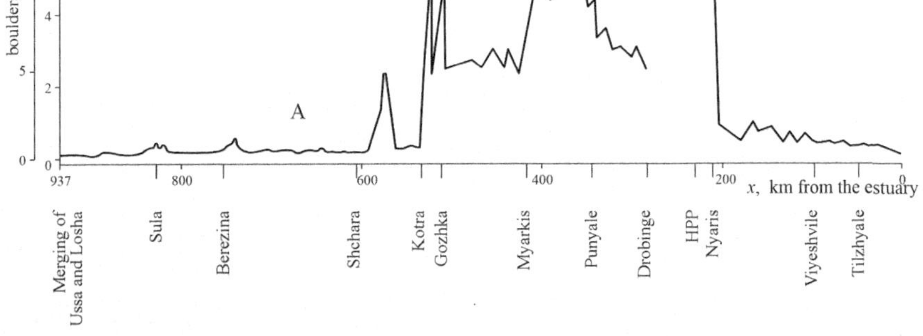

Fig. 3.11 Changes of the size d_{av} of bed material load (sand and gravel) (A) and boulders in the channel (B) along the Neman river

constrictions—gravelish-sandy sediments and gravel. From the confluence of the Kotra River to the mouth of the Gozhka River, the channel is dominated by pebbles and gravel, and in some places by pebbles and boulders. Below the estuary of the Gozhka River to the estuary of the Myarkis River, coarse-grained and gravel sands are the bed material load in the valley floor extensions; pebble-boulder material is found in the valley floor constrictions. Down to the Kaunas Reservoir (Drobinga River estuary), bed material load are represented by coarse-grained and gravel sands, with many "single" stones and rocky ripples in the channel. However, in areas of increased sediment thickness, only gravel sands and gravel are riverine sediments; pebbles and pebble-boulder are quaternary perluvium [according to Lamakin (1943)] that form the bedrock and the banks of the river. Here Neman crosses the Grodno Upland, the Baltic and South-Lithuanian moraine ridges and glacial glacial prolongations made by fluvioglacial deposits (Gudyalis 1973; Dvaretskas 1989). Here the incised channel prevails, and only at crossing of prolongations it becomes wide floodplain on short pieces (Darbutas and Chalov 1991).

Directly below the Kaunas hydroelectric power station and up to the point of confluence with the Nyaris River, the composition of the load here is determined by the conditions for the erosion of the Neman riverbed in the area near dam. Gravel deposits dominate here. Below the mouth of the Nyaris River down to the delta branch on the branches near the town of Sovetsk (below the mouth of the Tilzhyale River), the channel is sandy, and at first medium-grained sands prevail, and then fine medium-grained ones. Sources of load are moraine loams and sandy loams from the Valdai lake glacial complex and sandy loam in the delta zone. Against the background of a general decrease in the size of the load, its fluctuations are related to local features of the geological structure of the coastline up to 1.5 times: where the channel passes through floodplain banks or along sandy terraces, the particle size of the load decreases; when moraine loams saturated with pebbles and gravel or coarse fluvioglacial sediments are washed away, it increases.

The relationship of the bed material load size on the slopes of flat rivers is manifested differently, determined by the geological and geomorphological conditions of bed formation. On rivers with pebbly-boulder and sandy-pebble bed (Lena, Aldan) it is clearly visible if significant fluctuations of slopes are observed along the length of the river (Fig. 3.12), and the closeness of the bond is approximately the same for the material composing the ripple, and for the alluvial thickness lying under it. In rivers with large sandy and smaller sediment compositions, this connection can only be observed in the wide floodplain channel with a relatively constant valley floor width and homogeneous geological structure (the upper Ob, from the confluence of Biya and Katun to the town Kamen-na Obi). However, the presence of several sources of load (inflows, sandy terrace cliffs and the plateau composed of loess-like loam) is the reason for the scattering of points in the $d_{av} \sim$ I coordinate field (the correlation coefficient is 0.6–0.7) (Table 3.9). At the Ob, below the Novosibirsk hydroelectric power station, and at the lower Lena and Vychegda, due to the complex geological structure of the seabed and banks, the relationship $d_{av} = f(I)$ is absent, despite the sandy composition of the load and the wide floodplain bed.

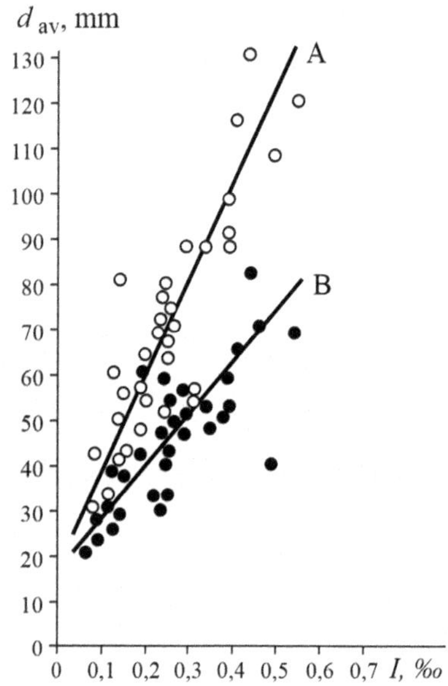

Fig. 3.12 Dependence of average size (d_{av}) of bed material load of a armouring layer (A) and alluvial layer (B) on slopes during the effective discharge passing on the Middle Lena river (Water ways … 1995)

Table 3.9 Changes in slope and particle size of bed material load on the upper Ob (Berkovich et al. 1990)

Distance from the confluence of Bia and Katun, km	I, ‰	$d_{average}$, mm	Distance from the confluence of Bia and Katun, km	I, ‰	$d_{average}$, mm
0–6	0.20	0.67 (3.7)[a]	94–106	0.17	0.46
6–12	0.17	0.74 (1.4)	106–114	0.16	0.42
12–24	0.18	0.52	114–135	0.12	0.39
24–36	0.16	0.45	135–150	0.11	0.45
36–45	0.20	0.56	150–175	0.13	0.40
45–56	0.20	0.51	175–200	0.09	0.35
56–72	0.16	0.47	200–250	0.08	0.35
72–82	0.17	0.44	250–320	0.06	0.24
82–94	0.15	0.48	320–360	0.07	0.34
			360–495	0.06	0.33

[a]In parentheses—including gravel admixture

In rivers characterized by fine sandy and sandy-silty loads (the Huang He, Yangtze, and Amu Darya) and very small bed material load loads and the absence of additional sources of their inflow, the length of the rivers shows a fairly clear relationship of their size on the slope. In the Huang He region, starting with the Xiaolandi HPP, the relationship of the median diameter d_{50} on the slope is particularly clear in comparison with the average river flow (Fig. 3.13), for which a large point range is observed in the graph; the load size here is more sensitive to

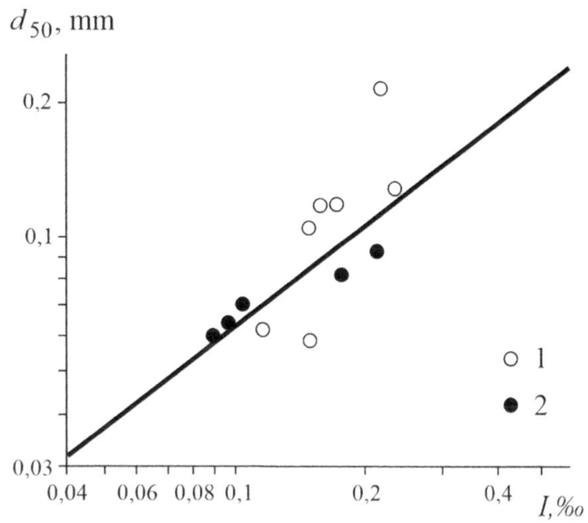

Fig. 3.13 Changes of bed material load size (d_{50}) in the middle (1) and lower (2) current of the Huang He river in dependence on the slope [according to Jiang Ning and Zhou Wen-Hao (1955)]

changes in slope, which is manifested in a decrease in the slope of the curve $d_{50} = f(I)$. However, in contrast to sandy and pebbly-boulder rivers, this relationship is logarithmic rather than linear. Such rivers are characterized by a close correspondence between the average diameter of the particles represented in suspension and the channel sediments of the d_{av} (see Fig. 2.10B). At the Huang He River, suspended sediment at a depth 0.031–0.022 mm in the lowland area of the Huang He River is reduced to 0.031–0.022 mm downstream. The average diameter of riverbed sediments is between 0.057 and 0.062 mm, practically unchanged along the length of the river. At the Yangtze River, the particle size of suspended load is basically the same, with a value of d_{50} almost identical to that of suspended load in Huang He waters; in the lower reaches (Nanjing city) it is noticeably smaller ($d_{50} = 0.012$ mm). Average diameter of channel sediments of the river. The Yangtze is a lot bigger than the Huang He, 0.156 mm. In the direction of the mouth of the river d_{50} decreases by almost 3 times. At such a ratio, even a relatively small change in the hydraulic characteristics of the flow leads to a mass transfer of particles from the bed sediments into the suspended matter and vice versa (Chalov et al. 2000).

On pebble-boulder rivers, due to the presence of armouring layer, the particle size of the load does not change in the long term with fluctuating water content. In years with increased runoff rates in the sandy bed, erosion occurs and an increase in the average particle size of bed material load is observed. The erosion products, accumulated downstream, lead to a decrease in the size of riverbed sediments. According to observations at the upper Ob, in an extremely high-water year (1984) below the confluence of Biya and Katun, the average diameter of sandy sediments (0.50–0.56 mm) did not change much after the first wave of flooding compared to previous years (1980–1981): only the size of pebbles had slightly increased, and the bed area occupied by them had doubled. In July, a powerful second wave of floods took place in the Ob, connected with the melting of snow in the mountains. After that, the average diameter of sandy load ranged from 0.58 to 0.64 mm, and the area of the bed occupied by pebbles was 39%. Downstream, where the accumulation of load was observed and the size of sand and pebbles decreased.

Such changes in sediment particle size are periodic in nature. In high water periods it increases slightly, while in dry periods it decreases. For example, in the upper Ob, in the dry 1980s, the average diameter of load was 0.4 mm, and in the high water 1984, 0.78 mm. At the same time, there was a sharp increase in the area of the channel occupied by coarse-grained sands.

A significant role in changing the composition of bed material load is played by the technogenic impact on rivers, thanks to which these changes take on a directional character with a clearly defined multi-year trend. On small flat rivers, siltation with soil erosion products in catchments leads to the replacement of sandy sediments with sandy loam and silt. On medium and large rivers, the size of bed material load is increasing due to the regulation of their flow by reservoirs, dredging and the development of channel quarries for construction materials.

An example of changes in the composition and distribution of bed material load under the influence of dredging is given by the upper Ob. Here it is shown at comparison of the data concerning the middle of 70-th–80-th years and the beginning of 60-th years of XX century (Lebedeva and Chalov 1966). In the early 1960s, medium and fine sands prevailed here, and the d_{av} varied from 0.42 to 0.24 mm. During this period, dredging was carried out in small volumes, sporadically and only at individual crossings to improve navigation conditions. In the 70–80-s, very large track works were carried out below the confluence of Bia and Katun, especially in the river section of about 70 km from the confluence. Intensive dredging has led to an increase in the particle size of the load by a factor of 1.5–2.

The anthropogenic loads in the bed material load of the Ob River below the Novosibirsk Hydroelectric Power Station have had an even greater impact. Its changes are connected with the directed channel erosion in the lower tailrace of the hydrounit, correction of the channel (development of capital dredging slots, construction of rectification structures) on a site with the length of almost 300 km. At the beginning of the 1960s, when the influence of hydroelectric power stations was observed only in a short −20 km—area near dam, the average particle size of river load $d_{average}$ varied from 0.73 to 0.50 mm in contractions to 0.30–0.40 mm in valley extensions (Berkovich et al. 1972). Local geological conditions had a significant impact on the $d_{average}$ value: rock outcrops and lower quaternary pebbles along the banks and in the channel. These factors explain the lack of connection between the Ob and the riverbed slope. At the end of the 1970s (Garrison 1979), the average diameter of bedrock deposits changed from 0.60 mm in expansions to 1.22 mm in narrowings in the river valley. Medium-large grained, coarse-grained and graveled sands began to predominate here. Their appearance is connected with the growth of dredging volumes, as their execution (development of slots) opens up layers of larger alluvium under the bed. The increase in the average diameter of the load is greatest where the bed is underlain by gravels or where there are rock outcrops (Fig. 3.14) and minimum with a homogeneous alluvial thickness.

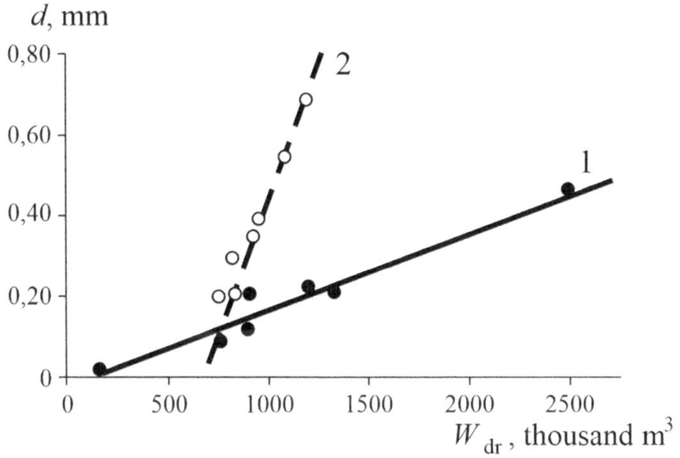

Fig. 3.14 Influence of dreding (W_{dr}) along Ob river (Novosibirsk HPP—Tom river estuary) on the size of channel sediments (Garisson 1979): 1—in sections with appearance of root rocks and pebble in the channel, 2—in sections of sandy alluvial channel

It is practically impossible to divide the influence of hydropower plants and dredging and correction works, as both factors act simultaneously and unidirectionally. No dredging has been carried out in some parts of the Ob, although the average diameter of the load has increased. For example, in the Umrevinsk channel, the runoff rates of which has not changed, it increased from 0.48 mm in 1968 to 0.96 mm in 1976, i.e. by 2 times. The underlying basal pebbles were uncovered when the wave of deep erosion from the dam of the Novosibirsk hydroelectric power station reached the Umrevinsk branching, and in spite of dredging, the coarseness in the navigable branch has increased little, because the bedrock there is deeply rooted.

Separate hoses serve as storage areas for spoil dumps, which, if washed away, result in an increase in the size of the sediment load. For example, in the right-bank arm near the Tumurchuk Ridgess to the Ob, the sediment load size increased from 0.37 mm in 1966 to 0.99 mm in 1976, while in the main arm this change was less —from 0.30 to 0.60 mm. In the Beloyarsk right-bank channel (776–783 km) the average diameter of load has increased from 0.20 to 0.99 mm, and in the navigable channel it has grown only 1.5 times. In general, the average diameter increment in the main branches is higher than in the non-navigational ones, except where this is due to the local peculiarities of the channel.

An increase in the size of bottom sediments under the influence of anthropogenic loads can be traced even in sandy muddy rivers. Table 3.10 shows that between the 1950s and 1960s and 1980s there was a significant (10–30%) increase in load on the Huang He River. The main reason for this is the flow regulation by the Huang

Table 3.10 Changes in the median diameter of d_{50} bed material load (mm) in the lower reaches of the Huang He River in time [by Qian Ning and Wenhao Zhou (1965) and "Studies of changes …" (1994)]

Hydrological station	d_{50}, mm	
	1950–1960s	1984
Huayuankou	0.092	0.102–0.200
Jiahetan	–	0.084–0.207
Gaotsun	0.057	0.089–0.136
Aishan	0.062	0.079–0.103
Lakou	–	0.083–0.112
Liczin	0.062	0.079–0.100

He Sangmenxian hydroelectric complex (built at the end of the 50-s), as well as mass construction of bank protection dams.

In the estuary areas of plain rivers, the sediment particle size along the main delta branches is gradually reduced from top to estuary, which is determined by a natural change in the hydraulic characteristics of the flow, the spreading of river jets on the estuary seashore, and the influence of surging events. Most of the load is siltstones and silt with particle diameters of less than 0.1 mm. Due to the uniformity of the composition and the small particle size of the alluvium, the influence of changes in the hydraulic characteristics of the flow in the distribution of load on the particle size is virtually non-existent. Sometimes there is some consolidation of channel load towards the estuary, which is explained by the fall in levels that occurs at the outlet of the flow during floods and runoffs (tides) into the sea. The current velocities increase, forming a erosion zone in the lower reaches of the delta branches.

3.3 Channel Stability

The degree of stability of the channel is an important characteristic of the channel, reflecting the nature of interaction between the channel flow and the soils composing the banks and bottom of the river, and in an integral form—the intensity of channel changes in the diversity of this phenomenon. The channel, which is stable in the dynamic sense, is characterized by the preservation of its main dimensions and position in the terrain over a long period of time under conditions of constant sediment transport rate along the length of the river (Grishanin 1974). The imbalance of sediment load is the cause of vertical channel changes. Since river runoff is uneven over time and the flow interacts with rocks and soils of different strength, the balance of sediment is constantly disturbed both in space and time. At the same time, lateral channel changes are carried out, manifesting themselves in the movement and constant change of channel shapes. Frequent changes in bottom marks are caused by movement of ripple forms of channel relief. Both occur at different speeds depending on the stability of the channel, which can be defined as the mechanical resistance of the channel to flow.

Using different indicators of channel stability, it is possible to compare rivers and individual river sections that differ in the intensity of channel change. At the same time, channel stability is assessed both by quantitative characteristics of channel changes (speed of riffle and sidebar shifting, intensity of bank erosion, etc.) and by qualitative descriptions based on the relationship of the channel shape on the degree of its mobility (all other things being equal, rivers with weakly stable channel more often branch into branches, they have no typical bend; on the contrary, the more stable the channel, the less likely it is to form branches of the channel and steeper bends.

The first experience of integrated assessment of channel stability using a special indicator belongs to Lokhtin (1897), who, calling the first fluvial processes driver the runoff rates of the flow, wrote that the river is created by a combination of runoff rates, "slope or steepness caused by the relief of the terrain crossed by the river, and greater or lesser dilution of the river bed, corresponding to the properties of its flow of land layers" (1948, p. 24). According to his ideas, the slope of the channel affects the flow rate and, consequently, its kinetic energy. Therefore, if the value of the latter is large enough "in comparison with the resistance of the river load … the channel will be subject to constant scouring, if not, it will be more or less constant" (1948, p. 41). Based on these considerations, the channel stability can be obtained from the partitioning between gravity expressed through the particle size of the river sediments (average diameter of their constituent particles d_{av}) and the square of the flow rate

$$k_s = \frac{d}{V^2}, \tag{3.15}$$

However, V.M. Lokhtin himself never wrote down the channel stability index in such a form (3.15), although in the literature it is sometimes possible to find a statement that this expression is the "Lokhtin's number" (Rzhanitsyn 1985). Actually, V.M. Lokhtin substituted its value in "expression for water pressure instead of its value for velocity $V\sqrt{2gH}$" (1948, p. 41) and substituting the square of velocity with kilometric fall N (m/km), at once received the expression

$$L = \frac{d}{H}, \tag{3.16}$$

which is known as Lokhtin's number and is widely used for comparative estimation of stability of channels or their separate sites (strictly speaking, in V.M. Lokhtin's $L = C_0\frac{d}{H}$ record, where C_0 is a multiplier uniting all constant values in expressions of both speed head and gravity of particles of the ground resisting shear by the flow).

The Lokhtin's number does not meet the requirements to the criteria, because in the numerator of the expression (3.16) d the value is dimensional, and the denominator H—the value is actually relative. As a result, the Lokhtin's number has a linear dimension and is subject to large-scale distortion (replacement of the kilometer drop of H by a slope of I in ‰ does not fundamentally change anything, and the Lokhtin's number remains measured). As a result, the value of the indicator is influenced by the size of the river. For small rivers, the Lokhtin's number is usually lower than for large rivers, which are less stable. Therefore, of Lokhtin's number allows us to judge about the relative stability of river channels with the same or similar size (order).

In order to remedy this shortcoming, a number of researchers have sought to give Lochtin's number a criterion form. Therefore, to characterize the degree of relative stability of the channel developed a large number of coefficients, summary and analysis of which are given in the works of Shataeva (1969), Chalov (1983a) and Rzhanitsyn (1985). A number of them are factor-integrated, evaluating the average particle weight partitioning between bed material load sediment and the parameter characterizing the transport capacity of the flow. One of the first modifications of the Lokhtin's number belongs to Velikanov (1948):

$$K_{\mathrm{V}} = \frac{gd}{V^2}, \tag{3.17}$$

in which the free fall acceleration value was used to bring it to a dimensionless state.

Makkaveev (1948) entered into the denominator of the expression (3.16) the depth of the flow h and replaced the kilometric drop with the dimensionless value of the slope I:

$$K_{\mathrm{M}} = \frac{d}{hI}. \tag{3.18}$$

In this case, the "sensitivity" (according to the expression of N.I. Makkaveev) of the indicator for most rivers decreases, because the depth of the flow increases its non-dilutive velocity. In this respect, the coefficient of M.A. Velikanov K_V (3.17) is more attractive, but it contains uncertainty in the purpose of the value of V, because the flow rate in real rivers is very variable in length, width and time.

Velikanov (1958) suggested that after multiplying the numerator and denominator in expression (3.17) by the acceleration of gravity, the members included in it should be replaced by the hydraulic particle size w and the dynamic velocity V_*, respectively, so that it took the form

$$K'_{\mathrm{V}} = \frac{\mathrm{w}}{V_*^2}. \tag{3.19}$$

In this formula, each member of the formula acquires a computational character, which makes the parameter itself practically unused, used only for purely theoretical constructions.

N.I. Makkaveev (Experimental … 1969) proposed the following variant of modification of the Lokhtin's number, which he called the channel stability coefficient

$$K_s = \frac{d}{Ib_{ch}} 1000, \quad (3.20)$$

, based on the fact that, firstly, the introduction of channel width into the denominator makes it possible to give the stability indicator a criterion look and, secondly, to take into account the influence of other factors of stability, which affect the relative width of the channel, reduce the value of K_s and affect the intensity of channel changes. According to N.I. Makkaveev, this makes the K_s index more universal than the number of Lokhtin. However, the channel width included in expression (3.20) depends on the flow rate in the river. In this connection, sections of small and large rivers with similar channel change rates and qualitative assessment of channel stability are characterized by different numerical values of K_s.

The basis of the given indicators (V.M. Lokhtin, M.A. Velikanov, V.M. Makkaveev, N.I. Makkaveev) is a condition of separation of soil particles (channel sediments) from the bottom and their driving, i.e., from the point of view of process physics, they characterize the stability of channel sediments or soils underlying the river flow with which it interacts.

The need to obtain a representative indicator of stability for solving practical problems (regulation of channels, ensuring their stabilization, creation of artificial channels—gravity-flowing channels, etc.) forced us to look for other ways, partially or completely different from the ideas of V.M. Lokhtin. Altunin (1962) proposed the criterion of channel width stability

$$A_A = \frac{b_{ch} I^{0.2}}{Q_f^{0.5}}, \quad (3.21)$$

where Q_f is the effective water discharge. In terms of its structure, it (3.21) represents the inverse value of channel stability, which is confirmed by the data of S.T. Altunin himself, who received the opposite course of values of A_A index and Lokhtin's number for Central Asia rivers.

Se Dzenhen (Sedimentation … 1993), using the inverse value of the Altunin $A' = \frac{Q_f^{0.5}}{b_{ch} I^{0.2}}$ criterion and a modification of the Lokhtin's number proposed by V.M. Makkaveev (3.18), obtained a complex index of channel stability F:

$$F = \frac{d}{hI}\left[\frac{Q_f^{0.5}}{b_{ch} I^{0.2}}\right]^2, \quad (3.22)$$

that takes into account the runoff rates of the river. According to the researches of Berkovich and Zlotina (2003) it correlates well with the coefficient of stability of the channel of N.I. Makkaveev K_s (3.20).

Qian Ning et al. (1987) proposed the index of the riverbed divagating, i.e. the variability of the channel position at the bottom of the valley, in the form of a formula:

$$\Theta = \left(\frac{\Delta Q}{0.5TQ_{av}}\right)\left(\frac{h_{l-w}I}{d_{35}}\right)^{0.6}\left(\frac{Q_{max}+Q_{min}}{Q_{max}-Q_{min}}\right)\left(\frac{b_{l-w}}{h_{l-w}}\right)\left(\frac{b_{max}}{b_{l-w}}\right)^{0.3}, \quad (3.23)$$

where $\frac{h_{l-w}I}{d_{35}}$ is the stability of the river channel (in fact, it is an indicator inverse to the sustainability criterion; if the numerator and denominator are swapped, the result is V.M. Makkaveev index): I—slope; b_{max}—channel width at channel-filling flow rate; b_{l-w} and h_{l-w}—width and depth of the channel during low-water, respectively; Q_{max} and Q_{min}—maximum and minimum water flow rate; Q_{av}—average annual water flow rate; ΔQ—amplitude of changes in water flow rate; T—duration of the intermittent period. For the wandering channel $\Theta < 0.5$, for the rest of the channel types (non wandering) $\Theta > 2$, the channels with $\Theta = 2 - 0.5$ are defined as transition channels.

Grishanin (1974) proposed an indicator of possible strain intensity

$$\frac{V^2}{\sqrt{gh}\cdot\omega} = \sqrt{Fr}\cdot\frac{V}{\omega}, \quad (3.24)$$

which represents a magnitude inverse of the stability of the channel. Unlike others, it allows predicting the variability of rolled products in different phases of the hydrological regime with changes in water discharge and other hydraulic flow characteristics. On rivers with different channel stability, determined by the degree of bed material load mobility (i.e. by the Lokhtin's number), a form of connection between the relative depth of the channel and the value of K.V. Grishanin's index (Fig. 3.15) is established: the greater the channel stability, the lower the value of the index of possible change intensity.

The use of such indicators (M.A. Velikanov, K.V. Grishanin) requires special data on the flow dynamics, which greatly limits the possibilities of their application. Similar difficulties also arise if the metrics include flow rate data. Such is, for example, the indicator proposed by Rzhanitsyn (1960):

$$K_R = \frac{db_{\mathrm{ch}}}{hI}\left(\frac{V_b^3}{V_{\mathrm{av}}^3} - 1\right) \quad (3.25)$$

where V_{av} and V_b are the average and bottom flow rates, respectively. These velocities vary depending on the water level in the river, the position of the measuring point in relation to the crest of the flood wave, the state of the bottom and other factors.

Another condition for the non-changing of the channel Grishanin (1972, p. 159) has received for rectilinear pool areas "with the shape of the channel close to the prismatic one. Their lengths usually range from 5 to 20 channel widths":

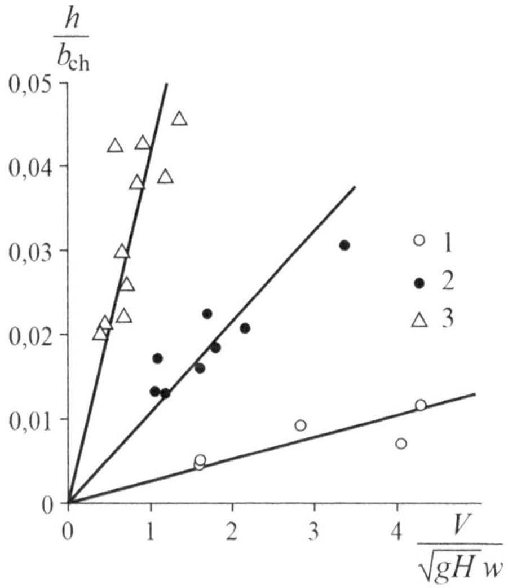

Fig. 3.15 Dependence of the relative depth of the flow from indexes of possible intensity of channel changes of K.V. Grishanin (3.24) and from channel stability: 1—Volga river, Cheboksary (high mobility of bottom particles, d_{av} = 0.25–0.40 mm); 2—Vym river, Polovinka village (fine movable bottom, d_{av} = 0.4–0.7 mm); 3—Assa river, Maimak village (semi-mountainous channel, d_{av} = 5–15 mm) (Grishanin 1974)

$$M = \frac{h(gb_{ch})^{0.25}}{Q^{0.5}}, \tag{3.26}$$

where Q is the average daily water discharge, m^3/s. For the channels folded by sand, M is on average equal to 0.9. At the same time, the dynamically stable state of the channel corresponds to the inequality

$$0.75 < M < 1.05. \tag{3.27}$$

If the value of M is high (>1.05), the river section is characterized by insufficient transport capacity of $W_{tr,}$ and within it there will be accumulation of sediment. On the contrary, at $M < 0.75$ W_{tr} exceeds the sediment load and the bed is washed out. However, in this case, we are not talking about the stability of the channel as an integral assessment of the intensity of its change, but about the correspondence of hydraulic and morphometric characteristics of the channel, which produces the flow in the linear moulded areas in the process of its long-term interaction with the channel. Accordingly, the criterion has different purposes than the Lochtin's number and its modifications.

Karasev (1975), combining in a single ratio the stability of the soil bed in the form of $V \leq V_n$ and the criterion of quasi homogeneity of the flow $\Theta = \frac{b_{ch}}{h}\sqrt{\lambda}$, recorded taking into account that $\lambda = \frac{2g}{C^2}$, received

$$\Theta_1 = \frac{b_{ch}}{h} \cdot \frac{\sqrt{2ghI}}{V_n} = \frac{1.41 V_* b_{ch}}{V_n h} \leq 4.5.$$

The value inverse of Θ_1, according to I.F. Karasev, is an erosion-morphometric index of dynamic stability of the channel, which in the case of substitution of V_n by the formula of B.I. Studenichenkov (4.49) has the form:

$$Y_{\text{e.m.}} = \frac{1}{\Theta_1} = \frac{h}{b_{\text{ch}}\sqrt{I}}\left(\frac{d}{h}\right)^{0.25} = \frac{V_n}{V_*}\cdot\frac{h}{b_{\text{ch}}} \geq 0.3, \tag{3.28}$$

Its structure is close to that of N.I. Makkaveev's channel stability coefficient (3.20), but includes not the absolute width of the channel, but the relative width of the channel, which makes it less dependent on the size of the river. Since $\frac{h}{b_{\text{ch}}} = f(N)$, where N is the order of the river (Rzhanitsyn 1960, 1985), I.F. Karasev's indicator retains the disadvantages inherent in other criteria. The meaning of $Y_{\text{e.m.}} \geq 0.3$ corresponds to absolutely stable, undeformable channels of medium and large rivers.

Some channel stability indicators are based on hydrological stations. In particular, the "coefficient of pool" proposed by H.M. Polin (Makkaveev 1949), takes into account the different steepness of the curve of the relationship between water discharge and river orders with a wide unstable and narrow relatively stable channel:

$$K_m = \frac{100\left(4\cdot\frac{Q_{\text{l-w}}}{Q'_{\text{l-w}}} + \Delta H\right)}{Q'_{\text{av}}}, \tag{3.29}$$

where Q_{l-w}—water discharge during two low-water months in the ice-free period; Q'_{1-w}—water discharge derivative by level for the low-water; ΔH—annual amplitude of levels; Q'_{av}—water discharge derivative by level within ΔH. Disadvantages of such indicators are related to the fact that, firstly, hydrological stations are located at a sufficient distance from each other, and therefore, according to their data it is impossible to judge about the variability of channel changes in the river sections stretching hundreds, and sometimes thousands of kilometers; Secondly, they are usually located where the river valley is narrowed and the channel is relatively straight and deformed, which does not allow extrapolation of parameters even to adjacent parts of the river, which differ in terms of channel morphometry and valley morphology. Nevertheless, Rabkova (1979), having analyzed the values of H.M. Polin's coefficient of mildew for a large number of navigable rivers, established that its ranges correspond to the formation of one of the channel patterns: incised and free bends, transition from meandering to branching, braided, etc. Rzhanitsyn (1985) established a close enough inverse exponential relationship between the K_m and the proposed by him "indicators of the channel regime" describing the changes in the shape of the living flow section with fluctuations in the runoff rates of the river. This is quite natural, since the initial data for its determination, as well as the K_m obtained from observations at hydrological stations. Polin's coefficient of pool was used by a number of researchers in the development of methods for calculating hydraulically maximum possible depths on navigable rivers in order to justify their radical correction (Chernyshov 1973; Rabkova 1979). In addition, either explicitly (3.25) or implicitly, through the relationship of the slope of the curve $Q = f(H)$ on the shape of the living flow section (3.29), both formulas include the value of the relative depth $\frac{b_{ch}}{h}$.

The fact that the shape of the living cross-section of the flow is determined by the stability of soil particles at its bottom, determines the existence of morphometric indicators based on the position: the less stable the channel, the less the partitioning between depth to width. This conclusion is based on the fact that the current velocity in the coastal zone attenuates towards the edge, and therefore the slope profile forms a curve, the average slope and curvature of which depends on the resistance of the soil particles to erosion. In essence, these considerations are similar to those used as a basis for the H.M. Polin's pool coefficient (3.29) and the index proposed by N.A. Rzhanitsyn (4.25). The first of a series of morphometric parameters was the parameter of Glushkov (1925)

$$A_G = \frac{\sqrt{b_{\text{ch}}}}{h}, \tag{3.30}$$

where A_G, in the author's opinion, depends on the erosion control stability of the soil: it is 1.4 for rocks and 5.5 for easily eroded rocks. Experimental studies (Makkaveev et al. 1961) have shown that erosion control stability of the river bed is not the only, but often the main factor determining the shape of the channel. In particular, the unevenness of runoff leads to a noticeable increase in the relative width of the channel. A certain influence on the value of the morphometric index of V.G. Glushkov A_G has the direction of vertical channel changes. The values obtained in the experiment range from 1.59 to 3.46 for preferential river incise and from 3.12 to 6.12 for sediment accumulation conditions. Calculations of the Glushkov A_G parameter for the Ob (Novosibirsk hydroelectric power station—the mouth of the Tom River) showed that they range from 3.7 to 126 (Channel mode … 1969). There is no clear connection between A_G and the alluvium size or the slopes of the channel. Apparently, in some cases, an increase in alluvium particle size at the locations of bedrock outcrops that narrow the bed, or, conversely, underfloor it at shallow depths, contribute to the development of lateral erosion and increase its width; in other cases, it is important to receive larger material from the slopes or the outcrops of ancient alluvial quaternary pebbles on the banks and at the bottom of the river.

The average value of the parameter A_G is within the limits of those values which were obtained in laboratory conditions (Makkaveev et al. 1961) for the rivers accumulating sediment, which was characteristic of the Ob below the city Novosibirsk in natural conditions (before the flow regulation of the Novosibirsk hydroelectric power station). At the 45-kilometer section of the river directly downstream of the Novosibirsk hydroelectric power station, a natural increase in A_G downstream (from 4.2 to 8.0) is observed. The inflow of the river below the dam of the hydropower plant affected the relative narrowing of the channel (this was also facilitated by the construction of dams covering the peripheral parts of the channel) and, as a consequence, a decrease in the value of A_G. Below the zone of intense deep erosion there is a "wave" of accumulation; the local values of this indicator here increase to 12.5.

Essentially, A_G reflects the inverse strain rate as a characteristic of the channel—its instability: the greater the A_G value, the wider and less stable the channel. In this respect, the parameter A_G has the same meaning as the criterion of stability of the channel width of S.T. Altunin (3.21). Therefore, to characterize the stability of the channel sometimes take the opposite value of the A_G criterion. So, for example, Berkovich and Zlotina (2003) have acted with criterion of S.T. Altunin, having written down it in the form

$$A' = \frac{Q_f^{0.5}}{b_{\mathrm{ch}} I^{0.2}}. \quad (3.31)$$

On the basis of their own researches and literary sources they have carried out comparison of stability indicators on 11 rivers (Table 3.11), having connected them with classification of rivers on degree of stability of channels (Chalov 1983a). As can be seen, all sustainability indicators, reflecting the overall trend, do not always coincide in terms of changes in values. Berkovich and Zlotina (2003) have revealed

Table 3.11 Comparison of different indicators of channel stability (Berkovich and Zlotina 2003)

River	Current section	Stability indicators					Characteristic
		K_S	K_M	A_G	A'	F	
Vakhsh	Upper	38.1	4.85	1.52	1.77	15.6	Stable
Yenisei	Middle	19.5	12.1	2.55	0.71	6.04	Stable
Oka	Upper	6.67	3.08	2.0	1.16	4.14	Relatively stable
	Middle	2.38	1.02	2.0	1.33	1.80	Relatively stable
Dniester	Middle	1.77	0.68	2.43	0.75	0.39	Relatively stable
Belaya	Lower	1.72	1.25	2.75	0.76	0.72	Relatively stable
Charysh	Lower	0.72	0.95	7.27	0.43	0.18	Relatively unstable
Ob	Middle	0.67	1.83	6.0	0.24	0.37	Relatively unstable
	Upper	0.52	1.93	11.7	0.30	0.18	Relatively unstable
Yangtze[a]	Lower	0.40	0.42	3.42	0.95	0.37	Relatively unstable
	Middle	0.30	0.36	2.66	0.91	0.29	Relatively unstable
Vistula[b]	Lower	0.34	0.42	5.21	0.42	0.17	Relatively unstable
Amu Darya	Lower	0.14	0.14	5.0	0.51	0.04	Unstable
Huang He	Lower	0.05	0.20	32.6	0.31	0.06	Unstable

[a]«Sedimentation» data … (1993)
[b]Data of Babiński (1997)

Table 3.12 Stability indicators for different channel patterns

Channel pattern	A_G	F	K_S	A'
Wandering	20–35	0.05–0.10	0.05–0.15	0.30–0.50
Braided	5–15	0.15–0.50	0.3–0.7	0.20–0.90
Meandering	2.0–3.5	0.30–12.0	1.5–6.0	0.90–1.80

a fairly close relationship between the channel stability coefficient of K_s N.I. Makkaveev (3.20) and the criterion of S.T. Altunin A' (3.31), the parameter of V.G. Glushkov A_G (3.30) and the stability index F (3.22). The correlation coefficient is 0.77 (for A') and 0.90 (for A_G and F). The connection between the rest of the indicators was weak.

For those indicators, between which there is a good correlation connection, Berkovich and Zlotina (2003) set the intervals of values, averaging them over the river sections corresponding to one or another channel patterns (Table 3.12). In addition to the commonplace, they have introduced a type of "wandering channel", which is often used in Chinese literature and differs from them by their extremely high mobility of the channel at the bottom of the valley, the effect of which bend or branchings have time to take hold.

The last modification of Lokhtin's number is the proposal of Obodovskiy (2001), which he introduced into it the relative width of the channel b_{ch}/h and the erosion coefficient A_{er}:

$$L_O = \frac{d}{H} \cdot \frac{b_p}{h} \cdot A_{er}. \tag{3.31}$$

Here, the A_{er} is a coefficient in the $R = A_{er}Q^m$ coupling equation called by Makkaveev (1955) as it reflects the conditions of erosion products inflow into the rivers; he also determined the A_{er} values for the rivers flowing in different natural conditions of the European part of the USSR. A.G. Obodovskiy obtained the average values of the A_{er} for the territory of Ukraine: in Polesye—0.0309, in the forest-steppe—0.0350, in the steppe—0.0530. Since the erosion coefficient of A_{er} has a dimension, its use provides a criterion character of the L_O number, which A.G. Obodovskiy called "erosion stability coefficient". However, the introduction of the Lokhtin ratio does not $\frac{b_p}{h}$ deprive it of the same disadvantage as, for example, that of V.M. Makkaveev (4.18). The value of b_{ch} changes in much larger limits than h, and its introduction into the numerator determines the "inverse" relationship between the characteristic of stability of the channel and the value of L_O: the more L_O, the less stability of the channel and the more intense channel change. Obodovsky (2001) explained this by the fact that the relationships he obtained for the conditions of free development of channel changes and are generally close to the index of the channel mode of N.A. Rzhanitsyn:

$$\frac{b_{\mathrm{ch}}}{h} = 2.35\left(\frac{Q_f}{Q_{\mathrm{l-w}}}\right) + 23.6, \tag{3.33}$$

where Q_f is the effective discharge in the floodplain edges; Q_{l-w} is the inter-stream flow rate. Since there is no relationship (3.33) in the conditions of limited channel change development, channel stabillity increases under higher L_O for respective rivers.

Shataeva (1969) proposed to determine the stability of the navigable river channels based on the materials of rolled strip surveys. Morphometric index of channel stability of S.G. Shataeva is the partitioning between increase of flow depth Δh at change of its width Δb_{ch}:

$$A_{\mathrm{M}} = \frac{\lg \Delta h}{\lg \Delta b_{\mathrm{ch}}}. \tag{3.34}$$

Unlike other indicators, it is applicable to the assessment of the stability of large forms of channel relief–riffles. It is determined by the cross-sections on the channel plans where the first (counting from the bottom) isobath is closed, i.e. on shallow waters, which usually correspond to the ropes. If the riffles follow the channel one by one, forming a riffle section, the average A_M parameter value characterizes the stability of this section as a whole. In rivers where the entire channel width is covered, i.e. in single-thread or poorly braided sections, its average value usually correlates well with the *Lokhtin's* number L and the channel stability coefficient of N.I. Makkaveev K_S. Changes in the stability of each crossing (A_M indicator) is consistent with changes in channel stability indicators as a whole (*Lokhtin's* number L and K_S). The exception is the rivers with a complex channel, where the size of the rolled channels is not comparable with the width of the whole channel, as well as the areas with single crossings, where this relationship is lost, and the average A_M value does not determine the overall stability of the channel.

S.G. Shataeva's method is also applicable to the pool valleys, allowing to assess the stability of channel sections practically devoid of rolling over. However, especially in the latter case, this is hindered by the lack of appropriate channel plans (they are prepared only for riffles and riffle sections).

Makkaveev (1955) was the first to carry out a mass determination of the Lokhtin's number for a number of rivers in the European part of the USSR and to establish a feedback between him and the rate of movement of the macroforms of the channel relief–riffle side-bars:

$$C_{\mathrm{s-b}} = 0.05\frac{Q_f}{L}, \tag{3.35}$$

where Q_f is the effective discharge. At Vychegda and Severnaya Dvina, the average shift rates of the riffles are 90–100 and 150–200 m/year, respectively; the Lokhtin's number at Severnaya Dvina does not exceed 3.5, while at Vychegda the average is 6.6. On the upper Ob at $L = 2.9$–4.7 riffles move at a speed of 300–350 m/year.

Berkovich and Zlotina (2003) have established close enough relationships of the rate of movement of the channel relief forms (side bars and mid bars) C_{s-b} and bank erosion U_b from the stability coefficient of N.I. Makkaveev K_S:

$$C_{\mathrm{s-b}} = 115.6\, K_{\mathrm{S}}^{-0.97}; \tag{3.36}$$

$$U_{\mathrm{b}} = 15.6 K_{\mathrm{S}}^{-0.65}, \tag{3.37}$$

for which correlation coefficients are 0.9 and 0.74.

Obodovskiy (2001) obtained the links of the modified Lokhtin's number with the rate of bank erosion

$$L_O = 7.66\, U_{\mathrm{b}} - 3.7 \tag{3.38a}$$

for free development of channel changes and

$$L_O = 12.4 - 3.99 U_{\mathrm{b}} \tag{3.38b}$$

for conditions of limited channel change development (correlation coefficient, respectively, 0.73 and −0.65).

The same (3.36) character is related to the movement velocity of the riffle side-bars movement and the Lokhtin's number (Fig. 3.16). Inverse, but at the same time linear, is the relationship between A_M indicator and the rate of side-bars movement. In the Middle Ob, this relationship looks like $A_M = 2.11 - 0.003\ \mathrm{C_{s-b}}$ (Berkovich et al. 1972). On other rivers, it differs in its free term and coefficient, the values of which are determined by the local conditions of formation of the channel.

The regional character, which is represented in the values of the coefficients, is the relationship of $C_{s-b} = f(K_S)$ and $U_b = f(K_S)$ for specific rivers or their sections that differ in runoff rates. For example, at the Lena River, the connection between the rate of floodplain bank erosion and the stability of the channel is expressed by the relationship $U_{\mathrm{b}} = aK_{\mathrm{S}}^{-0.8}$, and the closeness of the connection is quite high. The coefficient a is inversely related to the water flow rate. Obviously, the curvature of the channel in the high-water flow is smaller and, consequently, the acceleration of

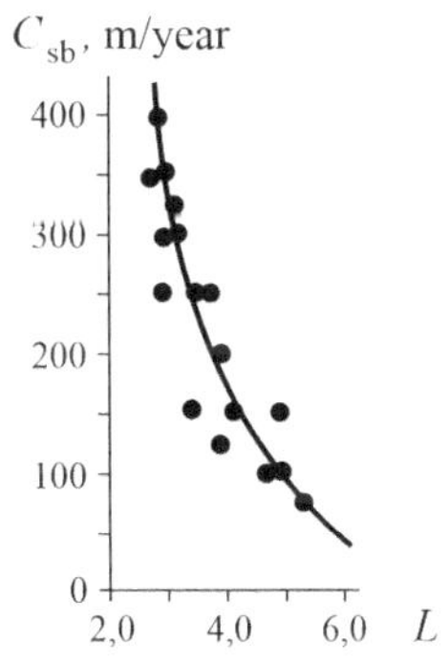

Fig. 3.16 Dependence of riffle side-bars shifting velocity (C_{rif}, m/year) on Lokhtin's code L on Vychegda river

the current and the circulation currents contributing to the erosion of the banks are less developed.

Table 3.13 shows the average values of channel stability in comparison with the rate of channel change on several rivers. A similar table for the rivers of Ukraine was compiled by Obodovskiy (2001).

Table 3.13 Channel stability indicators—Lokhtin's number *L*, stability coefficient of N.I. Makkaveev K_S and morphometric index of S.G. Shataeva A_M and their correlation to the rate of channel changes

The river, section	Q, m³/s	Stability indicators			Change rates		
		L	K_S	A_M	C_r m/year	U_b, m/year	T, years
Vychegda, downstream	1100	3.0–6.6	9.3	1.77	90–100	12–20	20–25
Ob the confluence of Biya and Katun—the mouth of the Charysh	1200	1.6–1.7	1.6–5.9	–	300–350	30–50	10–15
the mouth of the Charysh River—Barnaul	1500	1.9–5.5	1.3–6.6	–	200–250	20–30	–
Novosibirsk—Tom mouth	1800	6.3	9.0	1.06	150–200	15–20	35–10
Below the mouth of Tom	3800	6.7	8.5	–	150–200	15–20	8–15
Northern Dvina Vytchegda mouth—Vaga mouth	2000	2.8–3.5	5.4	1.50	150–200	10–20	15–20
Amu Darya, upper course	2400	0.7–2.1	0.5–1.9	1.04	500–1500	70–100	3–15
Viliui—lower course	1500	–	1.6–8.6	–	200–300	15–30	30–40
Lena Pokrovsk—Aldan mouth	8000	6.5	2.5–6.5	1.08	150–800	20–40	50–80
Viliui mouth—Zhigansk	13,000	7.7	1.5	1.76	200–600	20–40	80
Katun—the Katun mouth	630	12	18	–	40–50	4–5	–
Huang He—lower course	1500	0.13–0.3	–	–	2–460 m/day	130 m/day	0.5
Yangtze, middle and lower course	14,230–28,840	3.6–17.3	–	–	–	20–90	20–30
Dniester—lower course	330	–	22–490	–	stable	1–4	–

Notes Q—average annual water flow rate; C_r—speed of riffle movement; U_b—rate of bank erosion; T—period of the main stream of the river wandering along the branches

Table 3.14 Changes in the Amu Darya riverbed stability indicators above and below the tributary inflow—Surkhandarya river

Section	d, mm	I, ‰	b_{ch}, m	L	K_S
Above the inflow 0–3.5 km	0.17	0.08	1080	2.1	1.97
Below the confluence 0–2.5 km	0.21	0.29	1680	0.73	0.54
2.5–4 km	0.21	0.18	1740	1.7	0.67

The flat rivers of Central and South Asia, where flow velocities for most of the year are significantly higher than non-destructive ones, are characterized by the least stability and, accordingly, the highest intensity of channel changes. Even minor changes in slope or sediment size due to local conditions cause marked changes in channel stability, which affects the nature of channel changes. For example, in the Amu Darya, the inflow of the Surkhandarya, which carries a larger volume of material than the main river, causes changes in the slope and particle size of the load upstream and downstream of the mouth of the Amu Darya (Table 3.14). Directly below the inflow of tributary the channel has the lowest stability: the rate of movement of sand massifs here reaches 1500 m per year, which is almost 3 times higher than in the section above the mouth of the Surkhandarya, and 1.5–2 times higher than in the downstream section.

The most striking representation of the instability of the Amu Darya riverbed is the catastrophic erosion of the banks—"bank scour". The bankline retreat during bank scour is 1–1.5 m/min and the length of the scouring section is 1 km. A 300–500 m wide strip of land can be washed away in a year. Offset speeds of up to 1500 m/year are typical for both riffle side-bar movement and channel bend speeds. At the Huang He, whose channel is even less stable, the rate of movement of large accumulations of sediment into the high-water phase of the regime is measured in hundreds of meters per day; the position of the dynamic flow axis changes up to two times per year in a bandwidth of more than 2 km.

Rivers with stable channels are most often found in areas of limited channel change, where bed material load are represented by pebbles and boulders or even boulders and chunk at slopes typical of flat rivers (0.08–0.20‰). These channels are characterized by extremely low rates of change in their relief, which are established only when comparing channel surveys, often covering several decades, or they do not appear within the accuracy of measurements at all. At Aldan, Upper Lena, Angara, Upper Kama, Belaya, Chusovaya channel changes consist in periodic deepening and expansion of the ripples and minor changes in the configuration of the channel banks, which is associated with the movement of a stream of coarse chip material with relative stability of the channel shape and individual elements of the channel relief.

The stability of the channel also determines its main characteristics: width, degree of branching, radius of curvature of the bend, etc. At the upper Ob, the relationship of the degree of channel branching (by—the number of islands per 1 km of channel within the x km long sections) and the value of K_S correspond to

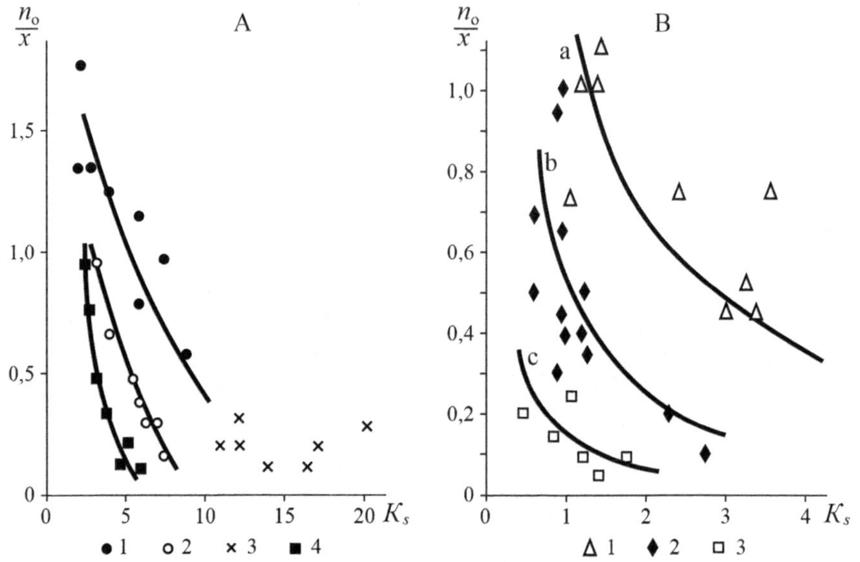

Fig. 3.17 Dependence of the braiding intensity n_{is}/x (n_{is}—number of islands on the 1 km of the channel in the sections with the length x km) on its stability K_s and channel pattern on the upper Ob (**A**): 1—parallel-branching channel, 2—alternate one-side branching, 3—one-side branching, 4—second- and third-order branching in a straight and meandering channel (Berkovich et al. 1990); and on total width of the braided channel on the Lower Lena (**B**): 1—complex-linked branching ($b_{\rho max}$ = 23 km); 2.3—parallel-branching channel (2—Sakham-Oton-Aryy, $b_{\rho max}$ = 16 km; 3—Kartylabyt-Khaastaabyt, $b_{\rho max}$ = 14 km) (Chalov, Berkovitch et al. 1976)

different types of braided channels, and for the single-thread or curved, where individual islands represent random formations, this connection is not traced (Fig. 3.17A). On the lower Lena, characterized by a complex braided channel, this connection, having the form

$$\frac{n_o}{x} = aK_S^{-1}, \tag{3.39}$$

is differentiated by three sections that differ in terms of the total (together with islands) channel width in each junction node: I − a = 2.6 at b_{ch} = 23 km; P − a = 1.0 at b_{ch} = 16 km; W − a = 0.3 at b_{ch} = 14 km (Fig. 3.17B). For the same sections, the relationships of the total width of the channel b_{ch} (together with the islands) on K_S were obtained (Fig. 3.18).

On meandering rivers, the bend sizes are inversely related to the value of channel stability coefficient K_S. For example, for Vychegda river (Chalov 1974):

$$r = 7.7K_S^{-0.96}, \tag{3.40}$$

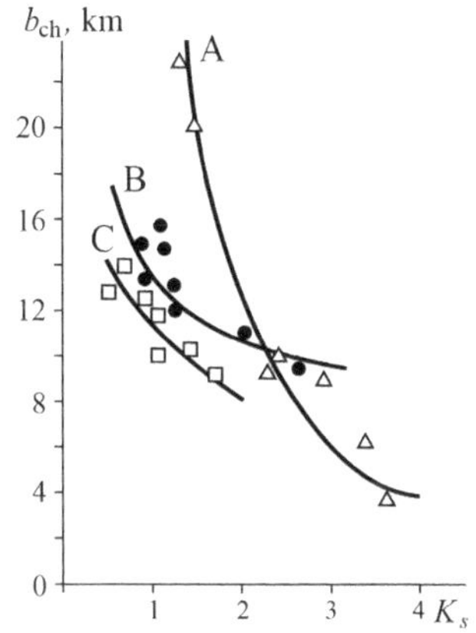

Fig. 3.18 Dependence of total width b_{ch} of Lower Lena channel on channel stability K_s: A—Monastyrsko-Cherpalskoye (complex-linked); B—Sakham-Oton-Aryy (parallel-branching); C—Kartylabyt-Khaastaabyt (parallel-brabching) (Chalov, Berkovitch et al. 1976)

where r is the radius of curvature of the bend (km). On rivers split into branches, on the contrary, the decrease in the value of Ks, which characterizes the stability of the entire channel, is accompanied by a decrease in the curvature radii of the main branches, in contrast to meandering rivers, the parameters of the bend of the main river flow among the islands are in direct relationship on the stability of the channel. In the Middle Lena, this relationship is expressed by the empirical formula

$$r = (1 - K_S)^2 \tag{3.41}$$

The higher the stability of the channel, the larger the area of the islands. The process of unification of the islands and, consequently, the reduction in the number of branches is most intensive in areas where the stability of the channel is relatively high. Therefore, increasing the stability of the channel leads to an increase in the runoff rates of the hoses and, due to the direct relationship between the radius of curvature r and water discharge $r = f(Q_f)$ (Makkaveev 1955), causes a greater radius of curvature.

The radius of curvature of the dynamic flow axis is determined by the stability of the ropes themselves, which is represented in its relationship on the morphometric index of S.G. Shataeva $r_{dfa} = f(A_M)$ (Fig. 3.19). On the Lower Lena, this relationship is differentiated by the types of braided channel (complex conjugate channel—a; parallel-branch—b, c) and the total width of the river (curves b, c); for the Middle Lena (curve d), the relationship is unified, not related to the channel pattern.

Changes in the depth of the channel at the crossings, determined by their seasonal changes, are related to the stability of the channel: the more stable the crossing is, the less the change in the height (washout) of its ripples during the year should be. Comparison of the morphometric index of S.G. Shataeva A_M with the volumes of dredging at Ob ropes shows that the relationship between them is expressed by two curves (Fig. 3.20), each of which corresponds to the riffles, timed to different parts of the channel branches, different in their genesis and, respectively,

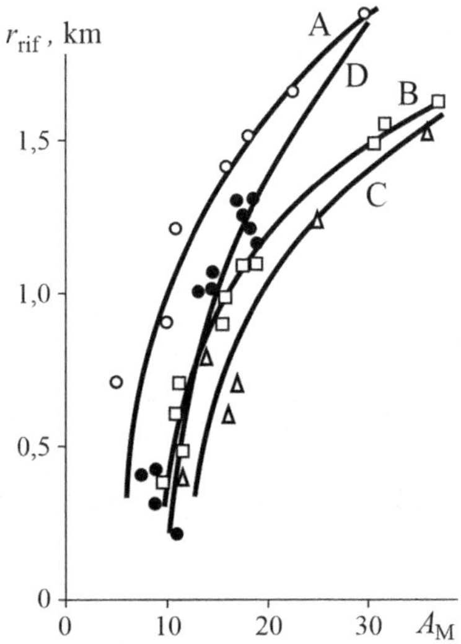

Fig. 3.19 Dependence of the curvature radius of the dynamic flow axis on riffles (r_{rif}) on its stability (A_M) on the Lower Lena river: A—complex-linked Monastyrsko-Cherpalskoye branching; B—parallel-branching channel Sakham-Oton-Aryy, C—the same Kartylabyt-Khaastaabyt; D—Middle Lena (Pokrovsk—Aldan estuary) (Chalov, Berkovitch et al. 1976)

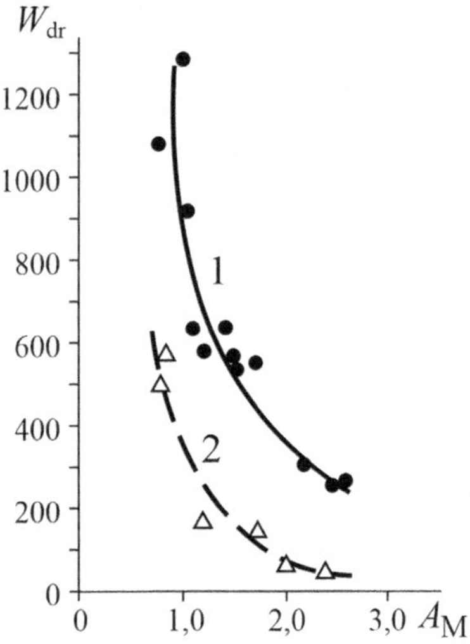

Fig. 3.20 Dependence of dredging volumes on riffles W_{dr} (thousands m^3/year) on morphometrical index of the stability A_M for the riffles on the braided Ob river channel: 1—near tip of the island of islands; 2—in downstream end of the island

the mode. The first characterizes the riffles formed at the tips of islands, the second one—in their ears (i.e. in the nodes of branching and merging of branches). All other things being equal, the transferability of riffles in island tips is greater than in their ears, which is explained by the greater deposition of sediment in the first case due to the dispersal of runoff and a general decrease in the transport capacity of the flow.

The stability of the channel within the riffle sections is characterized by the frequency of dredging operations per 1 km of the waterway, which is compared with the coefficient of channel stability of N.I. Makkaveev K_S in Table 3.15. The

Table 3.15 Channel stability coefficient K_S and frequency of dredging works on 1 km of waterway n_{dr} on the Ob River (50 s–70 s of XX century) (Berkovich et al. 1972)

Riffle section	K_S	n_{dr}
Bazikovsky–Dregunovsky	2.6	1.9
Krivoshchekovsky–Mochischensky	3.6	1.6
Mochishchensky–Bazikovsky	5.2	1.2
Kozhevnikovsky–Tagansky	5.4	1.0
Tasharinsky–Nizhnekameshkovsky	6.1	0.9
Nizhneelgansky–Kozhevnikovsky	7.0	0.8
Bogorodsky–Bazanakovskiy	10.0	0.7

higher the coefficient, i.e. the more stable the channel is, the less dredging work is carried out along the ship's route.

In the perrenial plan, the stability of the channel is also subject to certain changes. This is due to natural and anthropogenic changes in water and sediment yield. In the Ob, downstream of Novosibirsk, flow regulation by the reservoir (since 1959) and huge dredging works (in the 70–80 s) led to the fact that the indicators of channel stability (the *Lokhtin's* number L and the coefficient of stability of N.I. Makkaveev K_S increased by 2.6–3.6 times, and in the embankment section—almost 20 times. Equally significant changes in these indicators are observed in the lower reaches of the dams and other rivers (Table 3.16). These are consistent with changes in other sustainability indicators.

In 1962, the Lokhtin number in the upper Ob, where intensive dredging was carried out, was 2.4–2.6, and in the 1980s it increased to 2.7–4.5 (Berkovich et al. 1989). A more complex picture of changes in the stability is on Vychegda rive due to the use of different river length averages and the heterogeneity of the initial data on the composition of bed material load, although the overall trend of their growth is also evident (Table 3.17).

Table 3.16 Changes in the stability indicators at the sections near dams in the downstream waterworks (Berkovich and Zlotina 2003)

River	Hydrosystem (year of commissioning)	Year	K_S	F	A'
Ob	Novosibirsk (1959)	1954	0.67	0.37	0.24
		1964	1.08	0.50	0.61
		1987	12.7	8.00	0.61
Dniester	Dubosari (1949)	1950	1.77	0.39	0.75
		1983	9.52	3.12	0.89
		1987	39.6	8.89	0.85
Amu Darya	Tuyamuyunsk (1951)	1951	0.14	0.04	0.51
		1972	0.73	0.17	0.60
Vistula	Włocławek (1969)	1947	0.35	0.08	0.39
		1988	8.56	2.23	0.54

Table 3.17 Change in the Lokhtin's number (of channel stability) at Vychegda river

Plot, km from the estuary	Lokhtin's number, by year		
	1948 (Makkaveev 1955)	1959–1960 (Chalov 1979)	1994–1995
Syktyvkar–Chasovo (424–376)	3.4	3.8	–
Chasovo–Podgornaya (376–253)	4.7	4.7	–
Podgornaya–Yarensk (253–213)	5.6	4.9	5.5
Yarensk–Chakula (213–88)	4.1	5.0[a]	4.6
Chakula–Solvychegodsk (88–23)	3.5	10.0[a]	3.4
Solvychegodsk–estuary (23–0)	5.4[b]	1.6	3.1

[a]Data are overestimated due to the accounting for gravel and pebbles impurities in bed material load near primary banks
[b]Data refer to the area where the level decline curve was formed before merging with the Northern Dvina

Comparison of L and K_S values for rivers of close size and A_M with quantitative characteristics of channel change intensity allowed to give numerical gradations of indicators for assessment of river sections stability, average runoff rates (1000–3000 m^3/s) and individual crossings (Table 3.18) and on this basis to develop river channel classifications. However, all of them used only the gradations of the Lokhtin's number, and only Altunin (1962) gave them in comparison with the proposed criterion of stability of the channel width (3.21). The channel stability coefficient of N.I. Makkaveev K_S and the morphometric index of S.G. Shataeva for the first time for these purposes were used for the Ob channel (Berkovich et al. 1972) and later entered the classification of river channels on stability proposed by Chalov (1979, 1983a).

Obodovskiy (2001) proposed the following scheme for assessing the stability of beds by the L_O erosion coefficient for the conditions of free development of channel change: stable $L_O < 1$, relatively stable $L_O = 1 - 4$, relatively unstable $L_O = 4 - 7$, unstable $Lo > 7$; for the conditions of limited development of channel change L_O values are directly opposite. Unfortunately, A.G. Obodovskiy did not evaluate the links between L_O and L and other indicators of sustainability, which limits the possibilities of using the data provided by him.

In order to refine the gradations of L and K_S values corresponding to the transition from a weakly stable to an unstable channel, their values were compared with the ratio between the actual flow rate V_f and the erosion rate V_{er} for the upper Ob, at which the mass movement of bed material load begins (Zhaivoron and Chalov 1990). The analysis is based on the continuous maps of bed material load distribution and mass measurements of flow velocities at the site made in different years (1962–1987) and in different phases of the water regime. The critical share stress was determined by the C.E. Mirtskhulava formula (4.51) depending on the depth of flow for all types of bed material load lining the channel. The conditions of mass movement of bed material load were considered for two phases of the water regime: at a water discharge $Q_{l\text{–}w}$ = 1200 m^3/s, corresponding to the low water period, and

Table 3.18 Classification of river channels by degree of stability

Channel stability characteristic	The intensity of channel changes	Stability indicators			Characteristics of channel changes					
		L	K_S	$Y_{э.м.}$	C_{s-b}, m/year	$U_{b_{average}}$ m/year	$U_{b_{max}}$ m/year	T, years	I_b, length %	$\pm\Delta z$, cm/year
Absolutely unstable	Very high	<1.0	<2	–	>1000	>50	>100	<3	100	>20/ >5
Unstable	High	1.0–2.5	2–4	<0.05	500–1000	10–50	50	3–10	>80	10-20/ 2–5
Weakly stable	Increased	2.5–5	4–15	0.05	300–500	5–10	20	10–20	60–80	5–10/ 1–2
Relatively stable	Moderate	5–10	15–20	0.1–0.2	50–300	2–5	10	20–80	30–60	3–5/ 0.5–1
Stable	Low	10–50	20–100	0.2–0.3	10–50	<2	<5	>80	<20	<3/ <0.5
Absolutely stable	Absent	>50	>100	>0.3	<10	No erosion		Channel is stable		0

Notes C_{s-b}—the rate of movement of side-bars (mid-bars); $U_{b_{average}}$—the average rate of erosion (landfill) of banks; $U_{b_{max}}$—the maximum rate of erosion of banks; I_b—the length of the front of erosion of banks, in % of the length of the coastline in the area; $-\Delta z$—channel infeed (in the numerator—in the lower reaches of the waterworks, in the development of channel quarries; in the denominator—in the natural conditions); $+\Delta z$—sediment accumulation (in the numerator—in the zones of variable support and above reservoirs, in the denominator—in the natural conditions)

at a effective discharge ($Q_f = 3000$ m^3/s). Mass movement of sediment occurs in the part of the river channel where the actual flow velocities exceed the critical share stress. Maximum discharges are defined as a function of the average speed at the interval $V_{max} = f(V_{average})$ according to the data of actual measurements in 1985. This relationship is described by the equation $V_{max} = 1.4V_{average}$.

Comparison of the bed areas F_{ch}, where mass movement of bed material load occurs, with the stability indicator in the same areas (Table 3.19), makes it possible to clarify the assessment of the stability of the upper Ob riverbed. On the graph of the connection between the *Lokhtin's* number L and the area F_{ch} for the flood period (Fig. 3.21A) two areas are distinguished, the vertical boundary between which corresponds to the Lokhtin's number $L = 2.5$. For the left-hand region ($L < 2.5$), which characterizes those parts of the channel where the mass movement of sediment occurs on the area $F_{ch} = 0.8F$, a close connection with the Lokhtin's number can be traced. For the other area ($F_{ch} < 0.8F$) there is no connection between these parameters. For the conditions of the low level such division of the field on the communication chart $L = f(F_{ch})$ is less distinct (Fig. 3.21B). However, even in this case, an area with $L < 2.5$ with a linear relationship $L = f(F_{ch})$ and an area with $L > 2.5$, where there is no relationship between the variables, and the share of the bed area with mass movement of sediment varies over a wide range of values—from 20 to 50%. It is obvious that at $L < 2.5$ the channel is unstable, at $L > 2.5$—weakly stable.

Table 3.19 Relationship between Lokhtin's number L to the area of the upper Ob channel F_{ch}, %, with mass movement of bed material load

Branching	Distance from merging, km	The channel area F_{ch} with mass movement of bedloads when		L
		Qpr = 1200 m^3/s	Q_f = 300 m^3/s	
At the confluence of Biya and Katun	0–6	35	80	2.6
Fominsk	6–12	5	52	4.0
Ust-Peschansk	12–18	28	54	2.8
Ust-Aniuy	18–23	44	88	2.4
Akutikhinsk	36–43	70	90	2.1
Bystryi Istok	46–55	30	51	3.2
Rogachev	55–62	22	72	3.5
Zavyalov	62–72	33	70	3.2
Soldatov	72–77	26	78	3.6
Karpov-Shishevsk	77–93	53	53	3.7
Chekanov	93–100	66	96	2.2
Ust-Charysh	100–109	20	90	2.5

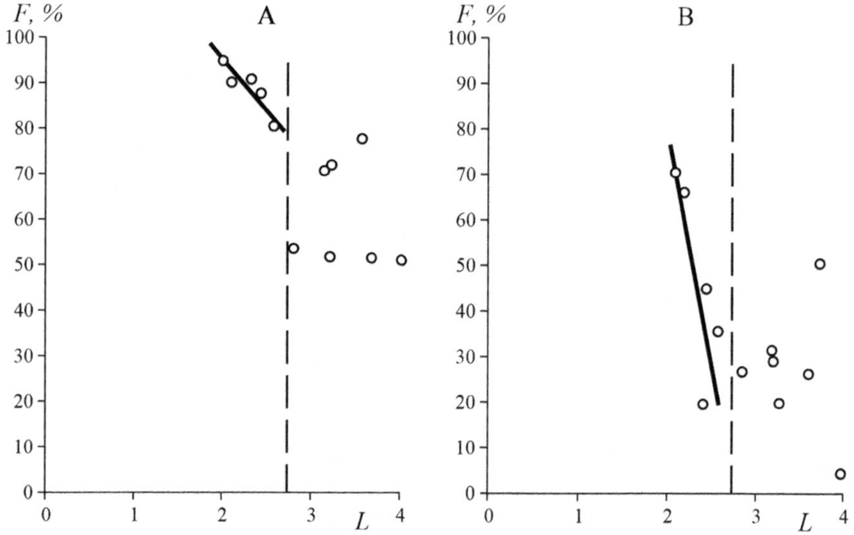

Fig. 3.21 Dependence of the upper Ob riverbed area F with the mass movement of bed material load (in % of total area of the section) at effective water discharge Q_f and discharge at design level $Q_{d.l.}$ on the Lokhtin's number L

Using the connection between the *Lokhtin's* number (*L*) and the sustainability indicator of N.I. Makkaveev (K_S), it is possible to determine the transition value of $K_S = 4$ from the unstable to the weakly stable channel. These boundary values of *L* and K_S are taken into account in Table 3.18, which specifies the previously proposed classification scheme (Chalov 1997; Chalov et al. 2000).

The areas with unstable channel ($L < 2.5$; $K_S < 4$) on the upper Ob include the merger of Biya and Katun (L = 2.0; K_S = 3.5), Ust-Aniuy (L = 2.0; Ks = 4.0), Akutikhinsk (L = 1.8; K_S = 3.8), Ust-Charysh (L = 1.7; K_S = 3.6). The remaining areas are not very stable.

The criterion character of the *Lokhtin's* number $L = 2.5$ is confirmed by the change of channel types, varieties of branching nodes, changes in channel width, velocities of movement of side-bars, mid-bars and spits, more or less complicated mode of reformation. Such branches as at the confluence of Biya and Katun, Ust'-Anuiy and Akutihin have been repeatedly described in the literature as examples of unstable channel with a small-island parallel-branch channel (up to 10 islands in the trunk) and very high intensity of changes (Malbert 1952; Popov 1962; Makkaveev and Chalov 1963a; Berkovich et al. 1989; Berkovich et al. 1990; Channel Processes and Waterways … 2001). The Fominskiy ($L = 4.0$) and Ust-Peschansk ($L = 2.8$) branching nodes are formed by one or three large islands separating two equally water-valuable branches. The transition from Akutikhin to Bystroistok and other branches, where $L > 2.5$, is accompanied by the separation of one high-water branch and a network of small channels between the coastal islands in each node located near the opposite banks of the river (alternating one-way branches).

The above classification cannot be extended to rivers with higher or lower runoff rates, as the adopted indicators are subject to various large-scale distortions, as they are either dimensional (*Lokhtin's* number *L*) or include a term (channel width), depending on the runoff rates of the river (K_S). In this connection, sections of small and large rivers with similar channel change rates and qualitative assessment of channel stability are characterized by different numerical values of K_S. Thus, on the middle Lena, where the average annual water discharge is 8000 m^3/s, the poorly stable channel is in the range of $K_S = 5$–10, and on the lower Lena ($Q = 13{,}000$ m^3/s)—$K_S = 3$–7, i.e., 1.5–2 times lower than on the rivers of average water content. Therefore, both the largest and the smallest ones have to develop "regional" classifications of river beds according to their degree of sustainability (Table 3.20). At the same time, due to the complex branching of the Lena channel and its large width, the morphometric index A_M reflects only the stability of the riffles, but does not characterize the stability of the channel as a whole, despite the riffle character of the sections of both the middle (Pokrovsk—Aldan river mouth) and lower (Aldan river mouth—Zhigansk) Lena. Riffles here form only a part of the morphological formation formed by several macroforms of the channel relief, or as single forms of relief are located in the branches, the runoff rates of which is not more than 50% of the total value of the river flow. Therefore, in the presence of separate relatively stable or stable riffles, the Lena channel itself (from Pokrovsk to Zhigansk) is weakly stable or unstable.

Table 3.20 Classification of sections of the Lena river channel and the Lena river riffles (middle and lower reaches) by channel stability

Stability characteristic	Sustainability indicators for channel sections				
	K_S		L		For riffles A_M
	Lower Lena	Middle Lena	Lower Lena	Middle Lena	
Unstable	<3	<5	<10	<9.5	<1.4
Weakly stable	3–7	5–10	10–17	9.5–16	1.4–1.7
Relatively stable	–	–	–	–	1.7–2.1
Stable	–	–	–	–	>2.1

Such a shortcoming can be eliminated by introducing a correction for the distribution of water flow rate on the hoses—"for flow rate". Due to the relationship between K_S and the relative aquifer capacity of the hoses, expressed as a percentage of the total river flow rate, a number of flow-corrected values K'_{S} are obtained. Further, we obtained a close relationship between the value of K_S and K'_{S}, that when substituting the expression K_S gave an equation:

$$K'_{\mathrm{S}} = \frac{1370d}{Ib_{\mathrm{ch}}} - 0.55. \qquad (3.42)$$

This Eq. (3.42) can only be used to compare the stability of the channels of a given river. It has allowed to receive, for example, much closer connection between radius of curvature of branches of the lower Lena and stability of their channels: $r = 39K'^{-1.92}_{\mathrm{S}}$ (correlation coefficient 0.93), whereas with application of K_S this relationship is less close (0.72).

Most indicators of stability do not take into account a number of important channel formation conditions in the specific natural environment. For example, rivers flowing in the permafrost zone are characterized by a mismatch between the intensity of some types of channel changes and the values of stability. On the lower Lena, which is characterized by an unstable channel, the speed of riffle movement is much lower than on the same rivers in the permafrost zone. In the deltas of the northern Siberian rivers, channel changes occur 3–5 times slower than in the southern rivers, which have the same stability coefficient (Korotaev et al. 1980).

In order to take into account the roughness of the channels, Makkaveev (1955) transformed the Lokhtin's number, replacing the drop in the level of the square speed and introducing a relative depth into it, and then substituted the expression of speed according to Chezy formula. The new indicator looks like this:

$$K''_{\mathrm{S}} = \frac{gd}{b_{\mathrm{ch}}C^2I}, \qquad (3.43)$$

where C is the Chezy's coefficient. However, this indicator has not yet been used for mass definitions, although it is free of large-scale distortion and, importantly, it can be used for loads of any size, from sandy loam to pebbles. Indeed, rivers with a high content of pebbles and boulders in the bed material load are characterized by increased values of stability, which do not reflect the true intensity of reformation. At the same time, the morphometric value of A_M does not depend on the composition of the sediment and can serve as a criterion for determining the gradations of other values corresponding to areas with different stability. Therefore, A_M values can be compared to the following numerical gradations of K_S stability coefficient for qualitative assessment of sand and pebble channel stability: unstable—<40, weakly stable—40 to 80, relatively stable—80 to 150, stable—>150. A_M index corresponds to Table 4.18; correlation coefficient depending on $K_S = f(A_M)$ is 0.92.

Plain rivers flowing in mountainous countries and mountain rivers with even larger alluvium (pebble, pebble-boulder, boulder-clump), with direct application to assess their stability of the Lokhtin's number L and the coefficient of stability of the channel of N.I. Makkaveev K_S are absolutely stable. With this characteristic, one should expect stability of channel forms and channel relief forms, and their morphology to remain unchanged over time. However, the very good pelletizing of boulders and their petrographic composition, different from the rocks that make up the slopes of the valleys and riverbanks, are evidence that they move over long distances. For example, on the upper Aldan, where limestone is widespread along the coastline, granites and granite gneisses, developed hundreds of kilometers upstream, dominate among boulders and large pebbles. Boulder accumulations form ripples like sandy ones. Zernov (2003), having carried out the analysis of the upper Lena river crossing plans, has shown that in spite of huge volumes of dredging works on the pebble-boulder riffles during 20 years, their condition practically has not changed, the depth increase has not occurred; the termination of these works in 90th years of increase of the bottom marks on the Lena river riffles by 40–50 cm, Aldan river—40 to 50 cm, Yana river—60 cm. In the branches of pebble-boulder channels on the branches there is a periodical development or dying out of the branches, erosion of the banks, movement of side-bars and mid-bars sometimes up to 30–40 m/year. Thus, the pebble-boulder channels are transformed more slowly and with certain specific features associated with the mechanism of transport of pebble-boulder material, peculiarities of formation of ripples, etc.

A special stability scale is required for comparing pebbles and boulders in terms of the intensity of changes. It can be constructed taking into account the shape of the channel cross-section on the basis of an empirical regularity: with the increase in the channel meltdown (relative width), i.e. distance from the most advantageous parabolic form, the mobility of channel forms increases and its stability decreases. The shape of the living section is explicitly included in the morphometric index of S.G. Shataeva A_M (3.34), which is the partitioning between depth increment of the flow when its width changes and does not depend on the size of channel sediments. This illustrates the graph (Fig. 3.22), which shows, using the Lena River as an example, that changes in the length of the river bed load particle size and gradient

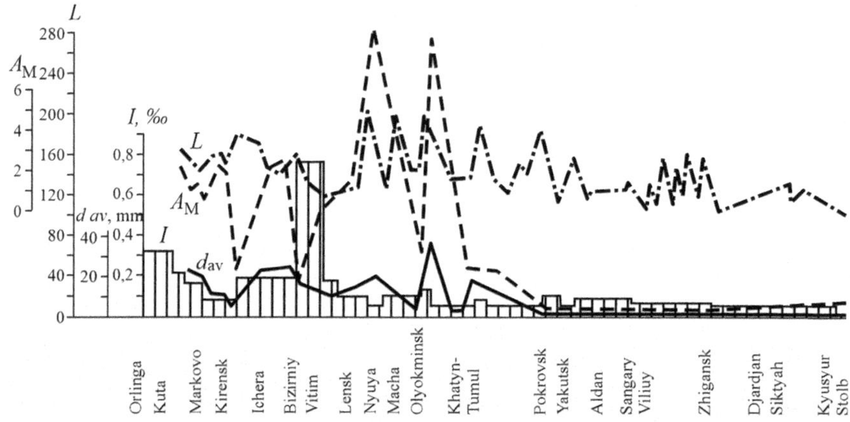

Fig. 3.22 Changes in average bed stability indexes—Lokhtin's number L and morphometric index of Shatayeva A_M, average particle size of bed material load d_{av} and longitudinal slopes of the river I on the Lena river (Ust-Kut—estuary) (Water ways … 1995)

are accompanied by corresponding changes in the Lokhtin's number. At the same time, there is a change in the absolutely stable channel on the upper and most part of the middle Lena to the slightly stable and unstable downstream channel. At the same time, A_M morphometric indicator shows only a weakly expressed downward trend only in the area from Pokrovsk to the mouth of the Vilyuy River, where the Lena channel is the least stable and differs from each other in terms of all the indicators by complex and intense changes.

The joint analysis of N.I. Makkaveev K_S stability coefficient and S.G. Shataeva A_M morphometric index for Aldan allowed to get links between them for sections of the incised channel with different pattern. For the bends

$$K_S = 556A_M - 180, \tag{3.44a}$$

for straight areas.

$$K_S = 338A_M - 550, \tag{3.44b}$$

for branching.

$$K_S = 634A_M + 220. \tag{3.44c}$$

The higher stability, all other things being equal, is characterized by the channel on the incised bends, the least—in the branches of the incised channel.

Table 3.21 Classification of pebbles-boulder channels and riffles by stability rates (example of the Aldan River)

Stability level	Stability indicators			
	For riffles A_M	For channel sections K_S		
		Bends	Straight sections	Branchings
Low	<1.4	<610	–	<310
Medium	1.4–1.7	610–780	110	310–330
High	1.7–2.1	780–950	110–230	330–350
Very high	2.1	>950	230	>350

The stability scale (Table 3.21), based on these links and field observations of channel dynamics, can be applied in the first approximation to other rivers with an incised channel and similar characteristics of pebbly-boulder sediments, slopes, etc. So far, it has a regional character. Each basin within a mountainous country or large flat river with a pebbly-boulder composition of bed material load needs to have its own ranges of variation in stability indicators. For example, for the lower Yana (Lower Yana … 1998) the following gradations of L and K_S are proposed for the degree of stability of the channel: low—$L < 180$, $K_S < 450$; medium—$L = 180$–200, $K_S = 450$–550; very high—$L > 250$, $K_S > 650$.

3.4 Effective Water Discharge

Since water and sediment load are linked by a quadratic or cubic relationship (2.19), the relative contribution of different water discharges in the transport of sediment and the formation of the channel is not uniform. On mountain rivers and rivers with a pebble-boulder composition, the sediment transport stops and the whole channel controls the flow; the water is completely clarified and the channel changes stop. During floods, when the slopes, depths and flow rates reach their highest values, the movement of pebbles and boulders is carried out along the entire channel, the flow is saturated with suspended material and the channel is actively reshaped. In easily washed out soils, the change of the channel does not stop completely in the sandy composition of bed material load. Complying with the law of nonlinearity of connections (Makkaveev 1976), they either cover the whole channel and sometimes the floodplain, or concentrate in the narrow core zone of the flow. In the first case, they cause changes in the shape of the channel as a whole, the accumulation of sediment on the floodplain, the development of floodplain ducts and the entire set of forms of channel relief. In the second case, they correspond to local erosion or accumulation of sediment, including in the ripples and basements of the ropes, causing changes in the bottom marks due to the movement of microforms of the channel relief, etc. If we take into account that in the low water level part of the river channel comes out of the water level, the difference in the results of the river fluvial processes driver activity in different phases of the regime is even more striking. A decrease of several orders of magnitude in the transport capacity of the flow at the flood recession and in the low water period creates the conditions under which the channel and forms of channel relief formed by the flood flow are relatively slow and only partially processed by the inter-flow, and this restructuring does not end by the next flood period. As a result, they act as a leading element in the flow-channel system in the lowlands.

The physical meaning of such temporal discreteness in the representation of fluvial processes is that the hydraulic characteristics of the flow are constantly changing over time, so that its transport capacity is periodically sufficient or insufficient to carry out channel changes. Even when the river is cut into rock, the flow during extreme floods produces the transport of boulders and clumps and changes the relief of the river channel. Due to the intra-annual flow irregularity and its long-term variability, the degree of water discharge contribution in the channel formation process is determined not only by their value, but also by their repeatability. Small water discharges with significant water content may eventually have no less impact on the channel than powerful but rarely repeated floods of low water content. Makkaveev (1955) proposed that the water discharge at which the maximum amount of sediment is transported (over a period of many years), and as a result their impact on channel formation, should be called fluvial processes driver.

Variability of flow and channel interaction conditions is the main reason for introducing the concept of effective discharge Q_f. However, the approaches to its definition, as well as the principle of their separation, are very different. In the majority of cases, they belong to the category of "appointed" based on the general understanding of the phase of the water regime on the river, in which the most significant changes in the channel and its relief are carried out. Analyzing similar approaches, Antropovsky (1969) and Lapshenkov (1979) noted that water discharges in floodplain logs, repeated once in 5–10 years from among the highest in the year, 1% exceedance probability average from the maximum annual water discharge, etc., are most often taken as such. At the same time, the functions prevailing in the formation of the channel are given to one of the highest costs and are denied by all other water discharges. The disagreement with this led Znamenskaya (1968) to the conclusion that "it would be more correct to abandon the term effective discharge' altogether, since each water discharge is to a greater or lesser extent effective, unless it leads to the transport of bottom sediments" (p. 84), although a few lines above it also substantiates the effective discharge "as some maximum flow rate, leading to the leveling of the velocity diagram in the channel part of the flow", and it is not identified by it with the flow rate within the floodplain logs (p. 84). This approach makes the effective discharge practically either uncertain, or at simplification and schematization of its definition it is reduced again to separation of a certain flow rate among the maximum, while fluvial processes driver activity of the flow in other phases of the regime is not considered. Also, it does not take into account the long-term recurrence of water discharge, which takes into account the importance of channel formation under different water discharge conditions, including those with high water content.

The opposite opinion was expressed by Leliavskiy (1893), who, studying mainly the conditions and mechanism of formation of riffles, proposed to take water flow rate in the low-water as a effective one for the river riffle sections (however, he has not used this term yet). But it is obvious that the formed bends of the beds on the riffles between their side-bars are forms of the second or third order compared to the bends (bendings) of the channel itself, and the relief of the channel on the bends mainly has a multiples structure. In the braided channel, the formation of branches takes place in the high-water phase of the water regime; in the low water, many of the branches become shallow, up to the point of drying up, and the mode of riffles in them and changes in the curvature of the flow are determined by the runoff rates of these branches.

Among the approaches that distinguish this or that phase of the water regime in the channel formation there are those that prefer the maximum water discharge (Karasev 1986). I.F. Karasev, speaking about "the active phase of the flood wave … characterizes the so-called dominant floods DF, which surpass all others in terms of the volume of sediment transport" (Karasev and Kovalenko 1992, p. 74). At the same time, the increased effective ability of water discharge in the range of their maximum values exceeding 0.9 Q_{max} over the years of average water content (the average multiyear maximum water discharge is assigned by I.F. Karasev as a standardizing value, and to determine the Q_f is considered typical for this river hydrograph for the years of average water content). These discharges are proposed to be called channel transformation discharges because they significantly, and sometimes radically, change the channel created by lower costs; when these costs are repeated, they accumulate a cumulative effect that affects channel morphology and, to some extent, manages fluvial processes in other phases of the regime. In the water flow range (0.6–0.7)Q_{max}, the flow affects large ripples, causing change of riffles, side bars, etc. I.F. Karasev calls these discharges «effective».

I.F. Karasev proposes to determine effective water discharge according to the curve

$$D = \delta_Q^2 \tau, \tag{3.45}$$

where D is a complex of effective transport capacity; $\delta_Q = \frac{Q_{in}}{Q_{max}}$—relative duration of Q_{in} flow intervals comparable in duration with the phase of the highest flow rate Q_{max}; τ—value characterizing the ratio $\frac{t_{in}}{T}$ (t_{in}—duration of flow intervals, T—duration of flood period).

The relationship of complex D on δQ (Fig. 3.23A–C) shows that the function has only one weakly expressed maximum at $\delta Q = 0.65$, which corresponds to a complete flooding of the side effects. The Q_f value is determined by the correspondence of the value of δQ to the average water consumption for this interval. With the increase Q after a weakly expressed maximum of D, the function $f(\tau)$ increases continuously, and thus the maximum flow rate plays a dominant role in

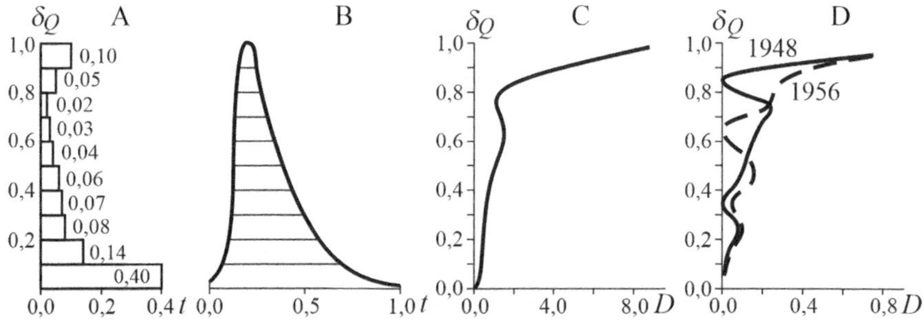

Fig. 3.23 Definition of effective water discharges accordind to Karaushevn (1986) (Karasev and Kovalenko 1992): A–C—by the author (**A** histogram of discharges range; **B** relative hydrograph Q_{in}/Q_{max}; **C** changing curve of D complex to determine Q_f); **D** by N.A. Mityukhina (changing curve of D complex for the Moscow river, Zvenigorod)

the formation of the channel itself. In fact, this is the second effective discharge equal to the average maximum or maximum for the average year in terms of runoff rates.

According to I.F. Karasev's proposed method, Q_f is determined for each of the selected average water years and, thus, characterizes the result of the flow for a particular year. At the same time, the average water years may differ significantly in the value of maximum water discharge. Thus, this method does not take into account the role of water discharge during relatively dry periods and its recurrence in the perrenial plan (by the way, these phases also accumulate cumulative effect, which cannot but affect channel changes during floods), as well as maximum water discharge in years of increased water content (above average), which contradicts the initial position of I.F. Karasev's method. In addition (and this applies to a very large number of calculation methods), I.F. Karasev has defined Q_f only for two hydrological posts (Don—Kalach, Volga—Nizhny Novgorod). Testing of the technique on other rivers carried out by N.A. Mityukhina has shown that in different years the function has two or three maximums, including those corresponding to 0.15–0.25 Q_{max} (Fig. 3.23D).

Antropovsky (1970) proposed to determine effective discharges by maximum curve

$$Q = f(N_{M_{sp}} P),, \tag{3.46}$$

where $N_{M_{sp}}$ is the specific flow rate related to the length of the river section; P—repeatability of the corresponding water discharge, which is taken from the generalized curve of the duration of daily water discharge. This approach does not take into account the fluvial processes driver activity of small flow rates, although the introduction of the P value into the design relationship (3.46) leads to the consideration of a complex hydrograph form and the appearance of the second maximum on the curve. At the same time, as the calculations have shown, it is likely that the emergence of such a shape of the curve, the function of which is deprived of $N_{M_{sp}}P$ any expressed maximums and, consequently, does not allow to determine Q_f.

Lapshenkov (1979) recommends to determine Q_f "by average water year (50% of runoff exceedance probability)". At the same time, for sandy channels it takes the highest water discharge, observed for 2–3 decades, for pebble channels—3 to 7 days a year. Later, Lapshenkov (1999) began to consider the river effective water discharge in the river corresponding to the water level in the floodplain edges. According to channel measurements and coastal levelling, the water level is set at the same level as the floodplain edge (i.e. at Q_f). Finally, the effective discharge is calculated by the following formula

$$Q_f = \omega C (h_{average} I)^{0.5}, \quad (3.47)$$

where ω is the area of the living cross-section of the channel in the floodplain edges; $C = \frac{h^y_{average}}{n}$—the Chezy coefficient; and n—parameter determined depending on the size of bed material load; I—the slope of the water surface.

These approaches to the definition of Q_f are based on general concepts of water flow performance and are not linked to the actual load flow and the duration of its impact on the channel in years of different water content. The same "designated" are the channel-controlled (channel-confined), effective and channel-controlling (channel-destroying) discharges released on mountain rivers (Talmaza and Kroshkin 1968; Artamonov et al. 1984), although it more reflects the specifics of pebble-boulder sediment transport. At the same time, on the basis of field data, the correspondence of water content for each allocated discharge to the channel state was established. Taking into account the complete cessation of transport of pebbles and boulders at the flood recession and in the low water period, Kamenskov (1986) proposed to subtract from the observed water discharge intervals taken for calculation (in the methods of I.F. Karasev, N.I. Makkaveev and V.I. Antropovsky), the discharge at which the movement of sediment does not occur. Obviously, this opinion makes sense only for pebbles and boulders, where it is consistent with the release of channel-controlled flow rate. Its practical use is complicated by the lack of information on the moment of termination of transport of bedloads.

The best known methods of determining effective discharges considered are based on specific experience and materials, limited to regional frameworks or specific channel forms. They include a`priori a number of provisions that do not take into account channel changes observed in different phases of the water regime. All this does not allow to establish connection between Q_f and various structural levels of representation of fluvial processes. In addition, they are not tested on a wide field definition of Q_f, and only the method of Lapshenkov (1999) is used in hydrotechnical calculations for the reclamation of small rivers.

Against this background, the method of determining effective discharges proposed by Makkaveev (1955) and Engineering of ship's passages … (1964) is more substantiated. It is based on the concept of the leading role of sediment in the formation of the channel and the duration of the impact on the channel of various water discharges. According to the first, the forms of representation of fluvial processes depend on the ratio between water and sediment flow, reflecting the physical and geographical conditions in river basins; the second determines the probabilistic nature of Q_f.

For the first time, a similar approach to the definition of Q_f was proposed by F. Schaffernack (Schaffernak 1950), who developed a graphical method based on the comparison of the curves of the frequency of occurrence of water levels and the bedload transport rate; Q_f corresponds to the maximum values of the product of the bedload transport rate for its repeatability, expressed in terms of time coordinates, i.e. the maximum sediment transport rate. Makkaveev (1955) noted that "the discharge (water—*R.Ch.*), which has a maximum mechanical impact on the river bed during the annual cycle, should leave the most significant traces in the relief of the channel compared to other flows", as "the sediment transported by the stream constantly exchange material with bed material alluvium and the greater the content of sediment in the stream, the (all other things being equal) with a more perceptible result in respect of the formation of accumulative forms this exchange takes place" (Makkaveev 1955, p. 183). However, in giving the maximum sediment transport rate a leading role in channel formation, N.I. Makkaveev considered it necessary to "warn against giving the concept of effective discharge' exceptional importance" and to identify it only with the maximum sediment transport rate at the maximum discharge of water. He noted that "even in the low inter-stream period, the flow does carry some sediment, the total volume of which can be quite significant due to the length of the intermittent period".

As a result, N.I. Makkaveev suggested that Q_f should be determined on the basis of the ratio (2.19) $W = A_{er}Q^mI$, where W is the total sediment transport rate (suspended and bed), Q is the corresponding discharge of water, I is the slope, and A_{er} is the erosion factor, the value of which depends on the peculiarities of erosion processes in the catchment area and its physical and geographical conditions. The value of the degree m depends on the particle size distribution of bed material load. Multiplication of the coupling coordinates $W = f(Q^mI)$ by the water discharge repeatability curve $P = f(Q)$, allows us to obtain, as with Schaffernack, several maximums of the product Q^mIP, which correspond to the largest volume of sediment load at a given value of Q over a long period of time. In addition, the term QI is an expression of the power of the flow, and thus this product actually takes into account in an integral form the probability of impact on the channel in the longitudinal section of the different phases of the water flow regime. N.I. Makkaveev suggested that water discharge corresponding to these maximums should be considered as effective discharge; the physical meaning of such an approach in the definition of Q_f is that their values correspond to the highest sediment discharge and, consequently, the most significant channel changes. Performed by Alekseevsky et al. (1983), Babich and others (1985) comparison of the results of calculations according to the method of N.I. Makkaveev with the data of calculations using direct measurements of water and sediment discharge, where Q_f was determined by maximum curves

$$Q = f(WP) \tag{3.48}$$

showed them a good convergence. Indeed, if you replace W with (2.19) in (3.47), you get the original $Q^m IP$ product. It should be borne in mind, however, that N.I. Makkaveev's method of calculation does not include the value of sediment load itself, but rather its relationship on water runoff, which is essentially the definition of Q_f, the water flow rate at which the greatest sediment load flows occur. This makes the methodology independent not only of the actual data on sediment load, but also of its real value and the partitioning between its suspended and bed load components.

N.I. Makkaveev suggested taking into account changes in the average flow rate and its transporting capacity during floodplain flooding, i.e. the kinematical effect (Zheleznyakov 1950; Baryshnikov 1984). For this purpose, the flow rate at which water starts to flow to the floodplain is established, and from this moment on, the coefficient σ, which depends on the width of the water spill in the high-water phase of the water regime, is introduced into the calculated product. However, a further increase in levels is accompanied by an increase in the depth of flow and a decrease in the relative magnitude of the roughness protrusions, which in turn may lead to a recovery and even greater increase in transport capacity and sediment load flow in floodplain conditions. Ultimately, Q_f is defined as the maximum expression

$$Q_{\mathrm{f}} = \max[f(\sigma Q^m IP], \tag{3.49}$$

where $\sigma = 1$ before water reaches the floodplain and $\sigma < 1$ when floodplain is flooded ($\sigma = 0.9$ when floodplain width is less than two bed widths and $\sigma = 0.5$ when floodplain width is more than ten bed widths); P—water discharge repeatability Q; I—water surface slope corresponding to it; m—water surface slope, depending on the size of the load and (by N.I. Makkaveev) assumed to be equal to 2 for the rivers with sandy, 2.5 for the rivers with sandy-pebble and 3 for the rivers with pebbles. At the same time, the time interval for which Q_f is determined should be not less than 20–25 years (Engeneering of ship passages … 1964).

For many hydrological posts, the connection $I = f(Q)$ is ambiguous, and in the Ob, the Fominskoye post is characterized by the absence of connection Q and I (Fig. 3.24) and a slight change in the width of the river in different phases of the regime before the release of water to the floodplain. In addition, actual data are often not sufficient to support the link $I = f(Q)$. Therefore, the right part of the expression for determining the effective discharge is simplified in practice—$\sigma Q^m P$, or the meaning of I is accepted as permanent.

A similar approach to the determination of effective water discharge was applied by some foreign researchers (Neff 1977; Andrem 1980; Nolan et al. 1987); the authorship of the methodology is attributed to M. Volman and J. Miller (Wolman and Miller 1960). Apparently, their proposal was made regardless of the previously

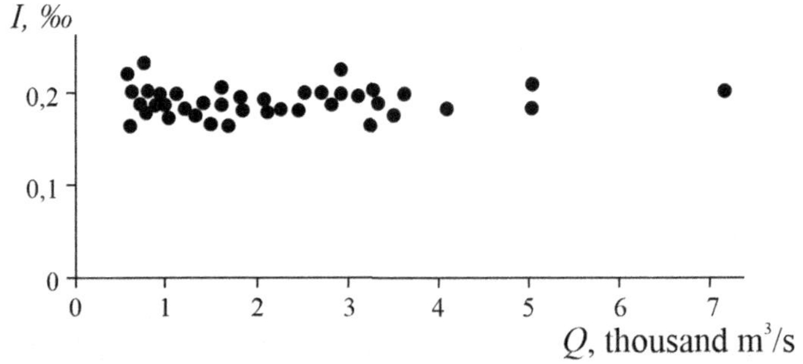

Fig. 3.24 Correlation between slopes free flow surface and water discharges (Ob river—Fominskoye)

published work of Makkaveev (1955), but it testifies, firstly, to the ignorance abroad of the Russian (then—Soviet) scientific literature and, secondly, to the coincidence of the direction of scientific research.

The application of N.I. Makkaveev's methodology makes it possible to calculate Q_f for rivers or their sections that are not provided with data on sediment discharge, i.e. to perform its determination on the basis of data from hydrological stations, where only water discharge is measured. However, according to Makkaveev (1955), *W* is the total suspended and tractional sediment transport rate, while network observations only determine the suspended sediment transport rate *R*. Nevertheless, using the data of the latter it is possible to take real values *m* from the relationship of $R = f(Q)$ instead of the recommended values ($m = 2$ on sandy rivers, $m = 2.5$ on sandy-pebbly rivers, $m = 3$ on pebbles and boulders) when calculating (Obodovskiy et al. 1987). This is possible because, as special studies have shown (Darbutas 1992; Nekos and Chalov 1997), *m* has one and the same value or it differs only in the second sign in the dependencies $W = f(Q)^m$ and $R = f(Q)^m$.

The inclusion of the sediment discharge formula in the Q_f methodology gives reason to consider Q_f as an integral characteristic of the influence of natural conditions on the formation and development of river channels (Chalov 1979). Indeed, sediment load depends on the physical and geographical characteristics of the river basins and is quite suitable for zoning, with the allocation of latitudinal zones, vertical belts and certain regions with local features of changes in sediment load caused by the geological-geomorphological structure or other natural factors. Similar conclusions can be drawn from the use of flow rate recurrence in calculated terms (3.49), which reflects the peculiarities of water runoff formation (hydrological regime of rivers).

A study of the connection between water discharge and suspended sediment load for different regions and basins has shown that the value of the degree *m* varies widely between rivers (up to 4.0). With the increase in the particle size of bed material load, the *m* parameter increases, which corresponds to the conclusions of Makkaveev (1955). However, each region (basin) has its own limits of change in the value of m. On the lowland rivers of Ukraine (Obodovskiy et al. 1987) they change from 1.12 to 2.60. The rivers of the Podolsk Upland, including the basins of the Southern Bug River and the Donetsk Ridge, are characterized by values of $m = 2.05–2.60$. In these areas, channel sediments are generally represented by gravel and pebbles. The rivers of Polesie and the left bank of the Dnieper have values $m < 1.6$. The lowest value of $m = 1.12$ was recorded on the Sula River near Lubna. Here bed material load are represented by fine sand and silt. Intermediate values ($m = 1.6–2.0$) are found in the rivers of the northern and eastern slopes of the Volyn-Podolsk Upland, the steppe regions of the Black Sea region, the basins of the rivers Samara, Severskiy Donets, middle and lower Dnieper.

In the Altai (the upper Ob basin), the value of m varies between 1.96–3.45 (Belotsky et al. 1995), and $m > 3.0$ is characteristic of large mountain rivers (Biya, Katun, Chuya, Charysh), m = 2.8–3.0—the rest of the mountain rivers. On the rivers of the foothills of the Altai and the upper Ob $m = 2.5–2.8$ (medium and coarse grained sands with an admixture of gravel and pebbles, sometimes sand and pebbles); small flat rivers of the Kulunda and Ob steppe plateau (fine and medium grained sands) have $m = 2.0–2.5$. The lowest values of m are characteristic of the rivers of the Biya-Katun interfluves in the foothill zone: here $m < 2$; these rivers are characterized by high sediment concentration and the contribution of suspended load in the bed material load composition, which results in increased sediment concentration of water in the lower reaches of the Katun.

In the Kama basin (Fig. 3.25), five groups of districts with specific intervals of the m indicator have been identified. In the basin of the Northern Dvina River, the value of m varies from 1.34 (the Maslyanaya river basin, the upper Sukhona River) to 2.34 (the Emtsa River); in the main rivers of the basin—the Sukhona, Vychegda, South and Northern Dvina—from 1.74 in the lower reaches of the Sukhona River to 2.31 (the lower reaches of the Northern Dvina River), increasing downstream in general (Reznikov and Chalov 2005). In the Ob River, m decreases from 2.40 at the confluence of Biya and Katun to 1.99 near Kolpashevo; at the Yangtze River, $m = 2.27$ at the city of Ichan and 1.98 at the city of Datun in the lower reaches of the river (Chalov et al. 2000). The average m value calculated for each selected area is assumed to be the calculated value in terms of effective discharges on rivers in the basin or region. Large and extended rivers are divided into sections for the same purpose, and the average value of m for each section is taken to calculate Q_m by formula (3.43).

Numerical values of indicator m sometimes significantly affect the value and exceedance probability of effective discharges and the form of “epure” Q = f ($\sigma Q^m IP$). Changing it may lead to the formation of a new one or the disappearance of one of the maximum calculated works, which naturally affects the reliability of the determination of effective discharges. For example, for a river Ros (Dnieper tributary) with different values of the degree m index obtained three variants of Q_f passing conditions (Obodovskiy et al. 1987). If we accept the recommendations of N.I. Makkaveev, then for the gravel and pebbles channel at $m = 2.5$ one effective discharge $Q_f = 900$ m^3/s exceedance probability of 0.07% is allocated, which takes place with flooded floodplain. At $m = 2.0$, there are three effective discharges, both

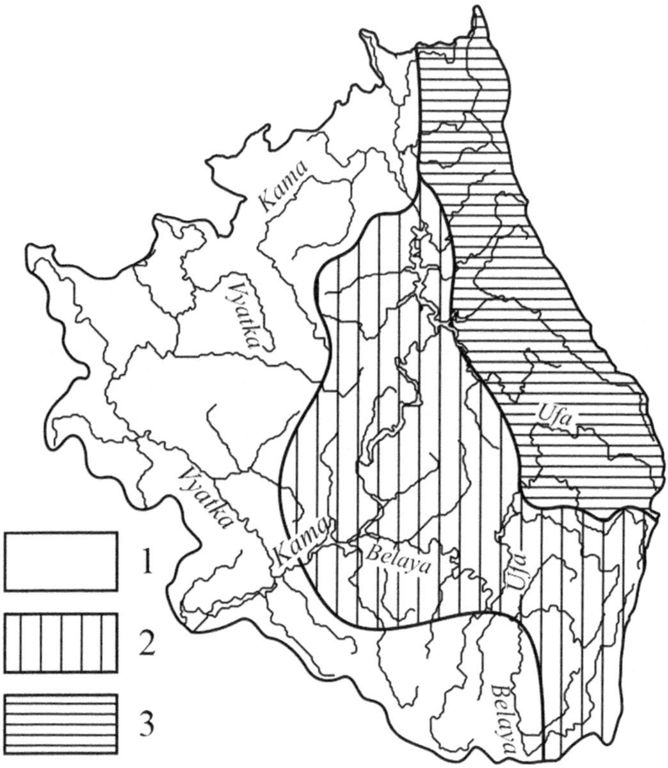

Fig. 3.25 Kama river basin zoning by the index of m degree in the dependence R = *f*(Q^m): 1—m_{av} = 1.7 (1.6 < m < 1.8); 2—m_{av} = 2.0 (1.9 < m < 2.1); 3—m_{av} = 2.3 (2.2 < m < 2.5) (Chalov, Shtankova 2000)

for floodplains and floodplain edges with water flow rates: 900 m^3/s, 725 m^3/s, 145 m^3/s, exceedance probability, 0.07%, 0.14%, 2.07%, respectively. Based on actual data m = 1.45. At the same time, two Q_f are distinguished, the absolute values of which are equal to 610 m^3/s and 78.0 m^3/s (exceedance probability, respectively, 0.22% and 4.51%). They pass both in the channel within the floodplain edges and in the floodplain.

Thus, increasing the m parameter increases the "weight" of the work $\sigma Q^m IP$ at high water consumption. This is due to the fact that, with the increase in the size of bed material load, the priority is given to the flow rates of floods and high floods, which have a greater impact on channel changes than the intermittent flow rates. The discharges of flooded floodplains are becoming increasingly important, and the role of water in the floodplains and the associated kinematical effects are diminishing. The latter through the coefficient σ, depending on the width of the floodplain, on the wide floodplain rivers can halve the value of the product σQ^mIP, which with small values of m leads to the "disappearance" of the upper Q^f.

An important element in the definition of Q_f by N.I. Makkaveev's method is to establish the moment when the coefficient σ should be entered. It is obvious that the influence of the floodplain on the conditions of riverbed formation affects when its flooding is accompanied by the formation of a transit floodplain flow, which has other than channel hydraulic characteristics. Its interaction with the channel flow determines the changes in speed and transport capacity of the latter (Baryshnikov 1984). Therefore, it is recommended to enter the coefficient σ when water levels are 0.5–1.0 m above the floodplain level. If the floodplain is single tier and flooding covers it across its entire width, then σ has a value corresponding to the partitioning between width of the floodplain B_{fl} and the width of the river b_{ch}. In the case of different heights of floodplains, the water spillage gradually extends to higher and higher levels. The value $\frac{B_{fl}}{b_{ch}}$ changes depending on this. This results in a consistent change of the coefficient σ from the one corresponding to the low floodplain flooding to the maximum flooding of the river.

When determining Qf, the time interval should be corrected according to the difference-integral curve of the river water content change. In this case, the repeatability of water discharge is determined for the period including the same number of dry, medium and wet years. If the calculation period covers mainly dry years, the results are underestimated, and if high water years are overestimated.

Depending on the height and duration of floods, the repeatability of each discharge interval, the degree of flooding, the nature of the floodplain surface and the composition of bed material load, effective discharge rates in rivers are observed in different ranges of levels and have different exceedance probability. In different natural conditions, there can be from one to three intervals of such discharges, which, according to their ratio with the water discharge corresponding to the beginning of complete floodplain flooding, are combined into four main varieties, which correspond to their "episodes" of the product (3.49) (Fig. 3.26). In some cases, the upper of the effective discharge intervals is observed at the levels corresponding to the highest floodplain flooding (A, B), in other cases, effective discharges take place only within floodplain edges (C, D)—the middle and lower intervals. There may be two intervals (B, D) before the floodplain release, the middle and lower ones, where the first one usually occurs during floods or high summer-autumn floods and corresponds to the levels immediately preceding the floodplain release, and the second one corresponds to the conditions of high low-water level, when only riverbed banks are under water, or one (A, C), which is either middle or lower.

The specific weight of each of the components of the work (3.49) in the formation of a particular type of "epure" is different. For example, the upper interval Q_f arises because of the large value of Q_{max}, due to which the influence of the coefficient σ does not affect so significantly (it corresponds to deep flooding of the floodplain), as well as at its high exceedance probability (flooding of the floodplain is lengthy). Almost always the average interval Q_f is associated with the phenomenon of kinematic effect at further increase of levels and flooding of the floodplain; the lower interval Q_f usually has a probabilistic origin due to its high

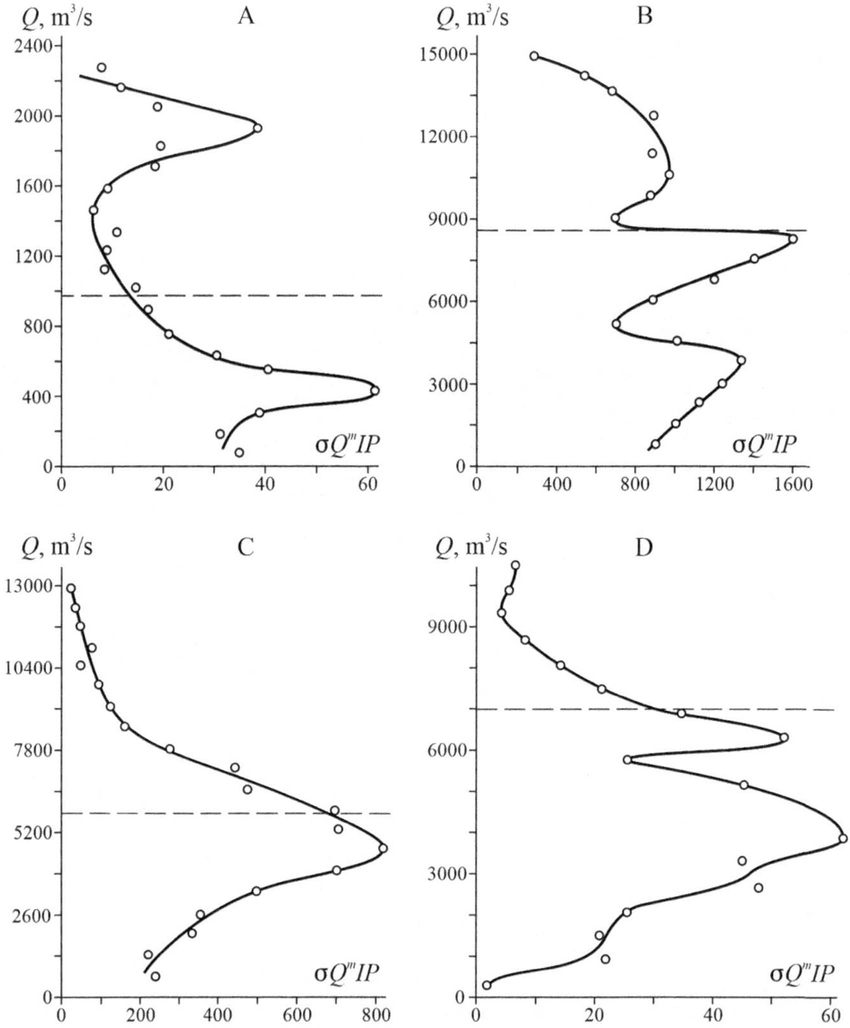

Fig. 3.26 Curve types for definition of effective discharges $Q_f = f[\max(\sigma Q^m IP]$: **A** Tobol river—Yalutorovsk; **B** Olenyok river—Olenyok; **C** Tom river—Tomsk; **D** Huang He river—Lijhing. Dotted line corresponds to the water appearance on floodplain

repeatability during the year and in the multi-year plan. On lowland rivers with sand alluvium, it is associated with erosion or shallowing of ropes in the low-water period; on rivers with pebbles and boulders, this interval is usually not represented.

Schoenberg (1989, 1990) has shown that the choice as an effective upper range is not well-founded, because its exceedance probability and value have high statistical instability and depend on the choice of time series used for calculations. Indeed, the analysis of the results of determination of effective water discharges on the rivers of

Northern Eurasia at 473 stations (Chalov and Bely 1975, 1984; Chalov 1979; Vlasov, and Chalov 1991) has shown that only in 4% of cases Q_f exceedance probability of the upper interval exceeds 5% and only in 20% more than 1%. Therefore, N.V. Schoenberg proposes to use not absolute ranges of change Q, but a more stable value—the average maximum water discharge in determining the design intervals. However, even in this variant of the methodology, the statistical instability of estimates of the frequency of large water discharge (at high levels) does not allow to accurately determine the value of Q_f *of* the upper range. Taking into account its importance for the assessment of channel changes in the high-water phase of the regime, it is necessary to make such assumptions and establish Q_f, which has low exceedance probability, correcting the results as the data are accumulated.

Definitions of Q_f, made practically on the basis of data of all hydrological posts on the rivers of the former USSR, which correspond to the requirements of N.I. Makkaveev's methodology, showed that the form of "epurs" depends on the type of water regime. For the majority of the rivers of the Russian Plain with the Eastern European type of water regime, two or three maximums of $\sigma Q^m IP$ have been set, of which the upper maximum corresponds to the flooded floodplain. The exception is the rivers or upper reaches of large rivers flowing from the main watershed of the ETR, which is the area of flow formation, as well as the extreme south of the plain —the area of insufficient moisture. Here, Q_f only passes through the floodplain edges, which are inundated only for a short time during floods. In the middle and lower Amur basin, the upper Q_f corresponds to the summer flood period associated with monsoon rains. At the same time, on the upper Amur and its tributaries, located mainly in the mountainous part of Baiksl region, Q_f pass before the water comes out to the floodplain. At Altai and Tienshan types of water regime (according to B.D. Zaikov) "epur" has trapezoidal form that reflects continuity of active channel reformations in the period of extended multi-peak flood. The genetic connection between the form of the "diagram" of effective discharges and the type of river water regime is illustrated by a good correspondence between the areas identified by the conditions of their passage (The work … 1987; Fluvial processes on the USSR rivers 1990) and the scheme of hydrological zoning of Kuzin and Babkin (1979). Thus, the effective discharge is an integral characteristic of water flow in rivers as a factor of fluvial processes, reflecting not only the runoff rates of the river, but also the peculiarities of its water regime.

Table 3.22 shows the results of calculation of effective water discharge for a number of large (and largest) rivers in Eurasia. At the same time, it is shown how Q_f conditions change in the flow regulation by reservoirs.

On the Northern Dvina before water releases to the floodplain, two Q_f intervals are distinguished, of which the lower Q_f has a very high exceedance probability (44%), creating conditions for complex and time-consuming rerouting.

On the lower Lena there is only one Q_f of the average interval before the water reaches the floodplain. Due to downward Q_f and water runoff rates increae, the exceedance probability decreases by 6 to 4 times. It is possible that this is affected by different positions of hydrological stations (Tabaga—in the section with wide floodplain channel, Kyusyur—with incised channel), as well as the impact of

Table 3.22 Effective water discharge and its exceedance probability (%) on some large rivers in Eurasia in natural (N) and regulated (R) conditions

River	Hydrological station		Effective water discharge, Q_f					
			Upper		Average		Lower	
			m³/s	%	m³/s	%	m³/s	%
Northern Dvina	Abramkovo		9500	3.7	1500	44.0	–	–
Lena	Tabaga		25,000	7.0	16,000	16.8	–	–
	Kyusur		115,000	1.2	75,000	4.9	–	–
Aldan	Verkhoyansky perevoz		39,000	0.4	27,000	2.9	19,000	7.4
Viliuy	Khatyryk-Homo	N	11,500	3.8	7000	8.9	–	–
		R	9750	1.0	4250	9.9	–	–
Ob	Fominskoye		–	–	2750	12.2	–	–
	Barnaul		–	–	3750	11.7	–	–
	Novosibirsk	N	–	–	4000	17.7	–	–
		R	–	–	3500	11.9	–	–
	Colashevo	N	–	–	12,500	10.0	4500	32.8
		R	–	–	11,500	7.3	3500	37.5
Chulym	Communarka		–	–	3000	6.7	800	36.1
Tom	Tomsk		–	–	4750	6.5	–	–
Yangtze	Ichan	N	–	–	28,500	13.27	–	–
		R	–	–	37,500	5.52	32,000	10.80
	Loshan	N	–	–	38,000	13.21	–	–
		R	–	–	37,500	13.42	25,500	35.07
	Hankou	N	–	–	38,000	18.42	–	–
		R	–	–	43,500	11.54	37,500	19.41
	Datun	N	82,000	0.34	50,000	11.29	42,000	25.09
		R	67,500	0.78	46,500	15.97	28,500	48.52
Huang He	Huayuankou	N	15,000	0.18	4000	11.27	–	–
		R	–	–	3000	11.75	–	–
	Jiahetan	N	13,000	0.19	3750	16.37	–	–
		R	–	–	3000	11.75	–	–
	Gaotsun	N	9000	0.72	3750	15.44	–	–
		R	–	–	3000	10.82	–	–
	Sunkou	N	8000	0.18	3600	14.84	–	–
		R	–	–	3000	10.22	–	–
	Aishan	N	8500	0.69	4500	8.82	–	–
		R	–	–	3000	10..13	–	–
	Lakow	N	6000	3.40	4200	10..57	–	–
		R	–	–	3000	9.67	–	–
	Liczin	N	6300	2.90	3900	12.38	–	–
		R	–	–	3000	8.13	–	–

tributaries, among which are large ones, such as Aldan and Vilyui. On Viliui, the conditions of Q_f coincide with those on Lena, but the exceedance probability of both their intervals is 2 times less. On the Aldan, the lower Q_f is absent on the main river, which reflects some peculiarity of its hydrological regime, which distinguishes it from the Lena and is associated with a great influence on the formation of its flow of Far Eastern cyclones and monsoon rains, causing the passage of several fairly high floods in summer and autumn and increased flow in the low-water. It can be assumed that the influence of Aldan on the Q_f conditions on the Lena River is represented in a decrease in the Q_f probability of the average interval characteristic of the main river. To the same effect leads to a low exceedance probability Q_f of the upper interval on Lena on Aldan.

Ob, in the upper and partly middle reaches, in contrast to the Northern Dvina and lower Lena, does not have an upper Q_f, which appears much downstream. Before water reaches the floodplain, there is only one Q_f of the mean interval, the absolute values and exceedance probability of which grow downstream. Only below the Tom estuary there is a noticeable (in natural conditions almost 3 times) decrease in its exceedance probability, which is represented by the influence of large right tributaries—the Tom and Chulym, which have a different area of flow formation (in the Ob such is the Altai, in the Tomi and Chulym—Salair and Sayan) and are characterized by lower exceedance probability Q_f (Chulym—also less its value). Besides, under the influence of Chulym the curve for determination of Q_f is transformed: on it (Kolpashevo), as well as on the curve for Chulym (Kommunarka settlement), there appears Q_f of the lower interval of high exceedance probability (over 30%).

At Huang He, in natural conditions (before the construction of the reservoirs, especially the Sanmenxia HPP), the upper Q_f is reduced by an absolute value downstream by 2.5 times (from 15,000 to 6000 m^3/s), and only at the lowest station, Liczin, there is a slight increase (up to 6300 m^3/s). This appears to be due to the location of flood dams directly along the riverbed, which creates an artificial analogue of the incised channel where the entire river stream is concentrated. Upstream, the dams will be located up to 10 km or more apart from each other, outside the channel wandering (or meandering) belt. The same fact leads to a decrease in the average Q_f downstream in the upper 2/3 of the section, after which it increases by 1.5 times among dams, reaching values higher than at the beginning of the lower reaches of the river, and decreases again to its mouth. The changes of Q_f exceedance probability in these conditions are more complicated: for example, in the case of the average Q_f, it first grows, then decreases and grows again with the maximum restriction of the flow by dams.

The curve for the determination of Q_f in the Huang He differs from other rivers in its significant completeness of the «epure», which consists of a large width of the distance between the upper and lower parts of the curve, which forms the maximum function (Fig. 3.27A). The completeness of the Qf «epure» is further enhanced for the conditions of an adjusted flow reservoir. In this case, at all hydrological stations in the lower reaches of the Huang He River (the figure shows, as an example, a curve for the determination of Q_f in Sunkou), a significant number of points lie on a weakly convex curve, on which the maximum is difficult to determine. In the given

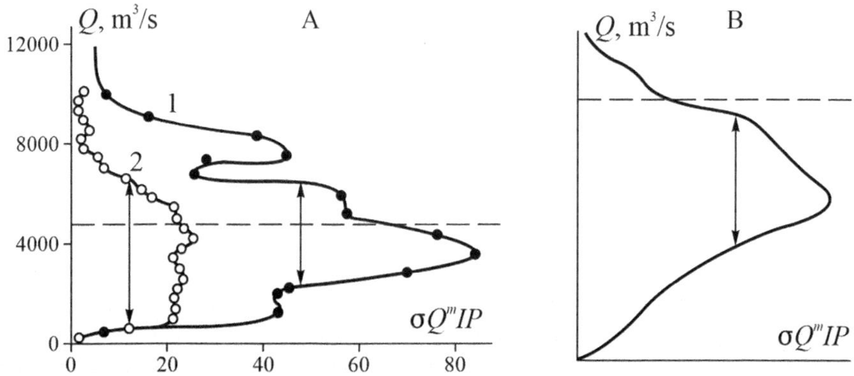

Fig. 3.27 «Epures» of maximum effective water discharges: **A** Huang He river (Sunkou station); 1—in natural conditions; 2—in regulated conditions; **B** Amu Daria river (Karamysh-Tash station). Arrows show the width of «epures», dotted line shows water appearance on floodplain

example (Sunkou station) after the creation of the Sanmensya HPP all water discharge in the interval from 1100 to 3400 m^3/s have essentially become effective, and only along the axis of the weakly convex curve the maximum function corresponding to Q_f *of* the average interval equal to 3000 m^3/s has been determined. Q_f exceedance probability at regulated flow at different hydrological stations changes mainly in the direction of its decrease. The main consequence of the flow regulation at the Huang He was the complete (along the whole downstream) disappearance (cutting) of the curve peak corresponding to the upper Q_f.

The completeness of "epure" Q_f brings the Huang He River closer to the plain rivers of Central Asia—the Amu Darya and the Syr Darya (Chalov and Belyi 1984). In this case, active channel changes are stretched for most of the year, while on rivers with a small fullness of the "epure" they are limited to a relatively short period of time. For the Amu Darya, such a stretch of the transformation stage was noted by many researchers of the river channel regime (Artamonov 1954; Altunin et al. 1975), and some of them saw this as a sign of chaotic channel changes. More complete Q_f episodes are usually typical of rivers with unstable channels and long floods or flood periods, if the flow velocities over the entire long period of high levels significantly exceed the non-destructive ones. In the first case, it determines the same or close to the maximum values of several flow intervals, so that effective discharges have a large range of exceedance probability. Floodplain inundation does not affect the transport capacity of the stream, and the upper parts of the single maximum may correspond to floodplain water levels. As a result, active channel changes occur continuously over a long period of time, stretching for most of the year or often repeated over the course of a year or a number of years. These are the conditions under which effective water discharges in the Amu Darya, the Huang He and other rivers are discharged, where this phenomenon is overlaid with absolute dominance of its suspended component in sediment load and its equal contribution in the formation of bed material load. In fact, the Amu Darya river effective water discharge is from 1 to 50%. This elongation of the intensive impact of the flow on the channel leads to the fact that a large amount of bottom sediments start to move even if the levels rise slightly, and the relief of the channel begins to adjust. Significant changes in the channel forms are observed in early spring, fading only by the end of autumn, when the low-water level occurs.

In mountainous and partly semi-mountainous rivers, effective value are often characterized by extremely high floods in addition to the main flow rates, the probability of which is tenth and hundredths of a percent. The emergence of another Q_f interval is a consequence of the turbulent flow during floods and the resulting specific forms of bed load transport. Sometimes only one effective discharge corresponding to the maximum level rise is observed under these conditions. Indeed, the mass movement of the clastic material is observed only during single catastrophic floods, when the largest blocks of 0.8–1.0 m and more in diameter come into motion. At the same time, the influence of the floodplain on the conditions of Qf passage has practically no effect on mountain rivers and all their intervals can be observed at its full flooding.

The presence of one or more effective discharges on rivers is reflected in the channel morphology, its change features and influence on the floodplain. In cases where Q_f passes before water reaches the floodplain, the channels have forms that are consistent with the floodplain bank configuration. On meandering rivers, steep bends develop, reaching a loopy or sinusoidal shape. The oxbows formed by their cutoff quickly turn into floodplain lakes or troughs, strongly curved in the plan. On the rivers with braided bed, the branches of the riverbed are laced off from the main bed at the source and estuary and gradually die off, remaining in the relief of the floodplains in the form of elongated hollows with lakes. The presence of one or two Q_f intervals within the floodplain logs determines a peculiar "stepwiseness" of the channel, in which a series of enclosed channel forms are formed, developing at different water levels. The point-bars on rivers with middle and lower Q_f interval form an intermediate step between the low-water channel and the floodplain. Q_f of the lower interval passes until these shoals are completely flooded and causes active changes of the riffles within their saddles. The mean interval is observed already at levels close to the height of floodplain edges, and it is associated with the reshaping of side-bars on riffles. In the presence of only an average interval of Q_f and a small height of shallows, the latter are separated from the banks by side channels, having the form of precipitates, sometimes moving quickly along the river. This usually contributes to the development of channel multithread. While its passing the most intensive development of the bend happens: the erosion of ones and the build-up of shallows near another banks, leading to the movement of the bend, the transformation of their shape and change in parameters. On rivers with one lower Q_f interval, the sides of the ropes are usually attached to the banks and inactive. Usually this interval is observed in the period of stable summer low-water period, when an important factor of channel changes is the duration of the flow on the channel. At this time, the bend and branches are complicated by secondary sinuosity or branching, the parameters of which are less than the parameters of the bend and channel branches themselves, correspond to the bends of the dynamic axis of the flow, enveloping the side-bars of the riffles.

When effective discharges are passed under floodplain conditions, i.e. in the presence of the upper Q_f interval, river floodplains are usually divided into separate massifs by branches that make up the anabranching channel system [according to Makkaveev (1955)]. On meandering rivers, it is the result of cutoff relatively gentle bends through the base of the floodplain segment, which is bounded by this bend. The main stream of the river moves into the cutoff bend hose, and the former channel is preserved as a floodplain channel. This type of anabranching channel is often identified with incomplete meandering (meander cutoffs) (Popov 1965; Kondratyev et al. 1982), i.e. with the formation of cutoffs. However, the anabranching channel system under such conditions of Q_f passage also appears as a consequence of incomplete dying of braided channel branches. Thus, the passage of Q_f upper interval, passing through the floodplain, creates favorable conditions for the cutoff of the free bend and the formation of "cutoff bend", contributes to the formation of braided channel and reshaping branches. Anabranching channel is (Chalov 1979) the most striking consequence of the passage of the upper Q_f. However, high sediment load may be one of the reasons why the river has no riverbed branches, despite other favourable conditions for its development—unstable riverbeds and accumulation of sediment. Apparently, the intensive straying of the channel and the filling of the emerging branches with sediment are more powerful factors preventing the development of branching.

With two or three Q_f on the river, each of them is a leader in the formation and development of various channel forms. This is represented, for example, in the different closeness of connection between the parameters of the channel forms (channel relief) and Q_f. Matveev (1985), for the sections of the Vilyuy, Aldan, Dniester, and Ponoy rivers with the meandering channel, obtained the highest values of the correlation coefficient for the effective discharge of the upper (for the given river) interval (in comparison with other Q_f), corresponding, according to his data, to the average monthly maximum water discharge. The greatest density of bonds is characteristic for sandy wide floodplain channels (correlation coefficient—0.87–0.95), the smallest (0.57–0.82) for incised bends. On other rivers, the role of each Q_f interval may be different. In the basin of the Don river (Nekos and Chalov 1997), many small rivers have higher correlation coefficients that are characteristic of the connection between the radius of curvature of the bend and the upper Q_f, which takes place in the floodplain. On the other hand, on medium and large rivers, the closeness of the connection is greater for Q_f of the lower interval, although other combinations are also observed in different parts of the basin.

Flow regulation leads to the transformation of the diagram of effective water discharge (Table 3.22). The upper Q_f interval usually disappears below the waterworks because the high floodplain is not flooded or is covered with a layer of shallow water for a short period of time. At the same time, there is a decrease in Q_f, which takes place at levels below the floodplain edges, an increase in its exceedance probability and a decrease in its corresponding level. Sometimes the average and lower Q_f intervals are combined into a single one, and it can have a fairly large range of changes in absolute values and exceedance probability, so its impact on the channel increases compared to natural conditions. Above the reservoirs, floodplain flooding increases and Q_f floodplain exceedance probability during spills, or a new, upper Q_f interval occurs during maximum flood levels. At Vilyui, before commissioning of the Vilyui hydroelectric power station, two Q_f intervals of 3.5 and 8% were observed, of which the upper interval was at the floodplain completely flooded, and the lower interval was observed at the water levels below the bottomland edges. With the flow regulated, two peaks are also expressed on the Q_f profile, but both of them correspond to lower levels, as the high floodplain is not flooded due to the cutoff of the flood peak. In the Ob, thanks to the construction of the Novosibirsk Hydroelectric Power Station, the Q_f value was reduced here at levels below the floodplain surface, its exceedance probability increased from 15 to 18–20% and the corresponding level decreased by 0.5–1 m. On the Volga River, comparison of effective discharge values calculated for the modern conditions of regulated flow with the data of Makkaveev (1955) obtained for the domestic regime (Table 3.23), shows that, despite the decrease in the height and duration of the flood, in the lower reaches of the hydroelectric complexes there was no sharp decrease in the supply of Q_f of the upper interval. However, its absolute value decreased on average by 1.5 times, although the level of passage remained above the floodplain edges. Exceedance probability of Q_f of the lower interval after flow regulation has increased from 27–30 to 52–88%, i.e. there was a significant increase in the role of low water level in channel changes due to both the increase in its runoff rates and the extension of the period of relatively high levels and the passage of pass waves in the zones of unsteady flow. In addition, Q_f of the average interval passing in floodplain edges and having exceedance probability of 4.5–8.0% was formed in conditions of the regulated flow (Vlasov and Chalov 1981).

These changes are related to both flow regulation and the strengthening of the Oka, the largest tributary of the Volga River, which remained unregulated under the new conditions. Previously, both rivers had different number of discharge intervals

Table 3.23 Effective discharges on mid- and lower Volga river before and after the creation of the reservoir cascade

Hydrological station	Effective discharge Q_f, m^3/s, and its exceedance probability, %					
	On a flooded floodplain		Upstream			
	Upper		Average		Lower	
	m^3/s	%	m^3/s	%	m^3/s	%
Natural runoff (Makkaveev 1955)						
Vyazoviye	28,000	0.5	–	–	2800	33.4
Samara	24,000	6	–	–	6000	35.8
Volgograd	36,000	2.4	–	–	8800	27.0
Regulated flow (Vlasov and Chalov 1981)						
Nizhny Novgorod	15,800	0.7	6580	8	3870	88
Kozmodemyansk	18,000	0.8	10,900	4.5	2920	76.3
Volgograd	24,500	3.2	20,800	6	5900	52

within the floodplain edges, which were of fluvial processes driver importance. After the Volga flow regulation below the Oka inflow, the shape of the curve $Q_f = f(\sigma Q^m IP)$ became completely the same as the diagram typical for the inflow. In other words, in the Volga riverbed mode, the middle course of the Volga river has become more distinct Oka features.

On the Yangtze (downstream) in natural conditions, before the river reached the Great Plain, there was only Q_f *of* one mean interval. In the lower reaches the Q_f were above the floodplain edges (upper interval), at riverbed filling to the floodplain edges (middle interval) and at the time when the riverbed banks dry up (lower interval). At the same time, the upper Q_f had the minimum exceedance probability —0.34%, and the lower Q_f—the largest on the rivers of China—25%. The absolute value of Q_f of the mean interval per Yangtze increased downstream by 1.7 times, but its exceedance probability at first significantly increased (from 13 to 18%), and then decreased to 11%. Under conditions of regulated flow in the lower reaches of the Q_f river the upper interval has been preserved, but its absolute value has decreased and exceedance probability has increased more than 2 times. Below the floodplain edges in the impregnated area (Ichan), the exceedance probability of Q_f became 2.5 times lower and its absolute value increased by more than 30%. Downstream changes are less significant for both indicators, but without any particular trend in the length of the river. At the same time, at all hydrological stations, a new Q_f of the lower range was formed, the absolute value and exceedance probability of which varies with different sign; the remaining lower Q_f in the lower reaches of the Yangtze (Datong city) is 1/3 less than in natural conditions, and almost twice as much exceedance probability (48.5% vs. 25%).

Significant changes in effective discharges occur above large reservoirs, where the river flow under a certain interval of levels is under pressure. Here, the floodplain flooding increases and the exceedance probability of Q_f during spills increases, or a new, upper Q_f interval is formed, corresponding to the maximum flood levels. At the Mologa River (Ustyuzhina town), it is equal to 1900 m^3/s, with a exceedance probability of 0.2%. Prior to the establishment of the Rybinsk water reservoir, this maximum did not appear at all. The values and exceedance probability of the other two Q_f intervals prior to the release of water to the floodplain have not changed in comparison with the domestic mode. On the Don above the Tsimlyansk reservoir in the zone of variable flow there are two intervals of exceedance probability 6–7 and 13–17%, which correspond to the full flooding of the floodplain before the creation of the reservoir the upper interval Q_f (under the floodplain) had exceedance probability of 1.5%, and the average passed below the floodplain logs, having exceedance probability of 17.5% (Makkaveev 1955).

In rivers with low sediment load, side effects are rejected and turned into aberrations as soon as the lower Q_f interval has passed. In natural conditions, such an axillary channel was typical of the Upper Volga and Don. On the Middle and Lower Volga, when only one lower Q_f was observed here, massive side-effects were inactive and confined to the banks. With the appearance of the average interval of Q_f flow regulation, their rejection from the banks and their transformation into mid-bars began. The branching of the main channel of the Volga River increased, which was also facilitated by a decrease in the height of the newly formed channel banks formed at reduced sediment load. At the embankment sections of the lower tailings of the hydroelectric power plant, the rejection of side effects and their transformation into a separations takes place as soon as Q_f of the lower interval passes. The formation of new mid-bars on the Volga River leads to the dispersal of the low-water flow and a decrease in its total transport capacity.

On the upper Volga, downstream of the Rybinsk reservoir, the average Q_f interval was "cut off". The riverbed has stabilized here, and its reshaping in the new conditions began to be limited only to the ripples of the ropes. There was also shallowing of branches, unification and joining to the bank of the islands, transformation of the mid-bar into riffle side-bars.

Since the design product (3.49) for determining effective discharge takes into account changes in the flow rate through the probability of water discharge, slopes, sediment particle size and floodplain width, this indicator reflects the impact on the channel regime of rivers in the catchment landscape, as well as the intensity of denudation processes, the products of which flow into the channel. As a result, effective discharges allow, according to their number, exceedance probability and ratio with floodplain height, to identify the most common spatial changes in the channel regime of rivers, and through them to establish regional features of morphology and channel dynamics. Since the complexity of the channel morphology, the nature and intensity of its transformation depend on the number, exceedance probability, the level of passage, the ratio with the floodplain height and the conditions of its flooding, the rivers of each region, which have their own form of Q_f "epure", are characterized by a specific set of channel forms, their parameters and changing regime. For example, where there is an upper interval of effective discharge, most rivers are distinguished by their anabranching nature, and their channels either form gentle bends or branch into branches; in other regions, the river channels form steep bends, and the floodplains are consolidated and not separated by ducts.

The method proposed by N.I. Makkaveev was included in the official guidelines for the calculation of corrective ship traffic routes on rivers (Engeneering guide … 1974); it was widely used in the study of fluvial processes in different natural conditions, including to detect their changes under the influence of flow regulation by reservoirs (Velikanov 1958; Degtyarev 1968; Sakharyuk 1980; Vlasov 1983; Rimkus 1984; Obodovsky et al. 1983). Characteristically, the number of Q_f and their exceedance probability, determined by the method of N.I. Makkaveev for semi-mountain rivers of Transcaucasia (Chalov 1969), were the same as calculated by Romashin (1972) at the rate of critical scouring rates to sediment particle size is confirmed by long-term stationary observations.

Comparison of the results of Q_f calculations according to N.I. Makkaveev's method with the calculations according to his proposed modifications or other methods leads either to relatively close results or to obtaining one Q_f out of three possible (according to N.I. Makkaveev's method). The latter gives reason to consider sometimes as "effective" water discharge averaged over a long period of time, which is close to one of the discharges having effective value. Thus, L. Lyudin (Antropovsky 1969) takes as such the average annual water discharge, believing that the effective one should correspond equally to all observed discharge values in the year. Leopold et al. (1957), Romashin (1968), Zavadsky and Chalov (1997, 2000) used the average of the annual maximum water discharge in the channel analysis. However, their values are significantly affected by the perrenial flow irregularity, which in some cases leads to significant deviations of these water discharges from the value of effective discharge. In addition, they may correspond to different Q_f intervals on different rivers, which introduces an element of uncertainty in the assessment of channel formation time conditions and the possibility of using them in relation to a particular structural level of fluvial processes.

Therefore, in most cases, the choice of a characteristic flow rate equivalent to the effective one is made depending on the research objectives. In particular, Zavadsky (2001) used water discharge of 30 days duration, which characterizes the runoff rates of the river in the most full-flowing phase of its water regime (flood or rain flood period, depending on the geographical location of the river basin). Its values are close to the effective discharge of the average interval (according to N.I. Makkaveev), in most cases slightly exceeding it (Table 3.24). Thus, the water flow rate of 30 days duration corresponds to the water levels prior to flooding of the floodplain, when the channel changes are most intensive.

The general disadvantage of using Q_f to identify conditions for channel development (regardless of the method of determination) is that they are calculated for the site of a particular hydrological station, and their magnitude and exceedance probability do not always characterize the conditions of Q_f passage at upstream and downstream sites. To neutralize this, it is necessary to correct the introduction of the coefficient σ into the calculated product according to the method of N.I. Makkaveev (3.49), taking it into account the partitioning between floodplain and channel widths in adjacent (typical) areas, if the hydrological station is located in the local narrowing of the valley, or to make additional hydraulic calculations to identify the peculiarities of the flow of the wide floodplain upstream and downstream. In most cases, this is the main reason for replacing Q_f with an equivalent characteristic water flow rate.

Table 3.24 Values of effective water discharge Q_f upper and average intervals (m^3/s) and 30-day water discharge duration Q_{30} (m^3/s) on some rivers (Chalov et al. 2004)

River	Hydrological station	Q_f		Q_{30}
		Upper	Average	
Vasyugan	Middle Vasyugan	610	530	586
Vah	Lobchynskoye	2800	1120	1440
Dniester	Oloneshty	361	–	255
Irtysh	Tobolsk	9000	4000	5890
Ket`	Maximkin Yar	1430	800	913
Conda	Bolchary	690	480	601
Pripyat	Turov	780	500	651
Seversky Donets	Belaya Kalitva	805	559	534
Chulym	Zyryanskoye	–	2600	2030
Yana	Jangky	6200	3600	3540

3.5 Flow Rate as a Condition of Riverbed Development

Analysis of the composition and distribution of bed material load on rivers, the stability of channels and effective discharges as the conditions in which the channel is formed, indicates the multifactorial nature of fluvial processes and the impossibility of identifying one or more of the leading ones, regardless of the specific natural conditions of channel formation. This conclusion confirms the criterion of fluvial processes types proposed by B.F. Snischenko (Kondratyev et al. 1982):

$$A_{\mathrm{S}} = \frac{I_{\mathrm{vb}}}{I} \cdot \frac{B_{\mathrm{vb}}}{b_{\mathrm{ch}}} \tag{3.50}$$

(here I—the slope of the channel; I_{vb}—the slope of the valley bottom; $\mathrm{b_{ch}}$—the width of the channel; B_{vb}—the width of the valley bottom). The $\mathrm{A_S}$ criterion is presented depending on the maximum water flow rate, which can be conventionally considered to be the upper fluvial process driver interval of the floodplain. The parallelism of the boundary lines indicates that the development of a particular type of channel (according to the SHI classification) does not depend on the importance of water content in the river. The determining role in these conditions is played by the width of the valley bottom and the partitioning between riverbed slopes to the valley bottom (Fig. 3.28). There is a natural change in the A_S criterion (3.50) from meandering to straight in accordance with the arrangement of the sequence of channel types, with some deviations of the same SHI classification scheme: free meandering → unfinished meandering → anabranching channel → limited meandering → side-bar type (straight channel) → channel multithread (in some classification variants the same as the side-bar type). Indeed, both the slope of the channel I, and the width of the bottom of the valley B_{vb} depend on the type of channel, i.e., the possibility of its movement (wandering) on the bottom of the valley: on the one hand, the more the curvature of the channel, the less its slope and greater ratio $\frac{I_{\mathrm{vb}}}{I}$; on the other hand, with other things being equal, the width of the bottom of the valley is maximum in the meandering channel and minimal in the rectilinear, single-thread (Makkaveev and Chalov 1986).

The assessment of conditions for the formation of different types of river channels by analyzing QI-diagrams is more promising, where Q—effective Q_f (or the average maximum $Q_{av.\ \mathrm{max}}$) water flow rate, I_{vb}—slope of the valley bottom. The choice of gradient of the valley, rather than the slope of the channel, is related to its inrelationship from the configuration and transformation of the channel (especially the wide floodplain one); it also determines the slope of the stream in the floodplain. The use of Q_f or Q_{max} is due to the fact that the former is often observed in floodplain globes, and then there is a discrepancy between it and the accepted slope I_{vb}; in the same cases, when Q_f passes with floodplain, its value is close to $Q_{av.\ \mathrm{max}}$.

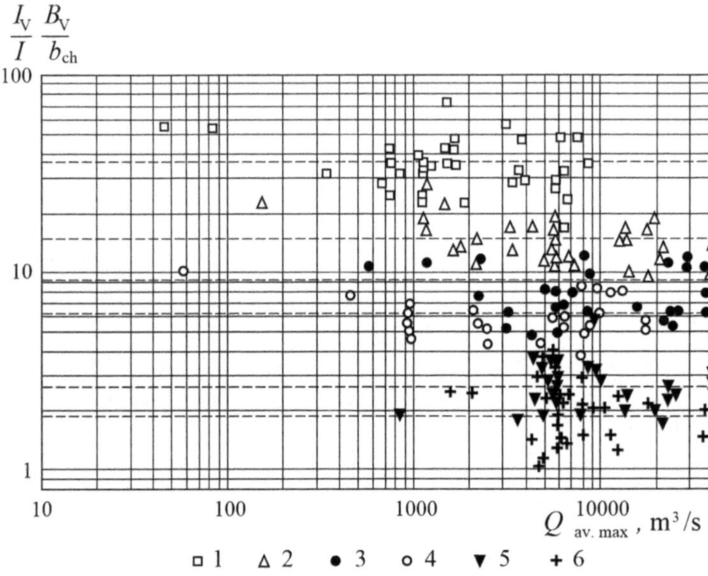

Fig. 3.28 Criterial graph of development conditions of different types of fluvial processes (according to SHI (Kondratiev et al. 1982): 1—free meandering; 2—incomplete meandering (meander cutoffs); 3—anabranching channel; 4—limited meandering (confined meandering channel); 5—side-bar type (straight single-thread channel); 6—channel multibranching

For the first time, *QI*-diagrams were used to analyze the conditions of formation of different types of channels by Leopold et al. (1957), who obtained a line of points corresponding to the meandering (meandering, with wide floodplain) channels occupying the lower part of the chart, and braided or straight, single-thread in the field of coordinates. As determining factors, they took the slope of channel I and water discharge in the floodplain edges Q_f, corresponding to the average interval of effective water discharge. Data on meandering and braided river channels of the USA and India were divided into two regions in these coordinates (Fig. 3.29), divided by the line

$$I_{cr} = 0.0125 Q_f^{-0.44}. \quad (3.51)$$

I_{cr} is called by them a critical meandering slope.

Subsequently, M. Carson (Carson 1984), having analyzed the data of L. Leopold and M. Walman, came to the conclusion that the relationship obtained by them primarily distinguishes between rivers with sandy and pebble bed. The slope above the critical one has pebble channels, most of which are braided in the initial data; the slope below the critical one has sandy channels, most of which meander. Therefore, Henderson (1961) introduced the coarseness of bed material load into the Leopold-Wolman formula:

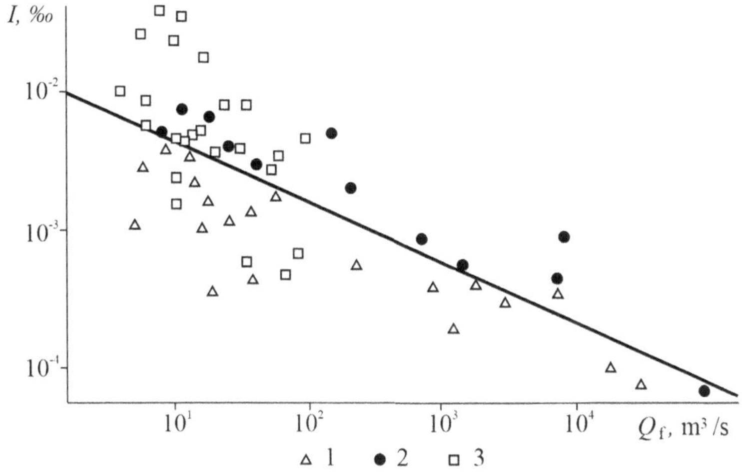

Fig. 3.29 Diagram «slope-discharge» I ~ Q for the rivers of USA and India (Leopold, Wolman 1957): 1—meandering channel; 2—braided channel; 3—straight channel

$$I_{\text{cr}} = 0.0196 d_{50}^{1.14} Q_f^{-0.44} \tag{3.52}$$

since the increase in the particle size of the load is usually accompanied by an increase in the slopes ($d \sim I$). Osterkamp (1978) according to the data of 76 rivers of the state of Kansas (USA) with sandy and sandy-pebble beds also showed that the ratio can be described by the family of equations with parameters d_{50}. D. Bray (1982), having analyzed the data on the pebble and pebble-boulder beds of Alberta (Canada), obtained that meandering beds with a degree of development l/L (l—the length of the bed on the bend, L—its pitch) more than 1.25 have a slope of less than

$$I_{\text{cr}} = 0.0367 Q_2^{-0.44}, \tag{3.53}$$

where Q_2 is the water flow rate for a two-year flood recurrence. Above this line (A) there are dots corresponding to weakly sinuous and braided channels (Fig. 3.30).

On the Fig. 3.30. The line corresponding to the Leopold-Wallman (B) Eq. (3.51) is also drawn. It can be seen from the comparison that the critical slope for rivers with coarsely clastic alluvium is considerably higher than the I_{cr} obtained without taking into account the particle size of the load. In general, the higher the l/L value, the more to the left and lower the chart shows the corresponding points. erguson (1984), according to the data for 89 sandy and pebble beds, obtained relationships for I_{cr} taking into account the particle size of the load:

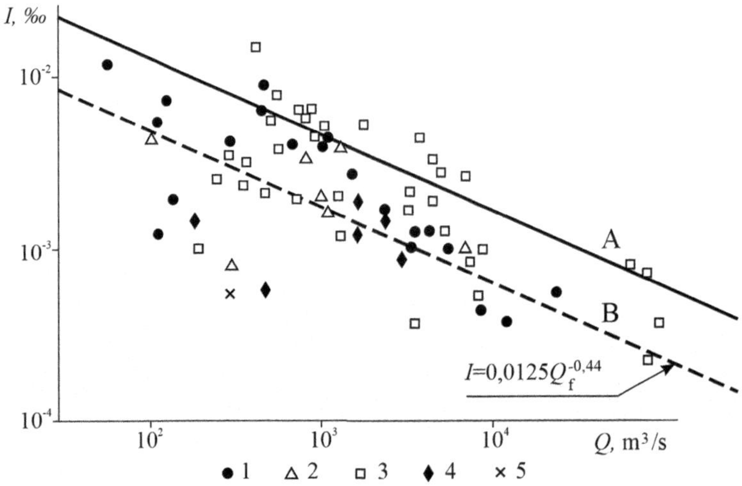

Fig. 3.30 Diagram «slope-discharge» I ~ Q for the pebble rivers of Alberta (Canada) with different degree of meandering (A) (Bray 1982); B—line, found by L. Leopold and M. Wolmal (Leopold and Wolmal 1957). Rivers with meandering coefficient (degree of development) l/L: 1—1.0 to 1.25; 2—1.25 to 1.55; 3—1.55 to 1.85; 4—1.85 to 2.15; 5—2.15 to 2.45

$$I_{cr} = 0.042 Q_f^{-0.49} d_{50}^{0.09}, \tag{3.54a}$$

$$I_{cr} = 0.017 Q_f^{-0.49} d_{90}^{0.27}. \tag{3.54b}$$

In this case, the Leopold-Wallman line is too high for sandy and too low for pebbles.

On a large actual interval along the rivers of the USSR Romashin (1968) has received for wide floodplain channels on the *QI*-diagram (Fig. 3.31) areas of meandering, braided and relatively straight, single-thread channels development. There is a transition area in which the meandering channels are dominated by channels with ruptured bends and braided channels are found. At the same time, the conditions for the development of braided channels coincide with the conditions for the development of the anabranching channel. The latter confirms the determinant role of Q_f of the upper interval (in the floodplain) both in the formation of channel branches and the preservation of floodplain branches, arising in the evolution of both meandering and braided channels. The same applies to the ruptured radiators, which are formed when Q_f passes above the floodplain edges. The boundary lines between the points corresponding to certain types of channel are described by equations:

$$I_{vb} = 1.35 Q_{av.max}^{-0.99} - \tag{3.55a}$$

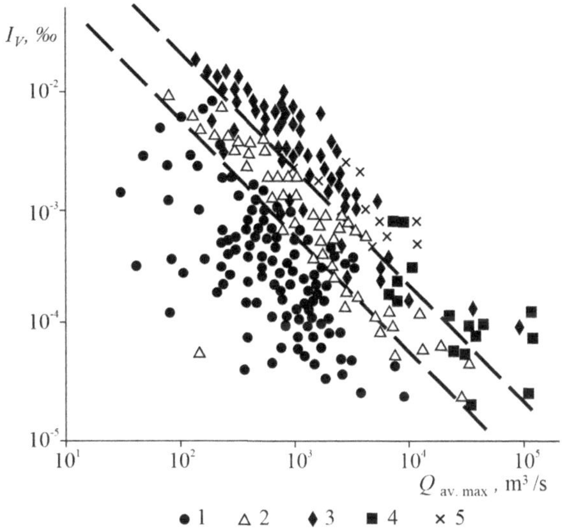

Fig. 3.31 Distribution of channel patterns of wide-floodplain channel in the field of QI-diagram [according to Romashin (1968)]: 1—bends; 2—meander cutoffs; 3—braided channel; 4—anabranching channel; 5—relatively straight, single-thread channel

Upper (between the branches, on the one hand, and the relatively straight, single-thread channels and transition area, on the other) and

$$I_{vb} = 0.334Q_{av.max}^{-0.99} - \quad (3.55b)$$

lower (between the transition area and meandering channels).

In China, the *QI-diagram* method was used to analyze the conditions of channel development throughout the Yangtze (Characteristics of the channel … 1985). At the same time, a single field of points corresponding to the upper, middle and lower reaches of the river was obtained, generally stretched along the line $I = 0.45Q^{-n}$. However, the selection of channel types has not been made and only the text refers to the predominance of points corresponding to the bends in the lower left part of the diagram and branching—in the upper right part. Asymmetrical branching with a hypertrophied bend of one of the branches ("duck's head") is referred by the authors of the book to the areas with a meandering channel.

Close results were obtained by Antropovsky (1969), who used data on 108 sections of 35 rivers of the USSR (Fig. 3.32): the points corresponding to the braided channel are located on the chart to the right and above, meandering—to the left and below; between them there is a strip in which the points belong to the cutoff (incomplete meandering (meander cutoffs), according to the classification of SHI). However, the points corresponding to the straight, single-thread turned out to be gentler in relation to the field lines of the points of braided channels, cutoffs and free meandering. V.I. Antropovsky points out the possibility of inaccuracies in the determination of channel types, because he used the data of R.A. Nezhikhovsky and schematic small-scale maps of Pinkovsky (1961, 1967), Kondratyev et al. (1959), compiled according to a simplified classification of the State Geological Survey.

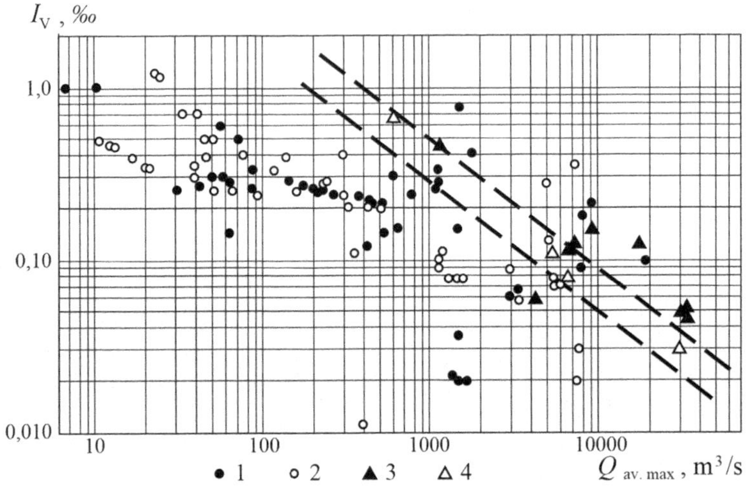

Fig. 3.32 Dependence of channel patterns on valley bottom slope I_{vb} and average maximum water discharge $Q_{av.max}$ (Antropovskiy 1969): 1—single-thread, straight channel; 2—free meandering channel; 3—braided channel; 4—cutoffs

A.M. Alabyan (1992) connects the physical meaning of *QI*-diagrams with changes in the specific power density of the flow (per unit of its length): the higher and to the right in the diagram field are the points, the greater the value of the flow power they correspond to. Indeed, the branching of river channels into the branches is characteristic of the largest rivers, which was noted by Velikanov (1958): the largest rivers with wide floodplain channel, they represent the dominant type of channel. At the same time, small and medium-sized rivers, as a rule, have freely meandering channels under similar conditions. This is consistent with the fairly clear relationship between the different forms of ripple motion of sediment and the flow rate (Fig. 3.33): the aisle points, which are the basis for the development of channel branches, are located in the upper right part of the QI diagram, while the lower left part of the diagram contains the points corresponding to the side-bars (the basis for the formation of bends) and the sand dunes.

The role of the power of the flow (its work, or "living force") in the formation of the channel was pointed out by Leliavsky (1893), who expressed it in the form of a work ρQH (*Q*—water flow rate; ρ—water density; *H*—kilometric drop). Approaching this question from the point of view of fluvial processes control to ensure normal conditions of navigation, he noted that "the desired correction of the channel is possible to achieve, mainly, not by restricting it, but by an expedient direction of the jets, and, of course, some restriction of living sections will follow" (1948, p. 107). In other words, in the self-regulating "stream-channel" system, the work of the stream is aimed at creating an optimal for its movement channel shape with the corresponding forms of a living section.

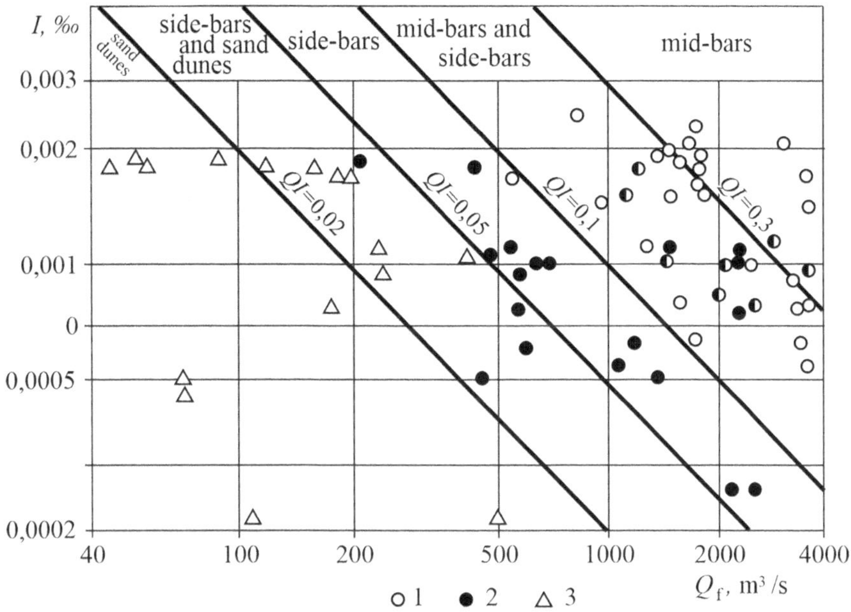

Fig. 3.33 Distribution of the macroforms of ripple movement of sediments (1—mid-bars; 2—side-bars; 3—sand dunes) in dependence on th effective discharge Q_f and channel slope during flood (Upper Ob river) (Alabyan 1992)

Antropovsky (1972) directly connected the conditions of formation of channels of different types with the partitioning between components in the equation of the specific energy of the flow section $h_{\text{average}} + \frac{V^2}{2g}$, where h_{average}—the average depth of the channel, numerically equal to the specific value of the potential energy of the flow, $\frac{V^2}{2g}$—the specific kinetic energy of the flow. According to its data (Fig. 3.34), the largest values of the specific energy of the flow corresponds to the rivers with braided channels, slightly smaller—cutoff and the smallest—freely meandering rivers.

The wide transition area between meandering and braided channels is explained by the fact that, although the power of the flow determines the formation of channels of one type or another, other conditions become essential in a particular natural environment. Also characteristic is the common field of points corresponding to the rectilinear, single-thread and braided or rectilinear, single-thread and meandering channels. In this case, the condition for the formation of the channel is the mobility of side-bars and mid-bars and the conditions for their fixation by vegetation. According to the data of A.M. Alabyan (1992), which was used to study the conditions of formation of river channels of different types of flow rate per unit of its length, the transition region is located in the range of flow rate from 4 to 15 kW/m ($0.4 < Q_f I < 1.5$). It may contain both braided and meandering channels, which is obviously connected with the diversity of channel formation

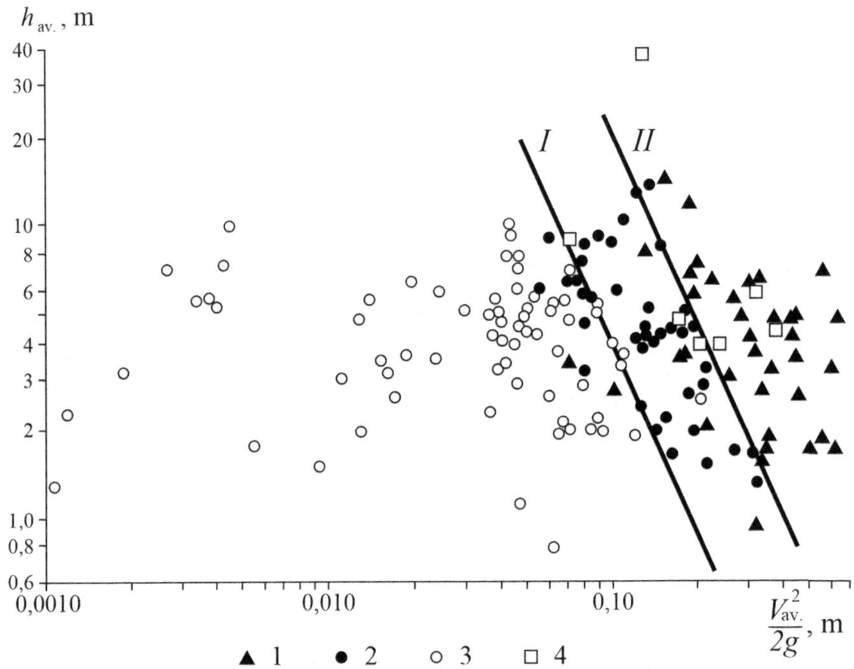

Fig. 3.34 Dependence of channel patterns on specific flow energy (h_{av}—depth; $V^2/2\ g$—velocity pressure). 1—braided channel; 2—incomplete meandering (meander cutoffs); 3—free meandering; 4—anabranching channel [according to Antropovsky (1972)]

conditions, including the erosion of the banks, the regime, the size and magnitude of bed material load, the stability of the channel, the unevenness of flow, ice phenomena (Alabyan 1992; Chalov et al. 1998), as well as the emergence in the flow of specific hydraulic phenomena associated with its uneven movement.

Begin (1981), having taken the tangent voltages as an indicator of specific energy or flow rate, calculated the value of bottom tangent voltage τ/τm (τm is an average of all data) for 126 highly sinuous (degree of sinuosity $l/L > 1.5$), 54 slightly sinuous ($l/L < 1.5$) and 82 braided channels. With the growth of τ/τm, the share of highly sinuous channels decreases and the share of braided channels increases; the maximum of weak sinuous channels is τ/τm = 2.0. In other words, an increase in the tangential stress corresponding to an increase in the flow rate is accompanied by an increase in the probability of branching of the channel. The mechanism of formation of channels of different types of Z. Begin connects with an increase in the rate of erosion of banks and transport of sediment with an increase in the specific energy of the flow. R. Ferguson applied a close approach (Ferguson 1973, 1984). From the expression for the specific power of the flow $N_{P_{sp}} = \rho g Q I b_{ch}^{-1}$ and from the statistically stable connection bp $\sim Q^{0.5}$, it follows that $N_{P_{sp}} \sim Q^{0.5}I$. The specific flow rate calculated in this way is approximately constant along the boundary line obtained by L. Leopold and M. Wolman (4.50) and is $N_{P_{cr}} = 50$ W/m^2. R. Ferguson concludes that the rivers with $N_{P_{cr}} > N_{P_{sp}}$ braided out, the rivers with $N_{P_{cr}} < N_{P_{sp}}$ meander. Many researchers have recorded an increase in the threshold combination of $I \sim Q$ (i.e. the flow rate) with an increase in the particle size of bed material load.

Along with the existence of a transition area between braided and meandering channels, which includes points both corresponding to both the straight and straight, single-thread channels and other types of main channel patterns, practically all *QI*—diagrams obtained by different researchers on their data sets, the positions of the section lines and their analytical expressions (coefficients and degree indicators) vary widely and, apparently, are largely determined by regional combinations of fluvial processes factors This is especially true for the differentiation of the main types of channel. Thus, for the Oka River basin (Barovskiy 2004), where only meandering and straight, unbroken channels are widespread, the obtained distribution of points on the *QI*-diagram (Fig. 3.35) is consistent with the above data: the points of the straightened unbroken channel are located in the area corresponding to a higher flow rate ($N_{P_{sp}} = \rho g Q I$). At lower flow rates, the wide floodplain channels meander. However, if usually, starting with the works of L. Leopold and M. In the field of *QI*-diagrams, in the field of *QI*-diagrams are drawn the lines of division between the points corresponding to the river sections with a certain type of channel, in this case, the separation of the benders by the degree of their development has led to the fact that the points belonging to the relatively straight channels, and the gentle bend with a degree of development $l/L = 1.15$–1.40 were in one upper part of the field, although with the shift in the graph to the right and upward the first begin to dominate. This gave rise to two $Q \sim I$ communication

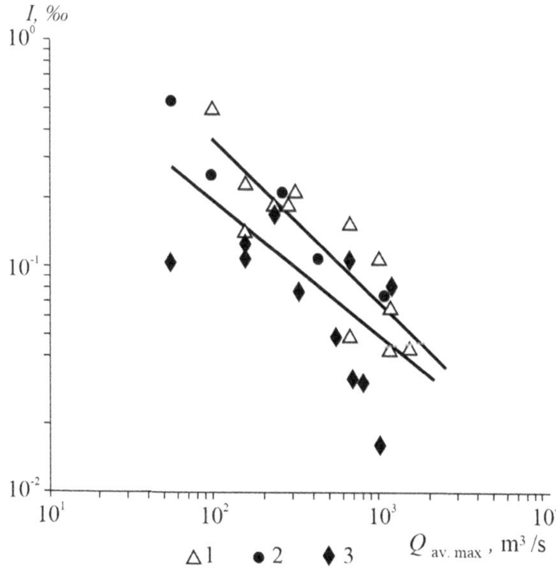

Fig. 3.35 QI-diagram for wide-floodplain channels on the rivers of Oka river basin: 1—straight channel; 2—gentle bends (*l/L* < *1.4*); 3—developed, segmental, sinusoidal bends (*l/L* > *1.4*) (Barovsky, Chalov 2004)

lines, corresponding to each of them, rather than to the division lines between the straight and meandering channels. The dependencies obtained in this way are approximated by equations: for straight channel—$I = 9.4Q^{-0.8}$ for straight channel —for meandering channel—$I = 2.9Q^{-0.6}$. The coefficients and degree indicators here have a regional meaning and differ from those obtained for rivers in other basins.

In addition, the multifactorial nature of fluvial processes determines the specificity of *QI*-diagrams for large and small, flat and mountainous, incised and wide floodplain rivers. You can see it in the Fig. 3.36. It presents *QI*-diagrams for the middle Amur, Northern Dvina, Ob, Lena with their largest tributaries—Chulym, Vilyuy, Aldan, Huang He, Yangtze (A), small and medium rivers of the Altai

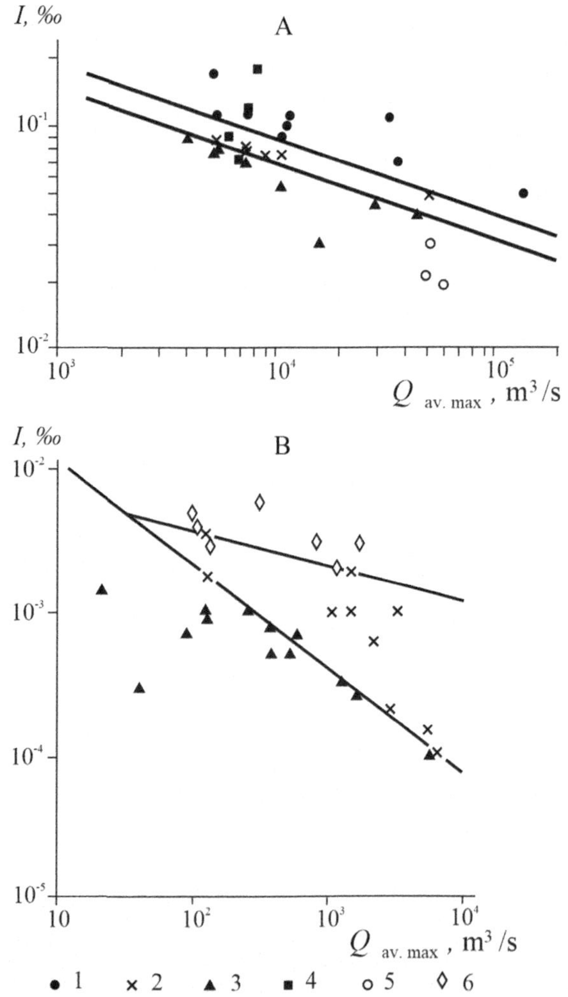

Fig. 3.36 Distribution of channel patterns in the field of QI-diagram: **A** for Middle Amur (Zavadsky et al. 2000), Northern Dvina, Ob, Lena with ducts, Huang He, Yangtze (Chalov et al. 2000); **B** small and medium rivers of Altai Region (Smirnova 2002). 1—braided channel; 2—straight, single-thread channel; 3—meandering channel; 4—Huang He river; 5—single asymmetric branching with tortuous meander in one branch (Yangtze river); 6—semi-mountainous incised straight channels of Altai Region rivers

region (B). For the first field of points on the chart, you can divide the points into three parts corresponding to a particular channel pattern. The points related to the free-meandering sections are located at the bottom of the *QI*-diagram; in the upper right corner of the chart there are points corresponding to the maximum flow rate and belonging to different types of braided channels; the points in the middle part of the chart correspond to a relatively straight channel. Section lines are approximated by equations

$$I = 5.2Q_{\text{av.max}}^{-0.48} \tag{3.56a}$$

for a straight and braided channel and

$$I = 2.1Q_{\text{av.max}}^{-0.37} \tag{3.56b}$$

for a straight and meandering channel.

A special position on the *QI*-diagram is occupied by rivers characterized by significantly higher runoff and intensive directed accumulation of sediment. The dots corresponding to the Yangtze River correspond to peculiar asymmetrical single branches, in which one of the branches looks like a swamped bend ("duck's head"—according to the literal translation of the Chinese name of this type of channel). They are located on the right side of the diagram and are generally lower than the other points of the meandering channel. Apparently, the formation of steep swamped bending radii of one of the branches is the leading process that brings this type of channel closer to meandering, resembling the formation of cutoffs by the mechanism.

Characteristically, below and to the left of the line (3.56b), there is a point of the incised rectilinear channel of the Yangtze River (Ichang) among the points corresponding to relatively straight, unbroken channels. The latter occupies a position close to the upper boundary of the section (upper right), which generally corresponds to the results obtained by Ivanov (1989) for rectilinear channels: on *QI*-diagrams, concerning only to this type of a channel, the points answering to the cut in channel, have settled down in the right top field of the chart.

The points corresponding to the channel of the Huang He River were spread out in such a way that in the upper field they belonged to the wandering varieties of the straight and meandering channel, in the central field—to the stable bend, in the lower field—to the stable rectilinear channel. However, the stability of the latter two is related to the anchoring of streambeds by collapse dams. This arrangement of points is quite logical: the wandering channel types are formed in conditions of high flow saturation of sediment, and their transport is carried out by the flow, characterized by the highest capacity. Therefore, the points corresponding to them occupy the highest position on the *QI*-diagram. The lowest position is occupied by the points belonging to the straight channel, which is completely bounded by dams and has the smallest slopes (lower reaches of the Huang He). Intermediate position is occupied by gentle bends, the development of which is also limited by dams (Chalov et al. 2000).

The location of the two uppermost points in the diagram field, corresponding to the braided channels of other rivers, is quite justified: with large sediment load, branching cannot be formed because of the rapid and complete introduction of ducts and branches by sediments and the attachment of the emerging islands to the bank; moreover, the wandering of the channel, with its absolute instability, leads either to such a rapid movement of the aisles that they do not have time to consolidate themselves by vegetation and turn into islands, or to their erosion, because they are composed of easily eroded riverbed sediments. It can be assumed that in the event of a decrease in river load (e.g. due to continued erosion control, bank stabilization and reservoir construction), channel branches will be formed in the upper half of the downstream section of the Huang He River.

The results obtained are generally consistent with the conclusions drawn from other rivers: the free-meandering channel is formed at a minimum flow rate, and the corresponding points are located in the lower left corner of the *QI*-diagram (Alabyan 1992; Antropovsky 1970). If we consider not only the free conditions for the development of channel changes on flat rivers, the maximum flow rate corresponds to the different types of incised channel (mainly incised bends), as it is observed when crossing the Amur mountain range Small Hingan, mountain and semi-mountain rivers. Panin (1991) has received a dividing line between the points corresponding to the incised bends formed in rocky (in the upper part of the diagram) and plastic (in the lower part) formations. Above this line there are also points corresponding to the branches of the incised channels. Indeed, the latter are more often found in local expansion of incised channels in the areas of rock distribution, while in the plastic ones mostly incised bends are formed.

For medium and small, including semi-mountainous, rivers of the Altai (Smirnova 2002), the *QI*-diagram shows a kind of "wedge" of braided channels, showing that small rivers are either freely meandering, or are relatively straight or form a incised bed, but are not divided into branches (Fig. 3.36B); the interface lines between braided and meandering channels are approximated by equations:

$$I = 0.066Q^{-0.74}, \tag{3.57a}$$

between braided and semi-mountainous relatively straight incised channels with a pebble composition of bed material load

$$I = 0.0138Q^{-0.28}. \tag{3.57b}$$

The points of the beginning of the "branching wedge" correspond to the boundary conditions for the formation of branching: the average annual water discharge of 30 m^3/s, the slope of the valley bottom—5.4‰; at lower values of water content and higher values of slope, braided channels are not formed, which can be accepted as a criterion for the formation of the latter for the region.

For a number of rivers in Eastern Europe, *QI*-diagrams (*I*—valley bottom slope, *Q*—effective discharge of the middle interval, i.e. filling the channel to the logs of the bottomland- channel-filling taking into account wide floodplain and incised types of channels. In all cases, lower water discharge with equal slopes of the valley bottom corresponds to meandering and straight channels under conditions of free development of channel changes. In the connection equation, $I_{vb} = \alpha Q_f^{-\beta}$ the degree $\beta = 0.72$ for all types of channel, while the coefficient α varies depending on the type of channel: for wide floodplain straight channels $\alpha = 0.0247$, for incised straight channels $\alpha = 0.0607$, for free bends $\alpha = 0.0188$, for incised bends $\alpha = 0.0478$.

Ivanov (1989), Chalov et al. (1998) on the *QI* diagram obtained differences in the conditions of formation of incised, confined and wide floodplain straight, single-thread channels (Fig. 3.37). The interface between the point fields corresponding to the incised and other channel types and the communication lines for the confined and wide floodplain channels is arranged sequentially in such a way that in the area of the highest flow rate (upper right side of the graph) there are incised channels with the lowest flow rate (lower left side)—straight channels with the widest floodplain ($B_{fp}/b_{ch} > 20$). The intermediate position is occupied by an confined channel and a channel with a smaller floodplain width. At the same time, the degree index in the connection equation for wide floodplain channels is the same and equals −0.421 (lines V-III). At the confined channels it is equal −0.456 (line P), and for a line of section (1 in Fig. 3.37) with the incised channels −0.445. In the field of incised straight channels the points corresponding to the channels formed in rocky (Eastern Siberia) and plastic (moraine loams, sea clays of the European part of Russia) rocks with the whole field, the former mainly correspond to the highest flow power, while the latter gravitate more to the dividing line. The values of the coefficient in the coupling equation increase consistently from the wide floodplain to the incised channels, making up 0.7 for line V ($B_{fp}/b_{ch} > 20$), 2.8 for line IV ($B_{fp}/b_{ch} = 10$–20), 2.8 for line III ($B_{fp}/b_{ch} = 2$–10)—2.8, for line II ($B_{fp}/b_{ch} = 1$–2)—5.3 and for I ($B_{fp}/b_{ch} < 1$)—9.5.

In order to clarify the conditions for the formation of different types of channel beds and to take into account the value of sediment load as a leading factor in fluvial processes, a graph was drawn up for a number of large rivers (Chalov et al. 2000), on which the coordinate axes are the slope *I* and sediment concentration *s* of

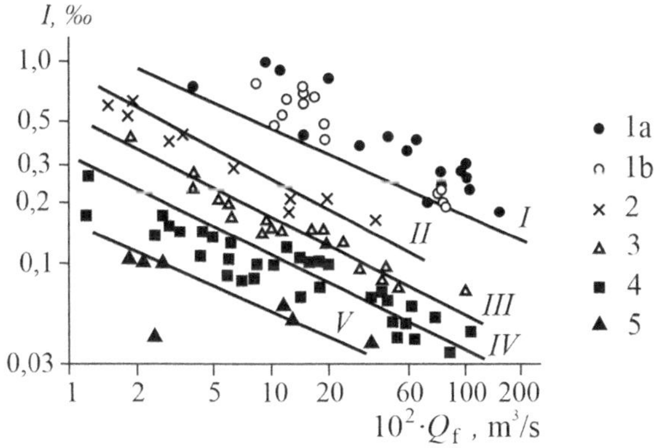

Fig. 3.37 QI-diagram for straight single-thread channels (Ivanov 1989): I—incised (a—in rocky, b—in plastic formations)—separation line; II—confined, III to V—wide-floodplain (III—at $B_{fl}/b_{ch} = 2$–10; IV—at $B_{fl}/b_{ch} = 10$–20; V—$B_{fl}/b_{ch} = >20$)—regression equations lines

the flow (Fig. 3.38A)—*sI*-diagram. In the upper left field of this chart there are points corresponding to the braided channels, to the right and below—all other types of channels. The boundary between them is 45°. The right part of the diagram is occupied by Yangtze points and, with a large gap, Huang He, and the lines of the interface between them can also be conventionally placed at an angle of 45°.

A close result is obtained when drawing a graph with coordinate axes—specific power of the *QI* flow and sediment concentration *s* of the flow—*QIs*-diagram (Fig. 3.38B). However, in one field to the right of the boundary of the division were the points corresponding to the rivers of Russia with an undeveloped channel and the Yangtze—with the meandering channel and single branches with a steep bend of one of the branches ("duck's head"). The extreme right position with a large gap

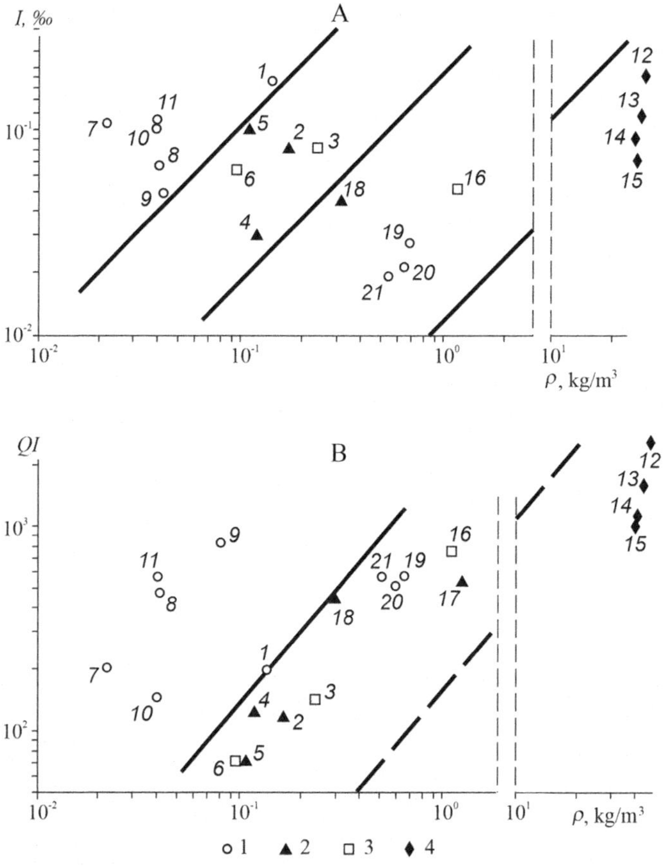

Fig. 3.38 Distribution of channel patterns in the field of *Is*-diagram (**A**) and in the field of the diagram specific power *QI*—turbidity *s* (**B**). 1—braided channel; 2—straight single-thread channel; 3—meandering channel; 4—Huang He river

is occupied by the points belonging to Huang He, and the boundary between them and the "central strip" can be drawn only conditionally.

The analysis of the diagrams (Fig. 3.38) shows that, with the same slope and flow rate, the increase in sediment load (sediment concentration of the flow) leads to a change in the braided channel, first single-thread, and then wandering (it can be assumed that the relatively stable bends and the rectilinear channel in the lower reaches of the Huang He are associated with an artificial restriction of their changes by dams). The smaller the sediment load, the lower the power and slope, the less the channel type changes. With low sediment load, no branching of the channel is formed at all. On the other hand, it is known that mountain rivers, characterized by low sediment concentration, have in the conditions of free development of channel changes (in the intermountain basins and at the exit to the foothills) difficult braided channel; the same is typical for pebble-boulder plain rivers flowing in mountain regions, but having a wide floodplain channel (Belyi et al. 1979; Berkovich et al. 1985). In the latter case, it was found that the branching of the channel increases in the narrowings of the valley bottom, where the slopes increase, although the channel remains wide floodplain, and decreases, up to the change of the channel type to a straight, single-thread, in the extensions within which the slopes are less.

A.M. Alabyan (1992), studying the relations between the positions of the points on the QI-diagram and the effective flow curves for the same points, proposed to divide the area of "epure" Q_f—communication graphs $Q = f(\sigma Q^m IP)$ into sections corresponding to one or another characteristic type of channel. This reception is carried out by means of horizontal lines in the field $Q \sim \sigma Q^m IP$, which secures the area of the "diagram" and corresponds to the values of water discharge, taken from the QI-diagram, at which: (1) water comes out to the floodplain [this line is carried out already during Q_f calculations and was first used by Chalov (1979)]—"floodplain line" (2) the value of QI is reached, which is sufficient for the formation of the channel of a pattern (this value is removed from the QI-diagram—"branching line", "straight channel line", corresponding to the critical slope of meandering [according to the terminology of L. Leopold and M. Wolman (1957)], or "meandering line". On the Fig. 3.39 several graphs are given to determine effective water discharge Q_f. At the middle Lena (Tabaga), the branching line is lower than both maximums of the function $\sigma Q^m IP$, which indicates that the transport capacity is not only in the high floodplain, when the floodplain is flooded, but also at the levels in the floodplain edges is quite sufficient for the formation of channel branches (Fig. 3.39A). The same result was obtained for the lower Lena (Kyusyur village). Indeed, the entire middle and lower floodplain channel of the middle and lower Lena is braided.

In the Northern Dvina (Abramkovo village), the "branching line" is located below the upper Q_f, which occurs when the floodplain is flooded, but above the lower Q_f, which has greater exceedance probability (Fig. 3.39B). From here, the braided channel of the Northern Dvina, as well as the anabranching channel is connected only with Q_f of the upper interval corresponding to the flood.

In the Ob (Kolpashevo), the "floodplain line" (Fig. 3.39B) is directly above the most pronounced maximum of the profile, which in this case is average; in such

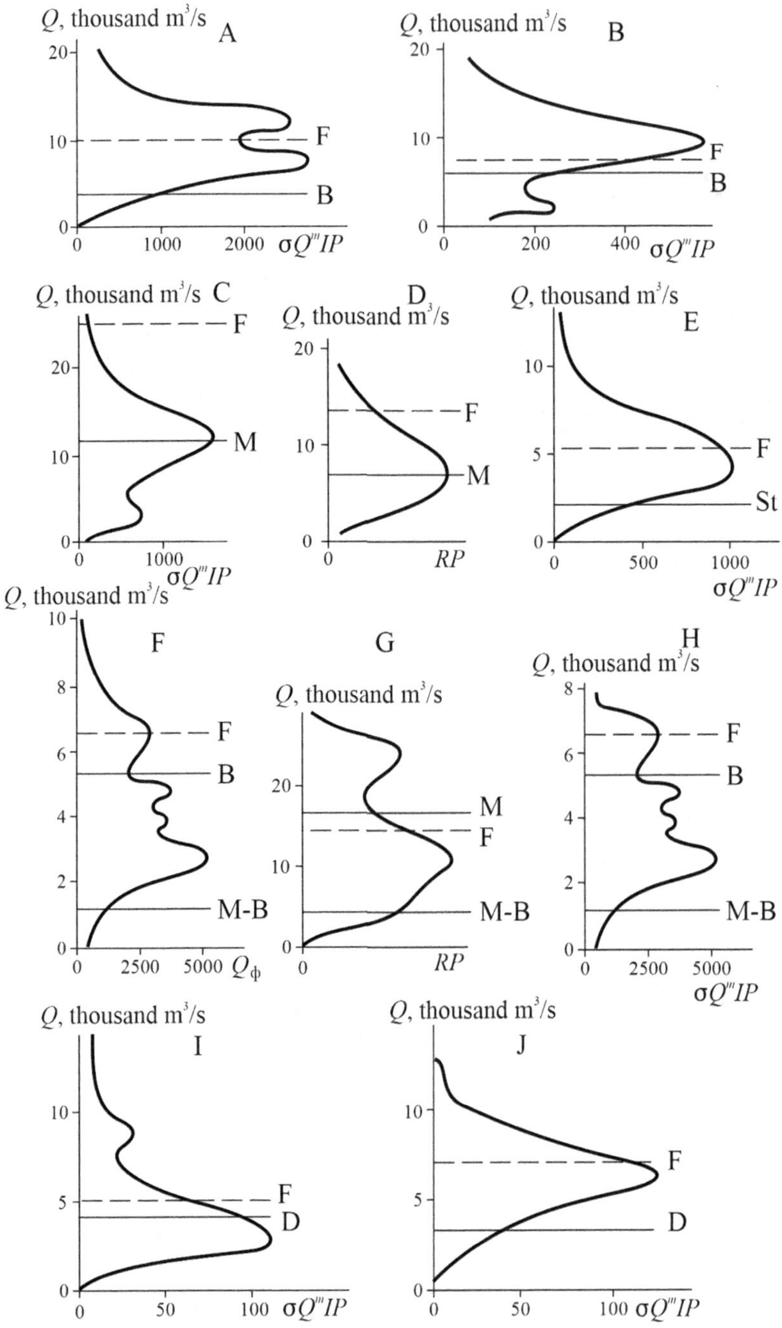
A
Q, thousand m³/s
20
10
0
F
B
0
1000
2000
σQ'''IP
B
Q, thousand m³/s
20
10
0
F
B
0
200
400
σQ'''IP
C
Q, thousand m³/s
20
10
0
F
M
0
1000
σQ'''IP
D
Q, thousand m³/s
20
10
0
F
M
0
RP
E
Q, thousand m³/s
10
5
0
F
St
0
500
1000
σQ'''IP
F
Q, thousand m³/s
10
8
6
4
2
0
F
B
M-B
0
2500
5000
Qф
G
Q, thousand m³/s
20
10
0
M
F
M-B
0
RP
H
Q, thousand m³/s
8
6
4
2
0
F
B
M-B
0
2500
5000
σQ'''IP
I
Q, thousand m³/s
10
5
0
F
D
0
50
100
σQ'''IP
J
Q, thousand m³/s
10
5
0
F
D
0
50
100
σQ'''IP

◂**Fig. 3.39** Epures of effective discharges and location on them of the lines of water discharges, corresponding to the flooding o the floodplain (F—dotted line), formation of different channel patterns (B—braided; M—meandering; St—straight, single-thread; W—wandering; Channel type —others; Mb—mid-bars). **A** Lena (Tabaga); **B** Northern Dvina (Abramkovo); **C** Ob (Kolpashevo); **D** Chulym (Zyryanskoye); **E** Ob (Novosibirsk); **F**—Ob (Fominskoye); **G** Ob (Salekhard); H—Huang He (Huayuankou); **I** Huang He (Gaotsun); **J** Yangtze (Hankou)

conditions, the anabranching channel does not develop. The channel type line (the Ob here meanders) is at the top of the epicure, and the whole epicure is located in a field corresponding to the development of the meandering channel. Thus, the average interval Q_f is responsible for the formation of bending in this area of Ob. A similar result (Fig. 4.39D) was obtained by A.M. Alabyan (1992) for the meandering Chulym channel (Ob tributary). Both rivers are characterized by similar conditions of Q_f passage (Chalov 1979). In Novosibirsk (Fig. 3.39E), on the contrary, the "straight channel line" lies below the peak of the chart and, therefore, the flow rate at effective discharges is sufficient to form a straight channel, typical for the Ob in this section of the river. On the upper Ob (below the confluence of Biya and Katun—the village of Fominskoye) "line of channel type" on the "epure" Q_f (Fig. 3.39F) is located below the peak, located in the floodplain's edges. Accordingly, the river is characterized by a braided channel. Its morphological complexity is promoted by the fact that the "line of mid-bars" (Alabyan 1992) lies in the lower part of the chart (below the peak corresponding to the lower Q_f, and their development is the main condition for the formation of numerous relatively small islands. At the same time, the "floodplain line" here is located above all peaks on the "epure" Q_f, which results in a consolidated, undivided floodplain ducts. At the lower Ob (Salekhard), the "floodplain line" and "branching line" lie below the upper Q_f, so the floodplain flow is here forming both the anabranching channel and channel branches.

Similar ratios are revealed for rivers in China. In the Huang He (Fig. 3.39G), the "wandering line" corresponding to the wandering straight channel (the Huayuankou station) runs in the lower part of the epure; thus, at all water flow rates, the power of the flow is sufficient for its intensive reformation and development of the corresponding channel type. The same ratio is characteristic for Lena, which confirms that the points of both rivers on the QI-diagram hit the field of the braided channel. At the Gaotsun hydrological station (wandering meandering channel), the channel type line on the "epure" is located between the peaks—upper and middle (Fig. 3.39H). Apparently, this determines the formation of a different type of channel here, the higher the current, but the fact that the corresponding point in the field of QI-diagram together with the braided channels is connected with the fact that this line is located below the upper Q_f (as in the Northern Dvina). On the Yangtze (Fig. 3.39I), the channel type line at all stations is located below the peak on the curve corresponding to the effective discharge, which distinguishes it from other rivers. This seems to be due to the very high runoff rates of the river and the high sediment load.

Joint analysis of QI-diagrams and curves for determination of effective discharges Q_f makes it possible not only to identify the conditions for the formation of a particular type of channel, but also to determine the flow rate, which is "responsible" for the development of the main forms of channel and forms of channel relief, as well as to predict its possible transformation in case of changes in the factors of fluvial processes.

Literature

Alabyan, A.M.: Types of the plain river beds and factors of their formation. Geomorphology (4), 37–42 (1992)

Alabyan, A.M., Alekseevsky, N.I., Berkovitch, K.M.: Sediment transport in the downstream area of Novosibirsk HES (in Russian). In: Proc. of WestSibNIGMI. Proc. 98, pp. 79–82 (1992)

Alekseevskiy, N.I., Vlasov, B.N., Chalov, R.S.: Some aspects of change of effective discharge of the Terek River. Water resources of the Terek Basin and their use, pp. 117–126. Yuzhgiprovodkhoz Publishing, Rostov-Na-Donu (1983)

Altunin, S.T.: Regulation of Riverbeds, p. 352. Selkhozizdat, Moscow (1962)

Altunin, V.S., Annaev, S.A., Ashirov, S.A.: Intensive Channel Changes of the Amudarya River, p. 103. Ylym, Ashkhabad (1975)

Andrem, E.D.: Effective and banxfull discharge of streams in the Vampa River basin, Colorado and Wyoming. J. Hydrol. **46**(3–4), 311–330 (1980)

Antropovsky, V.I.: Hydromorphological dependences and their further development (in Russian). In: Proc. of SHI. Op. 169, pp. 34–86 (1969)

Antropovsky, V.I.: Communication of types of fluvial process with determining factors. In: Proceedings of SHI. Op. 183, pp. 70–80 (1970)

Antropovsky, V.I.: Criterial dependences of types of fluvial process. In: Proceedings of SHI. Op. 190. pp. 5–18 (1972)

Artamonov, K.F.: Rucal deformations of the lower rivers. Hydrometeorology to help the national economy, p. 81. Publishing house of the Academy of Sciences of the Uzbek SSR, Tashkent (1954)

Artamonov, K.F., Kroshkin, A.N., Talmaza, V.F.: Basic principles of the theory of the fluvial process of stable mountain river channels. Dynamics and thermometry of rivers and reservoirs, pp. 139–145. Nauka, Moscow (1984)

Babich, D.B., Mikhailov, V.N., Morozov, V.N.: Estimation of river effective discharges in deltas of large rivers. Vest. Moscow. University. Ser. 5. Geography (in Russian). Vest. (1), 16–20 (1985)

Babinski, Z.: Procesy erozyjno-akumulacyjne poniiej stopnia wodnego Wiociawek, ich kosekwencje i wpiyw na morfodynamike planowanego zbioznika Nieszawa. PAN, Torun 37 s (1997)

Barovskiy, N.A.: Dynamics of the meandering channels of the Middle Ob inflows (in Russian). Erosion fluvial processes and hydroecology problems, p. 22 11. Geographer faculty MSU, Moscow (2004)

Baryshnikov N.B.: Morphology, Hydrology and Hydraulics of Floodplains, p. 280. Hydrometeoisdat, L. (1984)

Baryshnikov, N.B., Popov I.V.: Dynamics of Channel Flows and Fluvial Processes, p. 456. Hydrometeoizdat, L. (1988)

Begin Z.B.: The relationship between flow-shear stress and stream pattern. J. Hydrol. **52**, 307–319 (1981)
Belotsky, K.N., Ruleva, S.N., Chalov, R.S.: Principles of zoning of large regions by conditions of passage of effective discharges of water (on an example of the Altai rivers). Geography Nat. Resour. (3), 165–169 (1995)
Belyi, B.V., Berkovich, K.M., Borsuk, O.A., Zaitsev, A.A., Lodina, R.V., Chalov, R.S., Chernov, A.V.: Morphology, dynamics and regulation of the Kirenga riverbed in connection with transport development of BAM zone (in Russian). Soil erosion and fluvial processes, vol. 7, pp. 119–135. Moscow State University Publishing House, Moscow (1979)
Belyi, B.V., Ivanov, V.V., Nikitina, L.N., Chalov, R.S., Chernov, A.V.: Morphology and Dynamics of the Lower Vychegda Riverbed during Active Dredging and after their Termination. Soil Erosion and fluvial Processes, Issue 13, pp. 146–183. Moscow State University Publishing House, Moscow (2001)
Berkovich, K.M., Zlotina, L.V.: Calculation of the river channel stability under the anthropogenic load conditions (in Russian). Geography Nat. Resour. (4), 117—123 (2003)
Berkovich, K.M., Lodina, R.V., Chalov, R.S., Shutov, A.M.: Formation conditions, stability and mode of the middle Ob riverbed changes (in Russian). Soil erosion and fluvial processes. Ep. 2, pp. 186–210. MSU Publishing House, Moscow (1972)
Berkovich, K.M., Borsouk, O.A., Garrison, L.M., Kirik, O.M., Lodina, R.V., Rouleva, S.N., Chalov, R.S., Chernov, A.V.: Channel mode and regulation of the middle and lower Lena riverbed (in Russian). Soil erosion and fluvial processes, Issue 8, pp. 25–156. Moscow State University Publishing House, Moscow (1981)
Berkovich, K.M., Zaitsev, A.A., Lodina R.V., Chalov, R.S.: Fluvial processes on the large East Siberian rivers with a pebble-boulder alluvium and features of their regulation (in Russian). Vest. Moscow, Un. Ser 5. Geography (in Russian). Vest. (3), 35–41 (1985)
Berkovich, K.M., Rouleva, S.N., Chalov, R.C.: Upper Ob river regime (in Russian). Geography Nat. Resour. (4), 54–61 (1989)
Berkovich, K.M., Garrison, L.M., Roulev, S.N., Chalov, R.S.: Channel morphology and channel changes of the upper Ob (in Russian). Land and water resources: anti-erosion protection and regulation of channels, pp. 95–120. Moscow State University Publishing House, Moscow (1990)
Borsouk, O.A., Lodina, R.V.: Channel alluvium on the upper and middle Lena (in Russian). Soil erosion and fluvial processes, Ep. 4, pp. 149–154. MSU Publishing House, Moscow (1974)
Borsouk, O.A., Lodina, R.V.: Crushing of channel alluvium depending on the composition of rocks. Regularities of erosion and fluvial processes in different natural conditions, C. 355. MSU Publishing House, Moscow (1987)
Borsouk, O.A., Dobrovolskaya, N.G., Lodina, R.V., Chalov, R.S.: Morphology of channels and modern channel alluvium on the Western Tien Shan mountain rivers (in Russian). Geomorphology (4), 60–67 (1981)
Bray, D.I.: Regime equations for gravel-bed rivers. Gravel-bed rivers. Wiley & Sons, Chichtster (1982)
Budanov, V.I.: Methodology of expeditionary research of the coastal zone of the sea, p. 224. Nauka, Moscow (1964)
Carson, M.A.: The meandering-braided river threshold: a reappraisal. J. Hydrol. **73**(3–4), 315–334 (1984)
Chalov, R.S.: Morphology and dynamics of the mountain river beds (in Russian). In: Proceedings of Moscow. GS USSR, ed. 3. Geomorphology 8–10 (1969)
Chalov, R.S.: Emissions of the r. Vychegda. Soil erosion and fluvial processes, ed. 4, pp. 128–142. MSU Publishing House, Moscow. (1974)
Chalov, R.S.: Geographical Investigations of Fluvial Processes, p. 232. MSU Publishing House, Moscow (1979)

Chalov, R.S.: Indicators of channel stability, their use for estimation of intensity of channel changes and ways of their improvement. Dynamics of channel flows, pp. 46–53. LPI Publishing House, L. (1983a)
Chalov, R.S.: Factors of fluvial processes and hierarchy of channel forms. Geomorphology (2), 16–26 (1983b)
Chalov, R.S.: Channel Investigations, p. 106. MSU Publishing House, Moscow (1995)
Chalov, R.S.: General and Geographical Riverbed Science, p. 112. MSU Publishing House, Moscow (1997a)
Chalov, R.S.: Fluvial processes as hydrological phenomena and their geomorphological reflection. Hydrology and geomorphology of river systems, pp. 23–32. Irkutsk (1997b)
Chalov, R.S., Belyi, B.V.: Area of Siberia territory by character of the riverbed forming activity (in Russian). Meteorol. Hydrol. (12), 76–88 (1975)
Chalov, R.S., Belyi, B.V.: Regional features of the effective water discharge on the Central Asian rivers (in Russian). New. of the SA USSR. Ser. Geogr. (5), 107–116 (1984)
Chalov, R.S., Dobrovolskaya, N.G.: Biochemical weathering and channel alluvium of the Western Tien Shan mountain rivers (in Russian). Lesomelioration in Uzbekistan, pp. 21–25. UzNIILH, Tashkent (1993)
Chalov, R.S., Alabyan, A.M., Ivanov, V., Lodina, R.V., Panin, A.V.: Morphodynamics of the channels of plain rivers, p. 288. GEOS, M (1998)
Chalov, R.S., Liu, S., Alekseevskiy, N.I.: Sediment flow and fluvial processes on the big rivers of Russia and China, p. 216. MSU Publishing House, Moscow (2000)
Chalov, R.S., Zavadsky, A.S., Panin, A.V.: River Radiations, p. 371. MSU Publishing House, Moscow (2004)
Channel regime of Middle Ob River. M. 146 p. Dep. in VINITI № 1250–69 (1969)
Characteristics of channels and fluvial processes in the middle and lower reaches of the Yangtze River, p. 272. Science, Beijing (1985) (in Chinese)
Chernyshov, F.M.: Increase of Efficiency of Travelling Works on Multi-Branch Sections of Navigable Rivers, p. 324. NIIVT Publishing House, Novosibirsk (1973)
Darbutas, A.A.: Fluvial Processes of the R. Neman (Nemunas) and Influence of the Anthropogenic Factor on them: Abstract from the … phD Geogr. Sciences, p. 27. MSU Publishing House, Moscow (1992)
Darbutas, A.A., Chalov, R.S.: Morphodynamic types of the upper Neman channel, conditions of their formation and hydromorphological characteristics. Geomorphology (3), 72–79 (1991)
Darbutas, A.A., Lodina, R.V., Chalov, R.S.: Modern channel alluvium of Neman. Vest. Moscow, Un. Ser. 5. Geography. (5), 90–100 (1993)
Degtyarev, V.V.: Change of hydrological regime of Irtysh. River Transp. (12), 39–41 (1968)
Demin, A.G., Lodina, R.V., Ruleva, S.N., Chalov, R.S.: The role of geomorphological factor in changes of channel types and composition of bed material alluvium on large mountain rivers (by the example of Katun and Chuya). Geomorphology (4), 73–81 (1991)
Dobrovolskaya, N.G., Lodina, R.V., Chalov, R.S.: About a role of mechanical and biochemical weathering in formation of structure of channel alluvium. Geomorphology (1), 59–64 (1991)
Dvaretskas, V.V.: Structure and development of the river valleys of the edge zone of the last Scandinavian glaciation (by the example of the southern Baltic states): Author's thesis … Dr. Geogr. of Sciences. p. 52. MSU Publishing House, Moscow (1989)
Engineering of ship passages on free rivers. TsNIIEVT works, ed 36, p. 262. Transport, Moscow (1964)
Experimental Geomorphology, ed 2, p. 173. MSU Publishing House, Moscow (1969)
Ferguson, R.T.: The Threshold between meandering and braiding. In: Proceedings of International Conference "Channels and channel control structures", (6), pp. 15–29. Berlin e. a. (1984)
Fluvial processes on the rivers of the USSR. Scale 1:4000000. GUGGK USSR, 4 l (1990)
Fluvial processes and waterways in the rivers of the Ob basin, p. 300. RIPEL-plus, Novosibirsk (2001)

Garrison, L.M.: Influence of the hydrotechnical measures on the bottom sediment particle size (on the example of Ob) (in Russian). Vest. Moscow Un. Ser. 5. Geography (4), 47–50 (1979)
Glazovskaya, M.A., Dobrovolskaya, N.G.: Geochemical functions of microorganisms, p. 53. MSU Publishing House Moscow (1984)
Glushkov, V.G.: Morphology of the river channel. In: Proceedings of the 1st All-Russian Hydrological Congress, pp. 286–290. L. (1925)
Grishanin, K.V.: Theory of the River Fluvial Process, p. 216. Transport, Moscow (1972)
Grishanin, K.V.: Stability of River and Canal Beds, p. 144. Gidrometeoizdat, L. (1974)
Gudyalis, V.K.: Relief and Quaternary Deposits of the Baltic, p. 264. Mintis, Vilnius (1973)
Guidelines for designing a radical improvement of navigational conditions in the difficult areas of free rivers, p. 310. Transport, L. (1974)
Henderson, F.M.: Stability of alluvial channels. J. Hydr. Div. ASCE. **87**(6) (1961)
Ivanov, V.V.: Formation conditions, hydrological-morphometric dependences and changes in relation to the straight, unbranched channels: Author's … of Dis. PhD Geog. Sciences, p. 23. MSU Publishing House, Moscow (1989)
Kamenskov, Y.I.: To definition of the concept and value of effective water discharge. Meteorol. Hydrol. (7), 119–120 (1986)
Karasev, I.F.: Fluvial Processes at Transfer of the Flow, p. 288. Hydrometeoizdat, L. (1975)
Karasev, I.F.: Effective water discharges. Meteorol. Hydrol. (8), 94–98 (1986)
Karasev, I.F., Kovalenko, V.V.: Stochastic Methods of River Hydraulics and Hydrometry, p. 208. Hydrometeoizdat, SPb (1992)
Khakimov, S.K.: Fluvial processes on the mountain rivers of the Western Tien Shan: Author's thesis. PhD Geogr. Sciences, p. 29. MSU Publishing House, Moscow (1992)
Kondratyev, N.E., Lyapin, A.N., Popov, I.V., Pinkovsky, S.I., Fedorov, N.N., Yakunin, I.A.: Fluvial Process, p. 372. Hydrometeoizdat, L. (1959)
Korotayev, V.N., Sidorchuk, A.Y., Chalov, R.S.: Lithodynamics of the northern Siberian river deltas (in Russian). Water Resour. (1), 138–145 (1980)
Kondratyev, N.E., Popov, I.V., Snischenko, B.F.: Basics of Hydromorphological Theory of the Fluvial Process, p. 272. Gidrometeoizdat, L. (1982)
Kuzin, P.S., Babkin, V.P.: Geographical Laws of Hydrologic Regime of Rivers, p. 200. Gidrometeoizdat, L. (1979)
Lapshenkov, V.S.: Prediction of the Channel Changes in the River Junctions, p. 240. Gidrometeoizdat, L. (1979)
Lapshenkov, V.S.: River Hydraulic Engineering, p. 408. NGMA Publishing House, Novocherkassk (1999)
Lebedeva, I.V.: Change of particle size distribution of modern channel alluvium downstream of plain river (by the example of Vychegda river). Lithol. Miner. (5), 37–44 (1966)
Lebedeva, N.V., Chalov, R.S.: Experience of application of the map-metric method in studying of the river bed sediment granulometry (in Russian). MPEI Vest. Moscow Un. Ser 5. Geography (3), 92–94 (1966)
Leliavskiy, N.S.: On the river currents and the formation of a river bed. In: Procedings of the 2nd Congress of Hydraulic Engineers in 1893. St. Petersburg, 1893. Problems of hydraulic engineering of free rivers, pp. 18–136. Rechizdat, M. (1948)
Leopold, L.B., Wolman, M.G.: River channel patterns—braided, meandering and straight. US Geol. Sure Prof. Pap. 282-B, pp. 1–85 (1957)
Litvin, L.F.: Development of erosion processes in mountainous conditions (on the example of the southern slope of the Western Caucasus): Author's thesis … PhD Geogr. Sciences, p. 16. MSU Publishing House, Moscow (1971)
Lodina, R.V.: Sorting of alluvium in the rivers of Western Georgia (in Russian). Soil erosion and fluvial processes, ed. 4, pp. 143–148. MSU Publishing House, Moscow (1974)

Lodina, R.V., Rashutin, D.V., Sidorchuk, A.Y., Chalov, R.S.: Changes of channel morphology and bed material load from source to mouth (by the example of Terek River). Geomorphology (1), 86–94 (1987)

Lodina, R.V., Hakimov, S.K., Chalov, R.S.: Modern channel alluvium of the large mountain rivers of the Western Tien Shan and their tributaries (in Russian). Geomorphology (4), 70–76 (1994)

Lokhtin, V.I.M.: About the mechanism of the river bed. St. Petersburg, 1897. Problems of Hydraulic Engineering of Free Rivers, pp. 23–59. Rechizdat, M. (1948)

Lomakin, V.V.: On Residual River and General Residual Surface Deposits. New of the SA USSR. Ser. Geol. (2), 79–87 (1943)

Makkaveev, N.I.: Regressive transformations of river islands. Meteorol. Hydrol. (4), 44–50 (1948)

Makkaveev, N.I.: Channel Regime of Rivers and Tracing of Slits, p. 202. Rechizdat, Moscow (1949)

Makkaveev, N.I.: River Channel and Erosion in its Basin, p. 347. Published in the SA USSR, Moscow (1955)

Makkaveev, N.I.: General regularities of the erosion-fluvial processes (in Russian). In: Procedings of IV All-Union Hydrologic Congress. V. 10. Fluvial processes, pp. 8–12. Hydrometeoizdat, L. (1976)

Makkaveev, N.I., Chalov, R.S.: Methods of improvement of navigable conditions at Ob River. River Trans. (9), 45–47 (1963a)

Makkaveev, N.I., Chalov, R.S.: On morphological signs of modern accumulation in a river valley. New. of the SA USSR. Ser. Geogr. (3), 84–89 (1963b)

Makkaveev, N.I., Chalov, R.S.: Fluvial Processes, p. 264. MSU Publishing House, Moscow (1986)

Makkaveev, N.I., Khmeleva, N.V., Zaitov, I.R, Lebedeva, I.V.: Experimental Geomorphology, p. 196. MSU Publishing House, Moscow (1961)

Malbert, I.E.: To a question about the Siberian river channel processes (in Russian). In: Procedings of Transport Energetic Institute, West.-Sib. of. of SA USSR, ed. 3. Hydrol. pp. 73–85 (1952)

Matveev, B.V.: Morphology and geological-geomorphological factors of incised and free bends development. Autoref. diss. PhD of geography, p. 20. MSU Publishing House, Moscow (1985)

Neff, E.L.: Discharge frequency compared to long-term sediment yields. In: Procedings of Symposiun on River Morchology JANS, pp. 236–242. Pub. (1977)

Nekos, S.V., Chalov, R.S.: Sediment runoff and fluvial processes on rivers of the Don basin. Geomorphology (2), 60–71 (1997)

Nolan, K.M., Lisle, T.E., Kelsey, H.M.: Bankfull discharge and sediment transport in northwestern California. Erosion and Sedimentation in the Pacific Rim. Proc. Corvallis Simposirem. JANS. Pub. **165**, 439–448 (1987)

Obodovskiy, A.G.: Hydrologic-Ecological Galvanization of Fluvial Processes (at the Application Level of Ukraine), p. 276. Nica-Center, Kyiv (2001)

Obodovskiy, A.G., Zaitz, E.S., Potapenko, A.N.: Peculiarities of effective water discharge on rivers of forest-steppe zone of Ukraine. Physical geography and geomorphology, ed. 29, pp. 123–127. Higher School, Kyiv (1983)

Obodovskiy, A.G., Zaitz, E.S., Chalov, R.S.: Geographical substantiation of the determination technique of the water flow (by the example of the Ukrainian rivers). Vest. Moscow Un. Ser. 5. Geography (5), 67–72 (1987)

Osterkamp, W.P.: Gradient, discharge and particle-size relations of alluvial channels in Kansas, with observations on braiding. Ammer. J. Sci. **278**(9), 1253–1268 (1978)

Lower Yana: estuary and fluvial processes, p. 212. GEOS, M. (1998)

Panin, A.V.: Morphology and Dynamics of Incised pebble-boulder rivers. Autoref. dis. … PhD. Geogr. Sciences, p. 24. MSU Publishing House, Moscow (1991)

Pinkovsky, S.I.: Types of river channels of Middle and South Siberia. Proceedings of SHI, ed. 94, pp. 87–114 (1961)

Pinkovsky, S.I.: Types of riverbeds in the Soviet Far East (southern half) Proceedings of SGI, ed. 144, pp. 77–117 (1967)
Popov, I.V.: Application of morphological analysis to estimation of general channel changes of the p. Ob. In: Proceedings of SHI, ed. 94, pp. 22–86 (1962)
Popov, I.V.: River Bed Changes and Hydraulic Engineering, p. 328. Hydrometeoizdat, L. (1965)
Qian, N., Zhou, W.: Fluvial Processes in the Lower Reaches of the Huang He River, p. 224. Nauka, Beijing (1965) (in Chinese)
Qian, N., Zhang, Z., Zhou, Z.: Fluvial Process, p. 584. Science, Beijing (1987) (in Chinese)
Rabkova, E.K.: Methods of calculation of river beds and channels and their change on the basis of use of complex natural characteristics: Autoref. Dis. … Dr. Techn. Sc, p. 35. MGMI, Moscow (1979)
Razumikhin, N.V.: Experimental studies of evolution of rock debris pellets, p. 67. LSU Publishing House, L. (1965)
Reznikov, P.N., Chalov, R.S.: Sediment flow and conditions of channel formation on the rivers of the Basin of the Northern Dvina (in Russian). Geomorphology (2), 73–85 (2005)
Rimkus, Z.V.: Main channel-forming factors of Nemunas basin on the territory of Lithuanian SSR. Second All-Union Conference "Dynamics and terminology of rivers, reservoirs and estuaries". Theses of reports. V. II. M., pp. 28–29 (1984)
Romashin, V.V.: Types of fluvial process in connection with determining factors. In: Proceedings of SHI, ed. 155, pp. 56–63 (1968)
Romashin, V.V.: Estimation of a channel mode on morphological characteristics. In: Procedings of SHI (190), 48–56 (1972)
Rzhanitsyn, I.A.: Morphological and hydrological regularities of the river network structure, p. 240. Gidrometeoizdat, L. (1960)
Rzhanitsyn, N.A.: Channel-Forming Processes of the Rivers, p. 264. Gidrometeoizdat, L. (1985)
Sakharova, E.I., Lebedeva, N.V.: Change of average alluvium size along the length of a mountain river (on the example of the Mzymta River). Geomorphological and hydrological studies, pp. 24–30. MSU Publishing House, Moscow (1968)
Sakharyuk, N.P.: Determination of effective discharges for the rivers of the southern part of Primorye. In: Proceedings of DVNIGMI, ed. 84, pp. 94–96 (1980)
Schaffemak, F.: Grundriss dez FluBmozphologie und des Flussbaues. Spzingez, Wien (1950)
Sediment Yield, its Study of Geographical Distribution, p. 240. Gidrometeoizdat, L. (1977)
Sedimentation research in China, p. 250. China water and power press, Beijing (1993)
Shataeva, S.G.: Determination of dredging volumes by means of channel stability index. In: Proceedings of INIIEVT, ed. 68, pp. 84–121 (1969)
Shenberg, N.V.: Analysis of factors and methodology of calculation of effective water discharges. Water Resour. (4), 49–54 (1989)
Shenberg, N.V.: Use of the long-term characteristics of the flow regime for calculation of the effective water discharge (in Russian). Land and water resources: anti-erosion protection and regulation of the channels, pp. 69–76. MSU Publishing House, Moscow (1990)
Smirnova, V.G.: Hydrologic-morphological analysis of the branched riverbeds of the Altai region: Author's thesis … PhD Geogr. Sciences, p. 20. IG SB RAS, Irkutsk (2002)
Soviet Encyclopedic Dictionary, p. 1600. The Soviet Encyclopedia, Moscow (1982)
Studies of environmental changes in the Huang He River basin, water content and sediment transport. Jinan: Science and Technology (1994) (in Chinese)
Talmaza, I.F., Kroshkin, A.N.: Hydromorphometric Characteristics of Mountain Rivers, p. 204. Kyrgyzstan, Frunze (1968)
Velikanov, M.A.: Russian channel hydraulic engineering and its role in development of teaching about the fluvial processes (in Russian). Problems of hydraulic engineering of the free rivers, pp. 5–14. Rechizdat, Moscow (1948)
Velikanov, M.A.: Fluvial Process, p. 395. Gosfizmatizdat, Moscow (1958)

Vlasov, B.N.: Regionalization of UTS territory under conditions of effective discharges passing. Research of fluvial processes for practice of national economy, pp. 116–117. MSU Publishing House, Moscow (1983)

Vlasov, B.N., Chalov, R.S.: Effective discharges of the Volga R. New. of the USSR SA. Ser. Geograph (6), 75–81 (1981)

Vlasov, B.N., Chalov, R.S.: European territory zoning according to the conditions of effective water discharges on the rivers, pp. 32–42. Vest. MSU. Ser. 5. Geogr. №6 (1991)

Vozniachuk, L.N., Valchik, M.A.: Morphology, Structure and History of the Neman Valley Development in Neopleistocene and Holocene, p. 212. Science and technology, Minsk (1978)

Waterways of the Lena Basin, p. 600. MIKIS, M. (1995)

Wolman, M.G., Miller, J.P.: Magnitude and frequency of forces in geomorphic processes. J. Geol. **68**, 54–74 (1960)

Work of water flows, p. 194. MSU Publishing House, Moscow (1987)

Zavadsky, A.S.: Hydrologic and morphological analysis of free meandering of the plain river beds. Autoref. Diss. … PhD Geogr. Sciences, p. 27. MSU Publishing House, Moscow (2001)

Zavadsky, A.S., Chalov, R.S.: Regional analysis of the free meandering (in Russian). Vest. Moscow, Un. Ser 5. Geography (3), 32–36 (1997)

Zavadsky, A.S., Chalov, R.S.: Conditions of formation and morphology of free bends on the rivers of Northern Eurasia (in Russian). Geomorphology (4), 88–95 (2000)

Zernov, S.Ya.: Inland Waterways of the North-Eastern Region: Design, Construction, Operation, p. 124. Nauka, Novosibirsk (2003)

Zhayvoron, I.A., Chalov, R.S.: Stability of river channels and conditions of transport of bed material load (by the example of the upper Ob). Dynamics of channel flows and protection of natural waters, pp. 28–32. LGMI Publishing House, L. (1990)

Zheleznyakov, G.V.: Hydraulic Justification of River Hydrometry Methods, p. 163. SA USSR, Moscow; L. (1950)

Znamenskaya, N.S.: Ripple Movement of Sediment, p. 188 . Hydrometeoizdat, L. (1968)

Chapter 4
Classification of Fluvial Processes and Forms of Their Representation

4.1 Development of Ideas About the Classification of Fluvial Processes: Principles and Approaches to River Channel Classification

The issues of riverbed classification have been repeatedly considered in Russian and foreign literature. Accumulation of information on channel shapes, peculiarities of their changes, expansion of research geography, application of new methods of studying fluvial processes led to improvement of previously proposed classifications, their addition, development of new classification schemes. The first riverbed classifications appeared in the late 30-s of the twentieth century. Characteristically, even the founders of the doctrine of fluvial processes, V. M. Lokhtin and N. S. Leliavskiy, did not clearly distinguish the beds, which differ in their morphology, focusing on the formation of riffles. However, Leliavskiy (1893) noted the different structure of the flow at the bends and near the islands (in branches), which determine the differences in the mode of the riffles formed here and the methods of their deepening to improve navigation conditions.

More importance was given to channel morphology in the geomorphological literature devoted to the work of rivers as a relief-forming process. Thus, Krasnov (1897), describing the bend ("snake-like wriggles" or "curves") as the most characteristic form of channels and the mechanism of their development up to straightening, pointed out that in the middle and lower reaches of the river are often branched out to the branches. At the same time, "if a river, wriggling, at the same time, quickly deepens its channel, receiving more and more high banks, the gyrfalons lose the ability to movc, they are made permanent, fixed eingsecnkte Mäander (encaissé)" (p. 185). In fact, A. N. Krasnov for the first time singled out two types of bends—the gyrus formed "if the river flows through loose rock" (p. 179) and fixed meanders, which corresponds to the free and incised bends in modern terminology.

R. S. Chalov, *Fluvial Processes: Theory and Applications*,
https://doi.org/10.1007/978-3-030-66183-0_4

Later Shchukin (1933) singled out the primary bends (meanders)—the bends of the river associated with the relief of the earth's surface, and the secondary ones formed as a result of the work of the water flow itself. Later he specified that "the bend of the river on the floodplain plain is usually called wandering meanders" (Shchukin 1960, p. 248), and "directly fringed by steep and high sides of the valley, are called… incised meanders" (ibid., p. 236). At the same time, he singled out the rivers divided into branches by the islands, although the development of the river valley was associated solely with meandering of the channels. According to Edelstein (1947), "if the thalweg of the valley describes large bend shapes of arcs, often with very close ends, such areas are called meanders…. In the case when the whole valley is deeply cut into the earth's crust and meander, its meanders are called incised" (p. 90).

Thus, during the reporting period, the development of river channels was considered mainly in terms of meandering conditions, with two varieties of bendy, free and incised. For large rivers, multibranch channels were described, but only Krasnov (1897) singled out "four cases" of large branching and branching connected with small islets (after I.V. Mushketov he called them mid-bars), dividing the river into branches («volozhka» on the Volga river), connecting each other with "passages, ducts or lateral cannels" (pp. 197–198).

Existing river channel classifications are based on different attributes: (1) channel stability regardless of the type of channel morphology and type of its reformation; (2) development and shape of channel ripple relief; (3) riffle morphology; (4) flow kinematics, channel slope and form of sediment transport; (5) channel morphology and corresponding type of lateral channel changes.

The channel stability classifications are based on indicators that give an integral characteristic of channel change intensity. As a rule, they do not have any independent meaning, because they cannot be used without the channel characteristics according to their morphology and dynamics. However, one of the first riverbed classifications was based on the assessment of their sustainability. In the 30-s, Lvovich (1938a, b), using the Lokhtin' number (3.16), divided the river channels into stable and unstable ones. Several modifications of this classification were subsequently proposed. For example, Velikanov (1958) has already identified four types of channels according to the degree of stability: (1) rivers of high stability; (2) rivers of normal stability; (3) lowland rivers of low stability; (4) unstable rivers. A similar classification was proposed by Makkaveev (1971), who singled out the channels that are stable ($L > 15$), relatively stable (L = 8–15), slightly stable (L = 1.5–8) and unstable (L = 0.5–1.5). Chalov (1979), accepting this classification, substantiated the allocation of each channel type by intervals of the channel stability coefficient proposed by N. I. Makkaveev (3.20). Later on, this list of channel types with L and K_s values corresponding to each of them was supplemented first by absolutely stable (Chalov 1997a, b) and then by absolutely unstable channels (Chalov et al. 2000). At the same time, attempts were made to link channel stability with the development conditions in the rivers of certain channel patterns. However, the subjectivity of the approaches often led to contradictory results; in particular, the development of the bend (meandering) by different researchers correlated with different degrees of stability of the channels.

Other criteria for assessing riverbed stability were proposed by Altunin (1962). Grinberg (1950), Karaushev (1975), Makkaveev (1955, 1971). They proposed their own classification schemes, which not only determined the boundary values of the parameters for the channels of different types, selected by the intensity of changes, but also evaluated the conditions of development of various channel patterns, corresponding to certain ranges of stability parameters. Thus, according to Altunin (1962, p. 27), the roaming rivers (type I in terms of channel stability) "are divided into branches (channels) in some areas, and meanders are formed in others", the rivers are unstable and with low channel stability (type II) "form predominantly stable in terms of meanders"; at the same time, morphological characteristics are not given for stable channels (type III).

Empirical connections between channel stability (L, K_s) and bend parameters (radius of curvature r, step L) and branching were obtained by N. I. Makkaveev et al. (Experimental Geomorphology 1969), Chalov (1979). They have shown that free meandering and branching of the channel can develop on rivers with different channel stability, and only absolutely stable channels are characterized by incised channels.

In terms of the degree of development of ripple accumulative forms of channel relief (classification of the second group), the most common is the separation of riffle and pool areas on rivers, the formation of which is associated with changes in the forms of transport of bed material load, respectively, predominantly convex or suspended, during the passage of effective discharge rates. Makkaveev (1955, p. 301) divided the riffle sections of the riverbeds into two main types: (1) formed in areas of deep erosion, (2) formed in areas of lateral erosion. Within the limits of the first "riffles … are rare and their forms differ in comparative stability"; they are single formations, timed, for example, to the bends of the channel between adjacent bendings, knots of branching of the channel, local expansion of the channel, etc. With this spread, the riffles do not determine the morphological shape of the channel. Riffles within the riffle sections of the second type follow each other, are separated by short reaching valleys, usually move continuously and are constantly subject to long-term and seasonal reshaping. Makkaveev (1955) divided the riffle areas in the areas of lateral erosion into several varieties, which differ due to the reasons of formation of riffles. Boundaries of the reaches and riffles along the length of the river almost do not change, because at the beginning of the section to replace the riffles that replace down new ones are formed, and those that replace over the lower boundary of the stable pool section, on the contrary, are washed away. On rivers with a low stable channel, its division into reach and riffle areas is poor.

At present, there are a large number of classification schemes for riffles in which the types of riffles (or large ripple shapes comparable in width and height with the width and depth of the channel itself) are distinguished by different features (size, morphology, genesis, speed of movement along the channel, mode of transformation, seasonal changes in depth, etc.). In the classification of SHI (Kondratyev et al. 1982), the side-bar and sand dunes bed types essentially represent different forms of ripple formations that prevail in the bed and make up the entire riffle. Some classifications (Engineering of Ship's Passages… 1964; Chalov 1979) are based on the connection of riffle products with the shape of the channel itself (bendings, branches, straight sections), because the main features of their mode are determined by the hydraulic characteristics of the flow and its structure on the bendings, in the nodes of branching and merging of branches, etc.

The fourth group of classifications is based on the longitudinal slope of the river, which is the most important energy characteristic of the water flow, the degree of kineticity of the flow and the form of transport of bedloads. The hydraulic regime of the rivers, which varies widely depending on the gradient of the slope, determines the specificity of the interaction between the flow and the channel and, as a consequence, a great variety of conditions for the development of the channel relief and transport of sediment. Since from the river source to the estuary there is a natural change in the slopes, the main features of river fluvial process driver change in the same direction.

In flat areas, where the difference in the source and mouth of the main large river basin does not exceed the first hundreds of meters with its length of hundreds and thousands of kilometers, the change of slopes along the length of the river is insignificant and the water flow along the entire length of the river in most cases remains calm regime of the current. Under these conditions, changes in the fluvial process driver of rivers from upstream to downstream reaches are mainly determined by the transport capacity of the flow, which in turn depends on its runoff rates, as well as the relative role of other factors (development of aquatic vegetation, channel roughness, etc.). Changes in runoff rates create prerequisites for the formation of various channel forms, because its magnitude determines, all other things being equal, the width and depth of the channel, its curvature and other morphometric parameters. But even in the conditions of quiet flows, with naturally decreasing slope of the longitudinal profile, at first, in the upper reaches, fast enough and noticeable, and then very gradual, the flows are characterized by a different degree of kineticity. Therefore, the parameter of kineticity (Froude number) is included in many hydrologic-morphological relationships, reflecting the conditions of development of channel forms.

The most striking change in the fluvial process driver activity of rivers from top to bottom is seen in the rivers originating in the mountains. The natural decrease in their slopes and the generally concave longitudinal shape of their slopes reflect the geomorphological structure of the basins. The high gradients of the rivers in the mountains, which are several orders of magnitude greater than those of the lowland rivers, result in a high kinetic flow and specificity of the morphology and dynamics of the river channels, mechanisms and forms of sediment transport. For the first time V. M. Rodevich drew attention to the differences in the hydrological and hydraulic regimes of mountainous and lowland rivers (1931). Subsequently, Makkaveev (1955) and Altunin (1962) substantiated the necessity of separation of three classes of river channels—on mountainous, semi-mountainous and lowland rivers, which differ in kinetic water flows, different morphology of channels and forms of channel relief and principal difference of fluvial process driver activity of flows at large and small longitudinal slopes. Mountain rivers are characterised by rapid flows, step-pool cascade or with a smooth phase of sediment transport, and only at relatively low slopes for mountain rivers are ripples of sediment in the form of antidunes (Chalov 1968). The high gradients inherent in the first two types of beds, the structure-free transport of sediment and the structurally predetermined nature of river bedding make the beds of step-pool cascade and mountains with undeveloped alluvial forms (rippleless) relatively straight. The presence of ripples (side-bars) in channels with developed alluvial forms causes their sinuosity, but the active incision of mountain rivers with a stabilizing effect on the position of the flow of bedrock banks, leads to the development of incised bends, which are absolutely dominant in the mountains. Free bend formation is possible where rivers cross intermountain trenches; however, low (for pebble-boulder alluvium) stability of the channel within these trenches is usually accompanied by the destruction of the incipient bend and the formation of branches, the most complicated in their morphology and change regime. The same conditions of riverbed formation are characteristic of semi-mountain rivers, which are transition from mountain to plain in nature and form of fluvial processes. In limited conditions of channel changes development, incised bends are formed on them (in the low mountainous zone of the mountains); on the foothill plains, where there is usually a sharp decrease in the slope, the expansion of river valleys and the replacement of rocky sides by folded easily eroded sediments, branched channels are also formed, and only as far as the distance from the mountains may develop a free-meandering channel.

At the same time, a similar approach to the fluvial processes on mountain rivers was applied by Talmaza and Kroshkin (1968), who divided them into sections of the source, upper mountainous, middle and lower channels, each of which corresponds to a specific type of fluvial processes [according to the classification of Chalov (1979)]—step-pool cascade, mountainous with undeveloped alluvial forms, mountainous with developed alluvial forms and semi-mountainous. These sections of the channels are characterized by their straight (or slightly curved), meandering and scattered (decaying, according to A. N. Kroshkin, on branches and ducts); in terms of floodplain development, there are canyons, valleys, floodplains or wandering ones, with unstable channel among pebble deposits on the wide bottom of the valley; in terms of the nature of the soil composing the channel—rocky, transitional and alluvial; in terms of stability (intensity of changing)—the highest stability, stable and unstable. In a later version of Kroshkin (1970), this classification was presented in the form of a table (Table 4.1), in which he gave slightly different names to the sections of the channel, but justified their separation by hydromorphometric characteristics, a description of the movement of river load and the channel pattern. This makes it possible to compare the classification of A. N. Kroshkin with the classification of fluvial processes on mountain rivers, developed by Chalov (1968, 1979).

In lowland rivers with a predominant ripple form of sediment transport, beds of various channel patterns develop, among which bends predominate (meandering). Free bends, as well as branching, are most common on wide floodplain rivers under conditions of free development of channel change; where rivers cross structural,

Table 4.1 Types of mountain channels and associated hydromorphological characteristics (Kroshkin 1970, revised by author)

Section	*H/d*	*I* (‰)	*C*	*Fr*	Sediment movement	Channel type	
						by SHI (Kondratyev et al. 1959)	By Moscow State University (Chalov 1968, 1979)
Highland	<1	>100	<10	>1.2	Rippleless	Straight (with inclusion of foreign for the flow particles of debris by particle size)[b]	Step-pool cascade
Mountainous	1–3	100–20	10–20	1.2–0.8	Rippleless	Nonmeandering, alluvial	Mountainous plane channel
Mountainous-foothill	3–30	20–2	20–40	0.8–0.3	Rippleless transient (antidunes)[a]	Nonmeandering, limited meandering, scattered	Mountainous plane channel, mountainous riffle-pool channel
Foothill-plain	30	<2	>40	<0.3	Transient (antidunes)[a], ripple	Limited and free meandering, branching	Mountainous with developed alluvial forms, semi-mountainous

[a]Additions to Kroshkin (1970)
[b]Additions of the author

hilly-moraine and other uplands (limited conditions of channel change development), incised channels prevail.

The last fifth group unites the classifications of riverbeds by their shape (morphology) and their corresponding lateral changes. Such classifications are morphodynamic in nature (Chalov 1996) or hydrologic-morphodynamic, since the development of various forms of channel is determined by the runoff rates of the flow, runoff and the form of ripple movement of sediment.

The first morphodynamic classification of river channels belongs to Rossinsky and Kuzmin (1947, 1958), who singled out rectilinear (corresponding type of changes—periodic expansion), meandering (development of bends, meandering) and scattered (wandering) channels. Judging by the description, they belong to flat rivers, the channels of which are formed in the conditions of free development of channel changes, i.e. are wide floodplains. In this classification, a physically justified approach was used: the independent existence of three main types of channels, each of which is characterized by quite specific features of the dynamics and structure of the flow, the specificity of the channel relief and the nature of changes. In it, as in most subsequent morphodynamic classifications, the meandering branched and straight channels occupy an equal place. The reasons for the development of each type of changes K. I. Rossinsky and I. A. Kuzmin saw in a certain combination of geological-geomorphological factors and hydraulic characteristics of the flow. In their view, the periodic expansion of the channel is facilitated by the poor pliability of riverbank scouring combined with the uneven movement of bottom sediment accumulation, resulting in the process occurring in rivers with developed and shifting side-effects. Such an approach had a basis for criticism, because the described process gives a better idea of the dynamics of the riffles and their local transformation than of the development of the channel forms (Petrovsky and Sakharova 1958; Chalov 1979). However, it is important that K. I. Rossinsky and I A. Kuzmin were the first to consider the main types of river channels depending on the specifics of their formation and long-term changes.

Andreev and Yaroslavtsev (1958, p. 353) identified among the plain rivers the nonnon-navigating, "characterized by a side form of movement of sedimentary accumulations within the channel zone, which is slightly changing in time and in terms of the plan of the channel zone", the special case of nonnavigating rivers—with the island form of sediment transport, meandering with the "significant sinuosity of the channel zone", wandering, whose channels are constantly changing. The latter are common in places where rivers come out of the mountains into the foothills or plains, where there is a sharp decline in slopes. In this classification, branched, multi-branch channels are present only as a special case of non-controlling channels.

The two classifications considered are joined by the schemes proposed by Leopold and Wolman (1957), Grishanin (1972), and Rzhanitsyn (1985). Popov (1961), Kondratyev et al. (1959) typified the channels by the changes in the time of their outlines: (1) meandering or formation of bends; (2) movement of straight sections of the channel parallel to itself for a considerable length of time; (3) movement of the channel due to redistribution of runoff along the branches and formation of islands (channels of scattered type); (4) wandering channels with rapid movement of the flow from one arm to another in unstable soils. Among the free-meandering channels, they were distinguished by seven varieties that differ in the form of movement and transformation of the bend in the process of their development. Other types of channels are not separated, and branched in the future are characterized only in terms of meandering of river channels. At the same time, E. V. Popov distinguishes the rivers "with a meandering channel, but with minor planned changes …. (1) flowing in narrow meandering valleys with hard-to-flush slopes, with channel bends … repeating valley bends …; (2) rivers flowing through areas where upper, easily eroded material has already been removed …; (3) rivers of tundra and forest zone or flowing within ьфкыру forest areas, as well as in areas with permafrost; (4) rivers regulated by lakes and reservoirs, whose sinuous channel is the result of local obstacles, or inherited from the previous state of the river to its regulation" (Kondratyev et al. 1959, c. 85, 88). The first of them corresponds to the incised bends, the second—to the rocky bed, which bed is devoid of alluvium.

With the known inconsistency of classification features (along with incised and rocky channels—marshy, entrenched bend, adjusted, etc.) and a seemingly random set of selected varieties of channel, there is an attempt to cover in a single list of channel types their maximum possible diversity, including in rivers, where changes are relatively slow and manifest only over long periods of time.

This approach was also shared by N. I. Makkaveev (Engineering of Ship's Passages… 1964), who singled out meandering and nonmeandering channels, and among the latter—straight and braided, i.e. the sinuosity of the channel is accepted by him as the leading classification attribute. The legitimacy of such an approach is explained by the fact that in this case the criteria remain unambiguous, since the straight and meandering channels differ in their sinuosity, and the branching—in the presence of river branches (Chitale 1970), although the flowing of the island, as well as the side-bar in the straight channel, also causes the sinuosity of the dynamic axis of the flow in the branch. However, in earlier and later works Makkaveev (1955, 1971), without making a classification table, singled out bends, branched and straight channels, pointing out the differences in the flow structure and describing the change mode inherent in each of them. N. I. Makkaveev (Experimental Geomorphology 1969, 1971), also singled out incised, free, entrenched (or confined) bends. The first ones were called bends in areas where deep erosion prevails, while the second ones were in areas where lateral erosion prevails (Makkaveev 1955). He classified the latter by shape and morphology: segmental, sinusoidal, chest-like, omega-like, overwhelmed, and torn. Such a formal approach has at the same time a deep meaning, because the shape of the bend reflects the type and nature of their movement and the specifics of the reshaping. Comparison of this classification and the previously proposed (Makkaveev 1955) shows that segmental rays move downstream without changing the shape, the development of omega-like rays leads to the surviving necks of meanders and their subsequent "overeating" (the term of N. I. Makkaveev)—straightening due to the counter-blurring washout of the banks on the wings, the torn bends arise as a result of the breaking of the bends by the straightening duct, etc. According to N. I. Makkaveev, the chest-like bend is formed at the approach of the channel to the bedrock channel bank and the subsequent departure from it (in the first case —"oblique meanders" at the approach of the flow to the bedrock channel banks (Makkaveev 1955).

A similar method was used by Matveev (1970), who, considering the forms of the bend of the rivers near Moscow, singled out seven of their "geometric" types: (1) sinusoidal, (2) circular, (3) elliptical, (4) loopy, (5) complex, (6) asymmetric, (7) large bend shapes complicated by small meanders. Each shape of the bend is described by an analytical equation, but the conditions of the bend formation and changing were not evaluated, which reduces the importance of this classification.

In contrast to the approaches considered to the selection of classification features, which were based on the straightness or sinuosity of the channel, Rust (1981) suggested dividing the rivers into single- and multi-channel in terms of branching. In his opinion, single-channel, low meandering channels correspond to straight channels, while side-bar, малоизвилистые channels are referred to multi-branch ones. Characteristically, many researchers (Schumm and Khan 1972; Kellerhals et al. 1976; Miall 1977) consider the side-bar channels to be branched out. However, there is a shift in the classification traits according to the structural (discrete) levels of fluvial processes: side-bars are forms of ripple movement of sediment, branches are connected with the islands, and only on the basis of their formation are the side-bars fixed by vegetation, which changes their role in channel changes and the formation of morphological type of channel, in which the main thing is the fact that the channel flow is divided into several branches of the current, regardless of the causes that cause it. As a rule, this already applies to the conditions of the lowlands, when the high parts of the ripples, which are not adjacent to the bank and form side-bars, dry out, and in the beds with boulder-clump composition of bed material load—some of the largest boulders and blocks. Such a vertical structure for the classification of branches of flows, including point (elementary) branches due to the flow of coarse sediment particles, isseparated branches associated with dried up large ripples (macroforms), island branches (from single channel branches to bifurcated channels) and floodplain branches, was proposed by Chalov (2006) and Alekseevsky and Chalov (2006).

Consideration of branching has led to the creation of channel classifications based on a combination of morphological classification and sediment transport conditions. They were divided into single- and multi-branch channels (Schumm 1977; Miall 1977; Brotherton 1979). In the opinion of their authors, straight channels are characterized by a predominance of bedload in the total river load, high gradients and the melting of the bed; as these characteristics decrease, a transition to meandering and braided channels is observed. The difference between these classifications lies only in the fact that S. Schumm and D. Brotherton attributed the beds with apertures to the rectilinear divided all the beds into stable, cutoff and accumulating.

Taking into account the conditions of sediment transport or the transport capacity of the flow, new approaches to the classification of river beds emerged in the second half of the twentieth century. In the national literature, this was most clearly demonstrated in the classification of the State Institute of Geosciences, which was widely spread and included in the regulatory literature. The first version of this classification was proposed by Popov (1965); later it was developed by N. E. Kondratyev in 1968 (published in 2000), repeatedly published, reported and finally entered the monograph of Kondratyev et al. (1982). The channel types (fluvial processes) are arranged in a certain sequence (Fig. 4.1): (1) sand-dune; (2) side-bar; (3) limited meandering; (4) free meandering; (5) incomplete meandering; (6) anabranching channel; (7) mid-bar type or channel multi-branching. It considers the sinuosity of the channel as a leading classification feature, so that the meandering channels are represented by three types, and if we take into account that the development of a side dynamic axis of the flow in the lowlands is also twisting —four. The branched channels are very generalized, although the descriptions refer to the fact that they are very diverse. Straight channels do not stand out, although they can be correlated with the types of channel, identified by the shape of the channel relief (sand-dune, side-bar, mid-bar). Characteristically, N. E. Kondratyev himself (2000) wrote about the incompleteness of coverage in the classification of the entire diversity of river channels. In addition, the leading feature is broken in it —in one row there are channel forms of different order and mechanism of formation (side-bars, mid-bars, sand dunes—forms of channel relief, reflecting the ripple movement of sediment, and bend, channel and floodplain multifunctionality—

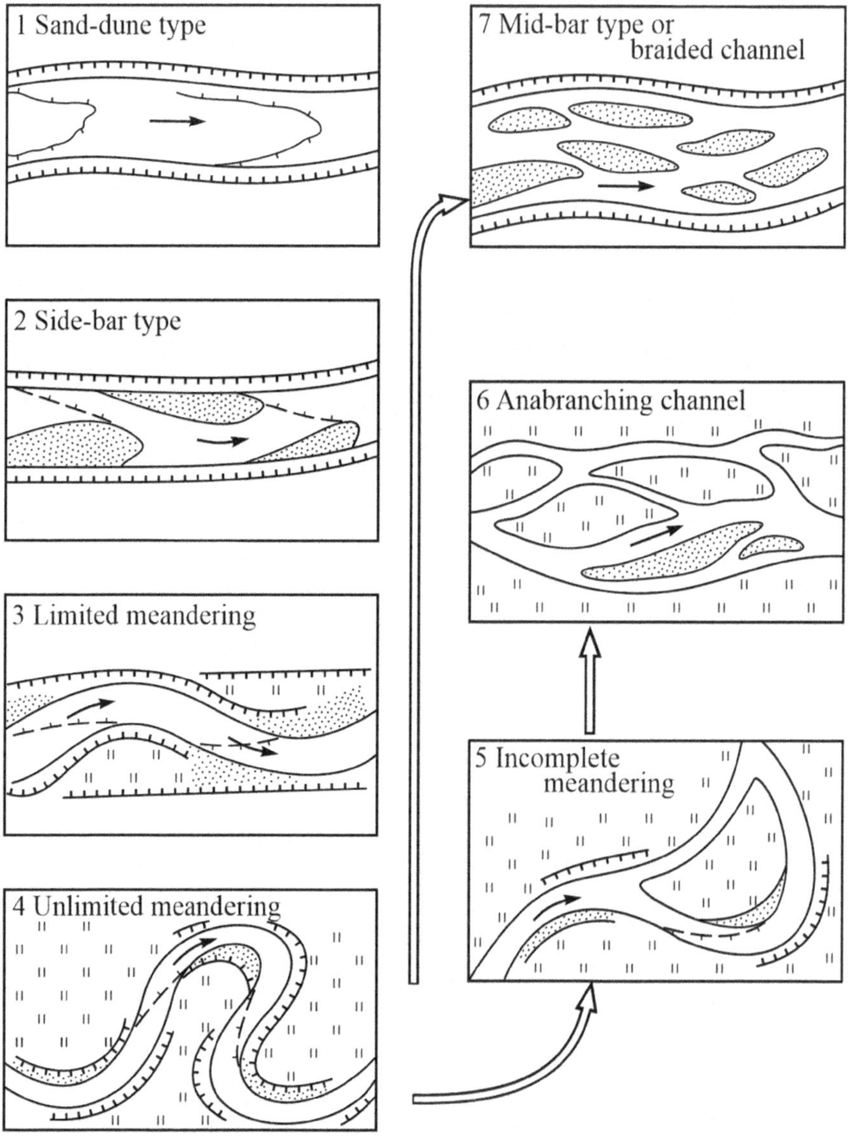

Fig. 4.1 SHI (Kondratiev et al. 1982) classification of river channels. Arrows direction corresponds to changes of flow transport capacity

forms of the channel itself). In order to avoid this shortcoming, Znamenskaya (1976, 1992) proposed to introduce "active" and "passive" processes into the classification of SHI. In her opinion, the forms of ripple motion of sediment that correspond to sand-dune one, side-bar and mid-bar processes are "active". The development and alternation of such processes are conditioned by the transformation of the ripple shapes in the event of a change in the partitioning between water discharge to bedload. Formation of different types of channels under reduction of water discharge in the lowland, which occurs on the basis of "active" processes due to fixation of large ripple forms of channel relief by vegetation, was attributed to "passive" processes. These proposals by N. S. Znamenskoy make it possible to separate the ripple transport of sediment from the channel types, i.e. to compare channel forms of different orders of magnitude, and to abandon the rigid sequence of channel type arrangement, which, according to the authors of the classification (Kondratyev et al. 1982), corresponds to an increase in the transport capacity of the flow.

Grishanin (1972), accepting the names of the channel types in accordance with the classification of the SHI, but refusing to arrange them in the proposed order, correlates them with the three main types of river beds of K. I. Rossinsky and I. A. Kuzmin, N. I. Makkaveev and others: (1) straight or slightly curved single branches with their chess-like sides; (2) sinuosity or meandering; (3) branching. In this case, the meandering channels are supplemented by entrenched bending of the channel, which occurs if the channel makes a sharp turn, approaching the bedrock channel bank, folded by bedrocks. Similar entrenched bends were also singled out by Makkaveev (1971), Experimental Geomorphology (1969) and Rzhanitsyn (1960, 1985).

Many foreign morphodynamic classifications are based mainly on the principles that formed the basis of classification of K. I. Rossinsky and I. A. Kuzmin (Leopold and Wolman 1957; Richards 1980; Schumm 1977), but they are supplemented by new features. Thus, Schumm (1977) correlates the type of channel with the form of sediment transport. It distinguishes between single- and multi-branch channels; the first type includes a slightly sinuous channel or a straight channel with spacing (with a larger share of bed load in total load), meandering (mixed composition of load) and steep meanders (with a smaller share of bed load). In another researching, S. Schumm (Shen et al. 1979) has its beds in the sequence of increasing its stability: (1) straight, stable; (2) straight, with meandering thalweg—usually stable, with shifting side-effects; (3a) meandering, stable, but small side-bars (beaches) may be rejected; (3b) meandering with extensive beaches near convex banks, relatively stable, with beaches detached, bend necks bursting, bend movement and growth; (4) transient from meandering to branching, unstable; (5) multi-branch channels, unstable.

Karasev (1975) singles out "the following types of fluvial processes" (p. 3, 37–38): (1) deep channel erosion in the absence of alluvial deposits; (2) sand dunes in the straight channel; (3) side-bars and asymmetrical ripples in the slightly sinuous channel; (4) bend development (meandering); (5) channel wandering and multi-branching; (6) irreversible accumulation of sediment near the general or local erosion base. Here, deep erosion and directed ("irreversible") accumulation of sediment has been placed on a par with channel patterns and ripple shapes, which disrupts the orderly and sequential classification, since in this case a variety of phenomena are classified, although the attempt to take into account both lateral and vertical changes in a single classification deserves attention.

A close approach was taken by Rzhanitsyn (1985), who singled out the "types of riverbed regimes" (described in the abbreviations, but with the preservation of the author's terminology): (1) deep incision, slightly sinuous channel, straight, incised in bedrock); (2) undeveloped meandering—transitional stage from plunging to free meandering (corresponds to the limited meandering in the classification of SHI—*R. Ch.*); (3) free meandering; (4) branching—accumulation of sediment; formation of relatively stable branching—bifurcation, when the branches have a large length and can be considered as independent flows, and multi-branching of the main channel; (5) wandering—accumulation and intense lateral erosion, continuous movement of the channel; (6) delta processes (Fig. 4.2). In both classifications (N. A. Rzhanitsyn and I. F. Karasev) different types of vertical changes are opposed, as the presence of alluvial deposits and floodplains is perceived as a sign of accumulation processes only. The advantage of N. A. Rzhanitsyn's classification is the partitioning between each type of rivers and flat and mountainous rivers, and its graphical representation shows the differences between them. A separate block in the classification gives the "types of fluvial process" by which N. A. Rzhanitsyn understands the different forms of ripple motion of sediment.

Table 4.2 shows the most widespread riverbed classifications in Russia. Most of them have three main channel patterns—meandering, multithread and relatively straight, not branched out, although other terms are often used to denote them (scattered nonmeandering, low meandering channel, channel multi-branching, etc.). Each of these types is independent, being defined in its formation by a complex combination of factors of fluvial processes. The exception is the classification of the SHI (Popov 1965; Kondratyev et al. 1982), in which the selected types of channel are arranged in a strictly defined sequence.

In most classifications, the classification of meandering river beds based on the peculiarities and mechanism of their formation is relatively well developed; at the same time, only general descriptions are given to the varieties of branched channels, and the signs of their separation are different: the length of the branches and the partitioning between branching nodes and the size of the channel, the number of islands forming them. Sometimes they are divided into two main types—side-bar and anabranching channel. At the same time, they represent different structural levels of representation of branching: the side-bar one is connected with large macro-form ripples drying up in the lowlands and separated from the banks by channels; the floodplain one is connected with the islands dividing the channel into branches. However, the side-bars while fastening by vegetation and their transformation into an elementary island underlie the branching of the channel. Further evolution of branching depends on the stability of the channel, the hydrological regime of the river and other factors. The result is either a system of small islands or large islands or island massifs composed of former elementary islands that have merged together. The first variety Popov (1965) and Grishanin (1972) are called the side-bar branching, and the second—the island branching.

Anabranching channel system is usually considered as a type of branched channels, and only Popov (1965) considered that this type of channel is an

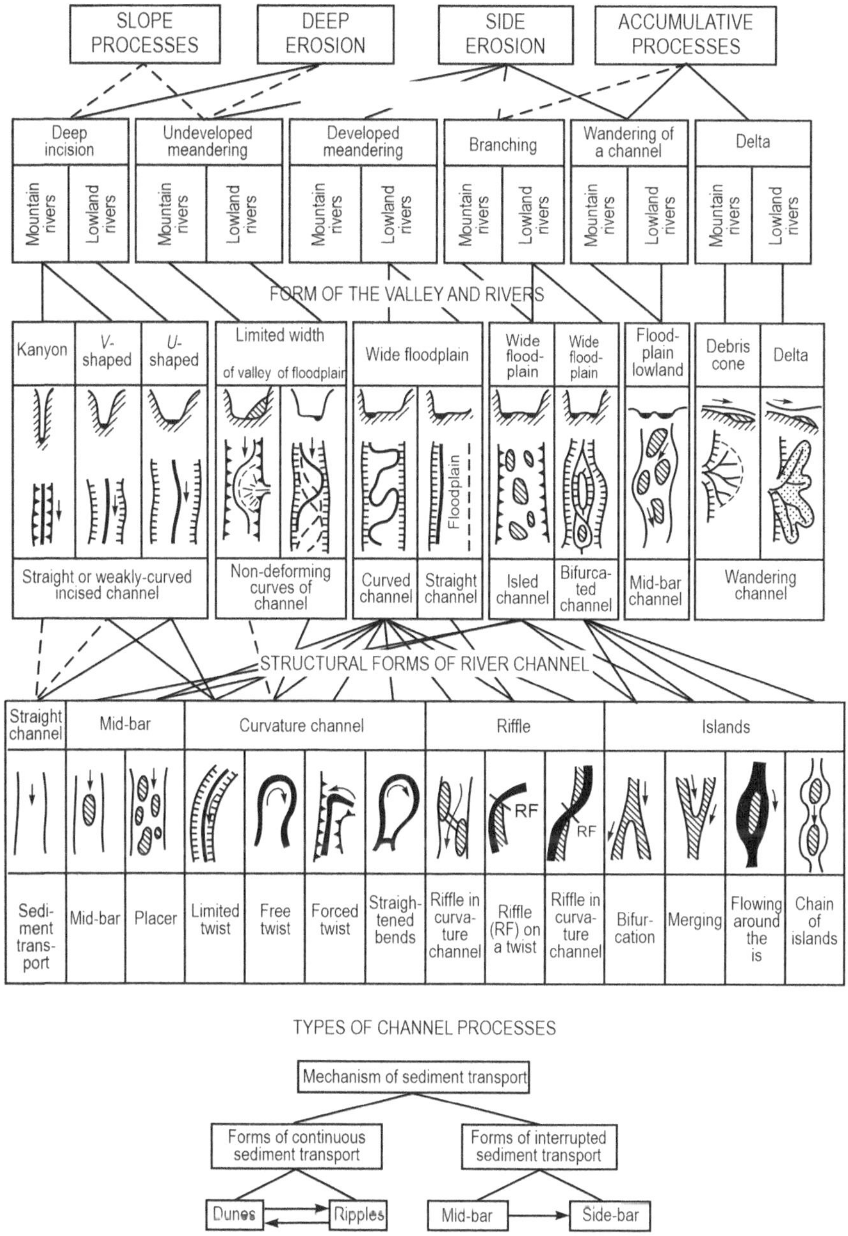

Fig. 4.2 Classification of fluvial processes and forms of riverbed configuration by Rzhanitsyn (1985)

Table 4.2 Comparison of the most widespread Soviet Union classifications of lowland river beds

Rossinsky and Kuzmin (1947)	Makkaveev (1955), Engineering of Ship's Passages… (1964)	Andreev and Yaroslavtsev (1958)	Popov (1965) (fluvial process types)	Grishanin (1972)
	Single-thread			
Straight, slightly curved (periodic widening)	(a) Low meandering and straight	Lowland, non-meandering (side-bar form of sediment transport)	1. Sand-dune	Straight, single branch
			2. Side-bar	
Meandering	(b) Meandering	Lowland meandering		Meandering, single branch
	(1) Incised bends			
	(2) Entrenched bends			(a) Entrenched bending of the channel
	(3) Free bends		3. Free meandering	(b) Free meandering
			4. Limited meandering	(c) Limited meandering
Spread out (wandering)	(c) Braided	Lowland, nonmeandering (island form of sediment transport)		
	(1) Floodplain channel		5. Incomplete meandering	(d) Incomplete meandering
				Branched out
			5a. Anabranching channel	(a) Floodplain branching
	(2) Bar-braided		6a. Mid-bar (channel multibranching)	(b) Channel branching (mid-bar)
	(3) Island-braided			(c) Channel branching (island)
Complex structure (riffles-placers, cohesive beds, floodplainless)		Wandering		

evolution of incomplete meandering. There is also a palliative opinion (Chalov 1979): the anabranching channel diversity is observed on both branched and meandering rivers as a result of the incomplete extinction of the oxbow lakes that were separated from the river by cutoff of bends and branches when the islands were attached to the bank.

Few classification schemes include entrenched and confined bend, although their change occurs according to other laws. Even rarer can be found in the classifications of incised channels or the division of channels by types of fluvial processes into mountainous and flat. The analysis of the considered river channel classifications shows: (1) the absence of a classification that includes the entire variety of channels of lowland and mountain rivers; (2) the difference in principles used by different researchers as the basis for the classification of channels, inconsistency of some and generalization of other classifications; (3) undeveloped classifications in terms of the nature of channel changes, which is primarily related to branched channels; (4) the lack of taking into account the conditions for the formation of channels, i.e. free and limited development (in the geomorphological sense) of channel changes.

In China, the existing channel patterns classifications rely on the scheme proposed by Rossinsky and Kuzmin (1947), singling out the meandering, straight and branched out channel as separate blocks, but making some additions to it, related to the specifics of fluvial processes on the rivers of China. The main thing is the selection of the wandering channel type. However, this term has a different meaning than in the classification of K. I. Rossinsky and I. A. Kuzmin. The last wandering channels were morphologically called anastomosing, characterizing them as "distinguished by the abundance of branches and ducts separated by side bars and islands" (Rossinsky and Kuzmin 1958, p. 10), and, consequently, identifying them with complex branched out. By wandering we mean the multistrained nature of the flow and the rapid movement of its dynamic axis along the numerous branches, their periodic development and shallowing. In Chinese literature, the wandering channel is characterized as very wide, and within it there is a constant (in the high-water phase of the regime and from year to year) and very intense change in the position of the line of greatest depths (its movement by hundreds and thousands of meters per year) among the vast, rapidly shifting shoals, with the relative straightness of the shape of the channel (in the floodplain or indigenous banks). The closest to this characteristic of the roaming channel is the block of "wandering rivers" in the Altunin's channel classification (1962), which contrasted the rivers with this type of channel in terms of the value of the stability index—the Lokhtin`s number (Syr Darya—$L = 2$–2.5, Amu Darya—$L = 1.0$) unstable ($L = 12$—Volga in the middle reaches, $L = 5$–6—Don in the lower reaches) and stable ($L > 30$) rivers.

In the Lin Chengquan riverbed classification published in the early 1960-s, meandering and branched channels are subdivided into stable and wandering bending or branching channels: the former develop if the channel topography stability is greater than that of the banks, the latter develop if the channel stability is less than that of the banks (Table 4.3). It also shows the character of the rhythmic soils forming riverbanks.

Many Chinese riverbed classifications, in contrast to Soviet and European ones, take into account a wide range of conditions for their formation. The latter are mainly designed for free conditions of channel changes on flat rivers. There are non-mountain and semi-mountain river channels, incised beds of plain rivers, formed in conditions of limited development of channel changes, occupy a

Table 4.3 Channel types and their characteristics, by Lin Chengquan (Chalov et al. 2000)

Channel types		The nature of the banks and their structure	The relief of the riverbed	Relative static channel characteristic: channel stability is greater (>) or less (<)than bank stability
Straight, gentle-curved		Bedrock banks with stable soils	Side bars along both sides of the river	<
Meandering	Stable bends	Dual or layer alternation of sandy and clayey soils	Convex bank with a high floodplain	>
	Wandering bends		Convex bank with low floodplain	<
Branched	Stable branchings	Sandy or dusty	Islands	>
	Wandering branchings		Islands	At low values of riverbed and bank stability

secondary position or are included as some additions. Later, Lin Chengquan proposed to classify the channels according to the conditions of formation (alluvial, semi-alluvial and non-alluvial, including incised) and only then according to their morphology (straight, meandering, branched) and the intensity of changes (e.g., wandering, stable meandering, free meandering). The incised channels (incised bends and rocky islands) are included in the classification of Sheng Yuchang and Gong Goyuan (Fig. 4.3). They are correlated with mountain rivers, where they are divided into normal and moving; the bend of flat rivers is divided into two types—simple and meandering. The branched channels are typified in more detail, which distinguishes this classification from the abovementioned ones. It tries to take into account the specifics of fluvial processes on mountain rivers. Single-thread channels are characterized by the degree of bend development, the partitioning between channel length and bend pitch *L*.

The most common classification in China is Qian et al. (1987), based on materials from large rivers of the world, including Chinese (Huang He, Yangtze, Weihe) (Table 4.4). In it, each channel pattern is accompanied by morphological characteristics, and for the bend of the channel—the assessment of channel changes and the characteristic of stability of the channel.

The last of the proposed Russian riverbed classifications is the morphodynamic one developed by Chalov (1996, 1997a, b), Chalov et al. (1998). It is a development of the scheme proposed by Makkaveev and Chalov (1986), which has been repeatedly specified and gained popularity in many publications using it as a classification of MSU. Its structure is multilevel, based on the principle of

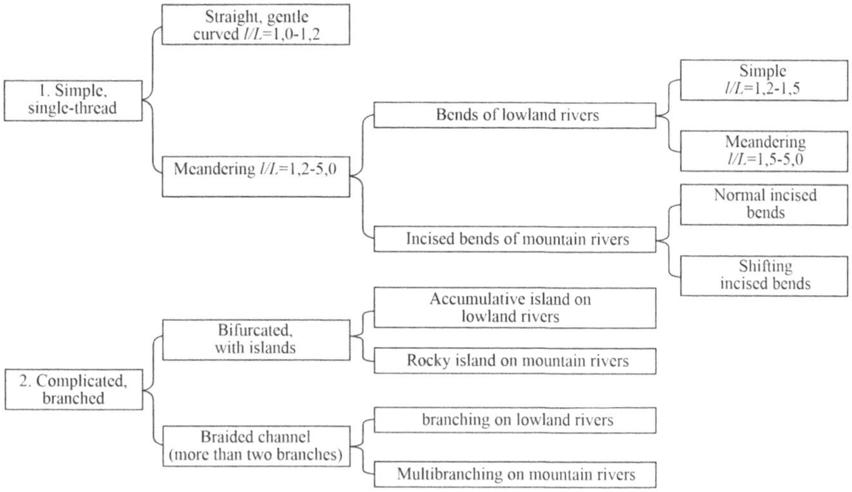

Fig. 4.3 Classification of river channels by Shen Uychan and Gun Goyuan (Chalov et al. 2000)

Table 4.4 Channel types and their characteristics, by Qian Ning (Chalov et al. 2000)

Channel types	Morphological characteristics	Changing	Stability characteristic	Specific examples
Wandering	A scattered multi-branch channel	Wandering	Very unstable	Lower course of Huang He, Brahmaputra
Branched	Branches	Periodic development of each branch	Unstable	The middle and lower reaches of the Yangtze, Niger.
Meandering	Meanders	Incised bend	Relatively stable	Yangtze middle course, Weihe, Mississippi middle course
		Free bends		
		Entrenched meanders		
Straight	Straight		Stable	The lower course of the Mississippi

inrelationship of the development of three main types of channels. It takes into account, on the one hand, the peculiarities of the interaction between the flow and the channel in the free (wide floodplain) and limited conditions of channel changes development, determined by the geological-geomorphological structure of river valleys. On the other hand, it takes into account the differences in the mechanism of fluvial processes (interaction of flow and channel, form of sediment transport) on mountainous, semi-mountainous and plain rivers. In this regard, for each of the main patterns (meandering, branched, relatively straight, single-thread) the channel can be wide floodplain, incised and confined, flat, semi-mountain and mountainous.

Each of them is subdivided into several varieties, differing in the form of the channel (for bending—the degree of their development l/L) and the nature of channel changes. Individual blocks in this classification take into account the composition of bed material load and the stability of the bed.

The classification proposed by Rosgen (1994) has a certain similarity to the classification of MSU, in which seven main types of channels are singled out in terms of their incising (the partitioning between floodplain width and the width of the channel B_f/b_{ch}), corresponding to the incised, confined and wide-floodplain channel, slopes and sinuosity (the partitioning between channel length to valley length, whereas in the MSU classification it is defined as l/L, where l is the length of the channel along the bend, L is its pitch) and the shape of the cross-section bch/h. Each type of channel, in turn, is divided into subtypes according to the predominant composition of bed material load. The combination of all traits allowed D. L. Rosgen to assign an index to each channel type and subtype (Table 4.5). The scheme accompanying the classification also shows the conditions of water effective discharge rates (Fig. 4.4): in floodplain and floodplain edges (bed-filling).

4.2 Types of Fluvial Processes

Principles of Fluvial Processes Classification

The mechanism and specificity of the interaction of the flow with the channel and the form of sediment transport, which are the essence of fluvial processes, vary widely depending on the runoff rates of the rivers, their water regime, the kinematics of the flow and the magnitude of the longitudinal slope, and the size and magnitude of bed material load. The longitudinal slope of the river is as important an energy characteristic of the flow as the flow rate, together with the flow rate. It also influences the morphology and dynamics of river channels not only through the stability and size of the effective discharge, but also mainly because it largely determines the degree of kineticity of the flow. The hydraulic regime of the rivers, which varies widely depending on changes in the longitudinal slope, results in a great variety of conditions for the development of the forms and transport of bed material load.

According to the definition of fluvial processes, their classification should be based on the differences in the mechanisms of interaction between flow and channel, bottom and bank erosion, sediment transport and accumulation. These differences depend primarily on the degree of kineticity of the flow, which determines the mechanism of the interaction between the flow and the channel itself, and, consequently, the conditions for its erosion and the form of transport of bed material load. Reflecting in channel morphology and specificity of channel change (development of bends, branches, straight channel, forms of channel relief), kinetic flow (rough or quiet), difference in gradients (fractions ‰ or their units and tens),

Table 4.5 Classification of river channels by Rosgen (1994)

	Single-thread														Branched			
Incising B_f/b_{ch}	Incised (<1.4)						Moderately incised (1.4–2.2) (confined)			Slightly incised (>22) (wide floodplain)[a]								
Width to depth ratio b_{ch}/h	Low <12				Medium-high >12		Average >12			Very low <12		Medium-high >12			Very high >40		Low <40	
Sinuosity	Small <1.2		Average >1.2		High >1.4		Average >1.2			Very high >1.5		High >1.4			Small		Small-high	
Type of channel	A		G		F		B			E		C			D		DA	
Slope (‰)	>1	0.4–0.99	0.2–0.39	<0.2	0.2–0.39	<,02	0.4–0.99	0.2–0.39	<0.2	0.2–0.39	<0.2	0.2–0.39	0.01–0.2	<0.01	0.2–0.39	0.01–0.2	<0.01	<0.05
Bedrock (1)	A1a +	A1	G1	G1c	F1b	F1	B1a	B1	B1c			C1b	C1	C1c-				
Boulder (2)	A2a +	A2	G2	G2c	F2b	F2	B2a	B2	B2c			C2b	C2	C2c-				
Cobble Pebble-boulder (3)	A3a +	A3	G3	G3c	F3b	F3	B3a	B3	B3c	E3b	E3	C3b	C3	C3c-	D3b	D3		
Gravel (4)	A4a +	A4	G4	G4c	F4b	F4	B4a	B4	B4c	E4b	E4	C4b	C4	C4c-	D4b	D4	D4c-	DA4
Sand (5)	A5a +	A5	G5	G5c	F5b	F5	B5a	B5	B5c	E5b	E5	C5b	C5	C5c-	D5b	D5	D5c-	DA5.
Silt/clay (6)	A6a +	A6	G6	G6c	F6b	F6	B6a	B6	B6c	E6b	E6	C6b	C6	C6c-	D6b	D6	D6c-	DA6

[a]*Note* Names corresponding to IYU classifications

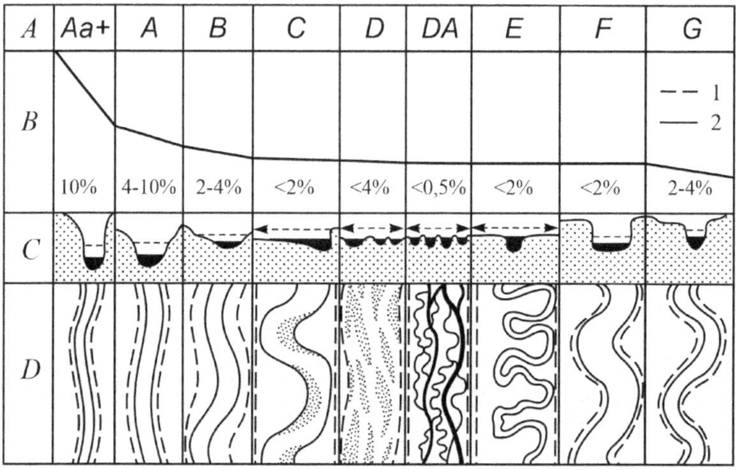

Fig. 4.4 Classification of river channels by Rosgen (1994): **A**—channel types (indexes); **B**—respective slope ranges and longitudinal profile; **C**—specific transverse profiles; **D**—plan outline. 1—boards of valley bottom; 2—river banks

predominantly sandy and sandy-pebbly or pebbly-boulder composition of bed material load, their ripple or unrimmed form of movement and a number of other features of the flow dynamics and, as a consequence, the representation of fluvial processes determines the separation of channels of lowland, semi-mountain and mountain rivers. Makkaveev (1955) proposed to use I_{kr} as the criterion of transition from lowland rivers to mountains the value of the critical slope of I_{cr}, because its ratio with the actual slope determines a calm ($I < I_{cr}$, $Fr < 1$), rough ($I > I_{cr}$, $Fr > 1$) *and* critical ($I = I_{cr}$, $Fr = 1$) state of the flow and, accordingly, the mechanism of interaction between the flow and the channel. In the given connection *Fr* is a kinetic parameter of the flow

$$Fr = \frac{V_{av}^2}{gh_{av}}, \tag{4.1}$$

where V_{av}—the average speed, h_{av}—the depth of the flow.

The stormy and calm currents differ in the nature of their free surface. The smooth surface of the calm flow is characterized by insignificant lowering of its marks (level reduction) over the individual obstacles. A stormy stream forms a splash over the obstacle—a standing wave. As a result, the surface of the turbulent flow is covered with a series of standing waves, creating an uneven surface. This difference is explained by the difference in energy losses between quiet and rough flows when crossing an obstacle. In a calm flow it is carried out at the expense of potential energy, i.e. it is accompanied by a decrease in depth *h* and a decrease in water level *H*. Kinetic energy is used in the stormy flow, so that the depth *h* increases, the flow rate *V* drops and the water surface level *H* rises sharply.

The stormy stream has a tendency to reflect from the banks. That's why he's changing direction on the way to the obstacle. At the same time, it flows around steep curves of the bank, without forming stagnant zones and whirlpools. If, however, a whirlpool is formed in the rough flow, then here, due to a sharp drop in pressure, according to the Bernoulli equation

$$z_1 + \frac{p_1}{\rho g} + \frac{V_1^2}{2g} = z_2 + \frac{p_2}{\rho g} + \frac{V_2^2}{2g} + \Delta H_{1-2}, \qquad (4.2)$$

(here z—specific position energy equal to the height of the level in the given site, p—pressure force, ρ—water density, $\frac{p}{\rho g}$—specific pressure energy, $\frac{V^2}{2g}$—specific kinetic energy of the flow, ΔH—pressure loss (flow energy consumption to overcome resistance, indexes 1 and 2 mean that the parameters belong to the position of the site) observed cold boiling fluid—cavitation, which determines the drilling of deep erosion boilers. The gentle flow of sediment flows smoothly around obstacles; below the capes it forms stagnant vortex zones in which sediment accumulation occurs. Thus, depending on the flow conditions the channel formation conditions change. In addition, the kinetic nature of the flow has a noticeable effect on the shape, intensity and direction of the ripple movement.

In natural channels, areas of calm and turbulent flow alternate. The mating of the rough and calm flow is known to take the form of a hydraulic jump. In full-scale conditions, hydraulic jumps are observed below waterfalls, step-pools, pebble-boulder ripples, some large fragments (boulders, blocks), where the depth of the flow is less than critical, and the flow becomes violent. In the jump zone, the entire stream is pulsating, which causes the erosion of the bottom ground, including due to the destruction of the strong rocks themselves. Downstream, $Fr<1$ conditions are restored, which corresponds to a calm state of the flow.

Hydraulic jump is represented in the form of a standing wave. The propagation velocity of a single wave on the flow surface is determined by the formula

$$u = \sqrt{gh}. \qquad (4.3)$$

A hydraulic jump in a rough flow occurs if

$$V_{av} \geq u, \qquad (4.4)$$

where u is the speed of a single wave. Hence, by substituting the formula (4.3) into the limiting equality (4.4) we can get the parameter of kineticity—the Froude number of (4.1).

Transition of the quiet flow into the stormy one is carried out in the form of a hydraulic waterfall (Fig. 4.5). The conditions conducive to the formation of hydraulic waterfalls, as well as hydraulic jumps, exist in the empty sections of mountain rivers.

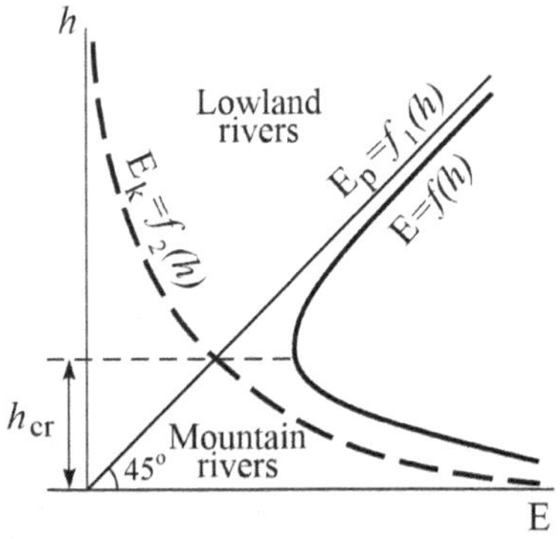

Fig. 4.5 Graph of the flow energy change in dependence on the depth and position of lowland and mountainous rivers on it

Equation (4.2) gives an expression for the estimation of pressure losses ΔH_{1-2}. If we accept infinitely small changes in position energy and consider the uniform flow rate at which $V_1 = V_2$, the reduction of the total pressure along the flow will be equal to

$$\Delta H_{1-2} = z_1 - z_2,$$

and its value related to the unit of flow length between the sections Δx, i.e. the hydraulic slope I_f

$$I_f = \frac{\Delta H_{1-2}}{\Delta x}, \tag{4.5}$$

i.e. the hydraulic slope is equal to the slope of the free flow surface. Hence, the noted differences in the character of the water surface of the stormy and calm streams.

To understand the physical essence of the concept of "critical slope" and to determine the conditions of transition from flat to mountain rivers, it is possible to use the structures known in river hydraulics. If we apply Bernoulli's equation (4.2) to a particular flow section, i.e. write it down relative to the lowest point of the river section, then the specific energy of the flow section

$$E = h + \frac{V^2}{2g} = h + \frac{Q^2}{2g\omega^2}. \tag{4.6}$$

In case of uniform movement of the liquid ($H = const$; $\omega = const$) the specific energy of the section is a constant value. But in rivers, the water flow is mostly uneven, so along the flow of energy section significantly changes, determining the local characteristics of fluvial processes.

It follows from the relationship (4.6) that the value E is a function of the depth of the channel (at $Q = const$). Figure 4.5 shows the change in all three values of equality (4.6) depending on the depth of flow. The change in the potential energy

h is represented by a straight line passing under the angle of 45° to the coordinate axes, since the specific potential energy of the section is equal to the depth of the flow. As the depth increases, the area of the living section increases, and the flow rate and, consequently, the kinetic energy of the E_{kin} flow decreases; the curve corresponding to E_{kin} asymptotically tends to get closer to the depth axis. A decrease in the depth and area of the living section leads to an increase in kinetic energy, and the curve corresponding to it approaches it asymptotically. Thus, when the depth changes, the equation of the specific energy of the flow section of the equation (4.6) changes in different directions. Therefore, the function $E = f(h)$ must have a minimum. The depth of filling of the channel corresponding to this minimum is called the critical depth V_{cr}. To determine it, we transform equation (4.6). The minimum of a function can be found by differentiating it and equating the derivative from function to zero:

$$\frac{\partial E}{\partial h} = 1 - \frac{Q^2}{g\omega^3} \cdot \frac{\partial \omega}{\partial h} = 1 - \frac{Q^2 b_{\text{ch}}}{g\omega^3}, \tag{4.7}$$

or

$$\frac{Q^2 b_{\text{cr}}}{g\omega_{\text{cr}}^3} = 1; \quad \frac{Q^2}{g} = \frac{\omega_{\text{cr}}^3}{b_{\text{cr}}} \tag{4.8}$$

where the index "cr" means that the flow characteristics are defined under the condition $h = h_{cr}$. For any shape of steering wheel, the critical depth is determined by the selection. In the sections of the rectangular form, h_{cr} is found by the formula arising from the expression (4.8):

$$h_{\text{cr}} = \sqrt[3]{\frac{Q^2}{g b_{\text{ch}}^2}} \quad \text{or} \quad h_{\text{cr}} = \sqrt[3]{\frac{q}{g}} \tag{4.9}$$

where q is the specific water discharge.

Changing the slope I at constant Q and h, as well as the corresponding change in the depth of the channel can lead to a situation where the normal depth is equal to the critical depth. The value of the slope at which the equality $h = h_{cr}$ is performed corresponds to the critical slope. Joint solution of the Shezy-Manning equations and (4.9)

$$Q = \omega C\sqrt{RI_{\text{f}}}$$

or for river flows

$$Q = \omega C\sqrt{hI}, \tag{4.10}$$

leads to a partitioning between

$$I_{\rm cr} = \frac{g\chi_{\rm cr}}{b_{\rm ch}C^2},\qquad (4.11)$$

$\chi_{cr} = \frac{\omega_{cr}}{h_{cr}}$ where there's a wet perimeter that's for natural channels $\chi_{cr} \to b_{ch}$, and then

$$I_{\rm cr} = \frac{g}{C^2}.\qquad (4.12)$$

It follows that the critical gradient of the riverbed can reach a significant value with a large roughness of the riverbed. In other words, the currents with a higher roughness of the channel can have very large slopes, while maintaining a calm flow. Therefore, the critical slope $I_{\rm cr}$, at which the transition from the lowland river channel to the mountain river channel is carried out, varies significantly (Table 4.6).

Critical slope $I_{\rm cr}$ can also be defined by solving equations (4.9) or the Shezy-Manning equation and the expression of the critical velocity of transition from calm to rough flow, which, based on the Froude number $V_{\rm cr} = \sqrt{gh}$

In this case

$$I_{\rm cr} = \frac{gn^2}{h^{1/3}}\qquad (4.13)$$

or by substituting in (4.13) the roughness coefficient as a function of sediment particle size d (e.g. Chang-Strickler) $n = k\,(d)^{1/6}$, we obtain

$$I_{\rm cr} = \frac{kgd^{1/3}}{h^{1/3}},\qquad (4.14)$$

i.e., the value of the critical slope is directly proportional to the size of alluvium, which determines the grained roughness of the channel, and is inversely proportional to the depth of the flow, depending on its runoff rates.

A close result was obtained by Kroshkin (1970) on the basis of analysis of hydromorphological characteristics of mountain rivers:

$$I_{\rm cr} = k\left(\frac{d_{\rm av}}{h}\right)^x \cdot \left(\frac{\rho_s - \rho}{1.65\rho}\right)^x,\qquad (4.15)$$

where ρ_s and ρ, respectively, are the densities of the sediments and the sediment-bearing flow.

Entering the depth value from the continuity equation $Q = \omega V$ into (4.9) and assuming that $\omega = b_{\rm ch}h^{2/3}$, we obtain

Table 4.6 Channel types and corresponding slope intervals, ‰, depending on their size (pool area and order)

Channel types	River slopes				Froude number, according to Kroshkin (1970) with the additions
	With a basin area (km^2)				
	>1000	1000–100	100–10	<10	
	Level (according to Scheidegger's scheme) (Pakhomova 2001)				
	>8.3	9.0–3.0	5.6–2.0	<2	
Lowland	<0.3–0.5	<0.5–0.7	<5–7	<10–15	<0.1
Semi-mountainous	0.3–7	0.5–7	5–10	10–20	<0.3
Mountainous					
(a) With developed alluvial forms	1.0–14	5–17	7–30	15–80	0.3–0.8
(b) With undeveloped alluvial forms	2.0–20	7–40	18–70	25–125	0.8–1.2
(c) Step-pool cascade	>4.0	>20	>25	>40	>1.2
(d) Mudflow	–	>15–30	>30–100	>70	>0.3
Rocky	All the slopes				All spectrum

$$h = \frac{Q}{b_{\text{ch}}^{2/3} V^{3/2}}. \tag{4.16}$$

Replacing in formula (4.14) the value of depths with the expression (3.15) and carrying out the necessary transformations, we obtain

$$I_{\text{cr}} = \frac{kgd^{1/3}B^{1/2}V^{1/2}}{Q^{1/2}}. \tag{4.17}$$

If we assume that $Q = f(F^m)$, where F is the catchment area and the degree index is on average 2/3 ($m = 1$ at $F < 50$ km^2, m = 0.84 at F = 50–100 km^2, m = 0.75 at F = 100–1000 km^2 and m = 0.56 at F > 1000 km^2) (Sokolovsky 1968), combining the coefficients and rounding, we can obtain

$$I_{\text{cr}} = kg\frac{d^{1/3}b_{\text{ch}}^{1/2}V^{1/2}}{F^{m/3}}. \tag{4.18}$$

Thus, the less the river's runoff rates and the area of its basin, the steeper the river remains calm. On the contrary, an increase in the flow rate, for example, as a result of its constriction under otherwise equal conditions, leads to an increase in the critical slope. This is in line with N. I. Makkaveev's ideas (1955), according to which the critical slope separating the flat and mountain rivers decreases with the increase of the river, as well as the conclusions of Charnomsky (1924) about the preservation of calm flow on small rivers with large slopes. At the same time, the increase in the particle size of the load causes an increase in the critical gradients (Chalov 1997a).

Using the connection between the area of the basin and the order of the river $F = k_1 e^{0.64N}$ and its aquifer capacity $Q = k_2 e^{0.60N}$, it is possible to obtain (Chalov et al. 1999; Pakhomova, 2001)

$$I_{\mathrm{cr}} = \frac{kg d_{\mathrm{av}}^{1/3} b_{ch}^{1/2} V^{1/2}}{e^{N \cdot m/3}}, \qquad (4.19)$$

where N is the level of the river, according to the scheme of Scheidegger (1964). Hence, the larger the size of the river (its runoff rates, area, level), the smaller the critical slope; on the other hand, the increase in the particle size of the river increases its value, which is in accordance with the relationship of the I_{cr} on the roughness of the channel (4.12). In many cases, mountain rivers are characterized by their comparable depths with the coarseness of bed material load represented by pebbly-boulder and boulder-clump material, i.e., $d \approx h$. Therefore, depending on the longitudinal gradient, the hydraulic regime of mountain rivers varies widely and, accordingly, the conditions for the development of channel forms, sediment transport and the movement of large debris change. The predominance of pebbles and boulder deposits on mountain rivers creates significant unevenness in the flow bed. Large fragments of rocks often form awards, which are partially protruding in the low water period. This leads to a large roughness of the riverbeds of mountain rivers and contributes to the formation of a rapid flow between the boulders of stagnant zones, which play the role of "water cushions" over which the upper layers slide, without involving the bottom layers in the general circulation of water. In these stagnant areas, smaller material usually accumulates. The coarser the bedding material, the more pronounced the velocity pulsation and the more heterogeneous the composition of alluvial sedimentation, the less sorted the bed material load.

All this results in a huge range of gradient changes (from the first tenths to hundreds of ‰) and a great variety of conditions for the formation of channels, the division of mountain rivers by type of fluvial processes into step-pool cascade, with undeveloped (rippleless) and developed (rippled) alluvial forms (Chalov 1968, 1979, 1997a, b). Transition from one type of fluvial processes on mountain rivers occurs according to the same laws as from flat to mountain rivers, described by formulas (4.16), 4.17), (4.18). The gradient ranges corresponding to the types of fluvial processes in mountain rivers are different for rivers of different sizes, differing in terms of the size of bed material load and the degree of congestion (width) of the channel. This determines the sliding scale of the slope ranges for rivers of different water bearing capacities (with different basin areas and order). Correspondingly, the critical gradients determining the transition from lowland rivers to semi-mountainous and from semi-mountainous to mountainous ones are different (Table 4.6).

Morphological differences between the beds of mountain rivers of different types are determined by the forms of transport by the rapid flows of bed material load. Mountainous channels with developed alluvial forms are characterized by ripple alluvial formations, i.e. antidunes commensurate with the width of the channel and developing during floods (Fig. 4.6). In the antidune zone of the channel are non-washed parts of the channel, where standing waves are formed by the flow of individual boulders. The riverbeds of this type in the lowland are an alternation of shallow water areas, i.e. riffles in the form of antidunes with rapid flow conditions and deep moulds with a calm flow regime (Fig. 4.7). This results in the fact that on mountain rivers with developed alluvial forms in the interstage waves on the surface of the flow are formed only on the ropes, where at low horizons the condition of preservation of a turbulent state for the zones of shallow relative depths is observed (Yakhontov 1960).

$$\frac{h}{\Delta} < 5 \tag{4.20}$$

where h is the depth of flow above the roughness projection of the Δ height determined by the bed material load particle size ($\Delta \sim d$). During floods, the flow becomes violent all over the river, and large standing waves occur on antidune riffles.

Mountainous channels with undeveloped alluvial forms are characterized by uniform distribution of depths along the flow length (Fig. 4.8), disturbed by large boulders and blocks. They form protrusions rising above the water level in the lowlands; their height is on average 0.7 d at least 20% of the total number of large boulders in the southeast (Talmaza and Kroshkin 1968). On rivers of this type, the ratio (4.20) is maintained for almost any point in the channel, which means that the flow here is generally turbulent. This change in channel shape is due to the

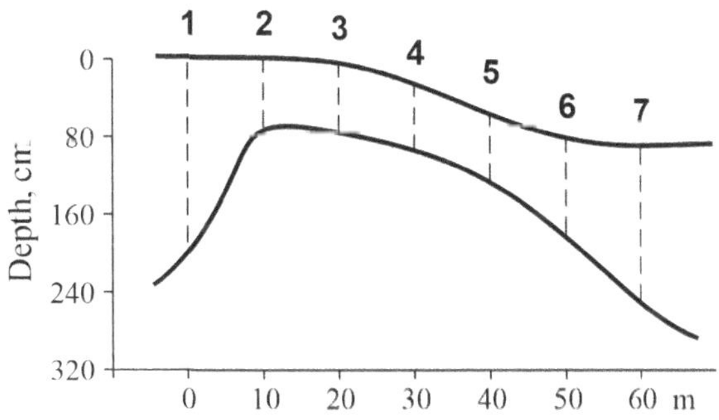

Fig. 4.6 Longitudinal profile of anti-dune ripple in a channel of mountainous river (data from the Kintrishi river near Kokhi village, Small Caucasus)

Fig. 4.7 Incised bends of mountainous river (Bashkaus river, Altai) with developed alluvial forms. Photo of the author

Fig. 4.8 Mountainous channel with undeveloped alluvial forms (Nachilova River, Kamchatka, photo by S.R. Chalolv). Photo of the author

increased kineticity of the flow and the Froude number, when the ripple structure of the bottom of the flow is lost and the "smooth phase" of sediment movement begins. This is consistent with the results of studies of the ripple motion of sediment under laboratory conditions (Znamenskaya 1968): when some critical values of the Froude number ($Fr > 2.5$–3.0) are reached, the ripple structure of the bottom of the flow is lost and the "smooth phase" of sediment motion begins.

Very high gradients of mountain rivers (Table 4.6) result in the formation of step-pool cascade (Fig. 4.9), the morphological appearance of which is determined by large boulders and blocks, the size of which depends on the quantity and nature of the receipt of clastic material from the slopes. There is an alternation of areas where the flow becomes ultra-drilling, up to the emergence of vertically falling hydraulic jets, and lake water reservoirs formed in erosion boilers.

Makkaveev (1955, p.82) considered that "the morphology of the mountainous and flat river bed has little in common". This conclusion is valid if we compare with flat and mountain rivers in their upper reaches, which have step-pool cascade channels. Together with the channels of lowland rivers, they form the extreme links in the chain of morphologically different types of channels, between which other types of mountain and semi-mountainous channels include an increasing number of features corresponding to or close to the channels of lowland rivers. The latter circumstance sometimes leads to the fact that the differences between the channel processes in the flat and mountain rivers are not made, and the beds of mountain rivers belong to the nonmeandering or rivers with the predominance of limiting

Fig. 4.9 Mountainous channel with step-pool cascade (Polovinka River, Kamchatka, photo by S. R. Chalov). Photo of the author

(according to SHI) factor (Kopaliani and Tskhadadze 1972). This, however, is related to lateral channel changes; the motion of the load is mainly assessed as un-rippled, spreading down to channels with slopes <2‰, i.e., semi-mountainous; the antidunes on mountain rivers are not considered to be a ripple form of the bottom topography. Sometimes, for the characteristics of mountain rivers, similar in appearance analogues of flat rivers are selected, additional definitions of the type of channel are allowed: for example, "mountain channel multibranch", etc. (Prokacheva et al. 1982).

The fluvial processes on semi-mountain rivers develop under condition of change in time of connections h h $\gtrless$ h_{cr}, I h $\gtrless$ I_{cr}. Therefore, semi-mountain rivers are characterized by a variable character of the current during the change of hydrological regime phases. In the lower reaches of the river bed, the flow is calm; in the shallow waters of the river bed, the flow becomes violent. In forests where the depth exceeds the height of flood waves, the average flow rate is less than that of a single wave and the current remains calm. This distinguishes the riverbeds of semi-mountain rivers from those of mountain rivers with developed alluvial forms, where the flow at the crossroads has a rough flow in the lowlands in all phases of the regime, and in the case of floods it becomes rough along the whole channel, regardless of its relief, as well as from plain rivers with a pebbly-boulder channel. The latter, due to their high depths, have a quiet flow regime despite relatively high velocities; in the low-water, at the riffles folded by boulders, which form indelible shallow waters with shallow depths, the longitudinal slopes increase sharply, and short stretches of stream flow appear on the river, which do not cover the entire width of the channel and are usually confined to the coastal zones. Therefore, the mechanism of the formation of the bed of semi-mountain rivers and its morphology bear the features of the beds of plains and mountain rivers, depending on the conditions in which these rivers flow—in the mountain valleys or in the foothills and intermountain basins. Correspondingly, asymmetrical ripple shapes with a steep lower slope (basement) can be found in the relief of semi-mountainous channels, which are identical to the ripples on flat rivers, but differ from them in their steepness, and antidunes corresponding to the riffles of mountain rivers. The lower the slope, the more similar the riverbeds of the semi-mountain plain rivers are. As the slope of semi-mountain rivers increases, their beds are increasingly similar to those of mountain rivers with developed alluvial forms. This resemblance is especially evident when the rivers become semi-mountainous as far back as in the mountainous area, where the rocky banks of deep cut river valleys cause a tightening of the flow, especially during floods, and the particle size of bed material load is increased in comparison with those in the foothills. In the foothills, where the bottom of the valleys sharply expands, the channels of semi-mountain rivers become more morphologically similar to the channels of lowland rivers. However, the smaller particle size of the load determines the presence of both large ripples of antidune form and ripples of smaller sizes (meso- and microforms), but predominantly asymmetrical form, on their surface.

Characteristically, on rivers with a rough flow, due to the large variability in the size of bed material pebbles and especially boulder-clump deposits, the flow rates averaged over the live section are less than the critical ones (Table 4.7). The trend towards a decrease in the number of Fruit trees is observed on the mountain rivers with the highest slopes, the largest and most heterogeneous bed material load and the empty waterfall channel. This is due to a number of features of mountain river beds. First of all, the ripples of roughness—clumps, boulders—contribute to the formation of stagnant zones at the bottom, which are most developed in the beds of the step-pool cascade type. Second, the coarser the bed material and the greater its heterogeneity, the greater the speed ripple and the greater the likelihood of hydraulic jumps and standing waves. Third, in the coastal part of the stream in the lowland and above the flooded boulder shallow, where the depths in the flood are smallest, the resistance to water flow provided by boulders and large pebbles increases, and therefore the peripheral parts of the mountain stream almost always have a quiet flow. The influence of channel roughness increases in the low-water when the Froude number is the lowest.

This was taken into account by Kroshkin (1970) when developing the classification of mountain river beds (Table 4.1) and is reflected in Table 4.6, in which the Froude number for the "mountain" and "mountainous-foothill" river sections is mainly less than 1 and only for the "highland" sections exceeds 1.2 with water discharge of 5–15% probability. Similar results were obtained by Kopaliani and Tskhadadze (1972), who calculated the Froude numbers for the rivers of the Western Transcaucasia at water discharge corresponding to the average maximum

Table 4.7 Values of the Froude number on some mountain rivers of the Western Transcaucasia with different types of channel (according to measurements of 1963–1965)

Channel type	River-station	Froude number (Chalov 1979)	
		During low-water	During flood
Semi-mountainous	Bzyb–Jirkhwa	0.02	0.15
	Musserka–10 km	0.00006	0.42
	Besleta–Sukhumi	0.01	0.21
	Majara–Machara	0.02	0.42
	Duab–Jali	0.05	0.17
	Okumi–Gudawa	0.011	0.42
Mountainous			
(a) With developed alluvial forms	Hipsta–the highway bridgc	0.05	1.05
	Gumista–Achadara	0.12	0.67
(b) With undeveloped forms	Kodori–Ganakhleba	0.08	0.59
	East Gumista–Sukhumi	0.20	0.71
	Chkhalta–Chkhalta	0.31	1.45
	Galidzga–Tavarcheli	0.10	0.82

[a]This is reprinted data from (Chalov 1979)

flood; according to their data, the value of the Froude number even in the flow zone is less than 1. Thus, at a slope of 275.2‰ (step-pool cascade bed) on the Geg River, $Fr_{av} = 0.53$, $Fr_{max} = 0.82$. Only on the Adzharis-Tskali river in the channel section with developed alluvial forms ($I = 5.9$‰) at $Fr_{av} = 0.84$ the value of $Fr_{max} = 1.50$, and in the channel section with undeveloped alluvial forms, where $I = 22.2$‰, $Fr_{av} = 1.12$, $Fr_{max} = 1.30$.

The results of the calculation of the Froude number in the lower reaches of the Zhoekwara River within the limits of the Gagra city, where the channel is transformed into a concrete channel, are indicative, there are no coarse sediment fractions, because they are accumulated by the upstream mud storage, and the slopes correspond to the semi-mountainous channels. Since there are no factors reducing the value of the Froude number, its fluctuations fully reflect the flow rate regime for semi-mountain rivers: in the low low-water period it is only 0.14–0.18, during floods—always more than 1. A special place in the classification of fluvial processes is occupied by mudflows of mountain rivers and rocky channels (Table 4.6). Mudflows are the channels through which mudflows pass, and the periodicity of mudflows is such that in the period between mudflows the channel flow does not have time to process mudflows and form a normal channel. On large mountain rivers (F >1000 km^2) mudflows do not develop, and they can only be receiving mudflows from tributaries, the outflows from which often determine the formation of channel sediments, change the shape of the longitudinal profile, the direction and intensity of channel changes, and sometimes change the type of fluvial processes. On rivers with a catchment area of 100–1000 km^2 mudflows occur only in areas corresponding to the slopes of step-pool cascade channels; on the smallest rivers (<10 km^2) mudflows extend beyond the mountains to the foothills, causing the replacement of semi-altitude channels with mudflows (Western Tien Shan, Zailiyskiy Alatau) (Borsuk et al. 1981; Kuznetsov and Chalov 1988). Intermediate position is occupied by rivers with catchment area of 10–100 km^2, on which mudflow activity affects the formation of all types of mountain channels.

Rocky channels are found on both mountainous and flat rivers in the entire range of slopes. They are formed at the river sections cut into rocky and plastic (semi-solid) rocks. Constriction of the bed by rocky banks and the bottom causes, in accordance with the underdevelopment equation, a sharp increase in speed, which ensures that the pebble-boulder (or sand) material coming to the areas with such a bed is taken out of their boundaries. In the case of high gradients in mountain rivers, gravity transport of sediment along steeply sloping rocky "trays" plays a major role.

Mountain rivers are never sandy in terms of the composition of bed material sediments, as the flow rate in them is several times higher than the incipient V_{in} for sand and even fine pebbles. According to Ts.E. Mirtskhulava (1988), at a flow depth of 1 m V_{in} for the largest gravel sands ($d_{av} = 0.9$–1.3 mm) varies within the range of 0.70–0.89 m/s depending on the degree of their sorting, whereas the actual velocities at the same depth, even at the minimum slopes typical of mountain rivers, are always greater than 1 m/s, reaching 3.7 m/s (Talmaza and Kroshkin 1968) and even 4.8 m/s (Kopaliani et al. 1992). Small boulders ($d = 150$ mm) begin to move at speeds greater than 1.60–1.98 m/s with a flow depth of 0.5 m and 2.14–2.52 m/s with a depth of 1.0 m. Smaller sediment fractions are suspended and are not bed material; they colmulate pebble-value deposits in the formation of shallows in the peripheral parts of the channel, where flow rates are relatively low.

Unlike mountain rivers, the plains can be either sandy or pebbly-boulder, depending on the conditions of their bed's formation. Large rivers flowing within the limits of orographically complex and occupying vast territories of mountainous countries (Altai, Sayans, mountains of Eastern Siberia and Central Asia), with high runoff rates (the area of basins more than 10,000 km^2) have slopes typical for semi-mountainous (Katun, Biya in Altai) and plain rivers (less than 0.30–0.50‰). For example, the upper Lena in the Baikal region mountains in the Kachug—Ust-Kut—Kirenga estuary section has slopes of 0.32–0.11‰ (Kachug F = 17.4 thousand km^2; below the Kirenga River estuary F = 140 thousand km^2; the average annual water discharge is 87.6 m^3/s for the Kachug River, 295 m^3/sec for the Ust-Kut and 950 m^3/sec for the section below Kirenga River estuary). These rivers, being flat in slope, kinetic flow and type of fluvial processes, have a pebble and pebble-boulder channel. These are not only the upper, but also the middle Lena, its large tributaries (Kirenga, Vitim, Olekma, Aldan), Yenisei, flowing along the border of the Central Siberian Plateau, Tom within the Kuzbass, the upper Amur. This is due to the direct inflow of coarse debris from mountain slopes into river channels and the removal of pebbly-boulder sediment from mountain tributaries. The pebble-boulder composition of bed material load on flat rivers in mountainous regions creates an external resemblance to mountain rivers, but they differ from them in terms of the specifics of bed formation and transport of sediments.

Pebble-boulder beds of lowland rivers are formed under conditions of limited development of channel changes predetermined by the location of river valleys or river crossings of uplands composed of bedrock, boulder-rich moraine and certain mountain ranges. This is the Lena riverbed in the lower reaches, where the river between the Central Yakutsk sandy lowland and the delta crosses the mountain massifs of the Kharaulakh mountain range of Cherkanovsky, Yana at the intersection of the Kularsky ridge, the middle and lower Amur—respectively, the Maliy Khingan and the Sikhote Alin spurs, etc. The pebble bed is typical for the upper Volga. The Neman, which has a sandy channel in the upper reaches, on average, crossing the Grodno moraine elevation, becomes a pebble-boulder.

Thus, the identification of flat, semi-altitude and mountain rivers corresponding to certain types of fluvial processes cannot be mixed with the definitions of rivers by their orographic position, i.e., with the concepts of "rivers in the mountains", "rivers in the foothills" and "rivers on the plains". As each term carries a certain conventionality in its definition, so each classification is to a certain extent conditional. N. I. Makkaveev wrote: "It is almost impossible to find any significant flat river, which would not have during its flow of "mountain" sections … many rivers, flat in the upper reaches, have in the middle or lower reaches, where the flow becomes stormy" (1955, pp. 79–80). Indeed, in relatively short river sections, the geological and geomorphological factor creates local conditions for a sharp increase in the longitudinal slope and the appearance of turbulent flows. These are the channels in the upper reaches of the rivers originating on the uplands (Central Russian,

Volyn-Podolsk, Volga, Northern Ridge, etc.). Here the slopes reach such values that during the flood (at effective discharge rate) they reach critical values and the flow becomes rough. Even such an extreme form of fluvial processes development on mountain rivers as mudflows can develop, under certain conditions (high depth of erosion), within the limits of the plains. For example, they occur in the channels of small rivers that divide the slopes of the middle Dniester valley, which is cut into 100–200 m in the Volyn-Podolsk Upland (Slastikhin 1987), in the streams that are the continuation of ravines that open into the valleys of the elevations of the European Part of Russia center, etc.

While the diversity of mountain river channels is determined by the wide range of changes in longitudinal slopes and their high value, which determine the diversity of sediment transport and morphological representations through their kinetic flow, the calm, flat rivers, which form the most common class of river channels, differ even more, but because of enormous differences in their runoff rates, water regime, sediment load size and composition: from streams and rivers of the 1st order with water discharge in the first cubic meters per second to the Amazon with an average annual water discharge of more than 170,000 m^3/s, from rivers with extremely low sediment concentration (≤ 10 g/m^3) to >1800 kg/m^3 during floods (Weihe River, a tributary of the Huang He River), from rivers with sandy mudflats to rivers with a pebbly-boulder composition of bed material load. They are united by their rippled form of sediment transport and calm flow conditions (kinetic flow conditions).

A typical form of riverbed relief in lowland rivers is a range of ripples of different sizes, with gentle upstream and downstream steep slopes. Moving downstream under the influence of the flow, they constitute the main form of sediment transport and form its runoff. Large ripples under changing water levels create the preconditions for the development of various forms of channel, as each of them, under certain conditions, can be mothballed and cause the transition of their component sediments into river sediments. The pebble-boulder composition of river load from lowland rivers results in a reduction in the number of ripple forms, up to and including the development of only macroforms that are commensurate with the width and depth of the river. In the case of silt and sand composition of bed material, the flow rates can be significantly higher than those of incipient sediments ($V > 2.5\ V_{in}$), in which the transport of sediments is carried out in a smooth phase.

Changes in the slope of the longitudinal profile along the length of the river cause appropriate changes in the conditions for the formation of riverbeds. This is especially evident on rivers that originate in the mountains and then enter the plain and have a concave longitudinal profile. Consecutive changes in the longitudinal profile of such mountainous-piedmont level rivers are accompanied by the change of one type of mountain channel to another, then to a semi-mountainous one and, finally, to a flat one. With the normal concave shape of the longitudinal profile, this determines the natural sequence in their distribution along the river length from foothills to highlands, which creates a vertical zonality in the spatial distribution of different types of fluvial processes. For example, on the territory of the Western Caucasus (Fig. 4.10), on all rivers flowing from the Greater Caucasus Range, the

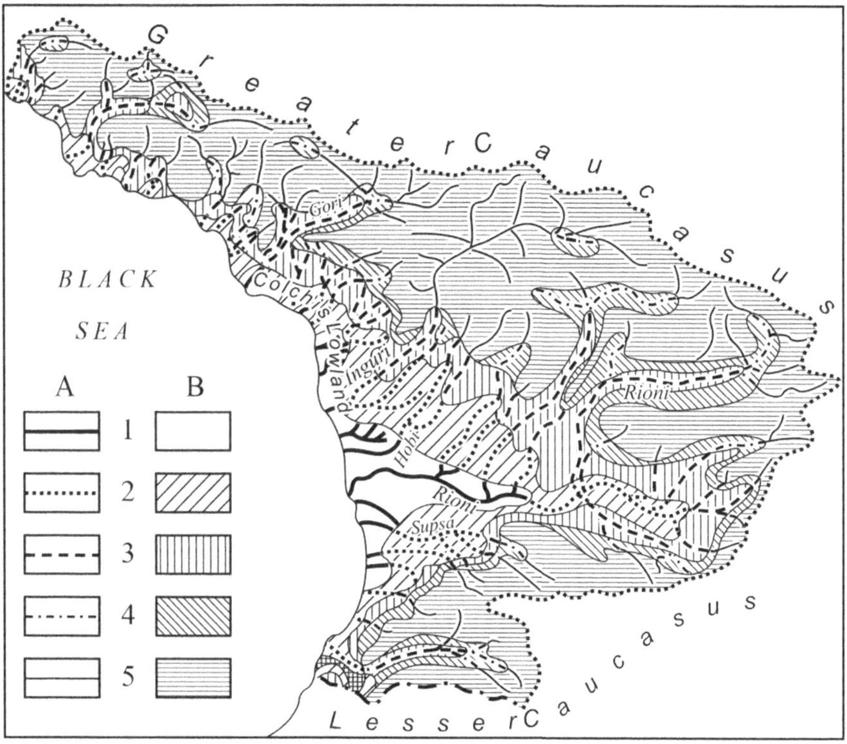

Fig. 4.10 Distribution of different types of fluvial processes on mountainous rivers of Western Caucasus. Channel types (**A**) and areas of its distribution (**B**): 1—lowland; 2—semi-mountainous; 3—mountainous with developed alluvial forms; 4—mountainous with undeveloped alluvial forms; 5—step-pool cascade

leading ripples of the Greater Caucasus, as well as from the ranges of the Lesser Caucasus, downstream successive change of step-pool cascade by mountain channels with undeveloped, and then with developed alluvial forms, semi-mountainous and, within the Colchis, flat beds takes place. The high and middle mountainous regions of the Greater and Lesser Caucasus are characterized by the development of step-pool cascade, which form the most significant belt in terms of area of distribution. Large rivers (Bzyb, Gumista, Kodori, Rioni, Kvirila, Adzharistskali) separate this belt, because along them and in the lower reaches of tributaries other types of channels corresponding to smaller slopes are developed. On these rivers, step-pool cascade are typical only of the upper reaches and their length is insignificant from the entire length of the rivers. As a result, the vertical zonality in the distribution of the types of mountain channels is represented both in general, from the watershed line of the main Caucasian range to the confluence of rivers into the Black Sea, and along the deep-cut valleys of large rivers, which form the local basis of erosion for their tributaries.

A similar picture is found on large rivers flowing through mountainous areas. Due to the concave longitudinal profiles, the deflection of the arrow to the upper reaches and the high runoff rates of the rivers within a large part of the upper, middle and lower reaches, they have slopes corresponding to the plain rivers and are, by the type of fluvial processes, such as in the mountains (the upper and middle Lena, its large right-bank tributaries—the Kirenga, Vitim, Aldan; the upper Amur and its components—the Shilka and Argun, etc.). In the intermountain basins and valley extensions, especially in the place of former lakes filled with powerful alluvium layers, where the local flattening of the longitudinal profile takes place, the step-pool cascade channel is replaced by both types of mountain alluvial channel, and then the channel with developed alluvial forms downstream passes into the channel with undeveloped forms, and then again into the step-pool cascade channel. This corresponds to the convex shape of the longitudinal profile below the troughs.

Repeated changes in channel types downstream can also be associated with lagging behind the process of formation of the developed longitudinal profile from the tectonic development of individual parts of the mountainous country. These are the rivers of the upper Amu Darya basin, which have a steep bend in the longitudinal profile on the border of the Eastern and Western Pamirs. In the Eastern Pamirs, the river has large slopes and waterfalls only in its upper reaches. Further downstream, they flow in low cut valleys, have small slopes and, mainly, beds with developed alluvial forms. When crossing the Western Pamirs, these rivers form deep gorges, their slopes increase tens of times, and the channels again become step-pool cascade.

On the rivers, from the source to the estuary, flowing through the plains, the concave longitudinal profile is not so pronounced; however, in case of the location of sources on the upstream slopes (1–4 order of watercourse), the formation of mountain-type channels, up to step-pool cascade ones, is conditioned here. Even within the limits of elevations such channels quickly become flat and are such throughout their entire length, if the smoothness of changes in the longitudinal profile is not violated by steps on the protrusions, folded by rocks (steps on the Dnieper, Yenisei). Constant alternation of mountainous, semi-mountainous and flat areas is typical for Onega and rivers of Karelia, which have a stepped longitudinal profile.

Even on such a typical flat river as the Neman in the upper reaches, where it flows first along the Stowbtsy plain and the Upper Neman lowland, down the river there is first a rapid and then a gradual decrease in slopes. Below the estuary of the Shchary River, at the intersection of the Grodno Upland and the Baltic Ridge, the slope increases, but below the Kaunas city it decreases again to the minimum values. Changes in the slopes are accompanied by a change in the wide floodplain channel in the upper reaches, mainly incised and confined in the middle and again wide floodplain in the lower reaches.

The consequence of the concave longitudinal profile is the predominance of large slopes in the upper reaches of mountain rivers and small ones in most of the middle and lower reaches. The first correspond to the step-pool cascade channels, the last to the mountain channels with developed alluvial forms and semi-mountainous ones. Between them, in a relatively short section, where the deflection of the longitudinal profile and the depth of the river incision reach the highest values, a mountain channel with undeveloped alluvial forms is formed. Thus, the relationship of the types of mountain channels on the value of longitudinal slope in the concave shape of the longitudinal profile determines a certain correspondence between their distribution of intensity of both the incision of the river itself and the entire complex of erosion and denudation processes within the basin. The latter is reflected depending on the value of specific effective water discharge on mountain rivers from the factors that determine the intensity of denudation processes in the catchment area: the absolute height of the mountains and slope steepness (Table 4.8). In addition, downstream of each river, i.e. with reduced slopes, sediment load size and flow kinetics, the specific formative flow rate decreases, i.e. as the river's "mountainousness" decreases, the most significant changes of the channel are carried out with less flow energy.

Fluvial Processes on Mountain and Semi-Mountain Rivers

Differences in the hydraulic flow regime of mountain rivers and, consequently, in the forms of sediment transport and the mechanisms of interaction between the flows and the river channels generally determine different conditions for the formation of the channels themselves, their morphology and dynamics. The geological-geomorphological factor is superimposed on it: at any type of fluvial processes of mountain rivers can be incised, confined or wide floodplain (it corresponds to conditions of the limited or free development of channel changes depending on a geological-geomorphological structure of territory), to be characterized by this or that morphology (a bend, branching, rather rectilinear channel and

Table 4.8 Relationship of specific effective discharges ($Q_{sp.f}$) in the rivers of the Western Caucasus on the intensity of denudation processes

A group of rivers	Rivers	Signs of denudation intensity	$Q_{sp.f}$ for each river	$Q_{sp.f}$ average for river groups
A	Inguri	Highland basin, the dominance steep slopes	6.4	5.0
	Chorokhi		3.4	
	Bzyb		5.2	
	Kodori		5.0	
B	Kintrishi	Mid-mountain basin, steep slopes predominance	2.9	1.9
	Gumista		1.6	
	Chakvis-Tskali		1.3	
C	Rioni	High-mountain basin, medium steep slopes	1.4	1.2
	Supsa		1.1	
D	Nathaneby	Low-mountain basin, gentle slopes	0.9	0.9

their variants) and features of reformations (lateral changes). Between mountainous rivers are predominant among the incised ones, although as the gradients decrease and the valley expands, as well as in the intermountain depressions and subsequent parts of the riverbeds they are wide floodplains. The latter dominate the exit of rivers into the foothills, but their intersection of different morphologically different sections here leads to the presence of incised and confined varieties of channels.

At the same time, the formation of a channel with a particular morphology occurs with a certain combination of hydraulic flow characteristics, particle size and sediment composition. For example, the formation of bends requires a combination of water discharge and slope (Makkaveev 1955). At small values of this ratio, which is typical for step-pool cascade and mountain channels with undeveloped alluvial forms, typical bend shapes are not formed. Makkaveev (1955, p. 279) explained this by the fact that in the turbulent flows of mountain rivers, "where the length of the curves of support sharply decreases and the formation of … transverse slope (necessary for the development of circulation) becomes impossible," "the formation of curvature is made impossible. Ruffles are usually straight or, in the wide floodplain channel, branched, and each branch can be step-pool cascade or have a channel with undeveloped alluvial forms. In the latter case, the channel is characterised by stable depth and unstructured (un-riddled) transport of sediment, the separating branches of the banks (drying out in the lowlands) are also unstructured accumulations of sediment formed between the flow rods in the deceleration zones. The rays of such channels are connected with structural and lithological conditions.

For mountainous channels with developed alluvial forms, as the river's runoff rates (annual average water discharge) increases downstream and gradients decrease, the likelihood of bend development increases. Gentle bends occur already at large for this type of slopes, and by the exit from the mountains to the foothills are usually formed a series of well-developed segmental bends. Only in cases when the river valley is relatively extended (crossing the zone of spreading of easily destructible rocks, tectonic depression, longitudinal depression between ripples, etc.), and the size of the bed material load sharply decreases, the river channel loses its stability and branches out on the branches. The same thing happens when a river comes out into the foothills, folded by loose or easily eroded rocks.

Incised mountain river beds are often rocky; the stream here looks like a rocky tray, the width of which is several times smaller than the adjacent alluvial areas (especially the wide floodplains). This leads to the transformation of the channel flow into a rapid flow (strawberry), the speed of which is much higher than that of the incipient for the largest pebbles and boulders, so that pebbles, boulders and even blocks are carried by the flow outside such areas.

These rivers, when entering troughs, valley extensions or foothills, tend to dramatically change the channel pattern. This is facilitated not only by the melting of the flow and the reduction in transport capacity, but also by the change in the form of sediment transport: if pebbles and even small boulders are transported in the suspended state in incised and rocky beds, then in the expansion of these fractions and even smaller pebbles and sand particles are transported only as conveying particles. The channels of these rivers branch into branches and are unstable. Floods usually change the position and number of ducts and branches, erode some channel shapes and form others. The floodplain does not have time to form, and the whole valley bottom in the lowlands is a wide and flat pebbly or pebbly-boulder field, among which are relatively small ducts. Morphological type—axerodic branching, and axerodic depending on the slope (and the type of mountainous channel) are non-structural sediment deposition (channel with undeveloped alluvial forms), ripples of the antidune (channel with developed alluvial forms) or a combination of antidunes and normal asymmetric ripples (semi-mountain channel). Accordingly, the low-water channel of each duct is characterized by depth stability along its length (undeveloped alluvial forms) or by the presence of riffles and pools (developed alluvial forms; semi-mountainous channel). Such pebble and boulder fields are higher than any large inflections of longitudinal profiles of mountain rivers, including those associated with mudflow removal cones from tributaries.

The morphological appearance of *step-pool cascade cannel* in the upper reaches of rivers is determined by a pronounced stepped shape of the longitudinal profile. Steps are formed on accumulations of boulder-boulder material (Fig. 4.9) and on the exits of relatively more stable rocks. Waterfalls prevail at the highest gradients, often exceeding 100‰. The steepness of the slope of a waterfall with a rocky base depends on the nature of rock formation (Fig. 4.11). Waterfalls are separated by deep moulds, which are formed above the fracture of the longitudinal profile in the zone of decline of levels and acceleration of the current, and erosion boilers below the waterfalls, where hydraulic jumps of wave type occur in the streams. Disorderly clusters of poorly sorted coarse pebbles and boulders are found beyond the banks, below the erosion boilers and in the moulds, without forming a continuous coating. The valley line of the channel here is usually incised in rocky bedrock and looks like a tray. Downstream, with relatively low slopes for this type of channel, the longitudinal profile is related to the presence of successive steps and micro waterfalls formed by continuous clusters of coarse chip material. Each step or microwaterfall is subject to the same hydraulic phenomena as those occurring on hard rock outcrops. The height of such microwaterfalls, depending on the size of the debris material composing the bed, ranges from 1–1.5 m to 10–20 cm. Such step-pool cascade is alluvial and microwaterfalls are formed due to the longitudinal sorting of pebble-boulder or boulder-clump (depending on the slope) material (Fig. 4.12) due to hydraulic phenomena in the rough flow (alternation of hydraulic jumps and hydraulic waterfalls).

The high longitudinal slope of the longitudinal profile on rivers with a waterfall and step channel determines a number of important specific features of the sediment transport mechanism. The tangential component of gravity affecting individual boulders or boulders at a steep slope of the bottom leads to a sharp increase in the probability of movement of the largest fragments with the least effort to bring them to a state of motion (Makkaveev 1973). The equation of critical velocity of erosion V_{cr} takes the form

Fig. 4.11 Small waterfall on mountainous river, connected with rocky formations outcrop (Kavavlya river, Kamchatka, photo by A.S. Chalova). Photo of the author

$$V_{\mathrm{cr}} = \kappa\sqrt{gd\frac{\rho_{\mathrm{s}} - \rho}{\rho}(k_{\mathrm{fr}}\cos\beta - \sin\beta)} + k_{\mathrm{c}}, \tag{4.21}$$

where ρ_{ns} and ρ—the density of sediment and water; g—gravity acceleration; d—diameter of blocks, boulders, pebbles; β—steepness of the bed of the channel in degrees; k_{f}—friction coefficient; k_c—coefficient depending on the force to be applied to tear off the chip from the bottom, if it is wedged between adjacent debris; k—coefficient depending on the magnitude of the slope of the channel. In the conditions of severe floods typical of mountain rivers, the movement of individual debris is facilitated by the effect of the wave impact on the obstacles that form blocks and boulders. Therefore, in many cases, the coefficient $k_c = 0$, and the value of V_{cr}, all other things being equal, is less than a slow increase in levels. As a result, even low-water mountain streams with high slopes and wave patterns can move very large debris. Stationary observations of the movement of large boulders on the rivers of the Western Caucasus with a step-pool cascade channel (Khmeleva et al. 2000) allowed us to establish that the rate of movement of debris up to 1.0 m in

Fig. 4.12 Riverhead if the Beliy Irkut river (Easter Sayany). Photo by N. L. Frolova

diameter at a maximum flow depth of 1.5 m/s is 0.2 m/s. After one heavy rainfall flood, a block of 0.8–1.0 m in diameter was carried out of the 300 m long section by a stream; larger 1.5–2.5 m of debris shifted to a shorter distance (the stream depth here reached 3.0 m).

For the transport of large fragments are important processes of washing out the fine material from under their base, which is facilitated by a lot of microwaterfalls that occur in the boulder-clump and pebble-boulder step-pool cascade channel and cause its local washout with the formation of small erosion boilers. This eventually leads to a loss of equilibrium and its movement downstream under the influence of gravitational forces. This phenomenon, called the "abluvial effect" (Experimental Geomorphology 1969), is one of the main reasons for the absence of a connection between the slope and the size of channel load on step-pool cascade rivers. For the first time, the abluvial effect was found on the eroded slopes of mountain valleys and in temporary waterways. For them, the distance x by which the debris moves depends on the thickness of the erosion layer h_{er} and slope steepness β (Makkaveev and Kalinin 1968; Makkaveev 1973).

$$x = h_{\text{er}}(9 \sin \beta + k_{\text{sl}}), \tag{4.22}$$

where the K_{sl} is a coefficient that depends on the shape of the slope. Formula (4.22) characterizes only the trends of distance change x. In the case of small rivers with a step-pool cascade channel, it can be brought to mind (Makkaveev et al. 1973).

$$x = kh_{\text{ch}} \sin \beta \tag{4.23}$$

The dependencies (4.22), (4.23) were obtained for experimental trays with a large longitudinal slope and depths commensurate with the size of the debris ($h \approx d$).

Mass movement of bed material load on rivers with step-pool cascade occurs only in some years, when floods are particularly severe. Usually, only a small number of fragments of relatively small size are moving. In any case, with the movement of a large piece of debris, a large number of smaller particles accumulated in the support area in front of them begin to move. Such a mass shift of fine material following the shift of a large block leads to a rapid propagation of a wave of regressive erosion in some part of the channel, which in turn favors the emergence of an abluvial effect for the upstream large boulders.

The presence of large boulders in mountain river channels has a turbulent effect on the water flow, which increases with the size of the debris and the degree of heterogeneity of the bed material. This also corresponds to the increase in the kineticity of the flow during the flood. The Froude numbers for rivers with step-pool cascade during floods reach 4–5. On the contrary, the roughness of the channel is so great that even at high gradients, the stream, divided into separate jets, covering numerous blocks and boulders, keeps a calm flow ($Fr \ll 1$). Both factors (additional turbulence and increased kineticity) contribute to an increase in the transport capacity of the flow on rivers with a step-pool cascade channel as compared to the flows flowing in alluvial channels. As a result, the volumes of sediment load on mountain rivers with the same runoff rates but with different types of channel differ by an order of magnitude (Table 4.9).

Depending on the geomorphological conditions, the upper reaches of rivers with step-pool cascade (mountain streams) are very diverse. On the slopes of glaciers, the channels of such streams are slightly cut into the slopes, representing steeply falling trays consisting of a continuous system of waterfalls and devoid of any alluvial deposits. In the well-designed valleys filled with slope material of ancient valleys with flat, steeply sloping, usually densely overgrown bottoms, the streams passively adapt their bed to the lithogenic base and vegetation cover, forming an erosionary shallow hollow with numerous microwaterfalls, microsteps and numerous bends.

In the zone of accumulation of fluvioglacial material below the glaciers ("Sandra") the bottom of mountain river valleys expands, becomes flat, steeply sloping, the river is divided into numerous channels with a step-pool cascade channel. At some distance from the glacier, the flow in the lowland is sometimes completely filtered into its own deposits and the channel dries up. The position of the duct changes after almost every flood.

Table 4.9 Sediment load from small rivers in the Caucasus with different types of channel

Type of channel	River	Sediment yield (t/year) (Mandych 1967)	
		Overall	With $d > 2$ cm
Step-pool cascade	Holodnaya river	3610	1530
	Cigiripsh	3610	1724
	Tsihreva	4500	1950
Alluvial	Azquara	890	309
	Psyrtkha	635	0
	Pshap II	1160	434

Mountain channels with undeveloped alluvial forms are characterized by the continuous attraction of bed material load over the entire living flow cross-section and are characterized by a trough-shaped cross-section and a shallow relative depth. Some clusters of boulders, which do not form ripple shapes, create only gently sloping platforms, timed to coincide with the peripheral zones of the riverbed at the outlet banks, below capes and large boulders that have fallen from the slopes. Mountain channels of this type usually occupy almost all the bottom of the valley, only rarely having a narrow floodplain. Only in mudflow unloading areas and valley extensions are the channels divided or multiple channels separated, including those between mudflows. In this case, each channel retains the morphological shape of the channel with undeveloped alluvial forms and is separated from the neighboring large longitudinal ramparts, which are accumulative mudflows. In the valley extensions between the branches there are vast, poorly vegetated pebbly-boulder fields formed during the river migration along the valley bottom. In a straight channel, the dried up pebble-boulder banks stretch along the banks of the riverbed, breaking off to it in small steps formed as the flood subsides. The flow itself is concentrated in the axial part of the trough channel, where the flow velocities at high gradients during the initial period of flood recession are still sufficient to move the boulder material. Therefore, in the coastal zone of the channel with the lowest depths, the sediment flow stops, while in the bedrock zone it is still preserved, providing some channel deepening relative to the banks (Fig. 4.13). Separate moulds emerge only when the current is piled to rocky banks.

In mountain channels with undeveloped alluvial forms, the abluvial effect is also a factor in the transport of sediment and channel changes. However, its effectiveness is much lower than that of waterfalls, as the microsteps associated with large boulders and boulders are relatively small in height. The most important factor in channel development is the comparability of individual boulder sizes and flow depths. Therefore, along with the continuing role of floods in sediment transport, the impact on individual boulders of not only high-speed but also hydrostatic pressure is of paramount importance. This is facilitated by the shape of the water surface near the projections of roughness—the rise of the level from the upper and

Fig. 4.13 Mountainous channel with undeveloped alluvial forms among pebble-boulder shoals (Bilshoy Azau river, tributary of Baksan river, Northern Caucasus). Photo by N. L. Frolova

the depression from the lower side (Makkaveev 1973). Under these conditions, the critical blur rate

$$V_{cr} = k\left(g\frac{\rho_s - \rho}{\rho}\right)^{1/2} d^{3/2}h^{-1} + k_c, \tag{4.24}$$

and weights P of the debris transported by the flow (Makkaveev et al. 1973)

$$P = (30 + 640I_b) \cdot 10^{-4}\rho V^2 h^2, \tag{4.25}$$

where I_b is the slope of the bottom; V is the current velocity, cm/sec; h is the depth of flow, cm; ρ is the density of water saturated with suspended sediments. Formulas (4.24), (4.25) were obtained by N. I. Makkaveev and his staff on the basis of processing of experimental research materials in laboratory trays using real small boulder-large pebble alluvium extracted from the bed of a mountain river in the Caucasus.

Thus, in mountainous beds with undeveloped alluvial forms, the movement of sediment and channel changes are carried out as a result of the movement of individual boulders and clumps under the influence of the full use of the specific energy of the flow section. The consequence is that the arrangement of large boulders and boulders in the channel is chaotic, and that they are transported a considerable distance from the point of entry into the channel. This is easily determined by the petrographic composition of boulders and clumps, which does not correspond to the prevailing complex of rocks forming the slopes of river valleys or their distance from the slopes to the channel. Active movement of the coarse-clastic material flow is observed most vividly during the passage of "sharp" floods (Mandych 1966), characterized by a rapid and significant rise in levels in a short time. The movement of each large piece of debris entails the involvement in the movement of the mass of small particles accumulated between the floods near the debris. As a result, severe floods usually have very high sediment concentration values, one or two orders of magnitude higher than the slow rise of levels (Table 4.10).

In addition to the hydrostatic pressure, the shear of large debris is also facilitated by the impact of smaller particles transported by the flow. The increase in sediment concentration and the increase in suspended solids content in the flow leads to an increase in water density and, consequently, a decrease in critical velocities (4.24) for fragments lying on the bottom (pebbles, boulders) (Makkaveev et al. 1970).

The most typical form of the relief of *mountain channels with developed alluvial forms* is the accumulation of antidune ripples, which are characterized by a continuous flow of their flow and the coincidence of the curves of the free surface of water in the flood with the profile of the ripple (Znamenskaya 1968). The channel itself forms the following incised segmental bends. Sediment accumulation is located at the bends of the stream, and the crests of the ripples are almost parallel to the concave bankline, which often forms a rocky scarp, near which a deep, calm pool forms. During a flood, the rapid flow is here, on a rocky cliff, in the conditions of support, thanks to which the base of the rapid decrease in speeds occurs and there is a zone of stable accumulation of sediment forming a riffle. Riffles are characterized by compression of the flow between the side streams, which ensures constant flow rate. Just below the roll in a deep pool valley, the flow is calm. The channel is expanding.

The constancy of the sediment accumulation zones determines the stable position of the riffle in the channel. Sediment transport is carried out by transferring individual particles from one ripple to another. Riffles in this case serve as a kind of "transshipment points" on the upper slopes of which the particles brought from above are deposited, and at the same time there is erosion of deposits and removal of particles from the bottom slope downstream. These changes occur only when the

Table 4.10 Influence of flood sharpness on sediment concentration of mountain river flow (r. Adzharis-Tskali, v. Makhinjauri, Minor Caucasus) (Mandych 1966)

Year	Date (month)	Water level (cm)	Flood sharpness	The sediment concentration of the water (g/m^3)
1952	31.03	111	0.28	140
	09.09	112	12.0	685
1955	18.03	109	0.60	8.9
	24.08	110	1.00	373

maximum water discharge, which is usually effective. At the flood recession and in the low flow rate, the transport capacity is insufficient for particle transport, channel change and channel relief forms stop, and the flow is almost completely clarified in the low flow rate.

Since the size of the ripples depends on the particle size of the load, only one generation of ripples develops on mountain rivers—commensurate with the width and depth of the channel. During the flood recession and in the low water period, when the movement of sediment stops, the transformation of the ripple in relation to the new phase of the regime does not occur. Side bars only sorting of pebbles and boulders consists in alternation of strips folded by larger boulders and smaller boulders and pebbles. When encountered on rivers with relatively low gradients for this type of channel, this banding seems to be a reflection of the structures of turbulent flow, leading to the formation of smaller ripples with a smaller alluvium composition of deposits.

The transformation of mountain channels with developed alluvial forms depends on the morphology of river valleys. In the gorges, where the meanders are developed, the predominant distribution of the meanders is undisturbed, strictly confined to the bends of the riverbed. According to observations on the rivers of the Caucasus—Bzyb for almost 30-year-old (Khmeleva et al. 2000) and Gumista for 15-year period (Vinogradova et al. 1979) the situation has not changed in terms of side effects. However, there are vertical changes on their surface; significant erosion or accumulation of sediment occurs along their outer edge. At the same time, the amount of change is associated with fluctuations in the runoff rates of rivers from year to year: during periods of increased runoff rates, the accumulation of sediment on their surface predominates, while in periods of increased runoff rates it is the accumulation of sediments on their surface that dominates (Khmeleva et al. 2000).

In some cases, mountain rivers have experienced a shift in the downstream movement of the ropes. For example, in Galidzga, Kodori and Bzyb (Western Caucasus), the riffle side-bars are shifting at an average speed of 10 m/year. In the lower reaches of Gumista (Vinogradova et al. 1979), significant changes in the channel forms are expressed in alternating addition to the side banks, their rejection and the emergence of new channels and branches. Riffle shifting on mountain rivers in the intermountain depressions and foothills are more intensive. Here the lateral movement rates reach 20–25 m/year (Galidzga, Duab). The same speeds are marked by Romashin (1967) in the lower reaches of the Psezuapse River.

The transgressive movement of large ripples (side-bars) in channels with developed alluvial forms contradicts the idea of regressive movement of ripples of antidune form, such as riffles, side-bars, and mid-bars on mountain rivers with this type of channel. The presence of a standing wave in the flood above such a ripple should be accompanied by accumulation of sediment on the upper slope of the ripple and erosion of the lower slope. The ripple itself has a symmetrical longitudinal profile with almost equal slopes or a slightly steeper upper slope. Such a movement of antidunes is typical for flows in laboratory trays (Znamenskaya 1968) and at the erosion of artificial bed in sandy soils with a rough flow at the initial high gradients (Makrinova 1953). In the latter case, channel infeed eventually leads to a decrease in slope, the rough flow is replaced by calm, and the antidunes are transformed into ripples with a normal asymmetrical profile.

The reason for the contradiction is not clear, which is explained by the unexplored mechanism of movement of pebble-boulder ripples at a rough state of flow, which is characteristic of mountain rivers with developed alluvial forms. The results of the flow of the coarse chip material are already being recorded, even in the longest and most thorough manner. In addition, these observations are carried out only for the dynamics of surface, drying in the lowland parts of the ripple (Romashin 1967; Khmeleva et al. 2000), which does not allow us to reliably assess the dynamics of the complete ripple profile. Also, there is no clarity on the mechanism of development of stable side-bars. They are essentially places of accumulation of mass pebbles and boulders in the form of a ripple-antidune.

In the foothills and intermountain trenches of mountain rivers, being branched out, in most cases they experience frequent and strong changes. For example, on the Kodori, Psou and some other rivers of the Western Caucasus directly below their exit from the mountains of the island in the channel are located chains, forming conjugate systems in which the activation of one arm in the upper link corresponds to the predominant development of another in the lower and vice versa. The factor determining the activation of one or another hose in the upper link is the movement of the hoses moving sequentially along the opposite banks of the side by side, their rejection from one bank to another and their joining to the other banks. V. V. On the basis of long-term visual and semi-instrumental observations in the lower reaches of the Psezuapse River, Romashin (1967) established the periodicity of the sequential development of one or the other branch in a similar conjugate system lasting 7–10 years.

In order to determine the conditions for the movement of sediment on mountain rivers with developed alluvial forms, formulas for critical velocity of erosion have been applied, the structure of which is identical to the formulas obtained for lowland rivers (Talmaza and Kroshkin 1968). The main feature of many of them is that the depth of flow is taken into account as a factor that makes it difficult for individual sediment particles to move. The values V_{cr} and h are directly related here, as the sizes of pebbles and boulders are often incomparably smaller than the depth of flow. A specific feature of the movement of bed material load on mountain rivers is their increased mobility with increased sediment concentration of the flow (Fig. 4.14). Similar to the transport of coarse-clastic sediment in step-pool cascade, Makkaveev et al. (1970) explained this by the fact that the stormy flow, carrying rather large sediment particles in suspended state, has a great force of dynamic influence on the debris forming the detachment. Destruction of the scaffolding causes additional fine material to flow into the stream, thus increasing the transport capacity in relation to coarse pebbles and boulders.

Semi-Mountainous rivers are mostly found in the foothills, but are less common in the intermountain basins; at the same time, this type of channel is dominant in large rivers in the middle-mountain and low-mountain regions of Eastern Siberia and the Far East. The morphological appearance of the semi-mountainous channels is different in the narrowings and extensions of the valleys. Rivers usually flow in

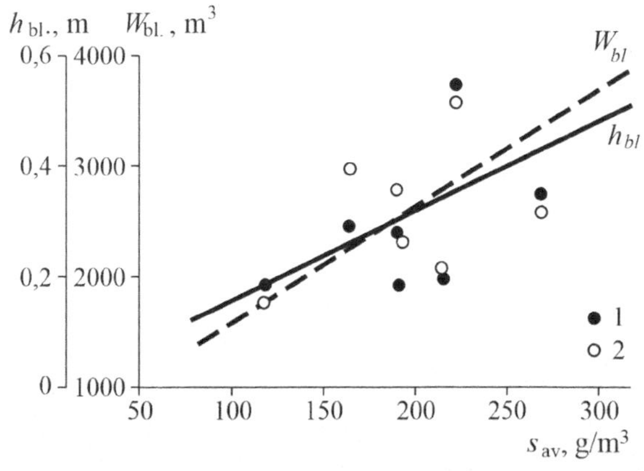

Fig. 4.14 Dependence of erosion volume (W_{er}) and average layer of erosion (h_{er}) on the surface of side-bar of Bzyb river near Girkhva HPP (Western Caucasus) on average flow turbidity (s_{av}): 1—volumes of erosion; 2—layer of erosion (Khmeleva et al. 2000)

constrictions in one slightly meandering channel, where pebbles and boulders are not uncommon, and pebbles are found. Extensions are dominated by multithread channels. Wide floodplain channels are usually characterised by anabranching channel flow rate due to the passage of floodplain bed water flow. Semi-hill river channels in valley extensions are often divided into a number of large branches forming complex floodplain-river branches, the regime of reformation of which is similar to the plain rivers with the same type of channel.

Riffles and sidebars of meandering semi-mountain rivers are characterized in average years by relative stability of their position on the river. In the gorges of the mountainous zone, they often have rocky ripples at their base, which form a kind of flowing ripples. Stationary observations (Vinogradova et al. 1979; Kopaliani and Tskhadadze 1972) have made it possible to establish that such crossings are transshipment points in the process of sediment transport: debris accumulates on their upper slopes, while debris is washed out and carried away at the lower slopes. With the overall stability of the riffle, its ripple shifts up or downstream by only 1–2 m during the flood, depending on the capacity of the blured or washed-out sediment layer. This changes the size of the debris located on the ripple. Sediment transport under these conditions has no ripple character, and channel changes consist of periodic partial erosion of the surface of side-bar. On average, over a long period of time, the flow of sediments in the side-shoots takes place within a certain range of the average thickness of their active layer. Upstream of the river, its size is decreasing. The reason for this is an increase in specific effective discharges (Chalov 1969) as the valley bottom narrows down and bed material load are consolidated in this direction even within a single channel type.

As the forms of the channel relief of semi-mountain rivers lose their connection with the relief of the main rock bed, which is more often observed in the foothill zone, the riffles becomes exclusively accumulative formations, which move in the same way as the riffles of lowland rivers. Their movement speed varies from 10–16 to 20–25 m/year.

For branched riverbeds of semi-mountainous rivers, a distinctive feature of the reshaping is their constant change due to the separation from the banks of riverbeds and the emergence of new channels and branches. Romashin (1967, 1975) established a link between the changes of such channels and the specificity of sediment transport. The various structural levels of sediment transport he has identified correspond to several orders of channel forms. The morphological intensity of these levels, as well as the degree of their mobility, is due to a certain range of water availability. As it grows and passes through some limits, the forms of transport of one level are changed to another, which is accompanied by the destruction of the channel forms of one generation, the emergence and development of others. The periodicity of branch development depends on the movement of side-bars, which periodically move to the source of the branch and cause a redistribution of runoff between them. As a result, the regularities of reformation are the same as in mountain rivers with developed alluvial forms in the conditions of free development of channel changes (wide floodplain channel).

In some cases, the rivers remain semi-hill in nature at considerable distances from the mountains and with sandy bed sediments. In the foothill plains of Central Asia, this is due to the systematic accumulation of sediment at the bottom of river valleys, which leads to increased slopes along the entire length of large transit rivers: for example, in the Syrdarya region within the Fergana Valley—4‰, at the beginning of the plain part of the basin—1.3‰; in the Amudarya region below the confluence of Vakhsh and Pyandz—0.4‰, etc. This determines the rapidity of the river flow into the flood when the bed of the channel is covered with antidune ripples. The stormy current is observed even at the end of the spring-summer flood recession within the riffle ripples.

Thus, the transport of sediment by mountainous and semi-mountainous (except in cases of sand alluvium) rivers is carried out only during floods and rain floods; in the low water period, when water discharge decreases by tens or hundreds of times, the sediment concentration of the water in the mountain river (and in the case of pebble and pebble-boulder alluvium) is close to zero, and the bottom of the riverbed becomes indissoluble, lined with alluvial detachment. Therefore, the development of channel changes only during floods is typical for all mountainous and most of the semi-altitude rivers. Depending on the local conditions, mainly on the peculiarities of the annual flow distribution, rivers in different mountainous areas differ in the number and security of effective discharges. For example, three effective discharge rates Q_f are usually observed on the rivers of the Tien Shan, of which two always occur when the floodplain is flooded within the extended areas of river valleys. Thanks to this, the wide floodplain sections of mountain rivers are characterized by a well-developed anabranching channel system, dissecting the entire valley bottom with long floodplain channels—branches to the complex system of islands. Combining with floodplain and riverbed branches, they create a complex labyrinth of branches, channels and branches at the bottom of the river valley.

Anabranching channel is generally a typical form of mountain channel in valley extensions, intermountain basins and foothills. This is due to the fact that effective floods are extremely rare, catastrophic and sometimes even hundreds of percentage points in number. Floodplain channels formed during the passage of such Q_f are well-defined if there is another flow interval on the river, which is of effective importance and is observed during flooding of the floodplain (the Tien Shan rivers). On rivers, where the second interval Q_f is not present, the branches dissecting floodplains are poorly developed, as they are overlapped by pebble-boulder banks formed in the main channel.

The rivers of the Western Transcaucasus are characterized by relatively low water level rise, which, with the prevalence of high floodplains or embankments, leads to the passage of Q_f before they are flooded, not counting Q_f, which corresponds to a catastrophic flood. Availability of this interval naturally decreases from lowland rivers to step-pool cascade in the alpine zone (Table 4.11).

The upper interval Q_f effective discharge corresponds to floods of rare recurrence. The remaining Q_f intervals take place in case of floods of higher security, causing transformation of different forms of channel or channel relief. Similar conclusions were also reached by Romashin (1967), who set water discharge values on the Psezuapse river (Western Caucasus), which lead to the transformation of

Table 4.11 Changes in effective discharge frequency on rivers of the Western Caucasus depending on the channel type

Channel type	River-station	Effective discharge frequency (%)
Plain	Inguri–Darcheli	28
	Supsa–Hidmagala	39
	Rioni–Sakachokidze	32
Semi-mountainous	Chorokhi–Erge	20
	Nathaneby–Nathaneby	24
Mountainous		
(a) With developed alluvial shapes	Kodori–Laty	15
	Kodori–Ganachleba	9
	Bzyb–Jirkhwa	10
	Ajaris-Tskali–Keda	16
	Gumista–Achadara	8.3
	Supsa–Chochatauri	16
	Kintrichi–Kohi	14
(b) With undeveloped alluvial forms	Chkhalta–Chkhalta	8.4
	Chakvi–Hala	7.0
	Rioni–Oni	8.3
(c) Step-pool cascade	Gega–Estuary	7.3
	Guangdra–Guangdra	1.0
	Ajaris-Tskali–Khulo	1.6
	Mirabis-Hevi–Shuahevi	1.1

secondary, development of some branches and dying out of other branches. At the same time, for the channel section with developed alluvial forms, this flow rate (Romashin, 1972) has a 15 % probability identical to the values given in Table 4.11 for effective discharges in other rivers of the Caucasus with the same type of channel.

Thus, each of the types of mountain river beds and semi-mountain river beds differs from each other in the prevailing form of sediment transport and, consequently, in the conditions in which they accumulate. The morphology of channels (except for step-pool cascade) can be very diverse depending on the conditions of their formation. In intermountain hollows and in the foothills of mountain and semi-mountain rivers, the channels of the mountain and semi-mountain rivers are branched into a complex system of branches due to the melting of the flow, the reduction of the specific effective discharge and the decrease in the transport capacity of the flow. In gorges, when the rivers cross the ripples across their direction, the river channels with undeveloped alluvial forms are usually low sinuous, often straight, the floodplain is either absent or fragmented. Rivers with developed alluvial forms form usually incised steep bends with relatively wide (for gorges) floodplains. The semi-mountainous channels are most often confined to the foothills, and sometimes also occur in the mountainous zone, where in narrow gorges they resemble mountain channels with developed alluvial forms, but the lower kineticity of their flow leads to the formation of bends and well-developed floodplains here. Semi-hill rivers are characterised by large, downstream ripple shapes (riffles) that cause changes of bends or branches similar to those in flat rivers.

Features of fluvial processes on large and small lowland rivers

In relation to fluvial processes, there is no classification similar to mountain rivers on lowland rivers. Usually, the types of channels to be selected are meandering, branching, etc.—correspond to channel patterns (Chalov 1979) [types of fluvial process, according to SHI (Kondratyev et al. 1982)], which correspond only to flat rivers; schemes of development (lateral changes) and methods of predicting long-term changes are proposed for them. Indeed, for lowland rivers with very low, in comparison with mountainous ones, values and the range of slopes, calm current and small values of the Froude number, the approach adopted for mountain rivers is obviously unacceptable. At the same time, the plain rivers differ in terms of the conditions of interaction between flow and channel, transport of bed material load, depending on their pebbly-boulder, sandy or sandy-silty composition, the partitioning between depth h and width b_{ch} of the channel, the shape of the living section and the degree of flow melting. The specific features of fluvial processes are characteristic of the river mouths and marshy rivers; with a pronounced deficit of load, rocky channels are formed; on the contrary, with very large sediment load flows exceeding the transport capacity of the rivers, bunded wandering channels are formed (Chalov et al. 2000). Many of these differences are not determined by the mechanism of interaction between flow and channel, but by the conditions of these processes (Fig. 4.15), which create the background against which this interaction, erosion, transport and accumulation of sediment takes place.

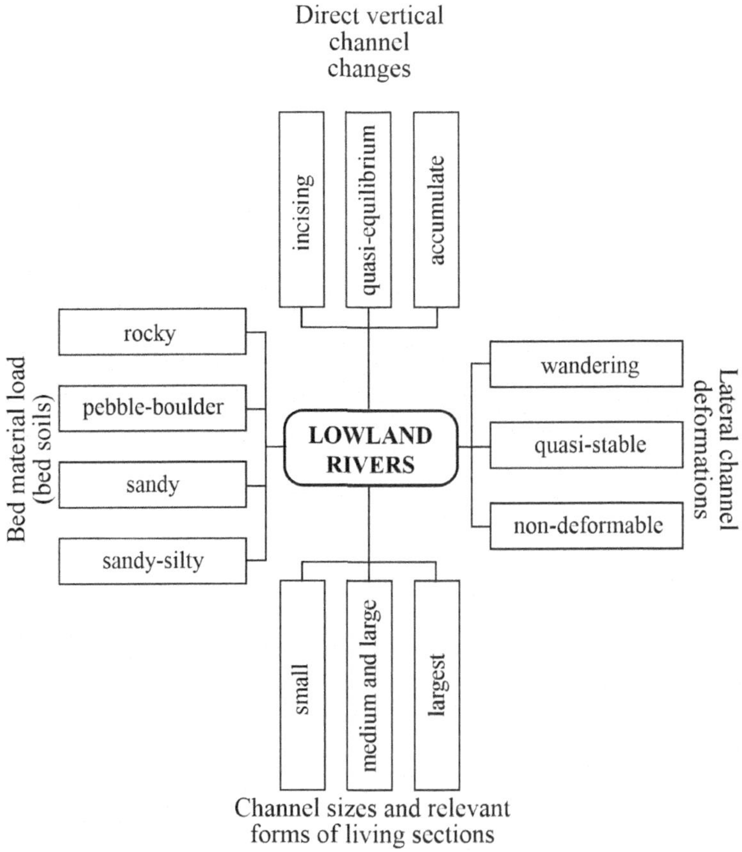

Fig. 4.15 Scheme of classification of fluvial processes on lowland rivers and forms of its representation

The type of bed material load determines the range of flow rates at which channel erosion, sediment transport or sedimentation occurs (Hjulstrom 1935). As you can see (Fig. 4.16), especially when extrapolating curves to larger particle sizes, particle shear and transport is carried out at velocities that are observed only in the high-water phases of the regime; in the low water period, the transport of sediment on such rivers ceases and their stoppage results in the formation of alluvial detachment from a material whose particle size exceeds the critical one for a given flow rate; smaller loads ($d < d_{cr}$) are carried to the lower parts of the channel, where they, in turn, form a detachment of smaller size. Thanks to this, there are only large ripples on pebble-boulder rivers that appear in the high-water phase of the regime; the transformation of the ripple relief of the channel with the formation of ripples of smaller size, corresponding to the hydraulic condition of the lowland, does not occur. On sandy rivers or where the channel becomes sandy, including by bringing in the relevant loads from upstream sections of the river, a hierarchy of ripple

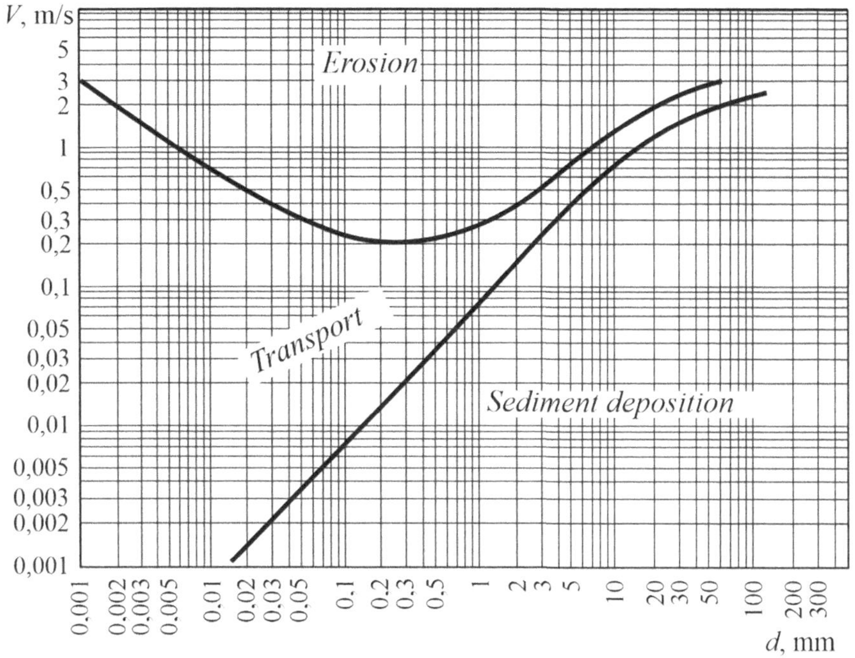

Fig. 4.16 Conditions of erosion, transport and deposition of bed material load of different grain size under different flow velocity (Hjulstrom 1935)

topography is formed—from macro- to micro- and smaller forms, or in the stretch zone of the flow there is a corresponding restructuring, leading to the stoppage (conservation) of macroforms and subsequent formation of ripple meso- and macroforms on their pressure slopes (Alekseevsky 1998). As a result, the transport of sediment on sandy rivers practically does not stop in the lowlands, but the size of the ripples, the movement of which determines their movement, changes, and the area of transport shrinks, because they dry up over a large area of the channel, or the flow velocity is less than that of incipient ones.

On pebble-boulder rivers, the movement of sediments and the movement of the ripples only begins when the armouring layer is destroyed. For the thickness below it, the actual flow rate is $V_{ac} \gg V_{cr}$ (V_{cr} is the critical velocity for particles of this size). As a result, the flow is rapidly saturated with fine material, which, like a similar phenomenon on mountain rivers, is accompanied by a decrease in critical velocities for coarse particles, including those that make up the detachment (Makkaveev et al. 1970; Rossinsky and Debolsky 1980). This is because the saturation of the water with suspended matter increases its density and, in the expression, $V_{cr} = f(d, g, h, \frac{\rho_s - \rho}{\rho})$ $\frac{\rho_s - \rho}{\rho}$ strives to 0.

Ripple movement velocities on pebble-boulder and sandy rivers differ by orders of magnitude, there are also differences in the all-zone mode of riffles. In sandy rivers, they are washed in the flood, washed in the low water level depending on the resulting load balance in a given river section; in the first case, the decrease in water levels ΔH partially, and the additional ΔH on the ripple due to the effect of overflowing partially or completely, is compensated by a decrease in the bottom marks at the riffles due to erosion of the ripple Δz (Fig. 4.17), i.e. $\Delta H + \Delta h \approx \Delta z$. On pebble-boulder rivers, the riffles turn into non-washed overflows, which, in turn, leads to the growth of velocities and an additional decrease (relative to the total) of levels on their ripples, not compensated by erosion.

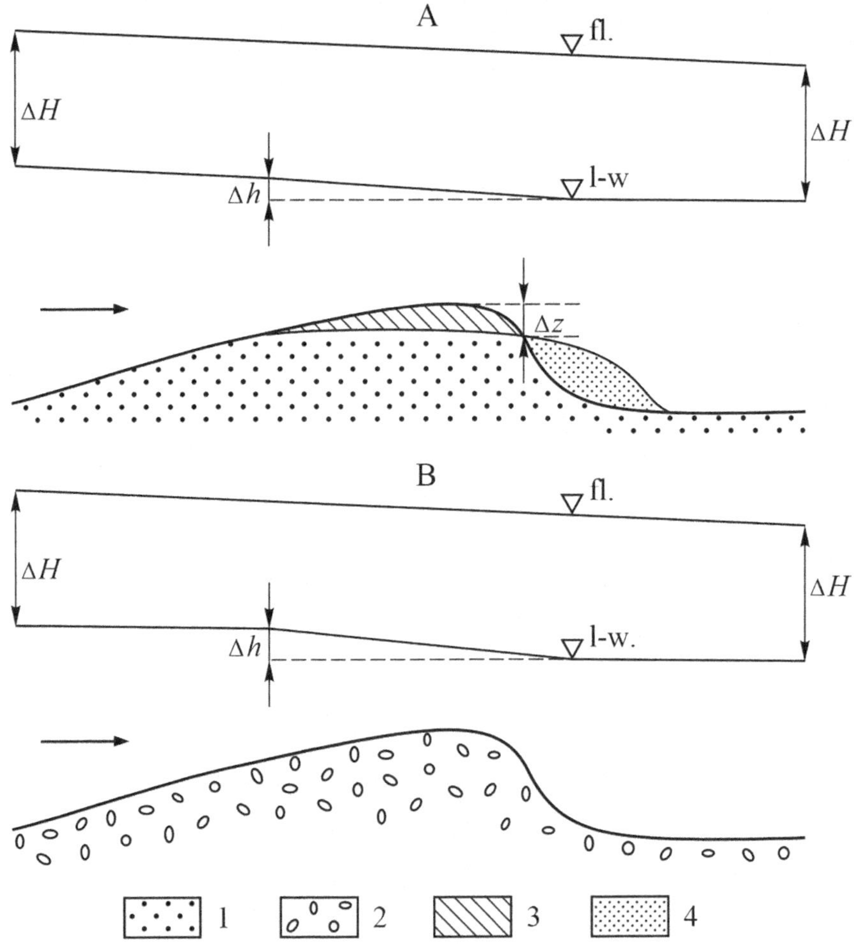

Fig. 4.17 Changes of the profiles of the free flow surface during flood and low-water and of river bottom with sandy (**A**) and pebble-boulder (**B**) composition of bed material load: 1—sand; 2—pebble and boulders; 3—erosion; 4—accumulation

On rivers with sandy-silty bed (not counting the estuaries, where the particle size of bed material load is related to their hydraulic sorting as slopes decrease, flow rates decrease on the estuary seashore, runoff and tidal phenomena, etc.), there is a possibility of un-rippled flow of sediments, as their small size provides a condition under which $V_f \geq 2.5V_n$ (Mirtskhulava 1988). Such a rippleless transport of sediment is characteristic of the Huang He and Amu Darya. Changes in the transport capacity of the flow during a change in the phases of the water regime lead to the formation of non-structural clusters of sediment, which form shallower areas—analogues of riffles in the low-water, their erosion and suspending of loads as runoff increases. On such rivers, the suspended component of the load is absolutely dominant, while the amount of bedload carried by the river is only 1% and less than the total amount of load transported by the river (Chalov et al. 2000). The transport capacity of the stream W_{tr} is realized by the suspended load of W_R, i.e. $W_R \sim W_{tr}$, while that of the sandy and pebbly-boulder river $W_R + W_G \sim W_{tr}$, where W_G is the suspended load transport rate.

Thus, the coarseness of bed material load determines the specificity of fluvial process driver, but their mechanism, which is determined by the interaction of the flow and channel, erosion, transport and accumulation of sediments, remains unchanged.

A special position on flat rivers is occupied by *rocky channels*, where the flow directly comes into contact with the bedrock. They are found at the intersection of mountain massifs by rivers or on the periphery of plain-platform areas (large rivers of Eastern Siberia, Yenisei, upper and middle—at the intersection of the Maly Hangan—Amur). Within the structural uplands, rocky riverbeds are also found on small rivers. They are characterized by a tightening of the flow by rocky banks and the bottom, which is accompanied by a significant increase in the velocities (up to the formation of fast currents, in which $V = 5$–7 m/s), providing a transit transport of load regardless of their size. The latter may form either scarce ripples or clusters in the form of short coastal ripples and "cobblestone bridges" on towpath (Lodina and Chalov 1994). Karasev (1975) has shown that channel constriction and the absence of channel sediments are also possible in cohesive soils, i.e. dense clays, whose effect on sediment flow and transport is similar to that of rock.

Karasev (1975) and Rzhanitsyn (1985), singling out channels in cohesive soils or cut into bedrock, identify them with the incision (deep erosion) of rivers, i.e. with vertical channel changes of the negative sign. The directed accumulation of load is usually represented in the lower reaches of lowland rivers due to their estuarine elongation or tectonic deflection of coastal lowlands, and in the inland areas with centuries-old fluctuations in the runoff rates of rivers. If the discharge is high, it somctimes leads to the formation of a bunded channel and its location even at intermittent levels, much higher than the adjacent areas. These are the channels of the middle and lower reaches of the Huang He River, raised to 7 m above the Great Chinese Plain, the Yangtze, the lower Amu Darya, the lower Terek, the lower reaches of the Kuban, the Rioni within the Colchis Khanate lowland, or in the lower reaches.

Depending on *the degree of stability of* lowland river beds, they differ in the intensity of lateral channel changes and movement of large ripple forms of channel relief (riffles, side-bars, spits, etc.). In extreme representation, they can be combined into three types according to the rates of changes (Fig. 4.15): non-deformable, quasi-stable, wandering. Most of the existing classifications, in which the channels are distinguished from unstable (L < 2) to stable (L > 10), deal with sandy and sandy-pebbly quasi-stable channels. They are characterized by a fairly large range of changing rates, but the very forms of the channel and the main (large) forms of the channel relief have dynamic stability, remain in basic features unchanged, although they undergo greater or lesser transformation as they develop and shift. Thus, depending on the stability of the riverbeds of the banks of the bend (respectively, and changes in the shape of the bend) occur on large rivers at a rate of 1–2 to 50–80 m/year, in branching with the same associated frequency of development and death of branches (1–2 years or tens and hundreds of years). In both examples, the channel shapes themselves (bend, branch) are preserved, experiencing only a change in their parameters.

Non-deformable channels can be called only conditionally, because their changing occurs due to the development of specific hydraulic phenomena and corrasses by moving particles (fragments). These include rocky channels, as well as pebble-boulder beds of large lowland rivers, which are characterized by the stability of their shapes in terms of plan and the speed of movement of coastal ripples no more than the first meters per year, which is almost impossible to fix. However, this is due not to the mechanism of fluvial processes, but to the big difference between the factual V_f and incipient V_{in}, the conditions of transportation of a stream of coarse chip material, and not its form.

The antipode of the non-deformable is the wandering channels, in which $V_f \gg V_{in(s)}$ and which it was suggested to refer to as absolutely unstable (L < 0.5). In this interpretation, the term "wandering" is given the meaning given to it by Chinese researchers (Liu et al. 2000): a complete change in the morphology of the channel during each flood due to the absolute dominance of a suspended component in sediment load. At the same time, wandering can be characteristic of any channel pattern, although almost instantaneous destruction of channel forms (bend, island) significantly limits the range of application of this term. Examples of such wanderings of the Huang He River bed (straight, single-thread and meandering) are shown in Fig. 4.18. It shows only some of the characteristic contours of the channel, and not its extreme positions during the course of the shift of channel shapes. In Russian literature, the term "wandering" is sometimes used (Rossiyskiy and Kuzmin 1958) to describe the intricate channels and in a number of classifications is identified with the "mid-bar multibranching" (Kondratyev et al. 1982; Grishanin 1972), which develops with the overgrowth of the aisles and their transformation into islands. Indeed, such channels are usually unstable or low stable in terms of their stability (the upper Ob is an example often cited in literature), but the islands that form such branches exist constantly, and there is only periodic, though with increased intensity, development of some and shallowing of other ducts and

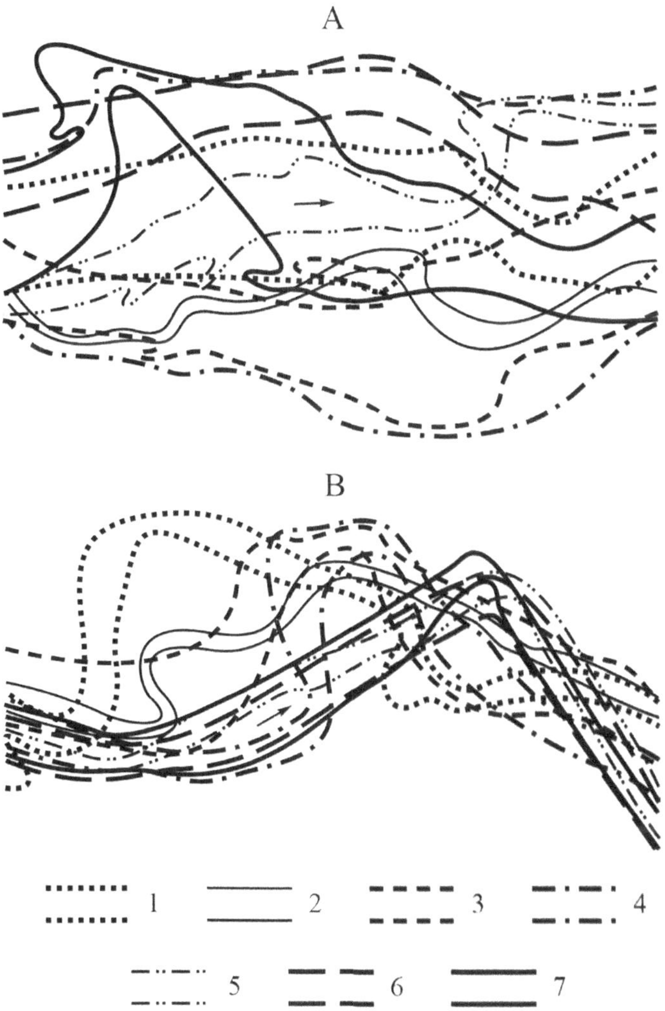

Fig. 4.18 Wandering of the absolutely unstable channel of the Huang He river in 1935–1982 [according to «Atlas of fluvial processes…» (1985)]: **A**—straight (Zhengzhou city–Jiahetan city); **B**—meandering (downstream Juanzen city). Channel position: 1—in 1935; 2—1956; 3—1960; 4—1964; 5—1972; 6—1967; 7—1982

branches between them with partial erosion of the islands themselves from the headboard or in their ears and changes in their configuration.

Thus, the stability of channels affects the intensity of their changes, which, in turn, to a certain extent represents itself in the prevalence of certain forms of channels corresponding to their different patterns.

From the point of view of fluvial processes, *rivers* were usually (or until recently) *singled out in terms of their size due to* mass siltation, overgrowth and degradation of small river channels due to excessive inflow of soil erosion products, respectively, the violation of condition $W \approx W_{tr}$ in the direction of exceeding the first ($W > W_{tr}$), while their eutrophication due to contamination of water by nutrients, and drying up of the rivers with a decrease in their runoff rates due to anthropogenic impact on catchments (Dedkov et al. 1995; Kovalchuk 1997; Chalov 2000). It is a question of the accumulation of sediment, which is almost directly fed into the channels of small rivers in the catchment area. The latter defines [according to Koronkevich et al. (1994, p. 14)] "the intermediate position (of small rivers—*R. Ch.*) between the catchment area, on the one hand, of the middle and large rivers, on the other hand". Such a buffer position of small rivers results in soil erosion products constituting up to 60–70% of the total sediment load. As the order of the rivers increases, the share of flushing products in sediment load progressively decreases. For example, in the Don basin, the annual accumulation of flush products from ploughed lands on rivers up to 25 km long averages 6–12 mm, and on rivers over 100 km long less than 1 mm.

However, small rivers, due to the rapid growth of their runoff rates from the source downstream, are heterogeneous in terms of flow cross-section and differ in this parameter from medium and even larger rivers. In turn, only large rivers are characterized by branching of channels; here this type of channel is the most widespread, forming long morphologically homogeneous sections of wide flood-plain rivers (under conditions of free development of channel changes). Already in medium-sized rivers, it is much less common, usually with the simplest forms, and is often confined to local conditions that cause the flow to be divided into separate branches (support from the confluence with the tributary, before the valley floor is narrowed or the river has to bend near the main bank; local expansion of the channel, etc.). On small rivers, branching occurs rarely, if there are local, often "non-stream" reasons (e.g., on forest rivers below logjams), and the less frequently, the smaller the river, and in the form of the simplest single forms, occupying a percentage of the total length of small rivers (Rzhanitsyn 1960). At the same time, rectilinear channels are most typical for the upper reaches of small rivers, which have 1–2 orders of magnitude in the steppe and forest-steppe zones. In the forest zone and especially in the taiga zone (in the areas of moraine loam distribution), where the deficit of sediment remains at a long distance from the source, the straightness of the channels is characteristic of rivers up to 5–6 orders of magnitude. The same reason explains the development of rectilinear channels on the rivers that connect the lakes or flow out of them. On large rivers of this type, channels of this type develop in the presence of additional conditions facilitating their formation (presence of the main leading bank, leveled along the length of the river; low stability of the channel, etc.) (Chalov 1979; Chalov et al. 1998). In addition, only small rivers have swampy and overgrowing beds. Only small rivers in vivo can dry up in the low-water.

Unlike small rivers, N. A. Rzhanitsyn drew attention for the first time to the difference between medium and large rivers in terms of their cross-sectional shape (1960), which showed that small rivers are characterized by a small ratio between the width b_{ch} and the depth h of the channel. That's why the shape of the channel cross-section is close to parabolic. At the same time, on rivers from 1 to 9–13 orders of magnitude (according to the scheme of Rzhanitsin (1985), the value of h/b_{ch} changes by 10–100 times. If the order is further increased, the h/b_{ch} decrease is relatively slow (Fig. 4.19).

If the nature of h/b_{ch} change is used to determine the "channel" boundary between small and medium rivers, then, according to Rzhanitsyn (1960), for rivers in the middle zone of the European Russia it corresponds to the order of 10–11, when a rapid change of h/b_{ch} is replaced by a slow one. In this case, the lower (counting from the source) limit of the length of the small flat river is 180–340 km, and in various natural conditions the "boundary" order of the river is less in humide and more—in arid areas. With the increased runoff rates of the river and natural flow regulation, the h/b_{ch} ratio decreases faster compared to dry rivers in arid regions, for which there is a pronounced intra-annual flow irregularity, up to drying up in the lowlands. Therefore, in arid conditions, the length of small rivers (from the source) is sometimes 500–800 km.

Other coding systems for the structure of the river network for the assessment of the boundary order of the river lead to similar results (Pakhomova 2001). As a result, as it was shown above, the order of the river (according to the scheme of A. Scheidegger) $N=10$ can be accepted as the lower boundary of small rivers, $N=14$—the upper boundary of large rivers.

Differences in the form of channel cross-sections determine the structure and dynamics of the flow, hydraulic resistances, and the relative roughness of river beds, which reflect the conditions of interaction between the flow and the channel. In the parabolic form of the cross-section, the width and depth of the channels of small rivers are comparable in size. On rivers of the first order $b_{ch} > 1.5$ h, and at N > 1 b_{ch}/h = 4–6. Therefore, the flow is equally affected by the bottom and the bank. In medium to large river channels, the influence of banks can already be neglected due to their extremely low contact with the flow compared to the bottom. In other words, the hydraulic resistances on small rivers are equally determined by the

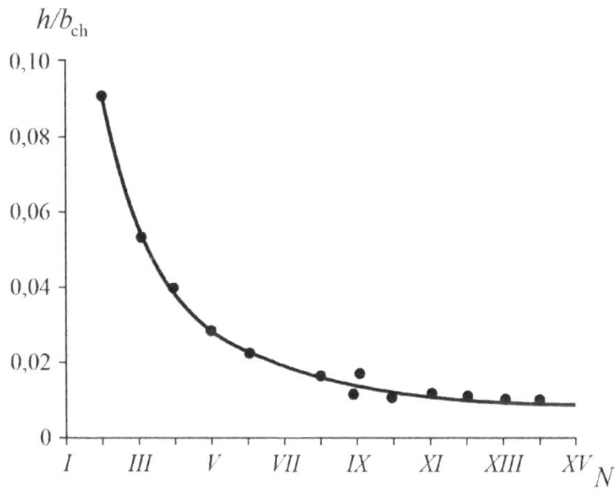

Fig. 4.19 Change of the relative channel depth h/b_{ch} (here h—depth of the pool valley; b_{ch}—channel width) with increasing of order of the river N (Rzhanitsyn 1985)

bottom and banks, while on medium and large rivers—almost exclusively—by the ripple or granular roughness of the bottom. The completeness of the cross-section (b_{ch} and h ratio) also determines the conditions for the development of the cross-sectional circulation in the channels. This is most clearly seen in the bends of the riverbed. It is known that the screw-shaped current on the benders appears in the channel, in which b_{ch} and h are comparable to each other. This was pointed out by Makkaveev (1955), who showed that the width of the screw-shaped current does not exceed 15–20 h. This conclusion was based, among other things, on Potapov's laboratory studies (1936), according to which artificial transverse circulation excitation as a method of channel deepening is possible only at $b_{ch} < 15$–20 h. In a wider direction, there is a divergence in the direction of the surface and bottom flow jets, or its division into several dynamic axes. On the basis of laboratory experiments, Losievsky (1934) has shown that the transverse circulation, typical for a narrow deep channel, is replaced by an inhomogeneous current with two or more dynamic axes in a wide, relatively shallow channel, and each branch of the flow can be characterized by its own screw-shaped current.

Karasev (1975) proposed to evaluate the flow conditions in line with the criterion of its quasi-homogeneity

$$\theta = \frac{b_{ch}}{h}\sqrt{\lambda} \tag{4.26}$$

where $\lambda = \frac{2g}{C}$ is the hydraulic resistance coefficient; C is the Shezy factor. Its analysis shows that in the channels of small rivers the value of θ reaches the lowest values (<4.5), which corresponds to the conditions of preservation by the flow of "the form of a compact jet having the size of the whole channel" (p. 100); on medium and large rivers the values of θ increase, and at θ > 9.5 in the flow two or more branches of the current are formed, between which the velocities are less and here are formed accumulative formations, thanks to which the channel, eventually, is divided into branches. The latter corresponds to the already large rivers, because the "internal branching of the flow" (according to I. F. Karasev) only begins on the middle rivers (4.5 < θ < 9.5) without formation of separate flow zones, i.e. there is a wide core zone without a clearly expressed dynamic flow axis. As a result, small, medium and large rivers differ in terms of the interaction between flow and channel and, consequently, the type of fluvial processes. These differences affect the distribution of meandering, branched and relatively straight, single-thread channels, and for meandering rivers—in a number of specific features of bend development on rivers of different sizes.

The rays of small rivers are steeper than those of medium and large rivers. They are dominated by loopy or steep segmental bens. The rapid change in the shape of the channel cross-section along the length of the river determines a natural change in the radii of curvature of the bend r depending on the average maximum or effective water discharge, and due to the large role of the cross-sectional circulation (screw-forming currents) in the development of the bend is a quadratic and uniform for all small rivers. For medium and large rivers, a similar relationship is linear in nature and is differentiated by river depending on the hydrological regime in different natural areas (Chalov et al. 1998, 2004).

At low completeness of the living cross-section of channels (i.e., h and b_{ch} ratio) and large hydraulic resistances, including those caused by a significant impact on the flow of banks, small rivers, all other things being equal, differ from the average smaller velocities of the current in the lowering to the minimum values. This is especially evident on the steep loopy bend shapes, which dominate on the small meandering rivers, with a pronounced asymmetry of the transverse profile in their tops, where deep moulds are formed. In the upper reaches of the Ug River, for example, at a channel width of 10–13 m and an average depth of 1.15 m, the maximum values of $b_{ch}/h = 3.5–4$, declining at the crossroads to 0.2–0.8. In the low flow rate between these moulds, the flow rate is reduced to 0.05 m/s and suspended sediment accumulation occurs. Overgrowth of the channel and its banks leads to stabilization of bend outlines, as vegetation prevents erosion during the period of increasing flow rates. Shallow riffle areas within other bend types are often composed of clays formed by compaction of muddy sediments. Most often they occur on small, drying up arid areas in the low watercourses.

In humide conditions, the increase in the size of small rivers and the b_{ch}/h ratio is accompanied by a decrease in the overgrowth of the channel, and then causes the disappearance of aquatic vegetation. The increased intensity of change results in a process of channel deflection that loses the characteristics of a small river.

A specific species is the swampy beds of small rivers. They are formed at slopes $I < 0.05‰$. Low flow velocities ($V < 0.1$ m/s) contribute to the development of aquatic vegetation. Riverbed sediments are dominated by silty-organogenic sediments. The boundaries of this channel are weak. The marshy floodplain passes into a marshy channel, along the length of which the swampy meadows (lake extensions) and narrow ducts alternate. The width of the swampy meadows can reach several tens of meters, and the width of the duct—3–5 m. The depth of water in the valleys varies from 6 to 1 m or less. Swampy beds often arise as a result of accumulation of soil flushing products from the surface of catchments in small river channels and overgrowth of shallow parts due to biogenic pollution of water and sediments. On the other hand, the presence of aquatic vegetation, increasing significantly the hydraulic resistance, cause an increase in water levels at a constant flow rate, which contributes to flooding of the coastline, the loss of clear shape of the channel.

Thus, small plain rivers are characterized by rectilinear, single-thread (formed on rivers of first order under conditions of sediment deficit), swampy, silty, marshy, meandering (with the steepest loopy shape) channels.

Based on the specifics of the interaction between the flow and the channel, the above considerations make it possible to type fluvial processes on flat rivers depending on the size of the latter, i.e. to identify the types of fluvial processes on small and medium-large rivers. Streams that correspond to watercourses of the 1st order are especially distinguished by fluvial processes. They are characterized by a great variety depending on the position of a constant source (groundwater outlet, erosion) on a slope, in a beam, at the bottom of a deep ravine cut to the waterproof or aquifer, etc.). On the other hand, the rocky channels are a reflection of a particular type of fluvial processes. All other approaches are either the identification of channel changes as a form of representation of fluvial processes, or reflect the specifics of their development associated with the size of load, the characteristics of their movement, depending on the partitioning between actual and incipient velocities, the presence of aquatic vegetation, etc. In the estuary areas of the rivers, this specificity is determined by the retention phenomena at runups and tides, and in the case of multibranch deltas, by the successive dispersal of water flow along the branches from the top of the delta to the estuary. Finally, in estuaries, the specific features of fluvial processes are determined not only by the influence of the sea factor, but also by the mountainous, semi-mountainous or flat type of river flowing into the receiving water body.

4.3 Morphodynamic Classification of River Channels

Despite the large number of already proposed and applied classification schemes, the issues of river channel classification are periodically raised in the Russian and foreign literature. This is due to the development of the theory of fluvial processes and, accordingly, the development of its hydromechanical foundations, and the expansion of the geography of channel studies, obtaining new information about the forms of representation of fluvial processes in various natural conditions. It is regional studies of fluvial processes that lead to the identification of new varieties of river channels characterized by a peculiar morphology and dynamics, clarification of previously proposed schemes of their development and, finally, the development of new classification schemes. Indeed, the mechanisms and specifics of the interaction between flow and channel, and the forms of sediment transport that make up the essence of fluvial processes, vary widely depending on the runoff rates of the rivers, their water regime, the kinematics of the flow and the magnitude of the longitudinal slope, and the size and magnitude of bed material load. This creates a great variety of conditions for the formation of river channels, determines the development of fluvial processes of different types and their representations in the form of river channels with different morphology and different modes of changes in different combinations of natural (and now—and anthropogenic) factors in different regions. However, according to the law of limited morphological complexes, which underlies the classification of fluvial processes, and the mutual control of the flow and channel leads to certain, most likely combinations between the morphometric characteristics of the channel and the hydraulic characteristics of the flow" (Velikanov 1958, p. 58). Earlier Velikanov (1948, p. 13) pointed out that "in nature, from the countless possible channel formations, only a limited number of relatively stable forms of such are carried out, which makes it possible to classify natural channel complexes by geomorphological means". According to Makkaveev (1976), the limitation of morphological complexes and the relative stability of the channel forms arising in the process of interaction of the channel forms is conditioned by the fact that the latter, being a derivative of the influence of the flow on the underlying surface of the flow, excite in the flow itself the phenomena and processes contributing to their renewal, i.e. predetermine the formation of one or another kinematic structure of the flow. Thus, in the bend of the channel and in the branching there are circulation currents and such a high-speed field of flow, which contribute to the preservation of them as forms of the channel, the growth of curvature of the bend, the development of the island in the middle of the channel, etc., the same applies to relatively straight, single-thread channels. In nature, however, one has to deal with different variations in channel shape corresponding to different stages of their development. Therefore, the classification of channels should be based not only on their appearance, but also on the entire set of possible changes in the shape of channels in the process of their evolution—from birth to death, ie, to link their morphological and dynamic characteristics. Hence, the morphological (by external signs) classification of channels is simultaneously dynamic, and any cell in the classification scheme, being called morphologically, corresponds to a certain scheme of channel changes; on the contrary, if such a cell is called a dynamic term, then this should be understood simultaneously as a well-defined morphological type of channel. For example, meandering channel and meandering, branched channel and branching of the channel. Classifications of this kind are morphodynamic, and their constituent elements are channel patterns.

Depending on the slopes and flow rate, its kineticity (calm or rough), the particle size of the load and the amount of it, various forms of sediment transport (rippled or un-rippled) and the mechanisms of interaction between the flow and the channel, against which the channel is already developing, morphology and reformation of a particular pattern arise. This determines the need to identify in the morphodynamic classification of channel channels the types of fluvial processes corresponding to flat, semi-mountain and mountain rivers (Fig. 4.20). Plain, semi-alpine and mountain rivers either have channel pattern (e.g., waterfalls and mudflows in mountain rivers), or the latter are developed in a specific way. Therefore, an additional characteristic is often added to the name of the channel type; usually it is done for the mountain river beds: for example, the mountain anabranching system (Kopaliani and Tskhadadze 1972), the mountain channel multi-branchingsystem (Prokacheva et al. 1982), etc.

In this approach to the classification of fluvial processes, meandering, branching and formation of relatively straight, single-thread channels are variations of lateral channel changes. Indeed, according to the definition of fluvial processes, their classification should be based on differences in the mechanisms of interaction between flow and channel, erosion of the river bed and banks (beds), sediment transport and accumulation. This corresponds to the separation of fluvial processes in mountainous, semi-mountainous and lowland rivers. Their differences determine the general background on which the channel changes take place and the channel of this or that pattern is formed.

The peculiarities of development of the channel of a pattern are different in different geological-geomorphological conditions. In this regard, they are highlighted (Fig. 4.21): A—incised (floodplainless) channels corresponding to the

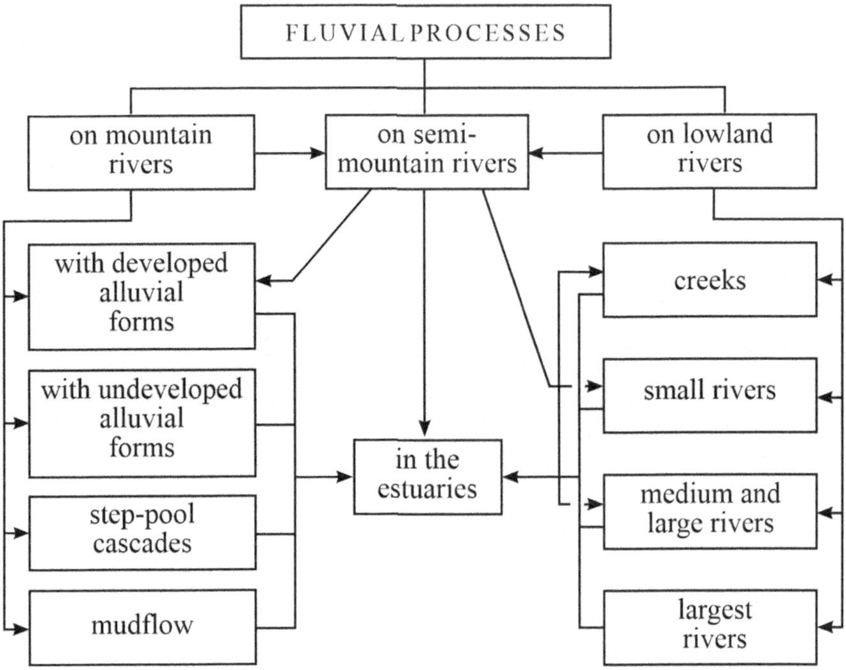

Fig. 4.20 Scheme of classification of fluvial processes (by conditions of the interaction of the flow and the channel and forms of sediment transport)

limited development of channel changes; among them there are two varieties associated with different lithology of bedrocks—in rocks and in plastic rocks; B—confined channels—intermediate type between incised and wide floodplain channels, developing in narrow valleys; C—wide floodplain channels corresponding to the free development of channel changes.

The incised channels are formed in conditions of limited development of channel changes. Often formed in a floodplain valley or with a small width of the floodplain (commensurate with the width of the channel), they have banks folded by bedrocks. If the latter are rocky, bed material load are usually made up of pebbles or boulders, which determines the high stability of the channel. Sediment transport only occurs during high floods and stops completely in the low-water. The coarse material forms side-bars and mid-bars, which are located in the core flow zone. Sediment movement stops abruptly, immediately after the flood wave has passed. The drop in velocity causes an equally abrupt stoppage of coarse material, which, by overlapping the thickness of heterogeneous deposits, forms a sink at the bottom of the stream. In the case of floods and smaller rises in water levels, such accumulations are not washed away, and the relief of the river bed completely reflects the conditions for the passage of effective discharges with the lowest exceedance probability, at which the flow velocities V_f exceed the dilutive Vs for the hours of

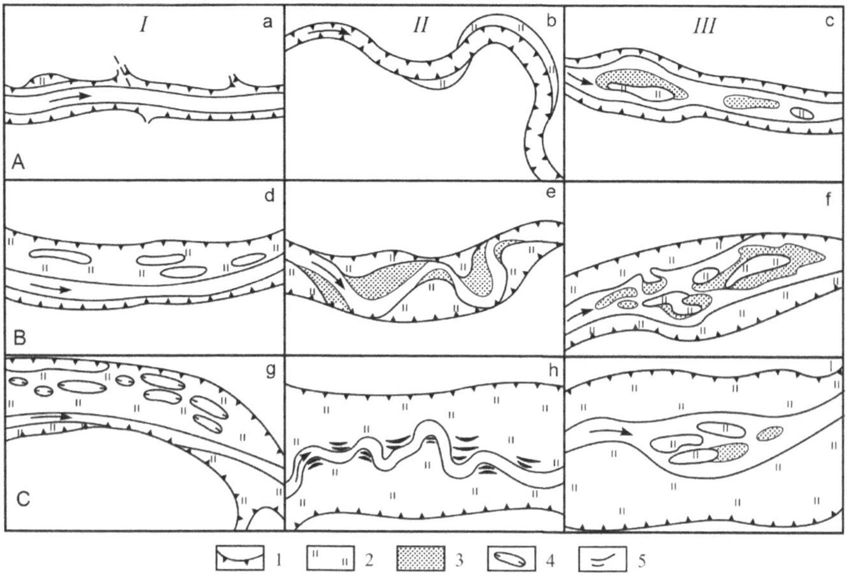

Fig. 4.21 Channel patterns in different geological-geomorphological conditions of the development of channel changes: **A**—incised; **B**—confined; **C**—wide-floodplain; I—relatively straight, single-thread; II—meandering; III—braided channel; a—Yug river near Vasilyevo village; b—Ponoy river near merging with Lopenyarka river; c—Sukhona river near Totma city; d—Don river near Kulyashovka village; e—Gauya river near Tsesis city; f—Northern Dvina upstream the merging with Vaga river; g—Oka river near Schurovo city; h—Zhizdra river near Pavlovo village; i—Pechora river near Velikanovysochnoye village. 1—valley boards; 2—floodplain; 3—large shoals drying up in low-water; 4—floodplain lakes; 5—swale floodplain topography

sediment of a given size. The ploughs on such rivers find themselves in the peripheral parts of the channel and are characterized by large depths; the amplitude of bottom markings in the system «riffle-pool-riffle» reaches large values.

The shape of the incised channels formed in the rocks is mainly determined by structural and lithological conditions. Usually straight channels follow large faults, areas of high fracturing or along the border of various lithological complexes. At the same time, large rivers sometimes stretching over dozens or even hundreds of kilometers do not form any significant bends. Differing resistance to weathering and erosion of rocks, as well as tectonic faults cause a well-defined beads in the bed. For example, when a rectilinear transverse fault is crossed by a rectilinear incised channel, a local fault fault widening occurs, within which the channel is divided into two branches (channels) by an island or an oblique one (Fig. 4.22A). Directly below such faults, as a rule, deep-water forest areas begin; on the contrary, upstream of the fault zone there is usually a rolling area, the depths in the channel decrease.

Another type of incised channels—incised meanders—is associated with the development of riverbeds of successively intersecting fault zones or zones of

Fig. 4.22 Incised channels in rocky formations: A—straight with the appearance of the mid-bar in local expanding at a break crossing on the Upper Aldan (photo of the author); B—Bend of the Upper Lena («Cheeks») with the whirlpool zone formation near the concave bank (photo by A. A. Zaitsev)

tectonic fracturing. In such cases, the tops of incised meanders adjacent to the fault crossings correspond to the expansion of the channel and vortex zones near concave banks. River turns in such cases can be very steep (up to 90°) and shoulders short (bend wings). These are, for example, bends of the upper Lena riverbed near the Lena «cheeks» (Figs. 3.22B).

A specific feature of the incised channels in the areas of crystal rocks is the rocky ledges on the bottom, which, creating a support, contribute to the formation of pebbles and boulders. They are associated with stony pearls, shivers, which narrow the flow and contribute to the emergence of areas with very high flow velocities (fast). Granite or granite-gneiss plates sometimes cross the channel from one bank to another, creating a shallow cofferdam that plays the role of a spillway in the low water level.

If the incised channel is formed in plastic rocks, its outlines (straight channel, bend) are the result of historical or geological development of the channel forms against the background of rather active river depth erosion. It is usually accompanied by narrow floodplains, occupying part of the meander neck bounded by the channel, or stretched in a narrow strip along one of the banks of the rectilinear channel. The channel sediments on such rivers are sandy; different stability of the channels is represented in the seasonal rerouting of riffle products, which occupy a certain position on the bend in accordance with the flow structure, in the rate of riffle products movement downstream in straight sections of the channel. In cases where moraine loams create limited conditions for channel change, their scour is associated with the release of large amounts of boulder material that is not transported by the flow and accumulates near the source of supply. Such deposits Lomakin (1943) proposed to call it pearljuice. They sometimes form boulder ripples similar to stony shivers in the channel or create an awning in the channel.

Sculptural branching is a kind of incised channel, which is formed if the river is cut into branches near an island channel. The latter is characterized by the presence of the main basement rising above the intercostal edge and the surface, which partially escaped from the flood level and is represented by high terraces. Along with sculptural branching, under favorable conditions, accumulative aisles can appear, which can be transformed into islands when fixed by vegetation.

Wide floodplain channels correspond to the conditions of free development of channel changes. The predominantly sandy composition of bed material load and the malleability of bank erosion are responsible for the greatest variety of channel forms corresponding to one or another of its patterns, their complex combinations with each other and with forms of channel relief, and the significant intensity of reformation.

In the most general form among the wide floodplain channels there are three main types—meandering (free bend), branched and single thread, relatively straight. They are accompanied by quite certain types of lateral channel changes, which allows us to judge by the shape of the channel in terms of general terms about the mode of its transformation: the meandering channels correspond to the process of bend development—free meandering, so they are also called meandering. Branched channels arise as a result of the formation of axes, their transformation into islands, the unification of islands and the division of the flow on the branches, which entails changes in the transport capacity of the flow. Sometimes this phenomenon is called channel furcation (Makkaveev 1971; Dynamic geomorphology 1992); Rossinsky and Kuzmin (1947, 1958) used the term "wandering". Straightened, single-thread channels are characterized by a shift of channel parallel to itself towards one of the banks, or by periodic local erosion or bank build-up, while maintaining the overall straightness of the channel in the longitudinal section.

Confined channels are formed in relatively narrow valleys, so that at floodplain spillage the share of water flowing in the channel remains greater or equal to the floodplain part, and its specific value Q_{max}/b_{ch} = 0.9–0.5 (Table 4.12). This distinguishes them from the wide floodplain channels, where in the high-water phase of the regime there is a smaller share of discharge, and its specific value is always less than 0.5, and incised, where water discharge is fully concentrated in the channel in all phases of the water regime. As a result, there are differences in the flow conveying capacity, which creates a specific change of the channels of each pattern. In particular, there is a partial control of their development on the bends by the bedrock channel banks, because the upper or lower wings of the bends, their top part is near the main bank. It transforms the shape of the bend, especially at the approach of the flow to the bedrock channel bank, resulting in a entrenched bend. Consistent arrangement of the bend near the opposite banks of the river causes the most complex interaction of channel and floodplain flows in connection with the periodic intersection of their axes (Baryshnikov 1984), the emergence of specific hydraulic phenomena when discharging clarified water from the floodplain into the channel directly under the bedrock channel bank, etc.

Taking into account the formation of bedding in alluvial sediments, rock or plastic rocks, the heterogeneous composition of bed material load (from sandy-silty to boulder-clump), different stability of beds and, consequently, different intensity of channel change, finally, the different anthropogenic (technogenic) transformation of channels, the orientation (sign) of vertical channel change, the development and diversity of the ripple relief of channels reflecting the movement of sediment, and other representations of fluvial processes, their classification and diversity of river channels can be presented in the form of a general scheme in Fig. 4.23. It shows the complex connection between the types of fluvial processes and all possible classification features that determine the forms of their representation, direction, rates and conditions under which channel changes are carried out.

In general, the morphodynamic classification of river channels is a system of blocks (Fig. 4.24), each of which corresponds to the level of development of fluvial processes and forms of their representations; fluvial processes of the previous block form the background on which the processes of channel formation occur, corresponding to each following block.

Table 4.12 Maximum water discharge ratio Q_{max} in channel and flooded floodplains with different geomorphological types of channel

Water discharge	Wide floodplain	Confined	Incised
Q_{max}, including	100%	100%	100%
$Q_{channel}$	≪50%	≥ 50%	90–100%
$Q_{floodplain}$	≫50%	≤ 50%	0–10%
$q_{sp} = \frac{Q_{max}}{b_{ch}}$	<0.5	0.5–0.9	0.9–1.0

Fig. 4.23 Pebble-boulder channel of the Neman river at the crossing of Grodno moraine highland (perluvial depositions). Photo of the author

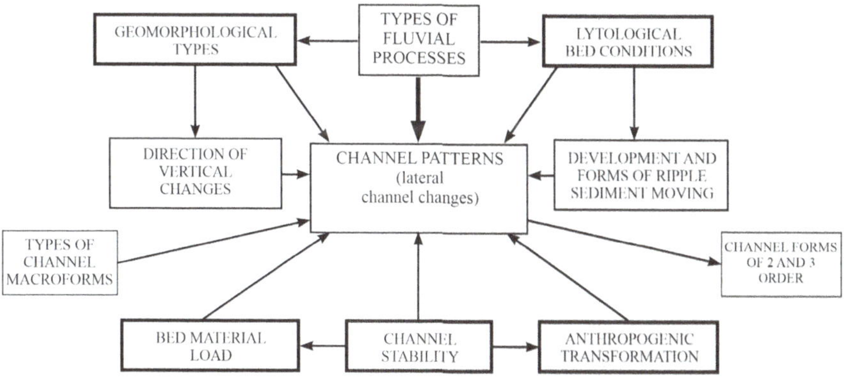

Fig. 4.24 Scheme of the overall composite classification of fluvial processes and river channels and relationships of separate classification signs

Block 1 comprises the types of fluvial processes corresponding to mountain, semi-mountain and plain rivers. At the same time, on mountain rivers there are channels with developed alluvial (ripple), with undeveloped alluvial (rippled) forms, alluvial threshold-waterfalls, as well as rocky (threshold-waterfalls and trays) and mudflows. For lowland rivers, as has been shown above, the following types of channel processes, i.e. the mechanism of interaction between flow and channel, erosion, transport and accumulation of sediment, can be distinguished: in the 1st order of watercourses—streams; on small rivers; on medium and large rivers (with the separation of sand, pebble-boulder and plain rocky rivers); in the estuaries of the rivers and on the estuaries of the river mouths. The former are characterized by a small parabolic form of the living section of the channel with a small partitioning between channel width to its depth b_{ch}/h (from 1:2 to 1:10) and its rapid change up to the rivers X–XI orders of magnitude in the humide regions (Rzhanitsin 1985). On medium and large rivers, this ratio becomes quasi-constant, amounting to about 0.001 and less; the channel is melted in rectilinear pool areas and looks like a trapeze with very small side sizes. Such differences in the shape of the living section and the b_{ch}/h ratio determine the different flow conditions and kinematical structure and, consequently, its interaction with the bottom and banks of rivers.

In the estuaries and at the estuaries of the estuaries, the specificity of fluvial processes is determined by the flow of water and the direct impact of marine factors on the flow of water (Mikhailov 1998).

Block II of the classification includes "geomorphological" types of channels—wide floodplain, confined and incised, the selection of which is based on geological-geomorphological factors of fluvial processes, each of them corresponds to a certain ratio between the width of the channel b_{ch} and the width of the floodplain B_{fl}. Makkaveev (1955) took these ratios as an indirect indicator of the prevalence of river erosion (incision) depth ($B_{fl} < b_{ch}$) or the predominant development of lateral erosion (lateral channel change) ($B_{fl} > 2/3b_{ch}$). In fact, the river is cut in if the sediment load is less than the transport capacity of the flow, i.e. there is a shortage of sediment. As a result, flow energy is used to scour the bottom; on the other hand, the shortage of sediment does not contribute to the development of ripple forms of riverbed relief and therefore reduces the possibility of lateral

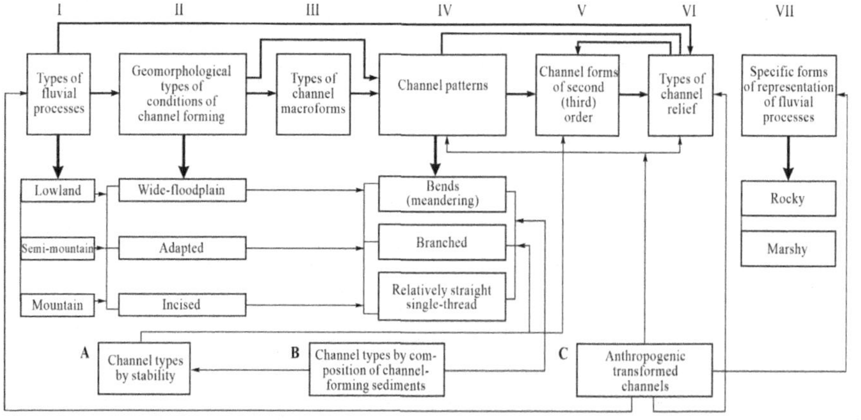

Fig. 4.25 Structure of the morphodynamical classification of river channels: I, II, …—main blocks; A, B, C—additional blocks

change, i.e., the expansion of the valley bottom. As a result, the greater the deficit of sediment, the more intensive the plunging, the narrower the bottom of the valley and the more likely the formation of the floodplain-free incised bed. This pattern is especially evident in the conditions of limited development of channel change.

In the absence of a shortage of sediment (the sediment load corresponds to the transport capacity of the stream) or its excess in the channel, a variety of ripple forms of relief (side-bars, mid-bars) are formed, which contribute to its wandering and formation of the wide floodplain channel. This is most clearly seen in the conditions of free development of channel changes, when the flow easily erodes the banks folded by loose rocks.

Confined channels (low deficit of sediment and still insufficient for the formation of thick riverbed sediments ($b_{ch} < B_{fl} < 2/3b_{ch}$) are usually found in border geomorphological areas (Vilyui between the Middle Siberian Plateau and the Central Yakutia Lowland, the Dniester River downstream of Dubossary, where the river crosses the slopes of the Volyn Podol Upland), in the valleys following the linearly elongated morphostructures, where their narrow bottom is enclosed between steeply sloping, folded bedrocks, limiting the possible range of lateral changes. Sediment load and the degree of development of ripple topography do not play a determining role, determining only the intensity of lateral channel changes.

Another criterion for determining the geomorphological types of the channel is the ratio between the width of the valley bottom B_v (floodplain + channel) and the channel wandering belt (meandering, branching) $B_{m(branching)}$. If $B_v = B_{m(branching)}$, then under condition $l \approx 1.6$ L the channel is confined (here l—the length of the channel, L—the step of the bend or branching node); if $Bv > B_{m(branching)}$—wide floodplain, $B_v < B_{m(branching)}$—incised.

Block III of the classification includes channel macroforms that are not directly related to lateral channel changes, exceed the limit dimensions of channel forms at the present-day water availability of the river, flow rate, sediment particle size, channel width, but at the same time influence their development (this block in the scheme—Fig. 4.25—is not deciphered). Panin (1991) proposed an approach similar to the separation of geomorphological types, i.e. with the subdivision of macro-emissions into free, confined (with one bedrock channel bank) and insised (Fig. 4.26) for the classification of macro-emissions—one of the types of channel macroforms. And all the macro bends can be: (1) inherited from the free

Fig. 4.26 Incised macro-bend of the Upper Aldan. Photo of the author

meandering channel; (2) inherited from the regular (normal) incised bend, corresponding to the parameters of the flow, but in its bedding and size under the influence of geological-geomorphological factors; (3) predetermined by geological and structural-tectonic conditions. The latter, in turn, are subdivided into the active tectonics formed under the influence of the active tectonics (for example, when the channel bending around the growing tectonic uplifts of the Samara bow type on the Volga River) and caused by passive adaptation of the channel to the relief and geological structure.

The incised macrobends develop due to the occurrence of hydraulic phenomena in the upper parts of the channel, which are inherent in the bending of the stream, the movement of the channel to the concave bank in accordance with the laws of development of channels with a one-side floodplain and the accumulation of morphological changes occurring under the influence of extreme floods. Macro-bends of wide floodplain channels (free) are considered to be relict formations or bends of the river meandering belt (Fig. 4.27A), which are developed during historical and geological periods of time through the consistent accumulation of morphological changes. This adds up or subtracts the effect of centrifugal forces on the bend of the channel and on the bend of the meandering belt, depending on the coincidence or mismatch of the signs of their bends.

Such an approach, obviously, can be applied to the classification of macroforms of branched and relatively straight, single-thread channel. Incised macro-branchings formed by the indigenous tall islands have dimensions commensurate with the incised macro-bends (e.g., Khortytsa island on the Dnieper). The analogue of the meandering belts for wide floodplain river branches is obviously the split channels (Smirnova 2002), i.e., the division of the channel into very long branches passing in opposite parts of the valley bottom, having a length (depending on the size of the river) of tens and hundreds of kilometers, exceeding the width of the channel by several orders of magnitude, and characterized by their peculiarities of the channel regime (Fig. 4.27B). The absence of meandering belt bending or bifurcated channel bend indicates the existence of a rectilinear wide floodplain bed macroform (Fig. 4.27B).

Block IV is represented by the channel patterns itself: bending (meandering), branching (channel multi-branching), relatively straight, single-thread channel with all its varieties. The channel of each of the types can be developed against the background of certain macroforms, be wide floodplain (for the benders the terms "free bend" or "free meandering" are widely used), confined (respectively, "entrenched", "confined" and "inscribed bend") or incised (for the bends—"incised bend"). At the same time, mountainous, semi-mountainous and plain rivers are characterized by specific features of development of channels of any pattern.

Each channel pattern is represented by different subtypes. In the classification scheme (Fig. 4.28) they correlate with free, confined or limited conditions of channel changes development, since the latter largely determine the specifics of the development of certain bendings, branches or straight, single-thread channel, their morphology, character and direction of lateral changes. They are most fully

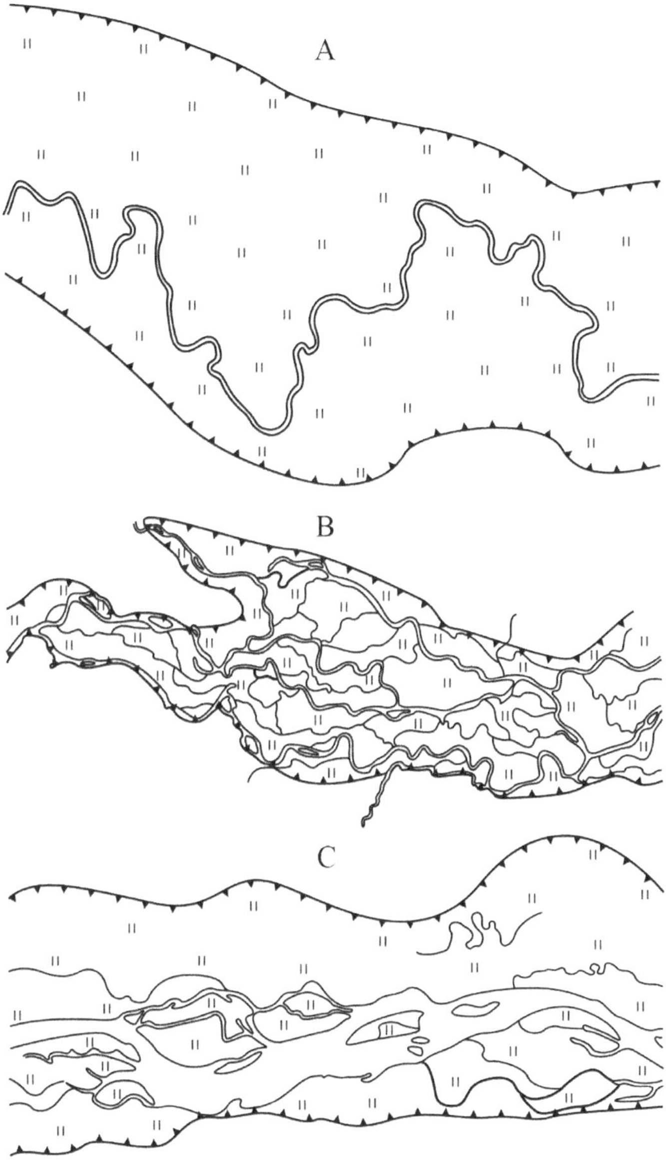

Fig. 4.27 Varieties of macroforms of wide-floodplain channels: **A**—bends of the meandering belt (Kamchatka river); **B**—bifurcated channel (Ob river); **C**—straight channel macroform with contemporary branching (Lena river)

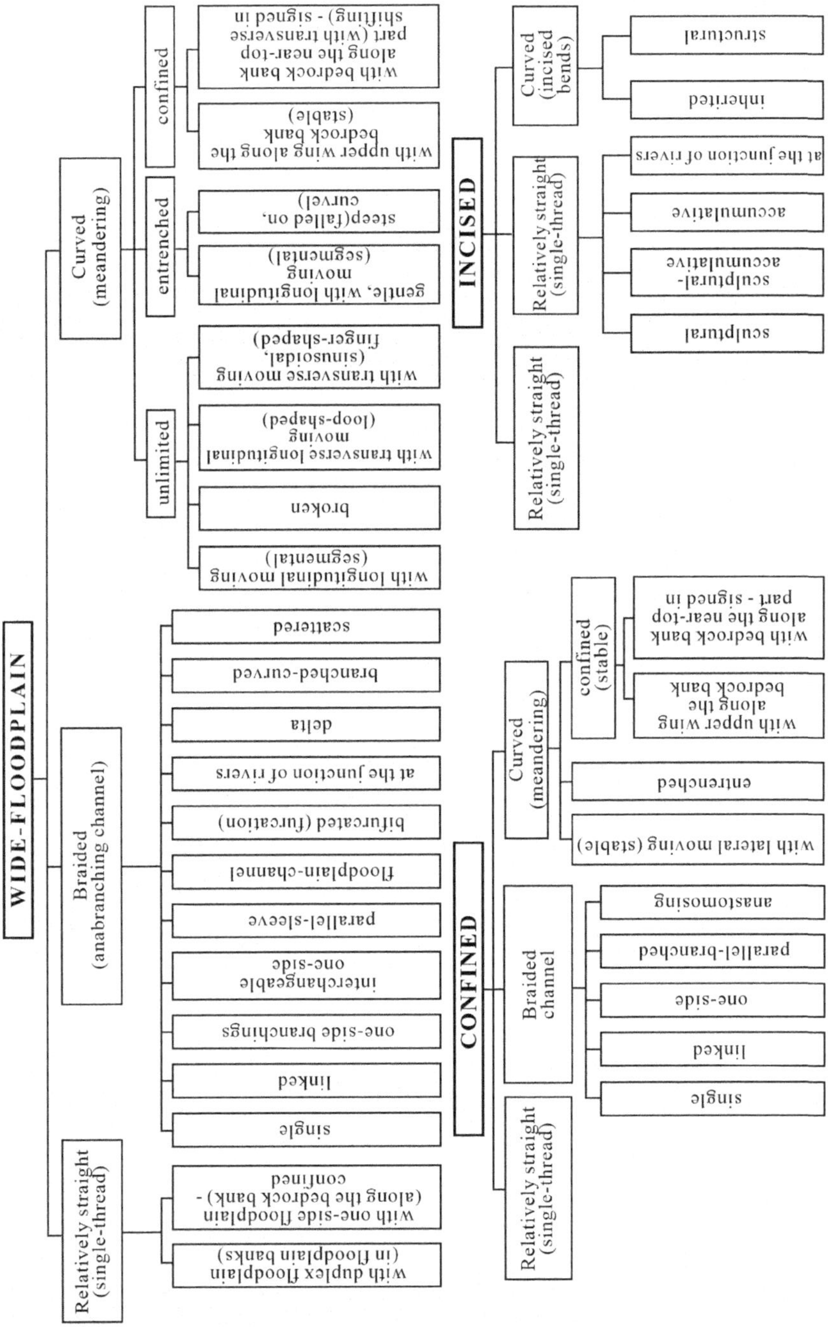

Fig. 4.28 Channel patterns of the low-land river channels and their varieties (block IV in Fig. 4.25 classification)

represented on flat rivers with wide floodplain channel. Some of them are not characteristic of mountain rivers, as well as only the latter are characteristic, for example, of step-pool cascades. Confined and incised channels have a smaller set of varieties of each pattern, as many of them do not have the conditions for their development in these conditions. Among the wide floodplain channels there are not only free but also confined and entrenched bends, which are found in the form of single forms where the meandering channel approaches the bedrock channel bank of the river (or deviates from it). Also confined is the wide floodplain rectilinear channel along the main bank. In a similar situation, there are also many varieties of branched channels, but they actually can only have one hose confined, while the other is between the coastal floodplain and the floodplain island. However, the presence of the indigenous coast has a significant impact on the reorganization of the entire branching.

Among the wide floodplain branched channels in the scheme there are "scattered channels". This is a prefabricated subtype that includes a large group of channels, poorly or practically unexplored with respect to their morphology and dynamics. It is from it that new varieties of complex branched channels are most often distinguished.

Block V of the classification includes channel shapes that complicate bending, branching or relatively straight, single-thread channel, and thus constitute the second and third order of formation. The less stable the channel, the greater the sediment load and the amount of effective discharge Q_f, the more likely it is that a hierarchy of nested channel forms will form, including those developed in different phases of the hydrological regime. In essence, the absence or presence of forms of the second or third order channel, creating their hierarchy, determines the morphodynamic structure of river channels, which, accordingly, can be simple or complex. On rivers with a freely meandering channel, the Q_f, with only one interval, produces the simplest bend shape, in which the dynamic axis of the flow as a whole follows the shape of the bend. If the Q_f of two or three intervals passes on the river, then there is an increasingly complex hierarchy of bends: complex bend consisting of several adjacent bend with a smaller pitch and radius of curvature, which, in turn, are complicated by the gyrus of the dynamic axis of the flow, enveloping the side-bars of the riffles.

In a branched channel, branchs can meander or, in turn, be divided by smaller islands into relatively short branches. In turn, each of these second order branches can be divided into even shorter branches by a small island, which forms a branching of the third order. Thus, there is a hierarchy of branches, and the drawing of the channel is sometimes a tangled labyrinth of branches of different sizes.

In a straight, single-thread channel, the formation of curvilines of the dynamic flow axis near the chess-like side-bars—analogues at the level of the ripple-macroform development is possible. Depending on the partitioning between channel parameters to islands, branching may have a second or third order, although the channel as a whole remains straight and single-thread.

In areas of limited channel changes, where channel straying is hindered by bedrocks or rocky formations forming riverbanks and beds, and bed material load is relatively small, the set of forms (hierarchy) of the second or third order is limited in the formation of incised channels. Often they have a non-stream genesis, being connected with the geological structure, or reflect the geological history of the river valley development, which is typical for the hierarchy of incised bends. More often than not, the forms of the second or third order channel are found in an incised branched or straight channel, where they are of accumulative origin.

Block VI of the classification comprises forms of channel relief and channel changes associated with the ripple movement of sediment, or sculptural changes caused by the unevenness of the river bed. In the first case, channels can be riffle (riffle sections), if its relief is determined by the developed macroforms of the channel relief. Following the suggestions of I. V. Popov (Popov 1965; Kondratyev et al. 1982), depending on the shape of the ripples, side-bars, aseparated and sand-dune beds are distinguished, in which side-bars, aseparated and sand dunes are independent forms of channel relief. In turn, the riffle products often represent single ripples comparable in width to the channel itself, in which the side-bars are part of them (Makkaveev 1949, 1955; Chalov 1979). If the flow with these hydraulic characteristics ensures the transit of sediment, a mould bed is formed in which the pool ripples do not define its main relief. Riffle channels are characterized by alternation of shallow waters (riffle) and pool cracks corresponding to inter-channel depressions of the channel bottom. The forest areas are deep-water, the channel is usually narrowed, and the ripples forming in the local expansions at the bends, at the channel branching or confluence points of the channels make up separate, unrelated increases in the bottom marks. Intermediate positions are occupied by the pool-riffle channels typical of meandering channels and some types of branching, in which the kinematic structure of the flow with high sediment load and weak stability of the channel provides for the alternation of moulds and riffles, the position of which is strictly confined to certain elements of the form of channel. This also includes the swampy channels of small rivers. This type of riverbed, as well as wetland rivers, can be considered as varieties of wide floodplain channels of small rivers.

Additional blocks classify the channels according to the indicators that determine the specifics of each channel pattern development. **Block A** gives an integral assessment of the intensity of reformation in terms of channel stability in accordance with the gradations of accepted indicators—for example, the Lokhtin`s number, channel stability coefficient of N. I. Makkaveev or others. **Block B** characterizes the channel in terms of the composition of bed material load, which determine the differences in morphology and riverbed dynamics through their stability and the development of the ripple structure of the channel relief. The first of these (A) affects channel development at the fourth, fifth and sixth levels (blocks) of the morphodynamic classification. The second (B) is one of the important conditions determining the type of fluvial processes.

The additional **Block B** takes into account the peculiarities of riverbed development during their infeeding or directed accumulation of sediment. The crossing of rivers is most clearly represented in the formation of incised channel types. However, even under wide floodplain conditions, the direction of vertical changes affects both morphology and channel dynamics. For example, in the branched channel the river crossing is first of all represented in more high-water channels, which leads to the unification of islands and their joining to the floodplain, and in the scales of long periods of time—to the reduction of the degree of branching of the channel (Makkaveev and Chalov 1964). The opposite trend is true if a river with a branched channel has a directional accumulation of sediment.

The fourth **Block D** is connected with the formation of the channel and channel relief in the conditions of their transformation by anthropogenic and other types of anthropogenic impacts up to the transformation of channels into channels, concreted trays, ploughed channels of small rivers, etc. This block system can be supplemented with blocks that take into account the drying up, freezing of rivers and other phenomena that also have an impact and cause the specificity of the representation of fluvial processes.

Similar approach to the development of river channel classifications Makkaveev (1976) compared with the principle of building mineralogical complexes, put forward by A.E. Fersman. It allows, on the one hand, to present in a single scheme the entire diversity of river channels and their changes, and on the other hand, to constantly include new elements (cells) in each of the blocks as the knowledge of the morphology of channels and the mode of their changes increases.

4.4 Types of Channel Changes

The morphodynamic classification of river channels is based, along with their morphology, on the type of channel changes, which characterizes changes in the channel pattern (contours) and its movement due to washout of some and other banks. This type of channel transformation represents lateral changes (Chalov 1979).

At the same time, during the interaction between the flow and the channel, during erosion, transport and accumulation of sediment, the bottom is washed out and its marks are lowered, or sediment accumulation and the channel bottom is increased, i.e. vertical channel changes are carried out. They accompany lateral changes due to simultaneous changes in the kinematic structure of the flow, the occurrence of circulating currents, the dispersal of flow along the forming hoses, etc. At the same time, vertical changes constitute the background on which lateral changes develop, imposing a certain imprint on the forms of their representation. At the same time, both lateral (development of channel shapes) and vertical (changes in channel bottom marks) changes are accompanied by the formation, movement, and spatial and temporal transformation of ripple forms of channel relief, which provide both transport of bed sediments and continuous changes in bottom marks. Therefore, the ripple movement of sediments can be considered as one of the independent types of channel changes.

Channel changes determine the development of various forms of river channels and channel relief, the mode of their seasonal, perennial and century-old changes, and determine the erosion of the river bed and banks, transport and accumulation of sediment, i.e., they are the main components of the fluvial processes. In such a context, channel changes are permanent channel changes that occur under the influence of the current water channel transformation. They can appear along the whole river, in the areas of greater or lesser length, within the limits of separate forms of the channel (bend, branching nodes, etc.) and channel relief (riffles, ripples of different sizes). Therefore, channel changes are the most obvious and, to a certain extent, the resulting form of representation of fluvial processes, and as a result they are often identified with a more general concept—"fluvial processes". The movement of sediment in channel flows is not only the content of fluvial processes, but is also a determining factor in the development of channel changes and river channel regimes.

Summarizing the existing ideas about the mechanism of formation of river channels in different conditions, all the variety of channel changes can be combined into several main groups (Table 4.13), differing from each other in their role in the implementation of soil erosion, the composition of the bottom and banks of rivers, transport and accumulation of sediment, spatial and temporal scale of development, the nature and complex of the factors that determine them. Depending on the development of channel changes in relation to the direction of gravity, the forms of their representations and the specifics of the movement of bed material load, channel changes are grouped into three main groups: (1) vertical, causing transformation of the longitudinal profile of the river and changes in channel bottom marks (incision or accumulation of sediment); (2) lateral, associated with the movement of the riverbed in the plan, scouring and reclamation (in case of overgrowth of emerging banks by vegetation) of the banks, leading to the formation of floodplains, and in case of erosion of primary or terrace banks—to the expansion of the valley bottom; (3) related to the ripple motion of sediments (formation, movement and erosion of alluvial ripples), which causes the formation of riffles, side-bars, mid-bars, spits and other accumulative forms of channel relief. Due to vertical changes, the bottom marks are lowered (washed out) or increased (due to accumulation of sediment), which affects the position of the free flow surface. Lateral changes correlate with the formation, development, evolution and extinction of various forms of river channels (bend cutoff, shallowing of channels and attachment of islands to the banks, etc.), corresponding to a particular channel pattern. The ripple movement of sediment (in terms of channel changes, it is obviously preferable to speak of the formation and movement of alluvial ripples) predetermines changes in the channel relief both vertically (the growth of riffle ripples in the flood, blurring them in the low water level and, conversely, eroding the gentle slopes of the ripples and the deposition of washed away material in their basement, etc.), as well as across the channel (laterally) with conjugation and simultaneity of erosion and accumulation processes on adjacent ripples, with their detachment by the flow from the banks or, conversely, joining them, etc.

The presented classification of channel changes is the only similar scheme, although the fact of existence of different types of channel changes is recognized by almost all researchers of fluvial processes. The approaches of Karasev (1975) and

Table 4.13 Types of channel changes and forms of their representation

Forms of representation	Types of changes		
	vertical	Lateral	Associated with the ripple motion of the sediment
(a) *By time*			
Directional	Transformations of the longitudinal profile of the river (incision, accumulation)	Shift of straight channels, meandering or branching belts towards one of the banks, development of incised bends	Vegetation of large ripples (macroforms) and their transformation into a floodplain
Periodic	Associated with the development of individual forms of channel (bend, branching nodes)	Transformations of individual bends, branching units, straight channel sections	Long-term and seasonal reshaping of riffles and other large ripples (macroforms)
Current	Changes in seabed markings due to seasonal fluctuations in transport capacity	Bankline scouring and build-up associated with movement and reshaping of sideline and sediments	Shifting of macro-, meso- and microforms of channel relief
(b) *In space (along the length of the river)*			
General (background)	Transformations of the longitudinal profile of the river (incision, accumulation), covering the whole river or parts of a long river	Covering areas of rivers with the same or predominant channel type (morphologically homogeneous)	Reshaping of riffle sections and moulded areas
Territorial	Associated with the development of individual forms of channel (bend, branching nodes)	Transformations of individual bends, branching units, straight channel sections	Formation and erosion of riffles and other macroforms
Local	Occurring within individual forms of riverbed terrain (riffles)	Shown within the elements of the channel shape (on the wings of the bend, in the branches, etc.); reshaping of side-bars and mid-bars and associated scouring and bank build-up	Formation and movement of individual ripples—meso- and microforms
(c) *In the form of spreading along the river in time*			
Transgressive	Incising propagation or accumulation downstream	Longitudinal bend movement; island tip erosion	Shifting of ripples downstream

(continued)

Table 4.13 (continued)

Forms of representation	Types of changes		
	vertical	Lateral	Associated with the ripple motion of the sediment
Regressive	Erosion or accumulation spreading upstream	Movement of bends, islands upstream	Building-up of side-bars, mid-bars, riffles, and other macroforms upstream enhanced by the accumulation of sediment on the upstream slopes
Regional	Simultaneous development across the whole river or on long stretches of land	Development of morphologically homogeneous areas	Riffle and pool areas development

Rzhanitsyn (1985) are the closest to it. Karasev singles out (1975, pp. 37–38: "(1) deep erosion of the channel in the absence of alluvial deposits; (2) sand dunes in the straight channel; (3) side-bars and asymmetrical ripples in the slightly sinuous channel; (4) bend development (meandering); (5) drifting and multi-branched channel; (6) irreversible accumulation of sediment near the general or local base level». Here, deep erosion and directed (irreversible) accumulation of sediment reflects vertical channel changes, morphological types of channel—lateral changes and ripple shapes—changes associated with the ripple motion of sediments.

N. A. Rzhanitsyn singles out two groups of "process types and channel forms": A. General orientation of fluvial process driver; B. Types of fluvial process driver. The latter corresponds to lateral, the first to vertical channel changes. In turn, it distinguishes between deep erosion, dynamic equilibrium and accumulation processes in terms of general orientation. In contrast to the above scheme, there is no "binding" of changes to the time of representation, their spatial characteristics and form of distribution along the river length. It can only be assumed that these are general directional changes. On the other hand, the separation of the dynamic equilibrium state obviously reflects only a very low rate of vertical changes, which leads to their erosion especially when other types of channel changes occur in a directional manner, and affects only long periods of time (historical and even geological).

Kondratyev et al. (1959) are also called directional vertical changes irreversible processes; lateral changes are identified with reversible processes.

The Chinese literature does not provide a classification of channel change species, but when considering the fluvial processes in the lower reaches of the Huang He and Yangtze, the main influence is given to the directional accumulation of sediment covering several hundred (up to a thousand or more) kilometers of riverbeds, i.e. vertical changes of the positive sign. An exception is the classification of the alluvial channel of the Weikhe River (Fluvial Processes…, 1984). It assesses the direction of vertical changes on the last large tributary of the Juanhe in terms of three gradations: stable bed, accumulation of sediment (very large sediment load) and incision of the bed (insufficient sediment load). At the same time, the conditions for the development of each type of change are set for different ratios of sandy and muddy load, bed material load and total sediment load; for stable channels (analogous to "dynamic equilibrium" or channels with very low rates of change), the values parameter of V. G. Glushkov $\frac{\sqrt{b_p}}{h}$, the degree of development of the bend l/L and the magnitude of the slopes I are given under these conditions; for the accumulation and incision channels, their qualitative characteristics are also given with respect to the selected conditions of sediment transport.

The vertical, lateral changes and changes associated with the movement of alluvial ripples reflect, in an integral form, the processes of separation, movement and settlement to the bottom of individual sediment particles. The latter are elementary channel changes that represent themselves at the level of individual solid particles, but together constitute the flow of sediment. According to Kondratyev (2000, p. 236), they represent the separation of a solid particle from the bottom, its

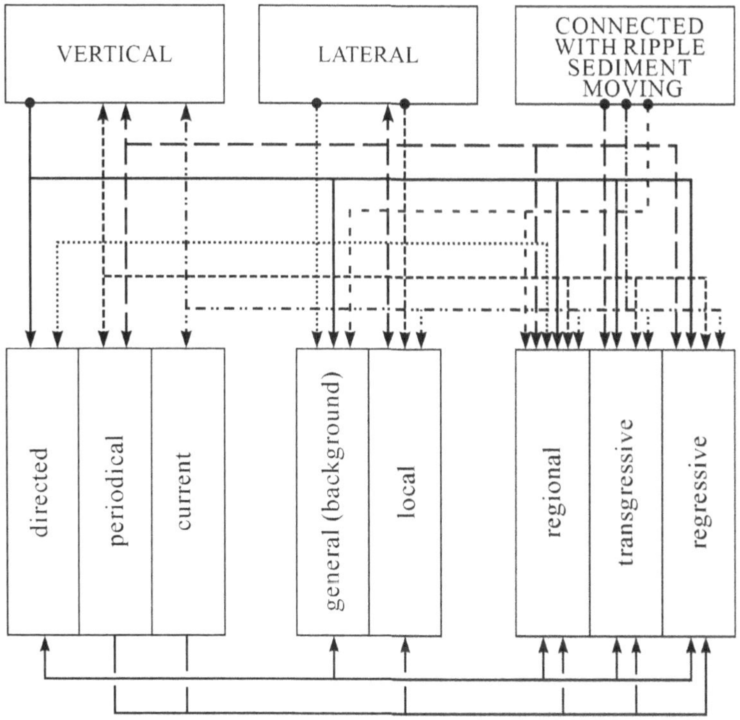

Fig. 4.29 Interrelationship of major types channel changes—vertical, lateral and connected with alluvial ripple moving

movement along the bottom (bottom sediments) or its transfer by the flow in suspended state (suspended sediments). "The movement of a single particle is the primary form of motion" (sediment—*R.Ch.*). Therefore, the impact of the flow on individual particles, their separation from the bottom, their subsequent movement, including the mass movement of such particles, and, finally, their stopping and transition to the category of sediments are considered independently of other types of channel changes, forming a practically independent area of research (Karasev 1975; Rossinsky and Debolsky 1980; Mirtskhulava 1988).

Each of the main groups of channel changes corresponds to different spatial and temporal scales of development: directed, periodic and current, on the one hand, and general (background), local on the other hand. For a given river cross-section, channel changes in terms of time of representation may be directional, periodic and current. The former are related to the most general conditions of riverbed formation and develop without changing the process sign during centuries, historical or geological periods of time. Periodic (alternating) changes are caused by seasonal or long-term fluctuations in the water availability of the river and the development of channel forms. Current changes are usually associated with the ripple movement of sediment, and only the form of their representation and the size of the ripple allow them to be attributed to lateral or vertical changes.

Channel changes of each species can occur along the river as a whole or over significant area of the river, representing general changes, or over limited lengths of the channel, which corresponds to local or local changes: in the former case, they occur in the pre-transition period.

The second corresponds to the elements of channel or channel relief shapes, meso- and microforms. Depending on the upstream or downstream distribution, channel changes are regressive (regressive erosion, regressive movement of islands; regressive movement of ripples—antidune) or transgressive (transgressive erosion, transgressive movement of bends, transgressive movement of ripples). If changes occur simultaneously along the whole river or over a sufficiently long section without a tendency to spread upstream or downstream, they can be called regional changes.

All types of changes interact with each other in a rather complex way, either by fully defining the development of one or more other species, or by exerting only indirect influence on them; some types of changes are connected by direct links, while others have clear, though not significant, secondary feedback (Fig. 4.29). For example, large ripples, falling into the zones of slowing down the flow on the bend or in branches of channels, into the places of bottom sedimentation by circulating currents in the stream (for example, in a straight line channel near the floodplain bank at a one-side floodplain), slow down their movement and are fixed by vegetation, turning into a floodplain and thus representing the most important condition for the development of channel shapes, directed and periodic lateral changes. In turn, directional vertical changes influence the development of channel and channel relief forms. All other things being equal, on accumulating rivers there is a higher probability of formation of branches of the channel, riffles and other ripple forms of the channel relief; single-thread channels and predominance of forest areas are more typical of incised rivers.

In rippleless mountain channels and on flat pebble-boulder rivers, channel changes are carried out in the form of movement of individual alluvium particles, i.e., they are elementary, although in the channels of mountain rivers of certain types and on flat rivers ripples are formed, which, depending on the amount of load, make up the macroforms of the channel relief or are scarce. Sediment transport begins with the destruction of the scaffolding, when the flow velocities exceed the incipient ones for the particles composing it. Sediments under the scaffold, consisting of particles of varying particle sizes, from sand to boulders, are involved in the movement. Such changes stop at the beginning of the flood recession and are therefore periodic in time of their representations. The formation and erosion of the macroforms of the channel relief, i.e. riffles, also belongs to the periodic ones.

The first group of channel changes united by the concept of "vertical changes" includes the processes of plunging or scouring the bottom, on the one hand, and the accumulation of sediment on the bottom, on the other hand. Depending on the spatial and temporal representations, they form a torus group: (1) Constant directional changes in the longitudinal profiles of rivers (deep erosion or systematic accumulation of sediment at the bottom of river valleys), the sign of which changes only when the general conditions of channel formation change; (2) local periodic (alternating) transformations of the longitudinal profile caused by the evolution of channel forms (e.g., caused by periodic cutoff and development of channel bends, long dying out and the development of short branches, etc.); (3) local bed scour or local accumulation of sediment associated with the development of ripple forms of sediment movement, seasonal and multi-year riffle mode, including the scour and movement of bottom sediments along the ripple's upper slope and their deposition in the basement, i.e., the movement of ripples downstream.

In the first case, vertical changes cover almost the entire length of the river, or parts of rivers of considerable length. Their development is influenced by natural conditions throughout the river basin. Such changes are common (background). The rate of development of deep erosion (plunging) or systematic accumulation (accumulation of alluvium in the channel) as their varieties only at anthropogenic genesis is the first centimeters per year, measured in natural conditions, usually in millimeters and fractions of millimeters. However, there are cases when the intensity of directed vertical changes in natural conditions is centimeters and the first tens of centimeters per year. Such rates of plunging are noted for mountain rivers of the Western Tien Shan—1–7 cm/year (Hakimov 1992). Maximal velocities (up to 32 cm/year) are fixed on the Alabuga river (Naryn tributary), which forms a canyon 50 to 130 m deep (Sidorchuk and Chalov 1996).

Intensive directed accumulation of sediment (also up to the first centimetres per year) has been observed in the lower Amu Darya as a consequence of high evaporation losses in the desert (3.5 cm/year in natural conditions, before runoff is disposed of for irrigation), The Rioni within the Colchis Depression due to its active tectonic deflection (1 cm/year), the Lower Terek (2–3 cm/year), where it is associated with fluctuations in the level of the Caspian Sea and estuary elongation of the river. For comparison: the average Amur load accumulation rate is estimated at 0.6–1.2 mm/year on average (Makhinov et al. 1994), and the highest rates of systematic accumulation of load are recorded at the Huang He River—more than 9 cm/year (Chalov et al. 2000).

Periodic vertical changes occur in short sections of the channel within the boundaries of one or more forms of the channel (bend, branching nodes), spreading upstream from the place of excitation (e.g., regressive plunging in the case of cutoff) at a relatively short distance. Being local in nature of distribution, they are connected with the evolution of channel forms, being their derivative. The local character of the riverbed is caused by its artificial congestion by dams, bridges, at the base of berthing walls and embankments, as well as at the embankment sections of the downstream waterworks and at the places of the riverbed quarries development. In the latter case, they spread from the quarry both transgressively (due to the interception of sediment load by the quarry) and regressively (due to the formation of a decline curve of levels upstream).

Current periodic vertical changes are represented in the fluctuations of channel bottom marks in the course of seasonal changes in the carrying capacity of the flow due to its different runoff rates in different phases of the hydrological regime. In particular, they include seasonal rearrangements of the riffles, which consist in the reclamation of their ripples in the flood and erosion in the low water level, or vice versa. The amplitude of the bottom marks is tens of centimeters—meters per year, depending on the size of both the river and the ripple itself. Thus, the rates of vertical changes are increasing from the periodic to the current, with the former being general, the latter local and the latter local.

When alluvial ripples move, vertical changes appear only within individual ripples. Large ripples, commensurate with the width and depth of the river, have an impact on the position of the free surface of the flow, also within the adjacent top of the pool or upstream ripple, contributing to the erosion of its ripple. On the contrary, when such a ripple is washed out, the sediment entering the stream accumulates on the lower ripple, causing an increase in the marks on the bottom of the channel. In all cases, vertical changes are due to the automatic levelling of the transporting capacity of the flow, the transformation of the specific energy of its living cross-section and the energy consumption along the length of the river for the flow of water, sediment transport and soil erosion of the bed and banks (Chalov 1986), and their direct cause is the change in the ratio between the value of sediment load W and the transporting capacity of the flow W_{tr}. With W_{tr} = const, an increase in sediment load is accompanied by accumulation ($W > W_{tr}$) and a decrease ($W < W_{tr}$) by bottom scour.

Lateral changes are either directional or periodic, depending on the conditions of channel formation. The directional ones consist in a constant movement of the straightened channel towards the main bank due to the appearance of such a kinematic structure of the flow at the effective discharge, which causes its erosion and accumulation of sediment along the floodplain (Chalov 1979). In the branched channel, for the same reason, the branches along the main banks of the river mainly develop, and when the river is located along them, the channel also tends to a directional moving over the years. A more common type of directional channel movement is the development of incised bends, where the bedrocks that make up the banks of the river prevent, and in the case of deep channel incisions and high bank heights, the possibility of cutoffs is excluded. Under the influence of such factors as Coriolis force (Baer`s law), stable direction of strong winds across the valley, tectonic distortions of the earth's surface (Shchukin 1933, 1960), can finally occur as a result of the millennial evolution of the channel, a consistent shift to one of the sides of the valley of the meandering belt or the riverbed branching belt. In such cases, the bottom of the valley acquires an asymmetrical transverse profile, and the belt of wandering the river on the bottom of the valley is much already across the floodplain.

Being present at long river sections, directional lateral changes are common. For example, the general shift of the Ob riverbed to the left bank between the mouths of the Charysha river near the village of Shelabolikha and to the right bank from the Novosibirsk to the mouth of the Tomi river or the Volga river (before the regulation of its flow) in the direction of the right high ("mountain") bank from the confluence with the Oka (Nizhny Novgorod) and almost to the top of the delta. Velocities of such shifts are insignificant: the Ob below the mouth of the river Charish, where the left bank is composed of loess-like loam up to 100 m high, is estimated at 0.5 m/year (Trepetsov 1963), on the Northern Dvina from the confluence of the Ug and Sukhona to the confluence with Vychegda—2–3 cm/year (here the retreating right bank, 30–40 m high, is composed of Upper Paleozoic mounds, marls, sandstones and clays) (Amalitsky 1931). However, on many rivers (or a significant length of their flow), the retreat of the main bank under the influence of the flow practically does not occur. The existing asymmetry of the transverse profile of the valley reflects the evolution of the channel over the entire history of its geological development and only in short areas, where the river is approached by the benches of floodplain terraces, there is observed erosion and local representations of directional movement of the channel. For example, the Ob river below Novosibirsk (see Fig. 2.7), located along the right side of the valley, washes away the benches of sandy terraces in the valley floor extensions, while in the narrowings to the river hollow slopes of riverside elevations descend. More often than not, the root slopes of the valleys directly approaching the river develop under the influence of landslide processes, which are provoked by the leaching of their foundations by the flow of water. These are, for example, the left bank of the Sysola before the confluence with Vychegda and Vycheda below the confluence (Ruleva and Chalov 2003), some parts of the Lower Dniester (Berkovich et al. 1992).

Periodic lateral changes consist, depending on the shape of the channel, in cutoff free bend, their formation and development to such a size that the channel is straightened again, in the alternating development of branches in the branched channel, the extinction of branches, the dying of branches, the division of the islands to the banks and the reformation of new islands that divide the channel into branches. Their analogue at the level of macroforms of the channel relief (riffles, side-bars, etc.) is the rejection of the sideways and braids, their addition to the bank, and the removal of sideways and other ripple forms of the channel relief.

Lateral changes develop as a consequence of the hydraulic structure of the flow, local changes in its living force and transporting capacity; they are associated with the formation of meandering, branched and relatively straight single-thread channels. They are most characterized by periodic (alternating) changes due to cutoff and development of bends, redistribution of runoff between branches and other types of channel wandering. Lateral changes are local within individual bend towers or junctions; if a channel type (and its corresponding lateral changes) are observed along the whole river or in areas of considerable length (morphologically homogeneous), they become general in nature.

Movement of alluvial ripples is the most universal form of representation of fluvial processes. It occurs, according to Makkaveev (1955), under conditions when the live force of the flow is insufficient for the transport of sediment of a given particle size in a suspended state; therefore, some of them are transferred to the composition of bedloads, the flow of which is carried out by the movement of their bottom congestions in the form of ripples. It is associated with continuous changes in the channel both vertically (increase or erosion of the rope ripples in different phases of the regime) and in the depth of the channel as it passes through the ripple or basement of the ripple as it moves by the flow) and laterally due to changes in the position of shallow and deep parts of the ripple and, consequently, the marks of the bottom across the channel. This type of change takes on a directional character when the ripples fall into the zones of slowing down the flow and accumulation of sediment and form powerful and high accumulations of sediment in the form of extensive side-bars or complex riverbed banks. Fixed with vegetation, they form part of the forming floodplain, in the structure of which they form a channel faction of alluvium. Large ripples are also characterised by periodic changes, as, depending on the position relative to the shape of the channel, their coastal parts (side-bars) are rejected from the banks and the sedimentary ripples are joining to the banks.

Local periodic representations of plunging or accumulation of sediment associated with the development of bends or branches of the channel are superimposed on the general directional vertical changes (Fig. 4.30). In some cases (when the signs of change coincide) they determine their local strengthening, in others—their weakening, and only sometimes—a temporary change of the sign of the direction of the process. In turn, the movement of large ripples due to the erosion of their upper (pressure) slope and the accumulation of sediment in the basement (on the lower steep slope) causes changes in the bottom marks, commensurate with the depth of flow—positive when the ripple is pushed on top of each other and negative (in this section) when the ripple moves downwards.

As a result, periodic changes in the intensity and direction of the processes occur against a general background of river incision or systematic accumulation of sediment. For example, the extension of the delta (estuary elongation of the river) causes a systematic accumulation of sediment in the lower reaches, spreading regressively upstream of the river. However, due to the restructuring of the hydrographic network in the delta, periodic development of newer, shorter delta branches, there is also a temporary development of an upstream regressively propagating channel incision, which is replaced by a new accumulation as the

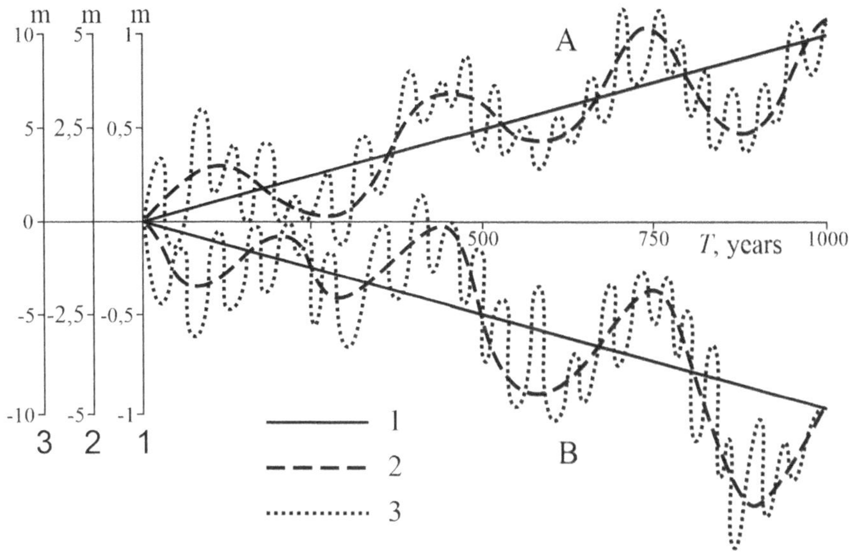

Fig. 4.30 Interrelationship over time T of total directed (1), periodical local vertical, connected with channel forms developing (2) and current alternating deformations, caused by the moving of ripple forms of channel relief (3): **A**—at directed sediment accumulation; **B**—at river incision. Each vertical scale (1, 2, 3)—change of the bottom marks relative to conventional zero at different types of vertical changes

estuary lengthening of the newly developing branch is taking place; against the background of the general process associated with the formation of the delta as a whole, local changes occur, which determine the dynamics of its hydrographic network. Similar occurrences can be observed in rivers that accumulate sediment when the bend is cutoff, causing local incision of the flow. Due to local incision of the riverbeds, in the course of cutoff, inundated areas located in the upper (including the current) parts of the floodplain segments are formed on the incising rivers (general vertical changes), which are small in area and are located in the highest parts of the floodplain segments (Chalov 1970). When a river with a branched channel is cut into a riverbed, such elevated (non-inundated in the high-water phase of the regime) sites appear on islands. Such local, periodic channel transformations do not lead to fundamental changes in the end result of overall directed vertical changes. On the other hand, general directional lateral changes appear only under certain geological-geomorphological conditions of channel formation depending on the nature of general vertical changes: incised bends of the channel are typical for the incising rivers in the conditions of limited development of channel changes, i.e., in the areas of distribution of stable ones; directional movement of the rectilinear channel towards one of the banks with the formation of a one-sided floodplain can take place both in the areas of channel incision and in the areas of sediment accumulation (first example is the upper Ob below the Charysh estuary, second—middle Ob below Novosibirsk).

The role of ripples in the development of channel changes of other species can be very diverse. The development of microforms in the bed depends only indirectly on the general orientation of vertical changes: in the conditions of sediment accumulation, all other things being equal, a greater amount of sediment is transported in the benthic area, which contributes to the formation of ripples. They have a more significant influence on the seasonal or multiyear transformation of large ripples comparable with the channel parameters, as they can be formed on the upper slopes of the latter and determine the portioned inflow of ripple marks, etc. Large ripples, falling into the zones of slowing down the flow on the bend or in the junctions of branching, in the places where the bottom sediments are brought in by the circulation currents in the stream to the coastal parts of the channel slow down their movement, are partially fixed by vegetation and turn into a floodplain, thus predetermining the directional or periodic lateral channel changes.

Due to different combinations of orientation and spatial and temporal representations of channel changes and their interrelation, each main type of channel changes corresponds to a certain stage in the morphological hierarchy of the forms of representation of fluvial processes: I—longitudinal profile of the river; II—forms of the channel; III—ripple forms of channel relief. Vertical changes in this hierarchy are reflected only as general directional changes in the form of the development of the longitudinal profile of the river; lateral changes are represented in the development of both general directional and local periodic channel forms; changes associated with the ripple movement of sediment correspond to all the spatial and temporal changes in the shape of the channel relief.

Directional vertical changes are carried out as a consequence of the entire complex of natural conditions in the catchment area, determined by fluctuations in the basis of erosion, climatic changes and tectonic movements. They are associated with the targeted development of longitudinal river profiles over long periods of time, covering historical and geological periods. The change in the river's plunging by accumulation of sediment, and vice versa, occurs when the natural conditions in the catchment area change (Makkaveev 1955; Makkaveev et al. 1961). Despite insignificant velocities, vertical changes nevertheless create a general background for the development of other types of channel changes, predetermining the formation of incised or wide floodplain channels, contributing to the strengthening or weakening of the wandering of the latter along the bottom of the valley, etc. Exceptions in terms of both time of representation and speed of development are transformations of the longitudinal profile of rivers associated with artificial flow regulation, agricultural development of catchment areas, and reclamation measures. Accordingly, transgressive deep erosion and regressive accumulation develops in the lower and upper reservoir banks of the dams, which during the first few decades after the construction of the dam cover areas of tens and hundreds of kilometers in length, and the intensity of their development, as a rule, exceeds the natural processes by an order of magnitude. Soil wash-off during plowing and deforestation leads to excessive inflow of sediment into rivers, causing siltation and even death of their upper reaches and sometimes affecting the shallowing of relatively large rivers.

Lateral changes are manifested within the limits of separate channel forms (bends, branching nodes) or relatively short parts of the river where these forms develop conjugate (series of bends, systems of conjugate branches). The rate of lateral change varies from a few centimeters to hundreds of meters per year, determined primarily by the stability of the channel. Only in the conditions of limited development of channel changes is the rate of development of incised forms of channel (bank retreat) comparable, and sometimes even less than the rate of incision of rivers. The morphological effect of lateral changes here is represented only in the geological scale of time. Its extreme view is the floodplainless channels, limited to the main sides of the valley. In this case, the channel shape is often predetermined by lithology, fracturing, and tectonic fragmentation of rocks, and in this case, it acts as a leading factor in fluvial processes. On the other hand, the type of lateral change and its reflection in the morphology of different types of channel depends on the direct influence of the hydrological regime of the river, sediment load, the peculiarities of the development of erosion and denudation processes in the catchment area, the geological structure of the bottom and banks, and the morphology of the valley. Therefore, the development of lateral changes in specific natural conditions, which differ in a certain combination of these or those natural factors, is accompanied by the emergence of their inherent set of channel forms of different orders, as well as the specifics of their development on specific rivers.

Channel changes occurring in the course of lateral channel changes are accompanied by local representations of vertical changes, which are excited by local transformations of specific energy of the flow section (e.g., when the bend is straightened or the channel is elongated when it turns into a loopy channel) and occur against the general background of the development of the longitudinal profile of the river.

Ripple formation is a phenomenon that is most autonomous in relation to natural factors of channel formation; the smaller the size of the ripple, the more mediated the connection between them becomes. Ripple microforms (sand waves, grooves), emerging as a reflection of the movable bottom of the kinematic structure of the turbulent flow, are mainly short-term formations. The only connection between them and the natural conditions is the amount and composition of the load coming from the catchment area and as a result of bank and bottom erosion. The largest forms of ripple relief (riffles, side-bars) are related to the shape of the channel and therefore among the entire hierarchy of ripple relief forms are most affected by the entire set of factors of fluvial processes. The rate of development of this type of channel change is extremely diverse. The degree of development of the hierarchy of ripple forms of the channel relief to a certain extent affects the development of vertical and lateral changes, determining in the first case the contact of the flow with the root bed, and in the second case, contributing to the wandering of the flow on the bottom of the valley. Feedback occurs between the types of channel changes: the formation of ripples (side-bars and mid-bars) is the basis for the development of a particular form of channel; the latter creates such a flow structure, which creates the conditions for the formation of ripples.

4.5 Elementary Channel Changes

Acting on the ground, the water flow washes them away, separates solid particles from the surface of the flow bed, which begin to move in the direction of the current. At the same hydraulic characteristics of the flow (water discharge, hydraulic slope, flow velocity), its ability to capture and carry solid particles can be different, because the intensity of soil erosion depends not only on the properties of the flow, but also on the characteristics of the lithogenic medium it affects. The degree of soil resistance is determined by the size of individual particles, their shape, cohesive forces between them, homogeneity of the soil, the speed and nature of rock weathering, etc. These properties of grounds (soils and rocks) ultimately determine their resistance to erosion.

The quantitative expression of soil erosion resistance is the critical average flow velocities: incipient velocity V_{in}—the limiting velocity at which the particles on the bottom still maintain a state of rest; critical velocity V_{cr}, which corresponds to the beginning of mass movement of particles, i.e., the state of continuous soil erosion (Karasev 1975; Mirtskhulava 1988). In addition, in order to fully characterize the regime of bottom erosion and sediment transport, Karasev (1975) singles out a non-silting velocity corresponding to the lower limit of velocity at which particles remain suspended and do not settle on the bottom of the flow.

The process of detachment of soil particles from the riverbed or banks means the beginning of change of the riverbed, as the surface of the bed is already transformed at this micro level and solids are transported. As a result of the separation of one particle, the conditions of the flow impact on the adjacent particles change, its structure changes (the scale of this change depends on the size of the particles and the depth of the flow), the contact of the flow with the particles, previously lying under the broken off and protected by it from the impact of the flow.

The particle's separation from the surface of the flow bed and its propulsion is mainly hindered by two factors: the particle's normal particle-gravity to the bottom surface and the bonding force between the particles. Besides, in each specific case the particle separation conditions are significantly affected by the density of loose soil, mutual jamming of particles, the degree of its moisture content, the presence of vegetation, physical properties of the soil, the angle between the direction of flow jets and the plane it affects (this is especially important for coastal slopes), magnetic

Table 4.14 Change of the surface area of particles at crushing of the cubic body (Makkaveev 1971)

Total volume (cm^3)	Cross section size (cm)	Number of particles	Total area of particle surface (cm^2)	The weight of the particles at their density in water ($\rho_s - \rho$)
1	1	1	6	2
1	0.1	10^3	60	0.002
1	0.01	10^6	600	0.000002
1	0.001	10^9	6000	0.000000002

and electrical phenomena in the system "liquid–solid body" in their mutual movement, and others. However, they do not define the mechanism of the erosion process.

Gravity has a predominant influence on the stability of the soil consisting of relatively large particles, as the particle's weight is proportional to the cube of its middle cross section. On the contrary, the bonding force is inversely related to the particle size, as it is proportional to the total surface area of the particles, which increases as they break up (Table 4.14). The adhesion force is noticeable in particles with a diameter of less than 0.1 mm if the soils composed of these particles were sufficiently cohesive.

On the Fig. 4.31 a graph showing the relative importance of gravity and grip in soil erosion is provided. Episodes of changes in the conditional values of erosion resistance under the influence of particle gravity and adhesion, depending on the particle size, intersect at a point corresponding to the average particle diameter d_{av} about 0.05 mm. The largest deflection of the total resistance curve corresponds to the same size. Thus, the minimum value of the total tear-off resistance is observed for particles with a diameter of about 0.05 mm. Such particle size has scaffolds, for which the incipient velocity is 0.07–0.20 m/s. To detach a particle of 0.001 mm from the bottom, a flow velocity of about 1.3 m/s is required, which is the same as for washing out fine pebbles with a cross section of 15 mm (Engeneering… 1964).

Soils with a particle size (d_{av} of more than 0.05 mm are classified as unconnected soils; soils with a particle size of less than 0.05 mm are connected (Mirtskhulava 1988). Depending on the nature of the bonding forces, they are divided into plastic (viscous) and rocky ones. In the latter case, erosion resistance is determined by rigid bonds that determine the structural bonding of rocky crystalline and cemented rocks. In crystalline rocks they have the character of ionic bonds having an electric nature; in cemented rocks (sandstones, mudstones) rigid bonds arise at their cementation in natural conditions by these or those substances. If the rock breaks rigid ties, they, like any solid body, do not recover.

The bonding of plastic rocks is different. The bonds between the individual particles are water-colloid and molecular and can be restored after the action of the

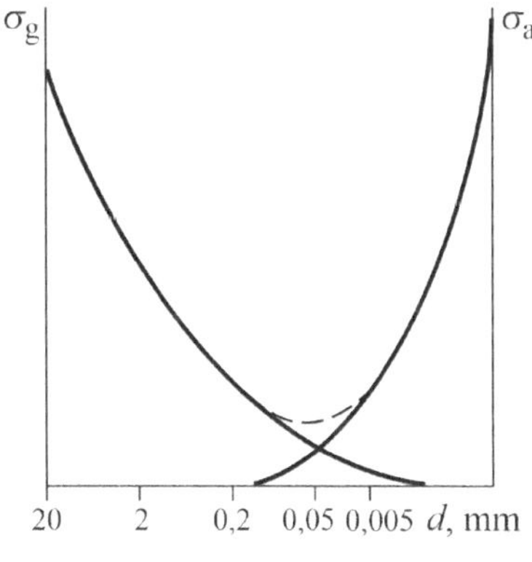

Fig. 4.31 Relative impact on soil erosion resistance of gravity force σ_g and adhesive force σ_{ad} of soil particles [according to N.I . Makkaveev (Design … 1964)]

factors that disturb them ceases. The strength of cohesion and, accordingly, erosion resistance of plastic rocks is determined by their dispersion.

In loose soils in a dry state there are no internal connections between particles. They appear and become stronger when the soil is moistened, and the finer the particles that make up the soil are.

The bonding forces between particles in cohesive and noncohesive soils F_{bond} can be expressed as a function of fatigue resistance to rupture of p_{rr}, the particle size d, the contact area ω_c and the thickness of the water layer δc (Karasev 1975):

$$F_{\mathrm{bond}} = f(d, \omega_c, \delta_c, p_{\mathrm{rr}}). \quad (4.27)$$

The thickness of the hydrate casings (water layer δc) and the contact area between the particles ω_c, in turn, depend on the damaging effect of the hydrostatic pressure. Accordingly, the adhesion force between the particles increases as the depth of flow increases and the particle size of the soil decreases.

The stability of individual soil particles against erosion is determined by the ratio between the hydrodynamic forces of the particle flow and the forces preventing them from being washed away by the flow. According to Karasev (1975), the hydrodynamic influence consists of three main components (Fig. 4.32A):

(1) the direct effect of the flow on the particle, i.e. the force of the velocity pressure exerting frontal pressure

$$F_{\mathrm{vp}} = k_{\mathrm{c-s}} \rho g d^2 \frac{V^2}{2g}, \quad (4.28)$$

where ρ—water density, $k_{\mathrm{c-s}}$—particle cross-sectional shape coefficient ($k_{\mathrm{c-s}} d^2$—midsectional section of the particle, directed to the elementary flow stream);

(2) difference in speed when the flow of a single particle flows, resulting in a pressure difference between the upper and lower surfaces of the particle. The pressure gradient is upwards and the resulting lifting force is

$$F_{\mathrm{l}} = k_{\mathrm{c-s}} \rho g d^2 \frac{V^2}{2g}. \quad (4.29)$$

The average lifting force is between 30 and 75% of the frontal pressure force. It decreases with increasing distance from the bottom of the flow, because in this direction there is a vertical alignment of speed, and ceases to operate at a distance of about several diameters of the particle. The increase in flow bottom roughness and flow rate increases the height where lifting power is generated;

(3) velocity ripples caused by turbulence of the flow. This results in a pressure difference between the front and back of the particle, below and above the particle. If there is no compensation due to additional factors, the particle is

either moved by a longitudinal pressure gradient or weighed by a vertical gradient.

Mirtskhulava (1988) has complicated this scheme (Fig.4.32B), taking into account the heterogeneity of the mass of particles composing the deposits at the bottom of the flow, and the emerging forces aimed at overcoming the resistance to shear, arising from frontal and lifting forces and under the influence of the water mass thickness, which has a damaging effect on the particles.

As can be seen from formulas (4.28) and (4.29), the frontal pressure force and the lifting force are proportional to the square of the speed, i.e. the pressure. Therefore, as a condition of the limiting equilibrium of the particle at the bottom of the flow in the order of first approximation, it is possible to take the equality of the velocity pressure force F_{vp} and the particle weight force F_P:

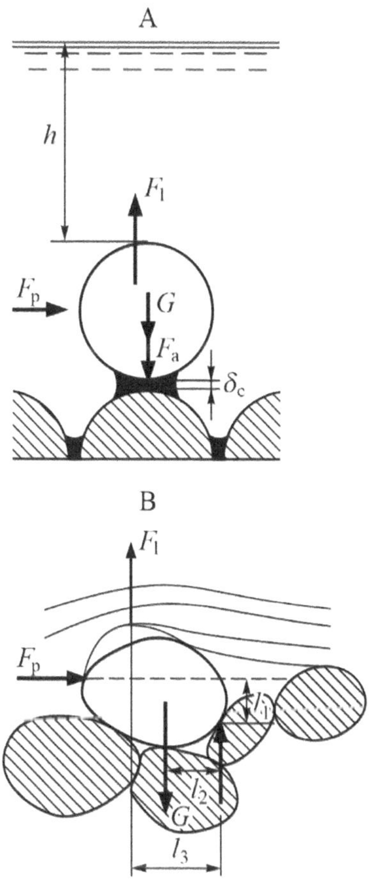

Fig. 4.32 Scheme of the forces acting on: **A**—particle of soil at the bottom of the stream: F_p—rapid pressure force; F_l—lifting force; G_B—particle gravity force; F_{ad}—adhesive force between soil particles; h—flow depth; δ_c—water column thickness between particles [according to Karasev (1975)]; **B**—on the soil particle, particle-surrounded. Vectors: of frontal force—l_1; lifting power—l_2; particle mass in water—l_3 [according to Ts.E. Mirtshulava (1988)]

$$F_{\mathrm{vp}} = F_{\mathrm{P}} \tag{4.30}$$

This state corresponds to a critical incipient flow velocity. If $F_{\mathrm{vp}} > \mathrm{F_P}$, the particle is shifted or weighed.

The gravity of the F_{P}, which prevents imbalance and counteracts flushing, can be imagined as

$$F_{\mathrm{P}} = W_s g(\rho_{\mathrm{s}} - \rho) = k_{\mathrm{sh}} d^3 g(\rho_{\mathrm{s}} - \rho), \tag{4.31}$$

where $\mathrm{W_s}$—particle volume; ρ_{s} and ρ—particle and water density, respectively; d—particle diameter in a characteristic cross section; *ksh*—particle shape coefficient. Substituting in equality (4.30) the values of the velocity force (4.28) and the particle weight in water (4.31)

$$k_{\mathrm{c-s}} \rho g d^2 \frac{V^2}{2g} = k_{\mathrm{sh}} d^3 g(\rho_{\mathrm{s}} - \rho), \tag{4.32}$$

if the flow reaches a incipient velocity, the value of the latter can be obtained

$$V_{\mathrm{in}} = k_0 \sqrt{gd \frac{\rho_{\mathrm{s}} - \rho}{\rho}} + K_{\mathrm{add}}, \tag{4.33}$$

where k_0 is a combined coefficient depending on the values of $k_{\mathrm{c-s}}$ and k_{sh}; K_{add} is an additional coefficient depending on the forces that need to be applied to tear a particle off the bottom, if it is connected with or stuck between adjacent particles by adhesive forces (Mirtskhulava 1988); frontal and lifting force coefficients are included directly in the expressions F_{vp} and F_{P}.

The obtained expression is valid for the conditions of fine-grained soils, when the partitioning between area of the midsection of a solid particle to the area of the living section of the flow is very small (i.e. with a certain degree of approximation $\frac{d}{\mathrm{h}} \to 0$), as well as for small longitudinal slopes. The latter determines the infinitesimal value of the tangential component of gravity on soil particles, which can therefore be neglected. On the other hand, the densities of alluvium and river water practically do not differ on different rivers. Therefore, the formula (4.33) can be simplified to allow sandy alluvial deposits to be washed away by river flows and relatively little suspension:

$$V_{\mathrm{in}} \approx 1.6 k_0 \sqrt{gd} \approx 5 k_0 \sqrt{d}. \tag{4.34}$$

At a time when the flow velocity exceeds a value equal to the incipient value, the particles composing the bottom of the flow shift. However, only a few particles come into motion. The mass movement of particles begins when the flow reaches the critical velocities, which, according to the research of Mirtskhulava (1988), are in the following ratio with incipient velocities:

$$V_{cr} = 1.4V_{in}. \tag{4.35}$$

The concept of particle motion continuity is used to characterize the erosion of the soil (movement of individual particles or their mass attraction along the bottom), which is [according to Karaushev (1977)] the partitioning between total volume of moving particles at the unit of the channel surface $\sum W_{mov}$ to the volume of the layer in which this motion occurs, i.e.$\sum W_{mov} + \sum W_{stat}$ where $\sum W_{stat}$ is the total volume of stationary particles:

$$\eta = \frac{W_{mov}}{\sum W_{mov} + \sum W_{stat}}. \tag{4.36}$$

In general, this indicator characterizes the degree of mobility of the surface layer of sediment that is directly affected by the bottom stream jets.

The particle density is determined by a number of factors: the shape of the particles; the degree of shading by the protruding particles of the bottom surface from the flow (i.e., by the protrusions at the bottom, which create stagnant zones in the flow; the duration of the excess of the instantaneous values of the pulsing velocity of the incipient values, the frequency of the bottom jets of the flow, in which the instantaneous pulsing velocity exceeds the incipient one. Thus, when the flow affects the rough surface of the bottom, only a fraction of the particles that compose it is directly affected by the frontal force—the force of the velocity of the head and the lifting force. A significant number of particles enter the stagnant zones. At value of average speed of a stream $V_{in} < V_{av} < V_{cr}$ the shift of particles occurs in a narrow range of the maximum pulsations where instantaneous speeds exceed not washing out. As the flow rate increases, the size of the stagnant zones decreases and more particles are subjected to shear and weighing forces. Therefore, the continuity of particle separation from the bottom of the flow with increasing speed increases and reaches its maximum at critical velocities. When the average flow rate is equal to the critical one ($V_{av} = V_{cr}$), the shift and weighing of particles is provided over the entire flow section, but only in the range of positive pulsations of instantaneous velocities. At $V_{av} > V_{cr}$, soil erosion also occurs in the negative ripple range.

Equality (4.32) also allows to obtain a critical size of soil particles for a given flow rate:

$$d_{cr} = \frac{V^2}{\rho_s - \rho}. \tag{4.37}$$

Erecting both parts of the equation (4.37) in a cube and multiplying them by $g(\rho_s - \rho)$, we obtain

$$P_{cr} = W_s g(\rho_s - \rho) \approx f(V^6), \tag{4.38}$$

where P_{cr} is the critical weight of the particle. This is an expression of Erie's famous law for the motion of bottom particles in the water flow: the weight of the particles carried by the flow is proportional to the sixth degree of flow velocity. The formula (3.38), in particular, shows that if the flow rate ratio is 1:2:4, the particle weight ratio will be 1:64:4096. Erie's law explains the fact that with relatively small differences in flow velocities, flows carry different material in size and weight. This difference is especially evident when comparing plain and mountain rivers: the former move only sand particles weighing thousands of times less than the weight of boulders and clumps moved by mountain streams, although the flow velocities of mountain rivers are only 3–4 times higher than those of plain rivers.

In addition, very large rock fragments often move in mountain river channels, weighing significantly more than the maximum weight of the particles that can tolerate the flow. Shifting of such fragments occurs as a result of abluvial effect (abluvio—buckwheat). As smaller particles are washed out near this debris, its position becomes less stable and even a small force that can be exerted by the flow is sufficient to disturb the balance. The rock chip is moved downstream for some distance and then the process of washing the base of the rock is repeated. The distance (x) by which the debris is moved depends on the capacity of the washed out fine soil (h_w) and the slope of the bed of the channel. The empirical formula obtained by Makkaveev and Kalinin (1968) looks like:

$$l = h_w\left(9 \sin \beta + K_{lp}\right), \tag{4.39}$$

where β—steepness of the channel bottom in degrees; K_{lp}—coefficient, the value of which varies from 0.3 to 0.5 depending on the shape of its longitudinal profile. As a result of the abluvial effect, rock fragments, regardless of their size and weight, can travel considerable distances on eroded slopes and in mountain river beds, even at relatively low gradients.

The tangential component of gravity acting on particles of soil at a steep slope of the bed of the channel (on a mountain river) significantly changes the results of

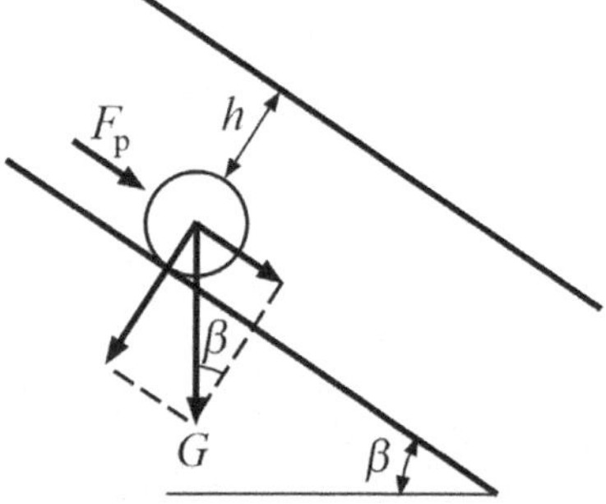

Fig. 4.33 Simplified scheme of forces acting on a particle of ground at the bottom of a stream with a large gradient: β—channel bed slope angle; other symbols—on the Fig. 2.32

definitions V_{in} by formula (4.33). The greater the slope of the bottom, the lower the component of the force of weight directed to the normal flow and the surface of the flow bed and the lower the force required to bring the particle into motion (Fig. 4.33). Indeed, the erosion-resistant force is proportional to cos β—the cosine of the angle composed by the lateral plane of the flow bottom, i.e. the slope of the flow bottom

$$F_P = W_s g(\rho_s - \rho)\cos\beta K_{fr} = k_{sh} d^3 g(\rho_s - \rho)\cos\beta K_{fr}, \tag{4.40}$$

where the K_{fr} is the friction coefficient.

The particle-movement force under these conditions will consist of the velocity force (4.28) and the particle weight component along the bottom slope, i.e.

$$F_{p-d} = k_{c-s}\rho g d^2 \frac{V^2}{2g} + k_{sh} d^3 g(\rho_s - \rho)\sin\beta. \tag{4.41}$$

Equating, as in the case (3.31), the forces shifting the particle (3.40) and forces resisting erosion (3.39), we obtain

$$V_{in} = k_0\sqrt{gd\frac{\rho_s - \rho}{\rho}(K_{fr}\cos\beta - \sin\beta) + K_{add}}. \tag{4.42}$$

When the $\beta \to 0$ two terms in brackets under the root sign are practically equal to 1, and equation (4.42) is reduced to equation (4.33). If $\alpha \to 45°$, then

$$V_{in} = k_0\sqrt{gd\frac{\rho_s - \rho}{\rho}\cdot\frac{\sqrt{2}}{2}(K_{fr} - 1) + K_{add}}, \tag{4.43}$$

i.e. at high gradients of the bottom of the flow, the incipient velocity tends to 0, and the movement of sediments becomes increasingly gravitational, and the effect of the velocity pressure is reduced.

In the cases considered, the potential energy of the flow has a relatively small impact on the value of the critical rate of erosion. The latter depends on the depth of flow (Goncharov 1962; Karasev 1975; Mirtskhulava 1988):

$$V_{cr} = f(h^n), \tag{4.44}$$

where the degree n value ranges from 0.15 to 0.35. Thus, the greater the depth, the higher the critical velocity required to bring the soil particle into motion. Hence, the structure of most calculation formulas for determining critical (incipient V_{in} or critical V_{cr}) velocities looks like

$$V_{\rm cr} = f\left(g^{0.5}, d^{0.5}, \mathrm{h}^{0.2}\right). \tag{4.45}$$

As an example, some formulas that are now widely used can be cited[1]:
Goncharov (1962)

$$V_{\rm in} = \lg\frac{8.8\mathrm{h}}{d}\sqrt{\frac{2\rho g(\rho_{\rm s}-\rho)d}{3.5\rho}} \quad \text{and} \tag{4.46a}$$

$$V_{\rm cr} = \lg\frac{8.8\mathrm{h}}{d}\sqrt{\frac{2\rho g(\rho_{\rm s}-\rho)d}{1.75\rho}}; \tag{4.46b}$$

Levi (1957)

$$V_{\rm cr} = 1.4\sqrt{gd}\,\lg\frac{12h}{d} \text{ at } \frac{\mathrm{h}}{d} > 60 \quad \text{and} \tag{4.47a}$$

$$V_{\rm cr} = 1.3\sqrt{gh}\left(0.8 + \frac{2}{3}\lg\frac{10h}{d}\right) \text{ at } 10 \ll \frac{\mathrm{h}}{d} 40; \tag{4.47b}$$

Shamov (1959)

$$V_H = 4.6\sqrt{d}\left(\frac{\mathrm{h}}{d}\right)^{\frac{1}{6}}; \tag{4.48}$$

Latyshenkov (1971)

$$V_{\rm in} = 1.6\sqrt{gd}\left(\frac{h}{d}\right)^{0.2}; \tag{4.49}$$

Studenichnikov (1964)

$$V_{\rm in} = 0.9\sqrt{\frac{g(\rho_{\rm s}-\rho)}{\rho}(hd)^{0.25}}; \tag{4.50}$$

Mirtskhulava (1988)

$$V_{\rm cr} = \lg\frac{8.8\mathrm{h}}{d}\sqrt{\frac{2m_1}{\alpha_1\rho m_2}\left[(\rho_{\rm H}-\rho)gd + \alpha_2 p_{\rm f.r.}k_{\rm gr}\right]}, \tag{4.51}$$

[1]A broader list of formulas, their analysis and evaluation is given in the work of Mirtskhulava (1988).

where m_1—coefficient, which assesses the strength of the soil (m_1 = 1 in sandy soil and clarified, without load flow; m_1 = 1.3 in the presence of load > 0.1 kg/m^3); m_2—coefficient that takes into account the pulsation nature of the velocity ($m_2 \approx 4$); $p_{f.r.}$—fatigue resistance of the soil to rupture, determined, in turn, for fine-grained, naturally densely packed soils according to the formula of Mirtskhulava (1988)

$$p_{\text{f.r.}} = 8\rho_s V(gV)^{1/3} d^{-1}, \tag{4.52}$$

where V is the flow rate; k_{gr} is the coefficient of homogeneity characterizing the variability of ground strength (k_{gr} = 1) for unbound soils; k_{gr} = 0.3–0.6 for clayey soils); α_1 and α_2 are dimensionless coefficients equal to those for sandy soils α_1 = 0.44, α_2 = 2, for clayey soils α_1 = 1.3, α_2 = 1.25. The C. E. Mirtskhulava formula (4.51) is the most universal one, as it allows to estimate erosion resistance of both unconnected and cohesive plastic soils. It is widely used for calculation of irrigation and main canals. Therefore, on its basis, C. E. Mirtskhulava developed special tables and a nomogram allowing determining the partitioning between incipient mean and bottom velocities depending on the diameter of particles and the depth of the flow. At the same time, they were offered approximate dependencies for particles with a diameter of more than 0.25 mm

$$V_{\text{in}} = \sqrt{4}g^2 d\text{h}, \tag{4.53}$$

where d and h are, respectively, the average particle diameter (mm) and depth of flow (m), and in the absence of data on them

$$V_{\text{in}} = k_{\text{cor}} Q^{0.2} \tag{4.54}$$

depending on the water flow rate Q (m^3/s) with regard to the correction factor k_{cor}.

The physical nature of the retardation effect of the flow depth on the intensity of soil erosion is determined by two aspects. Firstly, the adhesion forces between particles grow with increasing hydrostatic pressure, i.e. under otherwise equal conditions, the force to break off particles from the bottom should be greater in a deeper flow. Secondly, it is known that the greater the velocity gradient, the greater the pressure gradient in the flow, i.e.

$$\frac{\Delta V^2}{\Delta h} = \frac{\Delta P}{\Delta h} \tag{4.55}$$

Hencc, at the same average flow rate, the pressure gradient in the deep flow is less than in the shallow flow. Accordingly, the weighing force in the deep flow, which is determined by the difference in pressure between the bottom and top of the particle, will also be smaller; the value of the incipient velocity is greater. Table 4.15 shows K. M. Arbulieva (Rossinsky and Debolsky 1980) on the basis of experiments the value of the coefficient of lifting power at different Reynolds

Table 4.15 Influence of flow depth on the lifting power coefficient value [according to K. M. Arbulieva (Rossinsky and Debolsky 1980)]

Reynolds' number Re × 10^3	$\frac{h}{d}$	Lifting power coefficient
2	1.6	0.34
3.5	3.3	0.38
4.5	4.6	0.25
5.2	5.9	0.14
5.5	6.5	0.1
6	8.7	0.176
6.1	9.4	0.226
6.4	9.8	0.226

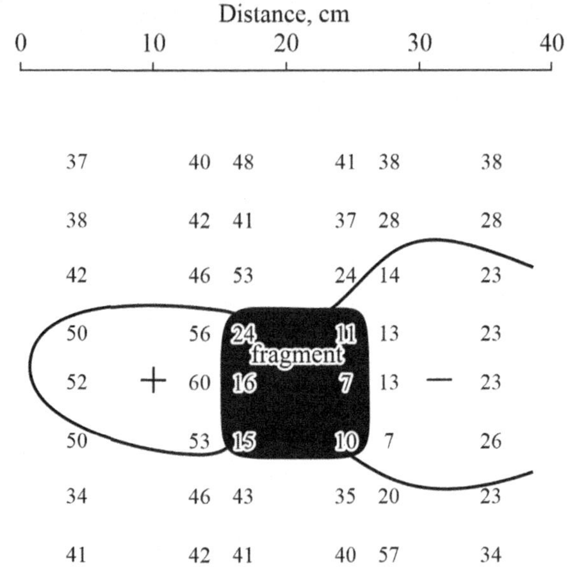

Fig. 4.34 Depth (mm) on the section of the lab channel near the rock fragment in the middle of the tray. Numbers on shaded area—water column thickness over the fragment; "+" and "–" signs mean elevation or depression of the water surface (Makkaveev and Kalinin 1968)

numbers and depths, taken in the ratio with the cross section of the particle. These data show that in a certain range of *Re* numbers the depth factor is decisive in the formation of the lifting force, but then a high degree of turbulence neutralizes the influence of depth growth, and the value of the coefficient of lifting force is even slightly increased.

Depth slowing down soil erosion is represented at low values of particle diameter to flow depth ratio. A different pattern is observed if the cross section of the alluvium particle and the flow depth are of the same order, which is typical for small mountain rivers, where the size of fragments (boulders, blocks) is comparable with the flow depth. Due to the steep slope of the bottom, such flows often move blocks, boulders or large pebbles, the cross section of which even exceeds the depth of the flow. Such large particles are brought into motion not only as a result of the effects of high-speed (kinetic) energy, but also as a result of hydrostatic pressure (potential energy). Each coarse particle causes a pronounced changes of the water surface, i.e. a level rise from the upper side and a level depression from the lower side, counting in the direction of the current (Fig. 4.34). Since the coarse particle moves at a slower rate than the velocity of water jets, it is a kind of mobile dam, continuously perceiving both high-speed and hydrostatic pressure. The latter (if the cross section of the particle is greater than the depth of flow) is approximately proportional to the square of the depth.

According to the experimental research data Makkaveev and Kalinin (1968); (Experimental Geomorphology 1969) have proposed the following expression for determination of critical incipient velocity for conditions of comparable depth of flow and size of debris particles:

$$V_{\text{in}} = K_{\text{p}}\left(g\frac{\rho_{\text{s}} - \rho}{\rho}\right)^{0.5} d^{1.5}\,\text{h}^{-1} + K_{\text{add}} \tag{4.56}$$

The value of the K_{p} coefficient here is a variable depending on the slope steepness. In the case of steep gradients of the flow bed, a complication similar to that of the derivation of formula (4.42) should be introduced into equation (4.54).

The comparison of expressions (4.46) and (4.56) shows that the role of depth in both cases is opposite. In the second case, the total flow energy is used to perform erosion and transport work much more fully than in the first case. Due to this feature, often small rivers, flowing into a large river, form in its channel a cone of discharge. The coarse-clastic material of these cones is not washed away by the flow of the main river, although its velocity is higher than that of the inflow that formed the cone.

The critical velocity relationship on alluvium particle diameter and depth can be expressed in the following way in the criterion form

$$\pi_1 = \frac{V_{\text{n}}^2}{gh}; \tag{4.57}$$

$$\pi_2 = V_{\text{n}}^2. \tag{4.58}$$

It follows from the above that the relationship between the two criteria (π_1 and π_2) may vary depending on the conditions that influence the use of the specific energy of the particle flow section.

The effect of the influence on the shear of debris particles of the total specific energy of the flux cross-section is amplified in the case of direct positive movement waves in the flux. The maximum velocities of the jets are observed in the leading edge of the wave, and in the rear part of the wave are reduced to values close to or even less than V_{in}. The greatest relative increase in the velocity of jets at the approach of the leading edge of the wave is observed in the near-bottom layers of the flow; the values of flow velocities in the surface and near-bottom layers are almost identical. As a result of the synchronous impact of the obstacle and the sudden increase in velocity in the bottom layers of the flow, the movement wave can shift relatively large alluvial particles, whose weight is 2–3 times and more than the maximum weight of the particles carried by the flow, having a depth and average flow rate similar to that of the movement wave when passing its crest. A coarse alluvium particle moves slower than the leading edge of the wave, where the depths are still significant because the bottom flow velocities are much lower than in the leading edge area. The particle continues to move for some time as a result of inertia, but then stops.

According to the observations of L. F. Litvin (Litvin and Chalov 1975) for the movement of marked fragments in the channels of the smallest rivers—tributaries of the Western Gumista river (Western Caucasus), there is no connection between the size of the fragments and the distance at which they were transported: larger fragments were moved by a greater distance than the fragments, smaller in size (Fig. 4.35). Blocks of the same petrographic composition often found themselves in the channel sections in the area of rock distribution, supplying smaller material to the channel. On the Pskem River (Western Tien Shan), below the inflow of the Urungachsai, where the river flows in the area of red clay distribution, there are blocks of granite in the channel (1.6 × 1.0 × 1.20 m), brought by the flow from the upstream areas (Lodina et al. 1994). Stationary observations of the movement of large boulders on the rivers of the Western Caucasus with a porous boulder bed allowed us to establish that the rate of movement of debris up to 10 m in diameter at a maximum flow depth of 1.5 m was 0.2 m/s (Khmeleva et al. 1970). After the catastrophic rainfall flood, large boulders with a diameter of 0.8–1.0 m were moved out of the 300 m long section and blocks of 1.5–2.5 m in size were moved by a small distance (the depth of the flow reached 3 m).

With the movement of a single large piece (lump), a large amount of fine material begins to move immediately for which the piece (or clusters of lumps) served as a barrier, creating a zone of support where the accumulation of sediment occurred. In their "shadows" there is also often the accumulation of smaller material represented by small boulders, pebbles and even gravel. For example, on the river Chatkal (Western Tien Shan) below the inflow of Hargush-Sai behind the huge blocks thrown out by mudflow from this tributary, there are small boulders with a diameter of less than 100 mm and pebbles, which can be transported by the flow at lower speeds and slopes. This provides a mass movement of fine material after the shift of a large boulder or block, leads to rapid propagation of a wave of regressive erosion in some part of the river and, in turn, favors the emergence of an abluvial effect in the bed of upstream boulders and clumps.

The mass movement and transport of sediment particles by flow is largely determined by the degree of heterogeneity of the soil it affects. Typically, the underlying surface (river bed) is represented by a wide variety of material in terms of size. The composition of bed sediments on mountain rivers is particularly heterogeneous, with boulders, pebbles and gravel of three to four orders of magnitude differing in size from one another within a small site (large boulders reach a cross section of 1 m, while gravel size is 1–10 mm). On flat rivers, the composition of the surface alluvium layer, which is directly affected by the flow, is much more homogeneous. Sands of various types, differing in content of individual fractions, dominate here. The particle size of the extreme sand fractions differs by a factor of 10. Fine sands are characterised by a good sorting and a low diversity of fractions. On the contrary, coarse sands are usually poorly sorted, consist of a large number of fractions and have an admixture of gravel or even fine pebbles.

The heterogeneity of the grain size of the soil causes a number of specific features of its erosion. At any given moment, the velocity (including ripple velocity) may be greater than incipient for some particles, and less than incipient for others, larger, less than incipient. This fact is taken into account, for example, in the formula of incipient flow rate proposed by Goncharov (1962) not for individual particles, but for a soil of heterogeneous composition:

$$V_{\text{in}} = 3.9\left(\frac{d_{\text{av}}}{d_{90\%}}\text{h}\right)^{0.2}(d_{\text{av}} + 0.0014)^{0.3}, \tag{4.59}$$

where d_{av} is the average particle diameter of the soil; $d_{90\%}$ is the particle diameter corresponding to 90% of the composition on the integral curve of the granulometric analysis.

Due to the different impact of the flow on the soil particles of different particle size and selective erosion of the bottom, a detachment is formed (Fig. 4.36). On pebble-boulder rivers, the decanter is also formed as a result of the release of particles from the state of motion at the flood recession or flood, when the hydraulic characteristics of the flow (slope, speed, depth) decrease. A decrease in speed after the flood peak causes the largest material to stop abruptly enough. The greater the heterogeneity of the soil, the more likely it is to form a scaffold. In the lower part of the riverbed, the erosion of the riverbed protected by the scaffold on the mountain river and the flat river with pebble-boulder alluvium stops. With increasing water flow rates and flow rates at the first stage of erosion, the heterogeneity of the soil increases the erosion ability of the flow due to the formation of eddies near the individual large particles, which increases its degree of turbulence. However, later on, however, the formation of the detachment reduces the intensity of erosion or prevents it altogether. As a result, the erosion of soils formed by inhomogeneous particles represents itself in a light-like stage.

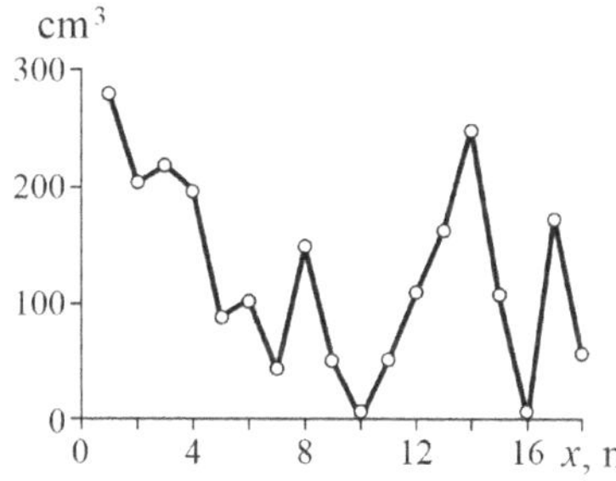

Fig. 4.35 Connection of average sizes of debris particles (cm³) with travel path length (x, m) in channels of small rivers in Western Gumista river valley (Western Caucasus) (Litvin and Chalov 1975)

Fig. 4.36 Pebble-boulder blind area on the shoal of the Middle Lena. Photo of the author

Disruption even of individual particles (wreckage) composing the decay leads to the impact of the flow on the underlying strata consisting of heterogeneous in composition and mainly smaller soil than the decay material. For it, $V_f > V_{in}$ and even $V_f \gg V_{in}$, which leads to a sharp activation of its erosion. In this case, as follows from the formulas for incipient velocities (4.33), (4.42), (4.46), (4.51), (4.56), the saturation of the flow with small fractions leads to a decrease in "n, which, in turn, facilitates the shift of larger particles. This is a phenomenon, the first time that N. I. Makkaveev and others have established (1970) for mountain rivers, later it led to the explanation of the mass transport of pebbly-boulder material by lowland rivers flowing in mountainous areas (Aldan, Vitim, upper and middle Lena, Yenisei, etc.) (Berkovich et al. 1985).

In the case of erosion of cohesive plastic soils, one of the main factors preventing erosion is the adhesion force between the individual particles. In addition, the soil compactness, the degree of its moisture content, the presence of soluble salts, the mechanical and mineralogical composition and other physical and chemical properties have a certain influence. The complex and often contradictory impact of these factors has led to the fact that there are still no reliable formulas for determining the critical rates of erosion of cohesive soils, although the search in this direction is very active. Above was given a universal formula of C. E. Mirtskhulava (4.51), in which fatigue strength and a number of special coefficients are introduced to take into account different properties of sandy and clayey soils.

Since the adhesion forces depend on the soil mechanical composition, Mirtskhulava (1988) proposed to divide the plastic soils into four classes according to their resistance to erosion, each of which was previously characterized by Mirtskhulava (1967) on the basis of laboratory and field experiments:

I soils of low resistance to erosion, having low cohesion, well aggregation: units are characterized by high mechanical strength and water resistance; V_{in} = 0.50–1.00 m/s;
II soils of medium resistance to erosion, have a noticeable bonding; V_{in} = 1.00–1.50 m/s;
III soils of high resistance to erosion, which have high cohesion in a solid state, but easily soaking; V_{in} = 1.5–2.00 m/s;
IV soils of especially high resistance to erosion, having high strength, cohesion and dense addition; V_{in} = 2.00–3.00 m/s.

In addition to the mechanical composition, the stability of cohesive soils, according to the descriptions given in the classification of C. E. Mirtskhulava, is greatly influenced by the saturation of the soil with moisture and the degree of compactness. Unsaturated bonded soil has little resistance to erosion. For example, the just-delayed sludge is washed out at a flow rate of 0.07 m/s. Gradual compaction leads to increased erosion resistance due to the formation of a structural skeleton. The latter is formed in the muds represented by hydro-muclides and kaolinite, for 15 days, and mormorillonites—for 20–30 days. The resulting clay is washed away at speeds 10 times higher than the freshly deposited silt—0.7 m/s. In the 9–10 months between the two clayey slots, the clay sludge can harden to such an extent that the incipient velocity for it will exceed 2 m/s.

The saturation of the soil with moisture has a dual effect on its resistance to erosion. On the one hand, it has been established that clays in a dry state are washed away at a speed 2–6 times less than in a state of complete water saturation. This is due to the increase in their plasticity as the humidity increases. Mirtskhulava (1967) gives the following example: incipient velocity of carbonate loams at humidity of 8–10% (hard consistency of the ground) is approximately 0.75 m/s, at humidity of 16–18% (low plastic consistency)—1.35 m/s, at humidity of 25–27% (plastic consistency)—2.6 m/s. When wet and ductile, carbonate loam withstands flow rates 3.5 times higher than when dry.

On the other hand, excessive saturation of the soil with moisture leads to soaking of the rock, because in this case the thickness of the hydrate cover reaches such values that the soil particles leave the sphere of molecular attraction. As a result, the washout of very dense clays does not stop completely, even at low flow rates; while under water, they are gradually liquefied and go into suspensions or colloidal solutions, which are easily removed by the flow.

Erosion of clayey soils is facilitated by the mineral salts and organic compounds they contain. They cement the rock dry, but weaken it when wet, causing swelling, soaking and leaching. Sodium and potassium compounds show the greatest activity in the process of diffuse leaching. Therefore, the content of these elements in clayey soils characterizes the reduced stability of soils against the chemical effects of water. There are data that the increase in the concentration of sodium, calcium and magnesium salts from 0 to 13% reduces the value of critical share stress by 1.5–3 times. One can judge about the intensity of hydrochemical erosion of clayey soils by the following data given by Karasev (1975) for the bed of the Bolshoi Yegorlyk river: the leaching rate reaches 36–38 g/day from 1 m^2, which corresponds to a daily bottom decrease of 0.9–1.4 mm.

The erosion of rocky soils is facilitated by the fracture of rocks, the occurrence of specific hydraulic phenomena in the flow, causing a sharp change in pressure at the border of liquid and solid media, the abrasive impact of the flow of bedloads, the joint effect of the flow on the rocks and their chemical and microbiological weathering.

All the rocks are characterized by cracked rocks, which causes them to be divided into different sizes. Mirtskhulava (1967) distinguishes two main groups of fractured rocks from each other in terms of their stability to erosion: the first group includes rocks crushed by cracks separately, between which there is practically no connection; the second group consists of rocks in which the cracks between the separations are filled with cementing material. For rocks in the first group, the resistance to erosion depends on the size of the separations, and thus appears to be in general terms similar to the resistance of unbound soils. In the case of rocks of the second group of rocks, the factor of connection between the separate compartments is added to it, and their destruction occurs only after overcoming the grip forces. The mechanism of rock bed erosion is that the flow causes the vibration of separations, which is stronger the greater the "angle of attack" formed by the direction of the jets and the rock surface; therefore, the greatest erosion capacity in relation to the crystalline rocks has a falling vertical water flow (waterfall, hydroelectric power plant spillway, etc.). As a result, the individual parts are loosened, the cracks are mechanically widened, the ties between the parts are weakened or completely destroyed. In these conditions, the excess of the incipient velocity of the flow leads to the failure of separations. In the presence of binder individual cement resistance to erosion increases significantly, because the bonding determines the direct relationship of the resistance forces on the fatigue resistance to rupture of the individual.

Mirtskhulava (1967) proposed an analytical formula for calculating the incipient velocity of rock fractured soil by a flowing stream of flow, which is not given here because of its complexity and limited application. It includes such indicators as the coefficients of working conditions of the individual for tensile and bending, the size of the individual in three axes, fatigue resistance of ties between the individual and a number of others.

The existence of cracks in rocks, their fragmentation into separate rocks can not always explain the erosion of rocks, as the size of the divisions often exceed the critical ones. The condition for the disruption of the compartments is their preliminary grinding. This is facilitated by the formation of cavitation zones in the flow and zones of transition of the turbulent flow into calm and vice versa. These phenomena can also act as independent factors of destruction and crushing of monolithic, crack-free rock. Accompanied by pressure pulsation, cavitation causes the destruction of the strongest crystalline rocks (basalts, quartzites, diabases, flints) on the steps and waterfalls.

One of the most significant factors of destruction by the water flow of rocks is the abrasive impact of solid particles (corrasion) transported by the flow. The greater the amount of suspended sediment in the stream and the larger it is, the more noticeable its mechanical impact on the rocky bed of the stream will be.

Rock erosion occurs in rivers as a result of the combined action of erosion and weathering. The flow breaks down individual rock fragments, which are weakened by the flow of water, carries out dissolution products and supplies more and more salts and acids, thus stimulating the process of dissolution and the processes of exchange reactions, which modify the rock. On the surface of the rock exposed to the action of aqueous solutions, there are almost separated from the monolithic grain minerals, for the separation of which from the rock is enough small mechanical impact of the flow.

The weathering of rocks under the river bed begins with cracks and separations. Cracks, which can be of primary origin (and not only as a result of weathering), allow the penetration of flowing water into the rock, accelerate the processes of its dissolution and biochemical weathering. In easily karsting rocks, chemical weathering is the main agent of the destruction of bedrock and, consequently, its erosion.

Particles of different particle sizes, being torn off from the soil composing the bottom and bank, are transported by the flow, unevenly distributed in its thickness. The nature of the flow of sediment and its distribution in the flow column is determined by the partitioning between its average velocity to its incipient velocity, the vertical component of the velocity and the hydraulic particle size of the load. Karaushev (1977), depending on these ratios, first of all on the partitioning between hydraulic grain size and the vertical component of the pulsation velocity, proposed the following classification of sediment load:

consisting of particles with a hydraulic particle size w less than the average value of the vertical component of the ripple velocity u_y, i.e. $w < u_y$; they are distributed evenly in the flow thickness and carried in suspended state without contact with the bottom (it corresponds to the vertical curve of distribution of the smallest sediment loads);
consisting of particles of hydraulic particle size within $u_y < w <, u_{y_{max}}$ where the $u_{y_{max}}$ maximum vertical component of velocity is achieved; they are transported both in the suspended flow thickness and by saltation in the bottom area;
sucking particles for which $w \leq u_{y_{max}}$, and $V_{in} < V_{av}$; they are transported in the bottom layer by rolling, drawing on the bottom and saltation.

This vertical distribution of load is called the sediment concentration profile, which is described by the V. M. Makkaveev formula (1967):

$$s = s_b \exp\left\{\frac{-2mC}{g} \cdot \frac{w}{V_{av}} \left(1 - \frac{y}{h}\right)\right\}, \tag{4.60}$$

where s is the concentration of sediment (turbidity) having a hydraulic particle size w at a depth y of the surface; s_b is the concentration of sediment (turbidity) at the bottom; V_{av} is the average vertical flow rate; C is the Shezy coefficient; m is a parameter that averages ~24 and depends on the Shezy coefficient; h is the flow depth.

The formation of a near-bottom layer of sediment loaded soil in the process of erosion reduces erosion, as it creates a kind of pillow between the flow and the stationary bottom. Removal of this layer leads to intensification of the soil erosion process, as the flow in this case comes into direct contact with it. Such a role is played by the circulation currents which cause the deflection of bottom, loaded jets towards water surface depressions. As a result, there is a deficit of sediment load saturation in the near-bottom area, and the bottom is washed away intensively.

However, the classification of sediment in one category or another, either suspended or bedload, according to the prevailing form of flow movement, is rather tentative. Changes in the hydraulic characteristics of the flow, both along the river and over time, in different phases of the hydrological regime and with fluctuating levels, result in the transfer of sediment particles from one form of movement to another. As the flow rate increases, bedloads begin to become suspended and, conversely, as the flow rate decreases, they are replenished with sediment that has previously been moved in suspension. In addition, there is a constant exchange of particles between the bottom and the main flow thickness, as in turbulent flow conditions, some particles settle on the bottom due to velocity pulsations, while others break off and are suspend in the water column. Sediment can therefore be categorised as suspended or outbound at a particular section of the river or even vertically at a crossing point and over a period of time to allow for uniform, steady flow conditions. The mechanism and basic patterns of suspended and tractional sediment transport are different. Extremely sized sediment fractions can only be moved in suspended or at-risk condition without ever moving from one category to another.

Between these two main forms of sediment transport, a transient form of saltation or particle spasm is distinguished. It occurs when the vertical component of the hydrodynamic force is significantly higher than the particle weight and the lifting force (4.28) is higher than the velocity force (4.27). As a result, the particle breaks away from the bottom and moves for some time in the near-bottom area of the flow.

The height of the jumps z_{jump} of individual particles is determined by the formula of Rossinsky and Debolsky (1980)

$$z_{\text{jump}} = \frac{u_0^2}{2j} \tag{4.61}$$

where u_0—vertical velocity of the particle at its separation from the bottom, associated with the bottom velocity of the flow and the value of the particle's frontal resistance; j—the value depending on u_0, the value and direction of the vertical pulsation component of the velocity and the time of the velocity increase to the value u_0. Altunin (1972) proposed to determine the thickness of the layer in which the sediment is transported in a steeply manner by the probability of the transfer of particles into the suspended sediment composition. Such a "critical" state can be determined by the average value of the product of longitudinal and vertical pulsation components of the velocity. In the case of mass particle failure, the maximum position of this product corresponds to a depth of 0.1 h. Below this boundary, the bulk of the load is carried by attraction along the bottom or in a jumpy manner, and above it forms suspended loads.

The jumps are made according to relatively correct parabolic trajectories. Particles rise with constant negative acceleration, i.e. evenly slowed down. At the increased turbulence of a stream of particles, having fallen to the bottom, do not stay, and again are picked up by ascending components of speed and make the following jump. The height and length of the jump, as well as the total number of particles separated from the bottom, depend on the flow rate and grow with its increase. On the other hand, at the same flow rate, small particles move slightly faster than large particles.

Bottom sediments move only under the influence of the force of the velocity head, which arises from the direct influence of the flow on the particle (4.28). If the dimensions of the load are comparable to the depth of flow, the movement of individual particles is not only due to the velocity of the head, but also due to hydrostatic pressure. In the case of large heterogeneity of the soil, the motion of large particles is associated with the abluvial effect, and in the case of significant slope of the bottom—with the tangential component of gravity. The passage of direct positive movement waves is also a factor that activates the movement of bottom sediments.

In general, the bed load is transported by dragging or rolling on the bottom. Their velocity is less than the flow rate by the value of incipient velocity:

$$U = \mathrm{V_b} - \mathrm{V_{in}}, \tag{4.62}$$

where $\mathrm{V_{cr}}$ is the flow rate at the bottom. Since the value of $\mathrm{V_{in}}$ depends on the particle size, the reduction of their size leads to an increase in the velocity of movement, which becomes closer to the flow rate. On the other hand, for the same particles, with increasing flow rate, their speed also increases. Rolling or dragging particles along the bottom alternates with short stops when instantaneous speeds are insufficient for particle movement.

Suspension of the movement of sediment particles is the final stage in elementary channel changes, which thus form a triad of "particle separation—transport (movement)—stop". It occurs at the recession of floods and rainfall floods as a consequence of reduction of hydraulic characteristics of the flow under the condition (Mirtskhulava 1988)

$$V_{ac} = 0.8V_{in}, \tag{4.63}$$

i.e., the flow rate corresponding to the beginning of accumulation is on average 20% less than the critical incipient flow velocity.

4.6 Ripple Relief of the Channel and Channel Changes, Associated with the Movement of the Ripples

Mass transportation of bed material load results in the formation of ripples that form a wavy surface of the bed of the channel flow. In a quiet flow, the longitudinal profile of each ripple is asymmetrical: towards the flow, the upper (pressure) slope is gentle, facing downstream (downstream slope)—steep (Fig. 4.37A). It has a steepness of a natural slope of 25°–30° and is called a basement (Makkaveev 1955; Znamenskaya 1968; Chalov 1979). However, the basement is often understood as a descent between the lower slope above the located ripple (at its sole) and the upper slope below it (Grishanin 1972). Ripples of almost symmetrical longitudinal profile are formed in the rough flow, or their top slope is even steeper than the bottom slope. Also the ripples were called antidunes (Fig. 4.37B). In the latter case, the shape of the bottom of the flow and the shape of its free surface coincide, because the high kineticity of the flow ($Fr > 1$) determines the formation of a splash over the ripple, so that the depth over its crest increases; on the contrary, in a calm flow "the ripple undulation and undulation of the free surface can … be in the opposite phase" (Kondratyev 1964), t.j. A decline curve of the levels above the ripple crest appears, the depth decreases and the current velocity increases.

There are many hypotheses about the origin of the ripples. Their selective analysis was made by A. N. Lyapin (Kondratyev et al. 1959), Mikhailova (1966), Znamenskaya (1968), Sidorchuk and Mikhinov (1985). Among them, the most common hypotheses are those proposed by M. A. Velikanov and V. M. Makkaveev. Velikanov (1955) related the reason for the formation of the ripple relief on the initially flat surface of the bottom, composed of easily eroded material, to the presence of macro-scale turbulence in the flow. Due to low-frequency velocity pulsations, the flat bottom experiences an uneven flow and

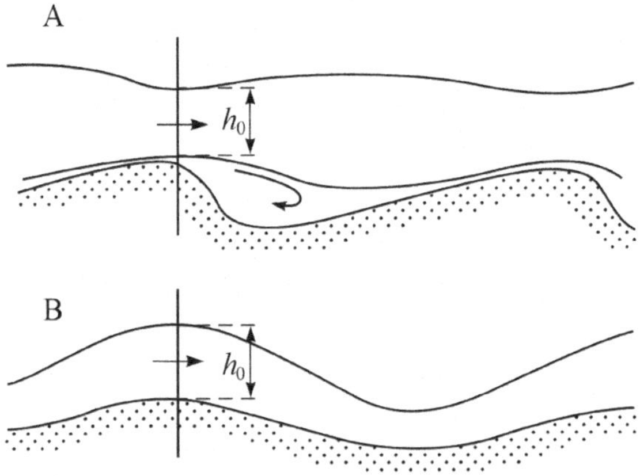

Fig. 4.37 Ripple form and its ratio with water surface profile: **A**—normal ripple asymmetry in a calm flow of the low-land river; **B**—symmetrical antidune-ripples in a turbulent flow of a mountainous river (h_0—flow depth above the ripple) (Kondratyev 1964)

acquires a waveform, which is then transformed into ripples. The assumptions of M. A. Velikanov were experimentally tested by Mikhailova (1966). It has established a correlation between the size of the ripples at the bottom of the flow and the size of the turbulent vortexes. These ideas were further developed in the works of Grishanin (1969, 1974). Sidorchuk (1992), having analyzed an extensive full-scale material, found that the complex structure of turbulent vortexes is reflected in the entire hierarchy of forms of channel relief—from large ripples, comparable to the width of the channel, to the smallest ripple formations that make up its thousands of shares.

Makkaveev (1948) proposed a hypothesis for the formation of a ripple based on the idea of the periodicity of hydrodynamic phenomena. It has received the equation of a wave surface of a bottom which included indicators of vibrational movements of a turbulent stream. Developing these representations, A. N. Lyapin (Kondratyev et al. 1959) explained the mechanism of formation of ripples by self-oscillatory movements of water flow and obtained an analytical expression for changing in time ripple profiles.

In many hypotheses, it is assumed that the initial impulse for the emergence of the bedrock ripple relief is perturbations in the flow that occur when an obstacle (a cape near the bank, a large projection on the bottom, etc.) is flowing. This, however, does not explain the existence and development of the ripples in all directions, although any obstacle can be considered as a special case of conditions for their formation.

Rossinsky and Debolsky (1980) have suggested that, since in the turbulent flow, the movement of bedloads is carried out at pulsating velocities, the result is change of the bottom surface, which takes on a wave-like (ripple) shape. In this case, each increase in the bottom has an asymmetrical shape, which leads to a roll in its basement—a vortex with a horizontal axis. This hypothesis does not allow to understand the reason of simultaneous coexistence of ripples of the different sizes making hierarchy of forms of a channel relief and does not explain the initial form of a profile of ripples which already is the reason of formation of a roller in a cellar of a ripple.

The following parameters are used to characterize the shape and dimensions of the ripples: height h_r, length l_r and slope, which is the ratio $\frac{h_r}{l_r}$. Besides, sometimes it is said about the width of the ripples b_r (especially for large ripples, comparable to the size of the channel) and its ratio with the width of the channel $\frac{b_r}{b_{ch}}$. The largest ripples have 1 $\frac{b_r}{b_{ch}}$, i.e. they cross the channel from one bank to another.

At very high flow velocities, the motion of bedloads in the rough flow ($Fr > 2$–3) is carried out in an even layer, without the formation of antidunes (Kennedy 1963). This corresponds to the development of a certain type of mountain channel—with undeveloped alluvial forms. On flat rivers with a sandy composition of sediments, the unriddled motion is carried out under the condition $V_f > 2.5V_{\text{in}}$ where V_{in} is a incipient velocity corresponding to the maximum resting state of the particles at the

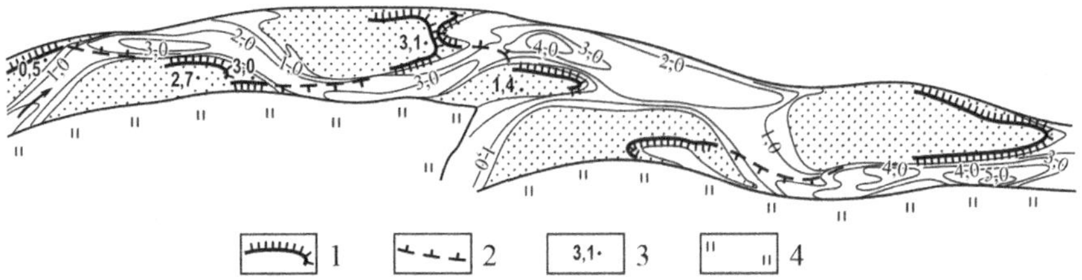

Fig. 4.38 Ripple relief (macroforms) in Vychegda river channel and channel plan in isobaths: 1—scour hole of ripple side-bars dried up during low-water, 2—scour hole of underwater parts of riffles, 3—marks of side-bars above the project (low-water) levels, 4—floodplain

bottom of the flow. Similar conditions are observed in nature with fine-sandy and muddy composition of bed material load (Amu Darya, Terek, Huang He, Yangtze).

The size of the ripple is very diverse. Large lowland rivers with sand-forming bedloads have several orders of magnitude, forming a complex hierarchy of channel relief forms. Their main classes (structural levels) are macro-, meso- and micro-forms of the riverbed relief. Macro-formes include riffles—large ripples crossing the channel from one bank to the other and causing support in the upstream section during low water levels (Makkaveev 1955), as well as ripples of 0.5–0.1 channel widths (independent side-bars, sand dunes, spits). The coastal parts of such ripples dry up at low levels, and their surface is clearly visible in the basement—the steep lower slope of the ripple (Fig. 4.38). Macroforms determine the main morphological shape of the channels (Fig. 4.39A) and cause changes in depth both along and across the river.

Mesoforms include large sand ripples, the sizes of which are smaller than the sizes of the riffles and sid-bars, arising on the upper slopes of the latter. The width of the mesoforms is the tenth and hundredths of the channel width. As a result, they are already three-dimensional, since the partitioning between ripple width to the channel depends on the degree of influence of the ripple on the structure of the flow; this distinguishes them from the two-dimensional ripple-macroforms, the width of which is comparable to the width of the channel, and their influence on the structure of the flow is determined only by the ratio $\frac{h_r}{l_r}$. The height of the ripple topography mesoforms is commensurate with the local depth of the channel. Therefore, they have mass distribution only at the channel banks (Fig. 4.39B) and in the coastal shallow parts of the channel. The formation of mesoforms is associated almost exclusively with the macroturbulence (hypothesis of M. A. Velikanov) of the channel flow. Their inverse effect on the flow is insignificant: thanks to them, whirlwinds are excited in the flow, which do not capture the entire thickness of the flow. Basically, mesoforms and the following microforms determine the roughness of the flow bottom and, consequently, the associated hydraulic resistances.

Absolute sizes of macro- and mesoforms of the bed topography vary very widely depending on the size of the rivers. On the largest rivers the length of the ripple reaches several kilometers with the height of 10–15 m and more. On small rivers, these sizes respectively decrease to the first tens of meters—the first tens of centimeters, i.e. they are equal to meso- and microforms on large rivers. Therefore, when determining the class of ripples, their relative rather than absolute dimensions should be taken into account.

Microforms of the channel relief are the smallest ripple formations at the bottom of the stream, the size of which (in terms of width and height) is incommensurably small compared to the size of the channel (Fig. 4.39C). These are small sand waves (including riffles), which create a peculiar flake or corrugated surface of the shallows. They are related to the flow rate fluctuations in the near-bottom area and have practically no effect on the flow structure. The height of such ripples varies from a few centimeters to 20–30 cm, and the length—from a few tens of centimeters to 10 m.

Nano- or ultramicroforms (Fig. 4.39D)—ripples no more than 2–3 cm high and up to 10–25 cm long are also distinguished. Ultramicroforms are miniature formations, which are sometimes called sand ripples or ripples signs in geological-geomorphological literature (Maslov 1933; Belostotsky 1940).

The greatest variety of shapes and sizes is characterized by macroforms; no more than two types of ripple formations are distinguished among microforms. The larger the relative size of the ripples, the more factors of fluvial processes are reflected in their development. The formation of microforms is influenced by a very limited set of factors (hydraulic structure of the flow and its changes in time and space, alluvium particle size).

Depending on the hydraulic flow characteristics and the particle size of the bed material load, the bed can be divided into classes or all major classes can be formed at the same time. On the upper slopes of the side slopes, smaller shapes are often formed. In such cases, the side banks that have come out from under the water level in the lower reaches of the riverbed consist of several independent ripples. In turn, the upper slopes of the meso- and microforms of the ripple relief are superimposed, respectively, on the microforms and ultramicroforms. As a result, there is a complex hierarchy of ripple channel formations. Rossinsky and Kuzmin (1947) observed on the Volga the following sequence in the subordination of the ripples: riffle side-bars 1200–1500 m long and 4–5 m high—mesoforms up to 200 m long and 0.7 m high —sandy ripples 5–12 m long and 0.1–0.2 m high—ultramicroforms 0.1–0.4 m long and 2–3 cm high. Korotaev (2003) in the Volga-Bakhtemir delta branch, on the basis of hydrolocation survey, found that the surface of the largest ripples, called by him "sand waves", the length and height of which decreases downstream, respectively, from 7 to 2 km and from 6 to 2 m, is complicated by successively superimposed "large ripples" 1–2 km long and 2–5 m high, "dunes" 100–150 m long and "riffles" 10–50 m long and 0.5–0.2 m high. Judging by the given sizes, the first two categories are macroforms or ripples of type A and B, according to N. I. Alekseevsky; then follow the mesoforms or ripples of type G—"dunes" and microforms or ripples of type D; apparently, mesoforms are ripples of type B or are absent or have not been selected during interpretation of local images.

In the moulded parts of rivers, where no micromorphic ripples are formed, the hierarchy of ripple relief is reduced: as a rule, micromorphic forms move along the bottom of moulded valleys, less frequently—mesoforms. Thus, Korotayev (2003) found in the branch of Kizan (Volga delta) 10–14 m wide ripples 10–50 m long and less than 1 m high and ultra-microform (the first ones he called "dunes", the second ones—"riffles").

On rivers with pebble-boulder alluvium, the hierarchy of ripple relief forms is simpler. Ripples of only one class usually appear here—macroforms. On flat pebble-boulder rivers these are normal ripples with gentle upstream and downstream steep slopes, on mountain rivers with a rough flow regime—antidunes. In addition, peculiar mesoforms in the form of coastal spits (Fig. 4.40) can be found on flat rivers. They are formed with a general deficit of sediment near the banks along which pebbles and boulders are transported during floods (e.g. at convex banks of bends).

In addition to the size of the ripples, they differ in the shape of their longitudinal profile, i.e. in the partitioning between height h_r to the length of l_r. Znamenskaya (1968) distinguishes the following varieties of ripples: (1) flat ripples with $\frac{h_r}{l_r}$ 0–0.04; (2) riffles $-\frac{h_r}{l_r}$ 0.04–0.15; (3) tilted ripples $-\frac{h_r}{l_r}$ 0.04–0.08; (4) steep ripples $-\frac{h_r}{l_r}$ 0.08–0.12; (5) very gentle ripples $-\frac{h_r}{l_r}$ 0.005–0; (6) Anti-dunes with steepness increasing with the increase in the Froude number and the partitioning between flow rate and hydraulic particle size, and then decreasing—the transition to a rippleless movement of sediment in the rapid flow begins. The same shape of the ripple appears under different hydraulic conditions, depending on the particle size of the load forming it. As a result, the ripples of each type occupy a certain position in the field of the communication diagram $Fr \sim V/w$ (Fig. 4.41). A special place belongs to the riffles, which on the chart have a closed area of development, and the parameter $-\frac{h_r}{l_r}$ coincides with the skewed and steep ripples, only in some cases slightly exceeding its maximum values. The difference between the riffles and other types of ripples is their two-dimensionality, so their appearance at the bottom of the flow determines its transverse (to the direction of the flow) banding like a "washing board" (Fig. 4.42). These are also the microforms in the Fig. 4.39B).

The classification given by N. S. Znamenskaya was developed for microforms of ripple channel relief. Applying it to the rest of the ripple types, it lowers the riffles, which, in her opinion, represent "ripples of mass distribution in the channel, the length of which is incommensurable with the flow width, … the riffles determine the roughness of the bottom … usually two-dimensional" (1968, p. 7). At the same time, following the approach of SHI, all the ripples are larger than the microforms, N. S. Znamenskaya considers mesoforms, because the category of macroforms includes either the forms of the channel itself (e.g., a bend) (Kondratyev et al. 1982), or, as it was presented in the first presentation of hydromorphological theory (Kondratyev et al. 1959)—complexes of simple forms (i.e. forms of many sizes), to which its authors include riffle and pool areas. In this version, the classification of N. S. Znamenskaya, obviously, can be extended to macroforms as well.

Fig. 4.39 Different size ripples, dried up during low-water: **A**—macroforms on the downstream Pechora river. Photo of the author; **B**—meso forms of Mezen river. Photo of the author; **C**—microforms on Usa river. Photo by N. N. Shtankova; **D**—ultra microforms on the Ob river in downstream end of the island of Medvezhiy island) downstream Novosibirsk city. Photo by A. M. Tarbeeva

Fig. 4.39 (continued)

Fig. 4.40 Riverside boulder spit—ripple (Aldan river). Photo of the author

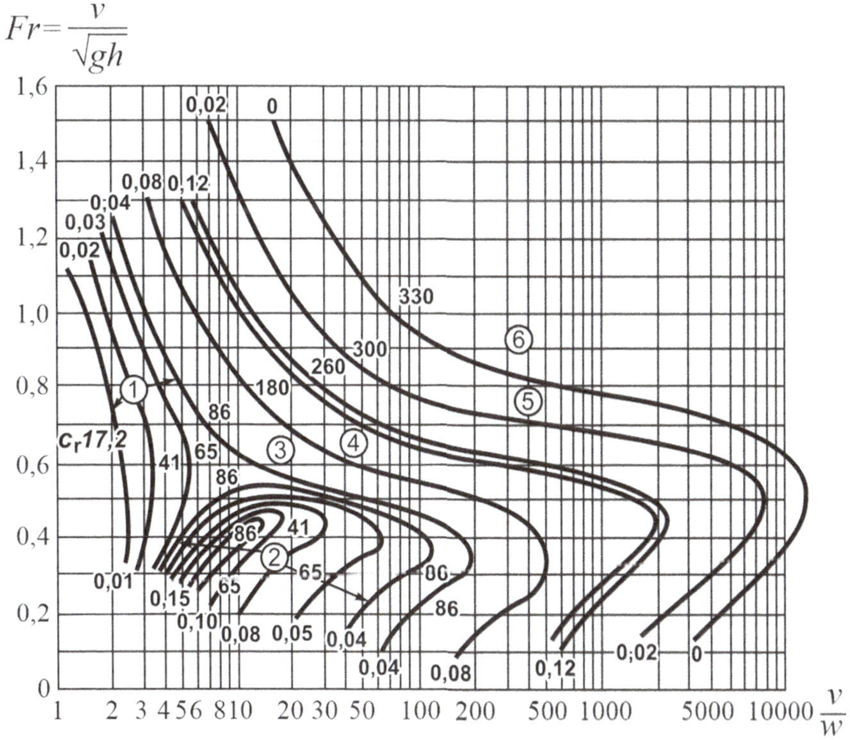

Fig. 4.41 Ripple classification of Znamenskaya (1968): 1—flat ripples; 2—riffles; 3—skewed ripples; 4—large ripples; between 4 and 5—eroding ripples; 5—very flat ripples; 6—anti-dunes

It is typical that Kondratyev (1959) along with micro-, meso- and macroforms (he also called them small and medium complex forms, respectively) singled out simple forms, a typical example of which he considered small sand riffles (analogue of ultramicroforms). However, in his other later work, published only in 2000, N. E. Kondratyev understood the riffles to mean all the ripples—micro- and mesoforms having a normal asymmetric profile, opposing them only to antidunes.

Sidorchuk (1992) applied a close approach to ripple classification, singled out: (1) ripples; (2) riffles (the smallest ripples), to which he referred, following Snishchenko (1983), the bottom forms of the order of depth of the flow and being two-dimensional in length (step); (3) small ripples, which include dunes—small ripples, the surface of which is modeled by riffles, and jets, the surface of which is complicated by dunes; (4) medium ripples—hulls and sand waves, from which on the surface of the first formed jets (it is characteristic that small and medium ripples A.Yu. Sidorchuk considers them in a single set and classifies them as three-dimensional); (5) large ripples with the size of the order of flow width; they include side edges, aseparations, and inter-row ripples; (6) the largest ripples, corresponding, in the author's opinion, to the most complex branching of channels (and he also correlates the bend with large ripples). Here, as in the GIS approach, when micro-, meso- and macroforms are singled out, the forms associated with the ripple motion of sediment and the forms of the channel itself are mixed. At the same time, following many foreign researchers who were the first to study the ripple relief of the riverbeds (Simons and Richardson 1961; Simons et. al. 1966; Bogardi 1974), A. Yu. Sidorchuk calls the ripples riffles and dunes, but at the same time he assigns Russian names to others, borrowing them from different sources: ripple is a term used by geologists (Maslov 1933; Belostotsky 1940), «zhastrugi» (Russian word for kind of ripples) and «shalygi» (Russian word for midchannel bars)—from the literature (Popkov 1962) and the regime of rifts (Sadovsky 1961) of navigable rivers, which have their roots in the Volga burlaks, and with the distortion of their original meaning.

Terminological variability, and often different approaches to the selection of ripples of different sizes and their names (almost all researchers agree only on the definition of the riffles), as well as the relationship of the ripple parameters on the size of the rivers (their runoff rates) can be considered quite acceptable to designate them at each hierarchical level as conventional indexes. Alekseevsky (1998) suggested that, in order to take into account all forms of the ripple movement of sediment, a morphological formation corresponding to sid-bars or riffles should be considered the largest ripple. Smaller channel forms, including those on its surface, are distinguished by the special character of changing the slope of the bottom (Fig. 4.43), and regardless of the size of the rivers they represent up to five types of ripples, not including ultra-microforms, which do not appear on the echograms due

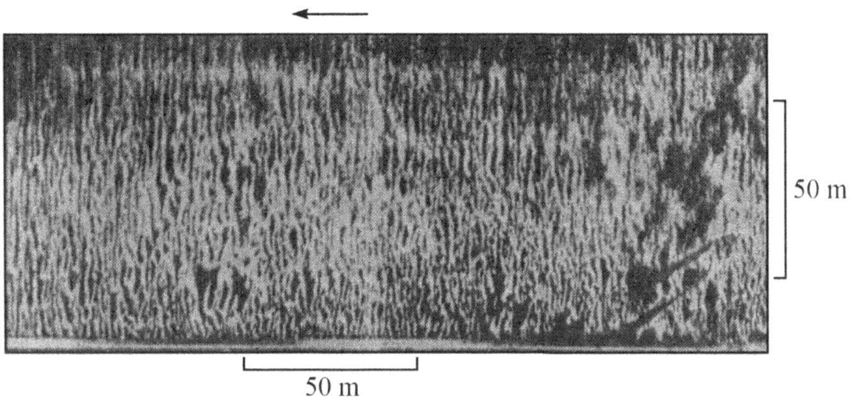

Fig. 4.42 Two-dimensional ripples—riffles in Lena channel near Mohsogollokh (localizer recording of lateral view by A. A. Zaitsev) (Sidorchuk 1992)

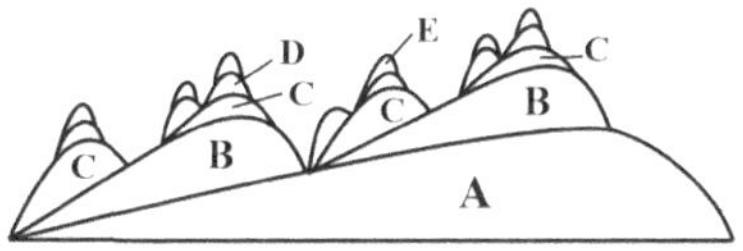

Fig. 4.43 Scheme of ripple channel relief: A, B, C, D and E—ripple types [according to Alekseevsky (1998)]

Table 4.16 Partitioning between some ripple form of channel relief classifications

By connection with flow structure and role in channel development (Makkaveev and Chalov 1986)	By transport conditions and role in bedload runoff (Alekseevsky 1998)	By morphology
Macroforms	A	Riffles, side-bars, mid-bars (Chalov 1979), large ripples (Sidorchuk 1992)
	B	
Mesoforms	C	Medium ripples (Sidorchuk 1992), sand dunes (Popov 1965)
	D	
Microforms	E	Small ripples (Sidorchuk 1992), sand waves (Mikhailova 1966)
Ultramicroforms (nanoforms)	–	Sand ripples (Belostotsky 1940), smallest ripples (Sidorchuk 1992)

to their small size (height). The largest of them (type A) are morphological formations, the length of which is proportional to the width of the channel. The smallest forms of channel relief (type D) are comparable in length to the depth of flow. The intermediate position is occupied by rows of type B, C and D, of which the rows of type B in size and ratio to the flow structure belong, as well as rows A, to the macroforms, rows C and D are mesoforms, rows D—microforms.

At present, there are at least three main approaches to the classification of ripple forms of channel relief (Table 4.16), but they all rely to some extent on the size of the ripple and their relationship with the size of the channel. The first of them is based on the degree of influence of the ripples on the flow structure and their role in the development of the channel itself (separation of macro-, meso- and micro-forms). The second approach is based on the partitioning between ripples of different sizes, the conditions of their movement in different phases of the water regime and the possibility of using them to determine the flow of bedload. The third approach is morphological; it is most diverse in terms of ripple terminology, size estimation and form descriptions.

The absolute dimensions of the micro- and mesoforms of the ripple relief, in addition to the hydraulic characteristics of the flow and the particle size of the bed material load, depend on the runoff of the latter. Therefore, the bedload transport rates G is included in the formula of B. F. Snischenko (Kondratyev et al. 1982), which determines the height of the ripple:

$$h_{\mathrm{r}} = \frac{G(gh)^{1.45}}{0.011V^{3.9}}. \tag{4.64}$$

The ripple parameters depend on the hydraulic flow characteristics. For sandy ripples, Snischenko (1980) obtained a nonlinear relationship between $\mathrm{h_r}$ and the flow depth h, which can be represented as two linear bonds depending on the flow depth (2.13): $h_{\mathrm{r}} = 0.25\,\mathrm{h}$ at $\mathrm{h} < 100$ cm and $h_{\mathrm{r}} = 0.20 + 0.1\,\mathrm{h}$ at $\mathrm{h} > 100$ cm. The length of the ripple is also related to the depth of flow and is determined by the formula of Grishanin (1962)

$$l_{\mathrm{r}} = \mathrm{h}\sqrt[3]{\frac{C^2}{g}}, \tag{4.65}$$

where h is the depth of flow; C is the Shezy factor.

N. I. Alekseevsky's approach is more universal (1998), according to which the total height of the ripples of different types of naturally changes depending on the size of the river and the phase of the water regime. The height of all types of channel formations, which are in accordance with the hydraulic regime, is proportional to the order of the river N, i.e.

$$\mathrm{h}_{r_i} = aN^m, \tag{4.66}$$

where the i index indicates that the ripple belongs to one of the five types of ripples (A to E).

A similar approach has been used to estimate the velocities of ripple shifting C_r. If B. F. Snischenko (Kondratyev et al. 1982) uses the relationship of the relative velocity on the Froude number of to estimate the C_r values

$$C_r = 0.021VFr^3 \tag{4.67}$$

or (Snishchenko and Kopaliani 1978) connects it with the ratios of actual fact and incipient velocity V_{in} (4.46a), sediment size d and height of the ripple h_r

$$C_r = 0.032(V_f - V_{in})\left(\frac{V_f}{V_{in}}\right)\left(\frac{d}{h_r}\right), \tag{4.68}$$

Alekseevsky (1998) proposes to apply a system of correlation relations $C_{r_i} \sim N^m$ when moving to seasonal peculiarities of the ripple movement:

$$C_{r_i} = V_{seas}N^m, \tag{4.69}$$

where S_{seas} is the average flow velocity of a river in a given season; m—*a, b, c,* d (including coefficients in formula 4.66)—empirical coefficients depending on the type of ripple and phase of hydrological regime in which the ripples of this or that type develop (Table 4.17).

Ripple movement is a type of channel change associated with the ripple movement of sediment. The form of the ripple movement is not the same in a calm and rough flow. In the quiet flow ($Fr < 1$), the normal ripple of the asymmetric profile (Fig. 4.44A) determines the alternation of areas of sudden expansion of the flow directly below its crest, where in the basement of the ripple formed a roller—a vortex with a horizontal axis, in which the bottom velocities are directed against the main flow, and compression of the flow jets over the upper (pressure) slope. The latter leads to an increase in the bottom velocity, which increases to the ripple ripple up to the vertical alignment of the velocity values and, if the ripple height is commensurate with the depth of the flow, to an increase in the average flow rate. This circumstance

Table 4.17 Coefficients in the height and the velocity of ripple movement formulas

Water regime phase	Coefficient *m*	Types of ripples				
		A	B	C	D	E
Floods	*a*	0.046	0.018	0.032	0.023	0.005
	в	1.95	2.00	1.5	1.4	1.9
	c	1.1×10^{-6}	3^{10}	3.5×10^{-4}	2.2×10^4	2.2×10^1
	d	5.4	−2.6	4.6	−3.5	−3.5
Low-water	*a*	–	–	–	0.005	0.003
	в	–	–	–	2.0	2.0
	c	–	–	–	600	600
	d	–	–	–	−2	−2

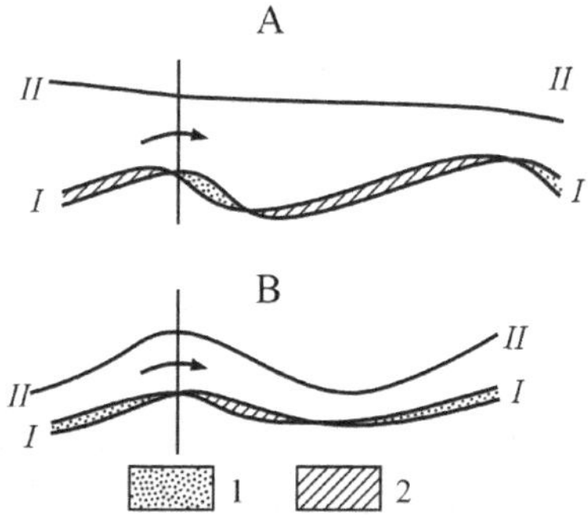

Fig. 4.44 Schemes of moving of normal asymmetrical ripples in a calm flow (**A**) and anti-dunes in turbulent flow (**B**): 1—sediment accumulation; 2—erosion of ripple surface. I—I ripple channel bed profiles; II—II—flow free surface profiles

is reflected in the form of a free surface profile of the flow, which acquires a steep slope in the area of the ripple crest and gentle—in the decrease between the ripples. As a result, on the gentle uphill slope the sediment particles are broken off and carried away by the flow to the crest of the ripple, i.e. the uphill (pressure) slope of the ripple is eroded, and the most intense erosion is observed approximately in the middle of the ripple. It attenuates to the ripple crest, as vertical velocity equalization leads to a decrease in the intensity of turbulent pulsation and vertical pressure gradient, which, in turn, contributes to a decrease in the erosion ability of the flow.

By passing the ripple crest, the sediment particles roll down into the basement, entering the roller development area here. Thanks to it, the major part of the particles are trapped on the downstream slope, and only a fraction of them are carried away downstream. The smallest sediment particles, which move mainly in a steeply way, can jump over the roller area to the top of the lower ripple. If the basement of the ripple is oriented at some angle to the direction of the current, the roller that appears here is transformed into a whirlwind current, which causes a part of the load to move along the basement. Thus, within the limits of each ripple there is an area of erosion corresponding to its upper slope and an accumulation area in the basement. As a result, the entire ripple moves downstream.

In a rough flow ($Fr > 1$), each ripple (antidune) causes a surge or standing wave on the free surface of the flow, which is accompanied by an increase in its depth to the crest of the ripple. The flow rate above the ripple is therefore dropping dramatically. In such a case—the coincidence of the bottom profiles and the free surface of the flow—the accumulation of sediment occurs on the upstream slope facing the flow of the ripple (in accordance with the decreasing speed here); the downstream slope, on the contrary, is eroded, and the sediment particles are transferred from here to the upper slope of the downstream ripple. As a result, the ripple (antidune) moves in the opposite direction to the movement of the particles, i.e. upstream (Fig. 4.44B).

Znamenskaya (1968) in the development of the classification (see Fig. 4.39) has found out the low rate of movement of the ripple along the contour lines $\frac{h_r}{l_r}$. This has given her reason to believe that each type of ripple has its own well-defined movement velocity, which is therefore dependent on the hydraulic characteristics of the flow and the particle size of the bed material load. Since there is a complex hierarchy of ripple forms of channel relief in the channel, the movement of smaller ripples along the surface of large ripples along with the movement of each ripple determines the complex nature of movement and reshaping of ripples, which, in turn, is complicated from small to large ripples. During the intermittent period, large ripples (macro- and mesoforms) move only due to the movement of shallower ripples on their surface, which cause a portioned and non-uniform across the width of the ripple to arrive in its basement.

The velocity of the ripples in accordance with the formulas (4.67) is directly proportional to the fourth degree of flow velocity and (4.68) is inversely proportional to the height of the ripple. The absolute velocity of the fastest moving ripples is usually hundreds of times less than that of the flow rate and tens of times less than that of individual sediment particles. Shamov (1959) describes the following observations of the ripple dynamics in the experimental tray: at a flow depth of h = 26 cm and d = 0.24 mm, the flow rate increased from 0.20 to 0.41 m/s; ripples with h_r = 1.21–2.12 cm high and l_r = 22.5–39.4 cm long were formed at the bottom of the flow; the flow rate of the ripple increased with an increase in the flow rate from

Table 4.18 Critical velocities C_r and ripple parameters (height h_r; length l_r) on different rivers[a]

River	Current velocity (m/s)	The depth of the river (m)	Ripple parameters		Critical velocities of the ripples (m/day[b])	Author
			height (m)	Length (m)		
Volga	0.50	5.0	0.80	24	0.56 (0.007 mm/s)	Shamov (1959)
Volga	0.50–0.75	1.0–4.0	0.30–1.30	8–35	0.3–1.7	Lopatin (1952)
Volga	0.97–1.06	4.8–5.8	0.17–0.90	13–60	1.68–8.88 (0.55–1.68 m/h)	Snishchenko (1966)
Nadym	0.434–0.62	0.32–0.50	0.52–0.90	–	0.27–2.10	Levashov (1973)
Luga	0.40–0.50	1.0–2.0	0.15–0.25	6–15	0.1–0.4	Lopatin (1952)
Polomet	1.42	1.9	0.4	–	117.1 (1.3 mm/s)	Korchoha (1968)
Polomet	0.90	6.0	0.75	75	8	Popov (1961)
Polomet	0.61–0.64	0.15–0.21	0.04–0.06	0.5–2.5	13.21–29.3 (0.55–1.68 m/h)	Snishchenko (1966)
Yangtze	0.73–1.34		0.76–1.64	–	19.8–26.0	Qian et al. (1987)
Chulym	0.50–0.60		0.2–0.4	2–5	9.6 (0.4 m/h)	Turykin (1997)

[a]A detailed summary of information on the ripple movement (macro- and mesoforms) is also made by B. F. Snischenko (Kondratyev et al. 1982)

[b]The speeds in the units of measurement accepted by the authors of the given data are indicated in brackets

0.062 to 0.768 mm/s, although the values of h_r and l_r varied unevenly. With the increase in sediment load uniformity, η (4.36) the speed of the ripple movement also increases. At the same time, the larger the ripples (in the hierarchy system on a given river), the slower they move. Therefore, smaller ripples overtake relatively larger ones by moving over their surface. As a result, in floods, when the active zone of ripple motion covers the entire channel, large ripples have their own movement, and indirectly, due to the movement of microforms on their surface.

Differences in flow rates in the bedrock zone and near the banks, depending on the shape of the channel (e.g., at concave and convex banks on the bend), in its contractions and expansions, causing alternation of areas of support and recession, compression of stream jets, etc., as well as heterogeneity in the composition of bed material load and the size of the ripples themselves cause changes in the velocities of the ripples of the classes of meso- and microforms of the channel relief (Table 4.18). The rates of movement of macroforms vary even further (from the first meters on pebble-boulder rivers) to hundreds of meters per year, depending on the shape of the channel, runoff rates of the high-water phase of the regime, alluvium particle size and other conditions.

During floods, bedrock macroforms of the ripple relief are formed and moved, which are complicated by meso- and microforms in their shallow peripheral parts, where flow rates are slow. When water flow rate is reduced, the ripples are reshaped, adapting to new hydraulic conditions; smaller ripples appear on the surface of the formed large ripples in the flow bar zone, corresponding to its new characteristics. The channel zone, which is active with respect to the movement of the ripple, is reduced, as the increased parts of the macroforms dry up, turning into extensive (sometimes taking up to half or more of the channel area in floodplain edges) channel banks or shallow waters along the periphery of the intertidal channel, where $V \leq V_{in}$ and the movement of sediment does not occur. Besides, in the period of the lowland, only micro- and partly mesoforms (D and E ripples, according to N. I. Alekseevsky) are actively moved.

However, the largest rows of macro-forms (such as A, according to N. I. Alekseevsky) are not always mobile formations, forming on bends between adjacent bending radii, in the nodes of branching and merging of branches, in the local expansion of the channel. Ripples of smaller sizes, moving along the surface of such macroforms, determine the seasonal and current changes in the relief of the channel, but getting into the basement, are washed away, the products of erosion or moved by the roller current diagonally to the direction of the current to the lower sideboard, increasing its size, or put in a pool valley and on the upper slope of the next ripple, where they form underwater meso- and microforms.

The degree of preservation of the ripple depends on the intensity of the decline in flood levels and the transition to the low water period. On rivers with short and high floods (rivers with East European type of water regime, according to B. D. Zaikov), the ripple relief does not have time to be reconstructed; in the lowland banks have flat surfaces of the upper slopes of the ripple-macroforms. Only in the flow retardation zones are the microforms of the ripple relief observed on them. In the case of the slow transition from flood to low water (rivers in the south of Western Siberia),

A

B

Fig. 4.45 Sideway lamination in sections of channel relief macroforms (side-bars, dried up during low-water): **A**—at personal ripple moving. Photo of the author. Sideway lamination in sections of channel relief macroforms (side-bars, dried up during low-water): **B**—during its moving on the surface of microform (Mezen river). Photo of the author

when the hydraulic characteristics of the flow change gradually, there is a transformation of the ripple movement of sediment. In these conditions, sandy micro- and ultramicroforms are well expressed on the surface of channel banks.

Such transformation of ripple forms of channel relief at change of hydraulic characteristics of the flow in time (from flood to low-water) and in space (from the core zone to the periphery of the channel) is typical only for rivers with sandy alluvium. On rivers with larger bed material load compositions, with reduced flow rates, alluvial sedimentation is formed, the movement of sediments is finished and, consequently, the ripple movement is finished. As a result, only macroforms are formed in river channels with a pebble-boulder composition of sediment. On rivers with gravel and pebble alluvium macro- and mesoforms develop, with the first ones, being formed by the flood flow, being antidunes, and mesoforms—asymmetric ripples in a steep lower slope.

The formation of macro-shaped ripple relief in the river channels is associated with the development of the forms of the channel itself (bends, branches) and the intensity of lateral channel changes—meandering, reshaping of branches, etc.

Ripples, influencing the flow structure, cause either bending of the dynamic flow axis or its splitting. The result is primary bending of the channel and scouring of the bank in front of the ripple (side-bar); elementary splitting of the flow is the basis for further development of the branches. At the same time, macroforms become floodplains when they are overgrown by elevated drying parts, as the vegetation on the surface of the shallow at its subsequent flooding causes a significant increase in the roughness of the underlying surface, a decrease in speed and an end to the washout of sediment particles; as a consequence, the movement of the ripple itself ceases, smaller ripples do not form on its surface and there is an accumulation of suspended sediment forming the floodplain alluvium facies.

In turn, in the process of moving sand ripples, the loads that have come into motion on a gentle upper slope settle on a steep lower slope (basement). Due to the rippling flow rate, this is where sediment of a larger or smaller size periodically enters. This is also facilitated by the movement of smaller ripples on the upper slope of a large ripple (up to ultra-microforms). At the same time, the swirl current in the basement of the ripple brings to the steep bottom slope only the deposits of the smallest fractions, which are buried under the larger ones, which get here due to the movement of the ripple. As a result, the oblique layers formed (Fig. 4.45), which have an inclination of about 25–300, differ from each other by the particle size. These layers are assembled in packs of different thicknesses depending on the size of the moving ripples, conditions of their safety in different parts of the channel (in the core or periphyric zones), or during the transformation of the ripple during the transition from the high-water phase of the water regime to the low water level. As the ripple shifts, these oblique sediments become eroded again; if the ripple is fixed and transformed into a fragment of the floodplain, they become alluvial sediments.

Slanting of deposits is poorly expressed in pebble-boulder beds, despite the asymmetrical ripple profile. This is explained by the low value of turbulent pulsations on the ripple crest of the ripple in comparison with the value of incipient velocity for coarse-clastic particles, as well as the absence of ripple micro- and sometimes even mesoforms. Also, there is no formation of oblique complexity of the deposits that compose the smallest ripples—ulramicroforms—because the swirl roll in the basement of such ripples does not appear (Kondratyev et al. 1959; Sidorchuk 1992); particles falling off their ripples are evenly placed in the basement, in the descent between the adjacent ripples and even at the beginning of the slope of the downstream ripple—ultramicroform. The layering of such ripples is wavy, finely lensed.

The transformation of different size ripples requires some time to change the hydraulic characteristics of the flow. At fast recession of a flood ripple basically do not have time to adapt to new hydraulic conditions, dry up and at fastening of channel shoals by vegetation form subsequently with cuts of a floodplain oblique layering which usually (Schanzer 1951) consider as the main attribute of channel facies of alluvial adjournment. With a slow decline in levels, the ripple movement of sediment is reduced to the movement of ultra-microforms, especially in the peripheral parts of the channel. The surface of the latter is covered with "scales" of small ripples—ultramicroforms, which later give a wavy, lenticular layering in the sections of the channel facies of alluvium. Thus, oblique alluvium layering is characteristic only of rivers with a hydrological regime similar to that of Eastern Europe; undulating, lenticular alluvium layering occurs on rivers with types of hydrological regimes, which are characterized by slow changes in levels and hydraulic characteristics of the flow from flood to lowwater.

4.7 Hierarchy of Channel Forms

The diversity and complexity of the form of the river hydrograph, the differences in sediment load and the peculiarities of its transport to different phases of the water regime, the development of ripple forms of the channel relief and their transformation from one phase of the regime to another—all this leads to the formation of a complex system (hierarchy) of channel forms nested in each other, the development of which occurs in different phases of the hydrological regime (Fig. 4.46). The influence of fluvial processes factors on the morphology and channel transformation is most fully evident on flat rivers flowing in the areas of free development of channel changes, having wide floodplain channels, where lateral changes are not limited to the geological structure of the banks and bottom. Under these conditions, the complexity of channel morphology and lateral change is determined, first, by the specificity of the active factors—the water regime and sediment load—which is represented integrally in the peculiarities of effective water discharge [in the understanding of Makkaveev (1955, 1971)], and, secondly, the degree of stability of the channel [as understood by Lokhtin (1897) and Makkaveev (1955)], whose parameters reflect the influence of geological-geomorphological factors (through the slope and particle size of bed material load) on the conditions of channel formation. The less stable the channel is against the flow, the more the channel controllability on the flow side is represented and, accordingly, the role of effective

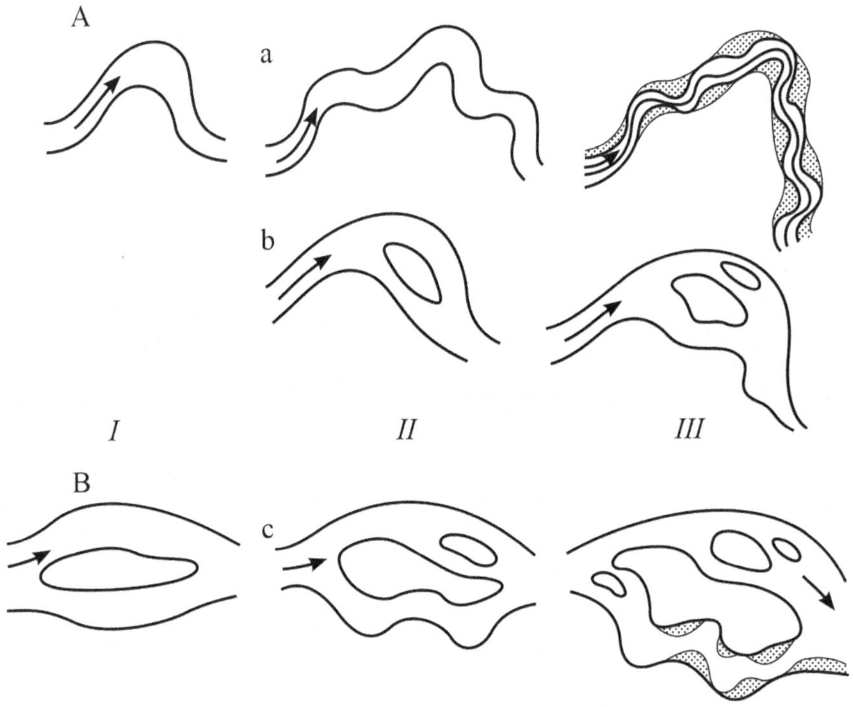

Fig. 4.46 River channel hierarchy on bends (**A**) and branching (**B**): I, II, III—channel forms orders; a, b, c—variety of combinations of channel forms of different orders

discharges in the development of the hierarchy of channel forms, the degree of complexity and intensity of reformation of each of its components.

Depending on the number of effective discharges in wide floodplain rivers with a low stable and unstable channel with sufficiently large bed material load, the forms of the channel of the 2nd and 3rd orders of magnitude arise, causing a complication of both the morphology of the channel as a whole and the mode of its transformation. Kondratyev (2000, p. 187) drew attention to this peculiarity, but only in relation to branched channels, in 1968, when he wrote that in each of the branches bending can occur as a form of the 2nd order and at the same time "each separate stream can be considered as a separate river and the fluvial process on it can develop in its own type". This was clearly illustrated by Klaven and Snishchenko (1978) by the example of the Selenga River, where in the branched channel the branches are characterized by the development of several bends; Antropovsky and Minaeva (1974) in the flows of the branched channel of the Viliuy River described the development of branched bends. Actually, the channel forms of the 2nd and 3rd orders of magnitude can occur in the channel of any pattern and together with it form complex combinations (Fig. 4.47).

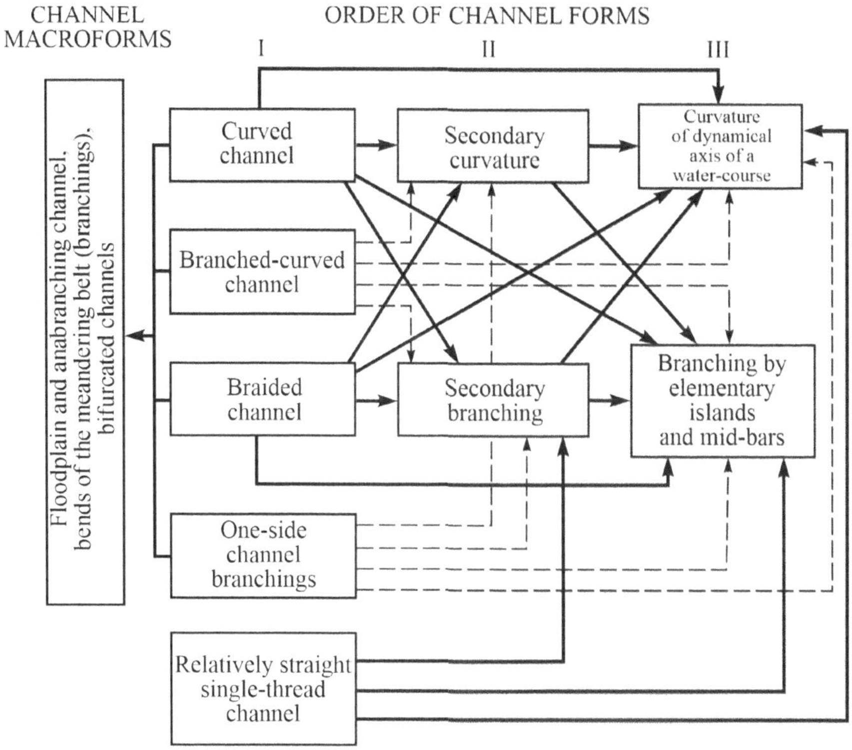

Fig. 4.47 Scheme of probable combinations of channel forms of different orders in the combinations (steepness) of channel forms

On rivers with free-meandering channel the number of effective discharge intervals and their correspondence to different water levels determine the degree of complexity of the bend configuration. If the Q_f on the river has one interval, the bend is characterized by the simplest form in which the dynamic axis of the flow as a whole follows the shape of the bend. When effective discharges are passed at two or three intervals, bend occur, each of which includes several (two or three or more) adjacent bend with smaller pitches and radius of curvature; the latter, in turn, are complicated by the gyrus of the dynamic flow axis enveloping the side-bars of the riffles. At the same time, the channel bend can be complicated only by the gyrus of the dynamic flow axis, i.e. the hierarchy of channel forms consists of the bend of the 1st and 3rd orders, or only the bend of the 2nd order. Makkaveev (1971) called such large bends of the channel complicated by smaller bends and complicated (Fig. 4.46A-a).

Zaitsev (1981), using the function of spectral density of its curvature for the analysis of the meandering channel shape, found the presence on the spectra of two or three extrema corresponding to the number of maxima on the diagram of effective discharge. Sidorchuk (1992) found that all structural links of channel forms (for example, bend-shaped) correspond to the number of turbulent vortices, which he called macro-, meso- and micro-vortices.

On branched rivers, the branches can meander themselves to form their own bend of branches. A common characteristic feature of branching, mainly relatively simple (single or conjugate links in the chain, etc.) is the formation of a bend of the flow enveloping each island and, consequently, the emergence in it of a hydraulic structure inherent in the bends of the channel (Makkaveev 1955; Chalov 1979). At the same time, there are essentially two bends near each island, the tops of which are located on opposite banks of the river (or are directed in different ways).

In addition to the main bend of the branches between the tip and the downstream end of the island separating them, they, with a fairly long length and a slight overall curvature, can meander themselves, forming a series of independent bend in each of the branches. In the presence of, besides, by-products, arranged in a chess order, there is a natural sequence of bends of branches of the branched channel: bends, which forms the channel of each branch, the gyrus of the dynamic axis of the flow. This ratio is analogous to the meandering channel bend hierarchy, and its structure is also determined by the number of effective discharge intervals. Like the development of bend-shaped bend-shaped arms in branched channels, the meandering channel can be characterized by the presence of islands and aisles on the wings and at the tops of the bend-shaped bend-shaped arms of the 2nd and 3rd orders. In cases when the channel meander neck consists of islands, the formation of which at the convex bank determines the process of meandering (or, that the same thing, in the process of development of the bend at their convex banks formed islands that make up the merger of the meander neck), the very type of channel is called branched-meandering, in other words, branched-meandering channel develops as a result of the formation of 2nd order islands that make up the meander neck (Fig. 4.48A) or formed in the lower wing of the bend (Fig. 4.48B).

It is not the sinuosity of the main branches of the branched channel that is more widespread, but their division into a network of relatively short branches of the 2nd order by smaller islands. Thanks to this, the pattern of the branched channel is often a tangled labyrinth of branches of different lengths and runoff rates. Genesis, morphology and the regime of branch changes of the 2nd order are generally identical to the main branches of the channel having the 1st order. Single branching of the 2nd order usually occurs at the beginning or end of the main branches; in the first case, the island is formed if the branch takes the main part of the sediment load, and the diversion of part of the water flow into the second main branch causes a decrease in its transport capacity; in the second case, the reason for the formation of the island is the pressure that the flow experiences when it merges with the second, more straight, main branch. Both islands—at the beginning and at the end of the branch—sometimes form a single chain with the islands developing inside the branch and forming with them an adjacent branching system of the 2nd order. In turn, each arm of the second order can be separated by a small, often elementary island or an isolated one into branches or channels of the third order. Thus, like a meandering channel, there is a hierarchy of branches, each order of which

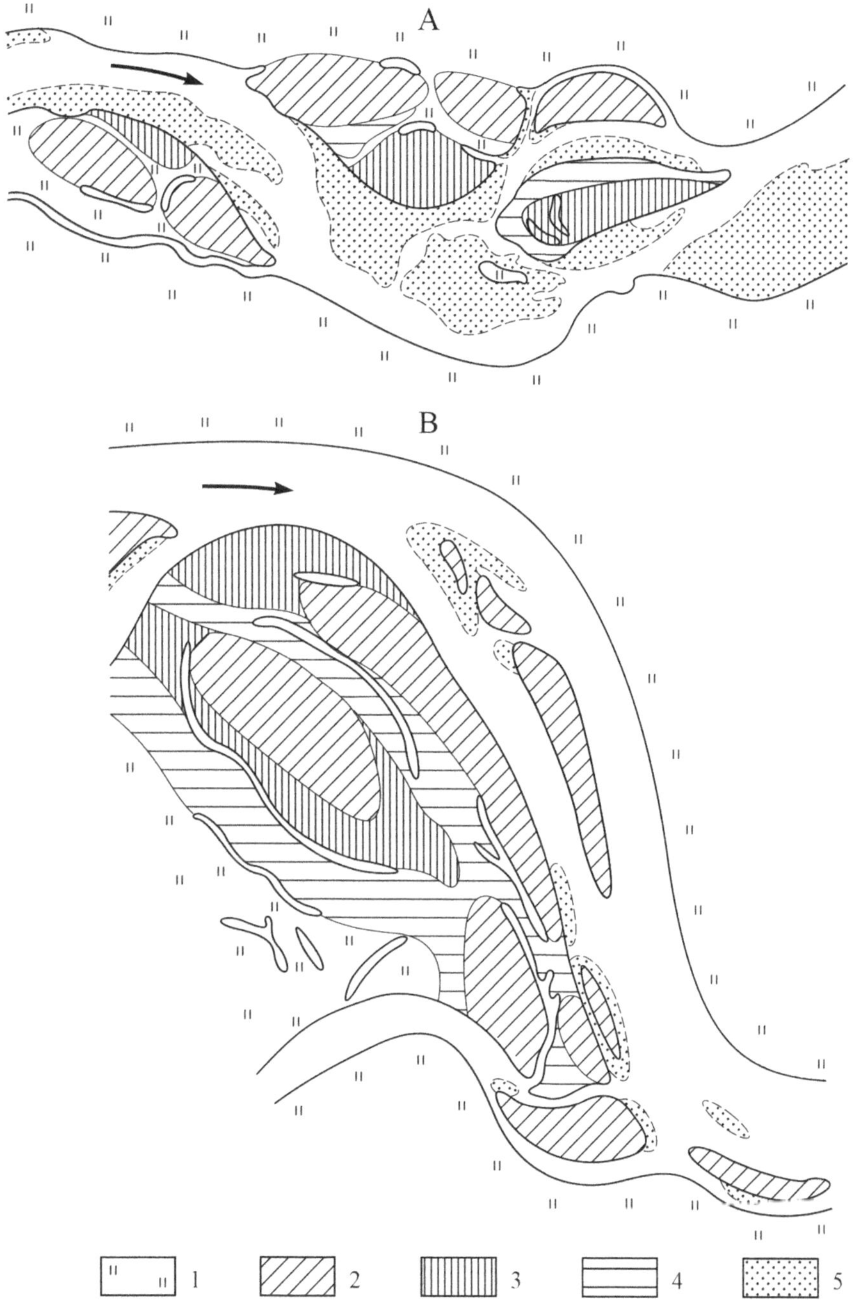

Fig. 4.48 Location of branching (islands) of II order on bends of braided-meandering channel: **A**—forming meander neck (Vychegda river, Solvychegodskaya bend); **B**—forming in the lower wing of the bend (Ob river, Taganskaya bend)

Fig. 4.49 Branching of the II order in the straight channel of Middle Lena (Tochilniy island). Photo of the author

corresponds to a certain interval of effective discharge. The most difficult branching is characteristic of rivers with Q_f in three intervals.

Relatively straight, single-thread channels (or channels with one-way branching as a transitional form between them and branched channels like branched and meandering channels, combining the features of both meandering and branched channels, can also be characterized by secondary branching under certain conditions. Depending on the partitioning between channel parameters to islands, the branching associated with them has the 2nd or 3rd order (Fig. 4.49), or the 2nd order, complicated by branching of the 3rd order. If we accept that in the branching of the second order the island has width $B_{is_{II}} < b_{br_I}$, in the branching of the third order, the island has width, in the branching of the channel, which has the third order $B_{is_{III}} < B_{br_{II}}$ (here indexes mean: is—island, br—branch), respectively, in a straight way the island, forming the branching of the third order, will have $B_{is_{III}} < 0.12b_{ch}$. These ratios are derived from the established Ivanov (1989), (Chalov et al. 1998) criteria for the separation of rectilinear channels, complicated by the 2nd or 3rd order islands, and the actual river channels, branched out into branches: $B_{is} < 0.4b_{ch}$.

The second order should also include branches that occur in a straight direction but are associated with islands of random origin; for example, separate long-distance coastal islands that occur directly below the outcrops of the original banks and do not create an overall morphological shape of the channel. In turn, in the branches of such branches can form mid-bars or side-bars, which, therefore, is already a form of channel of the 3rd order. Depending on the number of intervals and levels of effective discharges, there may be no second or third link in the bed material hierarchy. Then, within the bend, branching nodes and relatively straight single-thread areas, you can find the gyrus of the dynamic flow axis, enveloping side-bars, or the ducts formed by the emergence of axes in the channel.

In the scheme of hierarchical interconnection of channel forms (Fig. 4.45) an additional block—"channel forms"—is also selected. Its constituent elements (anabranching with its antipode—consolidated, non-dissected floodplain ducts; bends of the meandering belt or channel branching belt; bifurcated) are heterogeneous both in their origin and in their effect on channel formation processes. In the case of anabranching floodplains, a significant proportion of the flow goes into the floodplain branches, and the runoff rates of the flow in the main channel decreases even when the floodplain itself is already free from the floodplain water. The regulatory role of the consolidated floodplain is less in the high-water phase of the regime and fully ceases in the lowlands. The ramifications of the meandering (branching) belt play the same role as the bends of the river valley at the level of the modern floodplain with the difference that in the first case the regulating role of high floodplain is preserved, in the second case it is absent. Such macro bends (bends of meandering belts and channel branch belts) determine the direction of the floodplain flow, influencing the development of individual bendends and branches, the formation of ropes in the channel and their mode. For example, in a branched channel, the hose with the leading bank's shoulder flood flow is given priority development, under strong equal conditions. In the meandering channel there is a higher intensity of left bank erosion in comparison with the right bank or vice versa depending on the partitioning between bend elements with bends of the meandering belt.

In a bifurcated channel, the channels represent the division of the river into two practically independent, water-like streams, characterized by their channel regime different from that of the river above and below such macro-branchings.

In areas of limited channel change, where the lateral movement of channels is hindered by bedrocks or rocky formations that make up their banks, incised channels are formed with a minimal set of forms (hierarchy). They may have a non-channel genesis, being a predetermined geological structure. In this case, they are not related to the conditions of effective water discharge. However, the forms of channels of the 2nd or 3rd order, occurring in the incised channel, often have an accumulated origin, representing small islands, gentle confined bend, etc.

4.8 Rocky Beds, Sculptural and Accumulative Sculptural Forms of Channel Relief

The channel changes and morphology of channels are developed specifically if their banks and bottoms are composed of rocky formations. However, usually channel processes are considered only (or mainly) for the conditions of unconnected erosive soils. Thus, Grishanin (1972, p. 4), speaking of the interaction of water flow with the channel as the main essence of fluvial processes, emphasizes at the same time that we are talking about the impact of "the flow on the underlying solid (usually loose) environment". Such an approach is represented in the isolation of "limiting factors of fluvial processes" (Kondratyev et al. 1959, 1982).

In natural conditions, even large rivers with easily erodable channels are interrupted or interchanged with areas of greater or lesser length, where the flow is in direct contact with the bedrock and practically devoid of channel sediments. These are many mountain rivers, as well as semi-mountainous and flat rivers flowing in mountainous countries (for example, Katun in Altai, Vitim, rocky areas of upper Aldan); areas with a rocky bed are located on the upper and middle Lena, middle Yenisei (Kazachinsky and Osinovsky steps), middle Amur (Soyuznyi perekat), dominate the rivers of Karelia, and are found on the Onega and rivers flowing along the elevations of the East European Plain. Despite the absence of alluvium and the strength of the rocks with which the flows interact, fluvial processes in rocky channels are carried out, representing themselves in the incision of rivers, the transit transport of the pebbly-boulder-boulder deposits entering them, the formation of defibrillated ripples and the formation of stream-induced landslides, and the debris when the "presses" to the rocky banks.

In the upper reaches of mountain rivers and on small mountain rivers (first orders of magnitude), whose valleys separate the slopes of mountain ranges (basin area <10 km^2), rocky channels look like rectilinear steep slopes (Fig 4.50), sometimes stepped (Fig. 4.50) rocky trays, in which microwaterfalls (with a height difference of 0.5–1.0 m) and erosion boilers at their base continuously alternate. The hydraulics of the flows and the mechanism of sediment transport in such mountain rivers, which differ in large gradients (tens and hundreds of ‰), are similar to the temporary watercourses in the mountains (Litvin and Chalov 1975), which have such morphometric characteristics, but differ from them by the presence of a watercourse drying up in the most dry years during the dry season. With such high gradients, only the bed is washed out and downstream of its products and the material coming from the slopes, with the gravitational component contributing to the removal of sediment.

With the general reduction of the slope downstream of such rivers, the longitudinal profile acquires a more pronounced stepping stone, with large variations in the length of the river, up to and including waterfalls up to 10 m high and relatively gentle areas. According to measurements at the Chigirisha River (basin area 6.7 km^2) in the Western Transcaucasia (Khmeleva et al. 2000), the average gradients in the sections of the channel vary from 310 to 70‰ (Fig. 4.51). At the same time, even in the constrictions, in spite of rather large slopes (70‰) within their limits, there is a periodical accumulation and outflow of boulder-clump material, which transits here from the lying area, having a 4.5 times greater slope and a completely rocky bed. Vertical changes due to the accumulation and removal of sediment lead to fluctuations in the bottom marks from −3.5 m to +2.7 m (on average over the

Fig. 4.50 Channel rocky trays on small mountainous coinciding with the fall of rock strata (Cyprus island). Photo of the author

area from −0.9 m to +1.9 m). The same areas retain their rocky appearance when the bottom width remains the same, turning into trays with micro waterfalls when carrying out alluvium. The alternation of constrictions and dilatations, even with a generally homogeneous geological structure, is associated with lithological features and, consequently, erosion resistance of various bedrock formations. Already incised bends and sculptural side-effects are formed in the extensions near the convex banks (Khmeleva et al. 2000), arising as a result of more intensive destruction of the main bed due to the corrosive effect of sediments and incision of the bed in the core flow zone, where the main movement of coarse debris occurs. The formation of accumulations of bed material load is also possible here. Below the stages of the longitudinal profile (waterfalls) there is a phenomenon of cavitation, especially in floods, contributing to the formation of erosion boilers, the retreat of the stages upwards (regressively) and, at the same time, their greater intensity: if at the time of formation they are associated with different strength of rocks, now they are largely erosional. This is facilitated by the high kineticity of the flows: the number of the Fruit is almost always greater than 2.0 during floods, and in the steeply inclined tray (I = 310‰), which corresponds to the fall of the limestone bed surface, it reaches 10.8 (superturbulent flow); the flow rate in the latter case is 8.6 m/s.

On medium to large mountain rivers (basin area >10 km^2) in the gorges, rocky channels are sometimes significantly restricted by the bedrock channel banks and

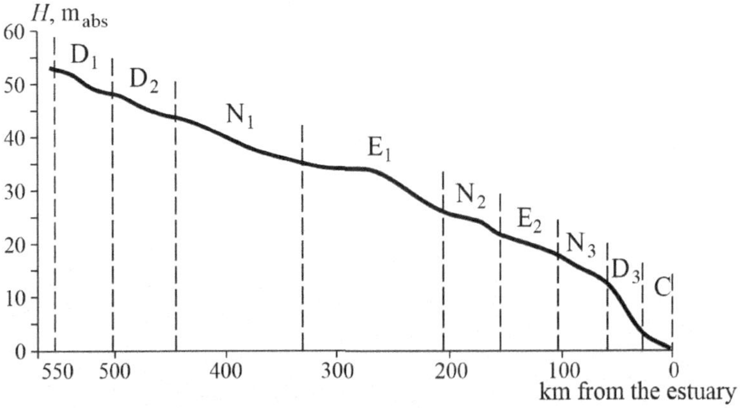

Fig. 4.51 Longitudinal profile of Chigiripsha river (Western Caucasus) downstream: bottom differences—steep sloping rocky trays; N—narrowing; E—expanding; C—debris cone. 1–3—numbers of sites [according to Khmeleva et al. (2000)]

bottom, and their width is several times smaller than in adjacent alluvial parts of the river. They correspond to the slopes of the channels with developed and undeveloped alluvial forms, which are upstream and downstream). Since $Q = \omega V$ (Q—water discharge; ω—living area of the flow; V—velocity), rapid currents with velocities of up to 5–7 m/s (instantaneous pulsation components reach 8–9 m/s) are formed here. This is facilitated by the very small size of the roughness of the bed, which is a rock polished by a stream (usually it is granite, gneisses, diabase, limestone, etc.). Such a significant increase in the average and instantaneous velocity with the same slope ensures that, in accordance with Erie's law (4.38), such rocky areas are taken out of the top of the bed, from tributaries or from the sides of the valley of coarse sediments, and that the channel is formed as a rocky tray. On Katun, the length of such rocky areas (Fig. 4.52) varies from the first kilometers to 70 km (in the lower reaches, before the river exits the mountains) with a longitudinal slope of less than 3–5‰. Separate accumulations of sediment can be found below rocky capes and islands, where local areas of silhouettes appear.

Boulders (pebbles, blocks) moved during floods corrode the bottom intensively; direct contact of the flow with the rock bed also provides active biochemical destruction of rocks (Dobrovolskaya 1980). Taking into account other factors, the incision of mountain riverbeds in rocky areas occurs at speeds of up to 1–7 cm/year (Hakimov 1992).

Rocky beds can be found on both semi-mountain rivers and flat rivers flowing in the mountains. The crowding of the flow causes high flow rates and intensive transport during floods of pebbles and boulders. It forms coastal ripples—big ripples and deficite ripples in other parts of the channel.

At the Vitim River in the lower reaches (downstream of Bodaibo), the proportion of rocky bed devoid of pebbles and boulders is 17.3% of the riverbed area at 300–171 km from the estuary and drops to 5.3% at 171–96 km (Zaitsev et al. 1998). In the

Middle Lena River, the riverbed from the Vitim estuary to Pokrovsk is characterized by the absence of a continuous cover of bottom sediments (Borsuk et al. 2001). Sediment deficit determines the insignificant thickness of alluvial deposits in the channel (Zaitsev and Savtsova 1994). The area of the channel devoid of riverbed sediments on the Nyuyskiy «riffle» is 10%, in the right-bank channel on the Tinskiy «riffle»—up to 40%, on the Zhedayskiy «riffle»—about 7%, and even in the lower part, for example, near the settlement of Nyuyskiy «riffle». Near Mohsogolloh it varies from 11 to 23% in different years (in the examples given here and below, the "riffles" on the Lena and other rivers is also called a shallow cofferdam in the channel, associated with the protrusions of the indigenous, rocky bed, causing difficulties for navigation, so they are quoted in the text). Lifting of the rock roofing causes local elevation of free surface slopes within their limits, especially in the lowlands, when such protrusions of the root bed begin to play the role of flooded weirs. Thus, according to the measurement data, the maximum water surface slopes were observed at low water horizons at the Elovsky riffle within the Nokhtuisko-Machinsky expansion (0.44‰) and at the Khatyn-Tumul "riffle" (0.55‰). The first is associated with diabase intrusion crossing the channel, expressed in its relief in the form of a projection with steep upstream and downstream slopes rising above the pool valleys by 5–6 m, and clearly visible in the hydrolocation image (Fig. 4.53); the second (Khatyn-Tumul) is determined by elevated limestone slabs.

A similar channel structure is typical of the Yenisei River below the Krasnoyarsk Hydroelectric Power Station. Here, at the riffles of Sr. Ladeyskoye, Nizhny Chastoostrovskoye, Sr. Shiverskoye, Atamanskiye stones and others, which are protrusions of rocky metamorphic rocks and granites, in which only depressions and cracks are filled with alluvium (Fig. 4.54A), the slope in the lowland increases in comparison with the flood by 15 times, reaching 0.35–0.54‰ at the Atalinskie stones "riffle" and 0.73‰ at the Berezovskie stones "riffle" (Berkovich et al. 2003).

On the upper Aldan, the local increase of slopes in the lowlands at the rocky outcrops (granites, gneisses) in the channel reaches 1.1 (Tungus "riffle")—1.7‰ (Sobolevsky "riffle"), and on the last one near the concave bank of the incised bend there are rocky aisles—boulders up to 1.5–2 m high above the low interlevel. On all the rivers with rocky bedrock, local bedrock projections, exerting a pressure on the flow, cause the formation of clusters of pebbles and boulders alluvium directly in front of them, and at the appropriate angle of encounter with the direction of the flow—and below, where it sometimes fills the erosion boilers. Such riffles are already sculpturally accumulative (Fig. 4.54B), in which pebbles and boulders are the result of changes in single rock projections of the hydraulic structure of the flow with the formation of a zone of support in front of them and "cross-shade" from the lower side. Similar riffles based on rocky foundations are widely spread on the upper and middle Lena, middle Yenisei, upper Aldan, Vitim and other rivers. With a lack of sediment load on these rivers, the thickness of riverbed sediments is insignificant, and in constrictions it is limited to the depth of the pool, which is underlain by the rocks. In the channel extensions, where the majority of riffles are formed, the thickness of the alluvium does not usually exceed 3 m. At the middle Lena River, the pebble-boulder alluvium thickness at the roffles ranges from 0.4 m

Fig. 4.52 Rocky channel of the Katun river (Altai) 5 km upstream (**A**) and 15–20 km downstream (**B**) of Elanda village. Photo of the author; To compare—mountainous channel of the Katun river with developed alluvial forms in the area of Chuya river estuary, 130 km upstream Elanda village (**C**). Photo of the author

c

Fig. 4.52 (continued)

to 1.0 m, increasing to the periphery of the channel, where side-bars of the riffles are formed, and in the local expansion of the channel.

Rocky ripples on the river bottom, being non-destructible, cause significant changes in the slope when changing the phases of the water regime. Zaitsev (1989) studied in detail this phenomenon at the "riffle" of Khatyn-Tumul on the Lena River (Fig. 4.55), which is a limestone slab with an uneven surface that crosses the channel in the section of 7 km long. The current velocity in the most submerged part of this plate reaches 2.0 m/s. At elevated levels, the longitudinal profile is aligned along the length of the section, and at the lowest interlinear levels, there is a pronounced bend corresponding to the lower end of the slab and the transition to the pool valley. This "play of slopes" [as Makkaveev put it (1955)] often leads to their ambiguous role in transport and accumulation of sediment in different phases of the water regime. On large flat rivers, the rocky areas represent a kind of spillway with a wide step. With large channel widths in the high-water phases of the slope mode, the slopes are reduced and the root bed is covered with an alluvial cloak. On the decline of levels and in the low speed, the depth drops, the deposited sediment is moved, washed away and carried downstream. These are the Union riffle on Amur, Olekminsk and Khatyn-Tumul on Lena. In a narrower channel, the rocky ledges are preserved in all phases of the regime, as the flow rates on them remain sufficient for the transit transport of sediment even when the slopes are reduced.

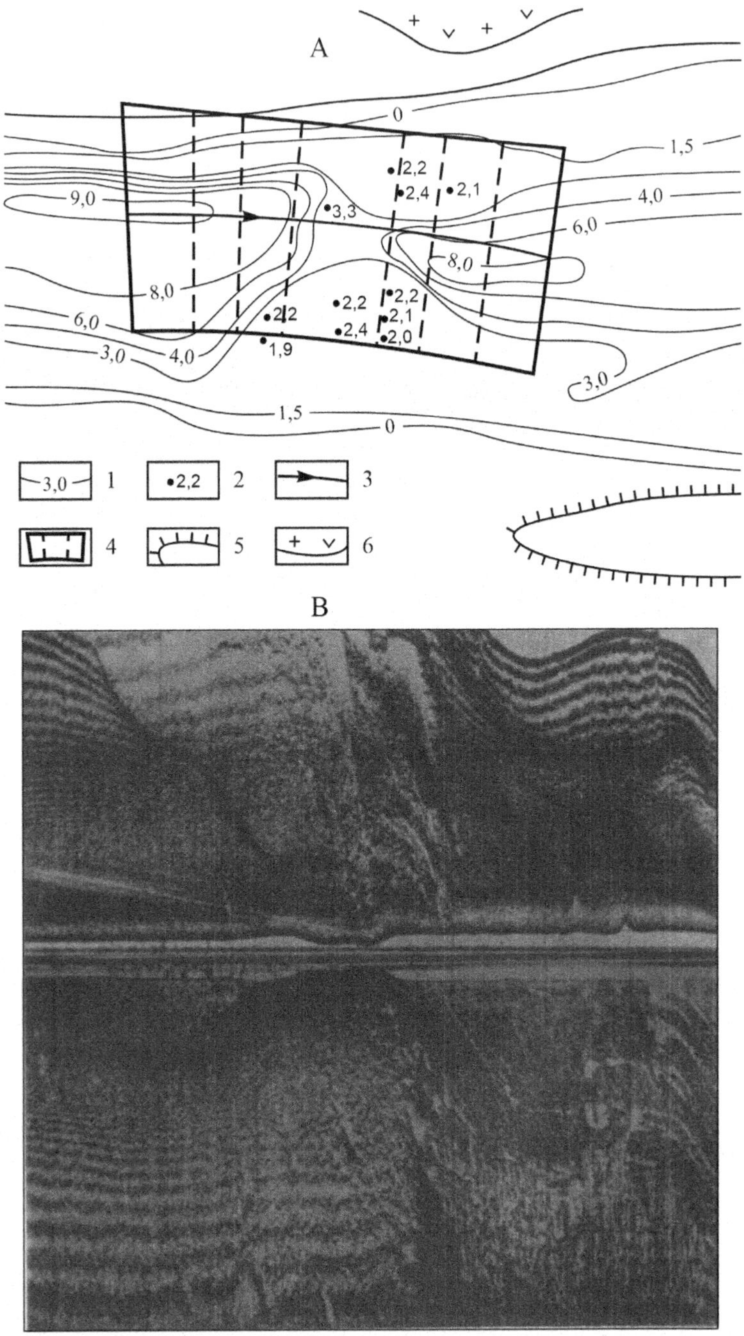

Fig. 4.53 Rocky outcrop on Elovsky riffle of Middle Lena. **A**—channel plan with isobaths (from the design level): 1—isobaths, m; 2—depth in separate points; 3—line of ship moving with the locator of lateral view; 4—sonar survey area; 5 -island edge; 6—rocky formation outcrops on the left bank. **B**—sonar image of rocky ledge (according to Zaitsev and Savtsova (1994)]

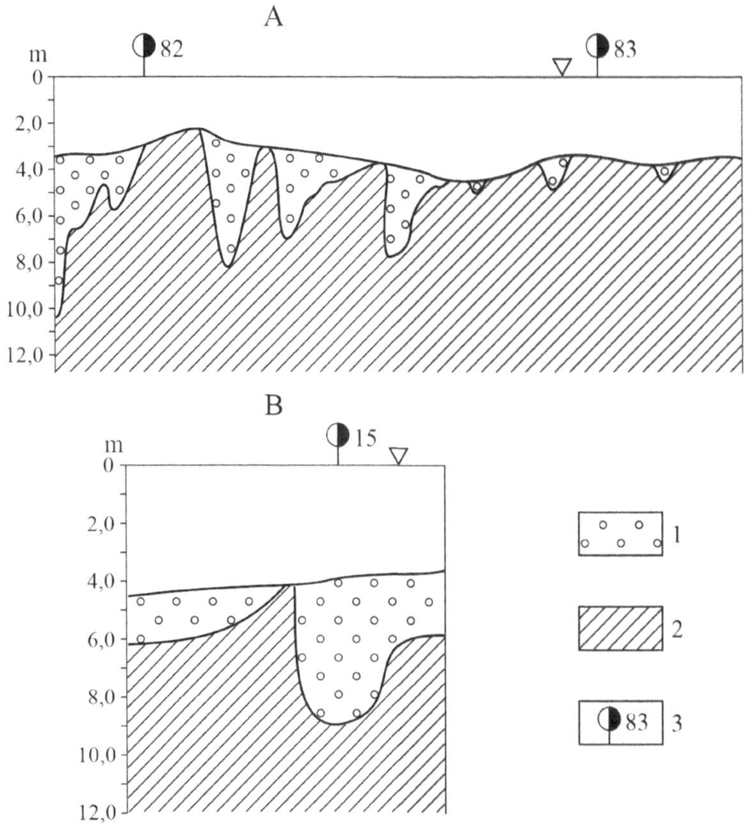

Fig. 4.54 Geological structure of the sculptural (**A**) and sculptural-accumulative (**B**) riffles on Yenisei river downstream Krasnoyarsk HPP (A—Atamanskiye stones riffle, B—Verkhniy Shaluninskiy riffle). 1—pebble alluvium, 2—rocky formations, 3—distance, km, along the ship's passage from HPP dam (Berkovitch et al. 2003)

A. V. Panin (Chalov et al. 1998) proposed an indicator called by him the K_R rocky coefficient as a partitioning between rocky areas to the total area of the channel. On the example of lower Vitim, he showed that downstream the bedrock ratio is decreasing, reflecting a decrease in the overall sediment deficit in this direction. The linear descending trend of rockiness is described by the relationship

$$K_R = \alpha x - \gamma, \tag{4.70}$$

where x is the distance from the mouth, km; $\alpha = 0.00111$; $\gamma = 0.084$ (for Vitim). Obviously, the values of α and γ on other rivers will be different depending on their slopes, other hydraulic characteristics and rock composition. A. V. Panin connected the directional decrease of the channel slope with the general decrease of the current velocity at the decrease of the channel slope, which is subject to the relationship V = 0.00125x +1.4. This leads to a reduction in the transit of particles off-site and an increase in the amount of bed load and sediments. In some local areas, the relationship between bedrock and current velocities is blurred by local factors. At the same time, sharp changes in geological-geomorphological conditions (Katun, Vitim) lead to an equally abrupt change in the rocky bed by alluvial ones and the disappearance of bed material load deficit.

The channel pattern has a certain influence on the degree of bed hardness. In Vitim, Middle Lena and Upper Aldan, the rockiest areas of the bed in the main banks of the river, where large depths of the bed prevail. Here, the cross-section is close to that of a parabolic cross-section, and the rocky bed is mainly confined to the lateral parts of the channel and rarely to its bedrock zone, where the main sediment load is concentrated. In this form of channel, the debris coming from the sides of the valley from the tributaries is quickly "rolled down" with gravitational forces to the middle of the channel, where the flow rate is sufficient for its transportation.

Within the branching usually the opposite picture: the rocky, alluvial-free bottom is confined to the central part of the branches. With a shortage of sediment in the wide channels of large lowland rivers, pebbles and boulders are transported in the coastal parts of the river, as transverse transport is difficult because of the high particle weight. This results in a more intensive depression of the bottom (due to corrosion) along the banks, and a rocky ripple forms in the middle of the channel,

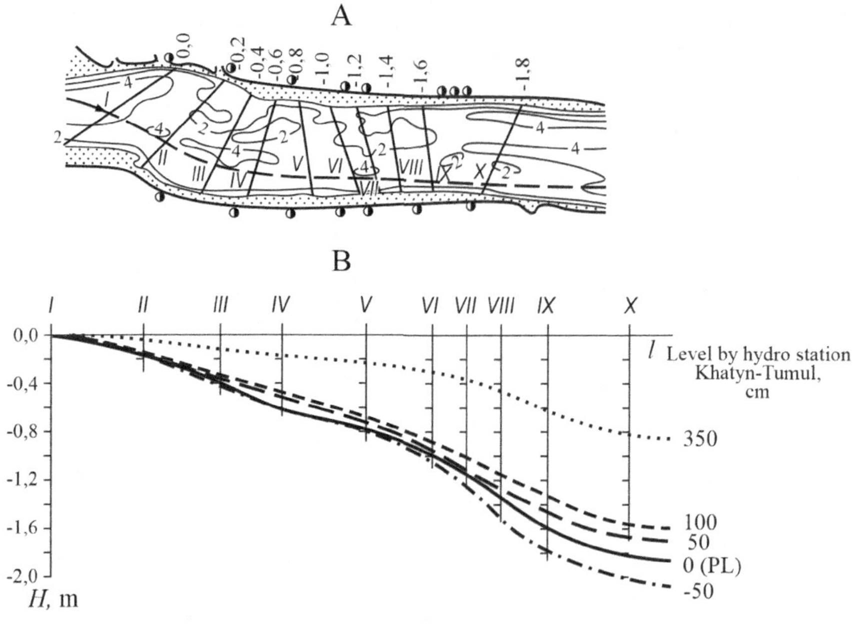

Fig. 4.55 Relief of water surface on Khatyn-Tumul riffle (Lena river) at the design level DL (**A**; isolines—m) and its decreasing along the main stream (axis of the ship`s passage) at different water levels from DL (**B**) (Zaitsev 1989)

which eventually turns into a basement island (Fig. 4.56). This determines the wide spread of rocky outcrops on the Lena, Aldan, Vitim, Yenisei and Angara islands. As they form, they affect the structure of the flow as they deepen the coastal parts of the channel, resulting in lower "velocity shadows" and higher levels of pebbles and boulders in the catchment area. As a result, the sculptural protrusions in the middle of the riverbed first turn into a sculptural and accumulative island. During the geological history of riverbeds development, the islands have also become floodplain islands on a significant part of their territory, sometimes towering over the river for tens of meters. These are, for example, numerous islands in the Angara, rising above the river up to 50–60 m, most of which are populated areas.

Separate rocky ledges are found on many lowland rivers, where they are linked to steps, sometimes waterfalls or, conversely, narrow gullies with abnormally large depths. They are timed to coincide with the intersection of hard to wash, usually rocky, crossed by rivers. Depending on the width of the strip of rock, these areas on the rivers are quite long or short. In doing so, they can be both narrowing the channel with large depths and rocky shallow waters forming empty areas. The first one is recessing to 35–40 m in the section of 1–1.5 km in the channel of the Severnaya Dvina river near the settlement of Orletsy, where the river crosses the area of distribution of the most dense limestone, narrowing to 200 m (against 1000–1500 m upstream at a maximum depth of 10 m); the river. Prior to its confluence with the Northern Dvina River, the Pinega runs for about 10 km between low limestone cliffs up to 15–20 m high, narrowing by more than 2 times compared to the upstream section and having a depth of up to 20 m, and the bottom of the river, as well as the banks, is rocky (upstream sandy channel is a complex riffle with a depth of about 1 m on the ropes (in the low-water) and up to 4–5 m in the pool valleys between them. The Yenisei channel in the 10 km "Cheeks" gorge has a similar appearance, but with incised bends, at the beginning of which it narrows down to 0.7 km (against 1.5 km in the single-thread section upstream). At the end of the gorge there are high rocky islands rising above the river by 30–40 m. The banks of the river are steep, rocky, the depth in "Cheeks" is up to 35 m; the bed is also rocky.

The same type of rocky but shallow water channels includes the former thresholds on the Dnieper River near the city of Zaporozhye. At the Tom River near the Tomsk city, the outcrop of rock at the Camp Garden forms a sharp fracture of the longitudinal profile with a slope of 0.8 ‰ in the 2 km section (with slopes above and below the current of 0.11 and 0.15‰, respectively). This fracture due to the regressive cutting of the channel due to the development of downstream quarries and the lowering of water levels in the river by 2.2–2.6 m has turned into a step that is the basis for erosion of the upstream section of the river (Berkovich et al. 1998). The same basis of erosion on the Dnieper was the steps in Zaporozhye before the construction of the Dnieper hydropower plant, and now are the Osinovsky and Kazachinsky steps on the Yenisei. The Kazachinsky step (see Fig. 2.3) represents two rocky ripples crossing the channel in a segment of about 4 km and creating a level difference of 3.8 m (flow slope—1‰). Between the rocks there is a narrow

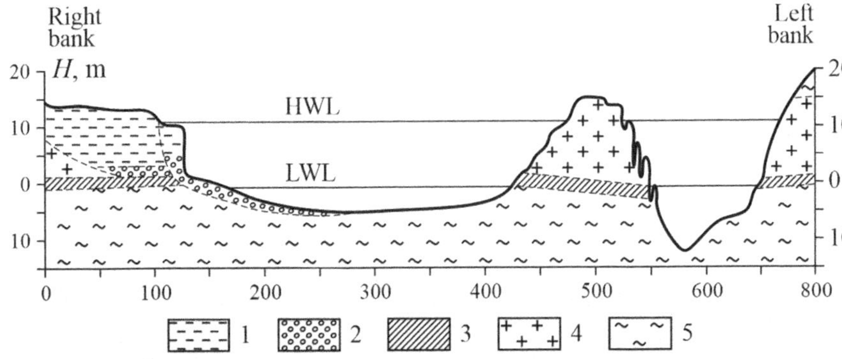

Fig. 4.56 Geological structure of the Vitim river channel near Kamenniy island (transverse profile): 1—loam; 2—pebble and boulder; 3—quartzites; 4—pegmatites; 5—gneisses and schists (according to Gurin and Lodina (1994)]

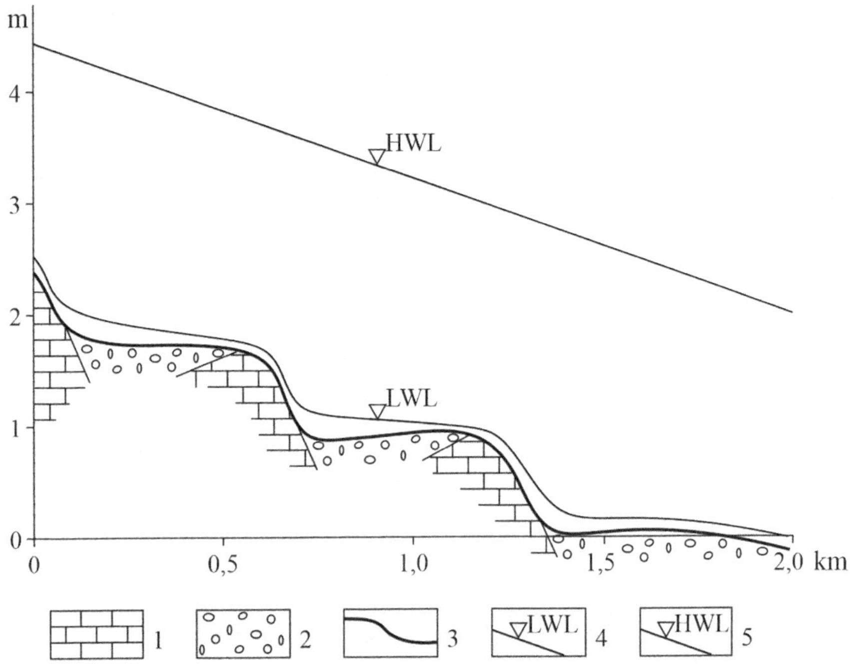

Fig. 4.57 Schematic longitudinal profiles of flow free surface during flood and low-water and of channel bottom of a small lowland river with steps and pools on the rock outcrops. 1—limestones; 2—pebble alluvium; 3—river bottom; 4—low-water level; 5—flood level

trough with a maximum depth of 3.2 m, on which the ship's passage is laid. Current velocities on the step reach 5 m/s.

The rocky channels are often accompanied by numerous protrusions and capes of banks, which have a directing effect on the flow, causing its local congestion, the formation of souvenirs and over-depressions. On the Aldan River in the section from Tommot to the Lower Kurum-Uryakh riffle outlets of granites and gneisses on the banks, hollowed under the edge, predetermine the shallow occurrence of rocky trays in the coastal parts of the channel and the presence of rocky ripples within the limits of the riffles at a small depth (Eloviy, Freimanovskiy, Sobolevskiy, Shestirucheiny riffles). When granite rocks are steeply cliff into the river, they are covered by alluvial boulder and pebble deposits within the channel.

Rocky capes on Yenisei, Amur and other Siberian rivers are called "bulls". On Yenisei the most outstanding cape of the right bank—Savinsky bull (311 km below Krasnoyarsk) blocks the riverbed by 700–800 m. Several rocky capes on the right bank of the Ob River below the Novosibirsk hydroelectric power station are connected with the flow of granite massifs by the river. As a rule, they have their continuation in the channel, forming numerous very local rocky outcrops scattered across the river and sometimes forming several consecutive coastal "shivers".

On small and sometimes middle plain rivers flowing along the structural uplands, rocky outcrops cause a stepwise progression of their longitudinal profiles. Each such stage consists of a step or microwaterfall, where the slopes are sometimes tens of times greater than in the areas between them (Fig. 4.57). During floods, the slope along the whole section with the stepped profile is levelled, but its total value is increased (for example, on the Tarusa river in the lower reaches it is 1.2‰), which causes the transfer of pebbles and small-boulder matherial by the flow.

Literature

Alekseevsky, N.I.: Formation and Movement of River Load, 203 p. Moscow State University Publishing House, Moscow (1998)

Alekseevsky, N.I., Chalov, S.R.: Hydrological Functions of Channel Branching. Hydrology, Hydrochemistry and Hydroecology. T II. Kiev: VGL "Obryi", pp. 53–59 (2006)

Altunin, V.S.: Channel Bed Changes, 120 p. Kolos, Moscow (1972)

Altunin, S.T.: Regulation of Riverbeds, 352 p. Selkhozizdat, Moscow (1962)

Amalitsky, V.P.: Diary of Observations of the Small Northern Dvina, 57 p. USSR Academy of Sciences, Leningrad (1931)

Andreev, O.V., Yaroslavtsev, I.A.: Channel Changes on the River Sections with the Bridge Crossings. Fluvial Processes, pp. 352–372. Publishing House of the USSR Academy of Sciences, Moscow (1958)

Antropovsky, V.I., Minaeva, G.I.: Background forecast of the riffle reformations on the river Vilyuy in the conditions of the regulated flow. Proc. SHI. Op. **216**, 52–63 (1974)

Baryshnikov, N.B. Morphology, Hydrology and Hydraulics of Floodplains, 280 p. Hydrometeoizdat, Leningrad (1984)

Belostotsky, I.I.: Observations over Ryabi signs. Izv. VGO. **2**, 225 (1940) (in Russian)

Berkovich, K.M., Zaitsev, A.A., Lodina, R.V., Chalov, R.S.: Fluvial processes on the large East Siberian rivers with a pebble-boulder alluvium and features of their regulation. Vest. Moscow Un. Ser 5. Geogr. Vest. (3), 35–41 (1985) (in Russian)

Berkovich, K.M., Zlotina, L.V., Ivanov, V.V., Nikitina, L.N., Ryazanov, P.I., Turykin, L.A., Chalov, R.S., Chernov, A.V.: Development of the lower and middle Dniester riverbed under the anthropogenic load. In: Ecological Problems of the Soil Erosion and Fluvial Processes, pp. 141–165. MSU Publications, Moscow (1992) (in Russian).

Berkovich, K.M., Rouleva, S.I., Surkov, V.V., Chalov, R.C.: Fluvial processes and anthropogenic reformations of the lower Tom riverbed and their influence on the floodplain landscapes. In: Soil Erosion and Fluvial Processes, vol. 11, pp. 215–240. MSU Publishing House, Moscow (1998) (in Russian)

Berkovich, K.M., Vinogradova, N.N., Ivanov, V.V., Chalov, R.C.: Yenisei channel reformation below Krasnoyarsk hydroelectric power station under intensive technogenic load. In: Soil Erosion and Fluvial Processes, vol. 14, pp. 144–160. Moscow State University Publishing House, Moscow (2003) (in Russian)

Borsuk, O.A., Dobrovolskaya, N.G., Lodina, R.V., Chalov, R.S.: Morphology of channels and modern channel alluvium on the Western Tien Shan mountain rivers. Geomorphology **4**, 60–67 (1981) (in Russian)

Borsuk, O.A., Zaitsev, A.A., Kirik, O.M., Lodina, R.V., Chalov, R.S.: Fluvial processes on the middle Lena (from Vitim to Pokrovsk). In: Soil Erosion and Fluvial Processes, vol. 13, pp. 203–228. MSU Publishing House, Moscow (2001)

Brotherton, D.L.: On the origin and characteristics of river channel. J. Hydrol. **44**(3–4), 211–230 (1979)

Chalov, R.S.: Some features of the mountain rivers channel regime. Meteorol. Hydrol. **4**, 70–74 (1968) (in Russian)

Chalov R. S. Morphology and dynamics of the mountain river beds. In: Proceedings of Moscow. GS USSR. Geomorphology, vol. 3, pp. 8–10 (1969) (in Russian)

Chalov, R.S.: Relief of floodplains. In: Soil Erosion and Fluvial Processes, vol. 1, pp. 192–204. MSU Publishing House, Moscow (1970)

Chalov, R.S.: Geographical Investigations of Fluvial Processes, 232 p. MSU Publishing House, Moscow (1979)

Chalov, R.S.: The main types of channel changes, forms of their manifestation and peculiarities of relations in different natural conditions. In: Proceedings of the Section of Fluvial Processes of Scientific Council "Integrated Water Resources Use and Protection". General Questions of Fluvial Processes Theory, vol. 1, pp. 100–112. Gidrometeoizdat, Leningrad (1986) (in Russian)

Chalov, R.S.: Types of fluvial processes and principles of morphodynamic classification of river channels. Geomorphology **1**, 25–36 (1996)

Chalov, R.S.: General and Geographical Riverbed Science, 112 p. MSU Publishing House, Moscow (1997a)

Chalov, R.S.: Fluvial processes as hydrological phenomena and their geomorphological reflection. In: Hydrology and Geomorphology of River Systems, pp. 23–32. Irkutsk (1997b)

Chalov, R.S.: Fluvial processes on small plain rivers as a special type of fluvial processes. In: Processes and Ecological Situation in Basins of Small Rivers, pp. 11–16. Udmurt Un, Izhevsk (2000)

Chalov, S.R.: Formation of the structure of channel branches. Geomorphology **1**, 92–102 (2006)

Chalov, R.S., Alabyan, A.M., Ivanov, V., Lodina, R.V., Panin, A.V.: Morphodynamics of the Channels of Plain Rivers, 288 p. GEOS, Moscow (1998)

Chalov, R.S., Zavadsky, A.S., Pakhomova, O.M.: Natural and anthropogenic manifestations of the fluvial processes in the different parts of the river network. In: Problems of Hydrology and Hydroecology, vol. 1, pp. 203–216. MSU Publishing House, Moscow (1999) (in Russian).

Chalov, R.S., Liu, S., Alekseevskiy, N.I.: Sediment Flow and Fluvial Processes on the Big Rivers of Russia and China, 216 p. MSU Publishing House, Moscow (2000)

Chalov, R.S., Zavadsky, A.S., Panin, A.V.: River Radiations, 371 p. MSU Publishing House, Moscow (2004)
Charnomsky, V.P.: About trays and channels of rectangular cross-section of the strengthened roughness in application to fish-passengers, fast-flows and platoon navigation. In: Materials on Investigation of the Volkhov River and its Basin, pp. 143–148. Leningrad (1924)
Chitale, S.V.: River channel patterns. J. Hydraul. Div. ASLE. **96**(HY-I), 201–222 (1970)
Dedkov, A.P., Kurbanova, S.G., Mozzherin, V.I.: About degradation of the river network in the Middle Volga region and its causes. In: Proceedings of the AVN, vol. 1, pp. 93–98. Moscow (1995)
Dobrovolskaya, N.G.: Features of subaqueous weathering on plain and mountain rivers (on the example of Lena and Kara-Bau, Western Tien Shan). Vest. Moscow Un. Ser. 5. Geogr. **2**, 71–76 (1980)
Dynamic Geomorphology, 448 p. MSU Publishing House, Moscow (1992)
Edelstein, Y.S.: Basics of Geomorphology, 400 p. Gostekhizdat, Moscow, Leningrad (1947)
Engineering of Ship Passages on Free Rivers/TsNIIEVT Works, vol. 36, 262 p. Transport, Moscow (1964)
Experimental Geomorphology, vol. 2, 173 p. MSU Publishing House, Moscow (1969)
Fluvial Processes in the Lower Reaches of the Weihe River. Nauka, Beijing (1984) (in Chinese).
Goncharov, V.N.: Dynamics of channel flows, 373 p. Hydrometeoisdat, Leningrad (1962)
Grinberg, Z.A.: About morphometric characteristics of rivers. Meteorology and hydrology. **4**, 43–45 (1950)
Grishanin, K.V.: Dynamics of Channel Flows, 488 p. Hydrometeoizdat, Leningrad (1969)
Grishanin, K.V.: Theory of the River Fluvial Process, 216 p. Transport, Moscow (1972)
Grishanin, K.V.: Stability of River and Canal Beds, 144 p. Gidrometeoizdat, Leningrad (1974)
Gurin, A.D., Lodina, R.V.: Peculiarities of fluvial processes and channel morphology of the Lower Vitim. Geomorphology. **1**, 87–93, (1994)
Hakimov, S.K.: Fluvial processes on the mountain rivers of the Western Tien Shan: Author's thesis. Ph.D. Geography Sciences, 29 p. MSU Publishing House, Moscow (1992)
Hjulstrom, F. (1935) Studies of the morphological activity of rivers as illustrated by the River Furis. Bull. Geol. Jnst. Upsala **35**, 35–58 (1935)
Ivanov, V.V.: Formation conditions, hydrological-morphometric dependences and changes in relation to the straight, unbranched channels: Author`s… of Dis. Ph.D. Geography Sciences, 23 p. MSU Publishing House, Moscow (1989)
Karasev, I.F.: Fluvial Processes at Transfer of the Flow, 288 p. Hydrometeoizdat, Leningrad (1975)
Karaushev, A.V.: River hydraulics, 416 p. Hydrometeoisdat, Leningrad (1975)
Karaushev, A.V.: Theory and Methods of Calculation of River Load, 272 p. Gidrometeoizdat, Leningrad (1977)
Kellerhals, R., Churh, M., Bray, D.J.: Classification and analysis of river processes. J. Hedraul. Div. Proc. Amer. Soc. Civ. Eng. **7**, 813–829 (1976)
Kennedy, J.F.: The mechanics of dunes and antidunes in erodible-bed channels. Fluid Mech. **16**, 521–544 (1963)
Khmeleva, N.V., Nikulin, F.V., Shevchenko, B.F.: Some results of stationary studies of erosion forms in the Northwest Caucasus. In: Soil Erosion and Fluvial Processes, vol. 1, pp. 142–151. MSU Publishing House, Moscow (1970)
Khmeleva, N.V., Vinogradova, N.N., Samoilova, A.A., Shevchenko, B.F.: Mountain River Basin and Exogenous Processes within it (Results of Stationary Studies), 187 p. MSU (2000)
Klaven, A.B., Snishchenko, B.F.: Basics of the feature of fluvial process in the lower course of Selenga R. Proc. of SHI **252**, 38–51 (1978)
Kondratyev, N.E.: Kinematic structure of a flow at a rippled structure of a bottom. Proc. of SHI **116**, 3–18 (1964) (in Russian)
Kondratyev, N.E.: River Fluvial Processes and Changes of Reservoir Banks, 258 p. Znak, SPb (2000)

Kondratyev, N.E., Lyapin, A.N., Popov, I.V., Pinkovsky, S.I., Fedorov, N.N., Yakunin, I.A.: Fluvial Process, 372 p. Hydrometeoizdat, Leningrad (1959)
Kondratyev, N.E., Popov, I.V., Snischenko, B.F.: Basics of Hydromorphological Theory of the Fluvial Process, 272 p. Gidrometeoizdat, Leningrad (1982)
Kopaliani, Z.D., Tvalavadze, O.A., Noselidze, L.V.: Hydraulic modeling of the fluvial process of the Anosovka river piedmont section on the bridge crossing. Newsletter of the Fluvial Processes Section of Scientific Council on the Problem "Integrated Use and Protection of Water Resources of SCST". Issue Paper. 3. Problems of Hydraulics and River Fluvial Process of the Mountain Rivers pp. 88–106. Gidrometeoizdat, SPb (1992) (in Russian)
Kopaliani, Z.D., Tskhadadze, V.S.: Types of river channels in Western Georgia. Proc. SHI **195**, 20–32 (1972)
Korchoha, Y.M.: Investigation of ridge motion of deposits on the river Polomet. Proc. SHI **161**, 98–119 (1968)
Koronkevich, N.I., Chernogayeva, G.M., Yasinskiy, S.V.: Conceptual Problems of the Solution of the Small Rivers Problem in Russia. Small Rivers of Russia, pp. 11–19. IG RAN-MC RGO, Moscow (1994)
Korotayev, V.N.: Morphodynamics of watercourses in the Western part of the Volga delta. Proc. AVN **9**, 123–139 (2003) (Moscow).
Kovalchuk, I.P.: Regional Ecological-Geomorphological Analysis, 439 p. Lvov, 1997
Krasnov, A.N.: Shapes of the land surface and figures who create them. In: Fundamentals of Land Science, vol. 3, 235 p. Kharkov (1897)
Kroshkin, A.N.: To determination of hydromorphological characteristics and average weight concentration of attracted load on mountain rivers. In: Movement of Load in Open Channels, pp. 111–119. Nauka, Moscow (1970) (in Russian)
Kuznetsov, K.L., Chalov, R.S.: Fluvial processes and morphology of the mountain river beds in conditions of active mudflow activity (on an example of the rivers of the northern slope of Zailijsky Alatau). Geomorphology **2**, 71–78 (1988)
Latyshenkov, A.M.: Basics of Hydraulics, 248 p. Hydrometeoizdat, Leningrad (1971)
Leliavskiy, N.S.: On the river currents and the formation of a river bed. In: Proceedings of the 2nd Congress of Hydraulic Engineers in 1893. St. Petersburg (1893) (Problems of Hydraulic Engineering of Free Rivers, pp. 18–136. Rechizdat, Moscow (1948)).
Leopold, L.B., Wolman, M.G.: River Channel Patterns—Braided, Meandering and Straight. US Geol. Sure Prof. Pap. **282-B**, 1–85 (1957)
Levashov, A.A.: Deformation rate determination of some characteristics of morphological formations in the Nadym river channel. Meteorol. Hydrol. **3**, 92–95 (1973)
Levi, I.I.: Dynamics of Streams, 250 p. Hydroenergoizdat, Moscow, Leningrad (1957)
Litvin, L.F., Chalov, R.S.: On the channel formation activity of temporary and permanent watercourses in the mountains. Geomorphology **1**, 81–86 (1975)
Liu, S., Chalov, R.S., Ding, J., Ivanov, V.V., Korotayev, V.N., Li, J., Yang, H., Yang, S.: Regional variations of the Asian rivers suspended sediment flow in estuaries. Vest. Moscow Univ. Ser 5. Geogr. **3**, 44–51 (2001) (in Russian)
Lodina, R.V., Chalov, R.S.: Cobblestone bridge on the big rivers. Nature **7**, 57–62 (1994) (in Russian)
Lodina, R.V., Hakimov, S.K., Chalov, R.S.: Modern channel alluvium of the large mountain rivers of the Western Tien Shan and their tributaries. Geomorphology **4**, 70–76 (1994) (in Russian)
Lokhtin, V.I.M.: About the Mechanism of the River Bed. St. Petersburg (1897) (Problems of Hydraulic Engineering of Free Rivers, pp. 23–59. Rechizdat, Moscow (1948)).
Lomakin, V.V.: On residual river and general residual surface deposits. New SA USSR. Ser. Geol. **2**, 79–87 (1943)
Lopatin, G.V.: Sediments of the USSR Rivers, 368 p. Geografgiz, Moscow (1952)
Losievsky, A.I.: Laboratory research of riffle formation processes. Proc. TsNIIVT **36**, 98 (1934)
Lvovich, M.I.: Experience of classification of the rivers of Russia. Proc. SHI **6**, 58–105 (1938a)
Lvovich, M.I.: Fluvial processes i gold content of a site of the river Sugor. In: Proceedings of trust "Zolotorazvedka" and Institute NIGRIZoloto. vol. 8. pp. 169–185 (1938b)

Makhinov, A.N., Chalov, R.S., Chernov, A.V.: Directed sediment accumulation and morphology of the lower Amur bed. Geomorphology **3**, 70–78 (1994) (in Russian)
Makkaveev N.I.: River Channel and Erosion in its Basin. Published in the SA USSR, Moscow (1955)
Makkaveev, V.M.: Some theoretical problems of the open thread dynamics. Proc. SHI **8**(62), 5–26 (1948) (in Russian)
Makkaveev, N.I.: Fluvial Regime of Rivers and Tracing of Slits, 202 p. Rechizdat, Moscow (1949)
Makkaveev, N.I.: Flow and Stream Processes, 116 p. MSU Publishing House, Moscow (1971)
Makkaveev, N.I.: Hydraulic typing of erosion process. In: Soil Erosion and Fluvial Processes, vol. 3, pp. 65–77. MSU Publishing House, Moscow (1973).
Makkaveev, N.I.: General regularities of the erosion-fluvial processes. In: Proceedings of IV All-Union Hydrologic Congress. Fluvial Processes, vol. 10, pp. 8–12. Hydrometeoizdat, Leningrad (1976) (in Russian)
Makkaveev, N.I., Kalinin, A.: On movement of large-fragment material in logs. Meteorology and Hydrology. **8**, 61–68 (1968)
Makkaveev, N.I., Chalov, R.S.: Fluvial Processes, 264 p. MSU Publishing House, Moscow (1986)
Makkaveev, N.I., Chalov, R.S.: On the development of the surface relief of river terraces and signs of deep erosion on the example of the upper Ob. New AS USSR. Ser. Geography. **4**, 120–125 (1964)
Makkaveev, N.I., Khmeleva, N.V., Zaitov, I.R., Lebedeva, I.V.: Experimental Geomorphology, 196 p. MSU Publishing House, Moscow (1961)
Makkaveev, N.I., Kalinin, A.M., Khmeleva, N.V.: Analysis of the material composition of piezoquartz placers for prediction of their bedrock sources. Geographycal questions, 92, pp. 143–154. Mysl, M (1973)
Makkaveev, N.I., Litvin, L.F., Khmeleva, N.V.: Using of the transport capacity of the river flow for practical purposes. Vest. MSU. Ser. 5. Geography. **2**, 82–89 (1970)
Makrinova, O.V.: Natural erosion of the diverter channel. Proc. SHI **40**(94), 29–43 (1953)
Mandych, A.F.: Influence of the mountain rivers flood waves on the sediment flow. Meteorol. Hydrol. **2**, 41–44 (1966) (in Russian)
Mandych, A.F.: The value of the sediment yield of the rivers of Western Georgia. Vest. MSU. Ser. 5. Geography. 41–44 (1967)
Maslov, V.P.: Some variants of the ripple river signs. Bill. MOIP. Depart. Geol. **XI**(2), 162–163 (1933) (in Russian)
Matveev, N.P.: Meandering of the Moscow region rivers. Uch. Zap. MOPI N.K. Krupskaya **267** (13), 3–24 (1970)
Miall, A.D.: A review of the braided-river depositional environment. Earth Sci. Rev. **13**, 1–62 (1977)
Mikhailov, V.N.: Flow and channel dynamics in non-tidal estuaries. Geoecology. **6**, 100–110 (1998)
Mikhailova, N.A.: Transfer of Solid Particles by Turbulent Water Flows, 236 p. Hydrometeoizdat, Leningrad (1966)
Mirtskhulava, C.E.: Erosion of channels and methods of their stability estimation, 180 p. Kolos, Moscow (1967)
Mirtskhulava, C.E.: Basics of Physics and Mechanics of Erosion of Channels, 304 p. Hydrometeoizdat, Leningrad (1988)
Pakhomova, O.M.: Hydrologic-morphological characteristics of river beds and orderly structure of the river network. Autoref. Diss. ... Ph.D. Geography Sciences, 28 p. MSU Publishing House, Moscow (2001)
Panin, A.V.: Morphology and dynamics of Incised pebble-boulder rivers. Autoref. dis. ... Ph.D. Geography Sciences, 24 p. MSU Publishing House, Moscow (1991)
Petrovsky, V. V., Sakharova, E.I.: About reformations of Krasnoborsk riffle (Northern Dvina). In: Channel Investigations to Improve Navigation Conditions, pp. 19–23. River Transport, Moscow (1958)
Popkov, I.F.: Total Inland Waterway Station, 279 p. River transport, Moscow (1962)

Popov, I.V.: Methodic Bases of the River Fluvial Process Research, 208 p. Gidrometeoizdat, Leningrad (1961)
Popov, I.V.: River Bed Changes and Hydraulic Engineering, 328 p. Hydrometeoizdat, Leningrad (1965)
Potapov, M.V.: Screw fluid motion in a direct open channel of a rectangular cross-section. In: Cross Circulation in an Open Flow and its Hydrotechnical Application, pp. 102–154. Agricultural Product, Moscow (1936) (in Russian)
Prokacheva, B.G., Snischenko, D.V., Usachev, V.F.: Remote Methods of Hydrological Study of BAM Zone, 224 p. Gidrometeoizdat, Leningrad (1982)
Richards, K.S.: The formation of ripples and dunes on an erodible bed. J. Fluid Meeh. **99**(3), 597–618 (1980)
Rodevich, V.M.: To the classification of rivers. New. SHI. **35**, 25–39 (1931)
Romashin, V.V.: Some features of the river fluvial process on the mountain river. Proc. SHI **144**, 66–76 (1967) (in Russian)
Romashin, V.V.: Estimation of a channel mode on morphological characteristics. Proc. SHI **190**, 48–56 (1972)
Romashin, V.V.: Properties of a channel wandering. Proc. SHI **225**, 5–15 (1975)
Rosgen, D.L.: A classification of natural rivers. CATENA **22**, 169–199 (1994)
Rossinsky, K.I., Kuzmin, I.A.: Some questions of the applied theory of the river channel formation. In: Problems of the River Flow Regulation, pp. 88–129. SA USSR, Moscow, Leningrad (1947) (in Russian)
Rossinsky, K.I., Kuzmin, I.A.: Regularities of river channel formation. In: Fluvial Processes, pp. 5–14. Publishing House of the SA USSR, Moscow (1958)
Rossinsky, K.I., Debolsky, V.K.: River Sediments, 216 p. Nauka, Moscow (1980)
Ruleva, S.N., Chalov, P.S.: Channel changes at the junction of the rivers Vychegda and Sysola and their influence on the infrastructure of Syktyvkar. Proc. AVN **9** 171–181 (2003) (Moscow).
Rust, B.R.: Sedimentation in arid-zone anastomosing fluvial system: cooper creek. Central Australia. J. Petrol. **51**(3), 745–755 (1981)
Rzhanitsyn, N.A.: Morphological and Hydrological Regularities of the River Network Structure, 240 p. Gidrometeoizdat, Leningrad (1960)
Rzhanitsyn, N.A.: Channel-Forming Processes of the Rivers, 264 p. Gidrometeoizdat, Leningrad (1985)
Qian, N., Zhang, Z., Zhou, Z.: Fluvial Process, 584 p. Science, Beijing (1987) (in Chinese)
Sadovsky, G.L.: Manual for Tracing Dredging Slots, 183 p. River Transport, Moscow (1961)
Schanzer, E.V.: Alluvium of plain rivers of temperate belt and its importance for learning the regularities of structure and formation of alluvial formations. Proc. Geol, Inst. SA USSR. Ser. Geol. **135**(55). 275 (1951)
Schumm, S.A.: The Fluvial System, 338 p (1977)
Schumm, S.A., Khan, H.R.: Experimental study of channel patterns. Geol. Soc. Am. Bull. **83**(6), 1755–1770 (1972)
Scheidegger, A.E.: Theoretical Geomorphology, 452 p. Progress, Moscow (1964)
Shamov, G.I.: River Sediment, 380 p. Hydrometeoizdat, Leningrad (1959)
Shchukin, I.S.: General Land Morphology. vol. 1, 366 p. ONTI, Moscow, Leningrad, Novosibirsk (1933)
Shchukin, I.P.: General Geomorphology, vol. 1, 616 p. MSU Publishing House, Moscow (1960)
Shen, H., IV., Schumm, S.Y., Doehring, D.O.: Stability of stream channel patterns. Transp. Res. Rec. **736**, 22–28 (1979)
Sidorchuk, A.Y., Mikhinov, A.E.: Morphology and dynamics of a channel relief. In: Results of a Science and Technology. Series "Land Hydrology", vol. 5, 163 p. VINITI, Moscow (1985)
Sidorchuk, A.Y.: The Structure of the Relief of the River Bed, 128 p. Hydrometeoizdat, SPb (1992)
Sidorchuk, A.Y., Chalov, R.S.: Incision of mountain rivers: speeds and causes. Nature **12**, 36–45 (1996)

Slastikhin, V.V.: Process of erosion on the selective watersheds in Moldova. In: Regularities of Erosion and Fluvial Processes in Different Natural Conditions, pp. 136–137. MSU Publishing House, Moscow (1987)
Smirnova, V.G.: Hydrologic-morphological analysis of the branched riverbeds of the Altai region: Author's thesis … Ph.D. Geography Sciences, 20 p. IG SB RAS, Irkutsk (2002)
Snischenko, B.F.: Movement of sand ripples in natural water flows. Proc. SHI **136**, 82–91 (1966)
Snishchenko, B.F., Kopaliani, Z.D.: About speed of a ridge movement in the rivers and laboratory conditions. Proc. SHI **252**, 30–37 (1978) (in Russian)
Snishchenko, B.F.: About the connection of the sand ripples height with the parameters of the river flow and channel. Meteorol. Hydrol. **6**, 84–91 (1980) (in Russian)
Snishchenko, B.F.: Pair connections of parameters of a ripple and characteristics of a stream and a channel. Proc. SHI **288**, 15–25 (1983)
Sokolovsky, D.L.: River Flow, 540 p. Hydrometeoisdat, Leningrad (1968)
Studenichnikov, B.I.: Erosion Ability of a Flow and Methods of Channel Calculations, 183 p. Gosstroyizdat, Moscow (1964)
Talmaza, I.F., Kroshkin, A.N.: Hydromorphometric Characteristics of Mountain Rivers, 204 p. Kyrgyzstan, Frunze (1968)
Trepetsov, E.V.: Washing out of the river banks in the steppe Altai and its engineering and geological characteristics. Bull. Sci. Techn. Inf. **8**(52), 3–19 (1963) (Gosgeoltekhizdat, Moscow)
Turykin, L.A.: Morphology and seasonal deformations of rolls on the rivers with unbranched channel: Authors' dis. Ph.D. Geography Sciences, 25 p. MSU Publishing House, Moscow (1997)
Velikanov, M.A.: Dynamics of channel flows. In: Flow Structure, vol. 1, 324 p. Gostekhizdat, Moscow (1954). Sediments and Channel, vol. 2. Gostekhizdat, Moscow (1955). 324 p
Velikanov, M.A.: Russian channel hydraulic engineering and its role in development of teaching about the fluvial processes. In: Problems of Hydraulic Engineering of the Free Rivers, pp. 5–14. Rechizdat, Moscow (1948) (in Russian)
Velikanov, M.A.: Fluvial Process, 395 p. Gosfizmatizdat, Moscow (1958)
Vinogradova, N.N., Samoilova, A.A., Khmeleva, N.V., Shevchenko, B.F.: Application of photo-theodolite method for study of channel forms on rivers of Abkhazia. In: Soil Erosion and Fluvial Processes, vol. 7, pp. 109–118. MSU Publishing House, Moscow (1979)
Yakhontov, S.A.: To estimation of resistance of rough flow in a zone of shallow relative depths. Hydrotech. Construct. **6**, 40–43 (1960)
Zaitsev, A.A.: Investigation of meandering process by spectral analysis method. In: Laws of Erosion and Fluvial Processes Manifestation in Different Natural Conditions, pp. 334–335. MSU Publishing House, Moscow (1981)
Zaitsev, A.A.: Mode of levels and slopes on boulder-pebble riffle and rocky ledges in the channels of large rivers. Meteorol. Hydrol. **12**, 103–106 (1989)
Zaitsev, A.A., Savtsova, T.M.: Sculptural and accumulative relief of the incised riverbeds of Eastern Siberia and features of its formation. Geomorphology **4**, 58–63 (1994)
Zaitsev, A.A., Kirik, O.M., Lodina, R.V., Panin, A.V., Chalov, R.S.: Hydromorphological characteristics and regulation of the lower Vitim riverbed in connection with its transport use. In: Soil Erosion and Fluvial Processes, vol. 12, pp. 189–214. MSU Publishing House (1998) (in Russian)
Znamenskaya, N.S.: Ripple Movement of Sediment, 188 p. Hydrometeoizdat, Leningrad (1968)
Znamenskaya, N.S.: Bottom Sediments and Fluvial Processes, 142 p. Hydrometeoisdat, Leningrad (1976)
Znamenskaya, N.S.: Hydraulic Modeling of Fluvial Processes, 240 p. Gidrometeoizdat, SPb (1992)

Chapter 5
Directional Vertical Channel Changes

5.1 Forms of Directional Vertical Changes and Their Morphological Signs

Directional vertical channel changes, due to the mismatch between the sediment load of the W_{G+R} and the transport capacity of the W_{tr}, i.e. $W_{G+R} \neq W_{tr}$, consist in the erosion of the bed bottom if $W_{G+R} < W_{tr,}$ or accumulation at the bottom of the sediment, their accumulation if $W_{G+R} > W_{tr.}$ In the first case, there is a decrease in the bottom marks and an overall decrease in the slope of the channel, in the second case (accumulation)—the increase in the bottom marks and longitudinal increase in the slope. Changes in channel bottom marks (negative $-\Delta z$ due to erosion or positive $+\Delta z$ due to accumulation of sediment) are accompanied by identical changes in water surface marks, especially at intermittent levels. Therefore, the combination of longitudinal profiles of the water surface according to its levelling data in different years at the same project (design) level or constructed on the bottom allows to obtain sufficiently accurate data from the intensity of vertical changes. In the latter case, the profiles should be built along the line of maximum depths of the pool valleys, as the bottom marks on the riffles and ripples are very variable, depending on the peculiarities of the ripple motion of deposits, long-term and seasonal changes of the riffles. Combining the longitudinal profiles of the bottom gives the expected result only when the time period is long enough, while the maximum depths can be used to judge about the vertical change of the channel in a relatively short period of time.

An example of the lowering of bottom marks and, accordingly, the reduction of the slope of the free flow surface with the general directional cut of the river is given by the upper Ob, including the lower reaches of its component rivers—Biya and Katun. According to the "Atlas of the Ob River …" (1897), the longitudinal profile at the end of the 19th century was characterized by a sharp bend in the area of the confluence of Biya and Katun (Table 5.1). The almost tenfold decrease in the

R. S. Chalov, *Fluvial Processes: Theory and Applications*,
https://doi.org/10.1007/978-3-030-66183-0_5

Table 5.1 Channel slopes in the lower reaches of Biya, Katun, on the upper Ob (in ‰) and their changes at river incising (Berkovich et al. 1990)

Distance from the merger of Biya and Katun, km	End of the 19th century	1966 year	1984 year
r. Katun			
31–20	3.6	–	0.45
20–9	2.9	–	0.45
9–0	1.5	–	0.18
r. Bia			
28–6	1.3	–	–
6–0	0.35	0.13	0.11
r. Ob			
0–35	0.16	0.16	0.16
35–55	0.12	0.18	0.17
55–95	0.21	0.16	0.16
95–135	0.11	0.16	0.14
135–155	0.16	0.12	0.12
155–240	0.10	0.12	0.10

downstream slope has led to the fact that sandy sediments carried upstream in suspended condition constitute the bulk of the bed material load, forming a complex branched channel.

By the end of the twentieth century, the point of inflection of the longitudinal profile due to the incision of the rivers moved upstream of the Biya and Katun rivers. A comparison of the Katun slopes measured at the end of the 19th century and in 1984 shows that the areas of higher values have decreased by a factor of 7 over a 90-year period, and by a factor of 3 in Biya. The directional change of slopes in the Ob with the continuing infeed of the lower Katun-Upper Ob system (Berkovich et al. 1990) is significantly less, expressing itself only in the form of a general trend. Here, at lower slopes, this phenomenon is masked by local variations in the length of the river and changes in the course of channel changes. Relatively elevated intermodal gradients are confined to complex branches and riffles; on relatively narrowed unbranched sections of the channel the slope is lower than average (0.09–0.13‰). The latter is a consequence of the fact that within these narrower areas, channel incisions are more obvious, affecting the local decline in longitudinal slopes. It can be assumed that the longitudinal bend of the longitudinal profile that existed at the end of the 19th century directly at the junction of Bia and Katun was regressively shifted to the lower reaches of Bia and Katun due to the incising of the riverbed in the 20th century.

On Fig. 5.1, A the longitudinal profiles of the bottom at the section of the Alabuga riverbed (right tributary of the Naryn river, Western Tien Shan), built according to the data of the Institute "Kirghizgiprovodkhoz" for 1977–1992, are given, which testify to the ongoing intensive incision of the river (Sidorchuk and

Chalov 1996). The overlap of the profiles of the Neman River bed near the city of Bridges over 50 years showed that the bottom marks (along the line of the deepest depths) decreased from 10 cm to 2 m (Fig. 5.1B), in other areas—to 1 m (Darbutas and Chalov 1993), although in some places the bottom line of 1989. Such deviations are connected with the formation of pool valleys at the riffles (lowering of the bottom marks up to 2 m and more) or, on the contrary, with the overturning of large ripples on the former pool valleys.

On Fig. 5.2 the combined longitudinal profiles of the water surface of the Belaya River between Ufa and Birsk are presented, reflecting its transformation during the period 1942–1995 due to the incision of the river, which spreads transgressively from the large channel quarries (Berkovich et al. 2001). In the 1940s and 1960s, the process of plunging and lowering the bottom marks covered a small part of the river downstream of Ufa, but by the end of the 1980s it had spread downstream to 40–50 km, and by 1995 to 70–80 km. Below the plunging zone there was an increase in levels (in 1940–1960s—in 70–90 km, in 1980–1990s—in 120–140 km below Ufa), which is explained by the partial accumulation of channel erosion products in the upper part of the channel. The decrease in the water surface marks over the period under consideration varied from −0.8 to −0.15 m along the length of the section covered by the incision and +0.2 m in the area of sediment accumulation.

The fact that changes in the bottom marks affect the position of the free flow surface levels determines the application of the analysis of water level and discharge curves over a multi-year period in order to obtain direct insight into the

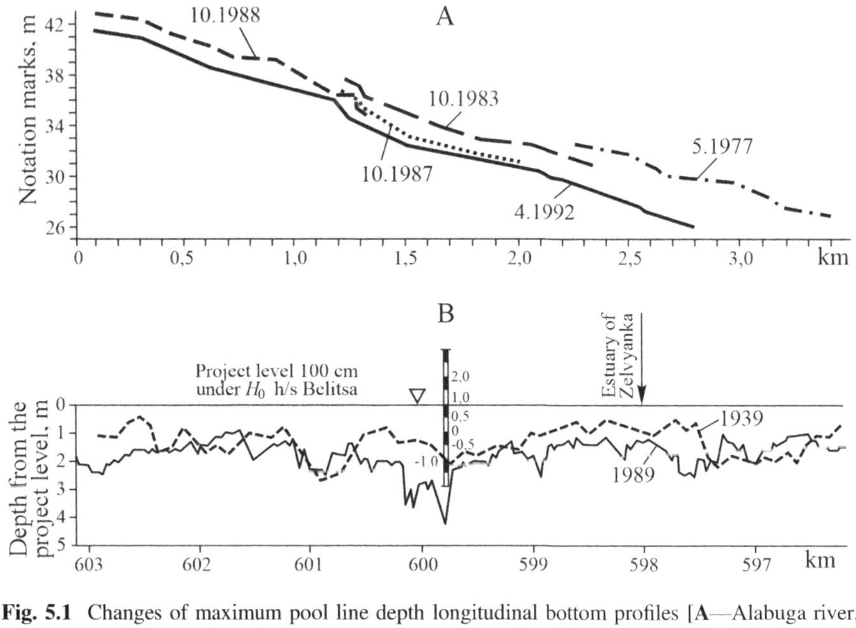

Fig. 5.1 Changes of maximum pool line depth longitudinal bottom profiles [**A**—Alabuga river, Western Tien Shan (Sidorchuk and Chalov 1996)] and profiles of measurement points along a line of maximum depths [**B**—Neman river near Mosty city (Darbutas and Chalov 1993)]

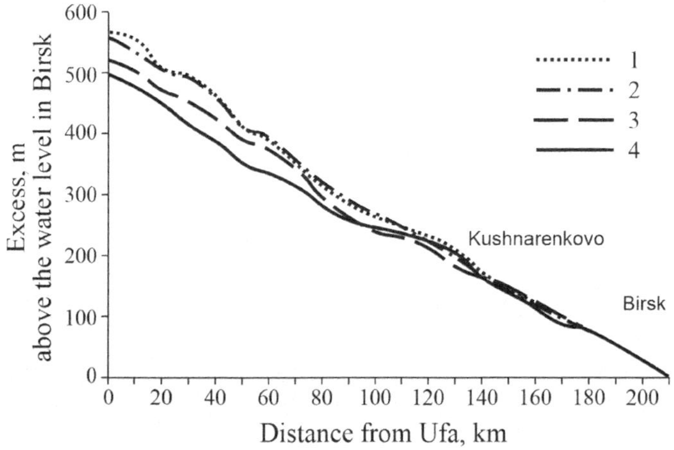

Fig. 5.2 Longitudinal Belaya river profile changes (Ufa city—Birsk city section): 1—1942; 2—1965; 3—1982; 4—1995 (Berkovitch et al. 2001)

direction of vertical changes and their rates. The shift of curves in the communication graphs from year to year up or down allows us to judge, respectively, about the accumulation of alluvium and increase in the river bed or its erosion and lowering of the river bed surface (Fig. 5.3). This method is based on regular observations of water levels and measurements of water discharge at hydrological stations. As the posts are usually located in the relatively narrower parts of the channel with a minimum (river specific) floodplain width, the result reflects the direction of the process in its purest form, as the stream is in direct contact with the river's bedrock or riverbed sediments. However, the same causes some overestimation of the plunging rate and underestimation of the rate of accumulation in comparison with the adjacent riffle areas of the wide floodplain channel, which are characterized by more complex morphology and the regime of perennial reformation. This can be seen from Table 5.2, which presents data on the direction and intensity of vertical channel change in some large rivers in Eurasia (Chalov et al. 2000), obtained from both the analysis of curves $Q = f(H)$ and the balance of sediment in the areas between hydrological stations. In addition, when analyzing curves $Q = f(H)$, it is important to select the reference level of the increase or decrease in the bottom marks. In the incised channel (Fig. 5.3A—r. Chatkal in the Northern Tien Shan; Fig. 5.3B—Amur River, Komsomolsk-na-Amure) with a parabolic or close to it form of a living section, the curves are located sub-parallel to each other, because the decrease (or increase) of the bottom leads to a similar decrease (increase) of water surface marks at any filling of the channel. In the wide floodplain branched channel there is a convergence of curves at high levels due to the successive flooding of the drying in the lowland shallows, low and then high floodplain (Fig. 5.3C—r. Ob, Barnaul); on the contrary, on mountain rivers with low floodplain but high abrupt floods, the curves diverge at high levels (Fig. 5.3D

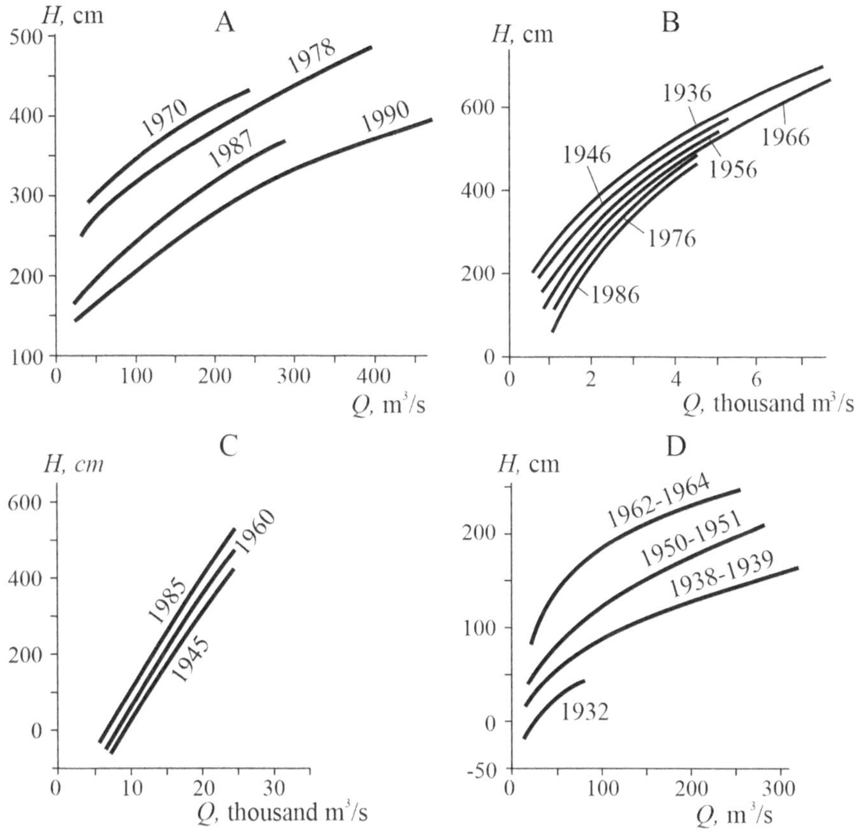

Fig. 5.3 Water flow and level connection curves Q = *f*(H), reflecting river incision processes (**A**—Chatkal river, Khudoidoi; **B**—Ob river, Barnaul) and sediment accumulation processes (**C**—Amur river, Komsomolsk-na-Amure; **D**—Kodori river, Ganakhleba) in incised (**A**, **C**) and wide-floodplain (**B**, **D**) channel

—Kodori r. in the Western Caucasus); the greater the width of the water mirror, the smaller the increment of +ΔH levels at the same increase in water discharge +ΔQ. Therefore, the plunging or accumulation value should be estimated according to the minimum levels. This gives a basis to determine the plunging/accumulation rate not from curves $Q = f(H)$, but from the graphs of changes in water levels at effective discharge over a multi-year period (Fig. 5.4). In the graph below, which indicates that the channel is incising and the levels are increasing, with the general trend towards a decrease in the latter, the years in which sediment accumulation and a temporary increase in levels have occurred are highlighted. In this case, this is due to the regime of releases from the Tsimlyansk reservoir, associated with changes in sediment discharge and accumulation/erosion ripples of riffles (Transport … 1972).

Table 5.2 Vertical channel change intensity on some large Eurasian rivers, determined by the curves $Q = f(H)$ and sediment balance (Chalov et al. 2000)

The river, the years	Hydrological station	Erosion layer (−) or accumulation layer (+), cm/year	
		By $Q = f(H)$ at Q_f (in the line of stations)	By sediment balance (between stations)
Hauang He 1950–1993	Huayuancou	+4.5	
			+3.5
	Jiahetan	+5.2	
			+5.7
	Gaotsun	+7.7	
			+6.6
	Sunkou	+7.4	
			+7.8
	Aishan	+7.3	
			+9.1
	Lakow	+7.7	
			+7.5
	Liczin	+6.0	
Yangtze 1950–1988	Ichan	–	
			+4.9
	Loshan	+3.7	
			.0.6
	Hankou	−1.7	
			−1.0
	Datun	−0.5	
Lena 1940–1985	Tabaga	−1.8	
			−0.4
	Kyusur	−0.7	
Northern Dvina 1939–1988	Abramkovo	−0.7	
			−0.01
	Zvoz	–	
			−0.38
	Ust-Pinega	–	
Ob 1953–1986	Fominskoye	−1.5	
			−0.22
1936–1986	Barnaul	−1.8	
			−0.71
	Kamen-na-Obi	–	
			+0.87
1936–1956	Novosibirsk	+0.8	
			+0.11
	Kolpashevo	+3.25	

(continued)

Table 5.2 (continued)

The river, the years	Hydrological station	Erosion layer (−) or accumulation layer (+), cm/year	
		By $Q = f(H)$ at Q_f (in the line of stations)	By sediment balance (between stations)
1956–1987	Novosibirsk	−2.3	
			−0.47
	Kolpashevo.	−1.99	

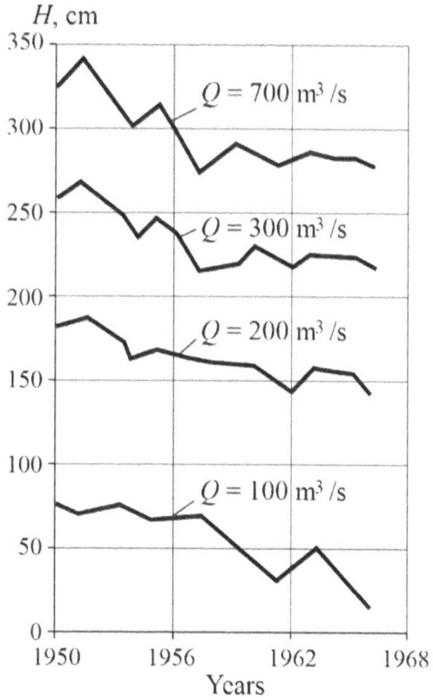

Fig. 5.4 Annual average water level H_{av} changes on the Don river (Nikolaevsky station) at water discharges 700, 50, 300 and 100 m^3/s (Transport … 1972)

Similar deviations from the general direction of movement are observed directly on the curves $Q = f(H)$ for individual years or periods of years, being associated with the expression of periodic vertical changes due to the development and cutoff of the bend, the movement of large ripples in the channel, the development of shorter and dying longer branches, and in the lower reaches of the river—the uneven extension of the delta and the redistribution of runoff between the delta branches. Therefore, a number of data for analysis should be as long as possible in order to avoid changes in the relationship curves caused by periodic phenomena. An example of such changes in the position of the curve $Q = f(H)$ for the Tola River (Mongolia, Ulan-Bator) is shown in Fig. 5.5. It first moved downwards (1945–1947), then moved to the highest position (1949), and then downwards again until 1954, reaching its minimum. Such a short-term (2–3 years) change in the general trend is obviously connected with the movement of a large ripple post across the yard (Natsag and Chalov 1978).

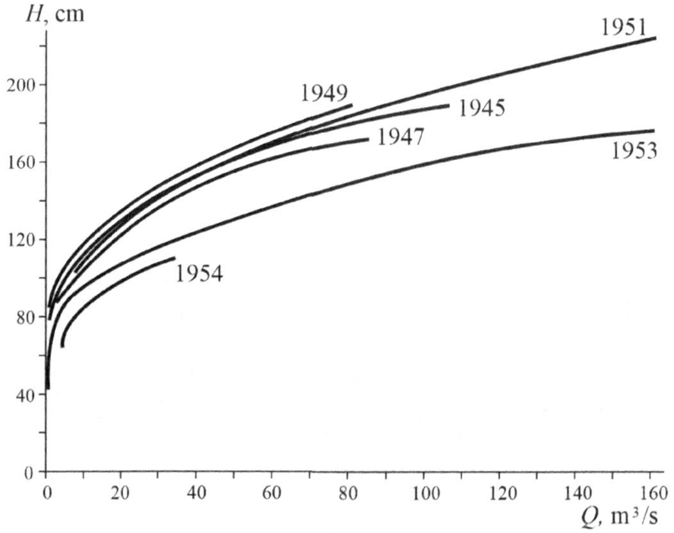

Fig. 5.5 Connection curves of discharges (Q) and water levels (H) on Tola river (Ulan-Bator city)

In the absence of measured water discharge data on channel incision or directed (systematic) accumulation of sediment can be judged by the change in the position of the corresponding level curves at adjacent gauging stations. On Fig. 5.6 the diagrams of the corresponding levels show changes in the direction of vertical changes in two sections between the adjacent gauging stations on the middle Amur between Blagoveshchensk (990 km from Khabarovsk) and Grodekovo (970 km) over a 30-year period (1955–1985), the curve of the corresponding levels has risen by 70–80 cm, which reflects an increase in the bottom marks due to the accumulation of sediment. The directional accumulation of sediment is also fixed above Khabarovsk (Fig. 5.6D), but the curve $H_1 = f(H_2)$ does not change its position in time when crossing the Malyi Hingan Massif (Fig. 5.6C). At the site Grodekovo—Konstantinovka (890 km) curve $H_G = f(N_K)$ reflects the directed incision of the river (Fig. 5.6B), but the graph gives overestimated results (60 cm for 30 years because of the increase in the levels in the site of Grodekovo: the actual decrease in levels was only 10–20 cm (Ivanov et al. 2000). Thus, using graphs $H_1 = f(H_2)$ to estimate the intensity of vertical changes, it should be borne in mind that the change in the position of the curves of the corresponding levels gives the average value of plunging or accumulation in the area between adjacent stations. Since different vertical change directions can be observed between the posts, especially those located at sufficient distance from each other, the method is applicable only to non-continuous sections and sections of rivers characterized by the same geological-geomorphological conditions. On the other hand, it obviously does not give the expected result when the river incise is compensated by the rise of the area, so that the corresponding levels remain constant.

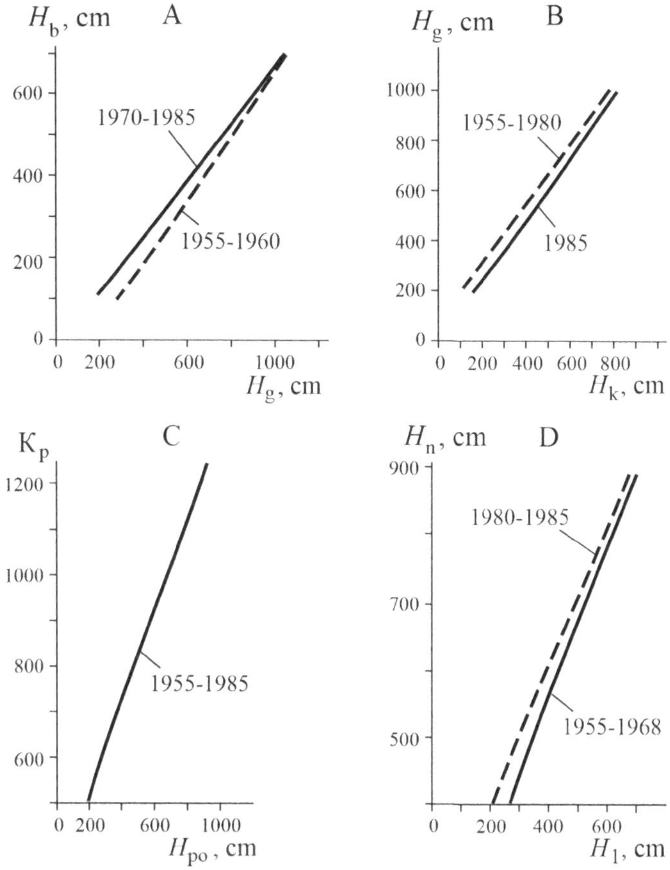

Fig. 5.6 Corresponding level relationship diagrams for adjacent Middle Amur hydrological posts for different periods of time: **A**—Blagoveshchensk (H_b) and Grodekovo (H_g); **B**—Grodekovo (H_g) and Konstantinovka (H_k); **C**—Pashkovo (H_p) and Pompeevka (H_{po}); **D**—Leninskoye (H_l) and Nagibovo (H_n) (Ivanov et al. 2000)

However, under certain conditions, bottom markings and free flow curves may differ from one to another. The above-mentioned water level decreases on the Belaya River are significantly lower than the bottom marks: −0.8 and −1.3 m, −0.4 and −1.1 m, −0.3 and −1.6 m, −0.15 and −0.35 m, respectively, in the sections (Berkovich et al. 2001). In the lower Neman (Darbutas and Chalov 1993), due to deep erosion in the lower reach of Kaunas HPP, construction of rectification facilities since the middle of the 19th century, which resulted in a two-fold contraction of the channel and the formation of a new coastline on both sides of the riverbed, and quarrying operations, the bottom marks were lowered to 2 m, while the decrease in water levels (Smalininkai) was at a water flow rate of 400–500 m^3/s only 35 cm. Such a discrepancy between the changes in the position of the bottom and the water surface is associated with the compensatory role of corrective structures, which provide an increase in water levels due to the tightening of the flow. In natural conditions, although on a much smaller scale, this may be a consequence of the concentration of the flow as it is cut into a narrower channel; on the contrary, when it is accumulated, the increase in levels may lag behind the increase in the bottom due to the flow spreading along the riverbed banks, which play an accumulating role (with respect to the water mass).

The most objective and reliable current direction of vertical change (riverbed incision or accumulation of sediment) is represented in changes in sediment load in a river section over a number of years. For the impenetrable section of the river,

$$W_1 - W_2 = \pm \Delta W \tag{5.1}$$

where W_1 is the volume of load entering the river section through the upstream section; W_2 is the volume of load removed from the section through the downstream section; ΔW is the resulting balance equal to the change in the volume of sediments (erosion/accumulation) within the section. A negative value of ΔW indicates the predominance within it of sediment accumulation (less than received, i.e. ΔW > 0), while a positive value indicates the predominance of erosion processes (more than received, i.e. ΔW < 0) (in Table 5.2 + and − reflect the change in sediment volume). Alekseevsky (1998), taking into account the inflow, proposes to define the resulting balance as

$$\Delta W \cong \left(W_{\text{er}} + \sum_{i=1}^{n} W_{\text{er}_i} \right) - W_{\text{ac}}, \tag{5.2}$$

where W_{er} and W_{ac} are the volumes of load entering the stream as a result of streambed scour or transferred to the channel sediments as a result of their accumulation W_{er_i};—the volume of load entering the river from tributaries; n—the number of inflows. In a more complete writing, he proposes to take into account the inflow of sediment from ravines W_{rav} as a result of the landslides W_{l}, screes W_{sc} mudslides W_{m} and solifluxation processes W_{s}, i.e.

$$\Delta W = \left(W_{\text{er}} + \sum_{i=1}^{n} W_{\text{er}_i} + W_{\text{rav}} + W_{\text{l}} + W_{\text{sc}} + W_{\text{m}} + W_{\text{s}} \right) - W_{\text{ac}}. \tag{5.3}$$

On the other hand, taking into account the forms of sediment transport and the different sources of the suspended and bed components of sediment load, the balance Eq. (5.1) can be presented as

$$W_{\text{G+R}} = (W_{\text{R}} + W_{\text{G}})_2 + \sum_{i=1}^{n} W_{\text{er}_i} \pm (\Delta W_{\text{R}} + \Delta W_{\text{G}}) \tag{5.4}$$

In other words, depending on whether the river cuts in or accumulates, the sediment load either increases or decreases. A first approximation of this can be made from the balance of suspended sediment (provided that it dominates total sediment load). For example, at the Ob site in Novosibirsk, the annual amount of suspended sediment carried by the river under natural conditions was 9 million m^3, and at the site near the village of Kozhevnikovo (230 km below Novosibirsk)—8.3 million m^3. Consequently, about 700,000 m^3 of sediment was deposited here every year. The area of the valley bottom (channel and floodplain) between these sites is about 1900 km^2. It follows from this that a layer of approximately 0.4 mm of sediment was formed annually on the bottom of the valley, which corresponds to 4 cm in a century. On the other hand, the balance of load in the upper Ob indicates the present-day incision of the river. Here, the average annual bedload transport rate in the upper section (Fominskoye village) is 160 kg/s, in the lower closing section (Barnaul)—276 kg/s. Sediment load from ravines and tributaries is estimated at 76 kg/s. Consequently, about 40 kg/s is removed from the site, which corresponds to an annual movement of 800,000 m^3 or an average layer of 0.8 mm per year of material removed from the river bed (Makkaveev and Chalov 1963b, 1964). Taking into account the bedload transport rates (Table 5.2), the accumulation of which or scouring of bed sediments mainly determines the condition of the riverbed, these data turned out to be much larger: the accumulation of sediment in the Novosibirsk-Kozhevnikovo section of the riverbed is 1.1 mm/year and the incision of the upper Ob is 2.2 mm/year. In addition, in the calculation of the change in suspended sediment load, the resulting distribution was not based on the area of the channel but on the entire valley bottom, while the balance sheet data in Table 5.2 was based on the area of the channel only.

Suspended sediments accumulate mainly within the floodplain, while their inflow during erosion of sediments in the river channel is insignificant. Sediment accumulation on the floodplain occurs regardless of the direction of vertical change in the river channel. The only exception is rivers with suspended sediment as their channel-forming source and their accumulation and flow due to scouring occurs in the channel itself. Therefore, taking into account the resulting load balance only in terms of the weighted component of the load, may lead to erroneous conclusions about the directionality and, even less so, the rate of vertical change.

However, the balance method, with long river sections and large inflows, cannot claim to be accurate either. In addition, the degree of hydrological study does not always make it possible to find out the nature of such links and ratios, reflecting the current trends in the development of the longitudinal profile of the river, or to extrapolate the data obtained to adjacent areas and even more so to other rivers. As a consequence, indirect morphological and dynamic features are required to assess the directionality of channel changes. However, not all of them allow us to judge it unequivocally, and for the correctness of the judgment, we should use other features, looking at them in a complex and involving analogues, which have these features confirmed by hydrological methods—the analysis of the links $Q = f(H)$, the corresponding levels, sediment balance, etc.). Thus, branching of the channel as a morphological form of representation of fluvial processes without taking into account their dynamics is sometimes considered (in the geomorphological literature) as a sign of directional accumulation of sediment (Dynamic geomorphology 1992, p. 155): "branching (channel—*R.Ch.*) shall be carried out when the stream, while accumulating sediment on the bottom, simultaneously moves a significant proportion of it to both banks ... Due to the simultaneous deposition of the same volume of sediment at the bottom of the channel and its sides, the latter moves upwards vertically ...". Based on such notions, Biske (1964), for example, argues that sediment accumulation occurs on the Lena River from Pokrovsk to Zhigansk, considering this process to be a representation of a strong branching of the channel and linking it to the tectonic depressions of the Nizhni-Aldan and Vilyuiysk depressions. However, the fact that the riverbed has branched does not in itself reflect the direction of development of the longitudinal profile of the river, determined by the volume of flow (the size of the river) and the degree of its unevenness, the size of sediment load, the stability of the channel, and, finally, the peculiarities of the dynamics of channel flows on large rivers (Losievsky 1934; Makkaveev 1955; Karasev 1975; Chalov 1979). The formation of islands only in isolated cases and under certain conditions can be a consequence of the accumulation stage of river development. For example, on the Middle and Lower Lena, the analysis of the curves $Q = f(H)$, the balance of load, and various morphological (floodplain stepness) and dynamic features (transgressive movement of the islands, their union and attachment to the coastal floodplain, and weak development of the anabranching channel during the passage of Q_f over the floodplain) testify to the inundation of the riverbed (Borsuk and Chalov 1973; Chalov et al. 2000). Similar "contradictions" are characteristic of the upper Ob below the confluence of Biya and Katun, which is distinguished by a very complex branched channel: here the river, as it was shown above, cuts in. It is characteristic that the branched channel of the Ob below the city of Novosibirsk accumulated sediment in natural conditions, but had a much simpler morphology (single and conjugate branches), while the upper Ob has the most complex types of branching, formed by groups of small islands. However, for the upper Ob it is typical to combine the islands, below Novosibirsk (in vivo)—their formation and regressive movement.

Even the incised channel is not always an absolute sign of the modern incision of the river, reflecting the history of its geological development. For example, the lower Amur River at the intersection of Sikhote-Alin (Komsomolsk-on-Amur—Bogorodskoe village) has an incised channel, but the curves of water discharge and $Q = f(H)$ levels indicate the directional accumulation of sediment (see Fig. 5.3B). It is reflected in the sub pond lakes (Fig. 5.7) at the mouths of tributaries, whose sediment load is disproportionately low compared to that of the main river (Makhinov et al. 1994). The formation of mouth-to-mouth spills on tributaries submerged by the main river in the form of lakes on the lower Amur is here the main sign of directional accumulation of sediment (Avaryaskin 1976). They are also found in areas of the middle Amur, where hydrological data show an accumulation of sediment and are very typical of the middle and lower reaches of the Yangtze River.

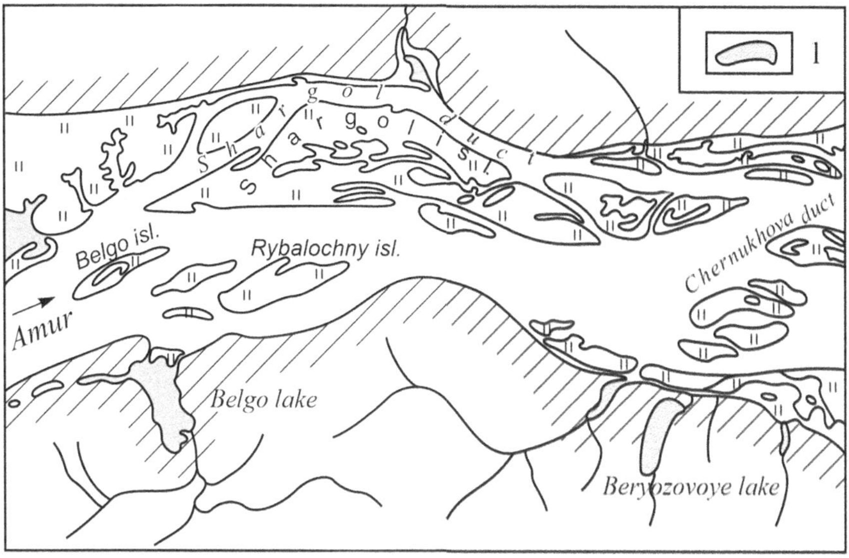

Fig. 5.7 Downstream pond lakes (1) on tributaries of Lower Amur on the section of braided channel

Deep erosion is evidenced by qualitative morphological, dynamic and other related features.

Gradual reduction in the number of branched bed branches. Comparison of the map of the «Atlas of the Ob…» (1897) with the modern survey of the channel allowed us to establish that the number of islands (and, accordingly, the number of branches) for 100 year decreased as a result of their joining the coasts or joining with neighboring islands by 14%. At the same time, the total width of the river tends to decrease (Makkaveev and Chalov 1964). The same phenomenon is noted in the branched wide floodplain channel of the middle and lower Lena (Borsuk and Chalov 1973) and is very typical for the lower water bodies, where river crossing occurs more intensively in comparison with natural conditions by an order of magnitude. The extreme representation of this process is the change of the channel pattern—branched out into a meandering one (this occurs at the Ob River 50 years after the construction of the Novosibirsk hydroelectric power plant).

Absence or weak development of anabranching channel on rivers, where the effective discharge Q_f takes place in the floodplain. As a result, the unification of the islands and their attachment to the banks is accompanied by the formation of a consolidated coastal floodplain not dissected or weakly dissected by ducts into separate massifs. On the average (from Pokrovsk to Aldan estuary) and lower Lena all floodplain branches are dry and are narrow channels; only in the floodplain they take in total up to 5–10% of the total water discharge in the river; in the low water period the majority of them dry up, and the total runoff rates of those still functioning is not more than 1–2%.

Both phenomena (unification of islands and extinction of floodplain branches) are determined by the unequal runoff rates of the branches and relationship on its transporting capacity of the flow $W_{tr} = f(Q^m)$, where m > 1, often m= 2–3. This process is more intensive in the high-water channels when the river is cut in, and when the bottom marks decrease, the low-water channels and floodplain branches become shallow and die, turning into floodplain hollows with lakes—the remains in the relief of the former channels' floodplains.

The transgressive nature of the islands' movement. As experimental studies have shown (Makkaveev et al. 1961) such (downstream) movement of islands is one of the signs that there is no systematic accumulation in the channel. However, here we are talking only about the islands, which form the main nodes of branching and are in the middle of the channel, creating a symmetrical division of the flow into branches. The tips of such islands are eroded, the shallows near them appear only when the top of a large ripple is pushed over, are not fixed by vegetation, and over time are either eroded or shifted into one or the other branch.

Deficiency of the ruptured bends on meandering rivers, despite the passage of Q_f in the floodplain. This is due to the high altitude of the rear, the oldest parts of the floodplain segments due to the gradual incision of the river and the increase in their marks relative to the water level. In such conditions, the cutoff of the bend is more often done by tearing off the top of the meander neck. If, however, the broken bends are formed, the floodplain multivariate as a consequence of their evolution also appears to be underdeveloped.

Prevalence of incised channel types on the river (Fig. 5.8A). If there are islands in the channel, the incision occurs in separate arms, so that the islands on the river are sculptural or sculpturally accumulative (Fig. 5.8B). Such islands are found on the upper and middle Lena (at Kirensk, below the mouth of the Chaya River, at the mouth of the Vitim River, above Lensk, near Olekminsk, above the Pokrovsk settlement), where they correspond in height I and II to the floodplain terraces (from 12–14 to 25–30 m) (Borsuk and Chalov 1973), on the upper Aldan, Vitim, Angara.

Stepness of the floodplain transverse profile (Fig. 5.8B). This is particularly evident in the unilateral spread of the floodplain and the lateral shifting of the channel towards the main bank. In this case, the oldest and most remote areas have the highest altitude. The differences in the heights of the high floodplain and the first floodplain terrace are the same as between the adjacent floodplain steps, and the terrace surface itself is often characterized by a well-preserved primary floodplain relief. The gradual decrease in the floodplain inundation and the withdrawal of elevated floodplain areas from the high water level during plunging leads to the transformation of floodplain landscapes (Surkov 1999): the meadow floodplain soils are replaced by the soils typical of the given zone and the increase of zonal elements in the vegetation cover as the transition from low, younger to higher, older stages takes place. The smoothness of the transition to zonal landscape soil and vegetation types indicates that the levels of spills in the river gradually decrease over time and that relatively elevated floodplains become part of a low floodplains terrace. This is also evidenced by the steppe meadows, which are widespread on floodplain ripples and whose area increases with time.

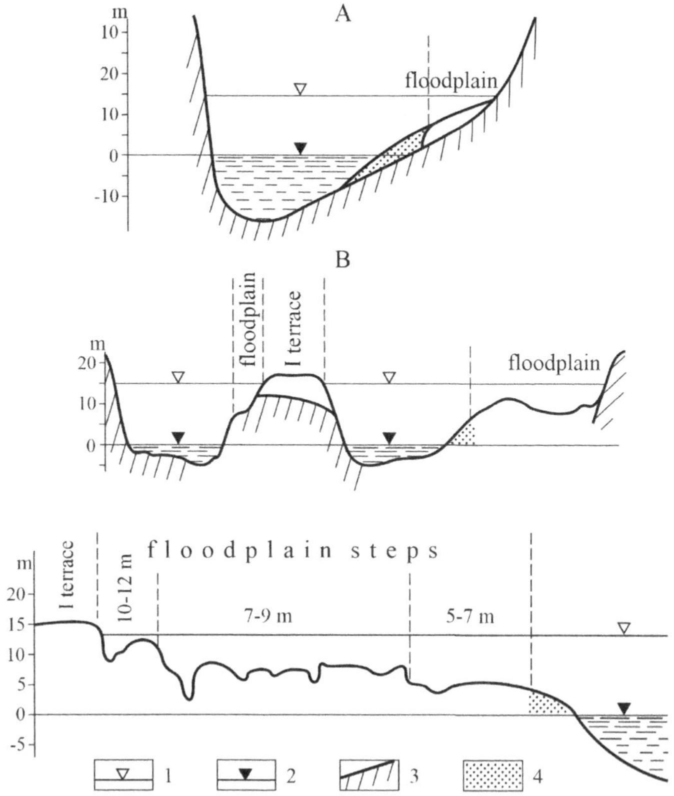

Fig. 5.8 Transverse profiles of the valley bottom of incisive rivers: **A**—incised channel (Upper Aldan); **B**—Incised channel with sculptural-accumulative branching: socle island with floodplain terrace (Lena river, downstream the Lensk city); **C**—wide-floodplain channel with stepped floodplain (Lena river downstream the Yakutsk city). 1—high flood level; 2—low-water level; 3—root banks and river bed; 4—bank shoals

If the channel constantly changes its position due to intensive lateral change (development, cutoff and redevelopment of the bend, reshaping of branches, formation and joining to the floodplain of the islands, etc.), the stepness of the floodplain acquires a peculiar pseudo-butte appearance: different in height areas (steps), up to the corresponding low floodplain terraces, are located in the middle of the floodplain, causing a mosaic of its altitudinal marks.

Character of longitudinal profile of floodplain steps. As the plunging intensity gradually decreases downstream, the longitudinal profile of the central floodplain surface is steeper than the interfluve water level in the river, i.e. the average relative height of the floodplain decreases downstream. In the upper Ob, below the confluence of Biya and Katun, there are significant floodplains, which still retain meadow soils, but which are not flooded even during very high spills and already have signs of transformation into zonal soils. In the area of the confluence of the

Anuy and Peschanaya Rivers, a solid strip of floodplain up to 4.5 km wide stretches out. Within its limits there are three high steps: the upper one reaches the height of 10–11 m above the interlevel, the middle one—7–7.5 m. Downstream, these steps are moved up to 3.5–4 m, i.e. up to the average lower floodplain level (Fig. 5.9A). On this high floodplain there are already signs of soil subsidence (Makkaveev and Chalov 1964).

On the middle Amur, in the junction zone of the Malyi Khingan and Middle-Amur lowland, a transition from incision of the river to the directed accumulation of sediment is observed. Decrease in intensity of plunging down the river is represented, as well as on the upper Ob, in successive decrease in the height of floodplain steps and their merging into a practically one-tier floodplain (Fig. 5.9B). Even further downstream in the area of directional accumulation further lowering of the surface of the floodplain terrace leads to the fact that it is here below the water levels of 1% security (Ivanov et al. 2000).

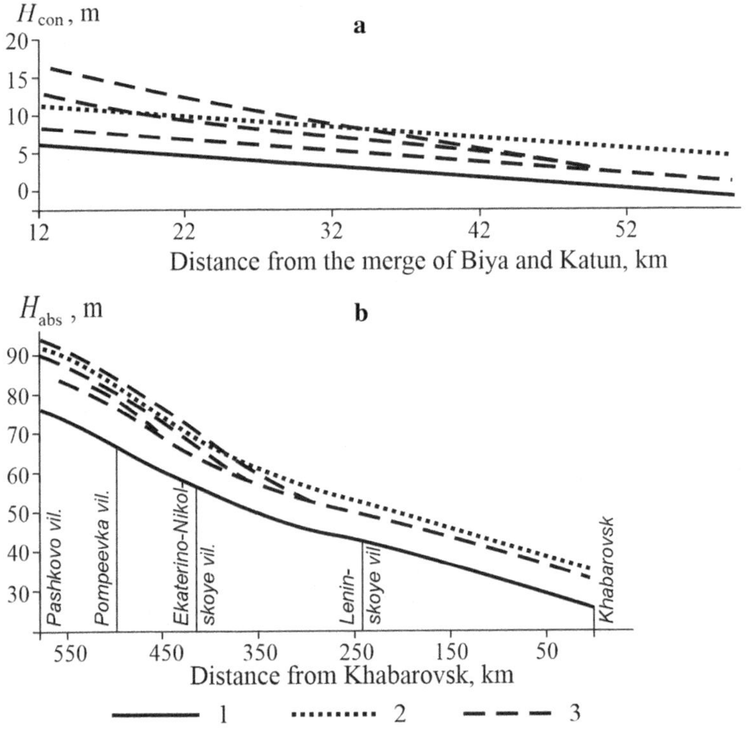

Fig. 5.9 Longitudinal profiles of water surface, floodplain steps and I floodplain terrace in conditions of increasing intensity of river incision downstream on Upper Ob (**A**) and transition from the incision to the directional accumulation on the Middle Amur (**B**). 1—minimal low-water level; 2—maximal flood level (flood on the Ob river, rainfall flood on the Amur river); 3—levels of floodplain steppes and of the I floodplain terrace

The difference between the above examples is that on the upper Ob two levels of high floodplain have moved out of the floodplain as a result of plunging (it was mentioned above about the shift of the inflection point of the longitudinal profile of the Ob in this area and the decrease of the longitudinal slope). On the middle Amur, the terrace is submerged under the floodplain level due to the change of accumulations downstream.

The presence of a process of directional accumulation of sediment is also confirmed by several morphological features.

An increase in the number of islands over time. In the Middle Ob, a comparison of multi-temporal channel maps, which have been regularly produced since the end of the 19th century (Atlas of the Ob … 1897), has shown that the total number of islands prior to the construction of the Novosibirsk Hydroelectric Power Station increased over a 60-year period (by the end of the 1950s, when the Novosibirsk Reservoir was created and the directed accumulation of sediment was replaced in the lower tailrace by the incision of the channel) by 6%, i.e., a trend towards the development of hoses was observed here.

Regressive growth of the islands. Near the heads of most of the islands that make up the main branches of the channel, riverbanks are regularly formed, which are then fixed by vegetation and attached to the island from the side of the capping or form independent islands separated from the main narrow inter-island channel (Fig. 5.10). According to experimental researchers (Makkaveev et al. 1961), such transformation of islands is one of the reliable signs of modern accumulation. The theoretical substantiation of the regressive movement of the islands as a phenomenon characterizing the process of directional accumulation of load at $W > W_{tr}$ was carried out by Makkaveev (1948).

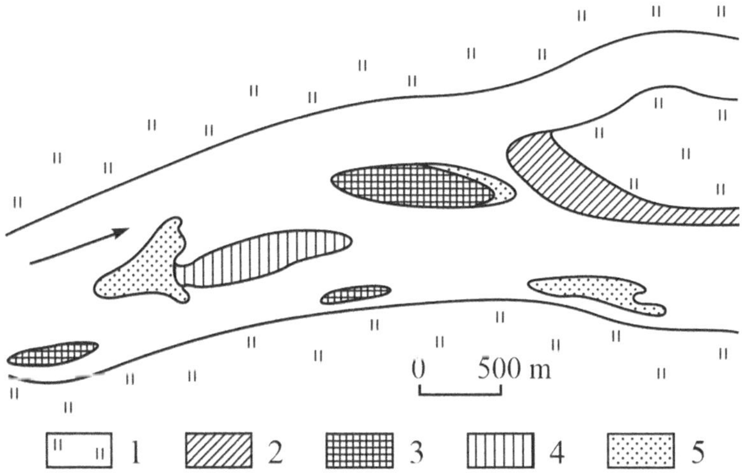

Fig. 5.10 Regressive island growing near the tip of the Eloviy island (Voronovskoye branching on the Ob river). Floodplain sections, developed: 1—before 1897; 2—1897–1927; 3—1927–1941; 4—1941–1957; 5—bank shoals, existed in 1962

Formation of ruptured bends on meandering rivers with wide floodplain channel. As the accumulation of sediment increases the marks on the bed bottom and water levels, the rear parts of the floodplain segments are relatively low, the floodplain inundating the floodplain is concentrated in them and the segmental bends are straightened along them. This is facilitated by the passage of Q_f with flooded floodplains.

Wide development of floodplain branches divided into numerous floodplain island massifs (Fig. 5.11). They are formed both as a result of the connection of old rivers in the torn bends and the preservation of the laced branches from the riverbeds at the incomplete division of the islands to the floodplain, concentrating the floodplain streams at the floodplain inundation. Together, these branches take up to 15–20% of the total water consumption in the river during the flood period and up to 5–10% during the low water period, some of them drying up. In navigable rivers, such floodplains are often used as off-road waterways to settlements remote from the main river channel.

The formation of a dammed channel. Under conditions of very high runoff and intensive accumulation of sediment, when $W \gg W_{tr}$, the river's intermodal water level is significantly higher than the adjacent areas where the developed areas, including cities, are located (Fig. 5.12A). If, during floods, water levels exceed bunding levels (natural dams are built up for flood protection), catastrophic floods occur. These are the channels of the Huang He, Yangtze, Amu Darya, Brahmaputra, Terek and Rioni rivers in their lower reaches.

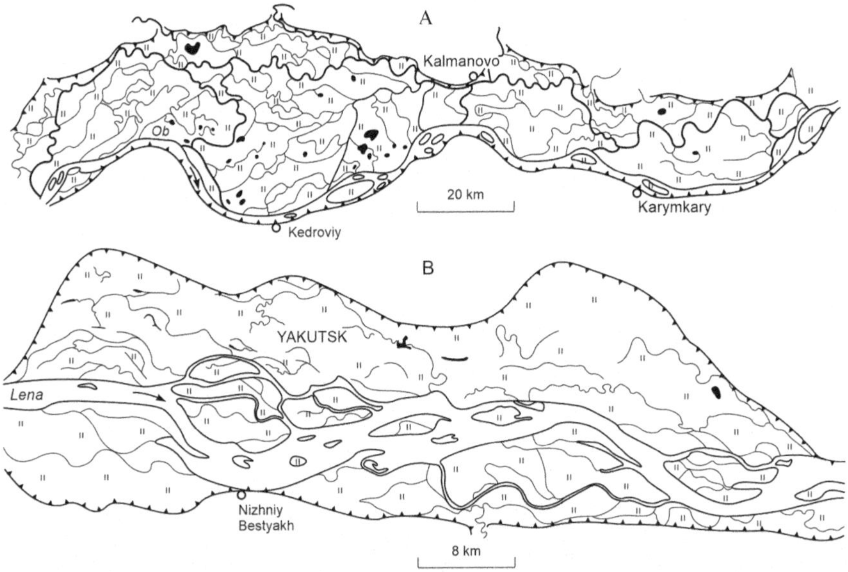

Fig. 5.11 Anabranching channel on accumulating (**A**, Lower Ob) and incisive (**B**, Lower Lena)

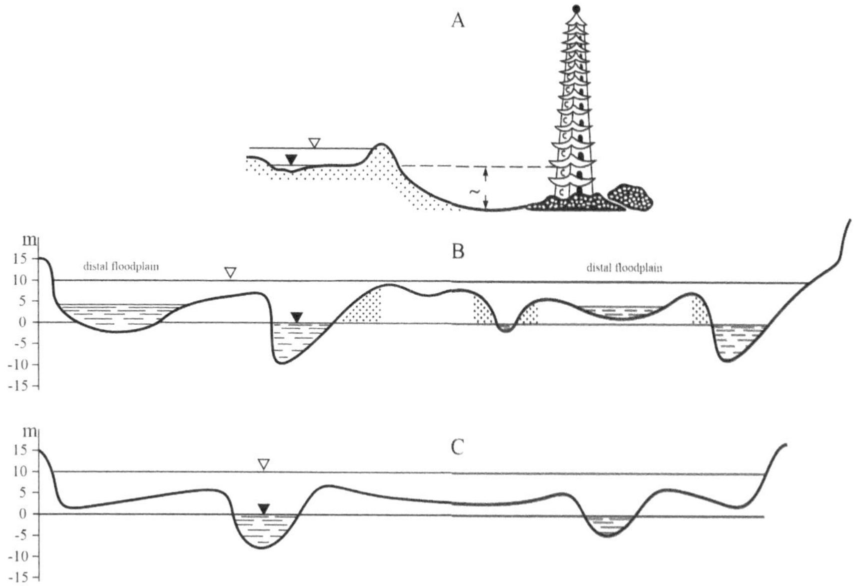

Fig. 5.12 Transverse profiles of river valleys bottom at different intensity of the directional accumulation of sediments: **A**—$W \gg W_{tr}$ (Huang He river near Keyfon city) (Series of picture albums … 1991); **B**—$W > W_{tr}$ (Lower Ob); C—$W \geq W_{tr}$ (Norther Dvina river)

Formation of one-tier floodplains and floodplains with elevated riverbed and lowered rear parts. At $W > W_{tr}$ and relatively low rates of accumulation of sediment under conditions of relatively low runoff and a significant share of the bedrock component in it, the bedrock parts of the floodplain turn out to be significantly higher than the central and rear parts because of the more active deposition of sediment here during floodplain flooding (Fig. 5.12B). The transverse profile of such a floodplain is an inclined line from the edge, which is several meters high above the dusk, to the rear seam, where the surface of the floodplain has a height of about 1 m and is often lower than the water level in the river. If the river is branched into branches, including floodplain branches, the lowered parts of the floodplain are confined to the central parts of the islands and island massifs. The rear parts of the floodplain, within the limits of which the accumulation of lime deposits is inconspicuous (already clarified waters penetrate here), lag behind the growth in the height of the central and even more so in the bedrock of the floodplain, become hypsometrically lower than the latter. These depressions are filled with water and turned into vast shallow lakes due to the fencing of the river or arm channels and branches by natural riverbed elevations, higher position of the groundwater mirror and heavier mechanical composition of floodplain sediments. Such lakes are widespread in the lower Ob, where they are called distal floodplains, and in the lower Amur.

On rivers with low intensity of $W \geq W_{tr}$ and accumulation processes, the floodplains are unicycled (Northern Dvina), lakes like ditches are not developed, but the islands and coastal massifs of the floodplain are characterized by the increase of riverbeds (Fig. 5.12B). As a result, the islands and island massifs have concave transverse profiles, and the coastal massifs have a distinctly sloping transverse slope from the river.

The presence of an overlayed floodplain. It is formed in those parts of valleys where the first low floodplain terrace or very rarely flooded high floodplain was widely developed before. As a result of the accumulation of spill levels, high floodplains are becoming more and more regularly covered with water, and the floodplains are being spread to the first terrace or high floodplains, which were previously rarely flooded. The thickness of the newest sediments composing the Ob floodplain from the surface, judging by the outcrops in the benches of the washed banks, is about 50 cm on average (Fig. 5.13). In the lower parts of the superimposed floodplain, the composition of clay sediments is clayey; in the upper parts of the floodplain, fine silt sands predominate. The young floodplain alluvium lies directly on the black earth zone or gray forest soil (50–70 cm thick). Thus, a characteristic feature of the superimposed floodplain is the burial of zonal or near-zonal soils, which has formed in the rarely flooded areas of the high floodplain or the first terrace to enter the river floodplain due to the increase in its levels of sediment accumulation.

Overlayed floodplains are characteristic of those river sections where, under the influence of centuries-long climate fluctuations and river runoff, incision is periodically replaced by accumulation. When plunging, a stepped floodplain is formed, and its high steps can turn into a floodplain terrace; when load is accumulated in a directional manner, the floodplain regime is restored to them, which results in a superimposed floodplain with a buried horizon of zonal soil. Often the surface of such a floodplain is transformed by the processes that took place in the period when it was a terrace.

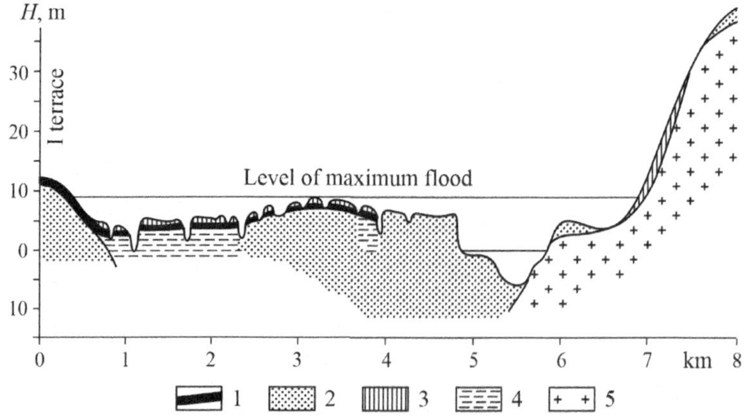

Fig. 5.13 Schematic transverse profile of the valley bottom with imposed floodplain (Ob river, Dubrovino village): 1—buried soil; 2—sand; 3—loam; 4—clay and loam; 5—rocky formations (granites) (Makkaveev and Chalov 1963b)

Formation of large sub-pond lakes at the mouths of small rivers—tributaries of big rivers (Avaryaskin 1976; Makhinov et al. 1994). The accumulation of sediment on the main river increases the erosion basis for its tributaries. With less sediment load than on the main river, they do not have time to accumulate sediment of the same thickness in their lower reaches and to raise their bottom and water levels. The lower reaches of the tributary are submerged by riverside banks or channel ramparts of the main river, a sub-pond lake is formed, into which the tributary flows and which connects with the river strait. Sometimes located in the rear parts of the floodplain, these lakes take over the floodplain branches and cover the estuaries of several tributaries. Such lakes are widespread in the middle and especially lower Amur, Yangtze, where they partially include lakes formed by the type of lakes (in the lower rear part of the floodplain) in the lower Ob.

5.2 Equilibrium Longitudinal Profile and Its Connection with the Directional Vertical Changes

Common directional vertical changes cause constant transformation of longitudinal river profiles. The latter can be considered in two forms: as a longitudinal profile of the river bed and as a profile of its water surface. The first one characterizes changes in the bottom marks reflecting, depending on the averaging and scale of the image, the ripple structure of the riverbed relief (large scale), or the basic roughness of any genesis and large size (small scale). Usually it is constructed in the coordinates of the flow depths from the accepted project (design) level (see Fig. 5.1B), and only when analyzed together with the water surface profile—in absolute (or conditional) bottom marks along the flow. From the point of view of the physics of the process and the mechanisms that determine the formation of the longitudinal profile of the river, it means the profile of the water surface. With the latter, it is possible to judge about the change of the hydraulic slope along the river flow—one of the main characteristics of the flow, reflecting the change in the length of the river flow energy losses to overcome the resistance. For their objective assessment, independent of multi-temporal changes in the water availability of the river, the longitudinal profile of the water surface of the natural channel flow is taken as the longitudinal profile of the river in the lowland, resulting in a cut-off level, when the distribution of slopes in the flow corresponds to the established flow. During floods, the flow is unsteady and its surface profile reflects the phases of wave propagation, sometimes overlapping with one water level or the other.

The value of the longitudinal profile of the water surface of the flow for the estimation of vertical channel changes follows from the equation of D. Bernoulli (4.2), recorded taking into account the pressure losses and the equality of the hydraulic slope to the slope of the free surface of the flow (4.5).

The starting (upper) point of the longitudinal profile of the river, which is the constant source of the river. The end (lower) point is the place where the river flows into the receiving water body, i.e. the water level in the sea or lake, or a smaller river into a larger one. This level of erosion is the basis for the flow formation and, consequently, at effective discharges, it is associated with hydraulic phenomena that determine channel scour or sediment accumulation, i.e. the direction of vertical changes in the lower reaches of the river. A good illustration of this is the graph of the amplitude of the levels (the maximum excess of the level in the flood above the low water level) on the Lower Volga before its transformation into a cascade of reservoirs (Fig. 5.14), given by Makkaveev (1955); according to it, the additional drop due to the spreading of the flood flow at the river inflow into the Caspian Sea increased from 1 cm per kilometer at the beginning of the decline zone to 4 cm/km on the lower 100 km before the mouth of the site. In the lowlands, the erosion base, like in the shallow water phase of the regime, is tied to the point where the current ceases to flow, and the river loads accumulate here. But if in a flood the flow gets additional fall, in the slope lowering period it gradually decreases up to the estuary where it becomes zero.

Along with the full melting of the flow at the mouths of the river, there are areas of partial melting where the flow is not interrupted, but its velocity increases and the depth of the flow sharply decreases. It happens on the steps and ripples of waterfalls. In significant valley and streambed expansion, in the places of inflow into the main river, which form local erosion bases, in the uncontrollable conditions a curve of decline in levels is formed, in the zone of distribution of which in the high-water phase of the regime longitudinal slopes increase (there is a curve of decline in levels), the rate of flow and erosion of the channel. Sometimes they cause the continuity of the longitudinal profile to break, creating its stages (waterfalls with large falls). Local erosion bases also include lakes (Makkaveev 1971), which are characterized by a variable regime of levels depending on the water availability phase of the rivers flowing into them. Riffles are a particular case of local erosion

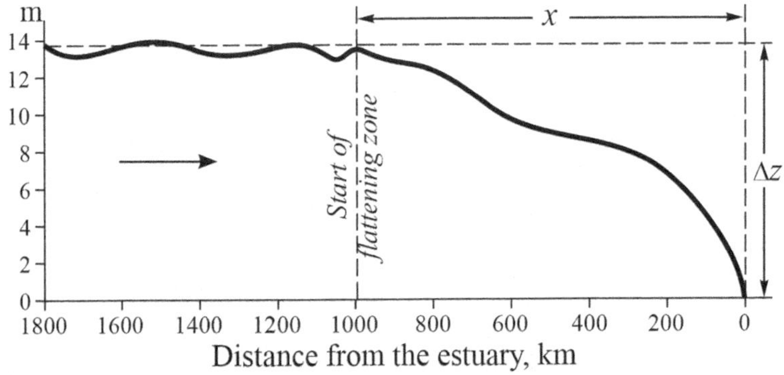

Fig. 5.14 Level amplitude on Lower Volga river in 1926 (Makkaveev 1955)

bases and also cause partial melting of the interflows (drop in crest levels) and their support within the overlying pool bed.

The distribution of private slopes of the free flow surface between local base levels along the river length determines the shape of the longitudinal profile. It consists of numerous bends that create a stepwise longitudinal profile of the river as a whole, and its steps have different sizes depending on the genesis of local base level.

Averaging of longitudinal profile marks leads to smoothing of its configuration, in which only large stages can be seen due to the geological-geomorphological structure of the territory along which the river flows (crossing of mountain ranges and structural uplands, steps, waterfalls). Therefore, the distribution of partial slopes along the length of the river makes it possible to distinguish four types of longitudinal profile forms: rectilinear (the value of the slope along the length of the river is constant), concave (the slope decreases from the source to the mouth), convex (the slope downstream increases), stepped (gentle sections alternate with steep ones) (Fig. 5.15). The latter, however, usually fits into one of the first three forms or characterizes their individual parts. The exception is some small rivers within structural uplands or plateaus composed of rocky, usually crystalline formation (e.g., Karelia rivers). The longitudinal profiles are also convex, with flat rivers flowing from lakes (Neva, Svir, Barnaulka); they are characterized by an increased slope from the source downstream.

The shape of the longitudinal profile and the distribution of slopes are constantly changing due to vertical changes. Incising of the riverbed in this part of the river leads to a decrease in the longitudinal slope, as the decrease in the bottom marks is accompanied by the same decrease in the water surface level. Propagating regressively or transgressively, plunging induces a consistent decrease in slope in the upper and lower slopes, respectively. At the same time, deep erosion can lead to the fact that at the moment the longitudinal profile from some site downstream has taken a convex form, because the transgressively spreading process has not yet reached here. At the same time, in the area where regressive erosion has not spread, a local increase in the water surface slope is observed. The accumulation of sediment, on the other hand, causes the bed of the channel to rise and the water level in the river to rise, increasing the gradient of the river. In relation to the upstream section, the accumulation of channel sediments leads to a decrease in gradient observed until regression accumulation and an increase in river gradient reaches the site. At the same time, accumulation can lead to smoothening of the longitudinal profile due to burial under the sediment layer of its steps, while incision, on the contrary, contributes to the fact that the root rock projections on the bottom of the river are dissected by the flow during the erosion of the bottom and begin to appear in the form of kinks or empty areas. An example of such a phenomenon is given by the Tom River in the lower reaches near the town of Tomsk (Berkovich et al. 1998), where the development of channel quarries caused incision, which, spreading regressively, led to the erosion of the alluvium layer and the exposure of the bottom at 73–75 km from the mouth of the rocky stepped river. The slope on the threshold increased with the further dropping of the slope levels, reaching 0.66‰ in the area

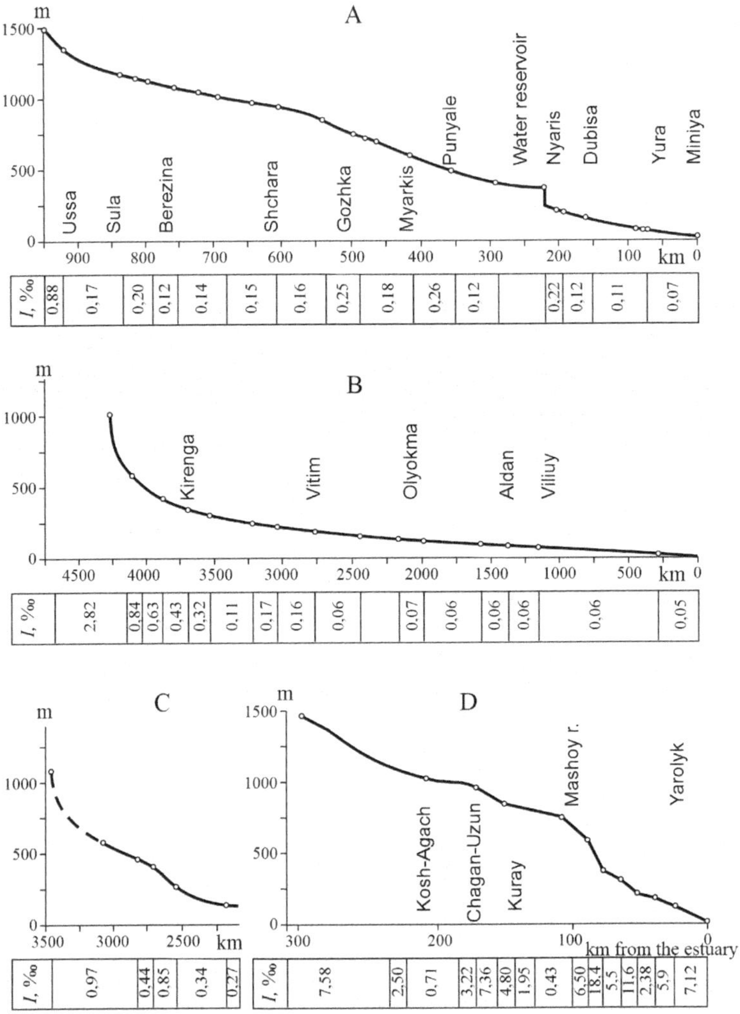

Fig. 5.15 Main types of rivers longitudinal profiles by their shape: **A**—straight (Neman river); **B**—concave (Lena river); **C**—convex (Olyokma river); **D**—stepped (Chuya river, Altai)

of 68–75 km in 1986 against 0.25‰ in 1983 and 0.19‰ in 1933 prior to the beginning of the quarrying operations, while the slopes were decreasing in the direction upstream and downstream (Table 5.3).

At the same time, incision of the river does not exclude the possibility of the development of accumulative forms of channel relief, the formation of which is associated with local changes in the transport capacity of the flow, the transition of sediment from suspended to bed-ridden), the magnitude of riverbed load and the ripple movement of the latter.

Table 5.3 Changes in the slopes (‰) of the Tom River in the area of Tomsk as a result of quarrying and incising of the riverbed

Year	Site, km from the estuary		
	0–68	68–75	75–82
1933	0.064	0.168	0.164
1963	0.043	0.220	–
1982	0.022	0.224	–
1983	0.014	0.258	0.133
1986	0.009	–	0.165

Makkaveev (1955) suggested that in order to assess the conditions for the development of general vertical channel changes, which lead to the transformation of the longitudinal profile, the slope should be considered as an indicator of the flow energy loss in a steady flow of water in the river

$$I = I_{\mathrm{fr}} + I_{\mathrm{er}} + I_{\mathrm{en}}, \tag{5.5}$$

where I_{fr}—corresponds to the energy consumption for the movement of the water itself, associated with overcoming the flow of friction on the bed, the viscosity of water and other obstacles, as well as the excitation of turbulent, circulating and wave motions; I_{er}—pressure loss associated with erosion and transport of sediment; I_{en}—increase (or decrease) of the slope due to changes in the kinetic energy of the flow. Energy losses I_{fr} flow of water are the main type of pressure losses. Total resistance to water flow (Baryshnikov and Popov 1988) is composed of resistance: (1) grain roughness determined by the size of bed material load; (2) ripple roughness (in sandy beds, the former can be neglected, as they are infinitely small compared to the latter; in pebble and especially pebble-boulder beds, on the contrary, the former is absolutely dominant); (3) channel shape; (4) additional resistance. Hence, member I_{fr} in formula (5.10) can be deployed and presented as

$$I_{\mathrm{fr}} = I_{\mathrm{r}} + I_{\mathrm{rid}} + I_{\mathrm{cs}} + I_{\mathrm{add}}, \tag{5.6}$$

where the indexes *r, rid, cs,* and *add* denote the affiliation of pressure losses to the influence of roughness, ripples, channel shape and causing additional resistance (turbulent viscosity, circulation currents, etc.). It is believed (Makkaveev 1971) that more than 80% of the energy of the flow is spent on the movement of water itself. When the discharges associated with the flow of water I_{fr} decrease in any part of the channel, there is a local increase in kinetic energy and an increase in the share of the head used for erosion and sediment transport, which ultimately causes the incision of the channel in this section and the local reduction of the slope. Indeed, the value of I_{fr} depends in turbulent flows on the flow rate, hydraulic radius, roughness of the channel, its shape in terms of plan and other hydraulic factors. Therefore, increasing the hydraulic radius, reducing the roughness, bringing the channel to the parabolic shape of the living section and the rectilinear appearance reduce this type of head loss. This is the case, for example, with the bend of the channel being straightened, which releases some of the energy that is used to deepen the local channel. As a result of the cutoff of numerous bends in a part of the upper Dniester in the early 1900s, the length of the river decreased by more than 2.5 times. This resulted in the river's inundation, which resulted in the lowering of the river's intertidal level by almost 3 m and the transformation of the floodplain into an inundated floodplain terrace. This process was most intensive during the first 15–20 years after the completion of melioration works, when the plunging speed averaged 20–30 cm per year (Afremov 1969). On the contrary, the riverbed elongation causes a loss of head and causes the accumulation of sediment spreading regressively up the river.

The energy losses of eroding and transporting sediment I_{er} can in some cases be significant. The addition of sediment to the stream, which is less than critical in size, causes a noticeable loss of head only at high suspended sediment concentrations. At the same time, the movement of bed material load significantly increases I_{er}. The amount of sediment that is transported by rivers in a semi-suspended and bed load state increases this part of the head loss almost twice as much as the same amount of sediment that is transported in suspension, although the increase in sediment concentration itself is accompanied by an increase in I_{er}. This phenomenon is especially pronounced in mountainous and semi-mountain rivers, where the range of changes in sediment concentration is particularly large. Available observations show that the flow velocities of mountain rivers at their saturation with sediments decrease in comparison with the velocity of clear water up to 67% (Pokrovsky 1940). At the same time, the value of I_{er} increases with a decrease in sediment load if, at the same time, the particle size of the load brought into the river from tributaries or slopes increases. That's why there are usually kinks in the longitudinal profile to such places. If there is a significant inflow into a river that carries more sediment than the main river, the slope of the latter below the junction point increases. If the inflow carries smaller amounts of sediment, the opposite is true: the slope of the main river is reduced. Makkaveev (1955) called this phenomenon "a game of tributaries".

The consequence of the interception of sediment load by reservoirs is the reduction of the I_{er} and the incising of river channels in their lower reaches. On the contrary, the inflow of additional sediment into the flow from any source causes the accumulation of sediment along the river. This phenomenon was described by Klaven et al. (2005) on the Zeya River as a result of the beginning of the White Mountains river washout, a sandy main slope of the valley; as the intensity of the washout decreased, the directional accumulation of sediment was replaced here by incision of the channel. Nevertheless, the accumulation wave, which was superimposed on the erosion products in the lower reach of the Zeya hydroelectric power station, reached the mouth of the river and represented itself in the 20 km section of the Amur River downstream (Ivanov et al. 2000).

The change in the partitioning between kinetic to potential energy of the flow cross-section (i.e., the loss of I_{en}) occurs along the length of each river. According to Makkaveev (1955), if as a measure of potential energy to take the depth of flow on the rolls and at the mouth of the river, each river is observed in the upper part of its channel system a gradual increase in potential energy (i.e., increase in the height of the water layer or depth of flow) due to the expenditure of some part of the kinetic energy. Starting from a certain point, the reverse phenomenon is observed in the lower reaches of the river at effective discharges—the flow thickness decreases due to the transfer of some potential energy into the kinetic one. Due to this phenomenon, the river flow moves the solid material to the estuary and actively extends the delta into the sea with minimal slopes and sometimes even negative slopes in the lower reaches of the river. Externally, the process of transition of the potential energy into the kinetic one is expressed in the reduction of the flood height towards the estuary, which becomes very small at the estuary bar. In general, the process of transformation of the flow energy contributes to some increase in the slope in the upper reaches of the river and decrease in the slope in its lower reaches, i.e., it is one of the factors contributing to the formation of the concave longitudinal shape of the river.

River incisings involve additional masses of sediment previously present in channel sediments or rock (soil) particles that are affected by the flow. The transport capacity of the flow rate W_{tr} should exceed the actual load flow: $W_{tr}>W$. On the contrary, the accumulation of sediment increases the volume of sediment at the bottom of the flow, which requires the opposite condition—$W_{\mathrm{tr}} < W$, *i.e.* the amount of sediment entering the flow exceeds its transport capacity. Since the runoff rates of the flow (including long-term, century-old, historical and geological scales of time), the inflow of sediment from the catchment area, the estuary elongation of the rivers due to the removal of sediment and the formation of deltas, the rise of land and the lowering of coastal areas, etc., are constantly changing, different ratios between W and W_{tr} appear along the length of the entire river or in its individual sections. Accordingly, if the carrying capacity of the stream is relatively high in a certain section of the river, i.e., $W_{tr} > W$, then the channel incisions or erosion of the bottom will occur more vigorously than in the neighboring sections, until the local slope (due to the erosion of the bottom) is reduced and the equal carrying capacity is achieved with the neighboring sections. Erosion or accumulation of changes in bed markings on a particular river section causes the same type of change, but gradually fading along the river length in adjacent areas as well. Such changes are carried out continuously, because the longitudinal profile tends to get the configuration corresponding to the condition $W_{\mathrm{tr}} = W$. The aspiration of the river from its source to its mouth to the formation of such a form of longitudinal profile, in which the length of the river is observed compliance of the transporting capacity of the flow of sediment load (the law of automatic leveling of the transport capacity of the flow), is the physical essence of continuous directional changes in the longitudinal profile of the river due to the ongoing processes of erosion, transport (movement) and accumulation of sediment.

Along the length of medium to large rivers crossing various geomorphological areas and natural zones, which integrate the influence of many factors, the transport of solid matter, the development of incisions and accumulation processes is carried out in a very complex form, with repeated alternations and temporal variability. Strictly speaking, this also applies to small rivers flowing into the sea. The larger the river, the greater the spatial and temporal fluctuations in sediment load and the alternation of solids transport conditions.

In the most general form, the intensity and direction of erosion and accumulation processes (vertical channel changes) is determined by the shape of the river's longitudinal profile and its correspondence to the condition of alignment of the transport capacity along the length of the river. Makkaveev (1955) suggested that the longitudinal profile, which has a relatively stable shape, corresponds to the leveled conveying capacity of the flow and a certain ratio between it and the slope, should be called developed. In other words, the developed longitudinal profile characterizes the stage of development of the channel with the established certain ratios between the sediment load and the transport capacity of the flow.

The mechanism of formation of the developed longitudinal profile of the river depends on its water-bearing capacity, the particle size of bed material load (alluvium) and their changes along the river length due to the intersection of different hydrological-geographical zones and geological-geomorphological regions. Therefore, the shape of the generated profile can be different: For example, it is generally concave in rivers with ever-increasing runoff rates and, consequently, increasing transport capacity; it is convex if downstream water flow decreases (arid conditions) and, consequently, the transport capacity decreases; The stepped shape of the profile is typical for the rivers crossing ridges and intramountain depressions in the mountains, uplands and lowlands on the plains, which is accompanied by changes in the amount and size of solid material entering them, etc.

The stability of the shape of the produced longitudinal profile is relative. Even if the elements of the geographical environment remain constant, it changes slowly but continuously. The length of the river increases due to the extension of its delta into the sea, which leads to the estuary elongation of the river, the loss of head and the development of regressive accumulation, covering, at low slopes, almost all lower reaches of the rivers (on large rivers, this corresponds to a length of thousands of kilometers). On the contrary, the erosion of the catchment areas is accompanied by a gradual levelling of the valley slopes, a decrease in the marks of watersheds and a decrease in the amount of solid material entering the river. This causes a reduction in the pressure loss on its movement and river incising which, in turn, contributes to an increase in the depth of river valleys and the activation of erosion and denudation processes; as a result, the inflow of sediment into the rivers increases again and slows down deep erosion (Borsuk and Chalov 1978). Constant climate change, changes in vegetation patterns, tectonic movements and economic activity cause general or local deviations in transport capacity from the "norm" (align), which causes a change in the direction of erosion and accumulation processes along the river length, the activation of incision in some areas and the accumulation of sediment in other parts of the river.

The model of the developed longitudinal profile proposed by Makkaveev (1955) is based on the analysis of the Shezy-Manning equation

$$I = \frac{n^2 V^2}{h^{4/3}}. \tag{5.7}$$

At the same time, Makkaveev accepts three assumptions: the first is that under the conditions of a longitudinal profile the current velocity along the river length is approximately equal; the second is that the roughness coefficient is constant along the river length; and the third is similarity in the length of the river of living sections of the channel. As a result, for the final expression received

$$I = kQ^{-\frac{2}{3}}, \tag{5.8}$$

for which Makkaveev himself limited the scope of application to flat rivers. On the other hand, the coefficient k depends on the roughness of the channel and the shape of its living section. But the latter makes sense only when comparing longitudinal profiles of different rivers, but does not allow to apply the formula (5.8) to the analysis of the crossing or stepwise profile of a given river. This is where the longitudinal profiles, whose shape deviates from the correct concave one, are referred to the category of undeveloped ones, although Makkaveev warned about the error of such a formal approach.

Acknowledging the limitations of the obtained formula (5.9), Makkaveev in further constructions connected with determination of the position of the point of maximum deflection of the concave longitudinal profile returned to the expression of relationship of its form on the roughness and cross-section of the channel:

$$x_1 = \frac{V^2 n^2 A_2}{k_3^{\frac{4}{3}} A_1^{\frac{2}{3}} I}, \tag{5.9}$$

where x_1 is the distance from the source to the point of greatest deflection (maximum incision of the river), V is the average flow rate at the effective discharge; n is the roughness coefficient, according to the Manning; k_3 is the coefficient in the equation of water depth and flow rate; A_2 is the coefficient in the equation of channel length and basin area; A_1 is the coefficient of flow intensity; I is the average river slope. Unfortunately, this expression is very rarely used (Makkaveev himself, except for his famous monographs (1955), did not use it anywhere else); however, it is only valid for comparing the profiles of different rivers, but it is applicable only to humid conditions, i.e. conditions of river water availability increase downstream.

In the real natural situation, the formation of river channels is such that the alternation of free and limited conditions for the development of channel changes occurs along the length of the river, mountainous channels in the upper reaches of the rivers are replaced by semi-mountainous and lowland ones (Chalov 1979), etc. All this is accompanied by changes in the living flow cross-section, flow velocities and the size of the bed load alluvium (from sand to boulders). To take these circumstances into account, the longitudinal profile equation can be obtained by solving the Shezy-Manning Eq. (5.7) and the continuity equation for the steady flow where $Q = \omega V = b_{\text{ch}} h V$; $V = Q/b_p h$, I is the slope; V is the average flow rate; h is the average flow depth; b_{ch} is its width; Q is the average flow rate; n is the channel roughness coefficient. By substituting the latter in (5.7) we get

$$I = \frac{n^2 Q^2}{b_{\text{ch}}^2 h^{\frac{19}{3}}}. \quad (5.10)$$

Taking, according to Grishanin (1974), for stable riverbeds (3.26), that $\frac{h(gB_{\text{ch}})^{1/4}}{Q^{1/2}} = M = const$; then

$$h = \frac{MQ^{1/2}}{(gb_{\text{ch}})^{1.4}} \quad (5.11)$$

and by uniting all the regulars, we get

$$I = kn^2 Q^{1/3} b_{\text{ch}}^{-7/6}. \quad (5.12)$$

If we accept from (4.21) that the width of the channel $b_{ch} = A\frac{Q^{1/2}}{I^{1/4\prime}}$, continuing to combine constants, we can get

$$I \cong k'' \frac{n^3}{Q^{1/2\prime}} \quad (5.13)$$

and replacing the value of n by Chang-Strickler, $n = k''d^{1/6}$:

$$IQ^{1/2} d^{-1/2} = const \quad (5.14)$$

Hence, all other things being equal, the decrease in longitudinal slope corresponds to the increase in downstream water discharge. On the contrary, the decrease in water availability should be accompanied by an increase in the slope, which reflects the development of accumulation processes. On the other hand, the increase or decrease in alluvium particle size reflects the increase or decrease in gradients; indeed, the increase in bed material load particle size is usually accompanied by an increase in gradients. For example, the sandy bed of the lower Neman is replaced by a pebbly-boulder bed in the middle reaches, where the river crosses the Grodno Upland; the slopes of the bed here increase by more than 1.5 times: from 0.15–0.16‰ to 0.26–0.26‰, and on the longitudinal profile, with its overall straightness, there is an inflection (Darbutas and Chalov 1993).

In lowland conditions on rivers with sandy alluvium (d_{av}< 1 mm) at fluctuations of its size in the areas of hundreds and thousands of kilometers in the range of 0.2–0.6 mm, this term can be neglected as a factor determining the roughness of the channel. The relationship takes the form

$$IQ^m = const \quad (5.15)$$

which corresponds to the formulas proposed by Makkaveev, Chizhov, Rybkin and others [their analysis is given by Makkaveev (1955)], in which the index of degree m varies within the range of $0.3 \div 0.67$ (maximum values in the formula of Makkaveev).

Taking into account that the degree values in the formula (5.15) may have a regional meaning, and that the value of n, according to the studies by Talmaza and Kroshkin (1968), depends on the partitioning between flow depth to the sediment particle size (e.g., with the comparability of these values, which is especially characteristic of mountain rivers, the value $n = 1/3$, if $2 < h/d < 10$), it is advisable to write it down in a more general way

$$IQ^m d^{-n} = const \tag{5.16}$$

A close result can be obtained by substituting in expression (5.7) the speed with its critical value from the Froude number, $Fr = 1 = \frac{V^2}{gh}$, i.e. $V^2 = gh$. This allows the obtained formula to be used for the analysis of longitudinal profiles of mountain rivers characterized by a rough flow and the above partitioning between its depths to the size of alluvium particles:

$$IQ^{5/12} d^{-1/3} g^{-1} = const \tag{5.17}$$

From here, rivers with the same runoff rates but with different sediment loads have different longitudinal slopes. In particular, the rivers of Eastern Siberia with pebble-boulder alluvium are characterized by 2–3 times greater slopes than the same rivers of Western Siberia or the East European Plain (respectively, 0.30–0.50 and 0.10–0.15‰).

According to Makkaveev (1955), the sediment concentration constancy of water along the length of the system should correspond to the transport capacity leveled along the length of the river as evidence of the depletion of the longitudinal profile

$$s = AQ^m I \tag{5.18}$$

that meets the condition $s = s_{tr}$. The second (along with the analytical conclusion) approach of Makkaveev to obtaining the formula of the developed longitudinal profile of the river is based on this circumstance.

If the condition of the transport capacity levelling is not complied, the longitudinal profile is in the stage of transformation, and significant fluctuations in the average annual sediment concentration of water occur along the length of the river. Such profiles are not developed. This is most clearly seen in the conditions of upward development of the relief, when the incision of the river as a whole lag behind the tectonic rise of the territory. In such cases, with an overall concave of the longitudinal profile, the point of deflection corresponds to the river section with the maximum intensity of incision. As a result, the midstreams of mountain rivers, which are generally subsequent, are characterized by the highest depth of the valley trenches and, as a consequence, by the spread of the steepest slopes in the adjacent part of the basin, the increased density and depth of the overall fragmentation of the territory by erosion forms (low-order river valleys, mountain ravines) and the maximum intensity of erosion and gravitational processes on the slopes. Intensive plunging and increased activity of erosion and denudation in the catchment area in the middle reaches of mountain rivers are integrated by the sediment concentration of water in the main river (or the river system, the order of which gradually decreases upstream), reaching the highest values here and decreasing both upstream and downstream (Makkaveev et al. 1968; Gadzhieva 1982). Data similar to those in Table 5.4 are obtained for the river Kodori (Western Caucasus), where the average annual sediment concentration of water is 2.5–3 times higher than the average annual sediment concentration of hydrological stations located in the upper reaches and close to the estuary (Mandych 1967).

Similar situations are typical for the rivers of the Greater Caucasus, a compact mountainous country with a relatively simple geomorphological structure originating in the high-mountainous zone (Rioni, Kodori, Mzymta, Bzyby, Kuban, Terek and others). At the same time, there are also conditions that determine the formation of large stages of the longitudinal profile of some rivers. They are associated with valley lakes formed by mountain precipices (Ritza, Amtkeli in the Caucasus, Sarezskoye in the Pamirs, etc.), with alternation of gorges along the conserved rivers at the intersection of forward ripples, which is accompanied by an increase in the particle size of alluvium, and intermountain troughs. The lakes intercept the entire runoff of pebble-boulder riverbed sediments. Vinogradova (1966), who studied the process of filling the Sukhumi reservoir with sediment, estimated that the total volume of sediment deposited in it over 16 years (1948–1964) was 52,000 m^3, and the average annual volume of sediment was 1,200 m^3, forming a layer of sandy-pebble sediment with a capacity of 1.5–6.0 m. It should be taken into account that mountain water reservoirs are regularly washed, where a part of the accumulated sediment is removed (carried to the downstream bank). Therefore, under natural conditions (in lakes of overburdened origin), the rate of sediment accumulation by rivers is much higher.

When lakes are filled with sediment, steps are formed in the longitudinal profile of the river, directly above which the accumulation of sediment predominates. After such a filling of the lake with sediment, the stream flows over the sediment that filled it within the valley extension and the natural dam, falling with it through a

Table 5.4 Changes in the average annual sediment concentration of water on the Rioni (Western Caucasus) (Makkaveev et al. 1968)

Hydrological station	Distance from the mouth	Slope, ‰	Average annual sediment concentration, g/m^3	
			1952 yr	1954 yr
Geby	302	20	490	536
Namokhvany	176	3.05	1020	1690
Sakochakidze	43	0.82	564	568

cascade of waterfalls. At the same time, there is a regressive plunging of the river into the debris strata and in the deposition of the former lake (Berkovich and Chalov 1969). This is the expansion of the Gegh valley (a tributary of the Bzybi River, Western Caucasus) above its confluence with the Yupshara River. Here, the formation of the lake now filled with sediment occurred as a result of the big collapse of limestone blocks, forming the slopes of the gorge during the earthquake in the zone of Dzyshra-Kugbaestinsky thrust. Within the limits of expansion—at the bottom of the former lake—the slope of the modern channel is equal to 14‰. Below the expansion, it increases by almost 15 times, reaching 200‰. The slope of the river is 40‰ above the expansion and 1 km below the bend of the longitudinal profile.

A similar phenomenon occurred on the Issyk River (Zailiyskiy Alatau) at the end of the 19th century. However, the resulting collapse did not have time to fill the lake with sediment, as it was lowered by a powerful mudflow on July 7, 1963 due to a flow breakthrough through the dam. The photos show the former lake hollow (Fig. 5.16A).

The rivers of more complex mountainous countries (Altai, Pamir, Tien Shan, Sayans, etc.) are characterized by a stepwise progression of longitudinal profiles and changes in the intensity and direction of vertical changes. In the Altai system in the Katun-Chui system, the crossing of large intermountain troughs ("steppes"—high-mountain valleys filled with moraines and spillage) and intensively rising ripples by rivers in the Altai causes a complex character of changes in slopes and alluvium particle size along the length of the rivers against the background of a clear relationship $d = f(I)$ and the total incision of the rivers (Demin et al. 1991). On the western Tien Hakimov (1992) set the speed of mountain rivers incision at the crossing of ripples in 1–7 cm/year. When sawing through bedrock formations by rivers and plunging them into the underlying ones, the velocities of deep erosion

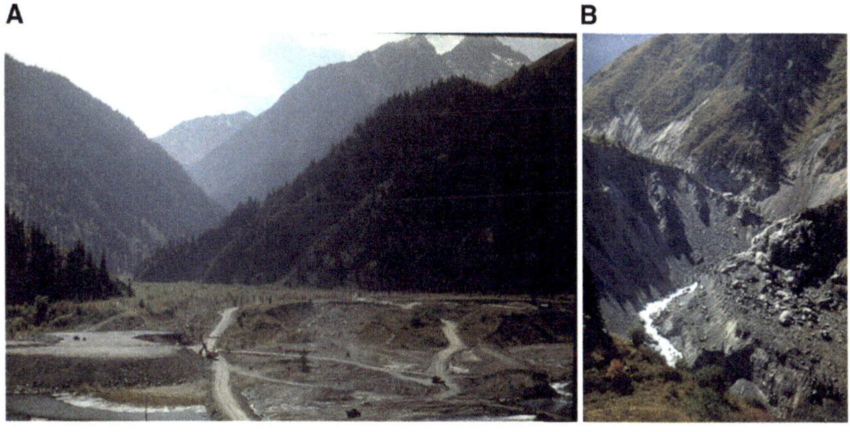

Fig. 5.16 Issyk river (Zailiyskiy Alatau) at the edge of a flat lake (**A**—view from below, **B**—top view) and downhill the rock-avalanche dam (**C**). Photo of the author

can reach 30 cm/year (Alabuga river, Central Tien Shan), which can be easily washed away as the bend (step) of the longitudinal profile is eliminated (Panin et al. 1990). However, as a rule, the speed of river intakes in the antecedent valleys dissecting rising ripples is no less or no less than the speed of ascent. Therefore, the intersection of the river systems of mountain ranges and their separating depressions often does not affect the direction and even the intensity of vertical changes, but is reflected in the bends of the longitudinal profile due to the local inflow from the slopes of the mountains of a larger debris material and an increase in the resulting loss of pressure on their movement and associated growth of grain roughness and hydraulic resistance.

The failure to develop the longitudinal profile of mountain rivers also express itself in the presence of areas with a predominant accumulation of sediment. For example, such a large river as the Yenisei accumulates sediment within the Minusinsk Depression (Bely et al. 2000). This process was interrupted by the construction of the Sayan-Shushensk HPP due to the cessation of sediment load in the minusinsk section of the Yenisei River. However, accumulation processes in pure mountainous conditions are quite rare (except for the above-mentioned river sections above, such as rock falls and lakes).

Thus, the undeveloped longitudinal profiles of mountain rivers are characteristic of the conditions of upward development of the relief, when the speed of the overall rise of the mountainous country is greater than the speed of incision of the rivers. Otherwise, as a result of river incision, the profile becomes depleted, the average annual sediment concentration along the length of the river is equalized, and the rate of vertical changes becomes minimal, determined by the general features of denudation of mountains and lowering the marks of watersheds. However, this is not related to the presence or absence of a longitudinal profile stepwise, which is determined by the size of the sediment load, which is of local origin.

In flat conditions, the longitudinal river profiles are usually developed, although due to the continuity of the processes of leveling the transport capacity (due to tectonic movements, including local rises and falls, changes in climatic conditions, including perennial and century-long fluctuations in the runoff rates of the rivers, etc.) in different parts of the river are incised or accumulated, and these processes may change over time. The longer the river is, the more likely it is to change the direction of vertical changes along its length.

In general, the river crossing the plain-platform area is slowly (fraction of a millimeter—millimeters-a-year) crashing, and only in the lower reaches is the accumulation of sediment as a result of the delta extension and estuary elongation of the river, which is often superimposed tectonic immersion of the coastal lowland. As a result, terrace stairs are formed and the floodplains of the river have a stepped transverse profile. In the lower reaches, where rivers accumulate sediment, vast alluvial-delta plains are formed, and the floodplains of the rivers are either single-tier or have a distinct excess of riverbeds over the rear ones, up to the point that the marks of the latter may be lower than the marks of the intertidal channel (lower Amur, Amu Darya, Ili, Terek, Huang He).

For example, on the Neman, the channel in the upper and middle reaches of the river incisions in natural conditions, and accumulates sediment in the lower reaches (Darbutas and Chalov 1993). The intensity of the incising processes varies along the river length from 1 cm per year in the upper reaches (here, however, during the time covered by the measurements, they were intensified under the influence of reclamation works in the river basin) to 3 mm per year in the middle reaches, where this process is limited by geological and geomorphological factors (crossing of the Grodno upland, South-Litovsk ridge, etc.). At the lower Neman, the bottom marks due to the accumulation increased at a rate of 0.5–1.2 cm/year, while the construction of Kaunas HPP, the continuous correction of the channel and the development of open pits did not change the sign of the direction of change.

Since changes in the runoff rates of rivers along their lengths are subject to the relationship (2.4) (Kositsky et al. 1999), i.e. they are rapidly growing in low order rivers and very few in large rivers, the longitudinal profile in humidal conditions is concave in shape, with the maximum deflection point (deflection point) shifted to the upper reaches of the rivers according to formulas (5.16, 5.17). Makkaveev (1955) came to this conclusion also on the basis of an analysis of changes in river length and sediment particle size, as well as the partitioning between channel depth to channel width. Nevertheless, the "portions" of water recharge in the river caused by the inflow of large (similar in order) inflows often leads to significant disruptions in the smoothness of changes in slopes of the longitudinal profile, even from large flat rivers. One of the main reasons for the formation of large levels of longitudinal profile, especially in the case of incising rivers, is also the heterogeneous lithology of the indigenous bottom. Strong rock outcrops, which are relatively resistant to erosion and weathering, produce steep changes in the longitudinal profile (steps, waterfalls), on which a significant part of the river fall often concentrates. For example, about 1/3 of the total fall of the Dnieper River was concentrated on the rapids of the Dnieper River.

On steep ledges of the longitudinal profile, the flow velocities and therefore the transport capacity of the flow are higher than on adjacent river sections. Thus, the process of leveling the conveying capacity gives the profile a smooth shape only if the geological structure of the valley bed is homogeneous.

The stepping of the longitudinal profile can also be caused by the fact that coarse material flows into the river unevenly along its length. The introduction of large deluvium from steep valley slopes or large alluvium from tributaries often causes a local increase in river slope. The indirect cause of the formation of the longitudinal profile steps is the local tectonic structures that cross the river. Usually the speed of tectonic movements is so inferior to the intensity of deep erosion or accumulation that the river has time to cut through the rising structure (if it is composed of rocks sufficiently malleable to erosion) or to fill the deflection formed when lowering the bottom. However, when elevated, strong rocks can be eroded, causing the formation of steps and kinks in the longitudinal profile; in addition, the arrival of coarse debris from steep valley slopes can also contribute to the local increase in slope. Local negative structures are usually marked by extended valley areas, where the river flows along the alluvial bed and has a relatively gentle slope.

On the background of the general transformations of the longitudinal profile local tectonic movements at low slopes and low rates of systematic accumulation lead to a local inversion of the directional vertical changes. Bylinskaya (1973) on the basis of analysis of water discharge curves and levels revealed a complex alternation of sections of incision and accumulation on the rivers of the center and south-west of the Central Russian Plain; the same conclusions were drawn with respect to the rivers of the north-western part of the Central Russian Plain, but using mainly geological data Meshcheryakov (1961).

On the middle Amur in its different parts (the Zeya-Bureya plain, the Maly Hanghan mountain system, and the western part of the Middle Amur lowland) there are five areas of directionality and intensity of vertical channel changes (Ivanov et al. 2000). On the first one, between Blagoveshchensk and Grodekovo (20 km), according to the curves $H_1 = f(H_2)$, accumulation of sediment is established (see Fig. 5.6A), which, however, does not correspond to the morphological features corresponding to the river incision. Obviously, this contradiction is influenced by the impact of sediment transport from the Zeya River, on which the Zeya Hydraulic unit was built during this period. Deep erosion in the lower tailrace of the hydroelectric power station has led to an increase in the flow of bedload, which has created a wave of accumulation on the Amur River directly below its confluence with the Zeya. As early as 30 km downstream (Grodekovo—Konstantinovka), the weak incision of the river is evidenced by both directional reduction of water levels at hydrological stations and morphological features. In the second section, up to the village of Pashkovo ("entrance" to the Khingan) in the Amur channel there is a directed accumulation of sediment, which is clearly manifested in the morphology of the valley and the Amur riverbed. The height of the floodplain is reduced, it becomes monolayer, and the riverbed is significantly higher than the rear part.

Changes in the direction and intensity of the vertical changes of the middle Amur River bed are reflected in the slopes of the water surface: the slopes of the river infeed are greater than those of the sediment accumulation, which indicates that the longitudinal profile of the river is not developed (Table 5.5).

The diversity of vertical channel changes and their different intensity within the Zeya-Bureya plain are caused by a number of factors, leading from which are modern tectonic movements. The Amur Valley passes here along the border of two tectonic structures—the Zeya-Bureya pain covering the left side of the basin, and the Maly Khingan—a mountain and folded structure lying on the right. Both structures have experienced a steady rise during the recent tectonic period, but it is sharply differentiated. The total newest tectonic elevation of the territory, where the valley section below the mouth of the Zeya River is located, is 100–200 m, and then it decreases to 50 m, and to the area of crossing the valley of the Maly Khingan spurs increases again to 150–200 m.

The decrease in the rate of elevation along the length of the pre-Hingan marginal deflection explains the change downstream of the river by the accumulation of sediment (before the impact of the Zeya HPP on these processes). The relative

Table 5.5 Slopes and average annual water discharge at sites of middle Amur with different directions of vertical change

No	Site	Length, km	Slope, ‰	$Q_{av.\ year}$, m^3/s
1	Zeya estuary—Bureya estuary	269	0.095	3520
2	Bureya estuary—"entrance" to Hingan (Pashkovo village)	136	0.085	4750
2a	Pre-Hingan site (Innokentyevka village—Pashkovo village)	108	0.092	4750
3	"Entrance" to Hingan—"exit" from Hingan	164	0.142	4750
3a	Maly Hingan (Pashkovo village—Pompeevka village)	81	0.099	4750
3б	Pompeyev Ridge (Pompeevka village—Ekaterino-Nikolskoe village)	83	0.185	4750
4	The "exit" from Hingan—Sungari estuary	146	0.071	4750
5	Sungari estuary—Khabarovsk	270	0.062	7450

reduction of slopes and, consequently, the transporting capacity of the flow ultimately leads to the accumulation of top-down sediment and an increase in the marks on the bed of the channel and the profile of the water surface.

Sediment accumulation in the channel immediately prior to Khingan is more intense than in the overlying areas. This is obviously due to the pressure conditions caused by the transition from free to limited development of channel changes. The wide floodplain channel is replaced by a cut one (the floodplain is wedged out) and narrowed by more than 1.5 times. Stress conditions are represented during the passage of effective water discharge when the floodplain is flooded; on the contrary, an increase in slope in front of the "entrance" to the Khingan Gorge is noted (Table 5.5, Sect. 2a). For the pre-Hingan section of the channel, the curve of the corresponding levels is unchanged (Fig. 5.6B). This is probably explained by the fact that the river crosses the mountain massif and the decrease occurs here at both stations with the same intensity.

Throughout the Hingan constriction, the Amur River has been cut into its bed. However, there are a number of indirect indications that local accumulation of sediment is currently taking place in the central part of the area where the river crosses the Sutarskaya Depression. This is indicated by the formation of mouth-water lakes on the tributaries of the Amur River (at the mouth of the Ulagakh River, etc.). Most of the deposits are formed by gravitational processes on slopes. The main role in the accumulation of in-stream sediment is played by the large pebble material, the main supplier of which is the embankment, which descends directly to the riverbed.

In the lower part of the Khinganskiy site at the maximum values of slopes, the river bed is either rocky or lined with a large pebble scaffold, the size of particles which exceed the critical for the shift of their flow. This is evidence of an intensive river intaglio. However, the corresponding level curves for posts in this section have constant coordinates over the entire observation period. Apparently, the rates of elevation of the mountain massif and the rates of plunging coincide. As a result, the Pompeevsk area with numerous bedrock outcrops on the bottom along its entire length creates an inflection of the longitudinal profile of the Amur, affecting its shape.

Below the Khingan constriction, deep erosion prevails in the short section (fourth, 40–50 km), characterized by low rates of erosion, followed by directed accumulation downstream. The river crossing is facilitated by the formation of a lowering curve regime in the high-water phase of the river at the exit of the river from the gorge to the valley with a wide floodplain and the change of the incised to the wide floodplain.

At the lower (fifth) 360 km section (up to Khabarovsk) there is a directed accumulation of sediment in the river, especially intense below the mouth of the Sungari River. Numerous estuarine lakes on tributaries (the rivers Bira, Vertoprashikha, Nongjiang, etc.) testify to the accumulation of sediment here. Above the confluence with the Sungari River, the accumulation on the Amur River is reinforced by a support on the supply side.

The directed accumulation of sediment covers the entire lower course from Khabarovsk (confluence with the Ussuri River) to the mouth of the Amur River (Makhinov et al. 1994). It is evidenced by the widespread distribution of estuarine lakes in almost all tributaries of the Amur River. At the same time, it is also observed within the incised channel at the intersection of the Sikhote Alin spurs between the Middle Amur and Lower Amur lowlands, where the estuary lakes are typical for small mountain rivers—tributaries of the Amur (see Fig. 5.7).

Thus, the process of levelling the transport capacity does not mean levelling the longitudinal profile line. The longitudinal profile can take a variety of forms and is characterized by a stepwise approach, determined by local changes in the energy density of the flow section and its transformations related to the local and general effects of changing environmental conditions on it.

Against the background of general vertical changes, there are periodic local short-term (periodic) changes in their direction. They are associated with the development of channel forms and cover small river sections. Cutoff of the bend, accompanied by reduction of channel length, leads to incision of the channel, which spreads some distance upstream from the place of cutoff (regressively) and slightly less—downstream (transgressively), where the products of erosion from the developing duct arrive (Fig. 5.17). As the new bend develops, the water surface marker gradually recovers. This process is superimposed on both the general infeed of the river, causing it to become more active locally, and the targeted accumulation of sediment, causing it to stop temporarily. This phenomenon is sometimes used to reduce the risk of flooding associated with a constant rise in water levels (Terek, Kuban). Cutoff the bend leads to a significant reduction in water levels. On the Yangtze River (Yang and Jichang 1999) within the Lianghu plain (Ichang city), it was 2 m in 40 years with an average annual flow rate of 14,230 m^3/s. Several successive bends were straightened out (Fig. 5.18).

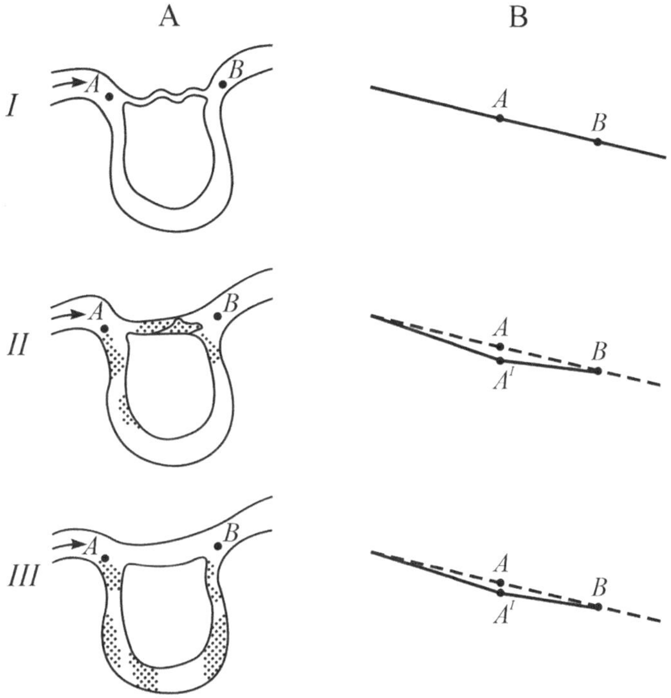

Fig. 5.17 Longitudinal profile development scheme at bend straightening (Gendelman and Popov 1987): **A**—channel plan; **B**—respective longitudinal water surface profiles. I, II, III—development stages of the straightening

In the lower reaches of rivers, the regressive accumulation of sediment associated with the extension of the delta is periodically interrupted by the straightening of the channel in the direction of the highest gradient. For example, in the lower reaches of the Huang He River such a concretion occurred in 1976. Mikhailova (1998) caused short-term regression erosion (a 0.5 m drop in levels at the Liczin station), which covered all the lower reaches of the river, but was replaced by accumulation since 1985. The layer of sediments was 130 cm long at the Liczin post in 8 years, and the new direction of the delta reached 13 km only in 1976–1980.

On the Black Sea coast of the Caucasus, the extension of the delta along some of the rivers also causes the accumulation of sediment, which is facilitated by the development of estuarine pebbles that deflect the estuary section of the river in the direction of longshore drift of sea sediments. On the Kodori River, the spit separating the riverbed from the sea reaches 1.5 km in length, almost completely encircling the outer edge of the delta. After reaching a certain critical length of the spit into the flood, the stream of the river breaks through the river flow in its root part, the channel straightens out and is cut into the channel, which is replaced by accumulation of the new mouth spit (Mandych and Chalov 1970).

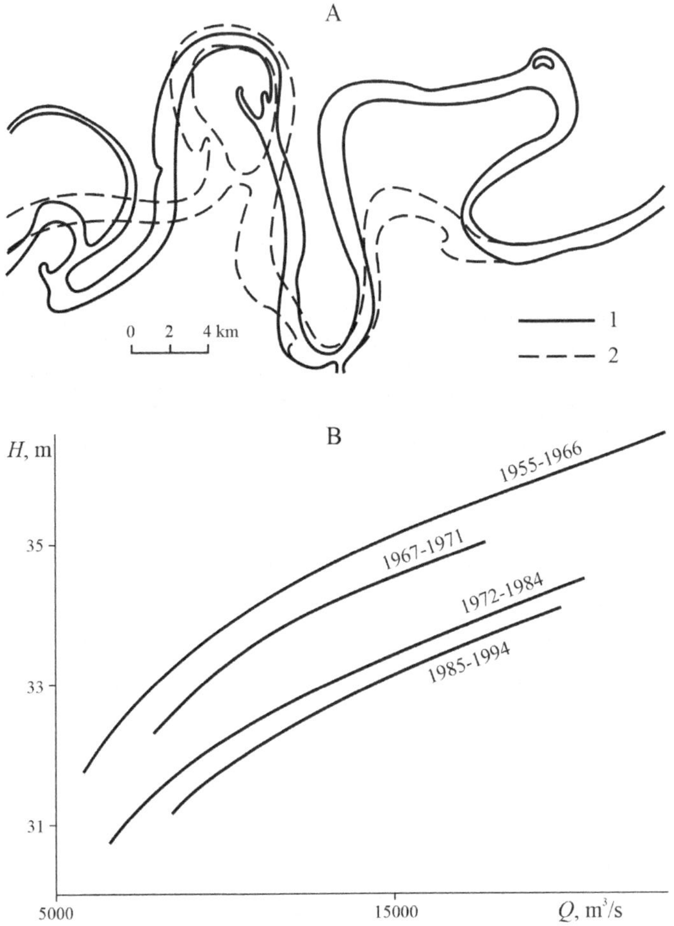

Fig. 5.18 Straightening of the bends of Yanghtze river (**A**) and connected decrease of curves Q = *f*(H) (**B**) (Yan amd Tan 1999): 1—position of the channel in 1940-s; 2—the same after straightening of bends (1960-s)

Overlapping of the general directed vertical change of the channel on the local, due, firstly, to the peculiarities of the geological-geomorphological structure of the territory, and, secondly, being periodic, connected with the evolution of the very forms of the channel, seasonal, perennial, secular and other changes in the factors of fluvial processes creates a rather complex and sometimes contradictory picture of the development of the longitudinal profile of the river. It can only be decoded if a number of observations are available, supplemented by certain morphological features.

Such changes in the slope and the concave shape of the longitudinal profile of the rivers are typical of humid conditions, when the growth of the catchment area corresponds to an increase in the runoff rates of the river. In arid conditions, from the moment the rivers reach the foothills, their flow is used for evaporation and is largely understood for irrigation. As a result, with the growth of the catchment area, the runoff rates of the rivers decreases, they accumulate sediment and form dry deltas, while the longitudinal profile becomes convex. However, the increase in gradients downstream of the gradually drying up river does not go beyond the gradients corresponding to the lowland rivers. An example of such rivers are the rivers of the peripheral margins of the Tien Shan, the catchment areas of which range from the first tens to the first hundreds of square kilometers, but their water availability growth is limited to the upper and middle reaches of the catchment area with the size of 100–10 km^2.

5.3 River Incision, Speeds and Causes

Transformations of longitudinal river profiles as a result of directed vertical changes of the negative sign channel—incision of the rivers causes a decrease in the marks of the channel bottom, which leads to a relative decrease in water levels at the same flow rate. Under these conditions the floodplain of the river ceases to be flooded and turns into a floodplain terrace. A series of terraces emerges in the long term (during historical and geological periods of time) when incision occurs.

River incision velocities are usually insignificant and differ by orders of magnitude from the possible velocities of lateral channel change (on rivers with wide floodplain channel). Therefore, valleys with wide floodplains and terraces are formed on the plains, where river crossings are accompanied by intensive bank erosion, which is tens and hundreds of times higher than the plunging speed. In mountains where bedrocks prevail, narrow valleys—canyons, gorges, often devoid of terraces—appear in the process of plunging; the latter are found and can even be well developed only in basins and at the exit of rivers into the foothills. The same is observed on flat rivers at the intersection of elevations formed by bedrocks, or mountain massifs, where incised channels are formed in narrow, often floodplain valleys. In most cases, all this is a consequence of a very long geological history, the evolution of the river valley for tens, hundreds of thousands, and sometimes tens of millions of years.

On flat rivers, speeds and plunging usually make up a fraction of a millimeter—the first millimeter and very rarely exceed 1 cm per year; maximum values, by orders of magnitude large, are characteristic of local and, as a rule, short-term expressions of deep erosion (e.g., by cutoff of individual bend, dredging the bottom of the concave bank of the bend, by shifting riffles and pools along the channel, etc.). These occurrences are of a periodic nature, are related to the development of channel forms or large ripples in the channel and do not affect channel morphology and the shape of the longitudinal profile of the river, although they are one of the main causes of accidents at pipeline crossings and other communications. Only in case of anthropogenic impact on the rivers (e.g. creation of reservoirs and large quarries intercepting sediment load, congestion of the channel by correction structures) can the speed of incision on flat rivers be the first centimeters per year. In this case, the incision spreads up or downstream by tens of kilometers and causes the bottom marks to drop by 0.5–2 m in 20–30 year.

Table 5.2 presents data on the incision velocities of a number of large rivers, as determined by the curves $Q = f(H)$ for effective discharge and the resulting sediment load balance along the length of the rivers referred to the bed area (Chalov et al. 2000). Characteristically, with unambiguous estimates of the direction of change, the actual values obtained by different methods differ by an order of magnitude (with some exceptions), and almost always the decrease (increase) of levels is greater than the decrease (increase) of the bottom marks. At low load rates, the shear/accumulation layer between the stations seems to integrate the influence of inflows below the mouths of which there is an increase in sediment load due to the removal of sediment from the inflows and an increase in the transport capacity of the flow due to an increase in the runoff rates of the river. In the presence of large inflows, the data on the sediment balance between hydrological stations provide a distorted view of the direction and rate of vertical change. Thus, in the Northern Dvina, the curves $Q = f(H)$ for the Abramkovo post recorded a decrease in the levels of 0.7 cm/year; in the sediment load balance between s. Abramkovo and s. Zvoz (above and below the inflow of the Vaga River) is 0.01 cm/year. The crossing of the river in the area of the Abramovo post is represented in the steepness of the floodplain and the transgressive movement of the channel forms. However, above the confluence with Vaga, a superimposed floodplain has been found, and the islands are characterized by regressive growth—obvious signs of sediment accumulation. Indeed, Table 5.2 does not take into account the fact that the Vaga (Ust-Syuma village) carries 0.42 million tons of load into the main river. The total sediment load of the Northern Dvina and Vaga (2.45 million tons) is greater than that determined at the site and at the village of Zvoz (2.38 million tons), which indicates a positive sign of vertical change in the lower reaches of the river. Sediment load from the Ust-Pinega post includes the outflows from the Pinega River, and therefore its increase also does not reflect the direction of vertical change, which is determined here by the support from the sea and the river's estuary elongation.

With large sediment loads from the main river, the impact of the outflow from the tributaries and the increased transport capacity of the river below its confluence with them is obviously not apparent. On the other hand, downstream of the rivers there is a change in plunging of accumulations. In the Northern Dvina, this occurs from the confluence with Vaga in the Ob (in its natural state, to the creation of the Novosibirsk reservoir)—from the city Kamen-na-Oby, covering all the middle and lower reaches of the river, and in the Amur—from the river's exit from the mountainous structures of the Maly Khingan to the limits of the Middle Amur lowland.

Sharp changes in the slope of the river from the mountains to the plain are typical for the Ob in the upper reaches. However, this possible cause of accumulation is veiled by age-old fluctuations in water availability, which lead to a shift in the boundaries of the vertical change sign change along the river: downstream in wet periods, upward in relatively dry periods (Makkaveev and Chalov 1970). At present, the upper Ob is crashing: the Fominskoye level of H at the post in 1953–1986 fell by almost 0.5 m, in Barnaul in 1936–1986—by almost 0.6 m (see Fig. 5.3B), i.e. the incision intensity was 1.5 and 1.8 cm/year, respectively. The high values of ΔH (at the Fominskoye post, 4.5 times higher than Δh, the plunging layer determined by the sediment load balance) are related to systematic dredging in the Fominskiy section of the Upper Ob and the development of channel quarries for construction materials in the area of Barnaul.

According to the Fominskoe post, until 1976, the value of ΔH = 1.1 cm/year; in subsequent years, it increased to 2.5 cm/year, which coincided with the growth of dredging at crossroads and radical correction of the channel. At the station of Barnaul until 1966, ΔH = 1.3 cm/year, in 1966–1976—3.0 cm/year (at this time there was active sand mining and dredging); in the following years, the decline in levels ceased, as regressive accumulation from the Novosibirsk reservoir reached here (Berkovich et al. 1990).

Below the city Kamen-na-Obi (before the Novosibirsk reservoir was established in the late 1950s), there was a directional accumulation of sediment. After the construction of the hydroelectric power station in the tailrace, deep erosion is developing, which represented itself in the decreasing of levels; its value (for 50 years at the Novosibirsk post) was more than 130 cm (2.5 cm/year). This process spread transgressively, leading to a change in the accumulation of erosion in the middle reaches of the river (Kolpashevo).

The crossing of rivers in the upper and middle reaches is usually replaced in the lower reaches, but at different distances from the estuary, up to the middle reaches, by the accumulation of sediment. At the same time, geomorphological conditions can lead to the development of river incisions up to the delta. This is the middle and lower Lena, the channel of which cuts into both the Central Yakutsk lowland (Tabaga settlement) and the Lena "pipe"—the river section before reaching the delta located between the mountain massifs (Kyusyur village), although the rate of lowering the levels is less than higher than in the current (Fig. 5.19).

The plunging rate changes along the length of the rivers. In the upper reaches of the Neman, the stepped floodplain is accompanied by a wide floodplain channel, which is cut in, according to the curves $Q = f(H)$ for the Stolbtsy station (Fig. 5.20A) at a rate of about 1 cm per year, as a result of which the water levels between 1960 and 1987 decreased by 20–25 cm. This occurred at a time when mass drainage was being carried out in the Upper Neman basin and their impact on river runoff led to some intensification of deep erosion. Downstream (Belitsa village, Fig. 5.20B), this trend is much less distinct, and due to constant oscillations of the curve it is impossible to quantify it. The position of the curve is probably greatly influenced by the movement of large ripples in the channel, which are characteristic of this section of the channel and cause the fluctuations of the bottom marks in the system of pool transition in 2–2.5 m.

The combination of longitudinal profiles in the river section at the beginning of its middle course (directly below the mouth of the Shchary River, where the channel still retains a wide floodplain character) has shown that over the last

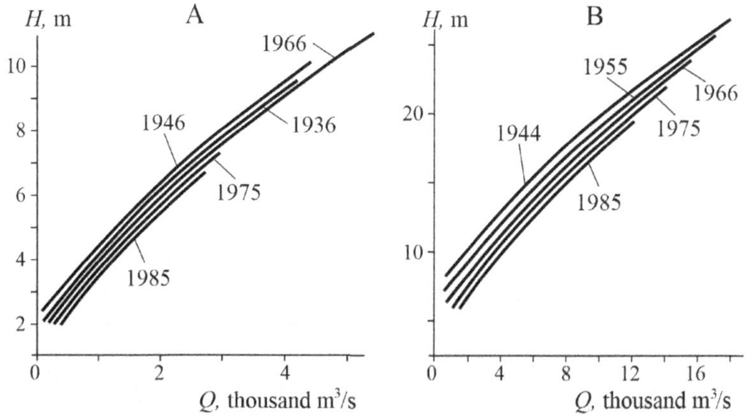

Fig. 5.19 Connection curves of water discharge and water levels Q = *f*(H) for hydrological stations on Lena river: **A**—Tabaga village (middle stream); **B**—Kyusur village (downstream)

50 years (1939–1989) the bottom marks have decreased from 10 cm to 1 m. This phenomenon is even more clearly observed in the incised channel near the town of Mosty (Mostovsky narrowness) (see Fig. 5.1B). In the incised channel of the middle course of the river (at the intersection of the Grodno Upland and the Baltic Ridge), the predominance of vertical changes of the negative sign is indicated by the incising nature of the channel itself. The data of M. J. Byatsonis also testify to the deep erosion of the middle Neman in the Lithuanian section of the river (Atlas of the Lithuanian SSR 1981). However, at Druskininkai post in 1956–1988 the curve $Q = f(H)$ is stable. Apparently, in such a short period of time (not more than 30 years) the geological structure of the channel, limiting the rate of incision, did not allow it to appear. At the same time, upstream (Grodno), the curve $Q = f(H)$ (Fig. 5.20B) shows a clearly marked decrease. During the period 1925–1987, the water levels between the two decreased by 0.5 m. Two stages with different rates of decreasing of levels are distinguished: before and after 1956. During the first 30 years, when the site was in a natural state, the decrease of levels was 10 cm (the rate of incision of the river is about 3 mm/year). After 1956 the district of Grodno became the arena for mass development of quarries of construction materials and construction of correction structures along the ship's route. As a result, the remaining 40 cm of reduction of levels fall on these 30 years.

Thus, in the natural conditions in the upper and middle reaches of the Neman riverbed is incise, and the intensity of the incision processes varies along the length of the river from 1 cm per year in the upper reaches to 3 mm per year on average, where this process is limited to geological and geomorphological conditions.

In the lower reaches of the Neman River, directional accumulation of sediment was developing in natural conditions, but the construction of the Kaunas hydroelectric power plant with a reservoir radically changed the direction of vertical changes. Deep erosion began to develop in the lower tailrace of the hydroelectric power plant, which together with the development of channel quarries led to a

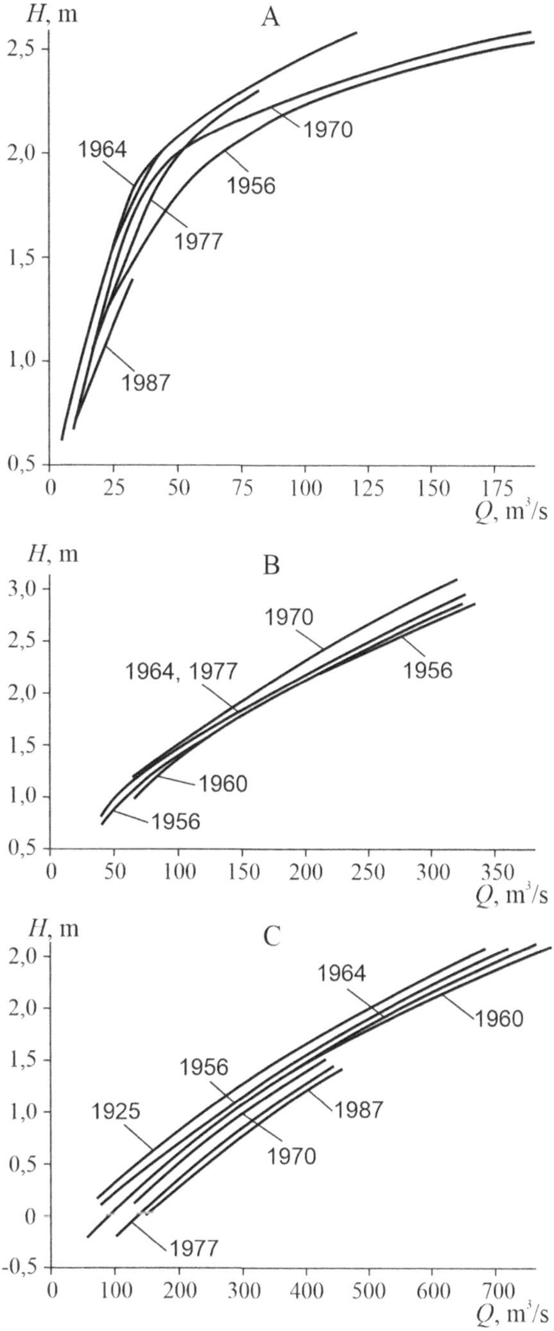

Fig. 5.20 Connection curves of water discharge Q and water levels H for hydrological stations on Neman river: **A**—Stolbtsy village, **B**—Belitsa village, **C**—Grodno city

160 cm drop in levels in the embankment section. The bottom marks at the area near dam were stabilized due to the formation of the scaffold (Malinauskas and Zdankus 1998). Below the mouth of the Nyaris River, the influence of the hydroelectric power plant is superimposed and coincides in the direction of vertical channel changes with the correction of the channel, dredging works on the rolled products and development of channel quarries. As a result, despite the stabilization of the bottom upstream, "leveling" continues here, but at the same time reducing the depth of the river.

It is believed that even in mountainous conditions the speed of river intakes is also relatively low, although it is higher than on flat rivers; nevertheless, the intakes are practically not felt for several generations of people and do not determine the service life of the construction on the river. Nikonov (1973) collected a large amount of data on the incision velocities of rivers in Central Asia, the Caucasus, the Carpathians, the Mongolian Plateau and other mountainous regions. It has shown that in the areas of modern active mountain formation the river incision speed does not exceed 0.5–7.0 mm/year. This is also confirmed by other researchers. For example, in California, in the junction zone of the three mobile lithospheric plates —Pacific, North American and Juan de Fuco—the tectonic elevation of the territory in some places is 2.5–3.0 mm/year, but the rivers crashed into the bedrock—shale in the Holocene at an average rate of 0.7–1.8 mm/year. In such actively rising mountains as the Himalayas, the incision velocity of the Indus River in the upper reaches is on average 2–12 mm/year in 7–60 thousand years; the maximum velocities are confined to the areas of the greatest tectonic changes (Burbank 1996; Howerd 1996). The current infiltration of Caucasian rivers within the mountainous area is clearly defined by the water discharge and level curve. At the Mzymta river (Western Caucasus), the decrease in absolute levels at the same flow rate in the area of the Ah-Tsu gorge corresponds to an annual deepening of the channel by 5.5 mm (Kozlovsky 1951). Nikonov (1973) estimates the plunging speed of the Greater Caucasus rivers at 0.5–1.0 mm/year. At the present stage, the rate of elevation is estimated to be more than 10 mm/year in the axial zone of the Greater Caucasus and about 4 mm/year in the Adjara-Trialets region (Lilienberg et al. 1972). This difference in the rate of tectonic rises and incisions of rivers determines the upward development of the relief and the overall increase in river gradients.

The crossing of rivers leads to the formation of deep mountain valleys, canyons, gorges, the edges of which are composed of practically non-destructible rocks, and the deep erosion itself occurs mainly in rocky soils. This is usually explained by the establishment of river valleys along fault lines or by the tectonic fracturing of rocks. In this case, however, debris that far exceeds its size limit for its movement by water flow enters the channel. Therefore, in addition to taking into account the tectonic factor, the role of chemical destruction of rocks by flowing water, as well as other weathering agents represented in subaquatic conditions, should be taken into account. The first factor was studied by Kochetov (1976), who established the average values of the flow modules of dissolved substances in the region of low mountains—33, middle mountains—105 and high mountains—55 m^3/km^2-years on the rivers of the Western Caucasus. These data are integral to both river erosion and erosion processes in the catchment area. Information about biochemical preparation of rocks for river flow erosion is provided by Dobrovolskaya (1980). It found that the effectiveness of biochemical weathering of erupted rocks is 100–1000 times greater than that of limestone. Due to the action of the constant water flow, the products of biochemical destruction are immediately washed away from the surface of the rock, which, thus, is constantly exposed and subjected to the activity of microorganisms. As a result of such peculiarities of preparation of the river bed for erosion in the mountains, where rocky rocks are prevalent, the main type of river erosion is deep. It also has a corrosive effect on the root floor of the debris transported by the stream. According to observations by Khmeleva et al. (1970), in some years, the Anahomsta River's (Western Caucasus) step-waterfall channel incisions into rocky limestone rocks at the passage of effective discharge (their water availability here is about 1%) reaches several centimeters, but in this case its value is determined by the intensity of weathering and, consequently, the preliminary preparation of the rock for erosion.

At such rates of change of river bottom marks (more than 1 cm—a few cm a year) incision can already affect functioning of economic objects and hydraulic engineering constructions on the rivers which service life usually is 50–100 years (decrease of marks will make at least 0.5–1 m). However, it is considered that intensive river intakes, which are significantly, by an order of magnitude greater than generally accepted, are rare and are usually associated with specific, extreme conditions of mountain river development. The existence of this opinion is explained by the poor development of the bottoms of mountain river valleys and the lack of the need to take into account the increased (in relation to the well-known) speeds of river intakes. Thus, Khakimov (1992) on changes of curves of water discharge and levels at hydrological stations for several decades established that the rivers of the Western Tien Shan (Pskem, Ugam, Chatkal, Akbulak, Ahangaran, etc.) are cut in with a speed of 1–7 cm/year (see Fig. 5.3A). The highest velocities are confined to narrow gorges, where the rivers are cramped by rocks (Fig. 5.21), and their channels are devoid of pebble-boulder alluvial cover. Due to the high current velocities during floods (up to 5 m/s and more), the load transits these river sections. The flow is in direct contact with the rocky bed, and the pebbles and boulders it moves corrode it, destroying the rock and speeding up the river's plunging. The same plunging speeds (5 cm/year) were established by Makhinov and Makhinova (2006) for the mountain rivers of Okhotsk region (Dzhugjur Ridge), linking them, among other things, with the corrosive effect of exposed sediments on the bedrock.

The Alabuga River, the left tributary of the Naryn River, is an even more striking example of incision, leading to catastrophic consequences during the development of the valley (Sidorchuk and Chalov 1996). A series of water intake stations for irrigation of agricultural lands in the river valley were designed here in the 70s. The project has been designed with the standard assumption that the riverbed marks are relatively constant on average over a long period of time. However, the operation of the Dzhergitalskaya pumping station, located 40 km from the river mouth, showed that over the past 10 years (1983–1992) without significant changes in external conditions the river hit 6.0 m. The main water intake structure dried up and had to

Fig. 5.21 Rocky channel oh the Chatkal river (Western Tien Shan). Photo of the author

Fig. 5.22 Bridge on the Alabuga river (Central Tien Shan), destroyed after intensive river incision. Photo of the author

be rebuilt twice in order to maintain the operation of the pumping station and supply the irrigated land with water. Above the water intake there is a road bridge. Its supports had to be constantly fortified due to intensive incision of the river, but the bridge was destroyed twice (since 1960) by the river (Fig. 5.22).

The Alabuga River, 180 km long in its lower reaches, runs along the Alabuga Depression for 85 km, crossing the spurs of the Western Ak-Shiyryak Ridge before flowing into Naryn. The bed material load are boulder-pebble, with an alluvium thickness of less than 1.5–2.0 m, which is less than the depth of the pool cracks; bed exits of bedrock are frequent, which indicates a shortage of bed material load; the narrow, stepped floodplain near the convex banks of the incised bend and even side-bars.

Within the limits of the Alabuga depression, the Alabuga valley is incised into the Neogene-Quaternary siltstones and the ancient pebble-boulder alluvium, which composes a relatively flat and wide main terrace. A modern canyon of Alabuga, 0.7–1.0 km wide and 50–130 m deep, is incised into this section; it has up to 6–8 local basement terraces covered with low-power (0.5–1.5 m) pebble-boulder alluvium (Fig. 5.23).

According to historical and archaeological evidence, up to the end of the 14th century there was no canyon-like incision of Alabuga, and on the surface of the main terrace, suitable for agriculture and nomadic cattle breeding, there was a branch of the Great Trade Route, which connected China with the countries of Europe and Asia Minor, located on different banks of the fortress of Chaldyvar and Shirdarbek, which served as caravanserais (History of the Kyrgyz SSR 1984). The riverbed was located at the level of the modern 19–22-m terrace, was melted and quite shallow, suitable for people and animal caravans for wading: it is impossible to cross the river with modern steep sides of the canyon and velocities of the current even in the lowlands of at least 2.5–3 m/s.

The time of formation of terraces up to 25 m high is determined by the thickness of the crust of desert tanning on pebbles (Panin et al. 1990)—up to 1300 years of terraces. The age of the floodplain and low terraces on the basis of the

Fig. 5.23 Incised bend of the Alabuga river (Central Tien Shan, tributary of the Naryn river) with the terrace cascade in its neck. Photo of the author

dendrochronological method is determined from 6 to 7 to 100 years. Fragments of the high floodplain (1988 data) are shown on large-scale maps of 1973 as riverbed shoals. Their relative excess over the current shallows and, accordingly, the size of the river incision for 15 years was determined by the levelling. These data and instrumental measurements of the process of lowering the bottom of the Alabuga river near the Djergital pumping station, carried out by specialists of the Institute "Kyrgyzgiprovodkhoz" in 1977–1982 (see Fig. 5.1A), showed that the maximum speed of river incision is 32 cm/year. At the confluence with Naryn, the incision speed is 0.4–0.9 cm/year, which obviously indicates that the channel of the main river has also been cut in. But these speeds already belong to the category of "normal", typical of most mountain rivers.

The Alabuga River within the Alabuga Depression is unique in many respects. The combination of mountainous terrain, large slopes of the river with easily eroded bedrock composing the valley, causes very high (catastrophic for the construction of engineering facilities) average speed of river incision, which apparently began only in the XV century. Before that, the Alabuga riverbed was located within the limits of the ancient alluvium thickness and had a small slope, but before entering Naryn it crossed the Ak-Shiyryak ridge, which created a step in the longitudinal profile, the steepness of which increased by an order of magnitude. The analogue of the Alabuga Valley at this stage is the modern valley of the Jamandavan River, a 64-km long tributary of the Alabuga River, which flows to the right of the mouth of the river 15 km. Its bottom is on the same level as the main terrace of Alabuga. Before flowing into it Jamandavan, it has an incised channel with a high slope (40‰).

It is obvious that at the turn of the XIV–XV centuries the stratum of bedrocks in the Ak-Shiyryak Ridge was cut through by the Alabuga River, and the river began to erode the alevrolites that were susceptible to erosion. As a result, a deep canyon was formed and the plunging speeds increased by 1–2 orders of magnitude. This made it impossible to continue using the valley, which was abandoned by man. Apparently, the same fate awaits the river Jamandavan, when she, like Alabuga, cuts through a layer of bedrocks.

Thus, in the conditions of mountainous terrain, the rapid incision of riverbeds occurs with a deficit of bed material load and low erosion resistance of the soils composing the river valley. The main signs of catastrophic plunging are the presence of main plinths in the side and a separated areas, shallow steps and basement of the floodplain and low terraces, good preservation of the riverbed relief on the terraces, canyon-like character of the valley and relative leveling of the surface of the main terrace into which the canyon is cut. Such canyons are widespread in the Tien Shan, Pamir-Alay and Altai mountains. In the foothills of the Fergana Valley, many rivers in the Syrdarya basin are incised into a flat surface formed by sand and pebble alluvium, forming canyons with vertical sides, narrow bottom and channel. In the valley of the Pskem River (Western Tien Shan), the channel is cut into Neogene "stone" clays, which form a base of high terraces in a narrow canyon, incised in a wide terraced surface, where settlements, highways, agricultural land are located. All the tributaries of the Pskem, which do not have time to follow the incision of the main river, have hanging estuaries, falling off with waterfalls formed on "stone clays". The Pskem channel occupies the bottom of the canyon, is free of sediment and is a tray made of clay. Similarly, the Katun River valley in the Altai above the Katun HPP and in the area of confluence with the Chui River, as well as the Chui River in the lower reaches.

Observational data and the structures of river valleys in the mountains indicate significant rates of river intakes in certain areas that significantly exceed the average characteristics of large regions as a whole. However, in all cases, the reasons for plunging may vary. But they are united by the representation of intensive plunging in relatively short sections (or on tributaries of large mountain rivers like the Alabuga) relative to the total length of the river system. For example, at the Katuni River in the area of the Chui estuary, intensive incision is obviously connected with the lowering of a large mountain lake, which is heavily loaded with sediment, when the threshold is sawn by a stream of water in the Aigulak Ridge, which washes the Ininsk Basin from below. It is also clear that the intensive incision phase at Katun has already ended and that modern deep erosion is occurring at a rate typical of most rivers in the region. Such mechanism of intensive river incision as a consequence of local tectonic elevation was studied on laboratory model by Makkaveev et al. (1961). According to them, if the rate of elevation is higher than the rate of incision or within it the river opens up an array of bedrocks, in the longitudinal profile there is a clear bend, or even there are thresholds and waterfalls. Under these conditions, the rising structure can support the river. At the same time, the slope below the structure axis (and the formation of bending) increases, and the erosion capacity of the flow increases. As a result, the river's infeed is intensified and the flow, if it does not meet bedrocks, quickly aligns the longitudinal profile.

The noted high incision speeds at the intersections of mountain ranges (Hakimov 1992) in comparison with the averaged parameters are due to the narrowing of the flow, its direct contact with the rocky bed and the corrosive effect of the transported pebble-boulder deposits. When such a river crosses the river, the energy of the flow is mainly used to scour the bottom. As a result, the valley and the channel are narrowing, which increases the specific effective discharge, which contributes to the strengthening of river incision.

The intersection of rivers, resulting in the transformation of their longitudinal profiles, can be caused by various reasons, which are combined into three main groups: (1) climate change and other physical-geographical conditions (vegetation cover, etc.), causing increase in water availability in rivers and increasing the degree of unevenness of runoff; (2) tectonic movements, causing distortions of the earth's surface; (3) decrease in the basis of erosion (the level of receiving water body) or changes in its position due to abrasion of the bank at the river mouth.

In all cases of vertical change, the law of factor relativity is more or less clear. Makkaveev (1955) has shown that in the upper links of the network erosion significantly changes its intensity with climate fluctuations, and in the lower links with tectonic movements. In general, channel flows with small catchments quickly and significantly change their regime even with relatively small changes in natural conditions in the catchment area. Large rivers with basins covering different landscape complexes are less susceptible to climate variability. At the same time, the erosion capacity of the flows in the lower reaches of the network (large rivers) reacts very sensitively even to minor changes in the slope; on the contrary, in the upper reaches of the network, the erosion intensity can only change significantly with large increments in the slope. The distribution of slopes along the length of the river with the longitudinal profile generated is consistent with the relationship (5.20, 5.21). Therefore, if the slope of the catchment surface increases by the same absolute value, the additional slope (and, consequently, the increase in erosion intensity) may be negligible for the flows in the upper reaches of the system and very significant for its lower reaches. This is particularly evident in the case of erosion baseline fluctuations, the impact of which on erosion and accumulation processes is, in most cases, limited to the lower reaches of the rivers.

The propagation of longitudinal profile changes from their initial representations along the channel system takes some time. The rate of change propagation is the higher the more flexible the soil is with respect to erosion, the more runoff rates of the flow and the greater the change itself. In some cases, there is a significant delay in the propagation of changes of the longitudinal profile, due to which the age of the resulting terrace can be different in different parts of the system.

Deep erosion spreads from the site of excitation regressively (upstream) or transgressively (towards the river flow), expresses itself simultaneously along the entire length of the channel system or develops in a limited area of the system. Regressive erosion is most often observed, for example, when the erosion base drops. In this case, the local increase in slope in the lower reaches of the river (caused by the retreat of the sea) disperses, spreading, gradually attenuating, to a more or less significant distance up the river. Transgressive erosion is common in cases where the transformation of the longitudinal profile of the river is caused by climatic causes. It is clearly expressed and has high rates of transgressive deep erosion below reservoirs.

Deep erosion can occur along the entire length of the channel system simultaneously with general changes in the water availability of the river and with tectonic movements causing a general change in the slope of the earth's surface. In local tectonic movements, deep erosion is limited only to the area where the river crosses the growing structure, and here incisions occur at higher velocities than in the background.

Since the transport capacity of the flow depends on the size of the flow, its changes resulting from general climate fluctuations (precipitation, evaporation), natural changes in the landscape in catchments or due to human activity (collector water discharge for irrigation, inter-basin redistribution of water resources, etc.), have a significant impact on the longitudinal profiles of the rivers. At the same time, the transport capacity of the river depends not only on the flow rate, but also on its regime. Therefore, one should take into account the possibility of influencing the longitudinal profile of changes in the flow regime towards its greater unevenness in the seasonal and perennial cross-section. Such a change may be due, for example, to the destruction of forests, which always causes an increase in the seasonal irregularity of flow, which is reflected in an increase in the capacity of floods and a decrease in the intermittent flow of water. Artificial changes in flow regime are also caused by the change of methods of agricultural machinery and regulation of flow by reservoirs.

In nature, all factors are interconnected and do not always clearly influence the development of rivers. For example, an increase in runoff caused by increased climate humidity is accompanied by a decrease in the concentration of sediment in the river water due to an increase in the density of the vegetation cover in the catchment area, etc. At the same time, the same change in the natural environment can lead to a contradictory impact of some of its components on fluvial processes: climate humidification is accompanied by an increase in runoff and, consequently, deep erosion on the rivers; the same factor, due to the development of vegetation, reduces the degree of irregularity of runoff and, thus, reduces the erosion ability of the flow. Therefore, in each case it is always important to determine which of the flow indicators was decisive in the current general vertical changes and transformations of the longitudinal profile of the river.

If the runoff layer increases throughout the catchment area and the effective discharge increases, the river deepens its channel. Since the transport capacity of the flow increases more than the flow rate of water (2.19), the plunging intensity increases downstream. However, in the lower reaches, the relative increase in water discharge is insignificant and decreases with the growth of water availability in the river and its order. Therefore, the intensity of the incision of the river here gradually diminishes. In addition, in the lower reaches of the river, the incision value is limited to the constant position of the erosion base, as well as in the upper reaches, the position of the river source is unchanged. As a result, the maximum intensity of incision is in the middle reaches of the river, closer to its upper reaches. As a result of plunging, the slope in the middle and lower reaches, based on the formula (5.16), is adjusted according to the relationship (Makkaveev 1971):

$$\frac{I_i}{I_{i+1}} = \left(\frac{Q_i}{Q_{i+1}}\right)^{2/3}. \tag{5.19}$$

In the upper reaches of the river, on the contrary, the slope of the river increases. The larger the runoff layer, the deeper the "deflection" of the longitudinal profile curve and the greater the value of river incision in the middle course of the river (Fig. 5.24A).

On Fig. 5.24, B shows the combined longitudinal profiles of the Oka and Don rivers with close basin areas. The sources and mouths of both rivers are located at approximately the same absolute levels. Eye flows through the forest steppe and the south of the forest zone, the average flow module from its basin is 5 l/s km^2, while the average flow module from the steppe basin of the Don is only 2.1 l/s km^2. The average slope of the compared rivers is almost the same; for the Don it is 0.096‰ and for the Oka it is 0.11‰. The degree of unevenness in the flow of both rivers is approximately the same: more than 50% of the annual volume of flow takes place during high but short-term floods. The comparison of profiles shows that the

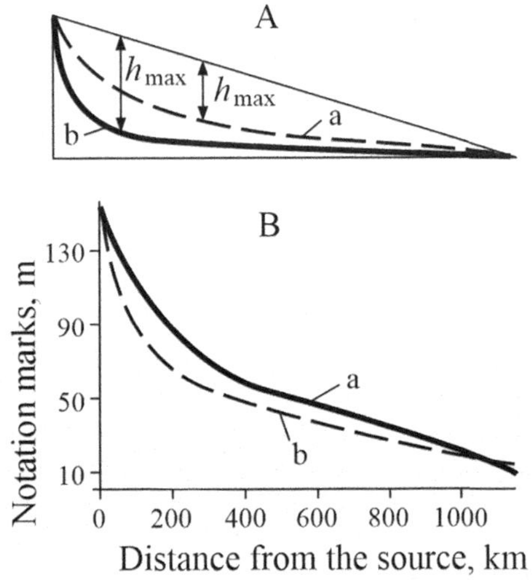

Fig. 5.24 Changes of the longitudinal river profile due to channel incision at increasing water content. **A**—theoretical scheme (a—longitudinal profile at low flow rate; b—longitudinal profile at high flow rate; h_{max}—deflection value); **B**—combination of longitudinal profiles of Don river (a) with small flow module and Oka river, (b) with large flow module (according to Makkaveev (1955)

longitudinal profile curve of the more high-water Oka is characterized by a relatively higher steepness in the upper and lower steepness in the middle reaches.

Deep erosion caused by increased river water availability can occur along the entire channel system simultaneously, if climate change is affecting the entire basin, as well as transgressively. The latter case is possible on those rivers that cross a number of natural zones, and climate change only covers the upper parts of them, gradually fading downstream. At the same time, the longitudinal slopes in the middle and lower reaches of the river are reduced during the plunging process. This results in a reduction in the particle size of bed material load, which in time can be an indicator for determining the causes of river incision.

When sediment load decreases (as a result of natural afforestation of the catchment area or "interception" by the reservoir), the channel is incised into the riverbed. In natural conditions, the maximum incision is observed in the middle and partially upper reaches of the river, gradually weakening both upstream and downstream. In the latter case, the decrease in the intensity of the incision is explained by the fact that in the lower reaches of the river a significant amount of energy is spent on the transport of sediment from the upstream sections of the river, where deep erosion is more intense. Thus, a decrease in the concentration of load in the river flow, as well as an increase in the runoff rates of the river or an increase in the unevenness of its flow, results in a predominant cut of the riverbed in the middle reaches, with a maximum at the point of deflection of the longitudinal profile curve.

Below the reservoirs, conditions occur due to the interception of sediment load $W_{tr} >>> W$; riverbed incision spreading along the length of the river is transgressive and fades out as the ratio Wtr~W is restored. The longitudinal profile becomes weakly convex, as the slope downstream increases as the energy consumption of the flow for the transport of sediments coming from the erosion of the riverbed increases. However, the erosion zones in the tail-water of the hydrosystems are characterized by an increase in the size of bed material load, since here, in the process of erosion, the basal horizon of alluvium is opened up, and due to the absence of top-down sediments, hydraulic sorting is not expressed in accordance with the change in slope.

River crossings in conditions of rising relief development are especially characteristic of mountainous regions, where the elevation of the territory is so great that the tectonic factor is the main factor in the transformation of longitudinal river profiles. Rivers originating in the mountains fall into the foothills and plains, where the speed of raising the area by an order of magnitude less or the elevation is replaced by a lowering. Therefore, within the limits of mountainous regions, almost all of the rivers flowing here, their channels cut into intensively; in the foothills and plains there is a complex alternation of areas with a continuing tendency to incision and areas of accumulation of sediment and increase in the bottom of the valleys, associated with changes in the length of the river transporting capacity of the flow, due to physical and geographical factors, the intersection of rivers of different in respect of the sign of tectonic movements of morphostructures, etc.

On rivers flowing entirely in flat conditions, the general rise of the territory creates a background on which vertical changes caused by changing climatic conditions occur. The tectonic rise of the continents in the river valleys has led to the creation of terrace staircases, although there have been numerous changes in the course of their geological history in the plunging of rivers by the systematic accumulation of sediment, which has resulted in the accumulation of powerful alluvial strata. In the subsequent phase of plunging, associated with increased water availability, uneven river flow or reduced sediment load, it occurs when the river's source rises above its erosion base, which increases the effect of deepening the valley. On the contrary, the same circumstance limits its representation in the phase of accumulation caused by climatic reasons. As a result, river valleys are not filled with alluvium.

In mountainous regions, under the conditions of general upward development of the relief, the infiltration of conserved rivers originating on the slopes of watershed ripples, corresponding to the axis of elevation, and crossing the mountainous area across the elevation in the direction of a decrease in its gradients, occurs due to the gradual, but quite intense increase in the slopes of the earth's surface on the wings of the elevation and the average slopes of the rivers. This determines the forms of representation of deep river erosion, which occurs when the territory is unevenly elevated. The upstream slope of the river is very steep, and its relative increment is only a few percent and cannot significantly affect the increase in transport and erosion capacity of the flow. In addition, the low runoff rates of the river headwaters causes less intensity of deep erosion compared to the average and downstream (before leaving the mountains) flow of the river, where high runoff rates and the increase in the energy of the river as a result of the rise create the preconditions for a relatively rapid incision of the river. However, the river crossing is limited either by the position of the erosion base, which is constant or changes by a small value and comparison with the total amplitude of the elevation, the change of the tectonic sign, or the river's exit to the sub lateral plain territory. Thus, in the source and on the border "mountain-plane" each river has "geometric" conditions limiting its incision, as well as the development of this process with increasing water availability.

Conflicting interaction of these factors leads to the fact that the maximum intensity of incision during the rise of the territory and, consequently, the deepening of the valley takes place in the middle course of the conserved river, where the relative increase in slopes is quite significant, the runoff rates is sufficient for a relatively rapid representation of erosion, and the "geometric" conditions created by the marks of the source and the basis of erosion (the place where the river leaves the mountains for the plain), allow the river to rather intensively cut into and form the deepest (relative to the line of the mountains) river valleys. At the same time, the growth of slopes above the rate of elevation of the territory over the inland watercourse leads to an increase in the flow rate (with the river's constant runoff rates) and a gradual intensification of the inland watercourse. These conditions are illustrated by the Fig. 5.25, which shows the profiles of some terraces of the r. Kodori (Western Caucasus), formed during the incision of the river during the rise of the mountainous country, and its modern channel. The maximum relative height of the plunging terraces is in the middle reaches, and the depth of the river plunging does not exceed 0.3–0.5 times the incremental value of the full river falls as a result of tectonic distortion of the basin surface. Therefore, the maximum depth of the river incision gives a significantly underestimated idea (2–3 times) of the amplitude of the uplift. The consequence of this is that the high terraces, formed at relatively low river gradients, are characterized by lower alluvium size than the lower terraces. Alluvium has the largest particle size in the floodplain and modern bed. The inverse relationship between the relative height of terraces in the valleys of the rivers of the Western Caucasus and the size of the boulder fraction in the basal layer

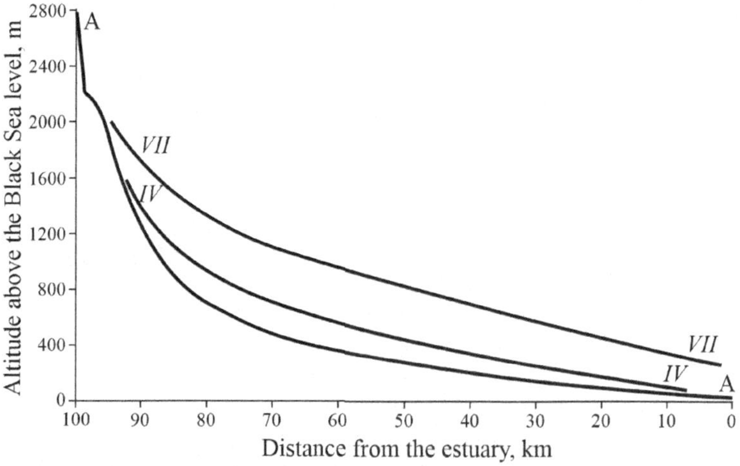

Fig. 5.25 Schematic longitudinal profile of the Kodori river (Western Caucasus) and its terraces: A—A modern channel; IV—fourth and VII—seventh terraces

of terrace alluvium for the first time drew the attention of Mikhailovskaya (1934). Obviously, it can be used as a sign that defines the tectonic nature of river intakes and terraces formation in conditions of upward topography development.

As a result of intensive incision of conserved mountain rivers, the presence of hanging mouths of their tributaries or a significant increase in the slopes of their beds before flowing into the main river (Table 5.6). The latter is due to the fact that most of the tributaries are oriented parallel to the uplift axis, and, consequently, the rise of the mountainous country does not have a direct impact on the fluvial processes; the transformation of their longitudinal profiles takes place only due to regressive erosion, spreading regressively up the tributary, starting from the main river bed (Makkaveev 1955). Lower runoff rates of the inflows and, consequently, lower flow rates and their erosion capacity also contribute to delays in plunging inflows.

In natural conditions, even some parts of the main river sometimes deviate from the normal to the uplift axis, and tributaries (and tributaries of tributaries) are oriented at different angles, up to the opposite one. Therefore, elevation causes very diverse changes in the transport and erosion capacity of flows in different parts of the basin and along individual rivers. Makkaveev (1955, 1971) and Makkaveev and Chalov (1986) for purely geometric reasons, obtained that the increase in the slope of the inflow of ΔIpr due to the elevation of the territory depends on the magnitude of the angle α between it and the direction of the main conserved river:

$$\Delta I_{trib} = \Delta I_{main} \cos \alpha, \tag{5.20}$$

where ΔI_{main} is the increase in the slope of the main river. The increment of the slope is zero for lines parallel to the uplift axis ($\alpha = 90°$) and negative for α greater than 90°.

The result of different ratios in the orientation of the main river and tributaries in relation to the axis of elevation is the restructuring of the longitudinal profile of each river, as some of the tributaries will strengthen its removal, others will weaken, and the slope of even the main river experiences different gradients in different

Table 5.6 Slopes of the estuarine sites of tributaries of some rivers of the Western Caucasus

Tributary	Orientation to the lifting axis	Main river	Slope, ‰	
			In middle current	Before estuary
Zheopse	In parallel	Sandripsh	20.0	23.2
Yupshara	Normally	Gega	7.5	30.1
Eastern Gumista	In parallel	West Gumista	28.3	34.9
Klych	Normally	Gvangdra	59.1	68.3
Bramba	In parallel	Kodori	31.6	54.5
Baramidze-Tskali	In parallel	Supsa	61.2	99.6
Skhalta	In parallel	Ajaris-Tskali	19.4	22.8

areas. As a result, local erosion basics and longitudinal profile steps in the form of scarps and lake extensions should be formed. The more intensive the tectonic movements and the more complicated the geological structure of the territory, the more significant the changes in the longitudinal profile will be. Some of the tributaries that do not receive a direct incremental slope are only cut in as a result of a reduction in the local erosion base (the receiving riverbed), while others may even be blocked by outbursts from their tributaries if the latter receive a positive incremental slope and the receiving river has a negative incremental slope. All this can also affect the profiles of terraces and lead to the formation of curves of their levels, which creates the false impression of the presence of local tectonic movements of different signs.

A river parallel to the lift axis should form a stepped longitudinal profile. Inflows from one of its banks, receiving a positive incremental slope, will increase the outflow of coarse debris into the main river; opposite inflows, on the contrary, receiving a negative incremental slope, will weaken the incision and outflow of material. This uneven elevation results in an alternation of steep and gentle longitudinal sections.

River valleys, and mountain valleys in particular, are characterized by alternation of narrowings and extensions, often caused by the peculiarities of lithology of rocks or tectonic structure. As a result, the stream periodically melts during the flood and collects again in a narrow gorge. As a result, the specific effective discharge is constantly changing. In constrictions, it increases, the transport and erosion capacity of the flow increases, and the river channel gets an additional impulse for incision. On the other hand, at the beginning of the expansion due to the flow spreading along the wide bottom of the valley at the effective discharge there is a curve of decline in water surface levels, and, consequently, there is a dilution and deepening of the channel. Only downstream, within the limits of expansion, the specific flow rate decreases and the intensity of the river's deep erosion decreases. All this is reflected in the form of a longitudinal profile, changes in the length of the river its slope and fluctuations in the power of the alluvial thickness.

The noted regularities of development of deep erosion are represented in a wide range of natural conditions. Even on such a large river as the Lena, the general uplift of the basin area is the main factor in the formation of the longitudinal profile, although in its middle course it leaves the middle mountains and flows along the territory of the Central Yakutsk Lowland for a considerable length of time (Borsuk and Chalov 1973). The analysis of the water level and discharge curves for the Lena hydrological stations shows a steady tendency to decrease the water level at the same flow rate, and the higher the flow rate, the more clearly this pattern appears. In the middle reaches of the Lena River, the intensity of the level decrease is about 0.5 mm/year. Modern plunging is reflected in the stepping of transverse profile of the floodplain, with the highest altitude being the oldest and farthest from the riverbed. The same is true of the transverse profile of the valley, with the excess of each successive terrace being greater than the previous one. Obviously, in the past, the plunging intensity was higher than it is now. If based on the results of determining the absolute age of a 40-m terrace (Kaplin et al. 1971), the plunging speed in the period since its formation, on average, is twice as high as that of the current channel, and at the first stages was at least 2 mm/year.

The position of the erosion base limits the processes of river intakes related to the general upwelling of the territory or increase in its runoff rates. However, the erosion basis itself is not constant, and its fluctuations determine the direction of vertical changes in the lower and sometimes middle reaches of rivers, where they propagate regressively.

When the base level is lowered, the slope of the surface above sea level due to the regression of the receiving basin may be greater or less than the slope of the downstream river. Therefore, different rivers react differently to the decrease in the erosion base, and the direction of development of vertical changes is influenced by the runoff rates of the flow. When the shelf has a slope greater than the average slope of the river downstream, the rivers are cut into any climatic conditions. The bigger the river, the faster and deeper its lower reaches run into it. Small rivers are "involved" in the erosion trenches of large rivers and become tributaries, increasing the runoff rates of the main river. Cutting spreads from the estuary regressively against the river flow. Thus, according to the law of factor relativity, speed of advance of front of regressive erosion and depth of plunging gradually decrease upwards on the river (Fig. 5.26), and plunging declines in that place where initial slope of a longitudinal profile is so significant that addition of its size caused by "wave" of regressive erosion, practically does not affect transporting and erosive ability of a stream. The average slope of the river becomes larger, which results in an increase in alluvium particle size compared to the pre-incision period. The rate of

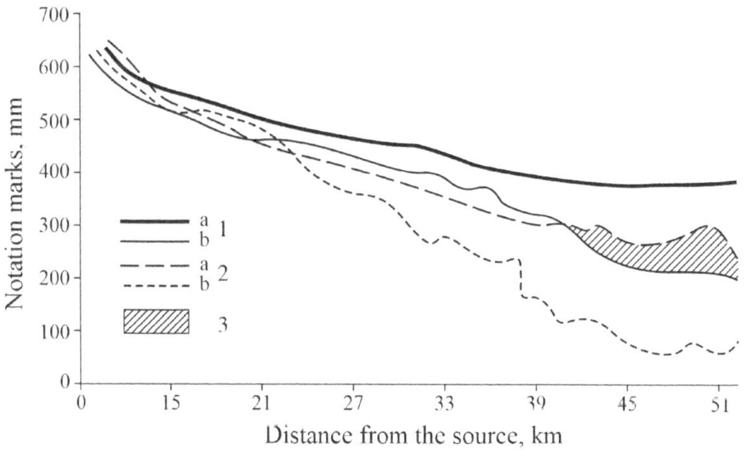

Fig. 5.26 Incision and transformation of longitudinal river profile downstream at base level lowering in steep slope conditions (regressive erosion): 1—water (a) and channel bottom (b) levels before base level lowering; 2—water (a) and channel bottom (b) levels in the next phases of incision; 3—sediment layer in delta, appeared in the first phase of incision (according to experimental data (Makkaveev et al. 1961)

erosion base decrease has a certain value. The bigger it is, the more likely it is that the river will be cut into the river even with small slopes of the newly formed channel on the trail.

Large rivers can also be crossed if the slope of the exposed shelf is slightly lower than the slope of the downstream river, as for them the river elongation is negligible compared to the size of the downstream river itself. Thus, the Volga, despite the shallow seashore, slightly crashed (by 25 cm during 1929–1955) with a decrease in the Caspian Sea level (by 2.2 m during the same period) (Makkaveev et al. 1958).

The depth and speed of plunging are also affected by the relative increase in flow resulting from the increase in the catchment area of the retreat of the sea and the conditions for the formation of the delta at the estuary. The higher the runoff increase, the more intensive the incision; the formation of the delta, on the contrary, slows down the incision and reduces its value.

The greater the average slope of the river, the shorter the section of the river (all other things being equal) to which the inset extends, caused by a decrease in the erosion base. On mountain rivers, terraces associated with fluctuations in the erosion base are seeped out at a shorter distance from the estuary than on flat ones.

Intensive incision of the Amu Darya channel in the lower reaches of the river resulted in a decrease in the level of the Aral Sea (Mikhailov 1997): during the period 1961–1979, the levels in 17 km from the sea (Temirbai station) decreased by 4.5 m, although the drop in sea level during the same period was 6.5 m. Differences in the size of the river infeed and sea level decrease are related to the influence of simultaneous river mouth elongation, which amounted to 17 km. The length of the plunging and lowering zone by 1980 was 50–60 km.

In the lower reaches of rivers with large multi-branch deltas, a number of side branches are dying out and the flow in the main, main branches is increasing (on the Amu Darya, the hydrographic network of the delta has practically degraded and the flow is concentrated in a single channel, the former Urdabai branch.

Intensive incision of the rivers flowing into Lake Sevan occurred with a sharp decline in its level in the 1950s (Ananyan 1960) and was recorded on the rivers flowing into lake Ilmen, and the river Volkhov as a consequence of the decrease in the level of Ilmen and Ladoga lakes (Bylinsky 1959).

The inbound lower reaches of the rivers of the Black Sea coast of the Caucasus are characterized by the transformation of high floodplain levels downstream into low floodplain terraces, i.e. the levels of young river terraces form an open fan downstream. According to research by Makkaveev et al. (1961), this shape of the longitudinal profile of the terraces indicates regressive deep erosion, which represents itself in the reduction of the erosion base. Apparently, river intakes in the lower reaches are connected with the periods of regression of the Black Sea, while the modern intakes are connected with the regression of the New Black Sea basin. This process was common to most of the Black Sea coast and apparently had a greater impact on the formation of river terraces than the negative tectonic movements of the tectonic sign, represented only in a narrow coastal strip.

In addition, for the most part, the Caucasian shores of the Black Sea are subject to intense abrasion, and the mouths of many rivers are located in the abrasion areas. Thus, the retreat of the sea coast near the mouth of the Gumista River is made with a speed of about 0.2 m/year (Zhivago 1952). Flushing the banks at the mouth of the river leads to a shortening of its channel length (Fig. 5.27), which in turn leads to a reduction in energy losses and, consequently, to an increase in the transport capacity of the flow. As a result, despite the tectonic lowering of the riverbed at the existing slope of 4.5‰ at the mouth of the river, the water surface slope is constantly increasing, which causes an annual deepening of the riverbed by 3 mm. Similar conditions prevailed at the estuary of the Galidzga River, where sea-bank erosion intensified after the construction of a breakwater that had intercepted longshore drift and reached a level of sediment in 1933–1948. 10–14 m/year (Zenkovich 1958). Shortening the length of the river by this value causes the slope to increase by 0,01‰ and deepening of the channel at the mouth by 5–6 mm per year. Thus, in the conditions of sinking foothills and transgression, the abrasion activity of the sea predetermined the preservation of the tendency to incision in the lower reaches of the rivers. River crossings in the lower reaches due to coastal erosion are typical of the rivers on the eastern slope of Sikhote Alin (these phenomena are described in V.K. Arseniev's book "Along the Ussurian Territory"), and the Kuril Islands up to the formation of waterfalls on abrasion cliffs (Korotkiy and Makarova 2006).

An analogue of coastal abrasion within the estuary is the straightening of the Terek Riverbed at its estuary with an artificial cut through the base of the Agrakhan Peninsula. Reduction of the estuary section length by 25 km caused a wave of regressive erosion, which spread regressively to 25–30 km (Alekseevskiy and Chalov 1997). This process was halted by the extension of the delta at the mouth of the slot and the beginning of the Caspian Sea level rise.

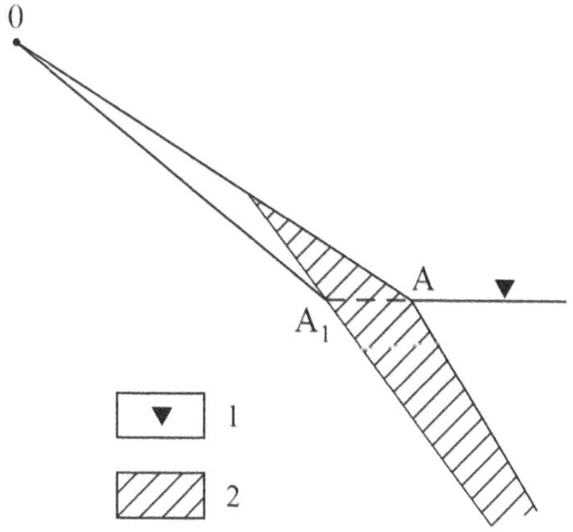

Fig. 5.27 River incision downstream at estuary location on the abrasion bank. 1—sea level; 2—abrasion section of the coastline. 0A and $0A_1$—position of the river profile before and after abrasion

5.4 Directional Accumulation of Sediment

Directed accumulation of sediment is a type of vertical channel change with a positive sign, representing itself in an increase in the marks on the bed of the riverbed along the length of the river. This applies to a wide range of phenomena—from the cessation of motion of individual particles (3.63) to the general transformation of the longitudinal profile, consisting in its movement to a higher hypsometric level. Essentially, accumulation is the transfer of sediment to channel sediments. Mass stoppage of particles means the accumulation of sediment at the bottom of the flow, which may re-emerge at increasing flow rates in the nearest flood or be involved in transport during the movement of ripples or channel shapes. All this together determines the temporal discreteness of the accumulation as a component of a single erosion-accumulation process. A prerequisite for the development of directional sediment accumulation and, consequently, the formation of channel sediments is the excess of total sediment load over the transport capacity of the $W > W_{tr}$. The intensity of the directed accumulation depends on the degree of severity of this inequality.

In general, the development of accumulation processes along a significant length of the river may be due to both tectonic movements of the negative sign, and climatic conditions, which are associated with high losses of runoff for evaporation and, consequently, a decrease in the transport capacity of the flow. An example of the latter is the Amu Darya downstream of Termez and to the estuary, which is characterized by a gradual decrease in flow and load downstream. Constant accumulation of sediment is characteristic of areas of high deflection, where it is compensatory in nature. It occurs in the foothill area because of a sharp decrease in slopes when rivers come out of the mountains to the plain, where it is favored by tectonic immersion in the area of edge deflections. In the lower reaches of the rivers, it develops as a result of fluctuations in the erosion base and the river's estuary elongation when the delta is extended to the sea. Local representations of directional accumulation on the rivers occur when they cross local negative tectonic structures, when there is a zone of support due to mountain collapses and above reservoirs. The cause of the disturbance of the balanced ratio W and W_{tr} may be the excessive inflow of sediment into the flow under changing natural conditions, including those caused by anthropogenic factors: deforestation, plowing of land and, consequently, the activation of soil erosion (the source of excessive inflow of sediment, primarily into the upper reaches of the river network) and the decrease in the runoff rates of rivers due to the lowering of groundwater levels.

All these factors overlap to some extent in many cases, causing extreme representation of the process. These are the rivers of China, which are distinguished by their huge sediment load and in their lower reaches are experiencing intensive directed accumulation (see Table 5.2). It is especially evident on the Huang He River, where its rates reach 7–9 cm/year and the channel rises above the surrounding territory by 3–10 m; in particular, the water level in the lowlands exceeds the level of the Great Chinese Plain, where the city Keifon is located (see

Fig. 5.12A), according to data for 1991 (Series of picture albums … 1991)—8 m, 2003. Li (2003)—at 13 m. A similar picture takes place in the lower reaches of the Amu Darya, Terek, Ili, Kuban, although with a smaller excess of the intertidal water level over the adjacent alluvial-delta plains.

The data in Table 5.2 are characterized not only by the unambiguous results obtained by different methods, but also by the close increases in bottom marks and levels (within the range of −35 +10%), which is characteristic of rivers with high sediment load. At the same time, at the Yangtze River section between the Loshan and Hankou hydrological posts, vertical changes with different sign and intensity were recorded along the curves $Q = f(H)$: at the first one—accumulation at a rate of 3.7 cm/year (Lianghu plain), at the second one—plunging at a rate of 1.7 cm/year (here the incised channel is formed at the crossing of mountain massifs). It appears that the plunging intensity (−0.6 cm/year) determined by the sediment balance between the two hydrological stations is underestimated, as the resulting balance includes the effects of two differently directed processes, which differ in intensity and cover different river sections. In the lower reaches of the Yangtze (the southern margin of the Great Chinese Plain), where there are no hydrological posts, the accumulation of channel change here is evidenced by its bunding, the widespread distribution of estuarine lakes on tributaries, and the constant threat of flooding (Muranov et al. 1959; Characteristics of the channel … 1985).

There are several reasons for the targeted accumulation in the lower reaches of the Huang He and Yangtze: sharp changes in gradients and, consequently, in flow rates at river outlet from mountainous areas to the plain, intensive river mouth elongation, mass water intake for irrigation (especially at the Yellow River), flow spreading over a wide floodplain during the flood period or insufficient transport capacity of the channel during the period of effective discharges when it is constrained by natural riverbed ramparts and artificial dams (to control floods). Modern tectonic sinking is also important, as it is characteristic of the lowlands of Eurasia adjacent to the Pacific Ocean. Thus, the accumulation of sediment in the lower reaches of the Yellow River and the Yangtze River is, to a certain extent, compensatory in nature.

In the lower reaches of the Huang He, the excess sediment transport rate in relation to the transporting capacity of the flow ($W >> Wtr$) is formed when the river crosses the middle reaches of the Loess Plateau (up to 60%), where soil washout reaches 3700 t/km^2 (Muranov 1957). At the same time, the accumulation of sediments is of a catastrophic nature, causing a constant threat of flooding due to the constant rise in water levels, the likelihood of overtopping of water through flood dams and their breach.

On Fig. 5.28 a photo-illustration is given, indicating an intensive accumulation of sediment on the river. The Weihei is the last tributary of the Huang He River, the confluence of which determines the boundary between the middle and lower reaches of the river (in Chinese scientific literature, the lower reaches of the river are usually considered to be from the Xiaoalandi hydroelectric power station, which is located near the river's outlet into the Great Plain of China); the bridge support indicates the levels of the bridge's position since the early 60s, with the lower mark

Fig. 5.28 Bridge downstream the Wei River (wright tributary of the Huang He river): on bridge support bridge levels are marked from the beginning of 1960s. Photo by Liu Shguang

corresponding to the level of the modern floodplain and riverbed banks. In 36 years, the bridge was rebuilt three times because of the accumulation of sediment and the increase in maximum water levels in the river, and eventually it was raised by 6 m. Thus, the accumulation rate here is 16.5 cm/year, which is twice as high as the accumulation rate at the lower reaches of the Huang He River (here the natural process is enhanced by regressive accumulation above the Sangmensya reservoir).

On Fig. 5.29 graphs of water level change at effective discharge of 3000 m^3/s for all hydrological posts of the lower reaches of the Huang He River for 1950–1993 are shown. The total increase in the level due to accumulation was about 2 m at the upstream downstream stations (Huayuankou and Jiayatan), and more than 3 m at the downstream stations (Liczin), where the levels rose by 2.5 m (alluvium accumulation rates, according to sediment balance data, increase from top to bottom from 3.5 to 9.0 cm/year; at Liczin, up to 7.5 cm/year).

Against this general background, the years 1960–1964 and 1980–1985 stand out when the river was incised and water levels decreased. The first of these periods corresponds to the lifetime of the Sangmensya Reservoir, which was filled in by 1960. Its interception of a significant part of the river load led to the incision of the river downstream of the hydropower plant at a rate of 0.39–0.17 m/year at the lower hydrological posts (except for the Lijin post, where the accumulation of sediment continued, albeit slowly). By 1964, the reservoir had been completely filled with sediment, the downstream sediment load had recovered and accumulation began to develop again. Over a 10-year period, the hydroelectric complex has been reconstructed twice, accompanied by the clearing of the reservoir and sediment transport rate into the tailrace. At this time, the intensity of accumulation reached 0.2–0.3 m per year, exceeding the average value for a 40-year period by 3–5 times.

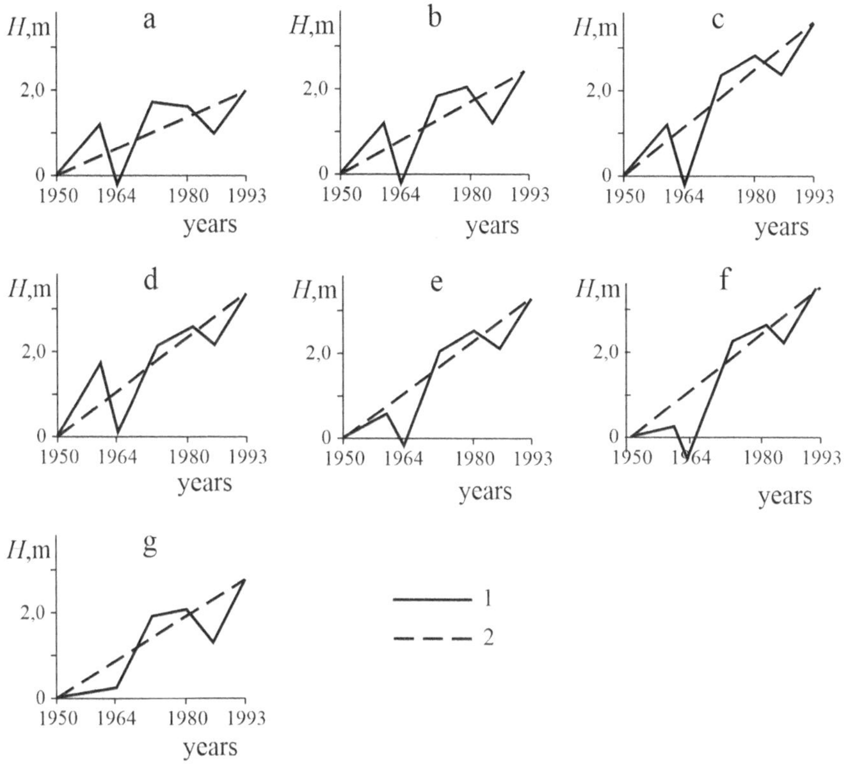

Fig. 5.29 Water level changes on the Huang He river (lower current) at the effective discharge 3000 m^3/s on hydrological stations: **a**—Huayuankou; **b**—Tzehetang; **c**—Gaotsun; **d**—Sunkou; **e**—Aishan; **f**—Lekou; **g**—Litzin; 1—Graph of the water level changes by characteristic years, 2—trend of changing levels

After the next reconstruction of the hydrosystem in the early 70s, the rate of accumulation decreased to 0.02–0.05 m/year, and at the nearest to the Huayuankou hydroelectric station there was some incision of the river. The superimposed high-water level with extremely high-water flow rate resulted in bed incision (0.06–0.14 m/year), which was observed for several years. In the lower reaches (Liczin), the artificial straightening of the channel before reaching the sea by 22 km with a change in the direction of the river flow, carried out in 1976, has also affected the short spread of regressive erosion and lowering of levels (Mikhailova 1998). Since 1985, the accumulation has re-emerged in the lower reaches of the Sangmensya HPP, with a layer of sediment (over 8 years) growing from 90 cm down the river at the Huayunkou post to 130 cm at the town of Liczinj. In the lower reaches of the delta, this was facilitated by the rapid advance of the delta into the sea: only in 4 years (1976–1980) the channel at the estuary lengthened by 13 km.

In the early 2000s, the accumulation process was again interrupted by the construction of the Xiaolandi HPP just before the river entered the Great Plain of

China. According to the available information (oral report of Professor Zhao Yeon) the infeed in the lower reach for 5 years was 3 m, and, spreading transgressively, it affected 100 km downstream (Huayuankou post near Zhengzhou).

The same directional accumulation, but with a lower intensity, is characteristic of the rivers of the lower Amur (Makhinov et al. 1994), although the beginning of its representation is timed to coincide with the release of the Amur from the mountain massif of Maly Khingan (Makhinov et al. 1994). It is confirmed by the analysis of changes in the length of the river load and the water discharge and level curves. The river carries 14.9 million tons of sediment into the Amur estuary annually. This is less than the annual sediment load in the Khabarovsk area, where it is 26.1 million tons. Downstream, to Komsomolsk-na-Amure, it is decreasing by 20%, amounting here to 20.7 million tons and reflecting the intensive accumulation of sediment within the Middle Amur lowland. Crossing the Sikhote-Alin spurs in the incised channel and accepting numerous mountain tributaries, the Amur increases the sediment load flow (up to 23.3 million tons near the village of Bogorodskoe), but then it decreases again as a consequence of the compensatory accumulation of sediment within the submerging Udyl Kizinskaya lowland (Nikolskaya 1972), and due to the tidal effects of the tides. At the same time, the water discharge curves $Q = f(H)$ show that in the Komsomolsk-na-Amure gauging station (the incised channel in the Sikhote-Alin section), there is a clear trend of increasing water levels (Fig. 5.3B). The tendency of increasing the levels is also characteristic of the lowest Bogorodskoe post, located at the lower end of the Udyl Kizinskaya lowland. Above Khabarovsk, the accumulation process is set according to the curves of corresponding levels (Ivanov et al. 2000). But in the site of Khabarovsk, thanks to dredging (along the route of the ship's passage) and quarrying in the channel prevails technogenic reduction of levels: with the consumption of water corresponding to the annual average—by almost 80 cm for the period from 1945 to 1985.

The total rate of accumulation of sediment at the Lower Amur is estimated at an average of 0.56–1.2 mm/year, which corresponds to an annual accumulation of 6 million tons of sediment at the bottom of the valley (in the channel and on the floodplain). Morphologically, this process is expressed in the formation of a powerful floodplain facies of alluvium, the bottom of which lies below the mean annual water level in the river; in the formation of subsurface estuarine lakes (lakes of flooding) on almost all tributaries of the Amur River, starting from the mouth of the Sungari River, including the sections of the incised channel, i.e. at the mouths of mountain rivers (Fig. 5.30) separated from the main river channel by riverbed ramparts (except for the mouths of the Ussuri and Sungari, large rivers whose own accumulation of sediment is comparable to the accumulation of sediment on the Amur River); in the relatively lower position of the rear part of the floodplain compared with the water level in the river, so that when inflows to the Amur River valley come out of the river, their water is flooded, forming complex lake spills in terms of shape. The largest lakes in the lower reaches of the Amur tributaries combine estuarine lakes and lakes formed as a result of flooding of the rear parts of the Amur floodplain.

Fig. 5.30 Estuary lake on the tributary of the Lower Amur (aerial photo)

The peculiarity of the development of directional accumulation on the lower Amur is its extension to the area of crossing the river lowland massif formed by the spurs of Sikhote Alin (on the right bank) and Mao Chan (from the left bank) and Chayatyn ridge (between Udyl Kizinsk and Lower Amur lowlands). Hills and small mountain ranges rising among the Amur floodplain within the Middle Amur lowland, evidently, indicate the involvement of mountain massifs in the general tectonic immersion of the territory. However, their crossing by the river causes increased inflow of solid material into the channel, which is greater than the amount of sediment accumulated in the channel. The consequence of this is likely to be an increase in sediment load between the Komsomolsk-na-Amure and Bogorodskoye gauging stations, despite the directional accumulation of sediment.

Intensive directional accumulation, covering most of the river length in the lowland part of the basin, is characteristic of the Amu Darya. The average value of the accumulation layer in the Amu Darya is 3.4 cm/year (Tsvetkova 1963). In the lower reaches, this leads to the formation of a bunded channel, the water level in which is higher than the floodplain surface (Artamonov 1954) (Fig. 5.31). Here,

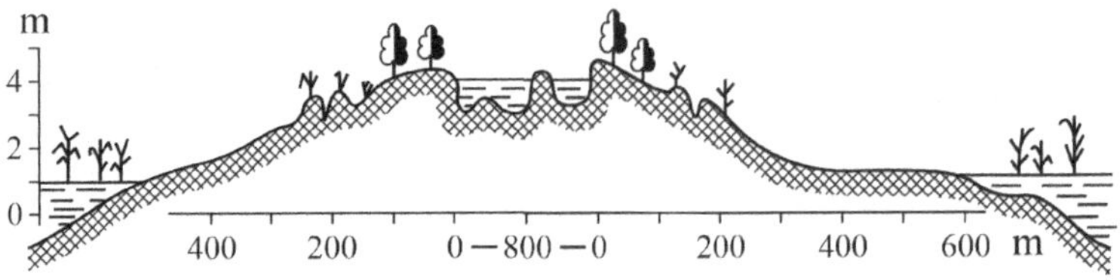

Fig. 5.31 Transverse channel profile of the Amu Daria river in the lower current (Neliubin and Amudaria 1963) and adjacent lowland

intensive accumulation of sediment is facilitated by evaporation losses and filtration of up to 20%. In addition, due to water withdrawal for irrigation, runoff is reduced to a minimum (only to the Karakum canal water consumption of up to 500 m^3/s and more), and in some years, the Amu Darya water does not reach the sea at all, starting from the 60s of XX century. As a result, downstream sediment load is noticeably reduced: in the site of Termez, it averages 478 million tons per year (Berkovich et al. 1973), in Kerka before the construction of water intakes and canals was 270 million tons, in the village of Ilchik—210, in the village of Chatly—133.5, in the city of Nukus—128 million tons.

Constant accumulation of sediment is characteristic of the Colchis rivers. The reasons for this are the intensive tectonic deflection of the Colchis Depression in the foothills, the advance of the Rioni, Okumi and other deltas into the sea, the eustatic rise of its level and the abrupt transition of the rivers (primarily the Rioni) from the intensively rising mountainous area, where they energetically cut into the foothills, which have a quiet tectonic regime (close to the stable in terms of the periphery lowland) and fall within the foothills of the bend. The combination of all factors results in an intensive accumulation of sediment, which leads to a systematic increase in the bedrock and an increase in the river slope. According to calculations (Mandych and Chalov 1970), the inherited sagging of Colchis (at its rate near Poti) is compensated, and the transport capacity is maintained if the annual accumulation of sediment is at least 6.5 mm. This process has been intensified by the regression of the New Black Sea basin, which has dried up an almost horizontal lowland with a width of about 7–8 km, and continues to be activated by the rapid advancement of the delta. According to Zenkovich's data (1958), at the historical time the sea coast in the area of the city of Poti (Rioni delta) was built up at a rate of 7–8 m/year, and the intensity of the delta extension sometimes reaches 140 m/year. As a result, the accumulation rate in the lower reaches of the Rioni is estimated at about 1 cm/year (Mandych and Chalov 1970). As a result, the Rioni and Khobi beds rise above the surrounding area, and only the presence of artificial dams limits the catastrophic flooding of rivers (in 1987, due to the extreme but intense snow melting in the mountains, flood waters broke through the dams on the right bank, and flooding occurred). In the lower reaches of the Okumi, Eris-Tskari, Ojogori and Gagida rivers, the process of accumulation caused a wide expression of bifurcation. Here, the coastal lowland is fenced with a system of multiple hoses, through which water is exchanged between these rivers in one direction and the other.

A number of other rivers in the Western Transcaucasus that accumulate sediment have a developed floodplain that can be traced back to the estuary and is as if "enclosed" in a low sea terrace; the latter is not flooded (Psou, Mchisht, Bzibi, Kodori). In the valleys of these rivers there is regressive movement of mid-bars and islands. Kochetov (1968) found buried soil horizons with a thickness of up to 55 cm lying at a depth of about 1.3 m in the floodplains of some rivers. Some rivers have the lowest terrace levels downstream, while the Psou and Bzybi rivers have their confluence with the floodplain, i.e., a fan of low river terraces converging with the mouth of the river. On these rivers the transformation of the low terrace into a floodplain is carried out in 6–8 km from the mouth of the mouth with the total length of the foothill section of 13–18 km. On Mchisht and Kodori, low terraces and floodplains did not merge, apparently due to a lower rate of accumulation. Comparison of aerial photos of the Psou riverbed and the analysis of the channel forms made it possible to determine approximately that a layer of sediment with a thickness of 2.5–3.0 cm is deposited annually in the riverbed. The analysis of the flow rate and water level curves at the Ganakhleb station on the Kodori River showed that the rate of accumulation here does not exceed 2.0–2.5 cm per year, and in some periods the curve of the connection shifts downwards, indicating a temporary incision of the channel (Mandych and Chalov 1970), associated with a periodic breakthrough of the river mouth of the spit in its root part. Indeed, the delta's maritime edge of the Black Sea coast of the Caucasus is being developed mainly through the development of spits that deviate in the direction of the longshore drift (southward in the northern part of the coast, northward in the southern part). As a result, the channel before flowing into the sea is located almost parallel to the shoreline. At the mouth of the Kodori River, the length of the spit at the mouth of the river (and the increase in the length of the channel) reaches 1.5 km. Upon reaching a certain critical length, the spit breaks through the stream during a high flood in the root part and a new short and straight channel is developed: the former spit in the delta relief remains in the form of a shaft separating the lagoon from the sea. The relief of the deltas, represented by a combination of such ramparts and separating their depressions, is typical for many rivers of the Black Sea coast of the Caucasus—Mzymta, Psou, Bzybi, etc. Such pulsation of delta formation processes is associated with intermittent accumulation in the river channel and periodic change of its some depression, during which non-inundated surfaces appear in the lower reaches, which exist until the incision is compensated by the accumulation accompanying the development of the next spit and the new stage of the delta extension.

The mechanism of influence of the river estuary elongation on the accumulation processes can be presented in the form of the following scheme. Most of the sediment transported by rivers does not reach the receiving basins, forming delta plains, deltas, estuaries, estuaries and other riverine accumulative forms. According to Korotayev (1991), there are between 25 and 50% of suspended sediment load and almost all exposed sediments within this range. The remainder of suspended sediment is transported to the open sea or bays (estuaries, lagoons, limans). For example, at least 24 million tons of sediment per year is transported to the top of the

Indigirka delta in a suspended state, but only half of this amount is transported to the estuary. The rest is accumulated in the delta. In addition, the delta build-up is due to longshore drift, which moves along the sea edge of the delta, accumulates between the estuary bars of the delta branches, blocks their way out into the sea by the estuary spits and levels the sea edge of the delta.

The estuary elongation and formation of primary channel forms leads to increased flow energy losses and the development of regressive accumulation covering the estuary area of the river and its lower reaches (Fig. 5.32). Sediment accumulation takes place at the river mouth in the form of a sediment cone (estuary bar, estuary delta) and its successive advancement into the receiving water body (sea, lake): the primary horizontal surface of the water in the reservoir ($I = 0$) is transformed into an inclined one ($I > 0$), and the area of accumulation and increase of the bottom marks spreads regressively (points O_2, O_3, O_4; at the same time, the development of the debris cone to the sea (transgressively) takes place, and its sea

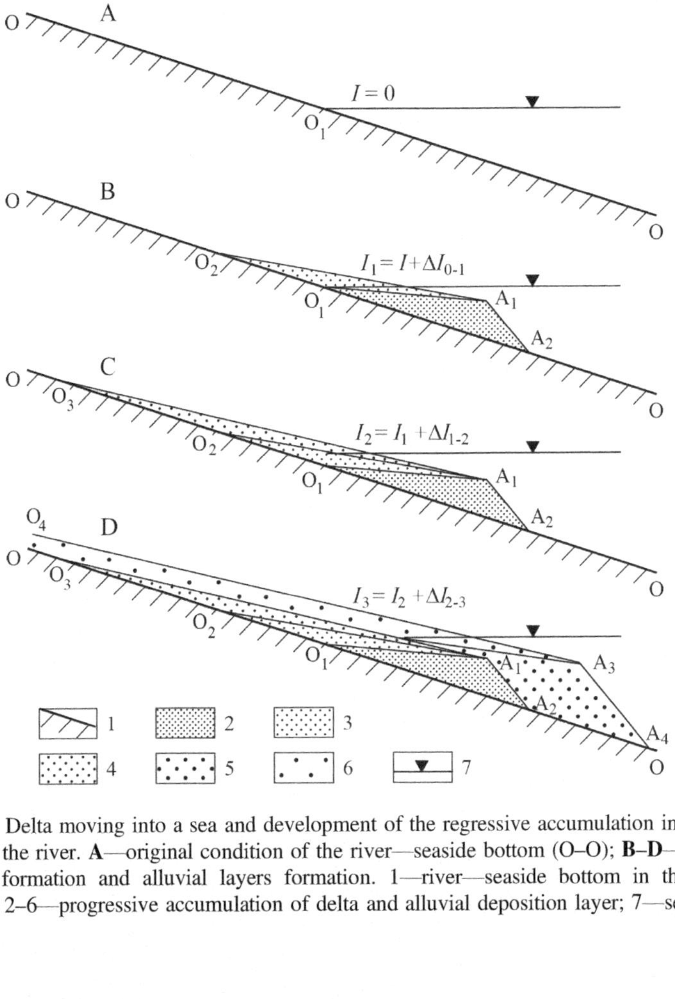

Fig. 5.32 Delta moving into a sea and development of the regressive accumulation in the lower current of the river. **A**—original condition of the river—seaside bottom (O–O); **B–D**—phases of the delta formation and alluvial layers formation. 1—river—seaside bottom in the original condition; 2–6—progressive accumulation of delta and alluvial deposition layer; 7—sea level

slope occupies the positions of A_3A_4. At each next stage, the channel slope increases within the former seaside by ΔI_{i+1}, where i is the number of the previous stage. In the river part, where regressive accumulation is associated with river mouth elongation, the local slope of the river can decrease ($I_1 < I_i$), and only a prolonged accumulation of sediment accompanied by an increase in channel marks above water body levels (absolute at the river's confluence with the sea) leads to an overall increase in the longitudinal slope. As a result, a decrease in sediment load is not only observed in the delta, but also upstream. For example, at the lower Ob (before the construction of the hydropower plant on the river and its tributaries), the annual sediment load was 20.1 million tons at the lower Ob (before the construction of the hydropower plant on the river and its tributaries) directly below the confluence with the Irtysh River (the village of Belogorye), and 13.6 million tons at the top of the delta (Salekhard) (Lopatin 1952). Thus, the accumulation of sediment at the confluence of rivers into water bodies is both transgressive (advancement of the delta into the sea and river mouth elongation) and regressive.

The rate and morphological representations of these processes depend on the magnitude of sediment load, the aquifer capacity of the rivers, the conditions under which they interact with the sea (discharge into open seashore or bays, shallow shelves or continental slopes, etc.) and, finally, the slope of the river itself, which determines the distance of regressive accumulation. For large rivers, the rate of delta outflow into the sea is 3–3.5 km per year on the Huang He River (Mikhailova 1998), 20 m/year on the Yenisei River (Korotayev 1991), and is virtually undefined during the period of observation (survey) at the mouth of other rivers in the north of Siberia, which, despite the high runoff rates, have a relatively small flow of sediment load, eroded by the cyclicity of the estuary elongation and the periodic development of various deltaic channels. When the delta is actively extended, the layer of sediment that has formed new parts of the delta and the thickness of alluvial deposits due to regressive accumulation is

$$\Delta h_{acc} = \Delta z \Delta x, \qquad (5.21)$$

where Δz is the average river fall in the delta region, m/km (or slope I, ‰); Δx is the increase in length (in km) over the calculated time period (Mandych and Chalov 1968). Assuming that the upper boundary of the regressive accumulation layer thickness is zero, and the decrease in the sediment accumulation capacity due to the accumulation of the upper river is linear in nature (Makkaveev et al. 1961), it is not difficult to determine the approximate volume of the accumulation body during a given period of time.

The extension of the delta and the river's estuary is the main condition for the accumulation of sediment and its regressive propagation during erosion. In this case, the gradient ratios between the individual links of the hydrographic network in the unflooded part of the territory do not change. Therefore, at a new level of river inflow into the receiving basin, conditions similar to those that preceded its increase are created.

At the same time, the estuarine decline in the flow into the water body, shifted upwards following the rise in the level, is even the cause of some incision in the new estuary of the river. However, since this also leads to a longer channel extension at the same time, the plunge cuts are quickly replaced by regressive accumulation. In the area where the flow backwater seeps out, the river begins to form a new estuary bar. The development of the bar, and then the general extension of the delta, results in the riverbed lengthening and increased energy losses along the length of the river. In the upstream section of the river, this results in back-watering, flow rates decreasing and accumulation of sediment. At the same time, the river bottom marks in the accumulation zone increase (Fig. 5.33A). Thus, the reason for the development of regressive accumulation in the course of the base level is not the fact of this rise, but the development of the delta estuary in a new position. This process can be superimposed on the accumulation of sediment that has penetrated here regressively from the previously formed delta at a lower erosion base position (Fig. 5.33B).

If, as the erosion and regression base of the sea drops, an area with a surface slope smaller than the slope of the downstream river is exposed, then in the lower reaches of the river there may also be accumulation of sediment, which will gradually spread upstream (Fig. 5.34). On the Black Sea coast of the Caucasus, during the regression of the New Black Sea basin at the mouths of the Mokva, Okumi and Chorokhi rivers, a shelf with slopes smaller than those in the lower reaches of these rivers was exposed. Thus, on the Okumi and Chorokhi the river

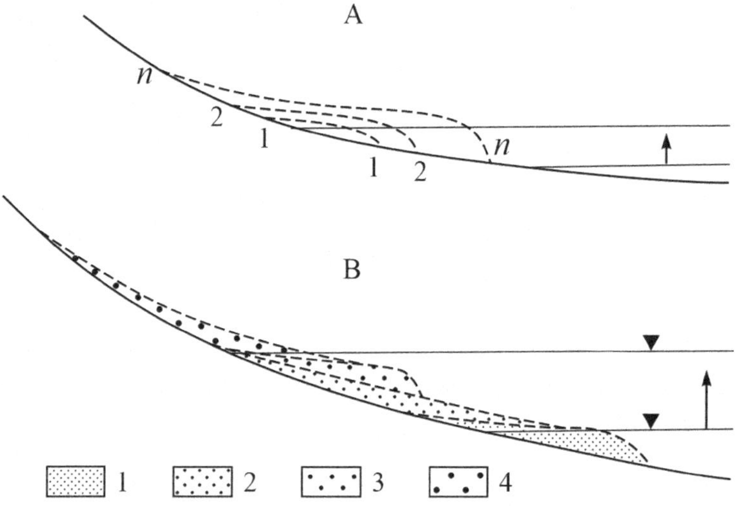

Fig. 5.33 Accumulation development downstream the river as a result of base level lifting. **A**—due to delta formation at a new sea level: 1, 2, n—consistent position of the deposition surface (Makkaveev 1971); **B**—at new delta overlaying on the alluvial depositions, formed as a result of the regressive accumulation at low position of the sea level: 1, 3—delta depositions; 2, 4—alluvial depositions before and after base level lifting

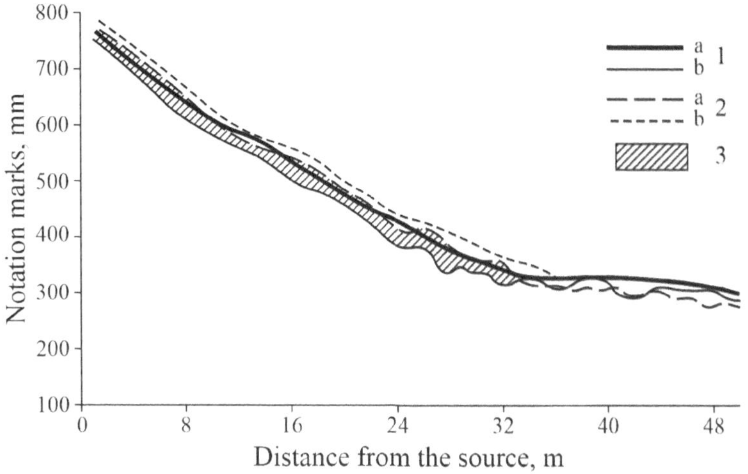

Fig. 5.34 Development of the regressive sediment accumulation at decrease of the base level in gentle shelf conditions: 1—water levels (a) and bottom levels (b) directly after base level decrease; 2—the same by the end of experience; 3—deposition layer (Makkaveev et al. 1961)

elongation only due to the lowering of the sea level was equal to 7.5–8 km, without taking into account the sinuosity of the channel (on the Okumi the length of the channel within the New Black Sea terrace is 18–20 km), and the slopes decreased by 4–7 times. Before reaching the young sea plain, the Okumi slopes were 1.9–4.2‰ and before the estuary they decreased to 0.48‰. At Mokva, the mouth elongation was 3.5 km (channel length 4.5 km), and the slopes decreased from 2.1 to 0.3‰.

The accumulation process is particularly intensive if the increase in the catchment area is not accompanied by an increase in runoff (in arid climates), and if the river flows into an area of an accumulative sea shore where its estuary is blocked by sediments brought by sea-side currents. This is also facilitated by the formation of the delta, which further reduces the longitudinal slope within the hollow shelf. For example, when the level of the Caspian Sea decreased in the 60 and 70s of the 20th century, some rivers "broke away" from the sea as a result of accumulation and formed blind mouths. The Emba River was shown on small-scale maps of the early 50s as far back as the Caspian shoreline and even having a small delta. Already on the maps of the early 90s the river began to appear as a drying dotted line, which does not reach the shoreline of the sea (Fig. 5.35).

Together, all of these reasons lead to a directional accumulation of sediment in the lower and sometimes middle course of lowland rivers. The distance of its distribution from the estuary is very different, depending on the size of the sediment load and, consequently, the intensity of the axial elongation, the slope of the channel and the tectonic regime. The slow rise of the lowland-platform areas covering their basins shows up only in the deltas and lower reaches of the rivers. These are the rivers of the East European Plain. In the Northern Dvina, for example,

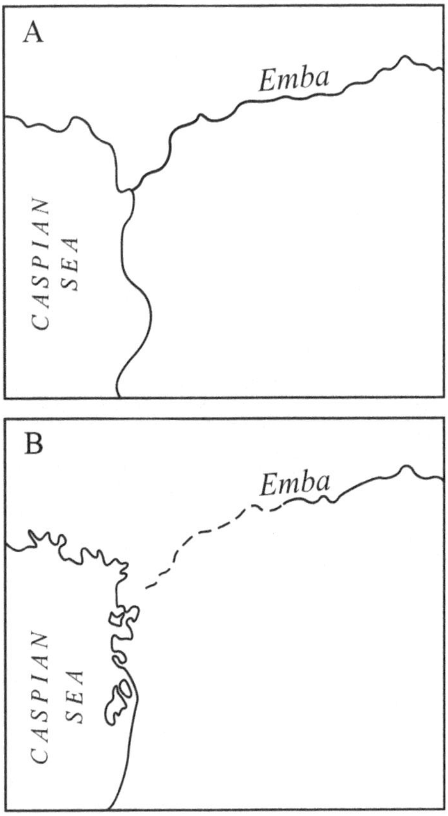

Fig. 5.35 Changes of the Emba river estuary positions at the Caspian Sea level decrease due to sediment accumulation in the gentle shelf conditions. Pictures of the estuary: **A**—1954; **B**—1990

the incision of the river in the upper and middle reaches of the river, alternating in the lower reaches with the accumulation of sediment, is characteristic of the Vaga estuary (Chalov et al. 2000). With low sediment load, the accumulation rate is relatively low, so that the bunded bed is not formed here, and the main morphological sign of accumulation is superimposed, or single tier, with increased lumpiness and marshing, floodplain. With large sediment load, a bunded channel is formed in the lower reaches of the rivers (the Huang He, the Kuban, the Amu Darya, and the Ili).

In the lower reaches of the Neman (downstream of the confluence with the Nyaris River), in its natural state, until 1960, when the Kaunas hydroelectric complex was built, the marks of the bed of the channel increased due to the process of systematic accumulation (Kolypaila et al. 1940; Zaikov 1954; Lasinskas et al. 1959). Kolypaila et al. (1940) identified this phenomenon from the differences in mean horizons between the periods 1812–1869 and 1870–1920 and from observations made each year between 1812 and 1920. According to Mingauda (Lasinskas et al. 1960), the Neman bottom mark at the Smalininkai post has risen by 0.75 m in 130 years and by 0.85 m near Kaunas in 70 years. The increase of the riverbed bed is confirmed by the analysis of the curve $Q = f(H)$: at H = 500 cm the water discharge for the period 1875–1932 was 1670 m^3/s, for the period 1949–1953—1540 m^3/s and for the period 1954–1958—1470 m^3/s (Jablonskis and Lasinskas 1962). Apparently, these data were the basis for M. I. Byatsonis (Atlas of the Lithuanian SSR 1981), who showed a "channel accumulation" on the whole of the lower Neman, starting from the mouth of the Nyaris River.

At the Ob, in its natural state (before the construction of the Novosibirsk hydroelectric power station, i.e. before the end of the 1950s), directed accumulation of load was observed up to the city of Kamen-na-Obi (upstream of the river), increasing in intensity downstream: while in Novosibirsk during the period 1936–1956 it was 0.8 cm/year, in Kolpashevo it was already 3.4 cm/year (Chalov et al. 2000). In the lower Ob, accumulation occurs as a result of the river mouth elongation when the Gulf of Ob is filled with sediment (regular overrunning is also important) and occurs throughout the entire period of floodplain formation (Riverbed regime … 1994). However, with relatively low sediment load (respectively, the river's estuary elongation rate is low), this process is not very intensive, is to a certain extent compensated for by sedimentation on the floodplain surface during the period of prolonged flooding in the floodplain and is represented in the development of the one-tier floodplain and the widespread distribution of lakes and reservoirs formed in the lower parts of the floodplain massifs as a result of the increase in water levels in the river (Lavrushin 1961). This process obviously extends to most of the sub meridional area of the Middle Ob (below the merger with the Irtysh River). Upstream the cause of directed accumulation is the removal of sediment from the mountainous part of the basin and from the incision section of the Upper Ob (confluence of Biya and Katun, Kamen-na-Obi), which are products of riverbed scouring, the amount of which in conditions of low slope (less than 0.10‰) is excessive in relation to the transport capacity of the flow. Both of these phenomena—regressive accumulation from the estuary of the river, and transgressive as a consequence of the inflow of sediment from the upper part of the basin—obviously close down somewhere below the town of Kolpashevo. This is evidenced by the fact that under the conditions of the Novosibirsk reservoir interception of the sediment flow from the upper Ob, the riverbed incision has spread to the middle Ob at least up to the town of Kolpashevo (mouth of the Ket river), where its rate of load balance is about 0.5 cm/year, and the curve $Q = f(H)$—1.9 cm/year (the latter is overestimated due to quarrying).

The conditions for the transgressive accumulation of sediment from the upper part of the basin are typical of the largest rivers in Siberia, the upper reaches of which are in the mountains, while most of the river and the basin are located within the vast lowland plain (for the Ob it is the Steppe Altai and the West Siberian lowland). This process is noticeably influenced by climate change, leading to centuries-old fluctuations in the water availability of rivers, which, in turn, determine the intermittent (in time) development of targeted accumulation. Therefore, a sharp change in the slope of the river exit from the mountains to the plain, characteristic of the Ob in the upper reaches, as the reason for the accumulation, is veiled by age-old fluctuations in water availability, which lead to a shift along the river boundary of the change of the sign of vertical changes: in wet periods—downstream, in relatively dry—upward (Makkaveev and Chalov 1970). Thus, changes in climatic conditions and water runoff may lead to different consequences in the development of the longitudinal profile in different parts of these rivers, and such periodic fluctuations in the rate of flow is particularly evident on the rivers originating in the mountains, crossing vast lowlands and only then flowing into the sea. In periods when the flow rate is relatively low, much of the solid material brought by the river from the mountainous part of its catchment area accumulates within the zone immediately below the river's outlet from the mountains to the plain. Within the second zone, which is far from the mountains, the decrease in water availability in the rivers is accompanied by a decrease in maximum water levels, a reduction in flood spills and a concentration of flow in the floodplain. Apparently, this fact, as well as the decrease in the amount of sediment coming from the upstream section, also causes some incision of the river. Both reasons—reduction of spills and plunging—lead to the transformation of the floodplain into a floodplain terrace or the development of zonal soil in its elevated areas, which are now flooded very rarely, in some years with extreme floods.

Subsequent increase in the flow rate within the upper zone leads to the development of deep erosion. Dilution products as well as the material brought from the mountains are accumulated within the second zone, which is located downstream and more stretched along the river length than the first one. As a result of the gradual increase of the bed bottom of the channel and the corresponding rise in the level of hollow water, a superimposed floodplain is formed here. Thus, the increase in water availability in one river section causes the riverbed to cut in, and in another, the accumulation of sediment brought in from the upstream section, which is excessive due to lower gradients.

In periods when climate change leads to a decrease in the runoff rates of the river, within the first zone (below the river's outlet from the mountains) deep erosion is replaced by accumulation, superimposed floodplain is formed in the river valley.

At the upper and partly middle Ob, such a periodicity in the development of directional accumulation and incision of the river is associated (Makkaveev and Chalov 1963a, 1964, 1970) with fluctuations in the runoff rates of the river, the step of which in time is approximately within the framework of the 1850-year period of changes in humidity, identified by Shnitnikov (1957). At the same time, in the upper Ob (below the confluence of Biya and Katun), where deep erosion is currently developing in all indicators and features, the horizon of the buried black earth soil is traced in the sections of the high floodplain; at the same time, it is formed as a modern one on its surface. Both of these features—buried soils at the previous stage of increased water availability and accumulation of sediment and modern black earth soil formation at river crossing—can be traced back to the tributaries of the Ob (Aleya, Charysh, Anui).

On the Ob river downstream of Novosibirsk, in addition to the horizon of recently covered by floodplain silt stock grey forest soils, there are two or three more buried soil horizons, which were formed in the previous climatic cycles of the

Holocene, indicating repeated changes in the value of river runoff and several stages of plunging and accumulation during the formation of the floodplain.

A similar location for the erosion and accumulation zone was established by Bakhtin (1957) on the Yenisei: from the source to the confluence with the Angara River, there is a constant increase in suspended sediment load and sediment concentration; downstream, there is a decrease. In a smaller scale (in relatively short areas), similar climatic shifts in the zone of deep erosion and directional accumulation occur in rivers flowing entirely through the plain areas. Thus, on the Northern Dvina the superimposed floodplain was found near the mouth of the Vaga River, i.e. on the border of the middle and lower reaches of the river. It is not found downstream, the floodplain here is one-tiered, because here regressive accumulation is developing due to the river mouth elongation; upstream (Abramkovo) the Northern Dvina and its tributaries (Vychegda, Sukhona, Ug) incise, and there is no superimposed floodplain.

Directed accumulation of sediment, caused by reduced runoff in arid conditions of climate change and representing itself over millennia and geologically over time, leads to ruptures in the river network, where inflows lose their connection to the main river. Makkaveev (1955) cites the p. Chu, which is currently not reaching the Syrdarya River, ending in a "blind" mouth, although the valleys of both rivers are joined. The same happened to the former tributaries of the Chu river—the rivers Sary-su, Talas and many others, lost in the sands, but in the historical past reached the main river. The tributaries of the Amu Darya were Tejen and Murghab, flowing down from the Kopet Dag, and the Zeravshan flowed into it still in the time of the Central Asian campaigns of Alexander the Great.

A peculiar expression of directional accumulation is the siltation of small rivers. It is a consequence of excessive inflow of material into the channels of small rivers, carried away from the slopes as a result of anthropogenic soil erosion during plowing and deforestation. The rate of sediment transport to rivers from flooded and forested slopes is tens to hundreds of times lower than from ploughed ones (Kovalchuk et al. 1996). In the forest zone in the east of the Russian Plain, the thickness of anthropogenic silt deposit was 1.5–2.5 m (Kurbanova and Petrenko 1990), which, with a simultaneous increase in the intra-annual irregularity of runoff, led to the fact that they began to be flooded and covered by the silt deposit of the surface of the first floodplain terraces. In the forest-steppe and steppe zones (the Don basin), the accumulation of flush products from ploughed land in the channels and floodplains of small rivers led to the complete burial of a part of the channels of small rivers under the sediment layer. According to Kovalchuk et al. (1996) over a 100-year period, about 25% of the 1st order rivers in the basins of small rivers in the Volyn-Podolsk Upland have ceased to exist, and the total length of the rivers (up to the 4th order) has decreased by more than 30 per cent. According to his calculations, from 0.3 to 1.1 billion m^3 of washed away material was received in the channels of these rivers. In the European part of Russia, the death of rivers due to siltation (taking into account some increase due to reclamation canals) was 44.4% in the steppe zone, 35.8% in the forest-steppe zone, and 1.64% (without taking into account the increase of about 3%) (Ivanova et al. 1996).

Dedkov and Mozzherin (1996) attach great importance to the reduction of water availability in the rivers at the time of deforestation and disappearance of their sources due to the ongoing decrease in the level of groundwater. Not excluding this reason for the degradation of small rivers, it is necessary to take into account the fact that the decrease in water availability must inevitably be accompanied by the accumulation of sediment, even if it does not change the way it flows into the channels from the catchment area. However, deforestation (fourfold reduction) was accompanied by a sixfold increase in sediment load, which, along with a decrease in the runoff rates of the rivers, caused siltation.

Thus, in the lower reaches of large rivers, as a rule, the accumulation of sediment occurs, and in the case of large sediment loads (the rivers of China) it is catastrophic in nature. When rivers cross mountain ranges, the accumulation is replaced by plunging. Under conditions of low sediment load, accumulation rates are low and the boundaries of sections with different directions of change migrate along the length of the river, depending on age-old changes in the runoff rates of the rivers. The alternation of dry and high-water years, with greater or lesser sediment load, results in a temporary reduction in the accumulation of rivers with a general trend towards higher levels due to riverbed incision. A special form of accumulation directions is formed by siltation of small rivers, connected with excess of load from slopes during plowing and forest data.

5.4.1 Geomorphological Effects of Directed Vertical Channel Changes

Directional vertical channel changes (river incision, systematic accumulation of sediment), due to their low rates of development, are usually not taken into account in engineering practice, as the time in which they occur and can affect the stability of structures, go far beyond the scope of their regulatory service. Exceptions in natural conditions are the intensive accumulation of sediment in the lower reaches of rivers with very high sediment load (this causes a constant threat of flooding due to the rise in water levels exceeding the level of branches), or isolated cases of catastrophic river tampering. Negative expressions are more frequent, mainly of river intakes in the lower tailings of hydrosystems, in the development of channel quarries of construction materials and, to a lesser extent, in the accumulation of sediment (in the zones of seepage and upstream of reservoirs, but only at high flow rates). Local representations of vertical changes, which are usually of a periodic nature, are characterized by higher intensity.

At the same time, the consequence of directional vertical changes is the fluvial relief of the land, the structure of river valleys and alluvial plains. Since this is related to the study of geomorphology of various territories, palaeogeographic reconstruction and the solution of many applied geological-geomorphological problems, the appropriate assessment is primarily important for geomorphology and dynamic geology.

As a result of plunging rivers into the land surface over long periods of time, comparable with geological epochs, river valleys with a complex of terraces are formed. In other words, terraces as a form of relief of river valleys arise in the process of incision of rivers and, consequently, the consistent movement of the water surface in the river (its level) to lower hyposometric levels. It is generally accepted that this process is represented until the river reaches such a state that the plunge fades, stops expressing itself and begins to dominate the wandering of the flow on the bottom of the already formed river valley, leading to its expansion. However, deep and lateral erosion as forms of representation of vertical channel changes develop synchronously with other things being equal, since they have different physical nature and mechanisms (Makkaveev 1955, Makkaveev and Chalov 1970; Makkaveev et al. 1961; Borsuk and Chalov 1978). Therefore, the identification of stages of river valley development on the basis of predominant development of one or another type of channel change is not always correct and legitimate. The accumulation of alluvium in the bottoms of river valleys, the increase of the river mirror to a higher level of gypsometric represents the opposite process of river development in the vertical plane. During the accumulation phase of river valley development, previously created terrace levels may be buried. The result of the accumulation is a gradual expansion of the river valley and, finally, its transition to a single alluvial plain, not separated into separate levels.

A characteristic feature of river valleys is the stepping of their transverse profile, i.e. the presence of terraces. Terraces in river valleys are evidence of the fact that during the entire period of formation of the valleys, the leading process of development has been river incision. Exceptions are relatively limited (relative to the total area of the continents) areas of some coastal lowlands experiencing tectonic immersion and arid zones (e.g., Central Asian plains), where climate aridization and reduction of river water availability have led to ruptures in the river network, within which vast alluvial plains are formed, as well as some intermountain basins and foothill deflections.

The wide spread of terraces in all main river valleys shows that the latter are formed under conditions of prevailing upward development of the relief (in mountainous countries, where tectonic recharge occurs more intensively than in the case of rivers) or in a balanced way, when the incision compensates for the rise and, with the persistence of other factors (runoff rates of the river, sediment load, the position of the erosion base), the longitudinal profile of the river does not change its hyposometric position.

The predominant rise of the continents and the ascendant development of the mountainous terrain, especially at the last (newest) stage of the transformation of the land surface, created the prerequisites for the dominance of the stages of plunging over the accumulation and formation of terraces in the valleys of the staircase. Even in cases where transit rivers cross large tectonic depressions, the river valleys have developed river terraces (the Lena River within the Viliui syneclise, Ob in the territory of West Siberian lowland, including Biysk-Barnaul, Kolpashevskaya and other depressions, Yenisei River within the Minusinskaya basin, etc.). The tectonic movements create a general background in which the climate plays a major role in the formation of river valleys and terraces. But depending on the scale of tectonic movements in mountainous countries and on plains, the importance of climate change varies. At intensive general tectonic movements of the positive sign in the mountains (upward development of the relief) climatic changes cause temporary oscillations of the process of terra formation, causing at certain stages the development of accumulation and, ultimately, the formation of accumulative terraces. In plain-platform areas, the general tectonic factor determines only the hypsometric position of river headwaters in relation to sea level (the base level), which at the rate of elevation in the proportion of a millimeter per year (10–20 cm for 1000 years) can be practically assumed to be constant, while the formation of river terraces on this general background of low elevation is associated with more effective in these conditions climate change. The process of climate-driven terra formation creates a picture against which local tectonic structures cause local deviations from the general scheme (Borsuk and Chalov 1978).

If the tectonic nature of terraces formation is determined only by a more intensive rise of the upper reaches of the river system in relation to the lower reaches or a stable position of the base level, the climatic factor (whether it creates a general background in the formation of terraces or determines local changes in the scheme in time or space terms) is associated with different phenomena in their nature. The same effect in the formation of a terrace complex can be achieved by changes in the water availability of the river, the degree of unevenness of flow, and an increase or decrease in the amount of sediment load entering the stream. The latter may be related not only to climatic conditions, but also to the overall fragmentation of the territory and the depth of the incision of the main river, which is often associated with tectonic movements of the positive sign.

As the river valley develops, the riverbed and valley bottom gradually decrease in relation to the watersheds during the incision process. At the same time, the depth of valleys is incomparably small in comparison with their width and seldom exceeds 100 m in flat floodplain rivers with the free development of channel changes, and against the background of plunging caused by the general rise of continents, there is a periodic change of process sign or activation of plunging associated with periodic fluctuations in the runoff rates of rivers (Makkaveev and Chalov 1970). In the conditions of limited development of channel changes, the sustainable incision of rivers leads to the formation of valleys, the depth of which reaches 1 km and significantly exceeds their width. These are gorges, gullies, canyons in mountainous countries. When lowland rivers are cut into bedrocks (moraine loams, marls, clays), the depth of the valley is the first tens of meters, but when crossing actively rising local structures, where the river forms the bed incision, the depth of the valleys can be significantly greater, and the width of the valley is also significantly reduced due to the limitation of lateral erosion (bank erosion) by hard rocky formations, their components.

With the developed longitudinal profile, the river slowly incisions are observed as a result of the development of its basin, involved, on the one hand, in the general tectonic uplift of the territory, and on the other hand—in the permanent denudation of the mainland territory. Therefore, under conditions of tectonic stability, the development of river valleys follows the same general scheme as under ascending topography development, and the process of incision of the river, deepening of the valley and formation of terraces can continue almost indefinitely, especially against the background of generally positive tectonic movements of the continents, but with the alternation of periods with different intensity of incision and even directed accumulation of sediment.

Due to the inundation and reduction of the maximum water level, the floodplains of the rivers are no longer flooded and turn into terraces. If the channel forms an incised bend, then during its longitudinal movement at a distance of a step the young floodplain at the convex bank will have time to turn into a floodplain terrace; in the incised rectilinear channel during this time the channel shifts by an amount equal to no more than one width of the channel. For the formation of incised bends as a consequence of vertical changes of the negative sign Panin (1991) received the expression

$$\frac{L_b}{U_b} > T = \int_{H_f^{min}}^{H_f^{max} f} \frac{dh}{U_{acc}(h) + U_{inc}}, \tag{5.22}$$

where L_b is the bend pitch; U_b is the speed of its longitudinal movement; and H_f^{max} *and* H_f^{min} is the maximum and minimum relative height of the floodplain; Ua_{cc} is the rate of sediment accumulation (sludge accumulation) on the floodplain surface; U_{inc} is the rate of river incision; h is the average relative height of the floodplain surface.

In the case of wide floodplain channels, the incision process takes place against the background of the river divagated along the bottom of the valley, which slows down the formation of each new terrace, but leads to the appearance of extensive terrace surfaces as the water level marks decrease. The reason for the slowdown is that during lateral channel change, riverbanks, including floodplains, are washed away and the transport capacity of the flow is realized not only through bottom erosion but also through lateral erosion. Since under the conditions of limited channel change and especially in the mountains, the bank erosion is minimal, the terraces formed during the incision process have greater differences in height than in the valleys of lowland rivers with a wide floodplain channel.

Regardless of the reasons of river intakes (tectonic, climatic), the terraces formed in this case have a chord character, increasing the relative height from the estuary upstream to some point in the middle reaches, but to the upper reaches there is a gradual decrease in the relative height of the terraces, as well as the merger of terraces of individual levels (see Fig. 5.35). In the valley of the Kodori River (Western Caucasus), which is formed under the conditions of rising relief development, in the foothill section there are five levels of floodplain terraces, the height of which grows upstream (Table 5.7). At 70–80 km from the estuary (middle reaches) the height of the terraces reaches the highest value; at the same time, the number of their levels increases: if 5–6 terrace levels are found in the foothills, then

Table 5.7 Changes in the heights of some of the Kodori Valley terraces (Berkovich and Chalov 1977)

Terrace	Terrace heights, m			
	Lower reaches	Borderline of foothills	Middle current	Upstream
I	4–6	13	40–44	5
II	10	28–30	70–80	9
III	25	70–80	120–140	27–29
IV	40–45	120	–	50
V	100	160	250	–

in the middle reaches they are already 10–11; in the upper reaches of the river the number of terrace levels decreases again to 4–5. A similar picture is observed in the valleys of other Caucasus rivers (Psou, Bzybi, Gumista, etc.). In the valley of the Psou River, the terrace, which has a height of 20 m in the foothills, rises to a midstream of 35 m. An increase in the relative height of the terraces from the lower reaches of the rivers to the middle reaches was found on the Rioni, Tskhonis-Tskali, Supse, Adzharis-Tskali, Chorokhi (Nemanishvili 1963, 1964; Astakhov 1961; Maruashvili 1960; Fedorov 1963). In the Rioni region, the height of the terraces from the Racha Basin (middle reaches) to the Onsk Basin (upper reaches) is clearly decreasing (Geomorphology of Georgia 1971).

The noted regularities of formation of terraces in the valleys at river crossings are expressed in a wide range of natural conditions. Even on such a large river as the Lena, the general upward development of the relief is the main factor in the development of the longitudinal profile, although in its middle course it leaves the middle mountains and for a considerable length of time flows through the territory of the Central Yakut lowland (Borsuk and Chalov 1973), within which only the rate of incision of the river is noticeably weakened, and along with the tectonic factor the leading role is played by the climate.

The peculiarity of most terraces of the Lena River is the width commensurate with the width of the floodplain. An exception is the so-called "Bestyakh terrace" (Timashev 1970), which has a height of 50–60 m near the city of Yakutsk and is composed of thin sands with a capacity of up to 60 m, which is 2.5–3 times higher than the capacity of modern alluvium in the riverbed. Its total width (on both sides of the valley) exceeds 60–80 km, while the cross-section of the valley bottom (floodplain and channel) does not exceed 25–30 km. The age of the "Bestyakh" terrace (Lungershausen 1961) is middle-quaternary. Therefore, from this time onwards, the plunging mode is constant during the development of the longitudinal profile and the formation of terraces. It is conditioned by the location of the upper reaches of the river in the North-Baikal mountainous country, part of the middle and lower reaches—within the limits of the Central Yakutsk lowland, and the lower reaches (Lena pipe)—between the ridge of Chekanovsk and Haraulakhsk mountains. The process of incision in the conditions of ascending development on the most part of the Lena basin leads to the formation of chorde type terraces

Table 5.8 Depth of the Lena River channel incision relative to the levels of eighth (VIII) and first (I) floodplain terraces

Points	Ust-Kut	Kirensk	Vitim	Olekminsk	Pokrovsk	Aldan river estuary
VIII	150–160	–	140–170	160–170	160–170	120–140
I	10–11	11–12	12–14	14–16	12–16	9–10
Points	Cartylabyt	Jarjan	Siktyakh—Govorovo	Kyusur	Chockurovka	–
VIII	~ 140	120–140	140–160	200	290–320	–
I	14–16	16	16–20	22–24	23–30	–

(Table 5.8). Same-aged terraces reach their maximum height on the middle Lena between the Vitim village and Olekminsk, decreasing both upstream towards Ust-Kut and downstream to Siktyakha village, experiencing local changes when crossing local structures. In the lower reaches, in the area of the Lena pipe, the sharpest rise of terrace levels is observed, which indicates the impact on the longitudinal profile of the river intensive local tectonic rise.

The climatic terrace of the chorde type is described in the valley of the Volga (Obedientova 1977), in the middle and upper reaches of the Dnieper (Bylinsky 1958), where their development was limited to the Dnieper steps, which created a local basis for erosion, the Dniester (Afremov 1978) and other rivers of the Russian Plain and Western Siberia.

Side erosion, which occurs simultaneously with incision, expands the river valley. However, as the river valley deepens over time, the rate of both inbound and outbound channel movement slows down, which in the case of the main and terrace banks of the river provides this expansion. In both cases, the decrease in the rate of deep and lateral erosion is due to an increase in the inflow of solid material from the slopes and the catchment area and, in the case of lateral erosion, also to the higher energy consumption for the removal of high bank erosion than in the case of low flow rates. As a result, under other equal conditions (including the homogeneous lithology of coastal rocks), the high bank retreats much slower than the scarp of a low terrace or floodplain. For example, on the upper Ob, a 100-m cliff of the Ob plateau, composed of easily washed out loess-like loam, retreats under the influence of the flow at a rate of no more than 0.5 m/year, while the intensity of erosion of the floodplain and sandy ledges of terraces on average is 5–10 m/year, sometimes reaching 30–40 m/year.

This phenomenon has two important geomorphological consequences. Firstly, since the river's overall inundation slows down as the valley deepens, all other things being equal, the vertical distance between the terraces that form will decrease over time. This also leads to a decrease in the longitudinal slope of the river as it incises, which is equivalent to maintaining constant water availability, reducing the transport capacity and, consequently, its transport capacity. Second, the increased depth of the valley and the general erosion of the area, accompanied by the arrival of more solid material in the rivers, leads to a simultaneous reduction in the extent of lateral wandering in the lateral plan, which, together with an increase in the height of the primary banks leads to a decrease in the probability of expansion of the valley bottom. As a result, the width of the valley bottom, all other things being equal, is reduced as the valley deepens (river intakes). An example of a simultaneous reduction in the width of the valley floor at different levels as the river tapers in and the width of the valley floor decreases at each of them gives the Alabuga river (Fig. 5.36, Table 5.9), the intensive incision of which began in the XIV century and led to the formation of a canyon with terraced sides (Panin et al. 1990).

Thus, the development of river valleys in the conditions of prevailing river intakes is accompanied by a narrowing of their bottoms, a reduction in the width of the channel wandering belt, and, as a consequence, an increase in the probability of preservation of higher terraces. The higher the level and the wider the surface of the terrace (i.e. the total channel wandering belt corresponding to the width of the future terrace), the more likely it is to destroy the previously formed terrace. On the contrary, with a narrower wandering belt corresponding to a deeper plaster terrace of the channel, the probability of preserving the wide and ancient terraces increases significantly. This is also facilitated by the increase in the height of the ancient terraces, as well as the formation of the indigenous basement, usually composed of bedrocks.

At the river valleys, where the incision is replaced by a directed accumulation from above downstream, the terraces first decrease in level, merge with each other and then bury them with younger sediments. Taller and older terraces are submerged under younger precipitation, the greater the older the terrace and the more

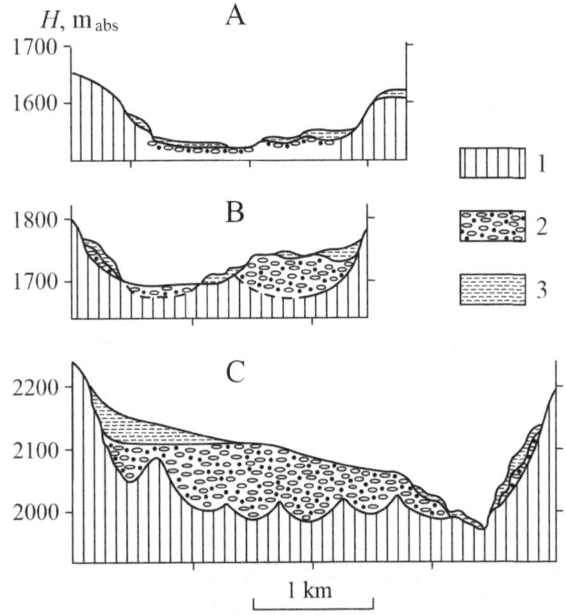

Fig. 5.36 Transverse profiles of the Alabuga river valley (downstream and midstream) at different distances from the estuary: **A**—0.2 km; **B**–19.5 km; **C**—54 km. 1—neogene lake siltstones; 2—ancient alluvium; 3—terrace alluvium; 4—slope depositions (Panin et al. 1990)

Table 5.9 Changes in terrace heights in the valley of the Alabuga River (Pani et al. 1990)

Distance from estuary, km	Valley width, km	Valley depth, m	Relational terrace heights, m	
			Upper complexes	Lower complexes
28.0	1.7	60	21.5; 32.0; 43.0	4.2; 5.6; 6.6; 7.0; 12.5; 14.1
36.0	1.2	70	28.5; 31.5; 38.0; 48.0	2.0; 3.5; 4.1; 4.6; 5.7; 9.5; 17.0
47.5	0.9	70	50.0; 60.0	4.0; 8.0; 10.0; 14.5; 31.0
54.0	0.8	90	56.0; 66.0	8.7; 11.0; 12.0; 12.5; 19.5; 22.0

intense the accumulation. Such crossing of terraces of different levels and younger deposits was called "scissors of Mirchink" (Makkaveev 1971), named after G. F. Mirchink, who first described this phenomenon in the Kuban valley. Their formation usually occurs when the foothill plain experiences sufficiently intensive tectonic movements of a negative sign or climatic conditions here are such that the river is characterized by large losses of runoff for evaporation and infiltration.

Thus, on the rivers crossing the central part of Colchis—Rioni and Khobi, flowing along the axis of the ancient deflection area, terrace "scissors" are very clearly expressed. The rivers filled an extensive sea bay over a long period of time, where marine quaternary sediments are buried under alluvial-deltaic sediments at depths starting from 300–400 m below the current sea level, and accumulation in this area occurred continuously, accompanied by an active extension of the delta. In the Rioni valley, the river terraces gradually decrease and in the central part of the lowland submerge under younger ones (Table 5.10).

The Amu Darya valley in the upper reaches, where, along with tectonic deflection, the main role in the accumulation of sediment is played by a longitudinal (downstream) decrease in flow due to evaporation losses in the absence of inflows, is characterized by a gradual decrease in the relative elevations of the terraces downstream of the river (Table 5.11), and then by the immersion of older levels under relatively young sediments near the settlement Kelif and downstream. A similar phenomenon is also observed on the middle Amur after its exit from the mountain structures of Maly Khingan (see Fig. 5.3B). Here, the transition from plunging the river to the directional accumulation of sediment within the Middle Amur lowland has been recorded in the confluence of floodplain steps and the floodplain surface, represented by the lake-alluvial plain, into a joint single-tier floodplain. Decrease in the level of the plain leads to the fact that it is below the water level of 1% of the water supply. In the upper Ob (see Fig. 5.9A) terrace "scissors" are not formed, but there is downstream unification of three or four steps of the floodplain, first in the one-tier floodplain (near the city of Barnaul), and then the appearance of the superimposed floodplain in the valley (below the city Kamen-na-Obi).

Table 5.10 Partitioning between terraces of the Rioni river and Colchis depression sediments ("scissors" of Mirchink)

Rioni Terraces (Geomorphology of Georgia 1971)					Colchis deposits near Poti	
No.	Relative height, m			Age	Laying depth of bottom below the sea level, m	Thickness, m
	Onsk hollow	Rachinsk hollow	Foothills (Kutaisi)			
VI	150–170	200–230	125–130	Chaudinsk	300–400	45–125
V	95–100	150	90–100			
IV	70	85–90	50–60	Ancient Euxiny	255–275	85–100
III	30–40	55–60	30–40	Usunlar	13–175	25–30
II	15–20	25–30	20–25	Karangatsk	105–145	85–100
I	5–10	10–15	5–10	New Black Sea	20–45	20–25
Floodplain	3	5–6	3	Nympea, modern	0–20	–

Table 5.11 Change in the relative heights of the upper Amu Darya terraces downstream (Central Asian materials of the Hydroproject Institute) (Berkovich et al. 1973)

Geomorphological level	Point and height above water level, m		
	Lower Panj	Termez	Kelif
Floodplain	5–6	4–5	3–5
I terrace	7–123	5–7	5–7
II terrace:			
Lower level	10–15	7–10	–
Upper level	40–45	12–13	10
Hilly watershed surface	50–100	20–50	–

If vertical changes in the lower reaches are associated with a reduction in the erosion base, this increases the accumulation of sediment and the effect of burial of terraces underneath the latest sediments in case of shelf erosion. In the case of incision in the lower reaches, the fan of the river terraces open towards the sea is formed Makkaveev 1955; Makkaveev et al. 1961), passing into one-age sea terraces. So, by Fedorov (1963) the Kodori terraces (Western Caucasus) within the seaside lowlands gradually move: I—to the New Black Sea, II—to the Late Karangatta, III—to the Early Karangatta, IV—to the Euxino-Uzuklar and V—to the Chaudin marine terraces; similar transitions are also observed in the lower reaches of other rivers of the Black Sea coast of the Caucasus—Psou, Gumista, Bzibi, Supsa, Natanebi.

Accumulation of sediment on rivers with wide floodplain channels leads to an increase in their bedding and an increase in flood levels. At first, high floodplains are more often flooded, then low terraces are covered with water, which turn into superimposed floodplains. The central and rear parts of the floodplains are below the intertidal level, separated from the rivers by channel ramparts (bunded floodplains), which are artificially built and transformed into flood dams. On a geological scale, the accumulation of sediment leads to the burial of terraces and even river valleys. As a result, accumulative alluvial plains are formed, the river network within which is intertwined, there is a phenomenon of bifurcation (the formation of branches connecting the neighboring rivers, on which the flow is carried out in one or another direction, depending on the height of the rise of levels on one or another river), consisting in their independent inflow into the sea, then in the transformation into a tributary of another river. Bifurcation is typical of the lower reaches of the Parana and Uruguay rivers in South America (Lukashova 1958), which flow together into the La Plata, Okumi and Eris-Tskari bays in the Colchis Depression (Mandych and Chalov 1970). In some cases, when the base level decreases, the incision leads to the "conservation" of such a furcation flow, which forms a peculiar transverse valley between the two independent rivers. Such is the connection between the Orinoco and Riu Negru rivers by the river Casikiara (Narskikh 1967). The Huang He, breaking through the riverbed ramparts and dams, periodically becomes a tributary of the Yangtze in the estuary area; repeatedly over the historical time, it changed its position in the band width of 600–800 km and flowed into the Yellow Sea, bypassing the mountainous Shandong peninsula from the north, as now, then from the south (Fig. 5.37). Similar reconstructions of the hydrographic network are typical for the lower Terek, which at the beginning of the 20th century directed its new channel—the Kargalynskiy breakthrough to the east, to the middle part of the Agrakhan bay; before that, from the middle of the 19th century from the station Kargalynskiy its main channel was the Talovka branch, which flowed into the Kizlyarskiy bay and had a north-north-eastern direction (Mikhailov 1997).

Lowering the base level and cutting rivers within such alluvial-deltaic plains can lead to the restructuring of the river network, as a result of which one river becomes a tributary of another, and its deltaic branches die off, and their continuation develops smaller independent river systems (Borsuk and Chalov 1975). The same effect is caused by the inflow of the delta branch of one river into the other, which is accompanied by an abrupt increase in water availability and erosion capacity of the latter. This phenomenon appears to have occurred in the lower reaches of the Northern Dvina. One of the branches, being more high-water, began to incise more intensively, which led to the rupture of the single bifurcation system of the Pinega-Northern Dvina; on the continuation of the former right branch of the Pinega, which independently flowed into the White Sea, formed an independent river network, while the left branch of the river. Pinega began to flow into the Northern Dvina. The formation of the Pinega-Severodvinsk delta plain led to the joining of the archipelago of the islands to the land, which now form an elevated watershed (the Yarga mountain has a height of 210 m) of the Northern Dvina and Kuloi rivers. At the same time, the upper reaches of the Kuloi and Pinega share the same floodplain, and part of the Pinega's flow goes to the Kuloi during floods.

Morphological evidence of this type of restructuring is the reduced alluvial watersheds between the sources of one river and the middle or lower reaches of another. Such is the decline in the Siberian Uvals at the Ob and Pura watershed,

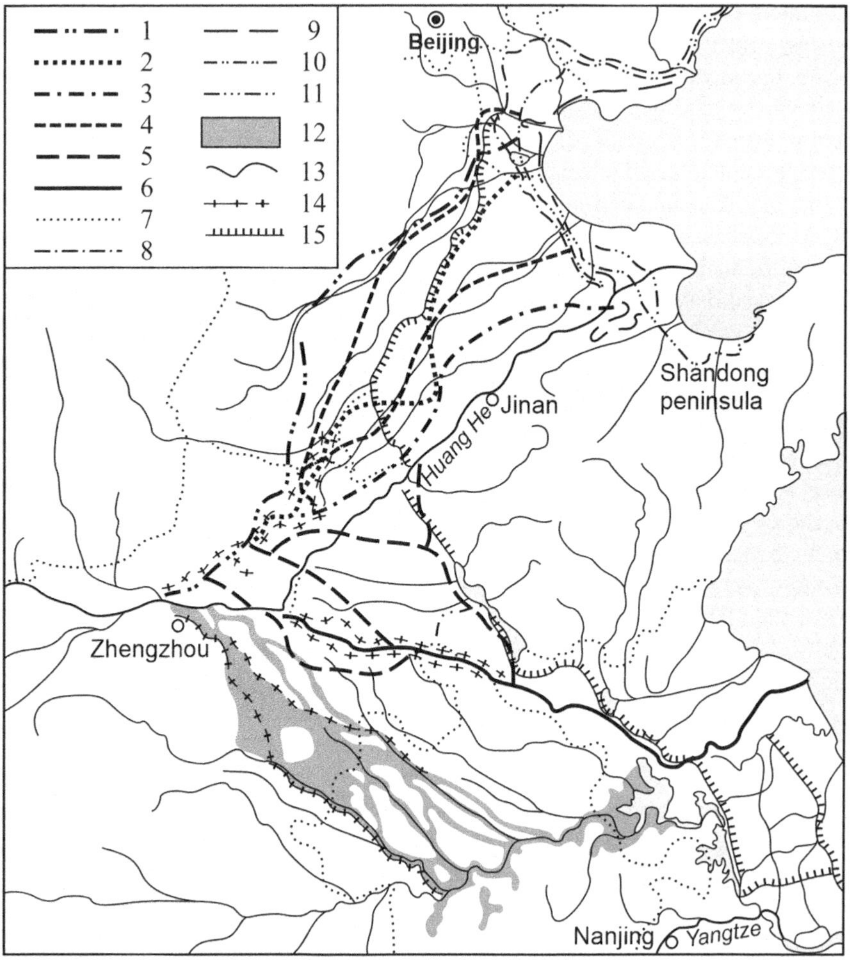

Fig. 5.37 Changes of the position of the Huang He river channel downstream beyond the historical period: 1—2278–602 B.C.; 2—602 B.C.–11 A.D; 3—11–1048; 4—1048–1128; 5—1128–1368; 6—1368–1855. Shoreline position: 7—602 B.C.; 8—11 A.D.; 9—1048; 10—1128; 11—1855; 12—flooded by Huang He river territory in 1938–1947; 13—modern channel of Huang He and Yangtze rivers and their tributaries; 14—flood control dams; 15—canals (Li 2003)

within which sandy lake-alluvial deposits were found (Arkhipov and Vdovin et al. 1970).

In the conditions of the formed river valleys, when there is no interaction of flows at the floodplain level, large regional changes in the structure of the hydro-network can be associated only with a powerful accumulation and increase of the river bed to the level of watersheds, i.e. with the restoration of the floodplain regime in large areas of land, similar to the coastal delta plains. In flat conditions, such transformations took place in the case of climate change towards dryness, often accompanied by ruptures in the river network and the formation of "blind" mouths of former tributaries of the main river. Thus, the Amu Darya changed its direction and began to flow not into the Caspian Sea, but into the Aral Sea, and later repeatedly changed the position of its channel.

In mountainous areas, the directional accumulation of sediment against the background of general tectonic movements of the positive sign leading to the filling of deep river valleys is usually associated with climate fluctuations or the local upwelling of rivers by local uplifts that affect them, similar to the formation of a rubble dam during earthquakes. In the valleys of large rivers in Central Asia, cyclic terraces are widespread, covering periods of formation including glaciation (filling the valley) and glaciation (incision) (Makarova et al. 1979). In the valley of the Alabuga river (see Fig. 5.36), the first one is evidenced by the Late Pleistocene pebble-boulder alluvium with a capacity of 50–150 m, in which the modern channel is incise. The next stage is characterized by a whole set of buried trenches, the position of which, like the current Alabuga canyon, does not coincide with the axis of the Middle Pleistocene valley. It is obvious that such an intra-length shift of trenches could have taken place when the valley was filled up to its main sides, after which the channel could have been located anywhere in the valley's wide bottom at a new stage of incision.

5.4.2 Geographical Analysis of Vertical Channel Deformation Development

The tectonic regime of flat-platform, coastal areas and mountainous countries, differences in sediment load, climatic conditions in different regions and water availability in rivers, different development of sloping lands and, as a consequence, the inflow of products of soil flushing into rivers, the presence of hydroelectric installations, the development of channel quarries of building materials and other types of technogenic impacts—all this determines a wide range of development of directed vertical change of river beds. At the same time, many of these factors are characterized by geographical regularities of their distribution, which determines the zonal, regional and local forms of representation of vertical channel changes (see Fig. 5.38).

In general, the rise of the continents determines the absolute dominance of river incisions within the inland areas. It is intensive in mountainous countries, where its speed is 1–7 cm/year. The largest of these values are those of the mountains of Middle and Central Asia, where the maximum rates of incision of the rivers (over 30 cm/year) are also recorded. The latter are created by a combination of incineration-friendly local conditions: the presence of easily eroded rocks and sediments (e.g., lake siltstones or ancient alluvium, filling the intermountain hollows) and the elimination of the local basis of erosion formed by the step created by a rockfall or formation of bedrocks.

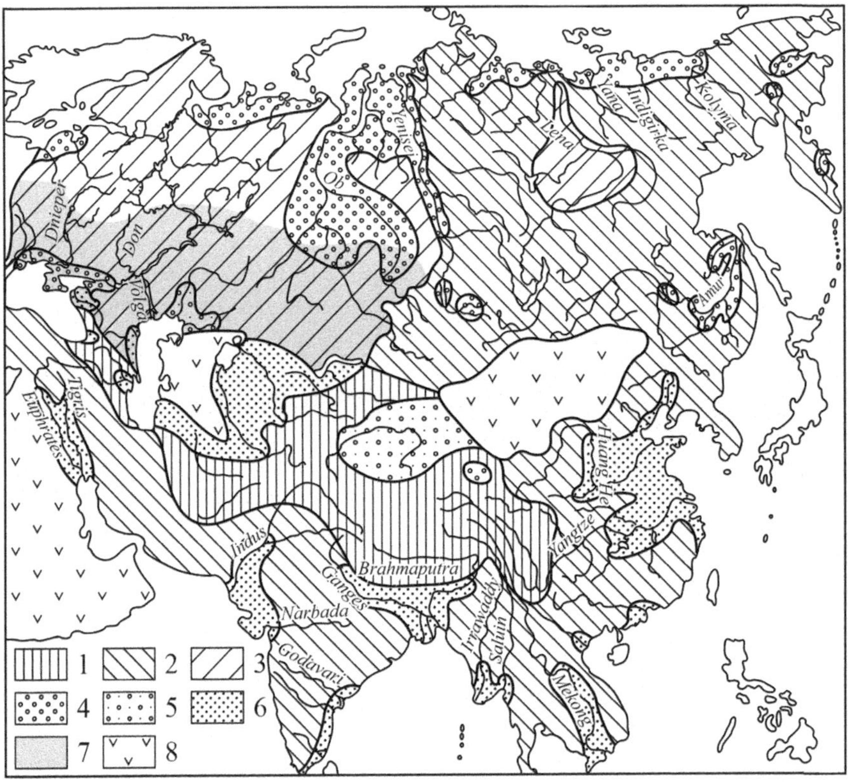

Fig. 5.38 Scheme of zoning of Eastern Europe and Asia (without island systems) at the orientation and speed of vertical channel changes on rivers. River incision (cm/year): 1—intensive (1–7 and more); 2—medium (0.5–1); 3—low (less than 0.5). Direct sediment accumulation (cm/year): 4—low (less than 0.5); 5—medium (0.5–1); 6—intensive (1–9); 7—small rivers siltation; 8—areas without the river runoff

Depending on the complexity of the mountainous terrain of the country, the incision velocities vary significantly along the length of the river systems, generally increasing from source to middle course and then decreasing the river flow from the mountains to the foothill zone. The presence of intermountain and mountain hollows, crossed by rivers, creates a complex mosaic of areas with different slopes and higher or lower plunging velocities, up to and including the change of plunging direction by the accumulation of sediment. The latter is represented along the periphery of depressions, where the rivers coming out of the mountain massif, due to a sharp change in slopes form almost merged debris cones (inner deltas). Some tectonic depressions experience submergence, and the rivers flowing through them accumulate sediment, constantly increasing bottom marks; in these cases, the directed accumulation is compensatory in nature. An example of such rivers is the upper Yenisei within the Minusinsk basin.

This is also the reason for the directional accumulation along the periphery of mountainous countries, where rivers run along or cross foothill boundary deflections with a steady tendency and descent. These are, for example, the rivers Colchis (Rioni, Khobi) and Pre-Caucasus (Kuban, Terek) (Gorelov 1958; Mandych and Chalov 197; Hydrology … 1993). Here, its speed reaches the first centimeters per year, and morphologically it is represented in the formation of a bunded channel, which is above the adjacent floodplain surface.

In the plain-platform regions, the river incision velocities make up the shares of millimeters—the first millimeters per year, increasing at the intersection of positive morphostructures and experiencing noticeable fluctuations in time depending on centuries-old and perennial changes in water availability: they are activated in humid years, cycles and periods and fade out in dry years. In many cases, river intakes are determined by morphological features only, since it is not possible to detect a movement of the curve $Q = f(H)$ within 1 cm from observations at hydrological stations over a period of even 100 years, and relatively small sediment load does not allow it to be detected by the sediment balance, which is within the limits of the accuracy of calculations. At the geological scale of time, major climatic changes significantly exceeded the impact of the tectonic factor, so that against the background of the general incision of rivers over their entire geological history there were large stages of accumulation and accumulation of thick layers of alluvial deposits.

In a number of cases, when local negative morphostructures intersect within the flat-platform regions, local contemporary accumulation of sediment is also noted (even from the curves of water discharge and levels) (Bylinskaya 1973).

Exceptions in the inland plain-platform areas are the areas of internal runoff, usually related to arid climatic conditions. The latter determine the absolute dominance of the directional accumulation on the flat rivers of Middle and Central Asia. It has led to the formation of a bunded channel (in the Amu Darya), ubiquitous ruptures in the hydrographic network and the formation of "blind" river mouths that do not reach the main waterways of the region and are lost in their own alluvial deposits.

Along the periphery of the continents, in the lower reaches of rivers flowing into marginal and inland seas and bays or directly into the oceans, directional accumulation of sediment predominates. Depending on the geomorphological structure of the territories flowing to the seas and oceans, it appears either only within the deltas of the rivers, or it infiltrates the largest rivers almost to the center of the continents, reaching the foothills. Such is the Ob, where in natural conditions the directed accumulation was observed from the middle of the upper reaches (the town of Kamen-na-Obi) up to the mouth, and in wet periods even the lower reaches of the Biya and Katun rivers were captured. At the Huang He River it is expressed along the entire middle and lower reaches corresponding to the meridional (south) and sub latitudinal (east and northeast) sections. On the Yangtze River the directed accumulation begins from the exit from the "Three Gorges", but at first it is intermittent (along the length of the river) due to the crossing of mountain ranges, and only from the city of Datuna goes down to the mouth. A similar picture is characteristic of the

Amur River, where the directed accumulation begins from its confluence of the Amur River with the Zeya river, where it first has an anthropogenic genesis (caused by increased outflows from the inflow of erosion products in the lower reach of the Zeya HPP), is then determined by pressure phenomena before the rise of the Maly Khingan, is replaced by plunging at the intersection of its mountain structures and then covers part of the middle and all the lower reaches up to the estuary, not stopping even in the incised channel at the intersection of the Sikhote-Alin spurs. Thus, the accumulation development of river channels is characteristic of the rivers of most of the West Siberian lowland, where it is manifested almost throughout the Ob and in the lower reaches of its tributaries, including the Irtysh, in the lower reaches of the Pur and Taz, the Yenisei, starting from the mouth of the Angara River (Bakhtin 1957), the Middle Amur and other lowlands, which flows through the lower Amur, the Great Chinese Plain. Directed accumulation of load is observed over a longer or shorter length of the rivers in southern Asia—the Pearl river, Mekong, Ganges and Brahmatutra, Indus, Tigris and Euphrates—covering those parts of their lower reaches where the rivers flow along the alluvial-deltaic plains (Gusarov 2002).

In the rivers of north-east Eurasia, the development of directional accumulation is limited to their deltas. On Lena and Yana, it is determined by their crossing in the very lower reaches, before reaching the delta, of the Kharaulakh mountains—the ridge of Chekanovsky (Lena) and Kularsky ridge (Yana). In other rivers in the region, the lack of accumulation is associated with low sediment load and the stability of the marine edge of the delta, i.e. the current lack of estuary elongation of these rivers. The estuarine lengthening of the rivers is one of the main reasons for the directional accumulation in the estuaries of the Yenisei, Taza and Ob, the deposits of which fill the estuarine bays, the Pechora, Mezen and Northern Dvina bays, forming the deltas of the Barents and White Seas, the Neman, which flows into the Curonian lagoon of the Baltic Sea, and the Dniester—into the Dniester estuary of the Black Sea.

The delta and estuary elongation of rivers on many rivers is superimposed on other factors causing the directional accumulation of sediment. The most important of them is the tectonic deflection of coastal areas, which sometimes corresponds to the exit to the foothills of the boundary deflections. This is the reason for the accumulation, along with the estuary elongation, in the plain part of the Kuban River and the Terek river. In the lower reaches of the latter, the Caspian Sea level fluctuations are also important: when the slope was lowered (before the development of the slope across the Agrakhan Peninsula), the slope was reduced due to the shallow surface of the estuarine seashore; when the flow was increased, the Caspian Sea level fluctuations also determined the accumulation of sediment at the mouths of the Ural River and, in particular, the Emba (Makkaveev et al. 1958). At the same time, on the Volga, due to the high flow rate, the lowering of the Caspian Sea level, despite gentle slope of the shelf, led to the development of deep erosion, and its current rise, due to the extent of the shallow seashore, which creates a kind of "buffer effect", practically did not affect the direction of vertical changes within the delta and began to express itself only 20 years later, when this effect was overcome, and the pressure spread upstream (Alekseevsky et al. 1988). At the same time, it is clear that the shortage of sediment that has arisen on the Volga River under the conditions of cascaded runoff reservoirs has played a certain role in this process. In the lower reaches of the Rioni, there is an effect of the estuary elongation, the compensatory filling of the tectonic deflection with sediments and the modern lowering of the sea level. On the middle and lower Amur, with relatively low sediment load and very low rates of estuary elongation, the main factor is the tectonic immersion, which covers lowlands and obviously involves the mountainous structures in themselves. The same is characteristic of the Great Plain of China and other coastal lowlands of south-east and south Asia. Starting with the Amur, despite its low rate (the first millimeters per year), and continuing southward, the directional accumulation of sediment leads to the formation of bunded channels and the excess of the surface of the vast surrounding area between the rivers by the water level. Almost the entire Great Plain of China is located below the water level on the Huang He and Yangtze, which has historically determined the repeated changes in the position of the Huang He channel within its width of more than 800. Thus, in the east, south-east and south of Asia, one can single out the Oceanic belt, within which intensive directed accumulation of load occurs on the rivers, at a rate quite high (more than 1 cm/year), sometimes reaching (at the Huang He river) catastrophic values (up to 8 cm/year). This is determined by the huge sediment load and the insignificant proportion of its component (0.1% on the Huang He river). The majority of suspended sediment (up to 80–90%) forms the channel. Suspended sediment load exceeds the transport capacity of the flows ($W_R >>> \mathrm{W}_{\mathrm{tr}}$), causing the development of directional accumulation.

It is small in the rivers of northern Eurasia, and only at the mouths of large rivers, with a significant proportion of bedload (Northern Dvina, Ob, Yenisei, Lena) it leads to river estuary elongation and regressive, albeit not intensive, accumulation, morphologically represented in the monolayer of floodplains, their high ash content and the formation of vast shallow lakes.

On the Yangtze river, a special place is occupied by a section of the river within the boundaries of the Lianghu Plain, which is a large local sinking morphological structure, occupied in historical time by a huge, relatively small lake. Nowadays, the lake is mainly filled with river sediment and only Dantinkhu and Poyankhu lakes have survived, connecting to the Yangtze, which flow into and out of the river. This dispersion of runoff increases the accumulation processes, the rate of which here is 3.7–4.9 cm/year (Chalov et al. 2000). This section of the Yangtze River is one of the most hazardous in terms of the probability of flooding.

On small rivers, due to accelerated (anthropogenic) soil erosion, accumulation processes cause siltation and degradation. At the same time, the relationship of soil erosion and gully erosion on natural conditions and the zonal nature of their development, including due to differences in the agricultural development of land in different natural zones, determined the zonal nature of these phenomena. They do not appear in the mountains and on small rivers of underdeveloped forest (taiga) areas in the European part of Russia, in Siberia and in the Far East. In the south direction, the degree of siltation of small rivers increases; if in the ploughed southern part of the forest and forest-steppe zones silted mainly streams (1st order rivers), and not everywhere, in the steppes of southern Russia and Ukraine siltation spreads to all small rivers up to the 4th order.

The generalization of available data on the direction and rates of vertical channel change made it possible to draw up a scheme of their development on the rivers of Eastern Europe and Asia (Fig. 5.38). Morphological features of both incision and directed accumulation of sediment have been taken into account, making it possible to extrapolate the data obtained to rivers and regions for which there are no accurate data based on network hydrometric measurements or sediment balance. The central part of the mainland within the alpine geosyncline belt is characterized by an absolute predominance of river incision and its highest rate (1–7 cm/year). In the mountainous structures, highlands and plateaus of south, east and northeast Asia, as well as south Siberia, which are very diverse in terms of tectonics, the incision of rivers is less intensive, obviously, due to the lower rate of tectonic rise, although it leads to the formation of incised, often rocky channels, devoid of alluvium. Deep erosion velocities here are millimeters per year. The rest, mainly northern Eurasia, belongs to the area of lowland rivers with predominantly weak riverbed incisions at first millimeter speeds, i.e. fractions of a millimeter per year. At the same time, such a slow incision covers almost all the rivers of the East European Plain (except for the lower reaches of some of them), the south of the West Siberian Lowland and Eastern Siberia, including those flowing through the Central Yakut Lowland. In particular, river Lena incises along its entire length, and the plunging velocities in the middle reaches (near Yakutsk) are 0.5 mm/year (Borsuk and Chalov 1973).

In the center of the mainland and along its periphery there are intermountain hollows, both forming areas of internal runoff and crossed by large rivers (Yenisei in the Tuva and Minusinsk hollows, the Amur above the Small Khingan). In these depressions, tectonic depressions cause a directed accumulation of sediment; in arid conditions (between the mountain structures of the Tien Shan and Kunlun), it increases due to a decrease in the runoff rates of the rivers downstream, from the mountains to the center of the depressions, where the discharge dries up at all.

The areas in which the rivers experience directional accumulation of sediment are confined to marginal coastal areas, usually experiencing tectonic depressions, and only within the West Siberian and Great Chinese Plains do they cover vast areas extending up the vast valleys and alluvial plains into the interior of the continent over distances of more than 1000 km. However, while in the north, due to low river load, the rate of accumulation is insignificant (fractions of a millimeter per year), in the south of the continent, where sediment load is colossal, it is becoming catastrophic. The latter is also characteristic of the internal flow areas in Central Asia (basins of the Syr Darya and Amu Darya rivers, Lake Balkhash, rivers flowing from the Kopet-Dag), as well as the rivers of the Pre-Caucasus and Transcaucasia flowing through foothill and intermountain deflections (Terek-Kuban, Kura-Rioni). The lower and partly middle Amur and its tributaries occupy an intermediate (in terms of tempo) position.

A special place in the scheme is occupied by the south of the forest, forest-steppe and steppe zones in the East European Plain and West Siberian Lowland, where the accumulation processes have covered small rivers, which are silted and degraded due to the excessive amount of soil erosion products in them.

Transformation (or strengthening) of the direction of vertical channel changes occurs as a result of the creation of reservoirs on rivers, which form the artificial basis of erosion for upstream river sections and create a deficit of sediment downstream. Even at the Huang He river after the construction of the Sangmensya hydropower plant, which spread all the way to the sea, but was then neutralized and compensated for by subsequent accumulation due to the continuing very intense river mouth elongation. A new wave of transgressive erosion occurred in the 2000s after the commissioning of the Xiaolandi HPP downstream. Sediment accumulation was replaced by riverbed incision in the lower reach of the Novosibirsk Hydroelectric Power Station in the Ob. The washout of the Zeya channel in the lower reach of the Zeya hydroelectric power station resulted in excess sediment entering the Amur channel, causing accumulation of sediment here (in natural conditions the channel below the confluence with the Zeya river slightly crashed). Above the reservoirs, due to the accumulation of sediment in the zone of variable support, regressive accumulation began to develop, expressing itself on rivers with sufficiently large sediment load (on the Ob above the Novosibirsk, on the Don above the Tsimlyansk reservoir, was recorded on the Dnieper above the Dnieper reservoir until it was in the cascade).

The development of quarries of building materials has led to similar consequences. On the Oka, the accumulation of sediment as a consequence of the transgressive spread of soil erosion products in the upper part of the basin, which was observed up to the 1940s, was replaced by a decrease in levels to 1.5–2 m. It was caused by both artificial deepening of the channel in the areas of non-refundable extraction sand and gravel extraction and by inbound river crossings associated with quarries due to the interception of sediment from upstream areas and the formation of level decline curves and an increase in the flow rate upstream of the river. Similar phenomena are observed on the rivers Belaya, Chulym, Tom, Neman and others, especially near large cities and urban agglomerations.

Literature

Afremov, D.N.: Anthropogenic influence on formation of longitudinal profile of the Dniester river. Izv. USSR Acad. Sci. Ser. Geogr. (3), 39–43 (1969)

Afremov, D.N.: Terrace spectrum and main stages of the Dniester Valley development. In: Longitudinal Profile of the Rivers and Their Terraces, pp. 70—82. MFGOSSR, M. (1978)

Alekseevsky, N.I.: Formation and Movement of River Load, p. 203. Moscow State University Publishing House, Moscow (1998)

Alekseevsky, N.I., Chalov, R.S.: Movement of Sediments and Fluvial Processes, p. 171. Moscow State University Publishing House, Moscow (1997)

Alekseevsky, N.I., Korotaev, V.N., Mikhailov, V.I.: Dynamics of the marine edge of the Volga Delta and the channel mode of its delta watercourses under Caspian Sea level fluctuations. In:

Soil Erosion and Fluvial Processes, Ep. 11, pp. 273–286. Moscow State University Publishing House, Moscow (1988)

Ananyan, A.K.: Methods of investigation of the process of change of the river beds flowing into Lake Sevan. In: Proceedings of III All-Union Hydrologic Congress, pp. 132–140. Hydrometeoizdat, T. V. L. (1960)

Arkhipov, S.A., Vdovin, V.V., Mizerov, B.V., Nikolaev, V.A.: West Siberian plain (in Russian). In: History of Siberia and Far East relief development, 280 p. Nauka, Moscow (1970)

Artamonov, K.F.: Rucal deformations of the lower rivers. In: Hydrometeorology to Help the National Economy, p. 81. Publishing house of the Academy of Sciences of the Uzbek SSR, Tashkent (1954)

Astakhov, N.E.: To geomorphology of a part of south-east Abkhazia. In: Proceedings of Institute of Geography Named After Vakhushti of the Academy of Sciences of GSR, T. XIV, pp. 19–27 (1961)

Atlas of the Lithuanian SSR, 215 l. GUGGK SSSR, M. (1981)

Atlas of the Ob River from Biysk to Yurt Tyaglovykh, 15 l. St. Petersburg (1897)

Avaryaskin, L.P.: Excess accumulation of Lower Amur sediment and its some hydrologic and morphological manifestations (in Russian). In: Problems of Far East Geography, pp. 11–20. T. 2. KhabSPI Publishing House, Khabarovsk (1976)

Bakhtin, N.P.: Some data on suspended runoff and erosion activities of the main rivers of the Krasnoyarsk Krai, vol. 8, pp. 127–133. Krasnoyarsk Krai University (1957)

Baryshnikov, N.B., Popov, I.V.: Dynamics of Channel Flows and Fluvial Processes, 456 p. Hydrometeoizdat, L. (1988)

Belyi, B.V., Vinogradova, N.N., Ivanov, V.V., Nikitina, L.N., Chalov, R.S., Chernov, A.V.: Morphology and deformations of the upper Yenisei riverbed between Sayano-Shushenskaya HPP and Krasnoyarsk reservoir (in Russian). In: Soil Erosion and Fluvial Processes, no. 12. MSU Publishing House, Moscow (2000)

Berkovich, K.M., Chalov, R.C.: Traces of the ancient lakes in Caucasus mountains (in Russian). Notes of Transbaik. br. of. GS USSR. vol. XXXI, pp. 48–53 (1969)

Berkovich, K.M., Lodina, R.V., Chalov, R.S.: Sediment runoff and regularities of the channel changes in the upper Amudarya flow (in Russian). In: Soil Erosion and Fluvial Processes, vol. 3, pp. 241–262. Moscow State University Publishing House, Moscow (1973)

Berkovich, K.M., Rouleva, S.I., Surkov, V.V., Chalov, R.C.: Fluvial processes and anthropogenic reformations of the lower Tom riverbed and their influence on the floodplain landscapes (in Russian). In: Soil Erosion and Fluvial Processes, Ep. 11, pp. 215–240. MSU Publishing House, Moscow (1998)

Berkovich, K.M., Zlotina, L.V., Turykin, L.A.: Anthropogenic changes of the Belaya River bed. In: Soil Erosion and Fluvial Processes, vol. 13, pp. 184–202. Moscow State University Publishing House, Moscow (2001)

Berkovich, K.M., Chalov, R.S.: Longitudinal profiles of the Western Transcaucasia rivers (in Russian). Izv. USSR Acad. Sci. Ser. Geogr (1), 65–70 (1977)

Berkovich, K.M., Garrison, L.M., Roulev, S.N., Chalov, R.S.: Channel morphology and channel changes of the upper Ob (in Russian). In: Land and water resources: anti-erosion protection and regulation of channels. Moscow State University Publishing House, Moscow, pp. 95–120 (1990)

Biske S. F. About the conditions of formation of the quaternary terraces of the Lena valley between the settlement Pokrovsk and the village Zhigansk. In: Proceedings of institute geology and geophysics, no. 8, pp. 5–39. SIB. Branch of the USSR SA (1964)

Borsouk, O.A., Chalov, R.S.: About the Incision of the Lena Riverbed. News VGS **105**(5), 452–456 (1973)

Borsouk, O.A., Chalov, R.S.: About the mechanism of the river network reorganization (in Russian). In: Problems of reorganization and interception of the river valleys, pp. 26–30. MF GS USSR, M. (1975)

Borsouk, O.A., Chalov, R.S.: About the mechanism of formation of river terraces. In: Longitudinal profile of the rivers and their terraces, pp. 13–18. MF GS USSR, MOSCOW (1978)

Burbank, D.: Nature **379**, 505–510 (1996)
Bylinskaya, L.N.: Fluvial processes and modern movements of the Earth's crust (by the example of the center and south-west of the Russian Plain). In: Modern movements of the Earth's crust, no. 5, pp. 484–490. Tartu (1973)
Bylinsky, E.N.: Influence of decrease in levels of Ilmen and Ladoga lakes on development of longitudinal profiles of inflows of Ilmen and Volkhov lakes. Vest. Moscow Un. Ser. Biol. Soil Sci. Geol. Geogr. (3), 221–231 (1959)
Bylinsky, E.N.: Influence of Dnieper thresholds on the development of longitudinal profile of the river. Dnieper in Quaternary Time, pp. 29–64. Collection of articles by post-graduates of MSU Ph. (1958)
Chalov, R.S.: Geographical Investigations of Fluvial Processes, p. 232. MSU Publishing House, Moscow (1979)
Chalov, R.S., Liu, S., Alekseevskiy, N.I.: Sediment Flow and Fluvial Processes on the Big Rivers of Russia and China, 216 p. MSU Publishing House, Moscow (2000)
Characteristics of channels and fluvial processes in the middle and lower reaches of the Yangtze River, 272 p. Science, Beijing (1985) (in Chinese)
Darbutas A.A., Chalov R.S. Vertical changes of the Neman riverbed and influence of anthropogenic factors on them. Geomorphology (3), 49–58 (1993)
Dedkov, A.P., Mozzherin, V.I.: Basic approaches to the study of changes in the flow regime and their geomorphological consequences. In: Causes and mechanism of drying of small rivers in Kazan: KazSU, pp. 9–26 (1996)
Demin, A.G., Lodina, R.V., Ruleva, S.N., Chalov, R.S.: The role of geomorphological factor in changes of channel types and composition of bed material alluvium on large mountain rivers (by the Example of Katun and Chuya). Geomorphology (4), 73–81 (1991)
Dobrovolskaya, N.G.: Features of subaqueous weathering on plain and mountain rivers (on the example of Lena and Kara-Bau, Western Tien Shan). Vest. Moscow Un. Ser. 5. Geogr. (2), 71–76, (1980)
Dynamic Geomorphology, 448 p. MSU Publishing House, Moscow (1992)
Fedorov, P.V.: Stratigraphy of quaternary sediments of the Crimean-Caucasian coast and some questions of geological history of the Black Sea. In: Proceedings of Geol, In-ta of the SA USSR, no. 88, 160 p. (1963)
Gadzhieva, Z.H.: Denudation Processes in Shale and Izvestnyakov Dagestan, 26 p. Abstract of a Dissertation phD Geographical Sciences. MSU Publishing House, Moscow (1982)
Geomorphology of Georgia, 611 p. Mezniereba, Tbilisi (1971)
Gorelov, S.K.: Regularities of the Structure of the River Floodplains of the Caucasus and Their Importance for the Analysis of the Young Tectonic Movements (in Russian), vol. 123, no. 6, pp. 1089–1091. FSA USSR. (1958)
Grishanin, K.V.: Stability of River and Canal Beds, 144 p. Gidrometeoizdat, L. (1974)
Gusarov, A.V.: Trends of erosion and suspended sediment flow in Asia in the second half of the XX century (in Russian). Geomorphology (4), 70–87 (2002)
Hakimov, S.K.: Fluvial Processes on the Mountain Rivers of the Western Tien Shan. Author's Thesis. phD Geographical Sciences, 29 p. MSU Publishing House, Moscow (1992)
History of the Kirghiz SSR, 798 p. V.I. Frunze, Kyrgyzstan (1984)
Howerd, A.: Nature **379**, 488–489 (1996)
Hydrology of the mouths of the Terek and Sulak rivers. In: Caspian Sea, 160 p. Nauka, Moscow (1993)
Ivanov, V.V., Makhinov, A.I., Chalov, R.S., Chernov, A.V.: Vertical channel changes on the middle Amur. Vest. Moscow Un. Ser 5. Geogr. (5), 32–38 (2000)
Ivanova, N.N., Golosov, V.N., Panin, A.V.: Agrarian development of territory and extinction of the rivers of the European part of Russia (in Russian). Geomorphology (4), 53–60 (1996)
Jablonskis, J., Lasinskas, M.: Lietuvos TSR upin, kadastras. Ill d, 640 p. Vilnius (1962)
Kaplin, P.A., Grekova, I.V., Parunin, O.B., Fyodorov, E.V., Shlyukov, A.I.: List of radiocarbon dating of laboratories of the geographical faculty of MSU and the Institute of Oceanology, SA USSR. Vest. Moscow. Un. Ser. 5. Geogr. (4), 104–108 (1971) (in Russian)

Karasev, I.F.: Fluvial Processes at Transfer of the Flow, 288 p. Hydrometeoizdat, L. (1975)

Khmeleva, N.V., Nikulin, F.V., Shevchenko, B.F.: Some results of stationary studies of erosion forms in the Northwest Caucasus. In: Soil Erosion and Fluvial Processes, Ed. I, pp. 142–151. MSU Publishing House, Moscow (1970)

Klaven, A.B., Vinogradov, V.A., Kostyuchenko, A.A.: Non-equilibrium processes in the formation of river channels. In: Makkaveevskie Readings—2004, pp. 8–25. MSU Publishing House, Moscow, (2005)

Kochetov, N.I.: Quantitative estimation of chemical denudation in the Western Caucasus on hydrochemical data. Geomorphology (3), 61–66 (1976)

Kochetov, N.I.: To the question of the modern process of sediment accumulation in the lower reaches of the Black Sea coast of the Caucasus and its connection with vertical movements. Rep. GS USSR **1**, 25–28 (1968)

Kolypaila, S.: Nemunas, 20 p. Kaunas (1940)

Korotayev, V.N.: Geomorphology of River Deltas, p. 224. MSU Publishing House, Moscow (1991)

Korotkiy, A.M., Makarova, T.R.: Main features of relief and exogenous geomorphological processes of the Kuril Islands (problem aspects). Geomorphology, (2), 82–92 (2006)

Kositskiy, A.G., Reteyum, K.F., Chutkina, L.P., Shenberg, N.V.: Scale effects of flow changes along the length of the river systems (in Russian). In: Problems of hydrology and hydroecology 1st Ed., pp. 141–156. MSU Publishing House, Moscow (1999)

Kovalchuk, I.P., Volos, S.I., Kholodko, L.P.: River Systems of the West of Ukraine: Scales and Tendencies of Transformation of Structure, Mechanism of Change of a Condition in XIX-XX Centuries And Reasons and Mechanism of drying of Small Rivers, pp. 43–56. KazSU, Kazan (1996)

Kozlovskiy, D.A.: Fluvial processes and the modern Earth crust movements (in Russian). In: Problems of Physical Geography, XVI Ed. Publishing House of the SA USSR, Moscow (1951)

Kurbanova, S.G., Petrenko, L.V.: Anthropogenic strengthening of accumulation of alluvium of the small rivers in east of Russian Plain (in Russian). In: Exogenous Processes and Environment, pp. 177–181. Nauka, Moscow (1990)

Lasinskas, M. Burnelkis, J.: Lietuvos TSR upin, kadastras, 206 p. II d. Vilnius (1960)

Lasinskas, M., Miceviclus, J., Jablonskis, J.: Lietuvos TSR upiu, kadastras, 214p. I d. Vilnius (1959)

Lavrushin, Y.A.: Types of Quaternary Alluvium of the Lower Yenisei. In: Proceedings of Geological Institute, SA USSR, 47 ed, 96 p. (1961)

Li, G.: Ponderation and Practice of the Yellow River Control. Yellow River Conservancy Press, 272 p (2003)

Lilienberg, D.A., Setunskaya, L.E., Blagovolin, N.S., Gorelov, S.K., Nikonov, A.A., Rozanov, L. L., Serebryanny, L.R., Filkin, V.A.: Morphological and structural analysis of the modern vertical movements of the European part of the USSR (in Russian). Geomorphology (1), 3–18 (1972)

Liu, S., Chalov, R.S., Ding, J., Ivanov, V.V., Korotayev, V.N., Li, J., Yang, H., Shaw, Y.: Regional variations of the Asian rivers suspended sediment flow in estuaries (in Russian). Vest. Moscow Univ. Ser 5. Geogr. (3), 44–51 (2001)

Lopatin, G.V.: Sediments of the USSR Rivers, p. 368. Geografgiz, Moscow (1952)

Losievsky, A.I.: Laboratory research of riffle formation processes. In: Proceedings of TsNIIVT, 36 ed., 98 p (1934)

Lukashova, E.N.: South America, p. 468. Uchpedgiz, Moscow (1958)

Lungershausen, G.F.: Geological history of the Middle Lena and some issues of stratigraphy of Quaternary sediments of Eastern Siberia. In: Proceedings of All-Union Meeting on the Study of the Quaternary Period, vol. III, pp. 209–217. Publishing House of the SA USSR, Moscow (1961)

Makarova, N.V., Makarov, V.I., Akinin, B.E.: Regularities of structure and development of river valleys in Central Asia in Quaternary period. History of river valley development and land reclamation problems, pp. 93–100. Western Siberia and Central Asia, Novosibirsk, Nauka (1979)
Makhinov, A.N., Chalov, R.S., Chernov, A.V.: Directed sediment accumulation and morphology of the lower Amur bed (in Russian). Geomorphology (3), pp. 70–78 (1994)
Makhinov, A.N., Makhinova, A.F.: Transformation of the anthropogenic relief in the area of alluvial deposits development (north of Khabarovsk Territory). Geomorphology (1), pp. 43–49 (2006)
Makkaveev, N. I.: Regressive Transformations of River Islands. Meteorol. Hydrol. (4), pp. 44–50 (1948)
Makkaveev, N.I., Belinovich, I.V., Vyshlov, K.P.: About the influence of the Caspian Sea level decline on the navigable conditions in the lower Volga and Ural. In: Fluvial Investigations to Improve Shipping Conditions. River Transport, Moscow, pp. 24–28 (1958)
Makkaveev, N.I., Chalov, R.S.: On the development of the surface topography of river terraces and signs of deep erosion by the example of the Upper Ob. New. Ser. geogr. **4**, 120–125 (1964)
Makkaveev, N.I.: Flow and Stream Processes, p. 116. MSU Publishing House, Moscow (1971)
Makkaveev, N.I.: River Channel and Erosion in its Basin, p. 347. Published in the SA USSR, Moscow (1955)
Makkaveev, N.I., Chalov, R.S.: Fluvial Processes, p. 264. MSU Publishing House, Moscow (1986)
Makkaveev, N.I., Chalov, R.S.: Methods of improvement of navigable conditions at Ob River. River Transp. no. 9. pp. 45–47 (1963a)
Makkaveev N.I., Chalov R. S. On Morphological Signs of Modern Accumulation in a River Valley. New. SA USSR. Ser. Geogr. (3), 84–89 (1963b)
Makkaveev, N.I., Chalov, R.S.: Some peculiarities of the big river valley floor connected with periodic changes of the flow norm (in Russian). Geography Issues, Col. 79. Geografgiz, Moscow,pp. 156–167 (1970)
Makkaveev, N.I., Kalinin, A.: On movement of large-fragment material in logs. Meteorol. Hydrol. (8), 61–68 (1968)
Makkaveev, N.I., Khmeleva, N.V., Zaitov, I.R, Lebedeva, I.V.: Experimental Geomorphology, 196 p. MSU Publishing House, Moscow (1961)
Mandych, A.F.: Value of the hard flow of the rivers of Western Georgia (in Russian). Vest. Moscow Un. Ser 5. Geogr., 134–137 (1967)
Mandych, A.F., Chalov, R.S.: Consideration of directionality of erosion-accumulative processes in calculations of solid runoff (on the example of the rivers of West Georgia). Problems of morpholithogenesis in river valleys, pp. 53–57. Transbaikal. branch GS USSR, Chita (1968)
Mandych, A.F., Chalov, R.S.: Direction of the modern erosion-accumulation processes in the lower reaches of the Western Caucasus rivers (in Russian). Vest. Moscow Un. Ser 5. Geogr. (1), 66–71 (1970)
Maruashvili, L.I.: Terraces of r. Kodor (Western Caucasus) as an example of terraforming in mountainous countries. Rep. SA USSR, vol. XXIV, no. 1, pp. 37–44 (1960)
Malinauskas, J., Zdankus, N.: Nemanas kasagi po Kainu. Mosrlfs Technika (8), 30 (1988)
Meshcheryakov, YuA: Young Tectonic Movements and Erosion-Accumulative Processes in the Northwestern Part of the Russian Plain, p. 90. Published in SA USSR, Moscow (1961)
Mikhailov, V.N.: Hydrological Processes in the River Mouths, p. 176. GEOS, Moscow (1997)
Mikhailova, M.V.: Hydrologic regime and dynamics of the Huang He estuary hydrographic network (in Russian). Water Resour. 28(1):105–117 (1998)
Mikhaylovskaya, O.I.: Quaternary terraces of Abkhazia. In: Proceedings of the First All-Union Geographer Congress, no. 3, pp. 82–94 (1934)
Muranov, A.P.: Huang He River, 88 p. Gidrometeoizdat, L. (1957)

Muranov, A.P.: Yangtze River. Hydrometeoizdat, L. 124 p (1959)
Narskikh, R.S.: Relief of South and Central America. In: Earth Relief, pp. 203–226. Nauka, Moscow (1967)
Natsag, J., Chalov, R.S.: Dynamics of River Flows and Channel Morphology on Rivers of Mongolia (Selenga River Basin). New. SA USSR. Ser. Geogr. (1), 110—116 (1978)
Neliubin, V.L., Amudaria, M.: Marine transport, p. 136. (1963)
Nemanishvili S. N. Terraces of the Rioni river valley above the city of Kutaisi. In: Proceedings of Geography Institute Named After Vakhushti, AS GSR, vol XX, pp. 36–38 (1964)
Nemanishvili, S.N.: Terraces of the Tskhenis-Tskali River Valley. In: Proceedings of Geography Institute Named After V.L. Vakhushti SA GSSR, vol XVIII, pp. 51–53 (1963)
Nikolskaya, V.V.: Morphosculpture of the Amur Basin, p. 296. Nauka, Moscow (1972)
Nikonov, A.A.: Determination of the river incision rate. Geomorphology (1), 24–35 (1973)
Obedientova, G.V.: Erosion Cycles and Formation of the Volga Valley, p. 240. Nauka, Moscow (1977)
Pokrovskiy, P.V.: On Roughness Coefficients of Mudflow Roughness. Meteorol. Hydrol. (7), 30–35 (1940)
Panin, A.V.: Morphology and dynamics of incised pebble-boulder rivers. Autoref. Dis. phD. Geogr. Sciences, 24. MSU Publishing House, Moscow (1991)
Pani, A.V., Sidorchuk, A.Y., Chalov, R.S.: Catastrophic speeds of a fluvial relief formation (in Russian). Geomorphology (2), 3–11 (1990)
Schnitnikov, A.V.: Variability of General Humidity of Materials in the Northern Hemisphere. Western GO USSR. vol. 16, 339 p. New series. M.; L.: Publishing house of SA USSR (1957)
Series of Picture Albums of the Yellow River, no 8, 21 p. Publication Center Yellow river conservancy commission, Beigangzi Street/Chongwen District, Beijing (1991)
Sidorchuk, A.Y., Chalov, R.S.: Incision of mountain rivers: speeds and causes. Nature (12), 36–45 (1996)
Surkov, V.V.: Dynamics of Floodplain Landscapes of the Upper and Middle Ob, p. 255. MSU Publishing House, Moscow (1999)
Talmaza, I.F., Kroshkin, A.N.: Hydromorphometric Characteristics of Mountain Rivers, p. 204. Kyrgyzstan, Frunze (1968)
Riverbed Regime in Northern Eurasia, 336 p. MSU Publishing House, Moscow (1994)
Timashev, I.E.: Inheritance of the Adjoining Section of the Lena River Valley. Vest. Moscow Un. Ser 5. Geogr. (1), 104–106 (1970)
Transport Use of Water Reservoirs, 224 p. Transport, Moscow (1972)
Tsvetkova, N.A.: Mode of Amudarya river sediments (in Russian). In: Questions of Hidrotechnik, no 13, pp. 5–86. Tashkent (1963)
Vinogradova, N.I.: About accuracy of forecast of siltation of reservoirs. Hydrotech. Constr. (9), 38–40 (1966)
Yang, H., Tan, Z.: Evolution of the Yangtze River bed in the Middle Reaches of the Tinjiang Region, 245 p. Water Management and Hydropower, Beijing (1999) (in Chinese)
Zaikov, B.D.: High Floods and Summer Floods on the USSR Rivers During Historical Time, 135 p. Hydrometeoizdat, L (1954)
Zenkovich, V.P.: Shore of the Black and Azov Seas, p. 376. Geografgiz, Moscow (1958)
Zhivago, A.V.: Some data on geomorphology and history of development of coast of Abkhazia, received by a method of comparative study of load. In: Proceedings in Geography Institute SA USSR, no. 51, pp. 12–32 (1952)

Literature

Afremov, D.N.: Anthropogenic influence on formation of longitudinal profile of the Dniester River. Izv. USSR Acad. Sci. Ser. Geogr. (3), 39–43 (1969)

Afremov, D.N.: Terrace spectrum and main stages of the Dniester Valley development. In: Longitudinal Profile of the Rivers and Their Terraces, pp. 70–82. MFGOSSR, M. (1978)

Alabyan, A.M.: Types of the plain river beds and factors of their formation. Geomorphology (4), pp. 37–42 (1992)

Alabyan, A.M., Alekseevsky, N.I., Berkovitch, K.M.: Sediment transport in the downstream area of Novosibirsk HES (in Russian). Proc. WestSibNIGMI 98, 79–82 (1992)

Alabyan, A.M., Alekseevsky, N.I., Evseeva, L.S., Zhuk, V.A., Ivanov, V.V., Surkov, V.V., Frolova, N.L., Chalov, R.S., Chernov, A. V.: Genetic analysis of the spring flooding causes of the Small Northern Dvina Valley in the area of Velikiy Ustyug city (in Russian). In: Soil Erosion and Fluvial Processes, Ep. 14, pp 104–130. Moscow State University Publishing House, Moscow (2003)

Alekseev, V.R.: Morphodynamic features of glacial sections of river valleys. In: Hydrology and Geomorphology of River Systems, pp. 207–208. SO RAS, Irkutsk (1997)

Alekseevskiy, N.I., Mel'nik, O.V.: Structure and peculiarities of the relief development of the gravel-pebble and boulder channels (in Russian). Geomorphology (3), 59–65 (1991)

Alekseevskiy, N.I., Sidorchuk, A.Y.: Accelerated erosion in the landscapes disrupted by mining work (by the Example of the Omoloy and Yana River Basin). In: Ecological Problems of Soil Erosion and Fluvial Processes, pp. 187–198. Moscow State University Publishing House, Moscow (1992)

Alekseevskiy, N.I., Vlasov, B.N., Chalov, R.P.: Some aspects of change of effective discharge of the Terek River. In: Water Resources of the Terek Basin and their use, pp. 117–126. Yuzhgiprovodkhoz Publishing, Rostov-Na-Donu (1983)

Alekseevskiy, N.I., Gaikovich, A.B.: To calculate the flow of incoming load on the unexplored rivers in the period between periods. Meteorol Hydrol (8), 96–102 (1987)

Alekseevskiy, N.I., Zaitsev, A.A., Chalov, R.S.: Balance of river load, channel changes and possibilities of regulation of the branched-out channel of the largest river (on an example of the river Lena near Yakutsk). In: Proceedings of AVN, no. 3, pp. 90–108 (1996)

Alekseevsky, N.I.: Formation and Movement of River Load, p. 203. Moscow State University Publishing House, Moscow (1998)

Alekseevsky, N.I., Chalov, R.C.: Movement of Sediments and Fluvial Processes, p. 171. Moscow State University Publishing House, Moscow (1997)

R. S. Chalov, *Fluvial Processes: Theory and Applications*,
https://doi.org/10.1007/978-3-030-66183-0

Alekseevsky, N.I., Korotaev, V.N., Mikhailov, V.I.: Dynamics of the marine edge of the Volga Delta and the channel mode of its delta watercourses under Caspian Sea level fluctuations. Soil erosion and fluvial processes, Ep. 11, pp. 273–286. Moscow State University Publishing House, Moscow (1998)

Alekseevsky, N.I., Chalov, S.R.: Hydrological functions of channel branching. In: Hydrology, Hydrochemistry and Hydroecology, T II. VGL "Obryi', pp. 53–59. Kiev (2006)

Altunin, S.T.: Regulation of Riverbeds, p. 352. Selkhozizdat, Moscow (1962)

Altunin, V.S.: Channel Bed Changes, p. 120. Kolos, Moscow (1972)

Altunin, V.S., Annaev, S.A., Ashirov, S.A.: Intensive Channel Changes of the Amudarya River, p. 103. Ylym, Ashkhabad (1975)

Amalitsky, V.P.: Diary of Observations of the Small Northern Dvina, 57 p. Published in the USSR Academy of Sciences, L. (1931)

Ananyan, A.K.: Methods of investigation of the process of change of the river beds flowing into Lake Sevan. In: Proceedings of III All-Union Hydrologic Congress. Hydrometeoizdat, T. V. L., pp. 132–140 (1960)

Andreev, O.V., Yaroslavtsev, I.A.: Channel changes on the river sections with the bridge crossings. In: Fluvial Processes, pp. 352–372. Publishing House of the USSR Academy of Sciences, Moscow (1958)

Andrem, E.D.: Effective and banxfull discharge of streams in the Vampa River basin, Colorado and Wyoming. J. Hydrol. **46**(3–4), 311–330 (1980)

Antropovsky, V.I.: Hydromorphological dependences and their further development (in Russian). Proc. SHI. Op. 169, 34–86 (1969)

Antropovsky, V.I.: Communication of types of fluvial process with determining factors. Proc. SHI. Op. 183, 70–80 (1970)

Antropovsky, V.I.: Criterial dependences of types of fluvial process. Proc. SHI. Op. 190, 5–18 (1972)

Antropovsky, V.I., Minaeva, G.I.: Background forecast of the riffle reformations on the river Vilyuy in the conditions of the regulated flow. Proc. SHI. Op. 216, 52–63 (1974)

Arkhipov, S.A., Vdovin, V.V., Mizerov, B.V., Nikolaev, V.A.: West Siberian Plain (in Russian). In: History of Siberia and Far East Relief Development, 280 p. Nauka, Moscow (1970)

Artamonov, K.F.: Rucal deformations of the lower rivers. In: Hydrometeorology to Help the National Economy, p. 81. Publishing house of the Academy of Sciences of the Uzbek SSR, Tashkent (1954)

Artamonov, K.F., Kroshkin, A.N.; Talmaza, V.F.: Basic principles of the river fluvial process theory of the stable mountain river beds (in Russian). In: Dynamics and Thermology of the Rivers and Reservoirs, pp. 139–145. Nauka, Moscow (1984)

Astakhov, N.E.: To geomorphology of a part of south-east Abkhazia. In: Proceedings of Institute of Geography Named After Vakhushti of the Academy of Sciences of GSR, T. XIV, pp. 19–27 (1961)

Atlas of the Lithuanian SSR, 215(l). GUGGK SSSR, M. (1981)

Atlas of the Ob River from Biysk to Yurt Tyaglovykh, vol. 15, no. l. St. Petersburg (1897)

Atlas of Fluvial Processes in the Lower Reaches of the Huang He River 114(l). Huang He River Water Management Commission, Zhengzhou (1985) (in Chinese)

Avaryaskin, L.P.: Excess accumulation of Lower Amur sediment and its some hydrologic and morphological manifestations (in Russian). In: Problems of Far East Geography, T. 2, pp. 11–20. KhabSPI Publishing House, Khabarovsk (1976)

Babak, Y.V., Turikin, L.A., Chalov, R.S.: Sediment flow, effective water discharge and channel patterns in the Oka basin. Geomorphology (4), 57–71 (2000)

Babich, D.B., Mikhailov, V.N., Morozov, V.N.: Estimation of river effective discharges in deltas of large rivers. Vest. Moscow. Univ. Ser. 5. Geogr. (1), 16–20. Vest (1985) (in Russian)

Babinski, Z.: Procesy korytowe Wisty ponizey Zaporu wodnei we Wloclawku. Dok. Geogr. Z(1–2), 94 s (1982)

Babinski, Z.: Procesy erozyjno-akumulacyjne poniiej stopnia wodnego Wiociawek, ich kosekwencje i wpiyw na morfodynamike planowanego zbioznika Nieszawa. 37 s. PAN, Torun (1997)
Babinski, Z.: Wplyw zapor na procesy kopytowe rzek abezialnych, 186 s. ydgoszcz (2002)
Bakhtin, N.P.: Some Data on Suspended Runoff and Erosion Activities of the Main Rivers of the Krasnoyarsk Krai, vol. 8, pp. 127–133. Krasnoyarsk Krai University (1957)
Barovskiy, N.A.: Dynamics of the meandering channels of the Middle Ob inflows (in Russian). In: Erosion Fluvial Processes and Hydroecology Problems, pp. 22–11. Geographer Faculty MSU, Moscow (2004)
Barovskiy, N.A., Chalov, R.S.: Hydrological-morphological analysis and the changes of the meandering beds of the Oka r. and its tributaries (in Russian). Vest. Moskow un. Ser 5. Geogr (2), 64–68. Vestnik Moskva (2004) (in Russian)
Baryshnikov, N.B., Samuseva, E.A.: Anthropogenic Influence on the Self-Regulating System of a Basin—River Stream—Channel, p. 220. Russian State Medical University, St. Petersburg (1999)
Baryshnikov, N.B.: River floodplains. Hydrometeoizdat, L., 152 p (1978)
Baryshnikov, N.B.: Morphology, Hydrology and Hydraulics of Floodplains, 280 p. Hydrometeoisdat, L. (1984)
Baryshnikov, N.B.: Influence of effect of interaction of floodplain and channel flows on transport capacity of channel flow. In: Erosion and Fluvial Processes, no. 4, pp. 141–151. MSU Publishing House, Moscow (2005)
Baryshnikov, N.B., Popov, I.V.: Dynamics of Channel Flows and Fluvial Processes, 456 p. Hydrometeoizdat, L. (1988)
Begin, Z.B.: The relationship between flow-shear stress and stream pattern. J. Hydrol. 52, 307–319 (1981)
Beliy B. V., Borsouk O. A., Ivanov V. V., Lodina R. V., Matveev B. V., Nikitina L. N., Chalov R. S., Chernov A. V. Conditions of formation, morphology, dynamics and regulation of the middle and lower Vilyui riverbeds (in Russian). In: Soil Erosion and Fluvial Processes, no. 9, pp. 146–175. MSU Publishing House, Moscow) (1983)
Belostotskiy, I.I.: Observations over Ryabi signs (in Russian). Izv. VGO (2), 225 (1940)
Belotserkovskiy, M.Y.: Erosion-ecological condition of the arable lands of the European territory of Russia (in Russian). In: Problems of the Russian Territory Ecological Tension Estimation: Factors, Zoning, Consequences, pp. 38–45. Moscow State University Publishing House, Moscow (1996)
Belotsky, K.N., Ruleva, S.N., Chalov, R.S.: Principles of zoning of large regions by conditions of passage of effective discharges of water (on an example of the Altai rivers). Geogr. Nat. Resour. (3), 165–169 (1995)
Belyi, B.V., Berkovich, K.M., Borsuk, O.A., Zaitsev, A.A., Lodina, R.V., Chalov, R.S., Chernov, A.V.: Morphology, dynamics and regulation of the Kirenga riverbed in connection with transport development of BAM zone (in Russian). In: Soil Erosion and Fluvial Processes, vol. 7, pp. 119–135. Moscow State University Publishing House, Moscow (1979)
Belyi, B.V., Vinogradova, N.N., Ivanov, V.V., Nikitina, L.N., Chalov, R.S., Chernov, A.V.: Morphology and deformations of the upper Yenisei riverbed between Sayano-Shushenskaya HPP and Krasnoyarsk reservoir (in Russian). In: Soil Erosion and Fluvial Processes, no. 12. MSU Publishing House, Moscow (2000)
Belyi, B.V., Ivanov, V.V., Nikitina, L.N., Chalov, R.S., Chernov, A.V.: Morphology and dynamics of the lower vychegda riverbed during active dredging and after their termination. In: Soil Erosion and Fluvial Processes, no. 13, pp. 146–183. Moscow State University Publishing House, Moscow (2001)
Berkovich, K.M.: Geographical Analysis of Anthropogenic Changes in Channel Processes, p. 164. GEOS, MOSCOW (2001)

Berkovich, K.M., Borsouk, O.A., Garrison, L.M., Kirik, O.M., Lodina, R.V., Rouleva, S.N., Chalov, R.S., Chernov, A.V.: Channel mode and regulation of the middle and lower Lena riverbed (in Russian). In: Soil Erosion and Fluvial Processes, no. 8, pp. 125–156. Moscow State University Publishing House, Moscow (1981)

Berkovich, K.M., Vinogradova N.N., Ivanov, V.V., Chalov, R.C.: Yenisei channel reformation below Krasnoyarsk hydroelectric power station under intensive technogenic load (in Russian). In: Soil Erosion and Fluvial Processes, no. 14, pp. 144–160. Moscow State University Publishing House, Moscow (2003)

Berkovich, K.M., Garrison, L.M., Roulev, S.N., Chalov, R.S.: Channel morphology and channel changes of the upper Ob (in Russian). In: Land and Water Resources: Anti-Erosion Protection and Regulation of Channels, pp. 95–120. Moscow State University Publishing House, Moscow (1990)

Berkovich, K.M., Zaitsev, A.A., Lodina, R.V., Chalov, R.S.: Fluvial processes on the large East Siberian rivers with a pebble-boulder alluvium and features of their regulation (in Russian). Vest. Moscow Un. Ser 5. Geogr. (in Russian), no. 3, 35–41. Vest (1985)

Berkovich, K.M., Zlotina, L.V.: Calculation of the river channel stability under the anthropogenic load conditions (in Russian). Geogr. Nat. Resour. (4), 117—123 (2003)

Berkovich, K.M., Zlotina, L.V., Ivanov, V.V., Nikitina, L.N., Ryazanov, P.I., Turykin, L.A., Chalov, R.S., Chernov, A.V.: Development of the lower and middle Dniester riverbed under the anthropogenic load (in Russian). In: Ecological Problems of the Soil Erosion and Fluvial Processes, pp. 141–165. MSU Publications, Moscow (1992)

Berkovich, K.M., Zlotina, L.V., Turykin, L.A.: Anthropogenic changes of the Belaya River bed. In: Soil Erosion and Fluvial Processes, vol. 13, pp. 184–202. Moscow State University Publishing House, Moscow (2001)

Berkovich, K.M., Zlotyna, L.V., Chalov, R.S.: Fluvial processes and urban industrial siltation of the Insar riverbed in Mordovia. Geogr. Nat. Resour. (2), 97–101 (1998)

Berkovich, K.M., Lodina, R.V., Chalov, R.S.: Sediment runoff and regularities of the channel changes in the upper Amudarya flow (in Russian). In: Soil Erosion and Fluvial Processes, vol. 3, pp. 241–262. Moscow State University Publishing House, Moscow (1973)

Berkovich, K.M., Lodina, R.V., Chalov, R.S., Shutov, A.M.: Formation conditions, stability and mode of the middle Ob riverbed changes (in Russian). Soil Erosion and Fluvial Processes, Ep. 2, pp. 186–210. MSU Publishing House, Moscow (1972)

Berkovich, K.M., Rouleva, S.I., Surkov, V.V., Chalov, R.C.: Fluvial processes and anthropogenic reformations of the lower Tom riverbed and their influence on the floodplain landscapes (in Russian). In: Soil Erosion and Fluvial Processes, Ep. 11, pp. 215–240. MSU Publishing House, Moscow (1998)

Berkovich, K.M., Rouleva, S.N., Chalov, R.C.: Upper Ob river regime (in Russian). Geogr. Nat. Resour. (4), 54–61 (1989)

Berkovich, K.M., Chalov, R.C.: Traces of the ancient lakes in Caucasus mountains (in Russian). Notes of Transbaik. br. of. GS USSR. vol. XXXI, pp. 48–53 (1969)

Berkovich, K.M., Chalov, R.S.: Longitudinal profiles of the Western Transcaucasia rivers (in Russian). Izv. USSR Acad. Sci. Ser. Geogr. (1), 65–70 (1977)

Berkovich, K.M., Chalov, R.S.: River channel mode and principles of its regulation at water transport development (in Russian). Geogr. Nat. Resour. (1), 10–17 (1993)

Berkovich, K.M., Chalov, R.S., Chernov, A.V.: Ecological Riverbed Science, 332 p. GEOS, M. (2000)

Berkovich K. M. Modern transformation of the upper Oka longitudinal profile (in Russian). Geomorphology, no. 1, pp. 43–49 (1993)

Biske, S.F.: About the conditions of formation of the quaternary terraces of the Lena valley between the settlement Pokrovsk and the village Zhigansk. In: Proceedings of Institute Geology and Geophysics, no. 8, pp. 5–39. SIB. BR. OF THE USSR SA (1964)

Bogardi, I.: Sediment Transport in Alluvial Channels, p. 826. Ahademia Kiado, Budapest (1974)

Bogomolov, A.L., Vinogradova, N.I., Vlasov, B.N., Krylenko, I.V., Chalov, R.C.: Influence of the catastrophic mudflows in Tyrnyauz on the Baksan River bed. Geomorphology (1), 65–74 (2002)

Bolysov, S.I.: Biogenic relief formation on land. Autoref. Diss. Dr. Geogr. Sciences, 41 p. MSU Publishing House, Moscow (2003)

Borovkov, V.S.: Fluvial Processes and Dynamics of the River Flows in the Urbanized Territories, 288 p. Gidrometeoizdat, L. (1989).

Borsouk, O.A., Gornak, A.A., Kirik, O.M., Chalov, R. S.: Eolian processes in the Lena r. valley. Bull. MPEI. Dep. Geol. (2), 151–152 (1975)

Borsouk, O.A., Dobrovolskaya, N.G., Lodina, R.V., Chalov, R.S.: Morphology of channels and modern channel alluvium on the Western Tien Shan mountain rivers (in Russian). Geomorphology (4), 60–67 (1981)

Borsouk, O.A., Zaitsev, A.A., Kirik, O.M., Lodina, R.V., Chalov, R.S.: Fluvial processes on the middle Lena (from Vitim to Pokrovsk). In: Soil Erosion and Fluvial Processes, no.13, pp. 203–228. MSU Publishing House, Moscow (2001)

Borsouk, O.A., Lodina, R.V.: Channel alluvium on the upper and middle Lena (in Russian). In: Soil Erosion and Fluvial Processes, Ep. 4, pp. 149–154. MSU Publishing House, Moscow (1974)

Borsouk, O.A., Lodina, R.V.: Grinding of channel alluvium depending on composition of rocks. In: Regularities of Erosive and Fluvial Processes in Different Natural Conditions, p. 355. MSU Publishing House, Moscow (1987)

Borsouk, O.A., Chalov, R.S.: About the Incision of the Lena Riverbed. News VGS 105 (5) 452–456(1973)

Borsouk, O.A., Chalov, R.S.: About the mechanism of the river network reorganization (in Russian). In: Problems of Reorganization and Interception of the River Valleys, pp. 26–30. MF GS USSR, M. (1975)

Borsouk, O.A., Chalov, R.S.: About the mechanism of formation of river terraces. In: Longitudinal Profile of the Rivers and Their Terraces, pp. 13–18. MF GS USSR, MOSCOW (1978)

Bray, D.I.: Regime equations for gravel-bed rivers. Gravel-Bed Rivers. Wiley & Sons, Chichtster (1982)

Brotherton, D.L.: On the origin and characteristics of river channel. J. Hydrol. **44**(3–4), 211–230 (1979)

Budanov, V.I.: Methodology of Expeditionary Research of the Coastal Zone of the Sea, p. 224. Nauka, Moscow (1964)

Burbank, D.: Nature 379, 505–510 (1996)

Butakov, A.N.: Hydraulics of the River Channel Mesoform Development, p. 216. RUDN Publishing House, Moscow (1999)

Bylinskaya, L.N.: Fluvial processes and modern movements of the Earth's crust (by the example of the center and south-west of the Russian Plain). Modern Movements of the Earth's Crust, no. 5, pp. 484–490. Tartu (1973)

Bylinsky, E.N.: Influence of Dnieper thresholds on the development of longitudinal profile of the river. In: Dnieper in Quaternary Time, pp. 29–64. Collection of articles by post-graduates of MSU Ph.

Bylinsky, E.N.: Influence of decrease in levels of Ilmen and Ladoga lakes on development of longitudinal profiles of inflows of Ilmen and Volkhov lakes. Vest. Moscow Un. Ser. Biology, Soil Sci. Geol. Geogr. (3), 221–231 (1959)

Carson, M.A.: The meandering-braided river threshold: a reappraisal. J. Hydrol. **73**(3–4), 315–334 (1984)

Chalov, R.S.: Geographical Investigations of Fluvial Processes, p. 232. MSU Publishing House, Moscow (1979)

Chalov, R.S.: Laws of Fluvial Geomorphology and Problems of Theoretical Geomorphology, pp. 111–121. Nauka, Moscow (1988)

Chalov, R.S.: Channel Investigations, p. 106. MSU Publishing House, Moscow (1995)

Chalov, R.S., Zavadsky, A.S., Panin, A.V.: River Radiations, p. 371. MSU Publishing House, Moscow (2004)

Chalov, R.S.: Determination of sediment discharge by the rate of displacement of accumulative forms of the river channel relief (in Russian). In: Quantitative Methods in Geography, pp. 19–22. MSU Publishing House, Moscow (1964)

Chalov, R.S.: Some features of the mountain rivers channel regime (in Russian). Meteorol. Hydrol. (4), 70–74 (1968)

Chalov, R.S.: Morphology and dynamics of the mountain river beds (in Russian). In: Proceeidngs of Moscow. GS USSR. Geomorphology, 3rd edn., pp. 8–10 (1969)

Chalov, R.S.: Relief of floodplains. Soil Erosion and Fluvial Processes, 1st edn., pp. 192–204. MSU Publishing House, Moscow (1970)

Chalov, R.S.: Emissions of the r. Vychegda. In: Soil Erosion and Fluvial Processes, 4th edn, pp. 128–142. MSU Publishing House, Moscow (1974)

Chalov, R.S.: Indicators of channel stability, their use for estimation of intensity of channel changes and ways of their improvement. In: Dynamics of Channel Flows, pp. 46–53. LPI Publishing House, L. (1983a)

Chalov, R.C.: Factors of fluvial processes and hierarchy of channel forms. Geomorphology (2), 16–26 (1983b)

Chalov, R.P.: Vertical Zoning in Development of the Fluvial Processes on the Mountain Rivers (in Russian). In: Study of Natural Conditions and its Applied Aspects, pp. 70–76. Nauka, Moscow (1985)

Chalov, R.S.: The main types of channel changes, forms of their manifestation and peculiarities of relations in different natural conditions (in Russian). In: Proceedings of the section of fluvial processes of Scientific Council "Integrated water resources use and protection', I edn, pp. 100–112. General questions of fluvial processes theory. Gidrometeoizdat, L. (1986)

Chalov, R.S.: General, geographical and engineering investigations: subject of research and positions in the system of sciences. Vest. Moscow. Un. Ser. 5. Geography (in Russian) (6), 10–16. Vest. (1992)

Chalov, R.S.: Types of fluvial processes and principles of morphodynamic classification of river channels. Geomorphology (1), 25–36 (1996)

Chalov, R.S.: General and Geographical Riverbed Science, 112 p. MSU Publishing House, Moscow (1997a)

Chalov, R.S.: Fluvial processes as hydrological phenomena and their geomorphological reflection. In: Hydrology and Geomorphology of River systems, pp. 23–32. Irkutsk (1997b)

Chalov, R.S.: Fluvial processes on small plain rivers as a special type of fluvial processes. In: Processes and Ecological Situation in Basins of Small Rivers, pp. 11–16. Udmurt Un, Izhevsk (2000)

Chalov, R.S.: Alpine rivers and rivers in mountains: longitudinal profile, morphology and channel dynamics. Geomorphology (3), 26–40 (2002a)

Chalov, R.S.: Erosion and fluvial processes in training courses for hydrologists and geomorphologists in state (classical) universities. In: Problems of Erosion and Fluvial Processes in Training Courses of Universities, pp. 6–11. AGPI Publishing House, Arzamas (2002b)

Chalov, R.S.: Rocky channels: spreading, morphology, specificity and conditions of deformations. In: Eighteenth Plenary Interuniversity Coordination Meeting on the Problem of Erosion, Fluvial and Etuary Processes, pp. 211–212. Kursk (2003)

Chalov, R.S., Alabyan, A.M., Ivanov, V., Lodina, R.V., Panin, A.V.: Morphodynamics of the Channels of Plain Rivers, 288 p. GEOS, M. (1998)

Chalov, R.S., Belyi, B.V.: Area of Siberia territory by character of the riverbed forming activity (in Russian). Meteorol. Hydrol. (12), 76–88 (1975)

Chalov, R.S., Belyi, B.V.: Regional features of the effective water discharge on the Central Asian rivers (in Russian). New. of the SA USSR. Ser. Geogr., no. 5. pp. 107–116 (1984)

Chalov, R.S., Berkovich, K.M., Borsuk, O.A., Kirik, O.M., Lodina, R.V., Paklina, L.M., Ruleva, S.N., Chernov, A.V.: Channel Mode of Middle and Lower Lena, 185 p. M., (1976) (Dep. in VINITI no. 2224–96)
Chalov, R.S. Dobrovolskaya, N.G.: Biochemical weathering and channel alluvium of the Western Tien Shan mountain rivers (in Russian). In: Lesomelioration in Uzbekistan, pp. 21–25. UzNIILH, Tashkent (1993)
Chalov, R.S., Zavadsky, A.S., Pakhomova, O.M.: Natural and anthropogenic manifestations of the fluvial processes in the different parts of the river network (in Russian). In: Problems of Hydrology and Hydroecology, 1st edn, pp. 203–216. MSU Publishing House, Moscow (1999)
Chalov, R.S., Kargapolova, I.N., Shtankova, N.N.: Study of fluvial processes on the Volga Basin rivers: history, condition, problems. In: Erosion and Accumulative Processes in the Upper and Middle Volga Basin, pp. 196–213. Udmurt Un, Izhevsk (2005)
Chalov, R.S., Liu, S., Alekseevskiy, N.I.: Sediment Flow and Fluvial Processes on the Big Rivers of Russia and China, 216 p. MSU Publishing House, Moscow (2000)
Chalov, R.S., Ruleva, S.N.: Changes of riverbeds and dangerous displays of fluvial processes on urbanized territories. In: Geography and Natural Resources, no. 4, pp. 17–23 (2001)
Chalov, R.S., Shtankova, N.N.: Sediment runoff, effective water discharge and morphodynamic types of river beds of the Kama basin. In: Problems of Physical Geography and Geoecology of the Urals, pp. 99–116. Perm State University, Perm (2000)
Chalov, S.R.: Fluvial processes on small rivers of Kamchatka and their technological changes. In: Soil Erosion and Fluvial Processes, 15 edn., pp. 164–175. MSU Publishing House, Moscow (2005)
Chalov, S.P.: Formation of the structure of channel branches. Geomorphology (1), 92–102 (2006)
Chalov R. S., Shtankova N.N. Sediment flow, the share of river load in it and their reflection in forms of manifestation of fluvial processes on the rivers of the Volga basin (in Russian). In: Proceedings of AVN, vol. 9, pp. 195–205 (2003)
Characteristics of Channels and Fluvial Processes in the Middle and Lower Reaches of the Yangtze River, 272 p. Science, Beijing (1985)(in Chinese).
Charnomsky, V.P.: About trays and channels of rectangular cross-section of the strengthened roughness in application to fish-passengers, fast-flows and platoon navigation. Materials on Investigation of the Volkhov River and Its Basin, pp. 143–148. L. (1924)
Chebotarev, A.I.: Hydrologic Dictionary, 224 p. Hydrometeoisdat, L. (1964)
Chemeris, M.P.: Estimation of the scale of anthropogenic transformations of water basins and channels of small rivers in Ukrainian Polesie. In: Problems of Erosion and Fluvial Processes, pp. 99–101. Izhevsk (1992)
Chernov, A.V.: Geomorphology of Floodplains of Plain Rivers, p. 198. MSU Publishing House, Moscow (1983)
Chernyshov, F.M.: Increase of Efficiency of Travelling Works on Multi-branch Sections of Navigable Rivers, p. 324. NIIVT Publishing House, Novosibirsk (1973)
Chistyakov, A.A.: Mountainous Alluvium, p. 288. Nedra, Moscow (1978)
Chitale, S.V.: River channel patterns. J. Hydraul. Div. ASLE. 96(HY-I.P.), 201–222 (1970)
Dal, V.I.: Explanatory dictionary of living Great Russian language, vol. 1, 700 p. Izd. of M.O. Wolf, St. Petersburg. M. (1880)
Darbutas, A.A.: Fluvial processes of the R. Neman (Nemunas) and Influence of the Anthropogenic Factor on them: Abstract from the … phD Geogr. Sciences, 27 p. MSU Publishing House, Moscow (1992)
Darbutas, A.A., Lodina, R.V., Chalov, R.S.: Modern channel alluvium of Neman. Vest. Moscow, Un. Ser. 5. Geography, (5), 90–100 (1993)
Darbutas, A.A., Chalov, R.S.: Morphodynamic types of the upper Neman channel, conditions of their formation and hydromorphological characteristics. Geomorphology (3), 72–79 (1991)
Darbutas, A.A., Chalov, R.S.: Vertical changes of the Neman riverbed and influence of anthropogenic factors on them. Geomorphology (3), 49–58 (1993)
Dasheng, Liu: Physical Geography of the World. Higher Education, Beijing (1986). (in Chinese)

Debolsky, V.K., Kotkov, V.M.: Peculiarities of the deficit forms dynamics in the incoming flows (in Russian). Meteorol. Hydrol. (10), 67–71 (1977)
Dedkov, A.P., Mozzherin, V.I.: Erosion and Sediment Sewage on Earth, p. 264. KazSU, Kazan (1984)
Dedkov, A.P., Kurbanova, S.G., Mozzherin, V.I.: About degradation of the river network in the Middle Volga region and its causes. In: Proceedings of AVN, 1s ed. pp. 93–98. M. (1995)
Dedkov, A.P., Mozzherin, V.I.: Basic approaches to the study of changes in the flow regime and their geomorphological consequences. In: Causes and Mechanism of Drying of Small Rivers in Kazan: KazSU, pp. 9–26 (1996)
Degtyarev, V.V., Degtyareva, V.V., Mukhin, MYu.: The Millennium of Dredging on Waterways (Review of Foreign and Domestic Periodicals), p. 28. NGAVT, Novosibirsk (2001)
Degtyarev, V.V.: Change of hydrological regime of Irtysh. River Trans. (12), 39–41 (1968)
Demin, A.G., Lodina, R.V., Ruleva, S.N., Chalov, R.S.: The Role of Geomorphological Factor in Changes of Channel Types and Composition of bed material Alluvium on Large Mountain Rivers (by the Example of Katun and Chuya). Geomorphology (4), pp. 73–81 (1991).
Design of Ship Passages on Free Rivers. In: TsNIIEVT Works. 36 edn., 262 p. Transport, Moscow (1964)
Dobrovolskaya, N.G.: Features of subaqueous weathering on plain and mountain rivers (on the example of Lena and Kara-Bau, Western Tien Shan). Vest. Moscow Un. Ser. 5. Geogr. (2), pp. 71–76 (1980)
Dobrovolskaya, N.G.: Microorganisms as a component of river geosystems. In: Soil Erosion and Fluvial Processes, 13 ed., pp. 133–144. MSU Publishing House, Moscow, (2001)
Dobrovolskaya, N.G.: Some features of microbial runoff of the small rivers (in Russian). In: XVIII Interuniversity Meeting on the Problem of Erosion, Fluvial and Estuary Processes, pp. 108–109. Kursk (2003)
Dobrovolskaya, N.G., Lodina, R.V., Chalov, R.S.: About a role of mechanical and biochemical weathering in formation of structure of channel alluvium. Geomorphology (1), 59–64 (1991)
Dvaretskas, V.V.: Structure and development of the river valleys of the edge zone of the last Scandinavian glaciation (by the example of the southern Baltic states): Author's Thesis … Dr. Geography of Sciences, 52 p. MSU Publishing House, Moscow (1989)
Dynamic Geomorphology, 448 p. MSU Publishing House, Moscow (1992)
Ecological Atlas of Russia, vol. 128, no. l. ZAO "Karta" (2002)
Ecology of Erosion and Fluvial Systems in Russia, 163 p. MSU Publishing House, Moscow (2002)
Edelstein, Ya.S.: Basics of Geomorphology, 400 p. Gosgeolizdat, M.; L, (1947)
Elenevskiy, R.A.: Questions of Study and Development of the River Floodplains, p. 100. VASKHNIL Publishing House, Moscow (1936)
Engels, F.: Dialectics of Nature, p. 260. Gosolitizdat, Moscow (1969)
Erosion Processes, 256 p. Thought, Moscow (1984).
Experimental Geomorphology, 2 edn., 173 p. MSU Publishing House, Moscow (1969)
Faktorovich, M.E.: Schematization of the channel formation processes and development of methods of calculation of channel changes (in Russian). In: Dynamics of Deposits in Open Channels, pp. 32–37. Nauka, Moscow (1970)
Fedorov, P.V.: Stratigraphy of Quaternary sediments of the Crimean-Caucasian coast and some questions of geological history of the Black Sea. In: Proceedings of Geol, In-ta of the SA USSR, 88th edn., 160 p (1963)
Fenixova, V.V.: Anthropogenic deposits of the South-East of Western Siberia. In: Problems of Geology and Paleogeography of Anthropogene, pp. 185–211. MSU Publishing House, Moscow (1966)
Ferguson, R.T.: The Threshold between meandering and braiding. In: Proceedings of International Conference on "Channels and channel control structures", no. 6, pp. 15–29. Berlin e. a. (1984)
Fluvial Processes in the Lower Reaches of the Weihe River. Nauka, Beijing (1984) (in Chinese)

Fluvial Processes and Waterways in the Rivers of the Ob Basin, 300 p. RIPEL-plus, Novosibirsk (2001)
Fluvial Processes on the Rivers of the Altai Territory. The Scale is 1:1000000, vol. 1, no. 1. GUGGK CSSR (1991)
Fluvial Processes on the Rivers of the Altai Region, 243 p. MSU Publishing House, Moscow (1996)
Fluvial Processes on the Rivers of the USSR. Scale 1:4000000, vol. 4 no. 1. GUGGK USSR (1990)
Gadzhieva, Z.H.: Denudation processes in Shale and Izvestnyakov Dagestan: Abstract of a dis… phD Geography Sciences, 26 p. MSU Publishing House, Moscow, (1982)
Geographical encyclopedic dictionary: Concepts and Terms, p. 432. Soviet Encyclopedia, Moscow (1988)
Geomorphology of Georgia, 611 p. Mezniereba, Tbilisi (1971)
Glazovskaya, M.A., Dobrovolskaya, N.G.: Geochemical Functions of Microorganisms, p. 153. MSU Publishing House, Moscow (1984)
Glushkov V.G. Morphology of the river channel. Proceedings of the 1st All-Russian Hydrological Congress, 286–290. L. (1925)
Golosov, V.N.: Erosion-accumulative processes in the upper links of a fluvial network of the mastered plains of a moderate belt. In: Proceedings of Intern. Dr. Geography of Sciences. MSU Publishing House, Moscow (2003)
Goncharov, V.N.: Dynamics of Channel Flows, 373 p. Hydrometeoisdat, L. (1962)
Gorelov, S.K.: Regularities of the structure of the river floodplains of the Caucasus and their importance for the analysis of the young tectonic movements (in Russian), vol. 123, No. 6, pp. 1089–1091. FSA USSR (1958)
Gorin, A.D., Lodina, R.V.: Peculiarities of the fluvial processes and morphology of the Lower Vitim Riverbed. Geomorphology (1), 87–93 (1994)
Gottlib, Y. L., Kuzmin, I.A., Razzorenov, F.F., Sokolnikov, N.M.: Natural Hydrological Studies in the Design of HPP, 268 p. Gidrometeoizdat, L. (1971)
Greenberg, Z.A.: About Morphometric Characteristics of Rivers. Meteorol. Hydrol (4), 43–45 (1950)
Grishanin, K.V.: Theory of the River Fluvial Process, p. 216. Transport, Moscow (1972)
Grishanin, K.V.: Dynamics of Channel Flows, 488 p. Hydrometeoizdat, L. (1969)
Grishanin, K.V.: Stability of River and Canal Beds, 144 p. Gidrometeoizdat, L. (1974)
Gudyalis, V.K.: Relief and Quaternary Deposits of the Baltic, p. 264. Mintis, Vilnius (1973)
Guidelines for Designing a Radical Improvement of Navigational Conditions in the Difficult Areas of Free Rivers, 310 p. Transport, L. (1974)
Gusarov, A.V.: Trends of erosion and suspended sediment flow in Asia in the second half of the XX century (in Russian). Geomorphology (4), 70–87 (2002)
Gusev, M.N.: Morphodynamics of the Bottom of the Upper Amur Valley, p. 220. Dalnauka, Vladivostok (2002)
Hakimov, S.K.: Fluvial Processes on the Mountain Rivers of the Western Tien Shan: Author's Thesis. phD Geogr. Sciences, 29 p. MSU Publishing House, Moscow (1992)
Handelman M.M., Popov I. V. About development of the river channel conjugations (in Russian) // Proc. of SHI. Ed. 307. 1987. p. 3-16.
Harrison, L.M.: Influence of the hydrotechnical measures on the bottom sediment particle size (on the example of Ob) (in Russian), (4), 47–50. Vest. Moscow Un. Ser. 5. Geography (1979)
Henderson, F.M.: Stability of alluvial channels. J. Hydr. Div. ASCE. 87(6) (1961)
History of the Kirghiz SSR, 798 p. Kyrgyzstan, V.I. Frunze (1984)
Hjulstrom, F.: Studies of the morphological activity of rivers as illustrated by the River Furis. Bull. Geol. Inst. Upsala. 35, 35–58 (1935)
Hmaladze, G.N.: Sediment Loads by the Rivers of the Black Sea Coast of Caucasus, 168 p. Gidrometeoizdat, L. (1978)
Howerd, A.: Nature 379, 488–489 (1996)

Hydrology of the mouths of the Terek and Sulak rivers. Caspian Sea, 160 p. Nauka, Moscow (1993)

Ivanov, V.V.: Formation Conditions, Hydrological-Morphometric Dependences and Changes in Relation to the Straight, Unbranched Channels: Author's… of Dis. phD Geog. Sciences, 23 p. MSU Publishing House, Moscow (1989)

Ivanov, V.V., Makhinov, A.I., Chalov, R.S., Chernov, A.V.: Vertical Channel Changes on the Middle Amur. Vest. Moscow, Un. Ser 5. Geography (5), 32–38 (2000)

Ivanova, N.N., Golosov, V.N., Panin, A.V.: Agrarian development of territory and extinction of the rivers of the European part of Russia (in Russian). Geomorphology (4), 53–60 (1996)

Jablonskis, J., Lasinskas, M.: Lietuvos TSR upin, kadastras. Ill d, 640 p. Vilnius (1962)

Jasmund, R.: Flissende Gewasser. Der Wasserbau. Bd. 111. Lerpzig (1911)

Kalinin, A.M.: Formation of dniester valley slopes and fluvial processes. In: Regularities of Erosion and Fluvial Processes in Various Natural Conditions, pp. 311–312. MSU Publishing House, Moscow (1987)

Kalinin, A.M., Kirik, L.S.: Quantitative estimation of a role of slope processes of a valley of the top Aldan as the factor of channel formation. Research of Fluvial Processes for Practice of a National Economy, pp. 229–230. MSU Publishing House, Moscow (1983)

Kamenskov, Y.I.: To definition of the concept and value of effective water discharge. Meteorol. Hydrol. (7), 119–120 (1986)

Kapitsa, P.L.: Experiment. Theory. Practice, 352 p. Nauka, Moscow (1977)

Kaplin, P.A., Grekova, I.V., Parunin, O.B., Fyodorov, E.V., Shlyukov, A.I.: List of radiocarbon dating of laboratories of the geographical faculty of MSU and the Institute of Oceanology, SA USSR. Vest. Moscow. Un. Ser. 5. Geography (4), 104–108 (in Russian) (1971)

Karasev, I.F., Kovalenko, V.V.: Stochastic Methods of River Hydraulics and Hydrometry, p. 208. Hydrometeoizdat, SPb (1992)

Karasev, I.F.: Fluvial Processes at Transfer of the Flow, 288 p. Hydrometeoizdat, L. (1975)

Karasev, I.F.: Effective water discharges. Meteorol. Hydrol. (8), 94–98 (1986)

Karaushev, A.V.: Problems of the Natural Water Flow Dynamics, 392 p. Gidrometeoizdat, L. (1960)

Karaushev, A.V.: River Hydraulics, 416 p. Hydrometeoisdat, L. (1975)

Karaushev, A.V.: Theory and Methods of Calculation of River Load, 272 p. Gidrometeoizdat, L. (1977)

Kellerhals, R., Churh, M., Bray, D.J.: Classification and analysis of river processes. J. Hedraul. Div. Proc. Amer. Soc. Civ. Eng. **7**, 813–829 (1976)

Kennedy, J.F.: The mechanics of dunes and antidunes in erodible-bd channels. Fluid Mech. **16**, 521–544 (1963)

Khmeleva, N.V., Vinogradova, N.N., Samoilova, A.A., Shevchenko, B.F.: Mountain River Basin and Exogenous processes Within it (results of stationary studies), 187 p. MSU (2000)

Khmeleva, N.V., Vinogradova, O.V., Sysoeva, S.M.: Influence of development of alluvial rashes on fluvial processes of alpine and one and a half rivers in Eastern Siberia. In: Soil Erosion and Fluvial Processes, 10th edn, pp. 121–132. MSU Publishing House, Moscow (1995)

Khmeleva, N.V., Nikulin, F.V., Shevchenko, B.F.: Some results of stationary studies of erosion forms in the Northwest Caucasus. In: Soil Erosion and Fluvial Processes, Ist edn, pp. 142–151. MSU Publishing House, Moscow (1970)

Klaven, A.B., Snishchenko, B.F.: Basics of the feature of fluvial process in the lower course of Selenga r. In: Proceedings of SHI., 252 ed., pp. 38–51 (1978)

Klaven A. B., Vinogradov V. A., Kostyuchenko A.A.: Non-equilibrium processes in the formation of river channels. In: Makkaveevskie readings—2004, pp. 8–25. Moscow: MSU Publishing House (2005)

Kochetov, N.I.: To the question of the modern process of sediment accumulation in the lower reaches of the Black Sea coast of the Caucasus and its connection with vertical movements 1st edn., pp. 25–28.Rep. GS USSR (1968)

Kochetov, N.I.: Quantitative estimation of chemical denudation in the Western Caucasus on hydrochemical data. Geomorphology (3), 61–66 (1976)
Kolypaila, S.: Nemunas. Kaunas, 20 p (1940)
Kondratyev, N.E.: River Fluvial Processes and Changes of Reservoir Banks, p. 258. Znak, SPb (2000)
Kondratyev, N.E.: About discreteness of the fluvial processes (in Russian). In: Problem of the Fluvial Processes, pp. 34–42. Hydrometeoizdat, L. (1953)
Kondratyev, N.E.: Kinematic structure of a flow at a rippled structure of a bottom (in Russian). In: Proc. SHI. 116 ed., pp. 3–18. (1964)
Kondratyev, N.E., Lyapin, A.N., Popov, I.V., Pinkovsky, S.I., Fedorov, N.N., Yakunin, I.A.: Fluvial Process 372 p. Hydrometeoizdat, L. (1959)
Kondratyev, N.E., Popov, I.V., Snischenko, B.F.: Basics of Hydromorphological Theory of the Fluvial Process, 272 p. Gidrometeoizdat, L. (1982)
Kopaliani, Z.D.: About the Ratio of Bottom and Suspended Sediment Discharge in Rivers. In: Hydrophysical Processes in Rivers and Reservoirs, pp. 143–147. Nauka, Moscow (1985)
Kopaliani, Z.D., Kostiuchenko, A.A.: Calculations of the bottom sediment discharge in the rivers. Coll. Works Hydrol. (27), 25–40 (2004)
Kopaliani, Z.D., Tvalavadze, O.A., Noselidze, L.V.: Hydraulic modeling of the fluvial process of the Anosovka river piedmont section on the bridge crossing (in Russian). In: Newsletter of the fluvial processes section of Scientific Council on the problem 'Integrated use and protection of water resources of SCST'', Issue paper. 3. Problems of hydraulics and river fluvial process of the mountain rivers, pp. 88–106. Gidrometeoizdat, SPb (1992)
Kopaliani, Z.D., Tskhadadze, V.S.: Types of river channels in Western Georgia. Proc. SHI. (195), 20–32 (1972)
Korchoha, Yu.M.: Investigation of ridge motion of deposits on the river Polomet. In: Proceedings of SHI, 161 edn., pp. 98–119 (1968)
Koronkevich, N.I.: Water Balance of the Russian Plain and Its Anthropogenic Changes. Scienc, 205 p. M. (1990)
Koronkevich, N.I., Chernogayeva, G.M., Yasinskiy, S.V.: Conceptual Problems of the Solution of the Small Rivers Problem in Russia. In: Small Rivers of Russia, pp. 11–19. IG RAN-MC RGO, Moscow (1994)
Koronovsky, N.V., Yakusheva, A.F.: Principles of Geology, p. 414. Higher School, Moscow (1991)
Korotayev, V.N.: Geomorphology of River Deltas, p. 224. MSU Publishing House, Moscow (1991)
Korotayev, V.N. Morphodynamics of Watercourses in the Western Part of the Volga Delta. In: Proceedings of AVN, 9 edn, pp. 123–139. M. (2003)
Korotayev, V.N., Lodina, R.V., Miloshevich, V.A. Sidorchuk, A.Yu, Chalov, R.S.: Formation of the Yana river delta and development of its estuary bars (in Russian). In: Soil Erosion and Fluvial Processes, 6 edn., pp. 123–159. MSU Publishing House, Moscow (1978)
Korotayev, V.N., Sidorchuk, A.Yu., Chalov, R.S.: Lithodynamics of the northern Siberian river deltas (in Russian). Water Resour, no. 1, pp. 138–145 (1980)
Korotkiy, A.M., Makarova, T.R.: Main features of relief and exogenous geomorphological processes of the Kuril Islands (problem aspects). Geomorphology (2), 82–92 (2006)
Kositskiy, A.G., Reteyum, K.F., Chutkina, L.P., Shenberg, N.V.: Scale effects of flow changes along the length of the river systems (in Russian) Problems of hydrology and hydroecology, 1st edn, pp. 141–156. MSU Publishing House, Moscow (1999)
Kostyaev, A.G.: Periglacial deposits and structure of the low terraces of the Walday age in the Northern Dvina valley (in Russian). In: Periglacial Phenomena on the Territory of the USSR, pp. 188–200. MSU Publishing House, Moscow (1960)

Kovalchuk, I.P., Volos, S.I., Kholodko, L.P.: River Systems of the West of Ukraine: Scales and Tendencies of Transformation of Structure, Mechanism of Change of a Condition in XIX–XX Centuries and Reasons and Mechanism of Drying of Small Rivers, pp. 43–56. KazSU, Kazan (1996)

Kovalchuk, I.: Regional Ecological-geomorphological Analysis, 439 p. Lvov (1997)

Kozlovskiy, D.A.: Fluvial Processes and the Modern Earth Crust Movements (in Russian). In: Problems of Physical Geography, XVI ed., pp. 79–102. Publishing House of the SA USSR, Moscow (1951)

Krasnov, A.N.: Shapes of the land surface and figures who create them. In: Fundamentals of land science II edn, 235 p. I. Kharkov (1897)

Kroshkin, A.N.: To determination of hydromorphological characteristics and average weight concentration of attracted load on mountain rivers (in Russian). In: Movement of load in open channels, pp. 111–119. Nauka, Moscow (1970)

Kurbanova, S.G., Petrenko, L.V.: Anthropogenic strengthening of accumulation of alluvium of the small rivers in east of Russian Plain (in Russian). In: Exogenous Processes and Environment, pp. 177–181. Nauka, Moscow (1990)

Kuzin, P.S., Babkin, V.P.: Geographical Laws of Hydrologic Regime of Rivers, 200 p. Gidrometeoizdat, L. (1979)

Kuznetsov, K.L., Chalov, R.S.: Fluvial processes and morphology of the mountain river beds in conditions of active mudflow activity (on an example of the rivers of the northern slope of Zailijsky Alatau). Geomorphology (2), 71–78 (1988)

Kzhemen, K., Helmitsky, V.: Application of numerical taxonomy to river bed typology (on the example of the Feshi River, Mount Cairngorm, Scotland). In: Soil Erosion and Fluvial Processes, 11 ed, pp. 177–188. MSU Publishing House, Moscow (1998)

Lapshenkov, V.S.: River Hydraulic Engineering, p. 408. NGMA Publishing House, Novocherkassk (1999)

Lapshenkov, V.S.: Prediction of the Channel Changes in the River Junctions, 240 p Gidrometeoizdat, L. (1979)

Lasinskas, M., Burnelkis, J.: Lietuvos TSR upin, kadastras. II d. Vilnius, 206 p (1960)

Lasinskas, M., Miceviclus, J., Jablonskis, J.: Lietuvos TSR upiu, kadastras. Id.Vilnius, 214 p (1959)

Latyshenkov, A.M.: Basics of Hydraulics, 248 p. Hydrometeoizdat, L. (1971)

Lavrushin, Yu.A.: Types of quaternary alluvium of the lower Yenisei. Proceedings of Geological Institute SA USSR. 47th edn., 96 p (1961)

Lebedeva, I.V.: Change of particle size distribution of modern channel alluvium downstream of plain river (by the example of Vychegda river). Lithol. Miner. (5), 37–44 (1966)

Lebedeva, N.V., Chalov, R.S.: Experience of application of the map-metric method in studying of the river bed sediment granulometry (in Russian). MPEI Vest. Moscow Un. Ser 5. Geogr. (3), 92–94 (1966)

Leliavskiy, N.S.: On the river currents and the formation of a river bed. In: Proceedings of the 2nd Congress of Hydraulic Engineers in 1893. St. Petersburg, 1893. Problems of Hydraulic Engineering of Free Rivers, pp. 18–136. Rechizdat, M. (1948)

Leliavsky, N.S.: On the study of the movement of sand spits near Alexandrovsk. In: Proceedings of the 3rd Congress of Russian Workers on Waterways in 1896. Part 2. St. Petersburg, 1896, pp. 146–156. Issues of hydraulic engineering of free rivers. Rechizdat, M. (1948)

Lelyavskiy, N.S.: About deepening of the big river niche (in Russian). In: 10th Congress of Russian Figures on Waterways. Kiev (1904) (Problems of Hydraulic Engineering of the Free Rivers, pp. 156–208. Rechizdat, M. (1948)).

Lenin, V.I:. Philosophical Notebooks/Selected Works, vol. 5, Part II, 672 p. Publishing House of thepolit, Moscow (1986)

Leopold, L.B., Wolman, M.G.: River channel patterns—braided, meandering and straight. US Geol. Sure Prof. Pap. 282-B, 1–85 (1957)

Levanidov, V.Ya.: On the hydrological regime of spawning grounds for chum salmon and pink salmon (in Russian). Izv. TIN-RO 64, 101–125 (1968)
Levashov, A.A.: Deformation rate determination of some characteristics of morphological formations in the Nadym river channel. Meteorol. Hydrol. (3), 92–95 (1973)
Levi, I.I.: Dynamics of Streams, 250 p. Hydroenergoizdat, Moscow, L. (1957)
Li, G.: Ponderation and Practice of the Yellow River Control, 272 p. Yellow River Conservancy Press (2003)
Lilienberg, D.A., Setunskaya, L.E., Blagovolin, N.S., Gorelov, S.K., Nikonov, A.A., Rozanov, L. L., Serebryanny, L.R., Filkin, V.A.: Morphological and structural analysis of the modern vertical movements of the European part of the USSR (in Russian). Geomorphology (1), 3–18 (1972)
Linguistic Aspect of Terminology Standardization, 126 p. Nauka, Moscow (1993)
Lisitsyn, A.P.: Processes of Territorial Sedimentation in Seas and Oceans, p. 272. Nauka, Moscow (1991)
Litvin, L.F.: Development of erosion processes in mountainous conditions (on the example of the southern slope of the Western Caucasus): Author's Thesis … phD Geogr. Sciences, 16 p. MSU Publishing House, Moscow (1971)
Litvin, L.F., Kiryukhina, Z.P.: Ecological aspects of soil erosion and contamination of surface waters with biogenic elements. In: Soil Erosion and Fluvial Processes, 10th edn, pp. 5–16 . MSU Publishing House, Moscow (1995)
Litvin, L.F., Chalov, R.S.: On the channel formation activity of temporary and permanent watercourses in the mountains. Geomorphology (1), 81–86 (1975)
Liu, S.: Comparative Analysis of Channel Processes on Large Rivers of Russia and China: Author's Disc. phD Geogr. Sciences, 23 p. MSU Publishing House, Moscow (1998)
Liu, S., Chalo, R.S., Ding, J., Ivanov, V.V., Korotayev, V.N., Li, J., Yang, H., Yang, S.: Regional variations of the Asian rivers suspended sediment flow in estuaries (in Russian). Vest. Moscow Univ. Ser 5. Geogr. (3), 44–51 (2001)
Lodina, R.V.: Sorting of alluvium in the rivers of Western Georgia (in Russian). In: Soil erosion and fluvial processes 4th edn., pp. 143–148. MSU Publishing House, Moscow (1974)
Lodina, R.V., Rashutin, D.V., Sidorchuk, A.Yu, Chalov, R.S.: Changes of channel morphology and bed material load from source to mouth (by the example of Terek River). Geomorphology (1), 86–94 (1987)
Lodina, R.V., Hakimov, S.K., Chalov, R.S.: Modern channel alluvium of the large mountain rivers of the Western Tien Shan and their tributaries (in Russian). Geomorphology (4), 70–76 (1994)
Lodina, R.V., Chalov, R.S.: Cobblestone bridge on the big rivers (in Russian). Nature (7), 57–62 (1994)
Lohtin, V.I.M: About the mechanism of the river bed. St. Petersburg, 1897. In: Problems of Hydraulic Engineering of Free Rivers, pp. 23–59. Rechizdat, M. (1948)
Lomakin, V.V.: On residual river and general residual surface deposits. New of the SA USSR. Ser. Geol. (2), 79–87 (1943)
Lopatin, G.V.: Sediments of the USSR Rivers, p. 368. Geografgiz, Moscow (1952)
Losev, K.C.: Water, 271 p. Hydrometeoizdat, L. (1989)
Losievsky, A.I.: Laboratory research of riffle formation processes. In: Proceedings of TsNIIVT, 36 edn., 98 p (1934)
Lower Yana: Estuary and Fluvial Processes, 212 p. GEOS, M. (1998)
Lukashova, E.N.: South America, p. 468. Uchpedgiz, Moscow (1958)
Lungershausen, G.F.: Geological history of the Middle Lena and some issues of stratigraphy of Quaternary sediments of Eastern Siberia. In: Proceedings of All-Union Meeting on the Study of the Quaternary Period, vol. III, pp. 209–217. Publishing House of the SA USSR, Moscow (1961)
Lvovich, M.I.: Experience of classification of the rivers of Russia // Proceedings of SHI, 6 edn., pp. 58–105 (1938a)

Lvovich, M.I.: Fluvial processes i gold content of a site of the river Sugor. In: Proceedings of trust 'Zolotorazvedka`` and Inst. NIGRIZoloto, no. 8, pp. 169–185 (1938b)

Lyamin, V.P.: Unity of integration and differentiation in modern geography (in Russian). In: Geography Methodology: Theory, Practice, Teaching, pp. 43–51. MSU Publishing House, Moscow (1986)

Makarova, I.V., Makarov, V.I., Akinin, B.E.: Regularities of structure and development of river valleys of Central Asia in the Quaternary period. In: History of Development of River Valleys and Problems of Land Reclamation. Western Siberia and Central Asia, pp. 93–100. Science, Novosibirsk (1979)

Makhinov, A.N., Makhinova, A.F.: Transformation of the anthropogenic relief in the area of alluvial deposits development (north of Khabarovsk Territory). Geomorphology (1), 43–49 (2006)

Makhinov, A.N., Chalov, R.S., Chernov, A.V.: Directed sediment accumulation and morphology of the lower Amur bed (in Russian). Geomorphology (3), 70–78 (1994)

Makkaveev, N.I.: Channel Regime of Rivers and Tracing of Slits, p. 202. Rechizdat, Moscow (1949)

Makkaveev, N.I.: River channel and erosion in its basin, p. 347. Published in the SA USSR, Moscow (1955)

Makkaveev, N.I.: Flow and Stream Processes, p. 116. MSU Publishing House, Moscow (1971)

Makkaveev, N.I., Chalov, R.S.: On the development of the surface topography of river terraces and signs of deep erosion by the example of the Upper Ob. New. Ser. Geogr. **4**, 120–125 (1964)

Makkaveev, N.I., Chalov, R.S.: Fluvial Processes, p. 264. MSU Publishing House, Moscow (1986)

Makkaveev, V.M.: Some theoretical problems of the open thread dynamics (in Russian). In: Proceedings of SHI, no. 62, 8th edn., pp. 5–26 (1948)

Makkaveev, V.M.: Transfer of the solid particles in the turbulent liquid flows (in Russian). In: Proceedings of Cord. 36th edn, 262 p. VNIIG Publishing House, L. (1967)

Makkaveev, N.I.: Regressive transformations of river islands. Meteorol. Hydrol (4), 44–50 (1948)

Makkaveev, N.I.: Mutual connection of the hydrological and geomorphological investigations (in Russian). In: Methods of Geographic Investigations, pp. 206–208. Geografgiz, Moscow (1960)

Makkaveev, N.I.: Fluvial processes and their reflection in relief. In: Modern Exogenous Processes of Relief Formation, pp. 96–102. Nauka, Moscow (1970)

Makkaveev, N.I.: Hydraulic Typing of Erosion Process. In: Soil Erosion and Fluvial Processes, 3rd edn., pp. 65–77, MSU Publishing House, Moscow (1973)

Makkaveev, N.I.: Forms of manifestation of fluvial processes. In: Educational Materials for the VI Session of 1974 International Hydrological Courses at MSU named after M.V. Lomonosov, pp. 17–21. MSU Publishing House, Moscow (1974)

Makkaveev, N.I.: General regularities of the erosion-fluvial processes (in Russian). Proceedings of IV All-Union hydrologic Congress, vol. 10, pp. 8–12. Fluvial processes. Hydrometeoizdat, L. (1976)

Makkaveev, N.I.: Denudation component of substance balance in the ocean-land system and its role in the formation of peneplates. Water Resour. (3), 147–155. River Transport, Moscow (1982)

Makkaveev, N.I., Belinovich, I.V., Vyshlov, K.P.: About the influence of the Caspian Sea level decline on the navigable conditions in the lower Volga and Ural. In: Fluvial Investigations to Improve Shipping Conditions, pp. 24–28. River Transport, Moscow (1958)

Makkaveev, N.I., Berkovich, K.M., Mandych, A.F., Chalov, R.S.: Classification of West Georgia river basins by signs that determine the Intensity of Denudation (in Russian). In: Geomorphological and Hydrological Research, p. 13–23. MSU Publishing House, Moscow (1968)

Makkaveev, N.I., Kalinin, A.: On Movement of Large-fragment Material in Logs. Meteorol. Hydrol. (8), 61–68 (1968)

Makkaveev, N.I., Kalinin, A. M., Khmeleva, N.V.: Analysis of the real composition of the piezoquartz spoon placers for prognosis of their primary sources (in Russian). In: Questions of Geography, col. 92, pp. 143–154. Mysl, Moscow (1973)
Makkaveev, N.I., Litvin, L.F., Khmeleva, N.V.: Use of Transport Capacity of a River Stream for Practical Purposes. Vest. Moscow Un. Ser. 5. Geogr. (2), 82–89
Makkaveev, N.I., Mandych, A.F., Chalov, R.S.: Influence of the upward relief development on the deep erosion and hard runoff of the Western Georgia rivers (in Russian). In: Vest. Moscow Un. Ser. 5. Geogr (4), 52–58 (1968)
Makkaveev, N.I., Sakharova, E.I., Chalov, R.S.: Modern aeolian processes in the upper Ob valley (in Russian). Vest. Moscow Un. Ser 5. Geogr (2), 49–53 (1966)
Makkaveev, N.I., Sovetov, V.S.: Trace of dredging slots on the riffles of the plain rivers of the European part of the USSR.In: Proceedings of the Central Research Institute of Russian Federation, 3rd edn, 60 p (1940)
Makkaveev, N.I., Khmeleva, N.V., Zaitov, I.R, Lebedeva, I.V.: Experimental Geomorphology, 196 p. MSU Publishing House, Moscow (1961)
Makkaveev, N.I., Chalov, R.S.: Methods of improvement of navigable conditions at Ob River. River Trans., (9), 45–47 (1963a)
Makkaveev, N.I., Chalov, R.S.: On Morphological Signs of Modern Accumulation in a River Valley. New. of the SA USSR. Ser. Geogr. (3), 84-89.
Makkaveev, N.I., Chalov, R.S.: Some peculiarities of the big river valley floor connected with periodic changes of the flow norm (in Russian). Geography, col. 79, . pp. 156–167. Geografgiz, Moscow (1970).
Makrinova, O.V.: Natural Blurring of the Diverter Channel. In:Proceedings of SHI 40th edn., no. 94, pp. 29–43. (1953)
Malbert, I.E.: To a question about the Siberian river channel processes (in Russian). Proceedings of Transport Energetic Institute, West.-Sib. of. of SA USSR, 3edn . Hydrol, pp. 73–85. (1952)
Mandych, A.F.: Influence of the Mountain Rivers Flood Waves on the SEDIMENT Flow (in Russian). Meteorol. Hydrol. (2), 41–44 (1966)
Mandych, A.F.: Value of the hard flow of the rivers of Western Georgia (in Russian). Vest. Moscow Un. Ser 5 Geogr. 134–137 (1967)
Mandych, A.F., Chalov, R.S.: Consideration of direction of erosive-accumulative processes at calculation of firm runoff (on an example of the rivers of Western Georgia). In: Problems of Morpholithogenesis in River Valleys, pp. 53–57. Zabaik. br. of. GS USSR, Chita (1968)
Mandych, A.F., Chalov, R. S. Direction of the modern erosion-accumulation processes in the lower reaches of the Western Caucasus rivers (in Russian). Vest. Moscow Un. Ser 5. Geogr (1), 66–71 (1970)
Map 'Erosion danger of agricultural lands of European part of USSR``. The scale is 1:2000000, 8 p. MSU Publishing House, Moscow (1987)
Maruashvili, L.I.: Terraces of r. Kodor (Western Caucasus) as an example of terraforming in mountainous countries. Rep. SA USSR.XXIV(1), 37–44 (1960)
Maslov, V.P.: Some variants of the ripple river signs (in Russian). Bill. MOIP. Dep. Geol. XI(2), 162–163 (1933)
Matinauskas, J., Zdankus, N.: Nemanas kasagi po Kainu. Mosrlfs Tech. (8), 30 (1988)
Matveev, B.V.: Influence of the geological-geomorphological factors on formation and morphology of the river bend (in Russian). Geomorphology, (3), 51–57 (1985a)
Matveev, B.V.: Morphology and Geological-geomorphological Factors of Incised and Free Radiation Development. Autoref. d. … phD Geogr. Sciences, 20 p. MSU Publishing House, Moscow (1987b)
Matveev, N.P.: Meandering of the Moscow Region Rivers. Uch. Zap. MOPI N.K. Krupskaya, vol. 267, 13 edn., pp. 3–24 (1970)
Matveevskaya, A. L., Ivanova, E.F.: Geological Structure of the Southern Part of the West Siberian Plain in Connection with Oil and Gas Bearing Issues, 264 p. Publishing house of the SA USSR, M.; L.

Maximov, S.P.: On study of river sediment movement (in Russian). In: Questions of River Life. St. Petersburg, pp. 243–254.
Meshcheryakov, YuA: Young Tectonic Movements and Erosion-Accumulative Processes in the Northwestern Part of the Russian Plain, p. 90. Published in SA USSR, Moscow (1961)
Miall, A.D.: A review of the braided-river depositional environment. Earth Sci. Rev. 13, 1–62 (1977)
Middle Ob River Regime, 146 p. M., (1969) (Dep. in VINITI № 1250-69)
Mikhailov, V.N.: Hydrological Processes in the River Mouths, p. 176. GEOS, Moscow (1997)
Mikhailov, V.N., Dobrovolsky, A.D.: General Hydrology, p. 368. Higher School, Moscow (1991)
Mikhailov, V.N.: Dynamics of flow and channeling in untidy estuaries. Proceedings of GOIN, 102 edn, 260 p. (1971)
Mikhailov, V.N.: River Mouths of RUSSIA and Neighboring Countries: Past, Present and Fuzture, 413 p. GEOS, M. (1976)
Mikhailov, V.M.: Talik development in the Kolyma river valley and the river water temperature (in Russian), (6), 100–110. Geoecology (1998)
Mikhailova, M.V.: Hydrologic regime and dynamics of the Huang He estuary hydrographic network (in Russian). Water Resour. 28(1), 105–117 (1998)
Mikhailova, N.A.: Transfer of Solid Particles by Turbulent Water Flows, 236 p. Hydrometeoizdat, L. (1966)
Mikhaylovskaya, O.I.: Quaternary terraces of Abkhazia. In: Proceedings of the First All-Union Geographer Congress, 3rd edn., pp. 82–94 (1934)
Milliman, J.D., Rutkowski, C., Meybeck, M.: River discharge to the sea. A Global River Index (GLORI), LOICZ Reports and Studies. Texel, The Netherlands (1995)
Mirtskhulava, C.E.: Washing out of Channels and Methods of Their Stability Estimation, p. 180. Kolos, Moscow (1967)
Mirtskhulava, C.E.: Basics of Physics and Mechanics of Erosion of Channels, 304 p. Hydrometeoizdat, L. (1988)
Mirtskhulava, C.E., Snishchenko, B.F.: Estimation of the fluvial and erosion processes at the complex use and protection of the water resources (in Russian). In: General Reports of the V All-Union Hydrologic Congress, vol. II, pp. 88–106. Hydrometeoizdat, L., (1986)
Molchanov, A.K. Research of Dependence of the Morphometric Characteristics of the Riverbeds on the Factors Determining Them: Authors' Dis… Phd Geogr. Sciences, 16 p. MSU Publishing House, Moscow
Morphology and river bed dynamics in the European part of Russia and neighboring countries. Sc. 1:2000000, vol. 4, no. l. Federal Geodesy and Cartography Service of the Russian Federation (1999)
Muranov, A.P.: Huang He River, 88 p. Gidrometeoizdat, L. (1957)
Muranov, A.P.: Yangtze River, 124 p. Hydrometeoizdat, L. (1959)
Narskikh, R.S.: Relief of South and Central America. Earth Relief, pp. 203–226. Nauka, Moscow (1967)
Natsag, J., Chalov, R.S.: Dynamics of River Flows and Channel Morphology on Rivers of Mongolia (Selenga River Basin). New. of the SA USSR. Ser. Geogr. (1), 110–116 (1978)
Nazarov, N.I., Yegorkina, S.S.: Rivers of Perm Territory: Lateral Channel Changes, p. 156. Star, Perm (2004)
Neff, E.L.: Discharge frequency compared to long-term sediment yields. In: Proceedings of Symposiun on River Morchology, pp. 236–242. JANS. Publ. (1977)
Nekos, S.V., Chalov, R.S.: Sediment runoff and fluvial processes on rivers of the Don basin. Geomorphology (2), 60–71 (1997)
Neliubin V.L. Amudarya. M: Marine transport, 1963. 136 p.
Nemanishvili, S.N.: Terraces of the Tskhenis-Tskali River Valley. In: Proceedings of Geography institute Named after V.L. Vakhushti SA GSSR, vol. XVIII, pp. 51–53 (1963)
Nemanishvili, S.N.: Terraces of the Rioni river valley above the city of Kutaisi. In: Proceedings of Geography institute Named After Vakhushti, AS GSR, vol. XX, pp. 36–38 (1964)

Nguyen, V.K.: Dynamics of the Estuary Areas of the Rivers of the Socialist Republic of Vietnam: Abstract of Diss. Doc. Geogr. Sciences, 53 p. MSU Publishing House, Moscow (1990)
Nikitina, N.A.: Branch development and erosion-accumulative processes in river junctions. Water Resour. (2), 67–71 (1997)
Nikolskaya, V.V.: Morphosculpture of the Amur Basin, p. 296. Nauka, Moscow (1972)
Nikonov, A.A.: Determination of The River Incision Rate. Geomorphology (1), 24–35 (1973)
Nolan, K.M., Lisle, T.E., Kelsey, H.M.: Bankfull discharge and sediment transport in northwestern California. In: Erosion and Sedimentation in the Pacific Rim. In: Proceedings of the Corvallis Simposirem, no. 165, pp. 439–448. JANS. Pub. (1987)
Obedientova, G.V.: Erosion Cycles and Formation of the Volga Valley, p. 240. Nauka, Moscow (1977)
Obodovskiy, A.G., Zaitz, E.S., Potapenko, A.N.: Peculiarities of effective water discharge on rivers of forest-steppe zone of Ukraine. In: Physical Geography and Geomorphology, 29 edn., pp. 123–127, Higher School, Kyiv (1983)
Obodovskiy, A.G., Zaitz, E.S., Chalov, R.S.: Geographical substantiation of the determination technique of the water flow (by the example of the Ukrainian rivers). Vest. Moscow Un. Ser. 5. Geogr. (5), 67–72 (1987)
Obodovsky, A.G.: Hydrologic-Ecological Galvanization of Fluvial Processes (at the Application Level of Ukraine), p. 276. Nica-Center, Kyiv (2001)
Osterkamp, W.P.: Gradient, discharge and particle-size relations of alluvial channels in Kansas, with observations on braiding. Ammer J. Sci. **278**(9), 1253–1268 (1978)
Pakhomova, O.M.: Hydrologic-Morphological Characteristics of River Beds and Orderly Structure of the River Network. Autoref. Diss. ... phD Geogr. Sciences, 28 p. MSU Publishing House, Moscow (2001)
Panin, A.V.: Morphology and Dynamics of Incised Pebble-Boulder rivers. Autoref. Dis. ... phD. Geogr. Sciences, 24 p. MSU Publishing House, Moscow (1991)
Panin, A.V., Sidorchuk, A.Yu., Chalov, R.S.: Catastrophic speeds of a fluvial relief formation (in Russian). Geomorphology (2), 3–11 (1990)
Petrovsky, V.V., Sakharova, E.I.: About reformations of Krasnoborsk riffle (Northern Dvina). In: Channel Investigations to Improve Navigation Conditions, pp. 19–23. River Transport, Moscow (1958)
Physical Geography of China: Land Waters. Science, Beijing (1981). (in Chinese)
Pinkovsky, S.I.: Types of river channels of Middle and South Siberia. In: Proceedings of SHI, 94 edn., pp. 87–114
Pinkovsky, S.I.: Types of riverbeds in the Soviet Far East (southern half). In: Proceedings of SGI, 144 edn., pp. 77–117 (1967)
Pinkovsky ,S.I.: Types of the riverbeds of the Caucasus and Central Asia. In: Proceedings of SGI, 136 edn., pp. 231–269 (1966)
Pokrovskiy, P.V.: On roughness coefficients of mudflow roughness. Meteorol. Hydrol. (7), 30–35 (1940)
Poliakov, B.V.: Hydrology of the Don River Basin, 328 p. Rostov-na-Donu, (1930)
Polyakov, B.V.: Changes in River Load and Riffles of the Volga River Due to Construction of Dams (in Russian). New Res. Inst. Hydrotech. **9**, 5–58 (1933)
Popkov, I.F.: Total Inland Waterway Station, p. 279. River transport, Moscow (1962)
Popov, I.V.: Methodic Bases of the River Fluvial Process Research, 208 p. Gidrometeoizdat, L. (1961)
Popov, I.V.: Application of morphological analysis to estimation of general channel changes of the p. Ob. In: Proceedings of SHI 94 edn., pp. 22–86 (1962)
Popov, I.V.: River Bed Changes and Hydraulic Engineering, 328 p. Hydrometeoizdat, L. (1965)
Potapov, M.V.: Screw fluid motion in a direct open channel of a rectangular cross-section (in Russian). In: Cross Circulation in an Open Flow and Its Hydrotechnical Application, pp. 102–154. Agricultural product, Moscow (1936)

Prokacheva, B.G., Snischenko, D.V., Usachev, V.F.: Remote Methods of Hydrological Study of BAM Zone, 224 p. Gidrometeoizdat, L. (1982)

Qian, N., Yi, Q., Zhou, W.: Changes in Water Flow and River Load of the Main Current of the Huang He River and Fluvial Processes. Beijing (1993) (in Chinese)

Qian, N., Zhang, Z., Zhou, Z. (1987) Fluvial Process, 584 p. Science, Beijing (1987) (in Chinese)

Qian, N., Zhou, W.: Fluvial Processes in the Lower Reaches of the Huang He River, 224 p. Nauka, Beijing (1965) (in Chinese)

Rabkova, E.K.: Methods of Calculation of River Beds and Channels and Their Change on the Basis of Use of Complex Natural Characteristics: Autoref. Dis. … Dr. Techn. Sci., 35 p MGMI, Moscow (1979)

Ravine Erosion, 168 p. MSU Publishing House, Moscow (1989)

Razumikhin, N.V.: Experimental Studies of Evolution of Rock Debris Pellets, 67 p. LSU Publishing House, L. (1965)

Reznikov, P.N., Chalov, R.S.: Sediment flow and conditions of channel formation on the rivers of the Basin of the Northern Dvina (in Russian). Geomorphology (2), 73–85 (2005)

Richards, K.S.: The formation of ripples and dunes on an erodible bed. J. Fluid Meeh. **99**(3), 597–618 (1980)

Rimkus, Z.V.: Main channel-forming factors of Nemunas basin on the territory of Lithuanian SSR. In: Second All-Union Conference 'Dynamics and Terminology of Rivers, Reservoirs and Estuaries``. Theses of reports. vol. II, pp. 28–29. M. (1984)

Rodevich, V.M.: To the Classification of Rivers. New. SHI. **35**, 25–39 (1931)

Romashin, V.V.: Some features of the river fluvial process on the mountain river (in Russian). In: Proceedings of SHI. 144 edn., pp. 66–76 (1967)

Romashin, V.V.: Types of fluvial process in connection with determining factors. In: Proceedings of SHI, 155 edn., pp. 56–63 (1968)

Romashin, V.V.: Estimation of a channel mode on morphological characteristics. In: Proceedings of SHI, no. 190, pp. 48–56 (1972)

Romashin, V.V.: Properties of a channel wandering. In: Proceedings of SHI., 225 edn., pp. 5–15 (1975)

Rosgen, D.L.: A classification of natural rivers. Catena **22**, 169–199 (1994)

Rossinskiy, K.I., Debolsky, V.K.: River Sediments, p. 216. Nauka, Moscow (1980)

Rossinskiy, K.I., Kuzmin, I.A.: Some questions of the applied theory of the river channel formation (in Russian). In: Problems of the River Flow Regulation, pp. 88–129. Published in SA USSR, Moscow; L. (1947).

Rossinskiy, K.I., Kuzmin, I.A.: Regularities of river channel formation. In: Fluvial Processes, pp. 5–14. Publishing House of the SA USSR, Moscow (1958)

Rossinskiy, K.I., Kuzmin, I.A.: Balance method for calculation of changes of the flow bottom. In: Proceedings of Hydroproject. col. 2, pp. 265–271 (1964)

Rubtsov, M.V.: Scientific Bases of the Forestry Measures in the Protective and Water-Protective Forests Along the Rivers of the European North: Author's Thesis … Dr. Techn. Sci, 37 p. L. (1982)

Rubtsov, M.V., Deryugin ,A.A.: Influence of forest on channel of banks of taiga rivers. In: Laws of Manifestation of Erosion and Fluvial Processes in Various Natural Conditions, pp. 264–265. MSU Publishing House, Moscow (1981)

Ruleva, S.N., Chalov, P.S.: Channel changes at the junction of the rivers Vychegda and Sysola and their influence on the infrastructure of Syktyvkar. In: Proceedings of AVN 9 edn., pp. 171–181. M. (2003)

Rust, B.R.: Sedimentation in arid-zone anastomosing fluvial system: Cooper creek. Central Australia. J. Petrol. **51**(3), 745–755 (1981)

Rysin, I.I., Petukhova, L.N.: Fluvial Processes on the Rivers of Udmurtia, p. 176. Scientific Book, Izhevsk (2006)

Rzhanitsyn, I.A.: Morphological and Hydrological Regularities of the River Network Structure, 240 p. Gidrometeoizdat, L. (1960)

Rzhanitsyn, N.A.: Channel-Forming Processes of the Rivers, 264 p. Gidrometeoizdat, L. (1985)
Sadovsky, G.L.: Manual for Tracing Dredging Slots, p. 183. River Transport, Moscow (1961)
Sakharova, E.I.: Geomorphological research in the study of fluvial processes in the Northern Dvina. In: Methods of Geographic Research, pp. 101–109. Geografgiz, Moscow (1960)
Sakharova, E.I., Lebedeva, N.V.: Change of average alluvium size along the length of a mountain river (on the example of the Mzymta River). In: Geomorphological and Hydrological Studies, pp. 24–30. MSU Publishing House, Moscow (1968)
Sakharyuk, N.P.: Determination of effective discharges for the rivers of the southern part of Primorye. In: Proceedings of DVNIGMI, 84 edn., pp. 94–96 (1980)
Saveliev, R.A., Zaitsev, A.A.: Definition of initial and boundary conditions of problems of computer modeling of fluvial processes. In: XIX Plenary Interuniversity Coordination Meeting on the Problem of Erosion, Fluvial and Estuary Processes, pp. 178–180. Belgorod (2004)
Schaffemak, F.: Grundriss dez FluBmozphologie und des Flussbaues. Spzingez, Wien (1950)
Schanzer, E.V.: Alluvium of plain rivers of temperate belt and its importance for learning the regularities of structure and formation of alluvial formations. In: Proceedings of Geological Institute of the SA USSR. Ser. Geology., 135 edn, no. 55, 275 p (1951)
Schnitnikov, A.V.: Variability of General Humidity of Materials in the Northern Hemisphere/ Western GO USSR, vol. 16, 339 p. New series. Publishing house of SA USSR, M.; L (1957)
Schumm, S.A., Khan, H.R.: Experimental Study of Channel Patterns. Geol. Soc. Am. Bull. **83**(6), 1755–1770 (1972)
Schumm, S.A.: The Fluvial System, 338 p (1977)
Sediment Flow, Its Study of Geographical Distribution, 240 p. Gidrometeoizdat, L. (1977)
Sedimentation Research in China, 250 p. China water and power press, Beijing (1993)
Series of Picture Albums of the Yellow River. Publication Center Yellow River Conservancy Commission, no. 8, 21 p. Beigangzi Street/Chongwen District. Beijing (1991)
Shamov, G.I.: River Sediment, 380 p. Hydrometeoizdat, L. (1959)
Shapiro, H.Sh.: Regulation of Solid Flow and Fluvial Processes at Water Intake in Irrigation Systems: Authors' Diss. Doct. Sci. (Techn.), 52 p. MGMI, M. (1973)
Shataeva, S.G.: Determination of dredging volumes by means of channel stability index. In: Proceedings of INIIEVT, 68 edn., pp. 84–121 (1969)
Shaydegger, A.E.: Theoretical Geomorphology, p. 452. Progress, Moscow (1964)
Shchukin, I.P.: General Geomorphology, vol. 1, p. 616. MSU Publishing House, Moscow (1960)
Shchukin, I.S.: General Land Morphology, vol. 1, 366 p. Novosibirsk: ONTI, M.; L. (1933)
Shen IV, H., Schumm, S.Y., Doehring, D.O.: Stability of stream channel patterns. Transp. Res. Rec. **736**, 22–28 (1979)
Shenberg, N.V.: Analysis of factors and methodology of calculation of effective water discharges. Water Resour. (4), 49–54 (1989)
Shenberg, N.V.: Use of the long-term characteristics of the flow regime for calculation of the effective water discharge (in Russian). Land and Water Resources: Anti-Erosion Protection and Regulation of the Channels, pp. 69–76. MSU Publishing House, Moscow (1990)
Shtankova, N.N.: Sediment flow on rivers of Volga basin (in Russian). In: Dynamics of Gully-Beam Forms and Fluvial Processes, pp. 112–119. MSU Publishing House, Moscow (2002)
Sidorchuk, A.Yu.: The Structure of the Relief of the River Bed, 128 p. Hydrometeoizdat, SPb (1992)
Sidorchuk, A.Yu., Mikhinov, A.E.: Morphology and dynamics of a channel relief. In: Results of a Science and Technology. Series 'Land Hydrology", vol. 5, 163 p. VINITI, M. (1985)
Sidorchuk, A.Yu., Chalov, R.S.: Incision of mountain rivers: speeds and causes. Nature (12), 36–45 (1996)
Simonov, YuG: Regional Geomorphological Analysis, p. 264. MSU Publishing House, Moscow (1972)
Simons, D.D., Richardson, E.V.: Forms of bed roughness in alluvial channels. ASCE. Proc. 87 (HY-3) (1961)

Simons, D.B., Richardson, E.V., Nordin, C.F.: Sedimentary structures generated by flow in alluvial channels. SEPM Div. AAPS Spec. Publ .(12), 48–67 (1966)
Sinitsyn, M.G., Rusakov, A.B.: Influence of river beaver activity on relief of valleys and river beds of small rivers of Vetluzhsko-Unzhenskoye Polesiye. Geomorphology (1), 85–91 (1990)
Sladkopevtsev, S.A.: Development of the River Valleys and Neotectonics, p. 184. Nedra, Moscow (1973)
Sladkopevtsev, S.A.: On the structure of forests in the valleys of the temporary waterways of Central Kazakhstan (in Russian). New. VGO. 95(5), 449–450 (1963)
Slastikhin, V.V.: Process of erosion on the selective watersheds in Moldova, In: Regularities of Erosion and Fluvial Processes in Different Natural Conditions, pp. 136–137. MSU Publishing House, Moscow (1987)
Small Rivers in the Volga Basin, p. 335. MSU Publishing House, Moscow (1998)
Smirnova, V.G.: Hydrologic-Morphological Analysis of the Branched Riverbeds of the Altai region: Author's Thesis … phD Geogr. Sciences, 20 p. IG SB RAS, Irkutsk (2002)
Snischenko, B.F.: Movement of sand ripples in natural water flows. In: Proceedings of SHI. 136 edn., pp. 82–91 (1966)
Snishchenko, B.F.: About the connection of the sand ripples height with the parameters of the river flow and channel (in Russian). Meteorol. Hydrol. (6), pp. 84–91 (1980)
Snishchenko, B.F.: Pair connections of parameters of a ripple and characteristics of a stream and a channel. In: Proceedings of SHI. 288 edn., pp. 15–25 (1983)
Snishchenko, B.F.: Development of theory of fluvial process in the XX c. In: Proceedings of SHI. 361 edn, pp. 5–26 (2002)
Snishchenko, B.F., Copaliani,Z. D.: About speed of a ridge movement in the rivers and laboratory conditions (in Russian). In: Proceedings of SHI. 252th edn., pp. 30–37 (1978)
Sokolov, Yu.I.: Transport capacity and changes of the river floodplains (in Russian). In: Proceedings of V All-Union Hydrological Congress, vol.10, Kn. 2, pp. 196–203. Hydrometeoizdat, L. (1988)
Sokolov, Yu.N.: Hydraulic Resistances on the Floodplains of Plains Rivers: Author's Disc. Dr. Sci. (Techn.), 39 p. YVP AN SSSR, M. (1991)
Sokolovsky, D.L.: River Flow, 540 p. Hydrometeoisdat, L. (1968)
Solovyev, I.A.: Fluvial Process and the Waterways of the LOWER Amur, p. 224. Dalknizhizdat, Vladivostok (1967)
Soviet Encyclopedic Dictionary, 1600 p. The Soviet Encyclopedia, Moscow (1983)
Sribnyi, M.F.: Formula of average speed of the rivers flow and their hydraulic classification by resistance to movement. In: Research and Complex Use of Water Resources, pp. 204–221. Publishing House of the SA USSR, Moscow (1960)
Studenichnikov, B.P.: Blurring Ability of a Flow and Methods of Channel Calculations, p. 183. Gosstroyizdat, Moscow (1964)
Studies of Environmental Changes in the Huang He River Basin, Water Availability and Sediment Transport. Science and Technology, Jinan (1994) (in Chinese)
Surkov, V.V.: Dynamics of Floodplain Landscapes of the Upper and Middle Ob, p. 255. MSU Publishing House, Moscow (1999)
Talmaza, I.F., Kroshkin, A.N.: Hydromorphometric Characteristics of Mountain Rivers, p. 204. Kyrgyzstan, Frunze (1968)
Tananaev, N.I.: Influence of perennially frozen soils on suspended sediment flow and fluvial processes. In: Dynamics of Ravine-Beam forms and Fluvial processes, pp. 107–111. MSU Publishing House, Moscow (2002)
Tananaev, N.I.: Seasonal and long-term freezing of the cryolithozone river beds and its influence on channel changes (on the example of average Lena). In: Erosional, Fluvial Processes and Hydroecology Problems, pp. 195–201. Geogr. ft. MSU, Moscow (2004)
The River Bed Regime in Northern Eurasia, 336 p. MSU Publishing House, Moscow (1994)
Timashev, I.E.: Inheritance of the adjoining section of the Lena river valley. Vest. Moscow Un. Ser 5. Geogr. (1), 104–106 (1970)

Transport Use of Water Reservoirs, 224 p. Transport, Moscow (1972)
Trepetsov, E.V.: Washing out of the river banks in the steppe Altai and its engineering and geological characteristics. In: Bul. Scientific-Technical, Information, no. 8(52), pp. 3–19. Gosgeoltekhizdat, Moscow (1963)
Tsvetkova, N.A.: Mode of Amudarya river sediments (in Russian). In: Questions of Hidrotechnik, 13th edn., pp. 5–86. Tashkent (1963)
Turykin, L.A.: Morphology and Seasonal Deformations of Rolls on the Rivers with Unbranched Channel: Authors' Dis. phD Geogr. Sciences, 25 p. MSU Publishing House, Moscow (1997)
Velikanov, M.A.: Fluvial Process, p. 395. Gosfizmatizdat, Moscow (1958)
Velikanov, M.A.: To statement of a problem of fluvial processes. Meteorol. Hydrol. (3), 36–46 (1946)
Velikanov, M.A.: Russian channel hydraulic engineering and its role in development of teaching about the fluvial processes (in Russian). In: Problems of Hydraulic Engineering of the Free Rivers, pp. 5–14. Rechizdat, Moscow (1948)
Velikanov, M.A.: Dynamic Structure of the Fluvial Process (in Russian). Izv. USSR SA. Ser. Geogr. Geophy. (6), 497–512 (1949)
Velikanov, M.A.: Dynamics of Channel Flows, vol. 1, 324 p. Flow Structure, vol. 2 (1954). Gostekhizdat, M. Sediments and Channel, 324 p. Gostekhizdat, M. (1955)
Vinogradova, N.I.: About accuracy of forecast of siltation of reservoirs . Hydrotech. Constr (9), 38–40. (1966)
Vinogradova, N.N., Samoilova, A.A., Khmeleva, N.V., Shevchenko, B.F.: Application of photo-theodolite method for study of channel forms on rivers of Abkhazia. In: Soil Erosion and Fluvial Processes, 7 ed., pp. 109–118 MSU Publishing House, Moscow (1979)
Vinogradovm, V.A., Klaven, A.B., Nikitin, V.I., Turutina, T.V.: Non-equilibrium processes in river bed formation. In: Proceedings of AVN, vol. 9, pp. 55–68 (2003).
Vlasov, B.N.: Regionalization of UTS territory under conditions of effective discharges passing. In: Research of Fluvial Processes for Practice of National Economy, pp. 116–117. MSU Publishing House, Moscow (1983)
Vlasov, B.N., Chalov, R.S.: Effective discharges of the Volga R. New. USSR SA. Ser. Geogr, (6), 75–81 (1981)
Vlasov, B.N., Chalov R.S. Regionalization of the European territory according to the conditions of effective discharge on the rivers.Vest. Moscow. Un. Ser. 5. Geography (in Russian), (6), 32–42. Vest (1991)
Vozniachuk, L.N., Valchik, M.A.: Morphology, Structure and History of the Neman Valley Development in Neopleistocene and Holocene, p. 212. Science and Technology, Minsk (1978)
Walling, D.E., Webb, B.W.: Material transport by the world's river. JANS Publ. (164), 313–329 (1987)
Waterways of the Lena Basin, 600 p. MIKIS, M. (1995)
Wexler, A.B., Donenberg, V.M.: Influence of the flow regulation on the Yenisei channel transformation (in Russian). Regularities of the Erosion and Fluvial Processes Manifestation in Different Natural Conditions, pp. 341–342. MSU Publishing House, Moscow (1981)
Wolman, M.G., Miller, J.P.: Magnitude and frequency of forces in geomorphic processes. J. Geol. 68, 54–74 (1960)
Work of Water Flows, 194 p. MSU Publishing House, Moscow (1987)
Yakhontov, S.A.: To estimation of resistance of rough flow in a zone of shallow relative depths. Hydrotech. Constr. (6), 40–43 (1960)
Yang, H., Tan, Z.: Evolution of the Yangtze River bed in the middle reaches of the Tinjiang region (245 p). Water Management and Hydropower, Beijing (1999) (in Chinese).
Yi, Q., Lo, Z., Sun, Z., Zhang, Y., Su, X., Cao, Y.: Geomorphology of the Bed in the Lower Reaches of the Huang He River. Nauka, Beijing (1990). (in Chinese)
Yi, Q., Ning, Q., Zhou, W.: The Main Characteristics of Fluvial Processes in the Lower Reaches of the Huang He r. Construction, Beijing (1993). (in Chinese)

Zaikov, B.D.: High Floods and Summer Floods on the USSR Rivers During Historical Time, 135 p. Hydrometeoizdat, L. (1954)

Zaitsev, A.A.: Formation of the Free Bends on the Plain Rivers: Author's thesis … phD Geogr. Sciences, 24 p. MSU Publishing House, Moscow (1979)

Zaitsev, A.A.: Investigation of meandering process by spectral analysis method. In: Laws of Erosion and Fluvial Processes Manifestation in Different Natural Conditions, pp. 334–335. MSU Publishing House, Moscow (1981)

Zaitsev, A.A.: Mode of levels and slopes on boulder-pebble riffle and rocky ledges in the channels of large rivers. Meteorol. Hydrol. (12), 103–106 (1989)

Zaitsev, A.A.: The role of ice phenomena in formation of beds and banks of channels of large rivers of Lena basin. In: Safety of Power Structures, Exhibit. 11, pp. 211–223. NIYEZ, M. (2003)

Zaitsev, A.A., Ivanov, V.V., Korotayev, V.N., Labutina, I.A., Lukyanova, S.A., Li J., Rimsky-Korsakov, I.A., Rychagov, G.I., Scroll, A.A., Sidorchuk, A.Y., Sychev, V.A., Chernov, A.V.: The Lower Volga: Geomorphology, Paleogeography and Channel Morphodynamics, 242 p. GEOS, M. (2002)

Zaitsev, A.A., Kirik, O.M., Lodina, R.V., Panin, A.V., Chalov, R.S.: Hydromorphological characteristics and regulation of the lower Vitim riverbed in connection with its transport use (in Russian). In: Soil Erosion and Fluvial Processes, 12 ed, pp. 189–214. MSU Publishing House (1998)

Zaitsev, A.A., Militeyev, A.I.: Ice jams and fluvial processes on the Lena river. Security of Power Structures, 11 ed., pp. 224–237. NIYEZ, Moscow (2003)

Zaitsev, A.A., Savtsova, T.M.: Sculptural and accumulative relief of the incised riverbeds of Eastern Siberia and features of its formation. Geomorphology (4), 58–63 (1994)

Zamarin, E.A.: Transport Capacity and Allowable Flow Speeds in Channels, 32 p. Gostransizdat, Moscow, L. (1951)

Zaslavsky, M.N.: Soil Erosion, p. 248. Mysl, Moscow (1979)

Zavadskiy, A.S., Chalov, R.S.: Regional analysis of the free meandering (in Russian). Vest. Moscow, Un. Ser 5. Geography (3), 32–36 (1997)

Zavadsky, A.S., Makhinov, A.N., Chalov, R.S.: Formation of the Middle Amur Channel and its Morphodynamic Types. Water Resour. **27**(2), 133–140 (2000)

Zavadsky, A.S., Chalov, R.S.: Conditions of formation and morphology of free bends on the rivers of Northern Eurasia (in Russian). Geomorphology, (4), 88–95 (2000)

Zavadsky A.S. Hydrologic and morphological analysis of free meandering of the plain river beds. Autoref. Diss. … phD Geogr. Sciences, 27 p. MSU Publishing House, Moscow (2001)

Zemtsov A. A. Peculiarities of fluvial processes development on taiga rivers. In: Regularities of Erosion and Fluvial Processes Manifestation in Different Natural Conditions, pp. 258–259. MSU Publishing House, Moscow (1976)

Zenkovich, V.P.: Bereg of the Black and Azov Seas, p. 376. Geografgiz, Moscow (1958)

Zernov, S.Y.: Inland Waterways of the North-Eastern Region: Design, Construction, Operation, p. 124. Nauka, Novosibirsk (2003)

Zhayvoron, I.A., Chalov, R.S.: Stability of river channels and conditions of transport of bed material load (by the example of the upper Ob). Dynamics of Channel Flows and Protection of Natural Waters, pp. 28–32. LGMI Publishing House, L. (1990)

Zheleznyakov, G.V.: Hydraulic Justification of River Hydrometry Methods, 163 p. SA USSR, Moscow; L. (1950)

Zheleznyakov, G.V., Baryshnikov, N.B., Altunin, V.C.: Impact of kinematical effect of the non-pressure flow on sediment transport (in Russian). In: Sediment Movement in Open Channels, pp. 19–24. Nauka, Moscow (1970)

Zhivago, A.V.: Some data on geomorphology and history of development of coast of Abkhazia, received by a method of comparative study of load. In: Proceedings in Geography Institute SA USSR, 51 ed, pp. 12–32 (1952)

Znamenskaya, N.S.: Hydraulic Modeling of Fluvial Processes, p. 240. Gidrometeoizdat, SPb (1992)
Znamenskaya, N.S.: Ripple Movement of Sediment, 188 p. Hydrometeoizdat, L. (1968)
Znamenskaya, N.S.: Bottom Sediments and Fluvial Processes, 142 p. Hydrometeoisdat, L. (1476)
Zorina, E.F.: Ravines, ravine formation and development potential. In: Soil erosion and fluvial processes, 12 ed., pp. 72–95. MSU Publishing House, Moscow (2000)
Zorina, E.F., Kosov, B.F., Prokhorova, S.D.: Experience of estimation of volume of ravine outcrops in the Don river basin. Vest. Moscow, Un. Ser. 5. Geography (3), 39–45 (1980)
Zorina, E.F., Prokhorova, S.D., Chalov, R.S.: The role of ravine erosion in formation of the river riffles (in Russian). Geomorphology (4), pp. 76–83 (2000)

Printed by Printforce, the Netherlands